D1483409

MONOLITHIC PHASE-LOCKED LOOPS AND CLOCK RECOVERY CIRCUITS

MONOLITHIC PHASE-LOCKED LOOPS AND CLOCK RECOVERY CIRCUITS
THEORY AND DESIGN

Edited by

Behzad Razavi
AT&T Bell Laboratories

A Selected Reprint Volume
IEEE Solid-State Circuits Council, *Sponsor*

The Institute of Electrical and Electronics Engineers, Inc., New York

This book may be purchased at a discount from the publisher when ordered in bulk quantities. For more information contact:

IEEE PRESS Marketing
Attn: Special Sales
P.O. Box 1331
445 Hoes Lane
Piscataway, NJ 08855-1331
Fax: (908) 981-9334

Printed in the United States of America

10   9   8   7   6   5   4   3   2   1

ISBN 0-7803-1149-3

IEEE Order Number: PC5620

Library of Congress Cataloging-in-Publication Data

Razavi, Behzad.
    Monolithic phase-locked loops and clock recovery circuits : theory
and design / Behzad Razavi.
        p.   cm.
    Includes bibliographical references and index.
    ISBN 0-7803-1149-3
    1. Phase-locked loops–Design and construction.   2. Timing
circuits–Design and construction.   3. Integrated circuits–Design
and construction.   I. Title.
TK7872.P38R39   1996
621.3815'364–dc20   96-6102
                                                                        CIP

Contents

PART 5   CLOCK AND DATA RECOVERY CIRCUITS   381

Contents

Preface

Phase-locked loops (PLLs) and clock recovery circuits (CRCs) find wide application in wireless and communication systems, disk drive electronics, high-speed digital circuits, and instrumentation. While the theory of operation of these circuits has long been developed and treated in a number of books (including two reprint collections published by IEEE Press in the 1970s and 1980s), design techniques for monolithic implementation of PLLs and CRCs in modern integrated-circuit (IC) technologies have appeared only recently, primarily in the form of conference and journal papers. For an IC designer entering this field, theoretical books often have too many details and IC-related publications too few.

This book deals with the analysis, design, simulation, and implementation of PLLs and CRCs in monolithic technologies. It has been organized with two goals in mind: first, to bridge the existing gap between the theory and the design of phase-locked systems, and second, to selectively present the published work on PLLs and CRCs in a coherent, unified volume, providing a self-contained reference for students and practicing engineers.

In studying phase-locked loops (PLLs), one encounters two difficulties. First, the basic PLL is deceptively simple, often tempting the reader to ignore the crucial details that determine the performance in a realistic environment. Second, the design of PLLs cannot be easily described using a straight "top-down" or "bottom-up" approach, because each level of abstraction entails issues strongly related to other levels as well.

In order to overcome the above difficulties, this book begins with a tutorial on phase-locked loops and clock recovery circuits. The goal is to develop an intuitive understanding of the underlying principles and describe the important issues, preparing the reader for more advanced topics in the following collection of papers. The tutorial itself is organized in a "bottom-up–top-down" sequence to gradually take the reader from basic components to the architecture level and back to the building blocks.

Following the tutorial are the selected papers, organized in five parts. Part I deals with theoretical aspects of PLLs, covering topics such as transient and frequency response, discrete-time analysis, frequency detection, and phase noise.

Part II is concerned with the design of PLL and CRC building blocks, in particular oscillators and phase/frequency detectors for both periodic and random binary waveforms.

Part III addresses the difficult issue of PLL modeling and simulation. SPICE macromodeling, nonlinear mathematical modeling, and high-level behavioral representation are described.

Parts IV and V present papers on monolithic implementation of PLLs and CRCs, respectively, demonstrating a wide range of speeds for various applications, from fiber optics to microprocessors.

This book has benefited from many contributions by many people. Stewart Tewksbury (West Virginia University and IEEE Solid-State Circuits Council) reviewed the original proposal and made useful suggestions with respect to the contents. Kameran Azadet (AT&T Bell Labs), Larry DeVito (Analog Devices), Ramin Farjad-Rad (Stanford University), Mark Johnson (Rambus), Marc Loinaz (AT&T Bell Labs), John Maneatis (Silicon Graphics), and Masoud Zargari (Stanford University) reviewed various parts of the manuscript, particularly the tutorial, and provided invaluable feedback. Despite his busy schedule, Larry DeVito has also contributed an original paper to this volume.

I would also like to acknowledge the support of the staff at IEEE Press and the sponsorship of the IEEE Solid-State Circuits Council for this book.

Behzad Razavi
Holmdel, New Jersey

Design of Monolithic Phase-Locked Loops and Clock Recovery Circuits—A Tutorial

Behzad Razavi

Abstract — **This paper describes the principles of phase-locked system design with emphasis on monolithic implementations. Following a brief review of basic concepts, we analyze the static and dynamic behavior of phase-locked loops and study the design of their building blocks in bipolar and CMOS technologies. Next, we describe charge-pump phase-locked loops, effect of noise, and the problem of clock recovery from random data. Finally, we present applications in communications, digital systems, and RF transceivers.**

1. INTRODUCTION

PHASE-LOCKED loops (PLLs) and clock recovery circuits (CRCs) find wide application in areas such as communications, wireless systems, digital circuits, and disk drive electronics. While the concept of phase locking has been in use for more than half a century, monolithic implementation of PLLs and CRCs has become possible only in the last twenty years and popular in the last ten years. Two factors account for this trend: the demand for higher performance and lower cost in electronic systems, and the advance of integrated-circuit (IC) technologies in terms of speed and complexity.

This tutorial deals with the analysis and design of monolithic PLLs and CRCs. Following a brief look in Section 2 at a number of design problems that can be solved using phase locking, we review some basic concepts in Section 3 to establish proper background as well as define the terminology. In Section 4, we introduce the phase-locked loop in a simple form, analyze its static and dynamic behavior, and formulate its limitations. In Section 5, we describe the design of circuit building blocks, and in Section 6, charge-pump PLLs. Sections 7 to 9 deal with phase noise, clock recovery circuits, and applications of phase-locked systems, respectively.

2. WHY PHASE-LOCK?

Phase locking is a powerful technique that can provide elegant solutions in many applications. In this section, we consider four design problems that can be efficiently solved with the aid of PLLs. We return to these problems after studying the principles of phase locking.

2.1 Jitter Reduction

Signals often experience timing jitter as they travel through a communication channel or as they are retrieved from a storage medium. Depicted in Figure 1, jitter manifests itself as variation of the period of a waveform, a type of corruption that cannot be removed by amplification and clipping even if the signal is binary.

A PLL can be used to reduce the jitter.

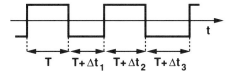

Fig. 1   Timing jitter.

2.2 Skew Suppression

Figure 2 illustrates a critical problem in high-speed digital systems. Here, a system clock, CK_S, enters a chip from a printed-circuit (PC) board and is buffered (in several stages) to sharpen its edges and drive the load capacitance with minimal delay. The principal difficulty in such an arrangement is that the on-chip clock, CK_C, typically drives several nanofarads of device and interconnect capacitance, exhibiting significant delay with respect to CK_S. The resulting skew reduces the timing budget for on-chip and inter-chip operations.

In order to lower the skew, the clock buffer can be placed in a phase-locked loop, thereby aligning CK_C with CK_S.

1

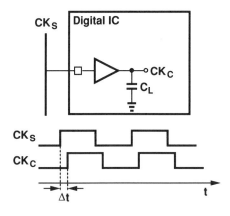

Fig. 2   Clock skew in a digital system.

2.3 Frequency Synthesis

Many applications require frequency multiplication of periodic signals. For example, in the digital system of Figure 2, the bandwidth limitation of PC boards constrains the frequency of CK_S, whereas the on-chip clock frequency may need to be much higher. As another example, wireless transceivers employ a local oscillator whose output frequency must be varied in small, precise steps, for example, from 900 MHz to 925 MHz in steps of 200 kHz.

These exemplify the problem of "frequency synthesis," a task performed efficiently using phase-locked systems.

2.4 Clock Recovery

In many systems, data is transmitted or retrieved without any additional timing reference. In optical communications, for example, a stream of data flows over a single fiber with no accompanying clock, but the receiver must eventually process the data *synchronously*. Thus, the timing information (e.g., the clock) must be recovered from the data at the receive end (Fig. 3). Most clock recovery circuits employ phase locking.

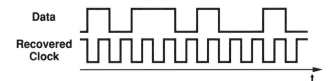

Fig. 3   Clock recovery from random data.

3. BASIC CONCEPTS

In studying PLLs and CRCs, we often need to draw upon concepts from the theory of signals and systems. This section provides a brief review of such concepts to the extent that they prove useful to IC designers.

3.1 Time- and Frequency-Domain Characteristics

Phase-locked systems exhibit nonlinear behavior at least during part of their operation (e.g., transients), thus requiring time-domain analysis in almost all applications. However, in the steady state and during slow transients, it is extremely helpful to study the response in the frequency domain as well, especially if the application imposes certain constraints on the purity of the output spectrum.

Most of the signals encountered in PLLs are either strictly periodic, for example, $x(t) = A \cos \omega_c t$, or phase-modulated, for example,

$$x(t) = A \cos[\omega_c t + \phi_n(t)] \tag{1}$$

We consider the second example as a more general case. The total phase of this signal is defined as $\phi_c(t) = \omega_c t + \phi_n(t)$ and the total frequency as $\Omega_c(t) = d\phi_c/dt = \omega_c + d[\phi_n(t)]/dt$.[1] PLLs usually operate on the "excess" components of ϕ_c and Ω_C, that is, $\phi_n(t)$ and $d[\phi_n(t)]/dt$, respectively.

In most cases of interest, $|\phi_n(t)| \ll 1$ radian, i.e., the difference between consecutive periods of the waveform is small and the signal only slightly deviates from a strictly periodic behavior. We call such a signal "almost periodic." This is illustrated in Figure 4, where two waveforms with equal "average" frequencies are shown. Note that while each period of the waveform in Figure 4b is very close to that in Figure 4a, the *phase difference* between the two can grow significantly because small phase deviations accumulate after every period.

---

[1] The quantity Ω_c is actually called the *angular frequency* and measured in rad/s, but for the sake of brevity we simply call it the *frequency*.

2

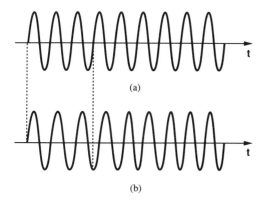

Fig. 4   (a) Periodic and (b) almost-periodic waveforms.

Let us consider examples of $\phi_n(t)$ in Eq. (1) that lead to the behavior depicted in Figure 4b. First, suppose $\phi_n(t) = \phi_m \sin \omega_m t$, where $\phi_m \ll 1$ radian. Then, (1) can be simplified as

$$x_1(t) = A \cos \omega_c t \cos(\phi_m \sin \omega_m t) - A \sin \omega_c t \sin(\phi_m \sin \omega_m t) \tag{2}$$

$$\approx A \cos \omega_c t - (A \sin \omega_c t)(\phi_m \sin \omega_m t) \tag{3}$$

$$= A \cos \omega_c t + \frac{A\phi_m}{2}[\cos(\omega_c + \omega_m)t - \cos(\omega_c - \omega_m)t] \tag{4}$$

Thus, the waveform has a strong component at $\omega = \omega_c$ and two small "sidebands" at $\omega = \omega_c \pm \omega_m$ (Fig. 5a). Second, suppose $\phi_n(t)$ is a stationary Gaussian random noise with a "low-pass" power spectral density:

$$P_\phi(\omega) = \frac{1}{1 + (\omega/\omega_0)^2} \tag{5}$$

Then, in a similar fashion, if $|\phi_n| \ll 1$, Eq. (1) can be expressed as

$$x_2(t) \approx A \cos \omega_c t + A\phi_n(t) \sin \omega_c(t) \tag{6}$$

The resulting spectrum exhibits noise "skirts" around the center frequency (Fig. 5b).

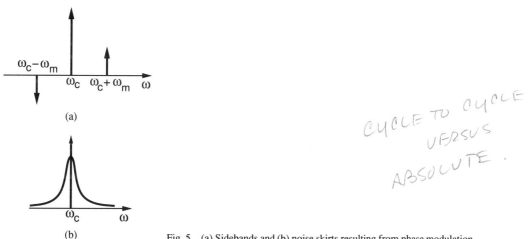

Fig. 5   (a) Sidebands and (b) noise skirts resulting from phase modulation.

From the above discussion, we can define a number of important parameters. "Cycle-to-cycle" jitter is the difference between every two consecutive periods of an almost-periodic waveform, and "absolute" jitter is the phase difference between the same waveform and a periodic signal having the same average frequency. If random, jitter is usually specified in terms of its root mean square (rms) and peak-to-peak values.

Frequency-domain counterparts of jitter are sidebands and "phase noise." As illustrated in Figure 5a, sidebands are deterministic components that do not have a harmonic relationship with the main component ω_c (also called the "carrier"); in most cases of interest, they are quite close to ω_c, i.e., $\omega_m \ll \omega_c$. Sidebands are specified with their frequency and their magnitude relative to that of the carrier.

In contrast to sidebands, phase noise arises from random frequency components (Fig. 5b). To quantify phase noise, we consider a unit bandwidth at a frequency offset $\Delta\omega$ with respect to ω_c, calculate the total noise power in this bandwidth, and divide the result by the power of the carrier (Fig. 6). Phase noise is expressed in terms of dBc/Hz, the letter "c" indicating the normalization of the noise power to the carrier power, and the unit Hz signifying the unity bandwidth used for the noise power.

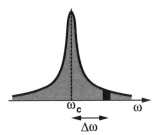

Fig. 6   Phase noise measurement.

It is interesting to note that although the phase modulation of $x_1(t)$ is deterministic and that of $x_2(t)$ random, their time-domain jitter may appear to be the same. In other words, if uniformity of zero crossings is critical, sidebands are as undesirable as random noise.

3.2 Voltage-Controlled Oscillator

An ideal voltage-controlled oscillator (VCO) generates a periodic output whose frequency is a linear function of a control voltage v_{cont}:

$$\omega_{out} = \omega_{FR} + K_{VCO} V_{cont} \tag{7}$$

where ω_{FR} is the "free-running" frequency and K_{VCO} is the "gain" of the VCO (specified in rad/s/V). Since phase is the time integral of frequency, the output of a sinusoidal VCO can be expressed as

$$y(t) = A \cos\left(\omega_{FR} t + K_{VCO} \int_{-\infty}^{t} V_{cont} dt\right) \tag{8}$$

In practical VCOs, K_{VCO} exhibits some dependence on the control voltage and eventually drops to zero as $|V_{cont}|$ increases. This is explained in Section 4.3.

It is interesting to note that if $V_{cont}(t) = V_m \cos \omega_m t$, then

$$y(t) = A \cos\left(\omega_{FR} t + \frac{K_{VCO}}{\omega_m} V_m \sin \omega_m t\right) \tag{9}$$

Called the *modulation index*, the quantity K_{VCO}/ω_m decreases as the modulating frequency ω_m increases, i.e., the VCO has a natural tendency to reject high-frequency components applied at its control input.

In studying PLLs, we usually consider a VCO as a linear time-invariant system, with the control voltage as the system's input and the *excess phase* of the output signal as the system's output. Since the excess phase,

$$\phi_{out}(t) = K_{VCO} \int V_{cont} dt \tag{10}$$

the input/output transfer function is

$$\frac{\Phi_{out}(s)}{V_{cont}(s)} = \frac{K_{VCO}}{s} \tag{11}$$

Equation (10) reveals an interesting property of VCOs: to change the output *phase*, we must first change the *frequency* and let the integration take place.[2] For example, suppose for $t < t_0$, a VCO oscillates at the same frequency as a reference but with a finite phase error (Fig. 7). To reduce the error, the control voltage, V_{cont}, is stepped by $+\Delta V$ at $t = t_0$, thereby increasing the VCO frequency and allowing the output to accumulate phase faster than the reference. At $t = t_1$, when the phase error has decreased to zero, V_{cont} returns to its initial value. Now, the two signals have equal frequencies and zero phase difference. Note also that the same goal can be accomplished by *lowering* the VCO frequency during this interval.

The above observation leads to another interesting result as well: the output phase of a VCO cannot be determined only from the *present* value of the control voltage, i.e., it depends on the history of V_{cont}. For this reason, we treat the output phase of VCOs as an independent initial condition (or state variable) in the time-domain analysis of PLLs.

In some cases, it is advantageous to control the output frequency of an oscillator by a *current*. Called a *current-controlled oscillator* (CCO), such a circuit exhibits the same behavior as a VCO.

---

[2]We assume the VCO has no other input to set its phase.

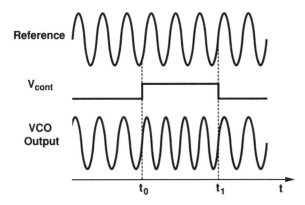

Fig. 7   Phase alignment of a VCO with a reference.

3.3 Phase Detector

An ideal phase detector (PD) produces an output signal whose dc value is linearly proportional to the difference between the phases of two periodic inputs (Fig. 8):

$$\overline{v_{\text{out}}} = K_{\text{PD}}\Delta\phi \tag{12}$$

where K_{PD} is the "gain" of the phase detector (specified in V/rad), and $\Delta\phi$ is the input phase difference. In practice, the characteristic may not be linear or even monotonic for large $\Delta\phi$. Furthermore, K_{PD} may depend on the amplitude or duty cycle of the inputs. These points are explained later.

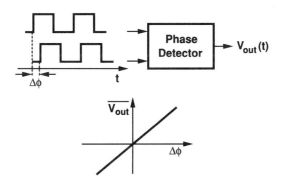

Fig. 8   Characteristic of an ideal phase detector.

Figure 9 illustrates a typical example, where the PD generates an output pulse whose width is equal to the time difference between consecutive zero crossings of the two inputs. Because the two frequencies are not equal, the phase difference exhibits a "beat" behavior with an average value of zero.

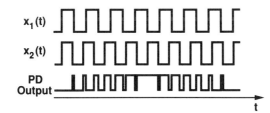

Fig. 9   Input and output waveforms of a PD.

A commonly used type of phase detector is a multiplier (also called a mixer or a sinusoidal PD). For two signals $x_1(t) = A_1 \cos\omega_1 t$ and $x_2(t) = A_2 \cos(\omega_2 t + \Delta\phi)$, a multiplier generates

$$y(t) = \alpha A_1 \cos\omega_1 t \cdot A_2 \cos(\omega_2 t + \Delta\phi) \tag{13}$$

$$= \frac{\alpha A_1 A_2}{2} \cos\left[(\omega_1 + \omega_2)t + \Delta\phi\right] + \frac{\alpha A_1 A_2}{2} \cos\left[(\omega_1 - \omega_2)t - \Delta\phi\right] \tag{14}$$

where α is a proportionality constant. Thus, for $\omega_1 = \omega_2$, the phase/voltage characteristic is given by

$$\overline{y(t)} = \frac{\alpha A_1 A_2}{2} \cos\Delta\phi \tag{15}$$

5

Plotted in Figure 10, this function exhibits a variable slope and nonmonotonicity, but it resembles that in Eq. (12) if $\Delta\phi$ is in the vicinity of $\pi/2$:

$$\overline{y(t)} \approx \frac{\alpha A_1 A_2}{2}\left(\frac{\pi}{2} - \Delta\phi\right) \tag{16}$$

yielding $K_{PD} = -\alpha A_1 A_2/2$. Note that the average output is zero if $\omega_1 \neq \omega_2$.

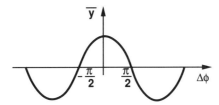

Fig. 10   Characteristic of a sinusoidal PD.

4. PHASE-LOCKED LOOP

4.1 Basic Topology

A phase-locked loop is a feedback system that operates on the *excess phase* of nominally periodic signals. This is in contrast to familiar feedback circuits where voltage and current amplitudes and their rate of change are of interest. Shown in Figure 11 is a simple PLL, consisting of a phase detector, a low-pass filter (LPF), and a VCO. The PD serves as an "error amplifier" in the feedback loop, thereby minimizing the phase difference, $\Delta\phi$, between $x(t)$ and $y(t)$. The loop is considered "locked" if $\Delta\phi$ is constant with time, a result of which is that the input and output frequencies are equal.

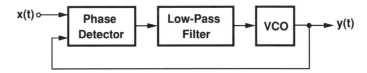

Fig. 11   Basic phase-locked loop.

In the locked condition, all the signals in the loop have reached a steady state and the PLL operates as follows. The phase detector produces an output whose dc value is proportional to $\Delta\phi$. The low-pass filter suppresses high-frequency components in the PD output, allowing the dc value to control the VCO frequency. The VCO then oscillates at a frequency equal to the input frequency and with a phase difference equal to $\Delta\phi$. Thus, the LPF generates the proper control voltage for the VCO.

It is instructive to examine the signals at various points in a PLL. Figure 12 shows a typical example. The input and output have equal frequencies but a finite phase difference, and the PD generates pulses whose widths are equal to the time difference between zero crossings of the input and output. These pulses are low-pass filtered to produce the dc voltage that sustains the VCO oscillation at the required frequency. As mentioned in Section 3.2, this voltage does not by itself determine the output *phase*. The VCO phase can be regarded as an initial condition of the system, independent of the initial conditions in the LPF.

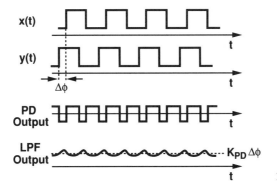

Fig. 12   Waveforms in a PLL.

Let us now study, qualitatively, the response of a PLL that is locked for $t < t_0$ and experiences a small, positive frequency step at the input at $t = t_0$ (Fig. 13). (For illustration purposes, the frequency step in this figure is only a few percent.) We note that because the input frequency, ω_{in}, is momentarily greater than the output frequency, ω_{out}, $x(t)$ accumulates phase faster than does $y(t)$ and the PD generates increasingly wider pulses. Each of these pulses creates an increasingly higher dc voltage at the output of the LPF, thereby increasing the VCO frequency. As the difference between ω_{in} and ω_{out} diminishes, the width of the phase comparison pulses decreases, eventually returning to slightly greater than its value before $t = t_0$.

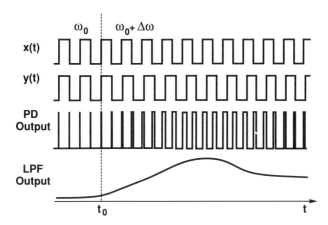

Fig. 13   Response of a PLL to a small frequency step.

The above analysis provides insight into the "tracking" capabilities of a PLL. If the input frequency changes slowly, its variation can be viewed as a succession of small narrow steps, during each of which the PLL behaves as in Figure 13.

It is important to note that in the above example the loop locks only after two conditions are satisfied: 1) ω_{out} has become equal to ω_{in}, and 2) the difference between ϕ_{in} and ϕ_{out} has settled to its proper value [1]. If the two frequencies become equal at a point in time but $\Delta\phi$ does not establish the required control voltage for the VCO, the loop must continue the transient, temporarily making the frequencies unequal again. In other words, both "frequency acquisition" and "phase acquisition" must be completed. This is, of course, to be expected because for lock to occur again, all the initial conditions of the system, including the VCO output phase, must be updated.

If the input to a PLL has a constant excess phase, i.e., is strictly periodic, but the input/output phase error, $\Delta\phi$, varies with time, we say the loop is "unlocked," an undesirable state because the output does not track the input or the relationship between the input and output is too complex to be useful. For example, if ω_{in} is sufficiently far from the VCO's free-running frequency, the loop may never lock. While the behavior of a PLL in the unlocked state is not important per se, whether and how it enters the locked state are both critical issues. Acquisition of lock is explained in Section 4.4.

Before studying PLLs in more detail, we make three important observations. First, because a PLL is a system with "memory," its output requires a finite time to respond to a change at its input, mandating a good understanding of the loop dynamics. Second, in a PLL, unlike many other feedback systems, the variable of interest changes dimension around the loop: it is converted from phase to voltage (or current) by the phase detector, processed by the LPF as such, and converted back to phase by the VCO. Third, in the lock condition, the input and output frequencies are *exactly* equal, regardless of the magnitude of the loop gain (although the phase error may not be zero). This is an extremely important property because many applications are intolerant of even small (systematic) differences between the input and output frequencies. Note that if the phase detector is replaced with only a frequency detector, this property vanishes.

While a PLL operates on phase, in many cases the parameter of interest is frequency. For example, we often need to know the response of the loop if 1) the input frequency is varied slowly, 2) the input frequency is varied rapidly, or 3) the input and output frequencies are not equal when the PLL is turned on. Therefore, the phase detector characteristic for unequal input frequencies plays an important role in the behavior of a phase-locked loop.

4.2 Loop Dynamics in Locked State

Transient response of phase-locked loops is generally a nonlinear process that cannot be formulated easily. Nevertheless, as with other feedback systems, a linear approximation can be used to gain intuition and understand trade-offs in PLL design.

Figure 14 shows a linear model of the PLL in lock along with the transfer function of each block. The model is to provide the overall transfer function for the phase, $\Phi_{out}(s)/\Phi_{in}(s)$; hence, the PD is represented by a subtractor. The LPF is assumed to have a voltage transfer function $G_{LPF}(s)$. The open-loop transfer function of the PLL is therefore equal to

$$H_O(s) = K_{PD}G_{LPF}(s)\frac{K_{VCO}}{s} \qquad (17)$$

yielding the following closed-loop transfer function:

$$H(s) = \frac{\Phi_{\text{out}}(s)}{\Phi_{\text{in}}(s)} \tag{18}$$

$$= \frac{K_{\text{PD}}K_{\text{VCO}}G_{\text{LPF}}(s)}{s + K_{\text{PD}}K_{\text{VCO}}G_{\text{LPF}}(s)} \tag{19}$$

In its simplest form, a low-pass filter is implemented as in Figure 15, with

$$G_{\text{LPF}}(s) = \frac{1}{1 + \dfrac{s}{\omega_{\text{LPF}}}} \tag{20}$$

where $\omega_{\text{LPF}} = 1/(RC)$. Equation (19) then reduces to:

$$H(s) = \frac{K_{\text{PD}}K_{\text{VCO}}}{\dfrac{s^2}{\omega_{\text{LPF}}} + s + K_{\text{PD}}K_{\text{VCO}}} \tag{21}$$

indicating that the system is of second order, with one pole contributed by the VCO and another by the LPF. The quantity $K = K_{\text{PD}}K_{\text{VCO}}$ is called the *loop gain* and expressed in rad/s.

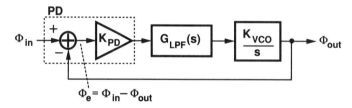

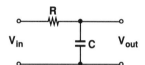

Fig. 14   Linear model of a PLL.

Fig. 15   Simple low-pass filter.

In order to understand the dynamic behavior of the PLL, we convert the denominator of Eq. (21) to the familiar form used in control theory: $s^2 + 2\zeta\omega_n s + \omega_n^2$, where ζ is the damping factor and ω_n is the natural frequency of the system.[3] Thus,

$$H(s) = \frac{\omega_n^2}{s^2 + 2\zeta\omega_n s + \omega_n^2} \tag{22}$$

where

$$\omega_n = \sqrt{\omega_{\text{LPF}}K} \tag{23}$$

$$\zeta = \frac{1}{2}\sqrt{\frac{\omega_{\text{LPF}}}{K}} \tag{24}$$

Note that ω_n is the geometric mean of the -3-dB bandwidth of the LPF and the loop gain, in a sense an indication of the gain-bandwidth product of the loop. Also, the damping factor is inversely proportional to the loop gain, an important and often undesirable trade-off that will be discussed.

In a well-designed second-order system, ζ is usually greater than 0.5 and preferably equal to $\sqrt{2}/2$ so as to provide an optimally flat frequency response. Therefore, K and ω_{LPF} cannot be chosen independently; for example, if $\zeta = \sqrt{2}/2$, then $K = \omega_{\text{LPF}}/2$. As explained in Sections 4.8 and 7.1, sideband or noise suppression issues typically impose an upper bound on ω_{LPF} and hence K. These limitations translate to significant phase error between the input and the output as well as a narrow capture range (Section 4.4).

The transfer function in Eq. (22) is that of a low-pass filter, suggesting that if the input excess phase varies slowly, then the output excess phase follows, and conversely, if the input excess phase varies rapidly, the output excess phase variation

[3]In a simple PLL, ω_n has no relation with the input and output frequencies.

will be small. In particular, if $s \to 0$, we note that $H(s) \to 1$; i.e., a *static* phase shift at the input is transferred to the output unchanged. This is because for phase quantities, the presence of integration in the VCO makes the open-loop gain approach infinity as $s \to 0$. To this end, we can examine the "phase error transfer function," defined as $H_e(s) = \Phi_e(s)/\Phi_{in}(s)$ in Figure 14:

$$H_e(s) = 1 - H(s) \tag{25}$$

$$= \frac{s^2 + 2\zeta\omega_n s}{s^2 + 2\zeta\omega_n s + \omega_n^2} \tag{26}$$

which drops to zero as $s \to 0$.

Since phase and frequency are related by a linear, time-invariant operation, the transfer functions in Eqs. (22) and (26) also apply to the input and output excess frequencies. For example, Eq. (22) indicates that if the input frequency varies rapidly, the instantaneous variation of the output frequency will be small.

It is instructive to repeat our previous analysis of the loop step response (Fig. 13) with the aid of Eq. (22). Suppose the input excess frequency is equal to $\Delta\omega u(t)$, where $u(t)$ is the unit step function. The output excess frequency then exhibits the typical step response of a second-order system, eventually settling to $\Delta\omega$ rad/s higher than its initial value (Fig. 16). The output excess *phase*, on the other hand, is given by

$$\Phi_{out}(s) = H(s)\Phi_{in}(s) \tag{27}$$

$$= \frac{\omega_n^2}{s^2 + 2\zeta\omega_n s + \omega_n^2} \frac{\Delta\omega}{s^2} \tag{28}$$

which is the response of a second-order system to a *ramp* input. More importantly, the phase error is

$$\Phi_e(s) = H_e(s)\Phi_{in}(s) \tag{29}$$

$$= \frac{s^2 + 2\zeta\omega_n s}{s^2 + 2\zeta\omega_n s + \omega_n^2} \frac{\Delta\omega}{s^2} \tag{30}$$

whose final value is given by

$$\phi_e(t = \infty) = \lim_{s \to 0} s\phi_e(s) \tag{31}$$

$$= \Delta\omega \frac{2\zeta}{\omega_n} \tag{32}$$

$$= \frac{\Delta\omega}{K} \tag{33}$$

Therefore, static changes in the input frequency are suppressed by a factor of K when they manifest themselves in the static phase error (Fig. 16). This is, of course, to be expected because for the VCO frequency to change by $\Delta\omega$, the control voltage must change by $\Delta\omega/K_{VCO}$ and the input to the PD by $\Delta\omega/(K_{VCO}K_{PD})$.

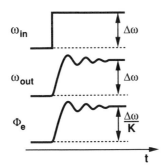

Fig. 16   Response of a PLL to a frequency step.

An important drawback of the PLL considered thus far is the direct relationship between ζ, ω_{LPF}, and K given by Eq. (24). For example, if the loop gain is increased to reduce the static phase error, then the settling behavior degrades. In order to allow independent choice of K and ω_{LPF}, a zero can be added to the low-pass filter as shown in Figure 17, modifying its transfer function to the following:

$$G_{LPF}(s) = \frac{R_2Cs + 1}{(R_1 + R_2)Cs + 1} \tag{34}$$

9

Thus, the PLL transfer function is

$$H(s) = \frac{K\left(\dfrac{s}{\omega_z} + 1\right)}{\dfrac{s^2}{\omega_p} + \left(\dfrac{K}{\omega_z} + 1\right)s + K} \tag{35}$$

$$= \frac{K\omega_p\left(\dfrac{s}{\omega_z} + 1\right)}{s^2 + \omega_p\left(\dfrac{K}{\omega_z} + 1\right)s + K\omega_p} \tag{36}$$

where $\omega_z = 1/(R_2 C)$ and $\omega_p = 1/[(R_1 + R_2)C]$. The damping factor is then equal to:

$$\zeta = \frac{1}{2}\sqrt{\frac{\omega_p}{K}}\left(\frac{K}{\omega_z} + 1\right) \tag{37}$$

subject to the constraint $\omega_z > \omega_p$. As an example, if $\zeta = \sqrt{2}/2$, then with $\omega_z = \infty$, the open-loop gain is $K = \omega_p/2$, but with $\omega_z \approx 4.57\omega_p$, $K = 32\omega_p$.

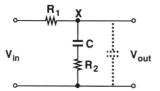

Fig. 17   LPF with a zero.

Note, however, that adding a zero to increase K has two side effects: 1) the -3-dB bandwidth of the system (roughly equal to $\omega_n = \sqrt{K\omega_p}$ for $\zeta = \sqrt{2}/2$) also increases, a trend that is desirable in some applications and undesirable in others; 2) the LPF attenuation of high-frequency signals is only $R_2/(R_1 + R_2)$, usually a troublesome drawback. To alleviate the latter effect, a second capacitor can be connected from node X in Figure 17 to ground so as to provide another pole beyond the zero (with some penalty in the settling time). (Because of the additional pole and zero, the optimum value of ζ may no longer be $\sqrt{2}/2$.)

In our discussion of PLLs thus far, we have assumed the feedback in the loop is negative. Interestingly, if the phase detector is realized with a multiplier, the polarity of feedback is unimportant. This is because the sinusoidal characteristic of the PD provides both negative and positive gains, allowing the loop to find a stable operating point by varying $\Delta\phi$. As illustrated in Figure 18, if $\Delta\phi$ begins around $+\pi/2$ and the feedback happens to be positive, the VCO frequency changes, driving $\Delta\phi$ toward $+3\pi/2$ or $-\pi/2$, where the PD gain has reverse polarity and the feedback is negative.

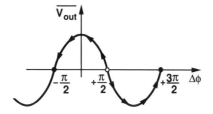

Fig. 18   Phase variation to provide negative feedback.

4.3 Tracking Behavior

The example of Figure 16 illustrates how a PLL can track the input frequency, indicating that in lock the input and output frequencies are equal but the phase error may not be zero. The natural question at this point is: How far can the PLL track the input frequency, i.e., what determines the "tracking range"[4] of the PLL? To answer this question, we consider two extreme cases: 1) the input frequency varies slowly (static tracking), and 2) the input frequency is changed abruptly (dynamic tracking). We will see that the tracking behavior is distinctly different in the two cases.

Suppose, starting from the VCO free-running frequency, the input frequency varies slowly such that the difference between ω_{in} and ω_{out} always remains much less than ω_{LPF} (or ω_p if the LPF contains a zero). Then, to allow tracking, the magnitude of the VCO control voltage, and hence the static phase error, must increase (Fig. 19). The PLL tracks as long as

---

[4]Also called the *lock range*.

the three parameters plotted in Figure 19 vary monotonically. In other words, the edge of the tracking range is reached at the point where the slope of one of the characteristics falls to zero or changes sign. This can occur only in the PD or the VCO (provided the LPF components are linear). Depicted in Figure 20 are examples of such behavior. The VCO frequency typically has a limited range, out of which its gain drops sharply. Also, in a typical phase detector, the characteristic becomes nonmonotonic for a sufficiently large input phase difference, at which point the PLL fails to maintain lock.

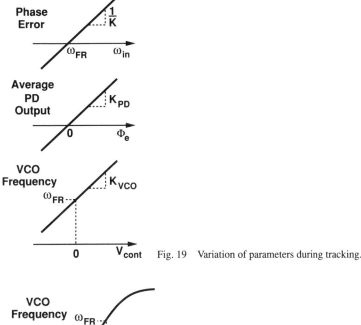

Fig. 19   Variation of parameters during tracking.

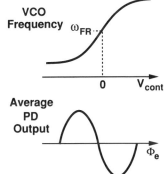

Fig. 20   Gain reduction in PD and VCO.

For a multiplier-type PD, as explained in Section 3.3, the gain, K_{PD}, changes sign if the input phase difference deviates from its center value by more than 90° (Fig. 10). Thus, the VCO output frequency can deviate from its free-running value by no more than

$$\Delta\omega_{tr} = K_{PD} \left(\sin \frac{\pi}{2} \right) K_{VCO} \qquad (38)$$

Therefore, the static tracking range of a PLL employing a sinusoidal PD is the smaller of K and half of the VCO output frequency range.

Now let us study the tracking behavior of PLLs when the input frequency is changed abruptly. Suppose the input frequency of a PLL that is initially operating at $\omega_{in} = \omega_{out} = \omega_{FR}$ is stepped by $\Delta\omega$. What is the maximum $\Delta\omega$ for which the loop locks again? Can $\Delta\omega$ be as large as $\Delta\omega_{tr}$ in the case of static tracking [Eq. (38)]?

To answer these questions, we first make an important observation. Strictly speaking, we note that for *any* input frequency step at its input, a PLL loses lock, at least momentarily. This is evident from the simplified analysis in Figure 13, where the loop requires a number of cycles to stabilize. During these cycles, the input-output phase difference varies, and the PLL can be considered unlocked. (Nevertheless, for small $\Delta\omega$, the loop locks quickly, and the transient can be viewed as one of tracking rather than locking.)

The key point resulting from the above observation is that the following two situations are similar: 1) a loop initially locked at ω_{FR} experiences a large input frequency step, $\Delta\omega$; and 2) a loop initially unlocked and free running ($\omega_{out} = \omega_{FR}$) must lock onto an input frequency given by $|\omega_{in} - \omega_{FR}| = \Delta\omega$. In both cases, the loop must *acquire* lock.

11

4.4 Acquisition of Lock

The second case mentioned above occurs, for example, when a PLL is turned on. If the initial conditions in the LPF are zero, the VCO begins to oscillate at ω_{FR}, whereas the input is at a different frequency, $\omega_{FR} + \Delta\omega$. The "acquisition range" (also called the *capture range*) is the maximum value of $\Delta\omega$ for which the loop locks. To understand how a PLL acquires lock, we study the response from two different perspectives, namely, in the frequency and time domains. For simplicity, we make the following assumptions: a) the PD is implemented with a multiplier; b) ω_{in} is within the VCO frequency range; c) the sum component at the output of the mixer [the first term in Eq. (14)] is attenuated by the LPF to negligible levels; d) the VCO output frequency increases as its control voltage becomes more positive.

Consider the PLL shown in Figure 21, where initially $\omega_{in} = \omega_{FR} + \Delta\omega$ and $\omega_{out} = \omega_{FR}$. We begin with the top two spectra and follow the signals around the loop. At first glance, it may seem that, because $\omega_{in} \neq \omega_{out}$, the average output of the PD is zero and the loop cannot be driven toward lock. The important point, however, is that the LPF does not completely suppress the component at $\omega_{in} - \omega_{out} = \Delta\omega$. Thus, the VCO control voltage, V_A, varies at a rate equal to $\Delta\omega$, thereby modulating the output frequency:

$$v_{out}(t) = A \cos\left[\omega_{FR}t + K_{VCO} \int A_m \cos(\Delta\omega t)dt\right] \tag{39}$$

$$= A \cos\left[\omega_{FR}t + \frac{K_{VCO}}{\Delta\omega} A_m \sin(\Delta\omega t)\right] \tag{40}$$

$$\approx A \cos\omega_{FR}t - \frac{K_{VCO}}{\Delta\omega} A_m \sin\omega_{FR}t \sin(\Delta\omega t) \tag{41}$$

where we have assumed $K_{VCO}A_m/\Delta\omega \ll 1$. As a result, the VCO output, V_B, exhibits sidebands at $\omega_{FR} \pm \Delta\omega$ in addition to the main component at ω_{FR}. When the PD multiplies the sideband at $\omega_{FR} + \Delta\omega$ by ω_{in}, a dc component appears at node A (Fig. 21), adjusting the VCO frequency toward lock [1]. The dc component may need to grow over a number of beat cycles before lock is achieved.

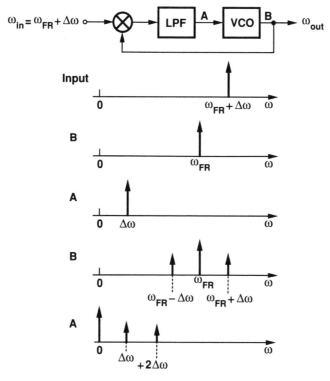

Fig. 21 Acquisition behavior in frequency domain.

From the above example, we note that the acquisition range depends on how much the LPF passes the component at $\Delta\omega$ and how strong the feedback dc component is, i.e., the acquisition range is a direct function of the loop gain at $\Delta\omega$. In other words, because the loop gain of a simple PLL drops as the difference between the input frequency and the VCO frequency increases, the acquisition range cannot be arbitrarily wide.

A second approach to analyzing the acquisition behavior is in the time domain. First, suppose the loop is opened at the VCO output and the feedback signal is replaced by a source oscillating at ω_{FR} (Fig. 22a) [2]. The output of the LPF is then a sinusoid at $\omega_{in} - \omega_{FR}$. As the sinusoid instantaneous amplitude increases, so does the VCO frequency and vice versa. Consequently, the difference between ω_{in} and ω_{out} reaches a maximum when the sinusoid is at a positive peak and a minimum when the sinusoid is at a negative peak. (This is simply the modulation phenomenon of Fig. 21 described in the time domain.) Now, if the loop is closed, the feedback signal has a *time-varying* frequency. When the LPF output goes through a positive excursion, ω_{out} approaches ω_{in} and the beat period increases. Conversely, when the LPF output becomes negative, ω_{out} moves away from ω_{in} and the beat period decreases. Shown in Fig. 22b, the resulting waveform at the LPF output exhibits longer positive cycles than negative ones, thus carrying a positive dc component and gradually shifting the average value of ω_{out} closer to ω_{in}.

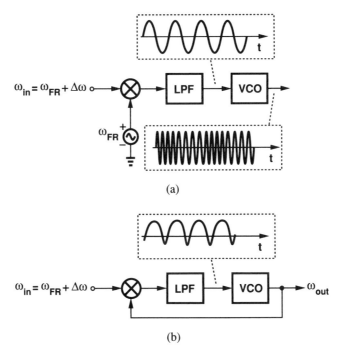

(a)

(b)

Fig. 22   Acquisition behavior in time domain.

The above time-domain analysis reveals two important points. First, if ω_{in} is sufficiently close to ω_{FR}, frequency acquisition is achieved at the first proper peak of the beat waveform (Fig. 23a). In this case, we say the PLL has locked with no "cycle slips." Second, if ω_{in} is sufficiently far from ω_{FR}, the beat waveform has little asymmetry and hence not enough dc voltage to drive the loop toward lock (Fig. 23b).

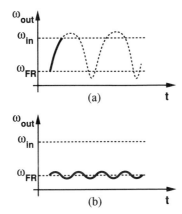

Fig. 23   (a) Capture with no cycle clips; (b) capture failure.

Figure 24 illustrates the simulated acquisition behavior of a 1-GHz PLL. Plotted here is the control voltage of the VCO as a function of time, exhibiting several cycle slips before the loop enters small-signal settling. While it is difficult to see in this figure, the peak of the beat cycles gradually becomes more positive and the period of each cycle slightly increases as the average value of ω_{out} comes closer to ω_{in}. Note, however, that cycle slips are observed only if ω_{in} is very close to the edge of the acquisition range.[5] Also, the number of cycle slips depends on the loop's initial conditions, i.e., those in the LPF as well as the initial phase of the VCO.

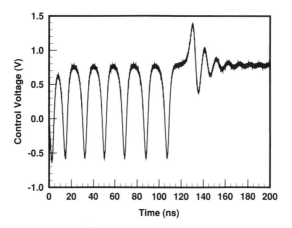

Fig. 24   Simulated capture behavior.

Acquisition range is a critical parameter because 1) it trades directly with the loop bandwidth; i.e., if an application requires a small loop bandwidth, the acquisition range will be proportionally small; 2) it determines the maximum frequency variation in the input or the VCO that can be accommodated. In monolithic implementations, the VCO free-running frequency can vary substantially with temperature and process, thereby requiring a wide acquisition range even if the input frequency is tightly controlled.

Unfortunately, it is difficult to calculate the acquisition range of PLLs analytically. To gain a better feeling about the limits, we consider a simplified case where the LPF output signal can be approximated as [2]

$$v_{\text{LPF}}(t) = K_{\text{PD}}|G_{\text{LPF}}(j\,\Delta\omega)|\sin(\Delta\omega t) \tag{42}$$

This signal modulates the VCO frequency, causing a maximum *deviation* of

$$(\omega_{\text{out}} - \omega_{\text{FR}})|_{\text{max}} = K_{\text{PD}}K_{\text{VCO}}|G_{\text{LPF}}(j\,\Delta\omega)| \tag{43}$$

As shown in Figure 23, if this deviation is equal to or greater than $\Delta\omega$, then the loop locks without cycle slips [3]:

$$\Delta\omega_{\text{acq}} = K_{\text{PD}}K_{\text{VCO}}|G_{\text{LPF}}(j\,\Delta\omega)| \tag{44}$$

With $G_{\text{LPF}}(s)$ known, $\Delta\omega$ can be calculated from this equation. For a simple low-pass filter:

$$\Delta\omega_{\text{acq}} = \left[\frac{\omega_{\text{LPF}}^2}{2} \left(-1 + \sqrt{1 + \frac{1}{4\zeta^4}} \right) \right]^{1/2} \tag{45}$$

which reduces to $\Delta\omega_{\text{acq}} \approx 0.46\omega_{\text{LPF}}$ if $\zeta = \sqrt{2}/2$.

The above derivation actually underestimates the capture range. This is because as ω_{in} is brought closer to ω_{FR}, the "average" frequency of the VCO also departs from ω_{FR} and comes closer to ω_{in} [4]. Thus, in the vicinity of lock, the difference between ω_{in} and ω_{VCO} is small and the LPF attenuation predicted by Eq. (42) too large. A more accurate expression is given in [5].

Most modern phase-locked systems incorporate additional means of frequency acquisition to significantly increase the capture range, often removing its dependence on K and ω_{LPF} and achieving limits equal to those of the VCO. This is discussed below.

4.5 Acquisition Time

The acquisition and settling times of PLLs are important in many applications. For example, if a PLL is used at the clock interface of a microprocessor (Fig. 2) and the system is powered down frequently to save energy, it becomes critical to know how long the system must remain idle after it is turned on to allow adequate phase alignment between the external and internal clocks. As another example, when a frequency synthesizer used in a wireless transceiver is switched to change its output frequency, the resulting loop transient causes "frequency spreading," in effect leaking noise into other channels. Furthermore, because channel spacing in such an environment is typically several orders of magnitude smaller than the VCO frequency range, precise settling of the loop (e.g., a few parts per million) is required.

---

[5]Or if the input is heavily corrupted by noise [1].

14

For a simple second-order system with $\zeta < 1$, the step response is expressed as

$$y(t) = \left[1 + \frac{1}{\sqrt{1 - \zeta^2}} \exp(-\zeta \omega_n t) \times \sin\left(\omega_n \sqrt{1 - \zeta^2} t - \psi \right) \right] u(t) \tag{46}$$

where $\psi = \sin^{-1} \sqrt{1 - \zeta^2}$. Thus, the decay time constant is

$$\tau_{\text{dec}} = \frac{1}{\zeta \omega_n} \tag{47}$$

$$= \frac{2}{\omega_{\text{LPF}}} \tag{48}$$

and the frequency of ringing equals $\omega_n \sqrt{1 - \zeta^2}$. For a frequency step at the PLL input, Eq. (46) can be used to calculate the time required for the output frequency to settle within a given error band around its final value.

Note that Eq. (46) assumes a linear system. In practice, nonlinearities in K_{PD} and K_{VCO} result in somewhat different settling characteristics, and simulations must be used to predict the lock time accurately. Nonetheless, this equation provides an initial guess that proves useful in early phases of the design.

4.6 Aided Acquisition

The acquisition behavior in Figure 23 and formulated by Eq. (45) indicates that the capture range of a simple, optimally stable phase-locked loop is roughly equal to $0.5\omega_{\text{LPF}}$, regardless of the magnitude of K. Since issues such as jitter and sideband suppression impose an upper bound on ω_{LPF}, the resulting capture range is often inadequate. Therefore, most practical PLLs employ additional techniques to aid the acquisition of frequency.

Shown in Figure 25 is a conceptual diagram of a PLL with aided frequency acquisition. Here, the system utilizes a frequency detector (FD) and a second low-pass filter, LPF$_2$, whose output is added to that of LPF$_1$. The FD produces an output having a dc value proportional to and with the same polarity as $\omega_{\text{in}} - \omega_{\text{out}}$. If the difference between ω_{in} and ω_{out} is large, the PD output has a negligible dc component and the VCO is driven by the dc output of the FD with negative feedback, thereby moving ω_{out} toward ω_{in}. As $|\omega_{\text{in}} - \omega_{\text{out}}|$ drops, the dc output of the FD decreases, whereas that of the PD increases. Thus, the frequency detection loop gradually relinquishes the acquisition to the phase-locked loop, becoming inactive when $\omega_{\text{in}} - \omega_{\text{out}} = 0$.

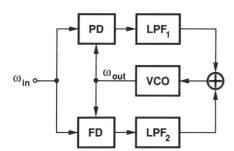

Fig. 25   Aided acquisition with a frequency detector.

It is important to note that in a frequency detection loop, the loop gain is relatively constant, independent of $|\omega_{\text{in}} - \omega_{\text{out}}|$, whereas in a simple phase-locked loop it drops if $|\omega_{\text{in}} - \omega_{\text{out}}|$ exceeds ω_{LPF}. For this reason, aided acquisition using FDs can substantially increase the capture range.

The configuration of Figure 25 is greatly simplified if a single circuit can perform both frequency and phase detection. This is discussed in Section 5.2.

4.7 Higher Order Loops

The generic phase-locked loop considered thus far is of second order. In principle, the low-pass filter can include more poles to achieve sharper cut-off characteristics, a desirable property in many applications. However, such systems are difficult to stabilize, especially when process and temperature variations are taken into account. On the other hand, in many cases the PLL inevitably has a third pole, for example, if a capacitor is connected in parallel with the LPF output port (Fig. 17) to suppress high frequencies. Thus, most practical PLLs can be considered as third-order topologies with the third pole being much farther from the origin than the other two.

4.8 Frequency Multiplication

Phase-locked loops are often used in applications where the output frequency must be a multiple of the input frequency. The design problem illustrated in Figure 2, for example, sets an upper bound on the external clock frequency because of signal distribution issues on PC boards, but it also requires higher internal clock frequencies for the processor.

A PLL can "amplify" a frequency in the same fashion as does a feedback amplifier. As shown in Figure 26a, to amplify the input, the output signal is divided down before it is fed back. Since the output quantity of interest in a PLL is the frequency, a frequency divider (e.g., a digital counter) must be inserted in the feedback loop (Fig. 26b). From another perspective, when the loop is locked, $\omega_F = \omega_{in}$, and hence $\omega_{out} = M\omega_{in}$.[6]

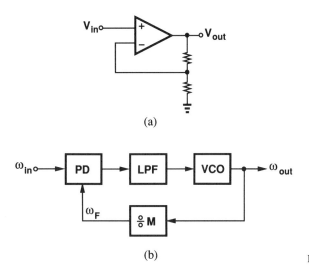

(a)

(b)

Fig. 26   Signal "amplification" in (a) a feedback amplifier; (b) a PLL.

The analogy depicted in Figure 26 also proves useful in studying the effect of the $\div M$ circuit upon the PLL behavior. As with the feedback amplifier, the loop gain is divided by M and hence the results of all of the previous static and dynamic analyses can be directly applied if K is replaced by K/M. Before examining the consequences of this change, we need to make an important observation.

Consider the system shown in Figure 26b with the phase detector implemented as a multiplier. When the loop is locked, the PD output consists of two components: one at $\omega_{in} - \omega_{out}/M = 0$ and another at $\omega_{in} + \omega_{out}/M = 2\omega_{in}$. Since the LPF has a finite stopband attenuation, the VCO control voltage contains a frequency component at $2\omega_{in}$, thereby modulating the output frequency and creating sidebands at $M\omega_{in} \pm 2\omega_{in}$. As mentioned in Section 3.1, the effect of these sidebands in the time domain can be viewed as jitter for most practical purposes. Furthermore, in applications such as wireless transceivers, the sidebands must be several orders of magnitude smaller than the carrier. Thus, ω_{LPF} must be chosen so as to sufficiently attenuate the component at $2\omega_{in}$. In other words, for a given ω_{out}, higher division ratios translate into lower ω_{LPF}. For simplicity, we assume the LPF cut-off frequency scales inversely with M.

With a divider in the loop and a fixed ω_{out}, the damping factor is:

$$\zeta = \frac{1}{2}\sqrt{\frac{\omega_{LPF}/M}{K/M}} \tag{49}$$

$$= \frac{1}{2}\sqrt{\frac{\omega_{LPF}}{K}} \tag{50}$$

and the natural frequency:

$$\omega_n = \sqrt{\frac{K}{M}\frac{\omega_{LPF}}{M}} \tag{51}$$

$$= \frac{1}{M}\sqrt{K\omega_{LPF}} \tag{52}$$

It follows from Eq. (52) that the settling is slowed down by a factor M.

It is interesting to note that frequency multiplication in Figure 26 also amplifies the input jitter. For example, jitter frequencies below the -3-dB bandwidth of the PLL are amplified by a factor M.

4.9 Delay-Locked Loops

A close relative of phase-locked loops is the delay-locked loop (DLL) [6, 7]. Shown in Figure 27a, a DLL replaces the VCO of a PLL with a voltage-controlled delay line (VCDL). The idea is that if a periodic input is delayed by an integer

---

[6]While the feedback amplifier suffers from gain error (due to the finite open-loop gain of the op amp), the PLL exhibits no such behavior. We leave the explanation as an exercise for the reader.

multiple of the period, T_{in}, then its phase shift can be considered zero. Thus, the phase detector drives the loop so that the phase difference between V_{in} and V_{out} approaches nT_{in}, where n is an integer (in most cases equal to unity). Note that the polarity of feedback must be negative.

The VCDL usually consists of a cascade of k identical gain stages with variable delay, as shown in simple form in Figure 27b. (Most of the VCO design issues described in Section 5.1 apply to VCDLs as well.) Note that, unlike oscillators, delay lines do not *generate* a signal, making it difficult to perform frequency multiplication in a DLL.

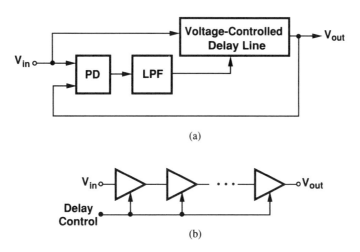

(a)

(b)

Fig. 27   (a) DLL block diagram; (b) delay line.

In addition to phase alignment, DLLs can provide precisely spaced timing edges even in the presence of temperature and process variations. To understand this, note that in Figure 27b, the delay between V_{in} and V_{out} is equal to nT_{in} when lock is achieved. Therefore, output signals of consecutive stages exhibit a phase difference of nT_{in}/k, a value independent of device parameters. In practice, mismatches between stages limit the edge delay accuracy.

DLLs have two important advantages over PLLs. First, because the delay line has no "memory," its transfer function is a constant, thereby yielding a *first-order* open-loop transfer function for the entire system (for a first-order LPF). Consequently, DLLs have much more relaxed trade-offs among gain, bandwidth, and stability. Second, delay lines typically introduce much less jitter than oscillators. Intuitively, this is because delaying a signal entails much less uncertainty than *generating* it. From another point of view, noise injected into a DLL disappears at the end of the delay line, whereas it is recirculated in an oscillator [8].

As DLLs can lock with a total delay of nT_{in}, some means must be provided to obtain the desired n. For $n = 1$, the maximum delay of the VCDL must remain less than $2T_{\text{in}}$ [7].

5. BUILDING BLOCKS OF PLLS

While the generic PLL architectures of Figures 11 and 25 have been extensively used with little modification, their building blocks have been implemented in many different forms. Innovations in the design of these building blocks have tremendously improved the speed, power dissipation, jitter, and capture range of phase-locked systems. In this section, we describe the design of voltage-controlled oscillators, phase and frequency detectors, and charge pumps.

5.1 Voltage-Controlled Oscillators

Perhaps the most critical part of PLLs and CRCs, oscillators have been the subject of numerous studies for more than half a century [9, 10], still defying an exact analysis. Most of these analyses have considered only "nearly sinusoidal" oscillations in conventional topologies (such as LC-based circuits), conditions difficult to create in monolithic circuits. For this reason, before our knowledge in this area advances sufficiently, we must rely on simulations to predict various parameters of oscillators (as far as simulations can go).

For a VCO that is to be used in a PLL, the following parameters are important. 1) Tuning range: i.e., the range between the minimum and maximum values of the VCO frequency. In this range, the *variation* of the output amplitude and jitter must be minimal. The tuning range must accommodate the PLL input frequency range as well as process- and temperature-induced variations in the VCO frequency range. The tuning range is typically at least $\pm 20\% \omega_{\text{FR}}$. 2) Jitter and phase noise: timing accuracy and spectral purity requirements in PLL applications impose an upper bound on the VCO jitter and phase noise. 3) Supply and substrate noise rejection: if integrated along with digital circuits, VCOs must be highly immune to supply and substrate noise. In the architecture of Fig. 26b, for example, the frequency divider can corrupt the

VCO output by injecting noise into the common substrate. Such effects become more prominent if a PLL shares the same substrate and package with a large digital processor. 4) Input/output characteristic linearity: variation of K_{VCO} across the tuning range is generally undesirable. If a PLL is used as an FM demodulator, the variation of K_{VCO} introduces harmonic distortion in the detected signal and must be below 1%. In other applications, this nonlinearity degrades the loop stability but it can be as high as several tens of percent.

Before describing VCO topologies, we must explain an extremely important point: in order to achieve high rejection of supply and substrate noise, both the signal path *and* the control path of a VCO must be fully differential. We also note that an oscillator in which signals exist in complementary form but have rail-to-rail swings is not considered differential because it exhibits poor supply rejection. Differential operation also yields a 50% duty cycle, an important requirement in timing applications, and is immune to the up-conversion of low-frequency noise components in the signal path [11].

A common oscillator topology in monolithic PLLs is the ring oscillator, shown in Figure 28. Here, a cascade of M gain stages with a total (dc) phase shift of 180° is placed in a feedback loop. It can be easily shown that the loop oscillates with a period equal to $2MT_d$, where T_d is the delay of each stage with a fanout of one. The oscillation can also be viewed as occurring at the frequency for which the total phase shift is zero and the loop gain is unity. Since in a typical IC technology the gate delay is monitored and controlled within the process corners, the oscillator frequency and its variation can be predicted with reasonable accuracy.

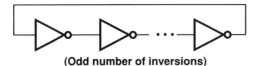

(Odd number of inversions)   Fig. 28   Ring oscillator.

The gain stages in a ring oscillator can be implemented in various forms, some of which are shown in Figure 29. Note that with the differential pairs of Figures 29b and c, the number of stages in the ring need not be odd; the total phase shift can be changed by 180° if the output signals of one of the stages are swapped. With an even number of stages, the oscillator can provide quadrature outputs, i.e., outputs that are 90° out of phase.

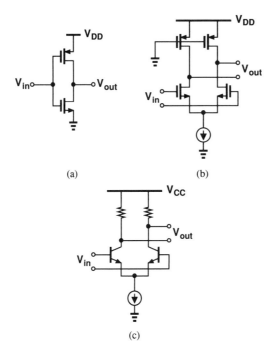

Fig. 29   Simple gain stages.

In order to vary the frequency of oscillation, one of the parameters in $2MT_d$ must change, i.e., the effective number of stages or the delay of each stage. The resulting techniques are conceptually illustrated in Figure 30. In the first approach, called "delay interpolation," a fast path and a slow path are used in parallel [12, 13]. The total delay is adjusted by increasing the gain of one path and decreasing that of the other, and hence is a weighted sum of the delays of the two paths. In the second approach, the delay of each stage in the ring is directly varied with negligible change in the gain or voltage swings.

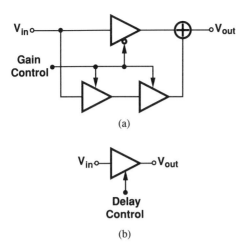

(a)

(b)

Fig. 30   Frequency variation techniques.

A simple implementation of delay interpolation is shown in Figure 31. The pair Q_1-Q_2 constitutes the fast path and the pairs Q_3-Q_4 and Q_5-Q_6, the slow path. The control input, V_{cont}, determines the gain of each path by steering I_{EE} from one to the other. The same topology can be used with CMOS devices.

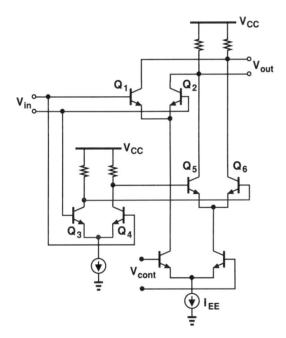

Fig. 31   Delay interpolation in a bipolar VCO.

In Figure 30b, the delay of each stage is tuned by the control input. This can be accomplished by varying the capacitance or the resistance seen at the output node of each stage. We first consider "capacitive tuning" to explain its drawbacks and lead to "resistive tuning" as the superior technique.

In order to vary the effective capacitance seen at a node, one of the two methods depicted in Figure 32 can be used. In Figure 32a, a voltage-dependent capacitor, for example, a reverse-biased pn junction diode, loads node X, and its value is adjusted by V_{cont}. The drawback is that the minimum value of C_1 (usually determined by the maximum range of V_{cont}) still loads the circuit, limiting the maximum frequency of operation. In Figure 32b, C_1 is constant but a MOS device, M_1, operates as a voltage-dependent resistor, thereby varying the "effective" capacitance seen at node X. The difficulty here is that K_{VCO} can experience substantial variation, especially if a wide tuning range is required. Additionally, both techniques use a single-ended control and are therefore susceptible to common-mode noise.

In contrast to capacitive tuning, resistive tuning provides a large, relatively uniform frequency variation and lends itself to differential control. Nonetheless, in addition to the effective load resistance, some other parameters of each stage must also be varied so that the voltage swings and/or the (dc) voltage gain remain relatively constant. To understand this issue, consider the gain stages shown in Figure 33. In Figure 33a, the load devices are biased in the triode region and their on-resistance is adjusted by V_{cont}. As V_{cont} decreases, the delay of the stage drops because the time constant at the output

19

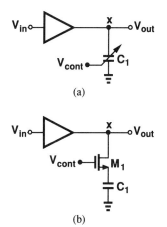

(a)

(b)

Fig. 32   Capacitive tuning.

nodes decreases. Note that the small-signal gain also decreases. The variable voltage swings present a non-optimal solution: when they are small, the signals are more corrupted by jitter, and when they are large, they require a higher supply voltage so as to preserve differential operation. Moreover, as the gain of each stage drops, the circuit eventually fails to oscillate because the total gain around the ring *at the frequency of oscillation* falls below unity. This is particularly problematic in CMOS oscillators if they employ a small number of stages and a low gain in each stage to achieve a high speed.

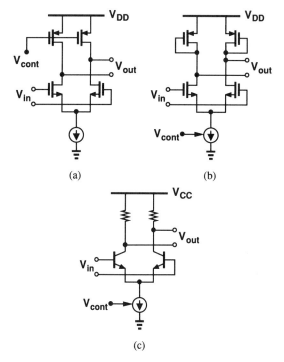

(a)   (b)

(c)

Fig. 33   Resistive tuning.

In Figure 33b, on the other hand, as V_{cont} adjusts the tail current, the small-signal impedance of the load devices varies accordingly, but the voltage gain remains relatively constant. Thus, the circuit may seem to be appropriate for a VCO stage. However, the large-signal output voltage swings still depend on the current because $I_D = \mu_p C_{ox} W (V_{GS} - V_{THP})^2/(2L)$.

It is also important to understand why a configuration such as that in Figure 33c is a poor choice for a variable-delay stage. Since the time constant at the collector of the bipolar transistors does not directly depend on the tail current, the tuning range of an oscillator employing this stage is very small. (In reality, the input capacitance of each stage increases with the tail current, but the time for which each stage is on decreases. Thus, the "average" input capacitance remains relatively constant.)

Another approach to varying the delay of the stages in a ring oscillator is to vary the effective resistance using local positive feedback. Feasible in both bipolar and CMOS technologies, this method is illustrated in a CMOS circuit in Figure 34. Here, the cross-coupled pair M_5-M_6 introduces a *negative* average resistance equal to $-2/g_m$, where g_m denotes the average transconductance of M_5 and M_6. This resistance partially cancels that seen at the drain of M_3 and M_4, increasing the effective output impedance and hence the delay. The key point is that the total current flowing from M_3 and M_4, and thus the voltage swings at X and Y remain constant as V_{cont} steers I_{SS} between M_1-M_2 and M_5-M_6.

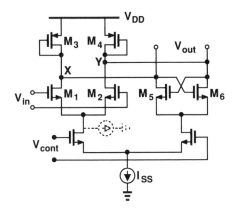

Fig. 34   Delay variation using local positive feedback.

In designing the circuit of Figure 34, two issues must be borne in mind. First, to avoid latch-up, the transconductance of M_5-M_6 must be less than that of M_3-M_4, a condition met if the latter devices are wider than the former and all four have equal lengths. Second, to ensure steady oscillations, the input pair, M_1-M_2, must have an additional constant bias current so that the stage has adequate gain even if a PLL transient steers all of I_{SS} to M_5-M_6.

While providing at least a two-to-one frequency range, this topology cannot operate from a 3-V supply if implemented in standard CMOS technology. This is due to the large gate-source voltages (including the body effect) of the stacked transistors. Figure 35 shows how two modifications can reduce the minimum supply voltage to approximately 2.5 V. First, the diode-connected PMOS loads are replaced with a composite PMOS/NMOS structure that exhibits approximately the same impedance but consumes less voltage headroom owing to the level shift provided by the NMOS source follower. Second, the control path is implemented as a folded PMOS stage to avoid stacking.

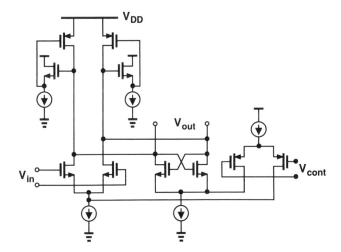

Fig. 35   Low-voltage variable-delay gain stage.

In bipolar technology, in addition to delay interpolation and local positive feedback, other techniques can be used to achieve a wide tuning range [14, 15].

An important issue in ring oscillator design is the minimum number of stages that can be used while attaining reliable operation. Since oscillation occurs at a frequency for which the total phase shift is zero and the loop gain is unity, as the number of stages decreases, the required phase shift and (dc) gain per stage increase. For example, for a three-stage oscillator, each stage must introduce a phase shift of 120° and a minimum dc gain of 2 [11]. While two-stage bipolar oscillators can be designed to achieve both sufficient phase shift and high speed [14, 15], simulations show that CMOS implementations with only two stages either do not operate reliably or, if they employ additional phase shift elements, oscillate no faster than three-stage configurations. Thus, CMOS VCOs typically utilize three or more stages.

5.2 Phase and Frequency Detectors

The PLL analysis in Section 4 indicates that many parameters of phase-locked systems, including tracking range, acquisition range, loop gain, and transient response depend on the properties of the phase and frequency detectors employed in such systems. Of particular interest are the following properties: 1) what is the input-phase difference range for which the characteristic is monotonic? 2) What is the response to unequal input frequencies? 3) How do the input amplitude and duty cycle affect the characteristic?

Gilbert Cell. A "combinational" phase detector often used in PLLs is the Gilbert cell (Fig. 36). For small signals applied to ports A and B, the circuit operates as an analog multiplier, with an average output given by Eq. (15). Note that as $\Delta\phi$ departs from $90°$, the slope of $\cos\Delta\phi$ and, hence, the equivalent K_{PD} decrease. For $\omega_1 \neq \omega_2$, the average output is zero; i.e., the circuit cannot be used as a frequency detector. Also, K_{PD} is a function of A and B, an undesirable attribute because a PLL employing such a phase detector exhibits amplitude-dependent static and dynamic behavior.

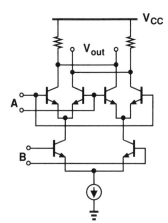

Fig. 36   Gilbert cell PD.

In the Gilbert cell of Figure 36, if the input signal amplitudes are much greater than kT/q, then all three differential pairs experience full switching and the circuit operates as an exclusive OR (XOR) gate. In this case, we consider the input waveforms to be triangular and examine the average output as the input phase difference varies (Fig. 37). If the two inputs are $90°$ out of phase (Fig. 37a), the output is a square wave with zero dc. As the phase difference deviates from $90°$, the output duty cycle is no longer 50%, providing a dc value proportional to the phase difference. Thus, in contrast to the small-signal multiplier, an XOR gate has a constant gain for $0 < |\Delta\phi| < 180°$ (Fig. 38). It can be easily shown that this translates to a factor of $\pi/2$ increase in the tracking range of PLLs. For this reason, and because K_{PD} is independent of the signal amplitudes, it is preferable to use a Gilbert cell with large signals.

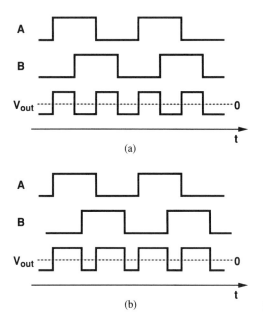

Fig. 37   Input and output waveforms of an XOR gate.

It is interesting to note that in a PLL with a large loop gain, a Gilbert-type PD forces the static phase difference between the input and output to remain close to $90°$ so that the dc output of the PD is small. In many applications, the $90°$ phase shift is unimportant or can be cancelled elsewhere in the system.

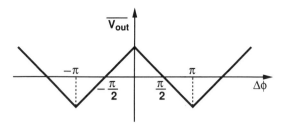

Fig. 38   Characteristic of an XOR PD.

In the Gilbert cell PD, the average output depends on the duty cycle of the inputs. Illustrated in Figure 39, this effect manifests itself as a static phase error in a PLL.

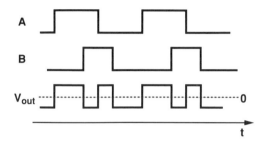

Fig. 39   Dependence of XOR output on input duty cycle.

R-S Latch.   Another topology that can be used for phase detection is an edge-triggered R-S latch, also called a two-state PD (Fig. 40). Here, the rising edge of *A* drives *Q* to *ONE* and that of *B* drives *Q* to *ZERO*. Thus, the differential output changes sign every time a rising edge at one input is followed by a rising edge at the other (Fig. 40b). Since this circuit changes state only on one edge of the inputs, its characteristics differ from those of an XOR in several respects: 1) the output frequency is the same as the input frequency; 2) the average output does not depend on the input duty cycle; 3) the input/output characteristic crosses zero when the inputs are 180° out of phase (Fig. 40b); 4) the monotonic range of the PD is ±180° around the center; 5) the shape of the characteristic is sawtooth rather than triangular. Among these, the first property is undesirable if the PLL performs frequency multiplication because it mandates a low −3-dB frequency in the LPF (Section 4.8).

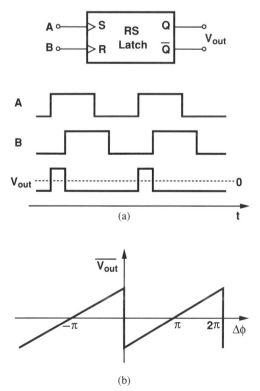

Fig. 40   Edge-triggered R-S latch as a PD.

It is also interesting to note that an R-S latch generates a nonzero dc output if one input frequency is an integer multiple of the other (Fig. 41), whereas an XOR gate exhibits no such behavior. Thus, a PLL employing an R-S latch as the PD may lock to a higher harmonic of the input if the VCO frequency range is sufficiently wide or the input spectrum has strong components at subharmonics of the VCO output.

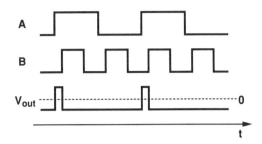

Fig. 41 Response of R-S latch to higher harmonics.

Phase/Frequency Detector. A circuit that can detect both phase and frequency difference proves extremely useful because it significantly increases the acquisition range and lock speed of PLLs.

Unlike XOR gates and R-S latches, sequential phase/frequency detectors (PFDs) generate two outputs that are *not* complementary. Illustrated in Figure 42, the operation of a typical PFD is as follows. If the frequency of input A, ω_A, is less than that of input B, ω_B, then the PFD produces positive pulses at Q_A, while Q_B remains at zero. Conversely, if $\omega_A > \omega_B$, then positive pulses appear at Q_B while $Q_A = 0$. If $\omega_A = \omega_B$, then the circuit generates pulses at either Q_A or Q_B with a width equal to the phase difference between the two inputs. (Note that, in principle, Q_A and Q_B are never high simultaneously.) Thus, the average value of $Q_A - Q_B$ is an indication of the frequency or phase difference between A and B. The outputs Q_A and Q_B are usually called the "UP" and "DOWN" signals.

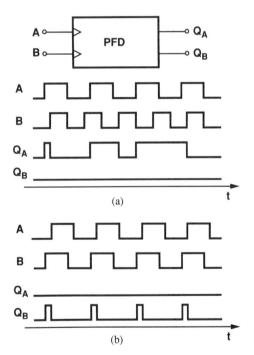

Fig. 42 Phase/frequency detector response with (a) $\omega_A < \omega_B$; (b) A lagging B.

To arrive at a circuit with the above behavior, we postulate that at least three logical states are required: $Q_A = Q_B = 0$; $Q_A = 0$, $Q_B = 1$; and $Q_A = 1$, $Q_B = 0$. Also, to avoid dependence of the output on the duty cycle of the inputs, the circuit should be implemented as an edge-triggered sequential machine. We assume the circuit can change state only on the rising transitions of A and B, and, for the sake of brevity, we will omit the adjective "rising" hereafter. Figure 43 shows a state diagram summarizing the operation. If the PFD is in the "ground" state, $Q_A = Q_B = 0$, then a transition on A takes it to State I, where $Q_A = 1$, $Q_B = 0$. The circuit remains in this state until a transition occurs on B, upon which the PFD returns to State 0. The switching sequence between States 0 and II is similar.

An important point in this state diagram is that if, for example, $\omega_A > \omega_B$, then there will be a time interval during which two transitions of A take place between two transitions of B. This ensures that, even if the PFD begins in State II, it will eventually leave that state and thereafter toggle between States 0 and I [16].

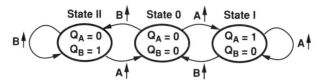

Fig. 43   PFD state diagram.

A possible implementation of the above PFD is shown in Figure 44 [16]. The circuit consists of two edge-triggered, resettable D flipflops with their D inputs connected to logical ONE. Signals A and B act as clock inputs of DFF_A and DFF_B, respectively. We note that if $Q_A = Q_B = 0$, a transition on A causes Q_A to go high. Subsequent transitions on A have no effect on Q_A, and when B goes high, the AND gate activates the reset of both flipflops. Thus, Q_A and Q_B are simultaneously high for a duration given by the total delay through the AND gate and the reset path of the flipflops. The implications of this overlap are explained later. Figure 45 shows the phase characteristic of the PFD.

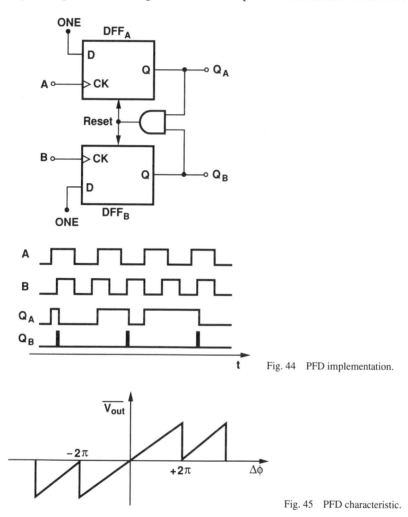

Fig. 44   PFD implementation.

Fig. 45   PFD characteristic.

The D flipflops in Figure 44 may employ different topologies in bipolar and CMOS implementations. In bipolar technology, a standard master-slave configuration with an additional reset input can be used. In CMOS technology, a simple circuit such as that in Figure 46 [17] proves adequate. Note that the D input is "hidden" here.

Other implementations of PFDs are described in [18, 19].

The output of a PFD can be converted to dc in different ways. One approach is to sense the difference between the two outputs by means of a differential amplifier and apply the result to a low-pass filter. Alternatively, the outputs can drive a three-state "charge pump."

5.3 Charge Pumps

In the low-pass filters considered thus far (Figs. 15 and 17), the average value of the PD output is obtained by depositing charge onto a capacitor during each phase comparison and allowing the charge to decay afterwards. In a charge pump, on the other hand, there is negligible decay of charge between phase comparison instants, leading to interesting consequences.

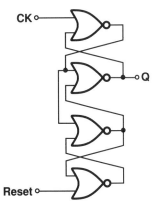

Fig. 46   Implementation of each *DFF* in Figure 44.

A three-state charge pump can be best studied in conjunction with a three-state phase/frequency detector (Fig. 47). The pump itself consists of two switched current sources driving a capacitor. (We assume herein that S_1 and I_1 are implemented with PMOSFETs and S_2 and I_2 with NMOSFETs.) Note that for a pulse of width T on Q_A, I_1 deposits a charge equal to IT on C_P. Thus, if $\omega_A > \omega_B$, or $\omega_A = \omega_B$ but A leads B, then positive charge accumulates on C_P steadily, yielding an *infinite* dc gain for the PFD. Similarly, if pulses appear on Q_B, I_2 removes charge from C_P on every phase comparison, driving V_{out} toward $-\infty$. In the third state, with $Q_A = Q_B = 0$, V_{out} remains constant. Since the steady-state gain is infinite, it is more meaningful to define the gain for one comparison instant, which is equal to $IT/(2\pi C_P)$.

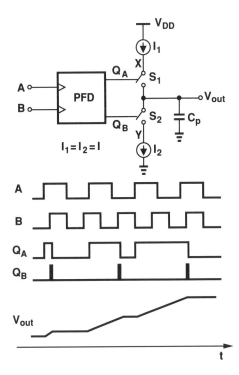

Fig. 47   PFD with charge pump.

An important conclusion to be drawn from the above observations is that, if offsets and mismatches are neglected, a PLL utilizing this arrangement locks such that the static phase difference between A and B is zero; even an infinitesimal phase error would result in an indefinite accumulation of charge on C_P.

The PFD/charge pump circuit of Figure 47 can potentially suffer from a "dead zone." To understand this effect, we let the phase difference between A and B approach zero and study the charge deposited on C_P. We also make two assumptions: 1) Q_A and Q_B exhibit relatively long transition times, for example, due to the capacitive loading of S_1 and S_2; 2) the delay from Q_A and Q_B through the AND gate and the reset path of the flipflops is small, i.e., when Q_A and Q_B exceed the threshold of the AND gate, the reset is immediately activated. The goal is to examine the increment in the charge deposited on C_P for an increment Δt in the delay between A and B (the small-signal gain).

Let us first consider a case where the delay between A and B is relatively large (Fig. 48a). We note that Q_A develops a full logical level and is reset when Q_B reaches the threshold of the AND gate. Thus, if the delay increments from T_1 to

$T_1 + \Delta t$, Q_A will be high for Δt seconds longer and the charge deposited on C_p will increase by $I\Delta t$. Now suppose, as shown in Figure 48b, $T_1 \approx 0$. What is the charge increment for a delay increment Δt? If B goes high Δt seconds after A does, then Q_B follows Q_A with the same delay. As soon as Q_B crosses the threshold of the AND gate, the reset signal is asserted, forcing Q_A and Q_B to return to zero. Thus, if Δt is small, Q_A does not reach a full logical level, failing to turn S_1 on or turning it on for an ill-defined length of time. In other words, the gain of the circuit drops as the delay between A and B becomes comparable with the transition time at Q_A and Q_B. This is illustrated in Figure 48c.

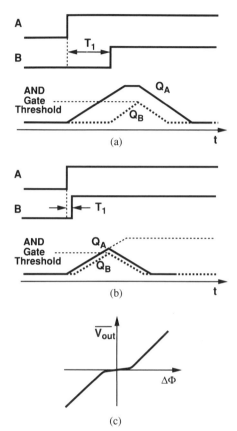

Fig. 48 Dead zone in a PFD with charge pump.

At this point, we must make two important observations. First, the dead zone is undesirable in a phase-locked system: If the phase difference between the input and output varies within the zone, the dc output of the charge pump does not change significantly and the loop fails to correct the resulting error. Consequently, a peak-to-peak jitter approximately equal to the width of the dead zone can arise in the output. Second, the PFD of Figure 44 is unlikely to suffer from a dead zone because its reset operation typically entails two or more gate delays (one due to the AND gate and one or more due to the flipflops). So long as the capacitive loading of S_1 and S_2 does not excessively slow down the output transitions, the reset delay ensures that Q_A and Q_B reach full logic levels. It is interesting to note that the dead zone existed in old implementations where the PFD output needed to drive an *external* charge pump and, hence, the capacitance associated with the IC pads and PC board traces. In those cases, additional inverters would be inserted at the output of the AND gate to increase the reset delay and eliminate the dead zone.

From the above discussion, we also infer that the dead zone disappears only if Q_A and Q_B can be simultaneously high for a sufficient amount of time. During this period, both S_1 and S_2 in Figure 47 are on, allowing the *difference* between I_1 and I_2 to vary the voltage stored on C_P. Since I_1 and I_2 typically have a few percent of mismatch, the output voltage varies even if the input phase difference is zero. Thus, a PLL employing this arrangement locks with a finite phase error so as to cancel the net charge deposited by I_1 and I_2 on C_p (Fig. 49). The important point is that the control voltage of the VCO is periodically disturbed, thereby modulating the VCO and introducing sidebands in the output spectrum.

Another error stems from mismatches between S_1 and S_2 in Figure 47. When these switches turn off, their charge-injection and feedthrough mismatch results in an error step at the output, changing the VCO frequency until the next phase comparison instant.

Our discussion thus far has assumed that I_1 and I_2 in Figure 47 are ideal. Since each current source requires a minimum voltage to maintain a relatively constant current, it is important that $V_{DD} - V_X$ and V_Y not drop below a certain

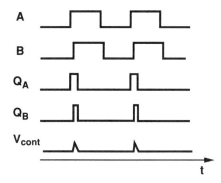

Fig. 49 Effect of mismatch in current sources of a charge pump.

level. If the extreme values of V_{cont} violate this condition, the current charging C_P varies and so does the overall gain [20], influencing the loop static and dynamic behavior.

Another related effect occurs when S_1 and S_2 are off: I_1 and I_2 pull nodes X and Y to V_{DD} and ground, respectively, causing charge-sharing between C_X, C_P, and C_Y when S_1 and S_2 turn on again (Fig. 50). If $V_{\mathrm{out}} = V_{DD}/2$, $I_1 = I_2$, and $C_X = C_Y$, then V_{out} is not disturbed, but because V_{out} determines the VCO frequency, it is generally not equal to $V_{DD}/2$, thus experiencing a jump when S_1 and S_2 turn on. This effect is also periodic and introduces sidebands at the output, but it can be suppressed if nodes X and Y are bootstrapped to the voltage stored on the capacitor [7].

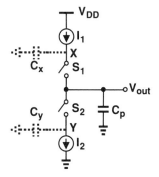

Fig. 50 Charge-sharing in a charge pump.

As noise immunity demands a differential control voltage for the VCO, the charge pump circuit of Figure 47 must be modified to provide a differential output. An additional advantage is that differential implementation alleviates the mismatch and charge-sharing problems as well. In a differential charge pump, the UP and DOWN signals activate only pull-down currents, and the pull-up currents are passive. Thus, when both UP and DOWN are low, a common-mode (CM) feedback circuit must counteract the pull-up currents to maintain a proper level.

Shown in Figure 51 is an example where differential pairs M_1-M_2 and M_3-M_4 are driven by the PFD and the network consisting of M_5-M_9 sets the output CM level at $V_{GS5} + V_{GS9}$. Because the CM level is momentarily disturbed at each phase comparison instant, it is important that CM transients not lead to *differential* settling components [21].

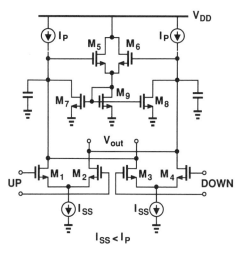

Fig. 51 Differential CMOS charge pump.

28

6. CHARGE-PUMP PHASE-LOCKED LOOPS

Charge-pump PLLs (CPPLLs) incorporate a PFD (or PD) and a charge pump (Fig. 52) instead of the combinational PD and the LPF in the generic architecture of Figure 11. As mentioned before, the combination of a PFD and a charge pump offers two important advantages over the XOR/LPF approach: 1) the capture range is only limited by the VCO output frequency range; 2) the static phase error is zero if mismatches and offsets are negligible. In this section, we study the characteristics of this type of PLL and make comparisons with the conventional type.

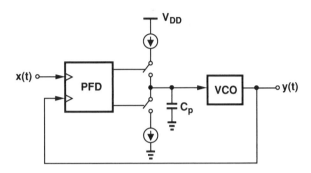

Fig. 52   Charge-pump PLL.

Charge pumps provide an infinite gain for a *static* phase difference at the input of the PFD. From another point of view, the response of a PFD/charge pump to a phase step is a linear ramp, indicating that the transfer function of the circuit contains a pole at *the origin*. With another such pole contributed by the VCO, a charge-pump PLL cannot remain stable. In fact, representing the transfer function of the PFD/charge pump with K_{PFD}/s, we note that the closed-loop transfer function of the PLL is

$$H(s) = \frac{\dfrac{K_{\text{PFD}}}{s}\dfrac{K_{\text{VCO}}}{s}}{1 + \dfrac{K_{\text{PFD}}}{s}\dfrac{K_{\text{VCO}}}{s}} \tag{53}$$

$$= \frac{K_{\text{PFD}}K_{\text{VCO}}}{s^2 + K_{\text{PFD}}K_{\text{VCO}}} \tag{54}$$

revealing two imaginary poles at $\omega = \pm j\sqrt{K_{\text{PFD}}K_{\text{VCO}}}$. To avoid instability, a zero must be added to the open-loop transfer function. This is in contrast to the case of a simple low-pass filter, where the loop is, in principle, stable even with no zero. The stabilizing zero in a CPPLL can be realized by placing a resistor in series with the charge-pump capacitor (Fig. 53).

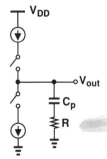

Fig. 53   Addition of a zero to a charge pump.

To perform a small-signal analysis, we note that the switching operation of the charge pump and the lack of a discharge path between phase comparison instants make the PLL a discrete-time system. However, if the loop bandwidth is much less than the input frequency, we can assume the state of the PLL changes by a small amount during each cycle of the input [20]. Using the "average" value of the discrete-time parameters, we can then study the loop as a continuous-time system [20].

Suppose the loop begins with a phase error $\phi_{\text{in}} - \phi_{\text{out}} = \phi_e$. Then, the average current charging the capacitor is given by $I\phi_e/(2\pi)$ and the average change in the control voltage of the VCO equals

$$V_{\text{cont}}(s) = \frac{I\phi_e}{2\pi}\left(R + \frac{1}{C_P s}\right) \tag{55}$$

Noting that $\Phi_{\text{out}}(s) = V_{\text{cont}}(s)K_{\text{VCO}}/s$, we obtain the following closed-loop transfer function:

$$H(s) = \frac{\dfrac{I}{2\pi C_P}(RC_P s + 1)K_{\text{VCO}}}{s^2 + \dfrac{I}{2\pi}K_{\text{VCO}}Rs + \dfrac{I}{2\pi C_P}K_{\text{VCO}}} \quad . \tag{56}$$

which has the same form as Eq. (36). Thus, the system is characterized by a zero at $\omega_z = -1/(RC_P)$ and

$$\omega_n = \sqrt{\frac{I}{2\pi C_P}K_{\text{VCO}}} \tag{57}$$

$$\zeta = \frac{R}{2}\sqrt{\frac{IC_P}{2\pi}K_{\text{VCO}}} \tag{58}$$

Note that ω_n is independent of R. If the loop includes frequency division, K_{VCO} must be divided by the division factor.

From Eq. (46), we note that the decay-time constant of the system is equal to $(\zeta\omega_n/2)^{-1} = (RIK_{\text{VCO}}/8)^{-1}$, a quantity independent of C_P.

In many applications, it is desirable to maximize the loop bandwidth, which is usually proportional to ω_n. While for a PLL with a sinusoidal PD (Section 3.3), ω_n and ζ cannot be maximized simultaneously, Eqs. (57) and (58) suggest that in a CPPLL both ω_n and ζ can be increased if I or K_{VCO} is increased. However, as the loop bandwidth becomes comparable with the input frequency, the continuous-time approximation used above breaks down, necessitating discrete-time analysis. Using such an analysis, Gardner has derived a stability limit [20] that can be reduced to

$$\omega_n^2 < \frac{\omega_{\text{in}}^2}{\pi(RC_P\omega_{\text{in}} + \pi)} \tag{59}$$

implying an upper bound on ω_n. This equation also indicates that R cannot be increased indefinitely [19]. (Typically, when the continuous-time approximation fails, the loop is unacceptably underdamped.)

In single-ended charge pumps, the resistor added in series with the capacitor can introduce "ripple" in the control voltage [20] even when the loop is locked. Since S_1 and S_2 turn on at every phase comparison instant, the mismatch between I_1 and I_2 flows through R, causing a step at the output. Furthermore, mismatch between overlap capacitance of S_1 and S_2 results in a net signal feedthrough to the output. Modulating the VCO frequency, this effect is especially undesirable in frequency synthesizers.

To suppress the ripple, a second capacitor can be connected from the output of the charge pump to ground. This modification introduces a third pole in the PLL, requiring further study of stability issues. Gardner provides criteria for the stability of such systems [20]. Note that fully differential charge pumps can reduce the magnitude of the ripple considerably.

The zero required in a charge-pump PLL can also be implemented using feedforward [22, 23]. This is accomplished by adding a fast signal path in parallel with the main charge pump. Illustrated in Figure 54 on page 31, this technique utilizes an auxiliary charge pump driving a wideband "dissipative" network, C_2 and R_2. The transfer function of the circuit is thus equal to

$$H_{CP}(s) = \frac{I_{P1}}{2\pi C_1 s} + \frac{I_{P2}R_2}{2\pi(R_2 C_2 s + 1)} \tag{60}$$

$$= \frac{(I_{P1}R_2 C_2 + I_{P2}R_2 C_1)s + I_{P1}}{2\pi C_1 s(R_2 C_2 s + 1)} \tag{61}$$

thereby providing a zero at

$$\omega_z = -\left(R_2 C_2 + R_2 C_1 \frac{I_{P2}}{I_{P1}}\right)^{-1} \tag{62}$$

Note that the addition of the zero inevitably introduces a pole at $\omega_P = -1/(R_2 C_2)$, making the PLL a third-order system. Moreover, the decay of the voltage across C_2 due to R_2 leads to ripple and, hence, frequency modulation of the VCO.

7. NOISE IN PHASE-LOCKED LOOPS

Since PLLs operate on the phase of signals, they are susceptible to phase noise or jitter. Within the scope of this tutorial, we consider phase noise as a random component in the excess phase, as exemplified by $\phi_n(t)$ in Eq. (1). For the sake of brevity, we use the term *noise* to mean *phase noise*.

30

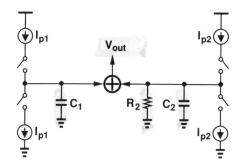

Fig. 54 Addition of zero by means of feedforward.

If the input signal or the building blocks of a PLL exhibit noise, then the output signal will also suffer from noise. In general, all the loop components, including the phase detector, the LPF, the VCO, and the frequency divider may contribute noise [24]. The goal is to understand how the spectrum of a given noise source is shaped as it propagates to the output.

In this tutorial, we examine two important cases: 1) the input signal contains noise, and 2) the VCO introduces noise; in each case, we find the transfer function from the noise source of interest to the PLL output. In monolithic implementations, the phase noise of the VCO is typically much more significant than that of other loop components.

7.1 Phase Noise at Input

Consider the PLL in Figure 55 where the input and output signals are $x(t) = A\sin[\omega_c t + \phi_{in}(t)]$ and $y(t) = B\sin[\omega_c t + \phi_{out}(t)]$. The transfer function $\Phi_{out}(s)/\Phi_{in}(s)$ is

$$H(s) = \frac{\omega_n^2}{s^2 + 2\zeta\omega_n s + \omega_n^2} \tag{63}$$

If the input (excess) phase, $\phi_{in}(t)$, does not vary with time, i.e., if the input to the PLL is a pure sinusoid, then $s = 0$ and $H(s) = 1$. Now, suppose $\phi_{in}(t)$ is varied so slowly that the denominator of Eq. (63) is still close to ω_n^2. Thus, $H(s)$ remains close to unity, indicating that the output phase (or frequency) follows the input phase (or frequency), a natural property of the PLL as a tracking system.

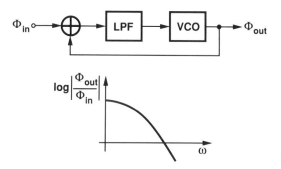

Fig. 55 Noise transfer function of a PLL from input to output.

What happens if $\phi_{in}(t)$ varies at an increasingly higher rate? Equation (63) shows that the output excess phase, $\phi_{out}(t)$, drops, eventually approaching zero and yielding $y(t) = B\sin\omega_c t$. In other words, for fast variations of the input excess phase or frequency, the PLL fails to track the input. Revisited in Section 9.1, this attribute of PLLs was the original reason for their widespread application in communications.

In summary, the input-phase noise spectrum of a PLL is shaped by the characteristic low-pass transfer function when it appears at the output. In order to minimize this noise, the loop bandwidth must be as small as possible, although it slows down the lock, limits the capture range, and degrades the stability.

7.2 Phase Noise of VCO

The phase noise of the VCO can be modeled as an additive component, ϕ_{VCO}, as shown in Figure 56. Assuming ϕ_{VCO} and ϕ_{in} are uncorrelated, we set ϕ_{in} to zero and compute the transfer function from ϕ_{VCO} to ϕ_{out}.[7] Note that $\phi_{in}(t) = 0$ means the *excess* phase of the input is zero, not the input signal itself; i.e., we must apply a strictly periodic signal at the input.

With $\phi_{in} = 0$ and a simple low-pass filter, we have

$$\frac{\Phi_{out}(s)}{\Phi_{VCO}(s)} = \frac{s(s + \omega_{LPF})}{s^2 + 2\zeta\omega_n s + \omega_n^2} \tag{64}$$

---

[7]Superposition holds for the *power* of uncorrelated sources.

31

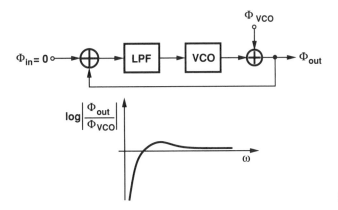

Fig. 56   Noise transfer function of a PLL from VCO to output.

As expected, this transfer function has the same poles as Eq. (63), but it also contains two zeros at $\omega_{z1} = 0$ and $\omega_{z2} = -\omega_{LPF}$, making the characteristic a *high-pass* filter.

The zero at the origin implies that, for slow variations in ϕ_{VCO}, ϕ_{out} is small. This is because, in lock, the phase variations in the VCO are converted to voltage by the PD and applied to the control input of the VCO so as to accumulate phase in the opposite direction. Since the VCO voltage/phase conversion has nearly infinite gain for a slowly varying V_{cont}, the negative feedback suppresses variations in the output phase.

From another perspective, the PLL can be simplified and redrawn as in Figures 57a and b. Because an ideal integrator placed in a negative feedback loop creates a "virtual" ground at its input, $\phi_{out} \approx 0$ for slow variations in ϕ_{VCO}.

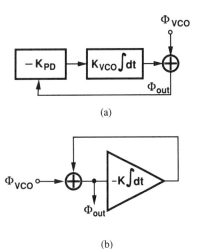

Fig. 57   Simplified model of PLL with VCO noise.

Now, suppose the rate of change of ϕ_{VCO} increases. Then, the magnitude of K_{VCO}/s and, hence, the loop gain decrease, allowing the virtual ground to experience significant variations. As the rate of change of ϕ_{in} approaches ω_{LPF}, the loop gain is reduced by the low-pass filter as well, an effect represented by the zero at $-\omega_{LPF}$.

From Eq. (64), we note that as $s \rightarrow \infty$, $\phi_{out} \rightarrow \phi_{VCO}$, which is to be expected because the feedback loop is essentially open for very fast changes in ϕ_{VCO}.

A common test of noise immunity in PLLs entails applying a small step to the power supply and finding the time required for the input-output phase difference to settle within a certain error band [18]. Since such a step predominantly affects the VCO output, we can use Eq. (64) to predict the circuit's behavior. For a phase step of height ϕ_1, the output assumes the following form:

$$\phi_{out}(t) = \phi_1 \left[\cos\sqrt{1-\zeta^2}\omega_n t + \frac{\zeta}{\sqrt{1-\zeta^2}} \sin\sqrt{1-\zeta^2}\omega_n t \right] \times \exp(-\zeta\omega_n t) \qquad (65)$$

Thus, the output initially jumps to ϕ_1 and subsequently decays to zero with a time constant $(\zeta\omega_n)^{-1}$. It is therefore desirable to maximize $\zeta\omega_n$ for fast recovery of the PLL.

From the above analysis, we conclude that to minimize the VCO phase-noise contribution, the loop bandwidth must be maximized, a requirement in conflict with that of the case where the PLL input contains noise. In applications where the input has negligible noise (e.g., because it is derived from a crystal oscillator), the loop bandwidth is maximized to reduce both the VCO phase noise and the lock time.

8. CLOCK RECOVERY CIRCUITS

As mentioned in Section 2.4, a clock recovery circuit produces a timing clock signal from a stream of binary data. In this section, we describe the design issues of CRCs and study various clock recovery techniques amenable to monolithic implementation.

8.1 Properties of NRZ Data

Binary data is commonly transmitted in the "nonreturn-to-zero" (NRZ) format. As shown in Figure 58a, in this format each bit has a duration of T_b ("bit period"), is equally likely to be ZERO or ONE, and is statistically independent of other bits. The quantity $r_b = 1/T_b$ is called the "bit rate" and measured in bits/s. The term "non-return-to-zero" distinguishes this type from another one called the "return-to-zero" (RZ) format, in which the signal goes to zero between consecutive bits (Fig. 58b). Since for a given bit rate, RZ data contains more transitions than NRZ data, the latter is preferable where channel or circuit bandwidth is costly. Note that, in general, data must be treated as a random waveform (with certain known statistical properties).

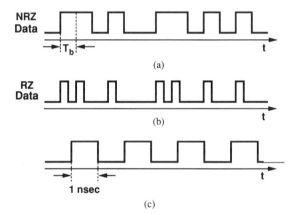

Fig. 58   (a) NRZ data; (b) RZ data; (c) fastest NRZ waveform with $r_b = 1$ Gb/s.

NRZ data has two attributes that make the task of clock recovery difficult. First, the data may exhibit long sequences of consecutive ONEs or ZEROs, demanding the CRC to "remember" the bit rate during such a period. This means that, in the absence of data transitions, the CRC should not only continue to produce the clock, but also incur negligible drift in the clock frequency. We return to this issue later.

Second, the spectrum of NRZ data has nulls at frequencies that are integer multiples of the bit rate; for example, if the data rate is 1 Gb/s, the spectrum has no energy at 1 GHz. To understand why, we note that the fastest waveform for a 1-Gb/s stream of data is obtained by alternating between ONE and ZERO every 1 ns (Fig. 58c). The result is a 500-MHz square wave, with all the even-order harmonics absent. From another point of view, if an NRZ sequence with rate r_b is multiplied by $A \sin(2\pi m r_b t)$, the result has a zero average for all integers m, indicating that the waveform contains no frequency components at $m r_b$.

It is also helpful to know the shape of the NRZ data spectrum. Since the autocorrelation function of a random binary sequence is [25]

$$R_x(\tau) = 1 - \frac{|\tau|}{T_b} \qquad |\tau| < T_b \tag{66}$$

$$= 0 \qquad |\tau| < T_b \tag{67}$$

the power spectral density equals

$$P_x(\omega) = T_b \left[\frac{\sin(\omega T_b/2)}{\omega T_b/2} \right]^2 \tag{68}$$

Plotted in Figure 59, this function vanishes at $\omega = 2m\pi/T_b$. In contrast, RZ data has finite power at such frequencies.

Due to the lack of a spectral component at the bit rate in the NRZ format, a clock recovery circuit may lock to spurious signals or simply not lock at all. Thus, NRZ data usually undergoes a nonlinear operation at the front end of the circuit so as to create a frequency component at r_b. A common approach is to detect each data transition and generate a corresponding pulse.

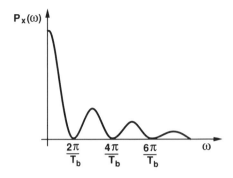

Fig. 59   Spectrum of NRZ data.

8.2 Edge Detection

Illustrated in Figure 60a, edge detection requires a means of sensing both positive and negative data transitions. In Figure 60b, an XOR gate with a delayed input performs this operation, whereas in Figure 60c, a differentiator produces impulses corresponding to each transition, and a squaring circuit or a full-wave rectifier converts the negative impulses to positive ones.

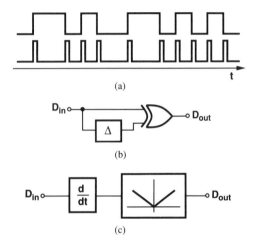

Fig. 60   Edge detection of NRZ data.

A third method of edge detection employs a flipflop operating on both rising and falling edges [26]. To understand this technique, we first note that, in a phase-locked CRC, the edge-detected data is multiplied by the output of a VCO (Fig. 61a). In effect, the data transition impulses "sample" points on the VCO output. This can also be accomplished using a master-slave flipflop consisting of two D latches: the data pulses drive the clock input while the VCO output is sensed by the D input (Fig. 61b). Since in this configuration the VCO output is sampled on either rising or falling transitions of the data, we modify the circuit such that both latches sample X_{VCO}, but on opposite transitions of D_{in}. Shown in Figure 61c, the resulting circuit samples the VCO output on every data transition and is therefore functionally equivalent to that in Figure 61a. We call this circuit a double-edge-triggered flipflop.

8.3 Clock Recovery Architectures

From the above observations, we note that clock recovery consists of two basic functions: 1) edge detection; 2) generation of a periodic output that settles to the input data rate but has negligible drift when some data transitions are absent. Illustrated in Figure 62a is a conceptual realization of these functions, where a high-Q oscillator is "synchronized" with the input transitions and oscillates freely in their absence. The synchronization can be achieved by means of phase-locking.

Figure 62b shows how a simple PLL can be used along with edge detection to perform clock recovery. First, suppose the input data is periodic with a frequency $1/T_b$. (The unit of $1/T_b$ is hertz rather than rad/s.) Then, the edge detector simply doubles the frequency, allowing the VCO to lock to $1/(2T_b)$. Now, assume some transitions are absent. During such an interval, the output of the multiplier is zero and the voltage stored in the low-pass filter decays, thereby making the VCO frequency drift. To minimize this effect, the time constant of the LPF must be sufficiently larger than the maximum allowable interval between consecutive transitions,[8] thereby resulting in a small loop bandwidth and, hence, a narrow capture range.

It follows from the above discussion that a PLL used for clock recovery must also employ frequency detection to ensure locking to the input despite process and temperature variations. This may suggest replacing the multiplier with the

---

[8] Most communication systems guarantee a certain upper bound on this interval by encoding the data.

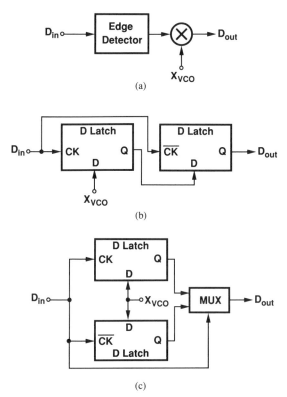

(a)

(b)

(c)

Fig. 61   Edge detection and sampling of NRZ data.

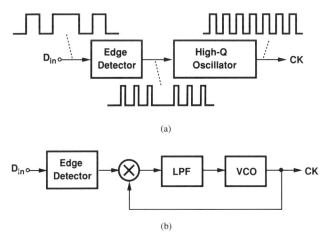

(a)

(b)

Fig. 62   (a) Conceptual realization of a CRC; (b) phase-locked CRC.

three-state PFD of Figure 42. However, the latter circuit produces an incorrect output if either of its input signals exhibits missing transitions. As depicted in Figure 63, in the absence of transitions on the main input, the PFD interprets the VCO frequency to be higher than the input frequency, driving the control voltage in such a direction as to correct the apparent difference. This occurs even if the VCO frequency is initially equal to the input data rate. Thus, the choice of phase and frequency detectors for random binary data requires careful examination of their response when some transitions are absent.

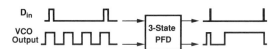

Fig. 63   Response of a three-state PFD to random data.

A clock recovery architecture that has been implemented in both analog and digital domains is the "quadricorrelator," introduced by Richman [27] and modified by Bellisio [28]. We first consider an analog representation of the architecture to describe its underlying principles. Shown in Figure 64, and bearing some resemblance to that in Figure 25, the quadricorrelator follows the edge detector with a combination of three loops sharing the same VCO. Loops I and II perform frequency detection and Loop III, phase detection. Note that the VCO generates quadrature outputs. The circuit operates as follows. Suppose the edge detector produces a frequency component at ω_1 while the VCO oscillates at ω_2. Mixing the VCO outputs with $\sin \omega_1 t$ and low-pass filtering the results, we obtain quadrature beat signals $\sin(\omega_1 - \omega_2)t$ and $\cos(\omega_1 - \omega_2)t$. Next, the latter signal is differentiated and mixed with the former, yielding $(\omega_1 - \omega_2) \cos^2(\omega_1 - \omega_2)t$ at node P. Representing both the polarity and the magnitude of the difference between ω_1 and ω_2, the average value of this signal drives the VCO with negative feedback so as to bring ω_2 closer to ω_1. As $|\omega_2 - \omega_1|$ drops, Loop III—a simple PLL—begins to generate an asymmetric signal at node M, assisting the lock process. For $\omega_2 \approx \omega_1$, the dc feedback signal produced by Loops I and II approaches zero and Loop III dominates, locking the VCO output to the input data.

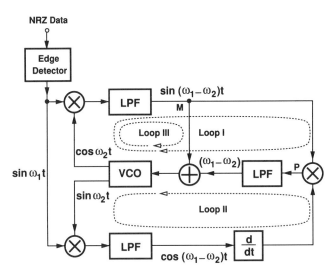

Fig. 64   Quadricorrelator.

The use of frequency detection in the quadricorrelator makes the capture range independent of the (locked) loop bandwidth, allowing a small cut-off frequency in the LPF of Loop III so as to minimize the VCO drift between data transitions.[9] Nevertheless, because the frequency detection circuit can respond to noise and spurious components, it is preferable to disable Loops I and II once phase lock is attained.

Further analysis of the quadricorrelator is given in [27, 29].

A drawback of the quadricorrelator in discrete technologies was that it required quadrature outputs of the VCO, a problem solved by passing the VCO output through a delay line to shift the phase by $90°$. The dependence of such a delay on frequency, temperature, and component values made the design difficult. In monolithic implementations, on the other hand, differential ring oscillators with an even number of stages easily provide quadrature outputs (Section 5.1).

The quadricorrelator can also be realized in digital form. Recall from Figure 61 that the combination of an edge detector and a mixer can be replaced with a double-edge-triggered flipflop. Thus, the architecture of Figure 64 can be "digitalized" as shown in Figure 65, which is notably similar to that in [26].

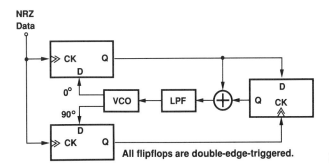

Fig. 65   Digital implementation of quadricorrelator.

Other types of phase and frequency detectors for NRZ data are described in [30, 31].

---

[9]It may be necessary to follow the mixer in Loop III with *two* LPFs: one with a wide band in Loop I and another with a narrow band in Loop III.

9. APPLICATIONS

9.1 Noise and Jitter Suppression in Communications

A common situation in communication systems is that a narrowband signal is corrupted by noise. This occurs, for example, in satellite transceivers where weak signals buried in noise must be detected by coherent demodulation. In order to achieve a high signal-to-noise ratio, the noise components around the carrier, ω_c, must be suppressed, implying the need for a narrowband filter. However, in most applications the required filter bandwidth is several orders of magnitude smaller than ω_c, thereby demanding filter Qs of greater than 1000.

A PLL can operate as a narrowband filter with an extremely high Q (Fig. 66). Recall that the input/output phase (or frequency) transfer function of a (continuous-time) PLL is that of a low-pass filter whose bandwidth is *independent* of the input frequency. Making the bandwidth of the PLL sufficiently small can therefore result in a very high equivalent Q. From another point of view, the PLL takes the average of the input frequency over a great many cycles, thus suppressing variations therein.

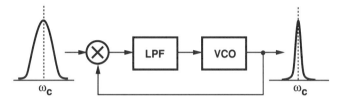

Fig. 66   PLL as a narrowband filter.

As mentioned in Section 7, with a small loop bandwidth, the VCO becomes the dominant source of output noise. Thus, in this application fully monolithic VCOs may not achieve sufficiently low phase noise, and external resonant devices such as inductors may be required.

In digital communications, transmitted or retrieved data often suffer from timing jitter, a problem similar to that illustrated in Figure 1. In order to lower the jitter, the data can be "regenerated" (or "retimed") with the aid of a phase-locked clock recovery circuit. Depicted in Figure 67, regeneration occurs when the data waveform is sampled at its peaks, farthest from zero-crossing points. Subsequent amplification then produces a logical ONE or ZERO according to the polarity of the sampled value.

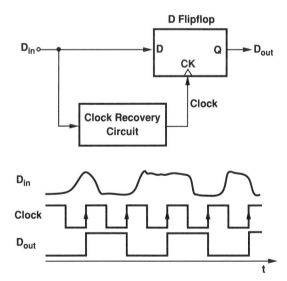

Fig. 67   Jitter suppression by data regeneration.

To appreciate the efficiency and robustness of this approach, we make two observations. First, as with the case shown in Figure 66, the phase-locked CRC rejects most of the noise that accompanies the data, generating a low-jitter clock. Second, peak sampling produces a correct output even if the data zero crossings deviate from their ideal instants by almost half a bit period (Fig. 68). Heavily deteriorated data can thus be recovered.

An important issue in clock and data recovery is "jitter peaking." In most PLLs, the input/output transfer function contains a zero, typically exhibiting a peaking in the frequency response. As a result, some high-frequency jitter components are actually amplified as they appear at the output. This issue is further discussed in [32, 33].

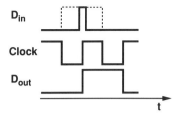

Fig. 68   Jitter margin for correct sampling of data.

9.2 Skew Suppression in Digital Systems

As described in Section 2.2, the interface between off-chip clocks and high-speed digital ICs typically introduces significant skew. This originates from the finite delay of the on-chip clock buffers used to drive the device and interconnect capacitance. Considering the delay as a phase shift, we postulate that its effect can be reduced if the buffers are embedded in a PLL with large loop gain. Figure 69 is an example of the on-chip buffered signal, CK_C, phase-locked to the off-chip clock, CK_S. With a three-state PFD and charge pump, the buffer delay is, in principle, completely cancelled.

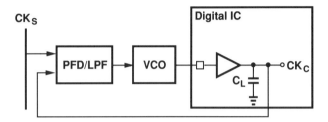

Fig. 69   Skew reduction using a PLL.

The above technique can be utilized more extensively in large clock networks. For example, if a processor clock is distributed on an "H" tree, the signal at the end of one of the branches can be locked to the input so as to null the delay associated with intermediate wires and local buffers [34].

An important issue in PLLs used in low-power systems is the lock time. A low-power processor is frequently powered down and up, requiring the PLL to perform phase alignment quickly. The load capacitance seen by the clock buffer changes when the processor is powered up. Thus, even if the PLL is always on, it still experiences a large transient. Another difficulty is that the lock time is a function of loop parameters and, hence, varies considerably with process and temperature.

9.3 Frequency Synthesis in RF Transceivers

RF systems usually require a high-frequency local oscillator (LO) whose frequency can be changed in small, precise steps. The ability of PLLs to multiply frequencies makes them attractive for synthesizing frequencies.

Figure 70 shows an example of a phase-locked synthesizer. The goal is to generate an output frequency that can be varied from 900 MHz to 925.4 MHz in steps of 200 kHz, covering 128 channels. The frequency divider in the loop is designed such that its division ratio is $M = NP + S$, where $NP = 4500$ and S can be programmed from zero to 127 by the "channel select" input. Thus, if $f_{REF} = 200$ kHz, then f_{out} can be varied from $f_{min} = (4500 + 0) \times 200$ kHz $= 900$ MHz to $f_{max} = (4500 + 127) \times 200$ kHz $= 925.4$ MHz.

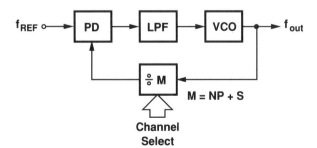

Fig. 70   Phase-locked synthesizer.

RF synthesizers typically impose severe constraints on the output phase noise and sidebands. In the example of Figure 70, phase noise is contributed mostly by the VCO because f_{REF} is provided by a low-noise crystal oscillator. Thus, it is desirable to maximize the loop bandwidth. However, as explained in Section 4.8, the second harmonic of f_{REF} at the PD output creates sidebands in the VCO, and can be suppressed only by lowering ω_{LPF}. This leads to a trade-off between phase noise and sideband magnitudes, making the architecture a viable choice only if the VCO phase noise is sufficiently small.

The lock time of RF synthesizers is also an important parameter. In "frequency-hopped" systems, for example, the channel is required to change in a short amount of time, a constraint in conflict with the small loop bandwidth needed to suppress the sidebands. Many architectures and circuit techniques have been devised to resolve these issues [35].

10. CONCLUSION

The concept of phase locking has proved essential in today's electronic systems. The ability of PLLs to control phase and frequency with extremely high precision provides efficient solutions to various design problems in communications, RF and wireless applications, disk drive electronics, and digital systems.

11. REFERENCES

[1] F. M. Gardner. *Phaselock Techniques*, Second Edition. New York: Wiley & Sons, 1979.

[2] A. B. Grebene, "The monolithic phase locked loop—A versatile building block," *IEEE Spectrum*, vol. 8, pp. 38–49, March 1971.

[3] R. E. Best. *Phase-Locked Loops*, Second Edition. New York: McGraw-Hill, 1993.

[4] G. S. Moschytz, "Miniaturized RC filters using phase-locked loop," *Bell Syst. Tech. J.*, vol. 44, pp. 823–870, May 1965.

[5] W. F. Egan. *Frequency Synthesis by Phase Lock*. New York: Wiley & Sons, 1981.

[6] M. Bazes, "A novel precision MOS synchronous delay line," *IEEE J. Solid-State Circuits*, vol. 20, pp. 1265–1271, December 1985.

[7] M. G. Johnson and E. L. Hudson, "A variable delay line PLL for CPU-coprocessor synchronization," *IEEE J. Solid-State Circuits*, vol. 23, pp. 1218–1223, October 1988.

[8] J. Sonntag and R. Leonowich, "A monolithic CMOS 10 MHz DPLL for burst-mode data retiming," *ISSCC Dig. Tech. Papers*, pp. 194–195, San Francisco, CA, February 1990.

[9] B. van der Pol, "The nonlinear theory of electric oscillators," *Proc. IRE*, vol. 22, pp. 1051–1086, September 1934.

[10] A. Borys, "Elementary deterministic theories of frequency and amplitude stability in feedback oscillators," *IEEE Trans. Circuits Syst.*, vol. 34, pp. 254–258, March 1987.

[11] B. Razavi, "Analysis, modeling, and simulation of phase noise in monolithic voltage-controlled oscillators," *Proc. CICC*, pp. 323–326, Santa Clara, CA, 1995.

[12] B. Lai and R. C. Walker, "A monolithic 622 Mb/sec clock extraction and data retiming circuit," *ISSCC Dig. Tech. Papers*, pp. 144–145, San Francisco, CA, February 1991.

[13] S. K. Enam and A. A. Abidi, "NMOS ICs for clock and data regeneration in gigabit-per-second optical-fiber receivers," *IEEE J. Solid-State Circuits*, vol. 27, pp. 1763–1774, December 1992.

[14] A. Pottbacker and U. Langmann, "An 8 GHz silicon bipolar clock recovery and data regenerator IC," *IEEE J. Solid-State Circuits*, vol. 29, pp. 1572–1576, December 1994.

[15] B. Razavi and J. Sung, "A 2.5-Gb/sec 15-mW BiCMOS clock recovery circuit," *Symp. VLSI Circuits Dig. Tech. Papers*, pp. 83–84, Kyoto, Japan, 1995.

[16] C. A. Sharpe, "A 3-state phase detector can improve your next PLL design," *EDN*, pp. 55–59, September 20, 1976.

[17] I. Shahriary et al., "GaAs monolithic phase/frequency discriminator," *IEEE GaAs IC Symp. Dig. Tech. Papers*, pp. 183–186, 1985.

[18] I. A. Young, J. K. Greason, and K. L. Wong, "A PLL clock generator with 5 to 110 MHz of lock range for microprocessors," *IEEE J. Solid-State Circuits*, vol. 27, pp. 1599–1607, November 1992.

[19] D. K. Jeong et al., "Design of PLL-based clock generation circuits," *IEEE J. Solid-State Circuits*, vol. 22, pp. 255–261, April 1987.

[20] F. M. Gardner, "Charge-pump phase-locked loops," *IEEE Trans. Comm.*, vol. COM-28, pp. 1849–1858, November 1980.

[21] B. Razavi. *Principles of Data Conversion System Design*. Piscataway, NJ: IEEE Press, 1995.

[22] D. Mijuskovic et al., "Cell-based fully integrated CMOS frequency synthesizers," *IEEE J. Solid-State Circuits*, vol. 29, pp. 271–279, March 1994.

[23] I. Novof et al., "Fully-integrated CMOS phase-locked loop with 15 to 240 MHz locking range and ±50 psec jitter," *ISSCC Dig. Tech. Papers*, pp. 112–113, San Francisco, CA, February 1995.

[24] V. F. Kroupa, "Noise properties of PLL systems," *IEEE Trans. Comm.*, vol. COM-30, pp. 2244–2252, October 1982.

[25] S. K. Shanmugam. *Digital and Analog Communication Systems*. New York: Wiley & Sons, 1979.

[26] A. Pottbacker, U. Langmann, and H. U. Schreiber, "A Si bipolar phase and frequency detector IC for clock extraction up to 8 Gb/s," *IEEE J. Solid-State Circuits*, vol. 27, pp. 1747–1751, December 1992.

[27] D. Richman, "Color-carrier reference phase synchronization accuracy in NTSC color television," *Proc. IRE*, vol. 42, pp. 106–133, January 1954.

[28] J. A. Bellisio, "A new phase-locked loop timing recovery method for digital regenerators," *IEEE Int. Communications Conf. Rec.*, vol. 1, June 1976, pp. 10-17–10-20.

[29] F. M. Gardner, "Properties of frequency difference detectors," *IEEE Trans. Comm.*, vol. COM-33, pp. 131–138, February 1985.

[30] D. G. Messerschmitt, "Frequency detectors for PLL acquisition in timing and carrier recovery," *IEEE Trans. Comm.*, vol. COM-27, pp. 1288–1295, September 1979.

[31] C. R. Hogge, "A self-correcting clock recovery circuit," *IEEE J. Lightwave Technology*, vol. LT-3, pp. 1312–1314, December 1985.

[32] L. De Vito et al., "A 52 MHz and 155 MHz clock recovery PLL," *ISSCC Dig. Tech. Papers*, pp. 142–143, San Francisco, CA, February 1991.

[33] T. H. Lee and J. F. Bulzacchelli, "A 155-MHz clock recovery delay- and phase-locked loop," *IEEE J. Solid-State Circuits*, vol. 27, pp. 1736–1746, December 1992.

[34] J. Alvarez et al., "A wide-bandwidth low-voltage PLL for PowerPC microprocessors," *IEEE J. Solid-State Circuits*, vol. 30, pp. 383–391, April 1995.

[35] J. A. Crawford. *Frequency Synthesizer Design Handbook*. Norwood, MA: Artech House, 1994.

Part 1
Basic Theory

IN this part, a number of papers dealing with theoretical aspects of phase-locked systems are presented. The first two papers are concerned with the basic PLL and its properties, including transient and frequency response, frequency detection, and capture behavior. The next three papers analyze PLLs with discrete-time behavior, an important topic in many of today's implementations. The four papers following are related to frequency detectors and clock recovery circuits.

The remaining papers in this part are devoted to the subject of phase noise: the first three in the architecture and the rest in the oscillator. While phase noise has always been of interest, it has found new importance in monolithic implementations. De-vice noise, supply and substrate noise, circuit nonlinearities, and the lack of integrated passive resonators have made the problem of phase noise a challenging one.

Additional References

W. R. Bennet, "Statistics of regenerative digital transmission," *Bell Syst. Tech. J.,* vol. 37, pp. 1501–1542, 1958.

D. L. Duttweiler, "The jitter performance of phase-locked loops extracting timing and baseband data waveforms," *Bell Syst. Tech. J.,* vol. 55, pp. 37–58, 1976.

F. M. Gardner, "Phase accuracy of charge pump PLL's," *IEEE Trans. Comm.,* vol. 30, pp. 2362–2363, October 1982.

K. Kurokawa, "Noise in synchronized oscillators," *IEEE Trans. Microwave Theory Techn.,* vol. 16, pp. 234–240, April 1968.

Theory of AFC Synchronization*

WOLF J. GRUEN†, MEMBER, IRE

Summary—The general solution for the important design parameters of an automatic frequency and phase-control system is presented. These parameters include the transient response, frequency response and noise bandwidth of the system, as well as the hold-in range and pull-in range of synchronization.

I. INTRODUCTION

AUTOMATIC FREQUENCY and phase-control systems have been used for a number of years for the horizontal-sweep synchronization in television receivers, and more recently have found application for the synchronization of the color subcarrier in the proposed NTSC color-television system. A block diagram of a general AFC system is shown in Fig. 1.

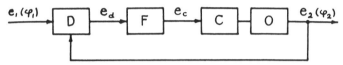

Fig. 1—Block diagram of A.F.C. loop.

The phase of the transmitted synchronizing signal e_1 is compared to the phase of a local oscillator signal e_2 in a phase discriminator D. The resulting discriminator output voltage is proportional to the phase difference of the two signals, and is fed through a control network F to a frequency-control stage C. This stage controls the frequency and phase of a local oscillator O in accordance with the synchronizing information, thereby keeping the two signals in perfect synchronism. Although in practice the transmitted reference signal is often pulsed and the oscillator comparison voltage non-sinusoidal, the analysis is carried out for sinusoidal signal voltages. The theory, however, can be extended for a particular problem by writing the applied voltages in terms of a Fourier series instead of the simple sine function. An AFC system is essentially a servomechanism, and the notation that will be used is the one followed by many workers in this field. An attempt will be made to present the response characteristics in dimensionless form in order to obtain a universal plot of the response curves.

II. DERIVATION OF THE BASIC EQUATION

If it is assumed that the discriminator is a balanced phase detector composed of peak-detecting diodes, the discriminator-output voltage can be derived from the vector diagram in Fig. 2. For sinusoidal variation with time, the synchronizing signal e_1 and the reference signal

* Decimal classification: R583.5. Original manuscript received by the Institute, August 21, 1952; revised manuscript received February 25, 1953.
† General Electric Co., Syracuse, N. Y.

e_2 can be written

$$e_1 = E_1 \cos \phi_1 \tag{1}$$

and

$$e_2 = E_2 \sin \phi_2. \tag{2}$$

ϕ_1 and ϕ_2 are functions of time and, for reasons of simplicity in the later development, it is arbitrarily assumed that ϕ_1 and ϕ_2 are in quadrature when the system is perfectly synchronized, that is when $\phi_1 = \phi_2$.

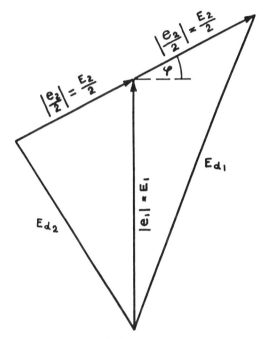

Fig. 2—Discriminator vector diagram.

While one of the discriminator diodes is fed with the sum of e_1 and $e_2/2$, the other is fed with the difference of these two vectors as shown in Fig. 2. The resulting rectified voltages E_{d1} and E_{d2} can be established by simple trigonometric relations. Defining a difference phase

$$\phi \equiv \phi_1 - \phi_2, \tag{3}$$

one obtains

$$E_{d1}^2 = E_1^2 + \frac{E_2^2}{4} + E_1 E_2 \sin \phi \tag{4}$$

and

$$E_{d2}^2 = E_1^2 + \frac{E_2^2}{4} - E_1 E_2 \sin \phi. \tag{5}$$

The discriminator output voltage e_d is equal to the difference of the two rectified voltages, so that

$$e_d = E_{d1} - E_{d2} = \frac{2E_1E_2}{E_{d1} + E_{d2}} \sin \phi. \quad (6)$$

If the amplitude E_1 of the synchronizing signal is larger than the amplitude E_2 of the reference signal, one obtains

$$E_{d1} + E_{d2} \cong 2E_1. \quad (7)$$

The discriminator output voltage then becomes

$$e_d = E_2 \sin \phi \quad (8)$$

and is independent of the amplitude E_1 of the synchronizing signal. As ϕ_1 and ϕ_2 are time-varying parameters, it should be kept in mind that the discriminator time constant ought to be shorter than the reciprocal of the highest difference frequency $d\phi/dt$, which is of importance for the operation of the system.

Denoting the transfer function of the control network F as $F(p)$, the oscillator control voltage becomes

$$e_c = F(p)E_2 \sin \phi. \quad (9)$$

Assuming furthermore that the oscillator has a linear-control characteristic of a slope S, and that the free-running oscillator frequency is ω_0, the actual oscillator frequency in operational notation becomes

$$p\phi_2 = \omega_0 + Se_c. \quad (10)$$

Substituting (3) and (9) into (10) then gives

$$p\phi + SE_2F(p) \sin \phi = p\phi_1 - \omega_0. \quad (11)$$

The product SE_2 repeats itself throughout this paper and shall be defined as the gain constant

$$K \equiv SE_2. \quad (12)$$

K represents the maximum frequency shift at the output of the system per radian phase shift at the input. It has the dimension of radians/second.

Equation (11) can be simplified further by measuring the phase angles in a coordinate system which moves at the free-running speed ω_0 of the local oscillator. One obtains

$$\boxed{p\phi + F(p)K \sin \phi = p\phi_1} \quad (13)$$

This equation represents the general differential equation of the AFC feedback loop. $p\phi$ is the instantaneous-difference frequency between the synchronizing signal and the controlled-oscillator signal and $p\phi_1$ is the instantaneous-difference frequency between the synchronizing signal and the free-running oscillator signal.

Equation (13) shows that all AFC systems with identical gain constants K and unity d.c. gain through the control network have the same steady-state solution, provided that the difference frequency $p\phi_1$ is constant. If this difference frequency is defined as

$$\Delta\omega \equiv p\phi_1 = \omega_1 - \omega_0, \quad (14)$$

the steady-state solution is

$$\sin \phi = \frac{\Delta\omega}{K}. \quad (15)$$

This means the system has a steady-state phase error which is proportional to the initial detuning $\Delta\omega$ and inversely proportional to the gain constant K. Since the maximum value of $\sin \phi$ in (15) is ± 1, the system will hold synchronism over a frequency range

$$|\Delta\omega_{\text{Hold-in}}| \leq K. \quad (16)$$

Equations (15) and (16) thus define the static performance limit of the system.

III. Linear Analysis

An AFC system, once it is synchronized, behaves like a low-pass filter. To study its performance it is permissible, for practical signal-to-noise ratios, to substitute the angle for the sine function in (13). Then, with the definition of (3), one obtains

$$p\phi_2 + KF(p)\phi_2 = KF(p)\phi_1. \quad (17)$$

This equation relates the output phase ϕ_2 of the synchronized system to the input phase ϕ_1. It permits an evaluation of the behavior of the system to small disturbances of the input phase, if the transfer function $F(p)$ of the control network is specified.

a. $F(p) = 1$

This is the simplest possible AFC system, and represents a direct connection between the discriminator output and the oscillator control stage. Equation (17) then becomes

$$p\phi_2 + K\phi_2 = K\phi_1. \quad (18)$$

If the initial detuning is zero, the transient response of the system to a sudden step of input phase $|\phi_1|$ is

$$\frac{\phi_2}{|\phi_1|}(t) = 1 - e^{-Kt}. \quad (19)$$

Likewise, the frequency response of the system to a sine-wave modulation of the input phase is

$$\frac{\phi_2}{\phi_1}(j\omega) = \frac{1}{1 + j\dfrac{\omega}{K}}. \quad (20)$$

The simple AFC system thus behaves like an RC-filter and has a cut-off frequency of

$$\omega_c = K \text{ [radians/sec]}. \quad (21)$$

George[1] has shown that the m.s. phase error of the system under the influence of random interference is proportional to the noise bandwidth, which is defined as

---

[1] T. S. George, "Synchronizing systems for dot interlaced color TV," Proc. I.R.E.; February, 1951.

$$B = \int_{-\infty}^{+\infty} \left| \frac{\phi_2}{\phi_1} (j\omega) \right|^2 d\omega. \tag{22}$$

The integration has to be carried out from $-\infty$ to $+\infty$ since the noise components on both sides of the carrier are demodulated. Inserting (20) into (22) then yields

$$B = \pi K \text{ [radians/sec]}. \tag{23}$$

It was shown in (15) that for small steady-state phase errors due to average frequency drift, the gain constant K has to be made as large as possible, while now for good noise immunity, i.e., narrow bandwidth, the gain constant has to be made as small as possible. A proper compromise of gain then must be found to insure adequate performance of the system for all requirements. This difficulty, however, can be overcome by the use of a more elaborate control network.

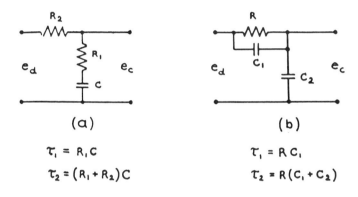

$$\tau_1 = R_1 C$$
$$\tau_2 = (R_1 + R_2)C$$

$$\tau_1 = R C_1$$
$$\tau_2 = R(C_1 + C_2)$$

$$\frac{e_c}{e_d}(p) = \frac{1 + \tau_1 p}{1 + \tau_2 p}$$

Fig. 3—Proportional plus integral control networks.

b. $F(p) = \dfrac{1 + \tau_1 p}{1 + \tau_2 p}$

Networks of this type are called proportional-plus integral-control networks[2] and typical network configurations are shown in Fig. 3. Inserting the above transfer function into (17) yields

$$p^2\phi_2 + \left(\frac{1}{\tau_2} + K\frac{\tau_1}{\tau_2} \right) p\phi_2 + \frac{K}{\tau_2}\phi_2$$
$$= K\frac{\tau_1}{\tau_2} p\phi_1 + \frac{K}{\tau_2}\phi_1. \tag{24}$$

ϕ_1 and ϕ_2 are again relative phase angles, measured in a coordinate system which moves at the free-running speed of the local oscillator. To integrate (24), it is convenient to introduce the following parameters

$$\omega_n^2 \equiv \frac{K}{\tau_2} \tag{25}$$

[2] G. S. Brown and D. P. Campbell, "Principles of Servomechanisms," John Wiley & Sons Publishing Co., New York, N. Y.; 1948.

and

$$2\zeta\omega_n \equiv \frac{1}{\tau_2} + K\frac{\tau_1}{\tau_2}. \tag{26}$$

ω_n is the resonance frequency of the system in the absence of any damping, and ζ is the ratio of actual-to-critical damping. In terms of the new parameters the time constants of the control network are

$$\tau_1 = \frac{2\zeta}{\omega_n} - \frac{1}{K} \tag{27}$$

and

$$\tau_2 = \frac{K}{\omega_n^2}. \tag{28}$$

With these definitions (24) becomes

$$p^2\phi_2 + 2\zeta\omega_n p\phi_2 + \omega_n^2\phi_2$$
$$= \left(2\zeta\omega_n - \frac{\omega_n^2}{K} \right) p\phi_1 + \omega_n^2\phi_1. \tag{29}$$

The transient response of the system to a sudden step of input phase $|\phi_1|$ is found by integration of (29) and the initial condition for the oscillator frequency is obtained from (10). The transient response then is

$$\frac{\phi_2}{|\phi_1|}(t) = 1 - e^{-\zeta\omega_n t}\left[\cos\sqrt{1-\zeta^2}\,\omega_n t - \frac{\zeta - \dfrac{\omega_n}{K}}{\sqrt{1-\zeta^2}} \sin\sqrt{1-\zeta^2}\,\omega_n t \right]. \tag{30}$$

For $\zeta < 1$ the system is underdamped (oscillatory), for $\zeta = 1$ critically damped and for $\zeta > 1$ overdamped (nonoscillatory). In order to avoid sluggishness of the system, a rule of thumb may be followed making. $4 < \zeta < 1^2$. The transient response of (30) can be plotted in dimensionless form if certain specifications are made for the ratio ω_n/K. As the time constant τ_1 of the control network must be positive or can at most be equal to zero, the maximum value for ω_n/K is found from (27), yielding

$$\frac{\omega_n}{K}\bigg|_{\max} = 2\zeta. \tag{31}$$

In this case the control network is reduced to a single time constant network ($\tau_1 = 0$). On the other hand, if for a fixed value of ω_n the gain of the system is increased towards infinity, the minimum value for ω_n/K becomes

$$\frac{\omega_n}{K}\bigg|_{\min} = 0. \tag{32}$$

Fig. 4 shows the transient response of the system for these two limits and for a damping ratio of $\zeta = 0.5$.

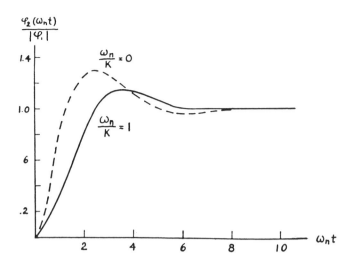

Fig. 4—Transient response for $\zeta = 0.5$.

The frequency response of the system is readily found from (24) and one obtains

$$\frac{\phi_2}{\phi_1}(j\omega) = \frac{1 + j2\zeta\frac{\omega}{\omega_n}\left(1 - \frac{\omega_n}{2\zeta K}\right)}{1 + j2\zeta\frac{\omega}{\omega_n} - \left(\frac{\omega}{\omega_n}\right)^2}. \quad (33)$$

Its magnitude is plotted in Fig. 5 for the two limit values of ω_n/K and for a damping ratio $\zeta = 0.5$. The curves show that the cut-off frequency of the system, for $\zeta = 0.5$, is approximately

$$\omega_c \cong \omega_n \text{ [radians/sec.]}. \quad (34)$$

If ϕ_1 and ϕ_2 in (33) are assumed to be the input and output voltage of a four-terminal low-pass filter, the frequency response leads to the equivalent circuit of Fig. 6.

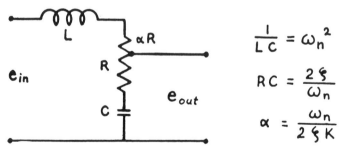

Fig. 6—Equivalent low-pass filter.

The noise bandwidth of the system is established by inserting (33) into (22) and one obtains

$$B = \omega_n \int_{-\infty}^{+\infty} \frac{1 + 4\zeta^2\left(\frac{\omega}{\omega_n}\right)^2\left[1 - \frac{\omega_n}{2\zeta K}\right]^2}{1 - (2 - 4\zeta^2)\left(\frac{\omega}{\omega_n}\right)^2 - \left(\frac{\omega}{\omega_n}\right)^4} d\left(\frac{\omega}{\omega_n}\right). \quad (35)$$

The integration, which can be carried out by partial fractions with the help of tables, yields

$$B = \frac{4\zeta^2 - 4\zeta\frac{\omega_n}{K} + \left(\frac{\omega_n}{K}\right)^2 + 1}{2\zeta}\pi\omega_n. \quad (36)$$

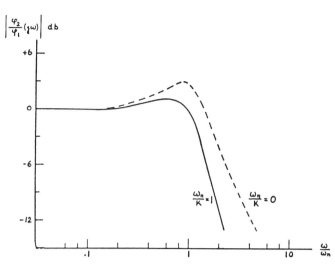

Fig. 5—Frequency response for $\zeta = 0.5$.

For small values of ω_n/K, it is readily established that this expression has a minimum when $\zeta = 0.5$. Hence, the noise bandwidths for the limit values of ω_n/K and $\zeta = 0.5$ become

$$B \big|_{(\omega_n/K) \to 1} = \pi\omega_n = \pi K \text{ [radians/sec]} \quad (37)$$

and

$$B \big|_{(\omega_n/K) \to 0} = 2\pi\omega_n \text{ [radians/sec]}. \quad (38)$$

The above derivations, as well as the response curves of Figs. 4 and 5, show that the bandwidth and the gain constant of the system can be adjusted independently if a double time-constant control network is employed.

c. Example

The theory is best illustrated by means of an example. Suppose an AFC system is to be designed, having a steady state phase error of not more than 3° and a noise bandwidth of 1,000 cps. The local oscillator drift shall be assumed with 1,500 cps.

The required gain constant is obtained from (15), yielding

$$K = \frac{\Delta\omega}{\sin\phi} = \frac{2\pi \cdot 1,500}{\sin 3°} = 180,000 \text{ radians/sec.}$$

Since K is large in comparison to the required bandwidth, the resonance frequency of the system is established from (38).

$$\omega_n = \frac{B}{2\pi} = \frac{2\pi \cdot 1,000}{2\pi} = 1,000 \text{ radians/sec.}$$

The two time constants of the control network, assuming a damping ratio of 0.5, are determined from (27) and (28) respectively

$$\tau_1 = \frac{2\zeta}{\omega_n} - \frac{1}{K} = \frac{1}{1,000} - \frac{1}{180,000} \cong 10^{-3} \text{ sec,}$$

and

$$\tau_2 = \frac{K}{\omega_n^2} = \frac{180,000}{1,000^2} = 0.18 \text{ sec.}$$

46

These values K, τ_1, and τ_2 completely define the AFC system. A proper choice of gain distribution and control-network impedance still has to be made to fit a particular design. For example, if the peak amplitude of the sinusoidal oscillator reference voltage is $E_2 = 6$ volts, the sensitivity of the oscillator control stage must be $S = 30{,}000$ radians/sec/volt to provide the necessary gain constant of 180,000 radians/sec. Furthermore, if the capacitor C for the control network of Fig. 3(a) is assumed to be 0.22 uf, the resistors R_1 and R_2 become 4.7 kΩ and 820 kΩ respectively, to yield the desired time constants.

IV. Non-Linear Analysis

While it was permissible to assume small phase angles for the study of the synchronized system, thereby linearizing the differential (13), this simplification cannot be made for the evaluation of the pull-in performance of the system. The pull-in range of synchronization is defined as the range of difference frequencies, $p\phi_1$, between the input signal and the free-running oscillator signal, over which the system can reach synchronism. Since the difference phase ϕ can vary over many radians during pull-in, it is necessary to integrate the nonlinear equation to establish the limit of synchronization.

Assuming that the frequency of the input signal is constant as defined by (14), (13) can be written

$$p\phi + F(p)K \sin \phi = \Delta\omega. \qquad (39)$$

Mathematically then, the pull-in range of synchronization is the maximum value of $\Delta\omega$ for which, irrespective of the initial condition of the system, the phase difference ϕ reaches a steady state value. To solve (39), the transfer function of the control network again must be defined.

a. $F(p) = 1$

The pull-in performance for this case has been treated in detail by Labin.[3] With $F(p) = 1$ (39) can be integrated by separation of the variables and it is readily found that the system synchronizes for all values of $|\Delta\omega| < K$. The condition for pull-in then is

$$|\Delta\omega|_{\text{Pull-in}} < K. \qquad (40)$$

Large pull-in range and narrow-noise bandwidth thus are incompatible requirements for this system.

b. $F(p) = \dfrac{1 + \tau_1 p}{1 + \tau_2 p}$

Inserting this transfer function into (39) and carrying out the differentiation yields

$$\frac{d^2\phi}{dt^2} + \left[\frac{1}{\tau_2} + K\frac{\tau_1}{\tau_2}\cos\phi\right]\frac{d\phi}{dt} + \frac{K}{\tau_2}\sin\phi = \frac{\Delta\omega}{\tau_2}. \qquad (41)$$

This equation can be simplified by inserting the coefficients defined in (25) and (26), and by dividing the resulting equation by $\omega_n{}^2$. This leads to the dimensionless equation.

$$\frac{d^2\phi}{\omega_n{}^2 dt^2} + \left[\frac{\omega_n}{K} + \left(2\zeta - \frac{\omega_n}{K}\right)\cos\phi\right]\frac{d\phi}{\omega_n dt} + \sin\phi = \frac{\Delta\omega}{K}. \qquad (42)$$

A further simplification is possible by defining a dimensionless difference frequency

$$y \equiv \frac{d\phi}{\omega_n dt} \qquad (43)$$

and one obtains a first order differential equation from which the dimensionless time $\omega_n t$ has been eliminated. It follows

$$\frac{dy}{d\phi} = \frac{\dfrac{\Delta\omega}{K} - \sin\phi}{y} - \frac{\omega_n}{K} - \left(2\zeta - \frac{\omega_n}{K}\right)\cos\phi. \qquad (44)$$

There is presently no analytical method available to solve this equation. However, the equation completely defines the slope of the solution curve $y(\phi)$ at all points of a $\phi - y$ plane, except for the points of stable and unstable equilibrium, $y = 0$; $\Delta\omega/K = \sin\phi$. The limit of synchronization can thus be found graphically by starting the system with an infinitesimal velocity Δy at a point of unstable equilibrium, $y = 0$; $\phi = \pi - \sin^{-1}\Delta\omega/K$, and finding the value of $\Delta\omega/K$ for which the solution curve just reaches the next point of unstable equilibrium located at $y = 0$; $\phi = 3\pi - \sin^{-1}\Delta\omega/K$. The method is discussed by Stoker[4] and has been used by Tellier and Preston[5] to find the pull-in range for a single time constant AFC system.

To establish the limit curve of synchronization for given values of ζ and ω_n/K, a number of solution curves have to be plotted with $\Delta\omega/K$ as parameter. The limit of pull-in range in terms of $\Delta\omega/K$ then can be interpolated to any desired degree of accuracy. The result, obtained in this manner, is shown in the dimensionless graph of Fig. 7, where $\Delta\omega/K$ is plotted as a function of ω_n/K for a damping ratio $\zeta = 0.5$. Since this curve represents the stability limit of synchronization for the system, the time required to reach synchronism is infinite when starting from any point on the limit curve. The same applies to any point on the $\Delta\omega/K$-axis, with exception of the point $\Delta\omega/K = 0$, since this axis describes a system having either infinite gain or zero bandwidth, and neither case has any real practical significance. The practical pull-in range of synchronization, therefore, lies inside the solid boundary. The individual points

[3] Edouard Labin, "Theorie de la synchronization par controle de phase," *Philips Res. Rep.*, (in French); August, 1941.

[4] J. J. Stoker, "Non-linear vibrations," *Interscience*; New York, 1950.

[5] G. W. Preston and J. C. Tellier, "The Lock-in Performance of an A.F.C. Circuit," Proc. I.R.E.; February, 1953.

entered in Fig. 7 represent the measured pull-in curve of a particular system for which the damping ratio was maintained at $\zeta = 0.5$. For small values of ω_n/K this

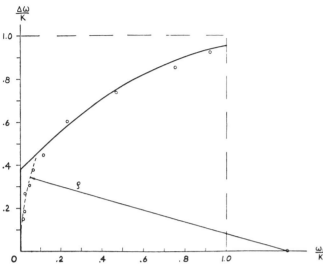

Fig. 7—Pull-in range of synchronization for $\zeta = 0.5$.

pull-in curve can be approximated by its circle of curvature which, as indicated by the dotted line, is tangent to the $\Delta\omega/K$-axis and whose center lies on the ω_n/K-axis. The pull-in range thus can be expressed analytically by the equation of the circle of curvature. If its radius is denoted by ζ, the circle is given by

$$\left(\frac{\omega_n}{K} - \zeta\right)^2 + \left(\frac{\Delta\omega}{K}\right)^2 = \zeta^2. \qquad (45)$$

Hence, for $(\omega_n/K) \to 0$, the pull-in range of synchronization is approximately

$$\left| \Delta\omega_{\text{Pull-in}} \right|_{(\omega_n/K)\to 0} < \sqrt{2\zeta\omega_n K - \omega_n^2} \cong \sqrt{2\zeta\omega_n K}. \qquad (46)$$

ζ can be interpreted as a constant of proportionality which depends on the particular design of the system, and which increases as the system gets closer to the theoretical limit of synchronization.

Equation (46) shows that the pull-in range for small values of ω_n/K is proportional to the square root of the product of the cut-off frequency ω_n and the gain constant K. Since the bandwidth of a double time constant AFC system can be adjusted independently of the gain constant, the pull-in range of such a system can exceed the noise bandwidth by any desired amount.

V. Conclusions

The performance of an AFC system can be described by three parameters. These are the gain constant K, the damping ratio ζ and the resonance or cut-off frequency ω_n. These parameters are specified by the requirements of a particular application and define the over-all design of the system. It has been shown that among the systems with zero, single and double time constant control networks, only the latter fulfills the requirement for achieving good noise immunity, small steady-state phase error and large pull-in range.

Color-Carrier Reference Phase Synchronization Accuracy in NTSC Color Television*

DONALD RICHMAN†, SENIOR MEMBER, IRE

Summary—The results of an evaluation of the capabilities of the NTSC color-carrier reference signal (the color burst) show this new color television synchronizing signal to be more than adequate; information inherent to the signal permits performance far in excess of that achieved by conventional circuits.

Phasing information inherent to the burst is considered first with particular regard to measures of accuracy, the required amount of integration, and the extent of the spectral region necessary to translate the burst information.

Properties of elementary passive and active circuits for using the burst in receivers are described along with a determination of the limits of burst synchronization performance for these circuits.

Fundamental considerations in the theory of synchronization show that better performance is obtainable with two-mode systems.

Properties of two-mode systems are considered and lead to an evaluation of the limits of synchronizing performance permitted by the color burst.

The mathematical derivations necessary to support the discussion are presented in the Appendixes.

NTSC COLOR television adds color to a monochrome picture by means of a narrow-band, frequency interleaved carrier color signal which carries one component of the color information in its phase, and another component in its amplitude. It is customary to provide a phase reference in the transmitted signal in order that receivers shall be able to measure the instantaneous phase angle of the carrier color signal so as to reproduce the desired color. This is accomplished by transmitting a short burst of oscillations at color subcarrier frequency during line retrace intervals,[1] at a reference phase which corresponds to the (*Y-B*) axis.[2]

The color burst carries phasing information. This paper shows how much phasing information is contained in the color burst, and how it may be used.

Analysis of the factors limiting performance shows that, even under extreme conditions of interference and of stabilization requirements, the burst contains adequate information to provide a reliable color-carrier reference signal; in fact, the amount of phasing information in the color signal appears adequate enough so that a customer-operated control relating to color sync should be unnecessary on NTSC color television receivers. Analysis shows that presently used sync instrumentation systems appear capable of meeting but not necessarily exceeding a reasonable measure of the above requirements. However, information existing in the *signal* permits substantially better performance.

The real limits of performance and sync systems which more fully utilize the signal information are discussed in this paper. Because of the excess of existing information, a variety of types of circuits can be used.

Several questions may be asked with regard to the amount of phasing information contained in the color burst and its application to provide a reference signal for color demodulation. These are: (a) How closely can the color-carrier reference signal be maintained to the true value, when signals are strong (and hence noise-free) and after transient effects have subsided? (b) How closely can the color-carrier reference signal be maintained in the presence of noise interference? (c) How long will a system or circuit designed to give satisfactory operation on (a) and (b) require to reach a stable mode of operation when stations are switched or a receiver is turned on? (d) How much performance is required in (a), (b), and (c)?

Of these questions, (d) is the most difficult to answer precisely; it depends on many subjective factors and may be obscured by temporary equipment difficulties. In order to provide a standard of comparison for use in this paper, a conservative (pessimistic) estimate has been made, based on past experience.

The answers to the questions are as follows:

(a) With a strong (clean) sync signal, the color-carrier reference signal may be maintained as closely accurate as desired, independent of other factors; in the presence of noise, the *average* phase may be maintained as closely as desired, independent of the required integration and transient characteristics; for example, designs presented later show how the static or average phase of the color-carrier reference signal may be controlled to within five degrees of the true value. Expressed as a time value this is an accuracy of approximately six mμsec. This phase accuracy implies a color fidelity probably substantially better than can be distinguished by the observer.[3]

(b) *The real limitation on performance* is thermal-noise interference, since this type of interference is the most difficult type to reject. It *is* rejected, however, *to any selected measure of reliability* by integration of the synchronization timing information over a suitably long period. Either of two basic types of integrators may be used. These are, one, passive integrators, and two, frequency-and-phase-locked self-oscillating integrators. The analysis presented in this paper shows that, under severe assumptions on the requirements of phase stability and signal-to-noise ratio, the required integration time for passive integrators is of the order of mag-

* Decimal classification: R583. NTSC Technical Monograph No. 7, reprinted by permission of the National Television System Committee from "Color System Analysis," report of NTSC Panel 12.
† Hazeltine Corp., Little Neck, N. Y.
[1] "Recent developments in color synchronization in the RCA color television system," RCA Labs. Report, Princeton, N. J.; Feb., 1950.
[2] Fig. 1 of "Minutes of the Meeting of Panel 14," NTSC; May 20, 1952.
[3] D. L. MacAdam, "Quality of color reproduction," PROC. I.R.E., vol. 39, pp. 468–485; May, 1951.

Reprinted with permission from *Proc. of IRE*, D. Richman, "Color-Carrier Reference Phase Synchronization Accuracy in NTSC Color Television," vol.42, pp. 106–133, January 1954. © Institution of Electrical Engineers.

nitude of 0.005 second, or less than a sixth of a frame period. Locked integrators on the same assumptions require 0.01 second for the integration to take place.

(c) The third requirement, of pull-in or stabilization time, is also limited by the signal-to-noise ratio and the requirement for integration. This may vary considerably with the method of instrumentation, but the limiting or optimum performance with regard to stabilization time is determined by the information carried in the signal; the limit imposed by signal information is found to be (for a reasonable measure of reliability) a few times the integration time discussed above. Later in this report this is shown to be approached under certain conditions by fairly simple passive integrators. It is also shown how locked integrators, characterized by some new forms of automatic frequency- and phase-control loops, may be made to achieve the upper limit of performance. Typical present APC (automatic phase control) circuits fall somewhat short of this limit, but when properly designed can be made to pull in quickly enough so as to appear virtually instantaneous, while permitting most of the burden of frequency stability to be borne by the transmitter.

These facts lead to the conclusion that there is adequate information in the color burst for completely automatic operation, without need for a customer control. The factors leading to this conclusion are presented in the following sequence:

Performance limitations for sync systems which are already synchronized are discussed first, in the section on "Synchronization Accuracy." The reliability of phase difference measurements, and factors relating to the integration time necessary to obtain a specified measure of reliability in the presence of noise are considered.

Then performance limitations of instrumentation systems are discussed with particular regard to the process of synchronization. The basic characteristics of passive and locked integrators are discussed in the section on "Elemental Sync Systems."

Evaluation of ultimate limitations for the signal, and factors leading to new sync systems capable of fully utilizing the signal information are presented in the section on "Theory of Synchronization." Factors of interest are mechanisms of pull-in, the reliability of frequency difference measurements, and the exchange of integration time for a specified measure of reliability in the presence of noise.

Effects of echoes and stability of the gate are briefly discussed.

The conclusions drawn regarding the adequacy of the signal are stated.

Mathematical derivations, which substantiate and illustrate the facts presented in this paper, are presented in several appendexes.

The NTSC Color Synchronizing Signal

Fig. 1 shows the NTSC color synchronizing signal in relation to the video and synchronizing wave form, in the vicinity of one line-retrace interval. It consists

of a burst of approximately 9 cycles of sinusoidal wave form at the color-carrier frequency of 3,579,545 ($\pm$.0003%) cps,[4] approximately centered on the portion of the line blanking pulse following each horizontal sync pulse. It is omitted during the nine lines in each field in which the field synchronizing information is transmitted.

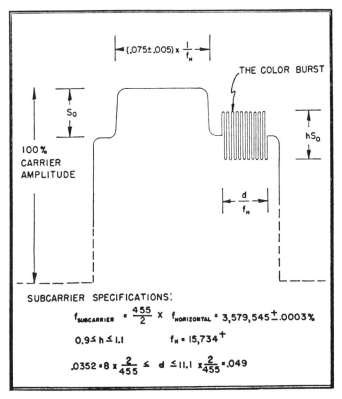

Fig. 1—Wave form during line retrace interval showing horizontal sync pulse and the NTSC burst reference signal.

Parameters of interest which are shown on the figure are:

S_0 = the amplitude of the line and field sync pulses, normally 25% of peak carrier amplitude measured in the video signal.
hS_0 = the peak-to-peak amplitude of the burst, measured in the video signal.
f_H = the line scanning frequency.
d = the duty cycle of the burst.

The color burst is used in the color television receiver to provide a control signal for the generation of a local continuous wave signal at the nominal burst frequency and locked to it in phase.

Synchronization Accuracy

Synchronizing Information

Any time-varying signal can carry timing information, the character of which depends on the distribution of signal energy throughout the frequency spectrum. In

---

[4] As specified by NTSC in February, 1953. The analysis is not critically dependent on the exact value of the color carrier frequency.

the case of a continuous sine wave, this timing information consists only of phase reference information because it is impossible to identify cycles of the carrier from each other. The same is essentially true of the pulse modulated sine wave which constitutes the burst; envelope information in the burst is not used. It is this phase reference information which is of interest with regard to color-carrier reference phase synchronization.

A signal which passes only through linear noiseless channels may be located in time (or phase) with theoretically unlimited precision. In the presence of noise the data obtained by a time (or *phase*) meter from the signal will fluctuate. This occurs because the timing information which can be extracted from the combination of signal-plus-noise in any specified interval is limited by the signal-to-noise ratio as well as by the statistical characteristics of the signal and noise.

Integration for Signal-to-Noise Ratio Improvement

The fluctuations in the phase data may be smoothed by integration. For example, the instantaneous output of the phase meter may represent the average of all data obtained over some preceding integration period T_M in duration.

Any measuring device which uses any form of integration or memory averages some effective number of independent measurements. One such integrator directly obtains a suitably weighted average (such as the least square error average) of all the data obtained in the preceding period T_M. Such an integrator provides a standard of comparison. Other forms of integrators may then be characterized by their effective integration times, T_M; several practical integrators are described later.

A section of a signal existing in an interval of duration T_M may be expressed as a sum of harmonics of the fundamental frequency $(1/T_M)$; the noise bandwidth associated with each component is equal to the spacing between components, or $1/T_M = f_N$. This means, for example, that if all of the timing information obtained in a period T_M from a signal consisting effectively of a signal sinusoidal component is averaged, that an improvement in reliability is obtained equivalent to that produced by passing the signal through a filter having a noise bandwidth of f_N.

Noise Interference

Noise is specified by its energy content and statistical characteristics. For a flat energy spectrum, taken as an example, impulse noise and white thermal noise represent opposite extremes, since for white thermal noise the relative phases of the several frequency components are completely random and incoherent; for impulse noise the relative phases of all components are related and are not random, although the time of occurrence of any impulse is a random variable.

Noise may be measured in terms of any convenient co-ordinate system into which the signal-plus-noise may be transformed, such as frequency, phase, amplitude, time of arrival, or more complex parameters.

Thermal noise is the most difficult to reject. It may be discriminated against only by averaging; this makes the effective error due to noise vary inversely as the square root of the number of measurements; hence, (for systems with fixed bandwidth) the error varies inversely as the square root of the integration time.

Impulse noise, or noise intermediate between thermal and impulse noise, may be rejected more easily than thermal noise since it represents a signal which can be recognized with a high measure of reliability and removed from the transmission channel.

A synchronizing system is a form of predictor which bases its estimates on past experience. When the input to the system has such a character (such as an improbable amplitude) that it is recognized with high reliability to be a disturbance, it is usually much better to use (at least approximately) the predicted signal as the input to the system for the duration of the disturbance. An equipment system for performing these operations is called an *aperture*. (Aperture systems are now widely used for line and field sync; the same principles are involved in the application to burst sync.)

Since thermal noise represents the most serious (as well as perhaps the most common) limitation to color synchronization performance, it is used in this paper as the measure of interference which must be overcome.

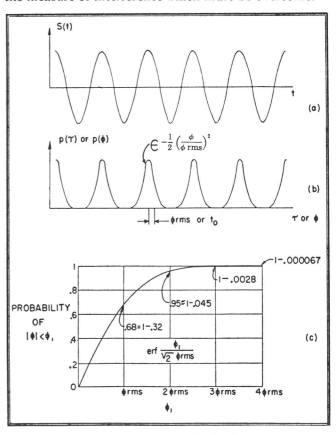

Fig. 2—Timing error distribution.

Measures of Reliability

A section of the burst reference signal is represented as $S(t)$ in Fig. 2(a). The time scale associated with the synchronizing signal may be identified with some representative point in a cycle which is selected as a reference.

51

The timing accuracy which is obtained for a given signal-to-noise ratio may be expressed in terms of a relative probability density function $p(\tau)$ such as is plotted in the curve of Fig. 2(b). The relative probability density curve permits the determination of the probability that the sync timing answer which results from a single measurement of the sync signal, using all of the information derived from the preceding period T_M, will occur within a specified time or phase interval. This probability is proportional to the area under the curve $p(\tau)$ or $p(\phi)$ within the specified interval. Due to the cyclic nature of the information, the time scale may be replaced by a phase scale. The curve for $p(\phi)$ defines the probability laws for the noise at the output of the synchronizing system. The curve is repetitive at the sync frequency. (The output noise from the sync measuring device has the same basic character from cycle to cycle.) For many signal energy distributions, and particularly for burst synchronization at the levels of output noise which give satisfactory performance, the curve $p(\tau)$ or $p(\phi)$ has very nearly the shape of a normal or Gaussian probability curve represented by the expression

$$\epsilon^{-1/2(t/t_0)^2} \quad \text{or} \quad \epsilon^{-1/2(\phi/\phi_{rms})^2}$$

in which case the phasing information may be completely described by the rms time error, t_0, or the rms phase error ϕ_{rms}, which may be expected for a specified set of measurement conditions.

For this case of the normal law the absolute probability that any measurement will yield an answer within a specified measure of the true answer may be represented in terms of the rms error. Fig. 2(c), which represents the integral of one lobe of the curve of Fig. 2(b) for the normal law, represents the probability that the magnitude of the phase error at any time is less than some selected phase error ϕ_1. ϕ_1 is measured in multiples of ϕ_{rms}. The curve illustrates that the probability is nearly unity only when ϕ_1 approaches $4\phi_{rms}$, which means that the effective peak value of Gaussian noise is near four times the rms value.[5]

The Sync Accuracy Equation

The parameters which determine the rms time error, seconds, for burst sync are:

The signal amplitude $\frac{1}{2}hS_0$ volts.
The duty cycle of the gated sine wave d as a fraction.
The rms noise (assumed flat over the band) N_W volts.
The video bandwidth occupied by the signal and noise f_W cycles per second.
The subcarrier frequency f_{SC} cycles per second.
The effective integration time T_M seconds.
The rms phase error ϕ_{rms} in degrees.
Equation (1) relates these parameters

[5] V. D. Landon, "The distribution of amplitude with time in fluctuation noise," Proc. I.R.E., vol. 29, pp. 50–55; Feb., 1941.

$$\frac{S_0}{N_W} = \frac{1}{\sqrt{df_W T_M}} \frac{1}{t_0 f_{SC}} \frac{1}{\pi h} \qquad (1)$$
$$= \frac{1}{\sqrt{df_W T_M}} \frac{360}{\phi_{rms}} \frac{1}{\pi h}$$

This equation is derived in Appendix A.[6] The physical significance of the several factors in (1) is as follows:

The factor S_0/N_W represents (for example) the smallest ratio of line sync amplitude to rms noise for which $t_0 f_{SC}$ will not exceed a selected arbitrary value. It may be visually estimated if the composite video signal is viewed with a wide band oscilloscope. When $S_0/N_W = 1$ the rms noise is equal to sync pulse amplitude. Since S_0 represents 25% carrier amplitude, and since the effective peak value of Gaussian noise is approximately four times the rms value, the condition $S_0/N_W = 1$ also corresponds to the "peak" noise being approximately equal to 100% of carrier amplitude.

The factor $t_0 f_{SC}$ represents the fraction of a cycle of phasing error at frequency f_{SC} corresponding to the timing error, t_0. Thus

$$t_0 f_{SC} = \frac{\text{rms phase error in degrees}}{360°} = \frac{\phi_{rms}}{360°}.$$

The factor $df_W T_M$ is the number of effectively independent measurements yielding phase information which may be made in the interval T_M on a signal which is present for only a fraction d of time, and which occupies portions of the bandwidth f_W. The signal is actually present for a period dT_M; the effect of integrating over the period T_M is therefore to reduce the rms error by

$$\sqrt{df_W T_M} = \sqrt{d\frac{f_W}{f_N}},$$

where $f_N = 1/T_M$ is the effective noise bandwidth.

The factor $1/\pi h$ is a constant.

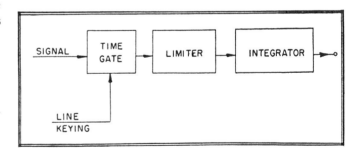

Fig. 3—Typical color-carrier phase reference generation system.

The Required Sync Accuracy

Equation (1) represents the theoretical upper limit of the phasing accuracy which may be derived from the subcarrier burst. A variety of circuits are available which can approach closely to this limit; these circuits are often of the form shown in Fig. 3. The composite

[6] D. Richman, "Theoretical limit to time difference measurements," Proc. NEC, vol. 5; pp. 203–210; 1949.

video signal is fed to a time-gate which is keyed from line flyback to select the burst, which is then amplitude limited and integrated. Practical integrators are described later.

The sync accuracy equation permits the determination of how much integration is required in order to obtain satisfactory performance under extreme conditions. However, due to the many subjective factors involved it is not possible to specify exactly what is the lowest level of signal-to-noise ratio which will be tolerable from a visual viewpoint;[7] it is equally difficult to specify exactly the largest value of rms phasing error which will not cause visible degradation of the picture. Accordingly, Fig. 4, which is a plot of (1), presents graphically the relations between the relevant factors over a range which probably includes the limiting case of interest. Fig. 4 is based on adverse tolerances presented below.

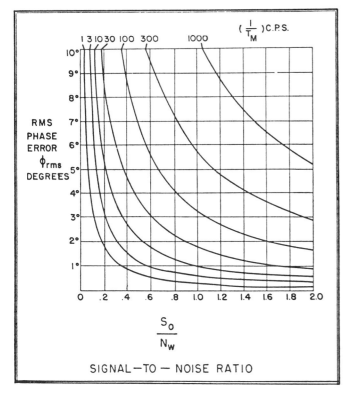

Fig. 4—Phasing accuracy relations for NTSC burst synchronization.

Fig. 4 presents the relation between the rms phase error, ϕ_{rms}, (in degrees) and the signal-to-noise ratio, S_0/N_W, with the integration time, T_M, (in seconds) as a parameter.

For the case corresponding to the most adverse tolerances, $h = .9$, $d = .0352$, and $f_W = 4.3$ mc. Equation (1) then reduces to

$$\phi_{\text{rms}} \frac{S_0}{N_W} = .33 \sqrt{\frac{1}{T_M}} = \frac{1}{3} \sqrt{\frac{1}{T_M}} \qquad (2)$$

which is shown graphically in Fig. 4.

[7] P. Mertz, A. D. Fowler and H. N. Christopher, "Quality rating of television images," PROC. I.R.E., vol. 38, pp. 1269–1283; Nov., 1950.

These curves show that any selected phase accuracy ϕ_{rms} can be obtained with decreasing signal-to-noise ratio S_0/N_W if more time T_M is taken for integration of the signal timing information; i.e., if more measurements are integrated in each complete measurement.

(The facts presented later in this paper with regard to the relations between noise integration and other properties of sync systems indicate that the conclusions reached regarding the reliability of the signal are *not* critically dependent upon the assumed values of S_0/N_W and ϕ_{rms}.)

System Efficiency and the Distribution of Timing Information

The relationships presented above describe the performance of the system when all of the information of the signal is applied usefully. Another parameter which needs to be introduced in order to determine the actual noise bandwidth required is the decoding efficiency, which represents the fraction of the timing information of the signal which is used. Systems with equal noise bandwidths but different decoding efficiencies will give different performance.

In the burst system practical considerations relating to tolerances and to the stability of the gate derived from horizontal sync may result in a gate width r times wider than the narrowest sync burst. Factors relating to this are described later. It results in a requirement of noise bandwidth and integration time such that

$$T_M = \sqrt{r}\, T_{M\text{LIMIT}}$$

$$f_N = \frac{1}{\sqrt{r}} f_{N\text{LIMIT}}$$

where

$(1/\sqrt{r})$ is a system efficiency

r = ratio of actual gate width to minimum burst width.

There is another cause of loss of decoding efficiency in sync systems which is of interest. This relates to the relative distribution of timing information in the frequency spectrum. For burst sync systems which are properly designed, effectively all of the information may be used; (common attainment in horizontal sync systems has not been so high).

Fig. 5(a) shows the relative distribution of timing information in the frequency spectrum occupied by the burst. The basis for this curve is discussed in Appendix A. The effective accuracy which can be obtained if only a portion of the information is used may be measured in terms of the ratio of the noise bandwidth required (at any specified signal-to-noise ratio) to the noise bandwidth required if all of the information is used. For example, a problem of interest in receiver design is the relationship between bandwidth in the burst amplification channel and efficiency. If a passband symmetrically tuned about subcarrier frequency is used in this channel then the system efficiency resulting is represented by the curve sketched in Fig. 5(b). The curve depends, of course, on the width of the burst. Even for the narrowest burst a total bandwidth of approximately 600 kc translates nearly all of the timing information.

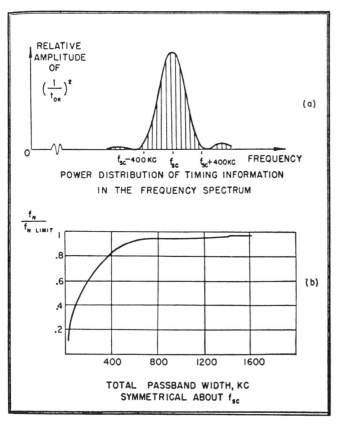

Fig. 5—Frequency distribution and system efficiency of burst sync timing information.

Example: As an illustration suppose the limiting parameter values of interest are approximately $\phi_{\text{rms}} = 5°$ and $S_0/N_W = 1$; these conditions correspond to the point in the center of Fig. 4; then from (2) $T_M \geq 0.0045$ second. *The required noise bandwidth for a gate width ratio $r = 1.2$ is then approximately $f_N = 200$ cycles per second. This figure is used as a basic design parameter for the practical forms of integrators which will be discussed in this paper.*

ELEMENTAL SYNC SYSTEMS

The function of combining signal information derived over an extended interval of time is accomplished by use of circuits which may broadly be classified as integrators. The performance characteristics of two basic forms of integrators are discussed below. The parameters of interest are:

1. The noise bandwidth and integration time of the system.

2. The static phase accuracy. In general, in systems involving feedback, this varies inversely with a circuit gain parameter and may be made nominally as small as desired.

3. The frequency pull-in range of the system. This is the maximum (single peak) frequency detuning for which the system will automatically achieve the desired final operating condition.

4. The stabilization time T_S; or the time required for all operating characteristics to reach effectively their stabilized conditions. This may consist of one or more definable segments.

5. The phase pull-in time T_ϕ; or the transient time required for the output phase of the system to reach some definable measure of its final conditions.

6. The frequency pull-in time T_F, applicable to systems in which a local signal oscillator must be controlled, or the time necessary for the oscillator frequency to be changed from its initial frequency to some selected reference frequency such as a frequency from which the net differential phase change between sync signal and reference oscillator will not exceed one whole cycle. This overlaps the phase pull-in time T_ϕ.

The first integration system discussed is the Passive Integrator. For this system stabilization consists effectively of a phase transient. The limitations of this system are: practical limitations on how high the circuit Q may be and the possibility of detuning.

These limitations are overcome in the second form of integrator called a Standard APC (Automatic Phase Control) System. In this system the signal is heterodyned against a local carrier at the same frequency permitting the desired filtering to be accomplished by means of a low-pass filter which thus effectively provides unlimited Q. The limitations of this system relate to the difficulties of obtaining synchronization and the long pull-in times which result when narrow noise bandwidths are required.

The real limitations imposed by the signal, and some system fundamentals related to using all of the information in the signal, are presented later.

Passive Integrator

The circuit of Fig. 6(a) shows one form of practical integrator. This is a passive integrator in which the required integration is obtained by use of a high-Q filter. The input signal to the filter consists of time-gated amplitude-limited bursts of sine waves at subcarrier frequency f_{SC}. Because of the gating and limiting, sidebands near f_{SC} (which are separated by integral multiples of f_H) as well as harmonics of f_{SC} which are generated in the preceding limiter, all have effectively the same phase modulation due to noise. The noise bandwidth of the filter needs to be less than or equal to the value of f_N which was computed above. If the filter is approximately equivalent to a single resonant circuit, the noise bandwidth is $f_N = (\pi/2)f_3$ where f_3 is the 3 db bandwidth. The bandwidth f_N is indicated in Fig. 6(b). Thus the filter bandwidth should be approximately $(2/\pi)(200) = 127$ cps between 3 db points. The Q desired is $f_{SC}/f_3 \approx 28,000$. This requires the use of a crystal filter. Practical crystals in the frequency range of the color subcarrier can achieve the required Q, but up to the present time apparently cannot exceed it by a large factor.[8],[9] The sum of transmitter frequency tolerance of ± 11 cps and the frequency tolerance of the crystal is comparable with the filter bandwidth. Fig. 6(c) shows how undesirably large static phase shift might result from normal detunings. This is prevented in the system shown in Fig. 6(a) by use of feedback for automatic

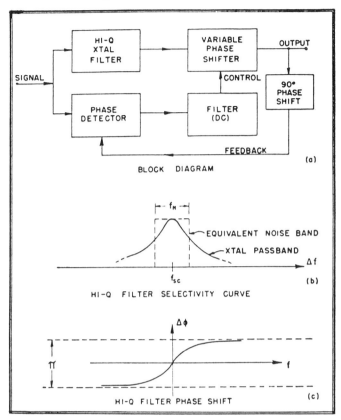

Fig. 6—Passive integrator.

static phase correction. The circuit includes in addition to the high-Q crystal filter a variable phase shifter, a phase detector (which has associated with it a 90° phase shift in one of the signal paths) and a low-pass (dc) filter in the feedback loop for correcting the average phase of the system. Other arrangements are possible; for example, a post-corrector might be used with the feedback signal derived directly from the output of the crystal filter, or a controllable reactance might be coupled to the crystal filter to insure optimum tuning.

The static phase may be maintained as closely accurate as desired by putting a suitably large amount of dc gain in the feedback loop. The signal-to-noise ratio at the output of the system will not be measurably changed if the dc filter is such that the bandwidth of the phase feedback loop is narrower than that of the crystal filter. Design considerations are discussed in Appendix B.

If the crystal stability is comparable to the transmitter frequency stability, the frequency error will be small enough so that rapid phase stabilization will occur when

[8] W. G. Cady, "Piezoelectricity," McGraw-Hill Book Co., Inc., New York, N. Y.; 1946.
[9] A. W. Warner, "High-frequency crystal units for primary frequency standards," PROC. I.R.E., vol. 40, pp. 1030–1033; Sept., 1952.

channels are switched. The switching transient is a phase transient and the stabilization time for small detunings will be or the order of a few times the transient time constant of the phase feedback loop. For the crystal bandwidth required, this time is essentially instantaneous. It may be noted however that if appreciable mistuning could occur the gain versus frequency characteristic of the high-Q filter would substantially reduce the amplitude of the correction signal, resulting in considerably increased stabilization time, and effectively reduced loop gain.

Standard Automatic Frequency and Phase Control Locked Integrator

Fig. 7(a) shows the block diagram of a standard automatic frequency and phase control loop. It includes a local reference oscillator, a phase detector which compares the relative phase difference between the sync signal and the oscillator, a filter which partly determines the transfer characteristic of the APC loop as an integrator, and a reactance tube for controlling the oscillator frequency. The loop gain for this system has the dimen-

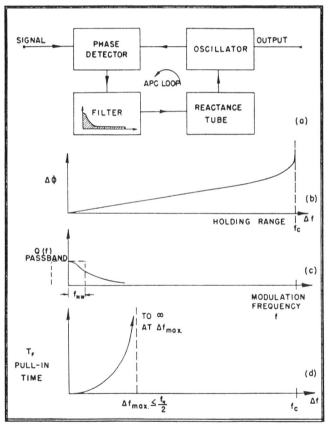

Fig. 7—Standard APC locked integrator.

sions of a frequency, f_c, which is equal to the frequency holding range of the APC system. Included in this characteristic is the dc transmission of the filter. Fig. 7(b) shows the relationship between the static phase error, $\Delta\phi$, and initial oscillator detuning, Δf. By making the holding range much larger than the normal operating range the static phase may be controlled as tightly as desired; here again the price of this control is high loop gain. Fig. 7(c) shows effective passband characteristic $Q(f)$ of the APC loop as a function of modulation frequency. This is determined largely by the ac transmission of the filter in conjunction with the feedback characteristics of the loop. The noise bandwidth f_{NN} is defined in the normal fashion and indicated on the figure. Since an APC loop phase detector is essentially a synchronous detector and does not distinguish between those noise components which are above or below the local oscillator frequency, then $f_{NN} = f_N/2$, and the effective integration time $T_M = 1/2f_{NN}$; the noise bandwidth of the APC loop should not exceed approximately 100 cps for equivalent performance with the high-Q filter.

Fig. 7(d) is a sketch of pull-in time for this loop as a function of Δf. The pull-in range cannot exceed half the gating frequency, i.e. $f_H/2$, and for many designs is substantially smaller. The pull-in mechanism of this loop is not the most efficient one possible. Pull-in times are particularly long near the limit of the pull-in range. The APC loop of Fig. 7(a) is of the same basic type[10] which has achieved essentially universal use in television receivers as an integrator for line frequency synchronizing information. A detailed analysis of the characteristics of this loop is presented in Appendix C and a derivation of the pull-in time relationships is presented in Appendix D.

The pull-in range and time are a function of some design parameters discussed later. It has been found that for optimum design there is a limit to the pull-in performance obtainable with this loop. For these limit designs the following performance is obtained:

(a) The static phase error $\Delta\phi$ may be as small as possible and in fact must be smaller than some specified number in order that pull-in time be minimized.

(b) The pull-in range is equal to $\pm(f_H/2)$.

(c) Except near the limit of pull-in range, the pull-in time and noise bandwidth are very nearly related to the frequency detuning, Δf, by (3)

$$T_F f_{NN} \approx 4\left(\frac{\Delta f}{f_{NN}}\right)^2. \qquad (3)$$

This has been used in Fig. 8 to plot the limit of pull-in performance for optimum design standard APC loops. Fig. 8 represents the pull-in time T_F in seconds as a function of the noise bandwidth f_{NN} in cycles per second. The range of f_{NN} in this log-log plot is from 10 to 1,000 cps with the approximate normal required bandwidth of 100 cps in the center of the graph. Pull-in times ranging from less than one-tenth to approximately one second appear instantaneous and may be characterized as "good." Pull-in times between 1 and 10 seconds are

acceptable but probably close to the limit of adequate performance and have been designated "fair." Pull-in times in excess of 10 seconds are definitely "poor."

The relationship between f_{NN} and T_F is shown for several values of Δf. For example an optimum design unit having a noise bandwidth of 100 cycles will require 4 seconds to pull in from 1,000 cycles detuning. This indicates that such a sync system should be adequate for completely automatic phase control but that it apparently does not have an excess of available performance; for example, if the noise bandwidth needed to be reduced to 50 cycles, then 32 seconds would be required to pull in 1 kc.

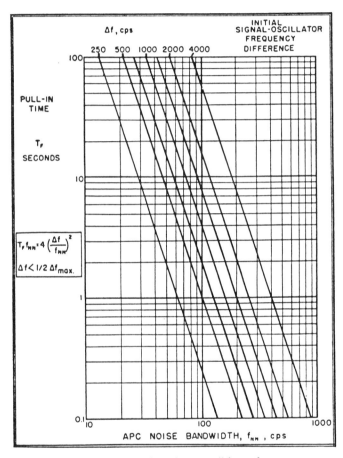

Fig. 8—Standard APC optimum pull-in performance.

Pull-In Performance Attainable with a Standard APC System

Not all designs of APC circuits will achieve the limits of performance discussed with respect to Fig. 8. In fact, partly due to economic limitations, the majority of past designs have fallen short of the limit. Accordingly, Figs. 9 and 10 are presented as a basis for demonstrating the pull-in limitations of the Standard APC System. The curves are expressed in terms of what are believed to be the parameters of interest to the user, specifically the noise bandwidth f_{NN}, the initial frequency difference Δf, and the frequency stabilization time T_F. The dimensionless parameters $T_F f_{NN}$, and $\Delta f/f_{NN}$, are used as ordinate

[10] K. R. Wendt and G. L. Fredendall, "Automatic frequency and phase control of synchronization in television receivers," Proc. I.R.E., vol. 31, pp. 7–15; Jan., 1943.

and abscissa. Two different parameters, designated m and K, which are discussed in Appendix C, appear. The parameter m varies inversely as the dc loop gain for fixed noise bandwidth. The figure shows that increased dc loop gain (smaller m) and hence tighter static phase control permit wider pull-in range and a closer approximation to the minimum pull-in time curve. The parameter K which is a damping coefficient (discussed in

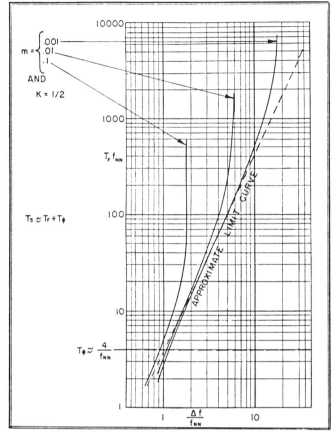

Fig. 9—Pull-in characteristics of standard APC loop.

Appendix C) determines the level of the limit curve as indicated in Fig. 10. Over part of its range of variation the parameter K permits an exchange of minimum pull-in time for pull-in range. The maximum increase, however, is limited to a 50% increase in frequency pull-in range, over designs which approach the optimum pull-in time limit curve.

The mathematics upon which these curves are based is presented in Appendexes C and D. Appendix C introduces and presents the relevant relations between the parameters of the Standard APC System. Derivation of the pull-in time equation and discussion of the pull-in phenomenon is presented in Appendix D.

THEORY OF SYNCHRONIZATION

Improved Sync Systems

The systems described thus far permit a level of performance which appears to satisfactorily meet the requirements for burst synchronization but do not appear to have a large excess of performance. The signal itself

permits substantially better performance.[11] This will be shown below by considering the limitations of the systems presented thus far and introducing the factors which lead to full utilization of the signal information. This leads to a sync system which appears capable of efficiently using all of the timing and synchronizing information in the signal. Then an implementation of this system is described which appears applicable to NTSC color television receivers to produce what may be ideal performance at no substantial cost increase.

Finally, the approximate upper limit of performance capability for the signal is evaluated numerically. The limitations on the previous system relate to the severe restrictions interrelating noise bandwidth and pull-in time. There appear to be a variety of new sync systems which can overcome this limitation. Several varieties have been instrumented and found practical. However, the potentialities of the NTSC burst sync system are perhaps most clearly demonstrated by examining what may be the upper limit of performance.

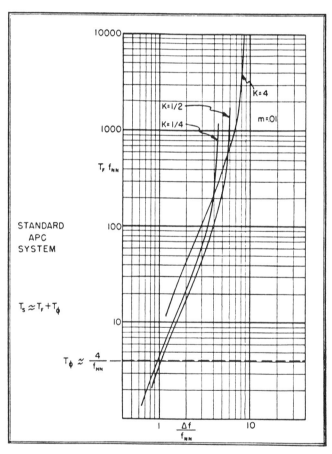

Fig. 10—Effect of variations in the parameter K.

Two Mode Systems

There are two separate and distinct modes of performance of sync systems. These relate to (a) the phase stability attainable after the system has achieved a stable synchronized operating condition, which has been

[11] D. Richman, "Theory of synchronization applied to NTSC color television," IRE CONVENTION RECORD, Part 4; 1953.

discussed in some detail earlier in this paper, and (b) the performance associated with the system achieving that final state. Each of these modes has fundamental physical restrictions and characteristics associated with it. The full measure of performance permitted by the signal can be achieved by a system which makes these two modes of operation as independent as possible of each other and of each other's limitations.

Some systems use the same mechanism for hold-in and pull-in. The Standard APC System falls into this category. It is inefficient in its use of signal information. Other types of systems use a multiplicity of mechanisms, usually two.[12] One mechanism is designed for stable performance after synchronization, the second mechanism is designed to produce synchronization. *Such a device must have within it the inherent ability to extract from the signal the necessary information with regard to the mode of performance which is required.* For example, it should not confuse noise which may be present when the system is synchronized with a beatnote indicative of a lack of synchronism.

Factors Relating to Frequency Pull-In

There are two basic factors which relate to frequency pull-in. The first problem is concerned with the mechanism whereby a frequency difference is recognized in the presence of strong signals and a control voltage generated which can be utilized for pull-in. The second problem relates to the ability of the mechanism associated with pull-in to discriminate against noise interference.

Frequency Recognition

This separation of the requirements of the system leads to the following principle. *The real limitation of a synchronization system with respect to frequency pull-in is the ability of the system when out of sync to recognize a frequency difference and distinguish it from noise.*

This sets the *real upper limit of performance.* If the frequency determination is effectively linear, then after a time delay which permits the frequency difference to be measured to within a suitable measure of reliability, the reference oscillator may be switched instantaneously by the proper amount to insure synchronization. A system for accomplishing this may be called an *ideal sync system.* Just as with phase measurements this reliability is obtained by integrating the frequency difference information for an adequately long period of time. The shortest stabilization time consistent with reliable performance is therefore determined by the integration time necessary to measure a frequency difference with a suitable measure of reliability.

The Pull-In Control Effect

Fig. 11 represents the generated control effect for pull-in for two important synchronization systems. Fig. 11(a) relates to the frequency pull-in characteristic of a

standard APC loop. The generated control voltage for pull-in is shown as a function of instantaneous applied frequency difference Δf. If the frequency is within a range roughly two-thirds that of the noise bandwidth, pull-in (as explained in Appendix D) is effectively instantaneous. The system never slips a cycle; a dc voltage for frequency control is generated which is proportional to the frequency difference. For larger values of Δf the

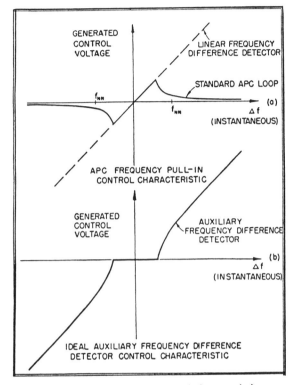

Fig. 11—Synchronization control characteristics.

system slips cycles but by virtue of the feedback in the APC loop generates a dc component of control voltage which varies in the inverse fashion with frequency difference indicated in Fig. 11(a). This inefficient control effect may be compensated for in this system by very high ratios of dc to ac loop gain ($1 \gg m$) but at the expense of the long pull-in times indicated by Fig. 9 and 10. An automatic frequency control system[13] containing a linear frequency difference detector[14] which generates a control voltage proportional to the frequency difference for all frequency differences of interest as indicated in Fig. 11(a) provides a more efficient indication of large frequency differences.

Improved performance may be achieved by supplementing the APC system with an "Ideal Auxiliary Frequency Difference Detector," the control characteristic of which is shown in Fig. 11(b). Such an auxiliary detector can provide a suitable control effect for nearly optimum pull-in performance and as indicated by the flat

[13] C. Travis, "Automatic frequency control," Proc. I.R.E., vol. 23, pp. 1125–1141; Oct., 1935.
[14] C. F. Shaeffer, "The zero-beat method of frequency discrimination," Proc. I.R.E., vol. 30, pp. 365–367; Aug., 1942.

[12] Fundamentals relating to systems analyzed here have been applied to automatic gain control circuits as well as to sync systems.

portion of the curve will *automatically turn itself off when synchronization has been achieved;* this occurs when the frequency difference is reduced to within the linear sloping portion of the curve of Fig. 11(a), within which range the standard APC loop can produce effectively instantaneous pull-in.

A Sync System which Efficiently Uses the Signal Information

Fig. 12 represents the block diagram for a sync system having the auxiliary frequency detection control characteristic described with regard to Fig. 11(b). It includes a Standard APC System such as was shown earlier in Fig. 7 and in addition an auxiliary frequency difference detector which supplements the pull-in performance of the APC system.

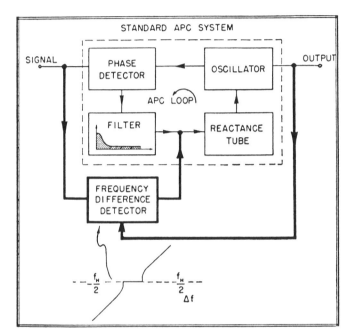

Fig. 12—A synchronization system capable of using total signal information at maximum efficiency.

The idealized upper limit performance described earlier under "Frequency Recognition" may be achieved by means of a suitable interconnection circuit. However, with the stepped characteristic of Fig. 11(b) an essentially direct connection is feasible. The composite system functions as a form of automatic frequency control system when out of sync and as an automatic phase control system when in sync; the auxiliary frequency difference detector turns itself off automatically by virtue of the shape of its control characteristic.

The ideal switched system has pull-in time equal for all frequency differences.

AFC systems normally require high loop gain and are characterized by a pull-in time constant. In some instrumentations of the system of Fig. 12 a loop gain of approximately unity (or a little more for tolerance purposes) may be adequate if the frequency difference detector includes a small amount of delay in its output. As

soon as the oscillator is brought near the frequency of the sync signal, the high-gain APC system becomes operative, and the frequency difference detector is automatically inactivated.

The Quadricorrelator: A Frequency Difference Detector

In order to illustrate in more detail the problems and characteristics associated with the achievement of effectively upper limit performance, a form of circuit arrangement is introduced here which appears capable of using elements already present in color television receivers operable on NTSC standards to achieve the ideal frequency difference detection described above. This form of circuit will be called a quadricorrelator in this paper. Analysis of the performance characteristics of the quadricorrelator presented in Appendix E shows that when preceded by a limiter it comes within a few db in signal-to-noise ratio of using all of the signal information for signal-to-noise ratios of interest here. When the limiter is omitted from the system, the quadricorrelator is an efficient frequency detector; the extra noise due to amplitude modulation disappears after pull-in.[15] It is a true frequency difference detector since it is not subject to tuning errors. The excess of available over required noise discrimination suggests that the limiter can be omitted.

There is no real purpose to accomplishing pull-in much more rapidly than perhaps a few tenths of a second. The simple quadricorrelator instrumentations appear (on this basis) to give effectively optimum performance.

A block diagram of a basic form of a quadricorrelator is shown in Fig. 13. Its elements are a pair of synchronous detectors which are fed with reference signals in quadrature with each other so that the phase detector outputs represent "in phase" and "quadrature" components of the applied sync signal. These output signals are then limited in maximum frequency to (for example) $f_H/2$ by filters as indicated in Fig. 13(a). The output of one of the synchronous detectors goes through a differentiating circuit which provides a 90° phase shift through the passband. The two signals are then heterodyned in another synchronous detector, and the output is integrated in a narrow band filter; a low-pass filter is shown. This filter exchanges brevity of integration time for reliability of frequency measurement. The resulting output signal is proportional to the frequency difference (as explained below) and is applied through an interconnection circuit to the controlled oscillator of the APC system of the receiver.

The mechanism by which the frequency difference is determined may be explained as follows: Assume that a frequency difference Δf exists between the sync signal and the local reference oscillator. The input noise may

[15] J. G. Chaffee, "The application of negative feedback to frequency modulation systems," Proc. I.R.E., vol. 27, pp. 317–331; May, 1939.

be considered for simplicity as the sum of two noise-modulated signals in quadrature with each other at the *oscillator* frequency. The output from one synchronous detector will contain a sine beatnote (see Fig. 13(b)) and the noise along one reference axis. The output of the other synchronous detector will contain a cosine beatnote (see Fig. 13(c)) and the noise along an axis in quadrature with the first reference axis. These two *noise* voltages are completely independent of each other.

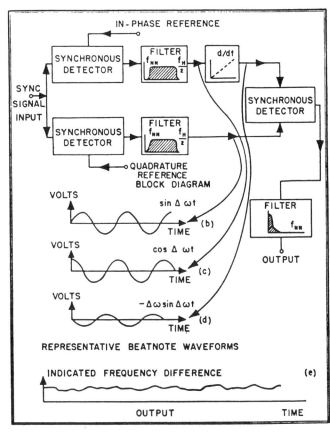

Fig. 13—Basic quadricorrelator.

The cosine beatnote is converted by differentiation to a sine beatnote having an amplitude which is proportional to its frequency, as indicated in Fig. 13(d); its associated noise is differentiated but the two noise voltages are still independent of each other. The output of the cross-multiplying synchronous detector will contain *a dc term proportional to and polarized according to the frequency difference.* In addition the output contains random noise; this noise output is discussed in Appendix E.

The quadricorrelator provides a convenient means for measuring a frequency difference with any selected measure of reliability in the presence of noise by integration of the frequency difference information.

An Illustrative Receiver Operable on NTSC Standards

Fig. 14 shows a partial block diagram of an NTSC color television receiver which includes color difference synchronous detectors, a Standard APC System, and a quadricorrelator for frequency difference detection. The composite video signal is fed to a pair of synchronous detectors for deriving the color difference video signal. The $R - Y$ synchronous detector output may be fed through an amplifier gated during line retrace to a filter, a reactance tube, and an oscillator, the output of which is fed back in the normal fashion to both synchronous detectors. These elements comprise a Standard APC System as described earlier. The gated outputs of both synchronous detectors are fed to a pair of filters as indicated. These may be bandpass filters having low-frequency cutoffs near the noise bandwidth of the APC system and having high frequency cutoffs not higher than half line frequency, as shown in Fig. 14. The differentiating circuit may be included in either beatnote translation path. The third synchronous detector and filter as indicated complete the elements of the quadricorrelator, the output of which is fed to the reactance tube. *The low-frequency attenuation characteristics of the filters in the two channels make the quadricorrelator have an essentially zero transmission characteristic for small beatnote frequency differences.*

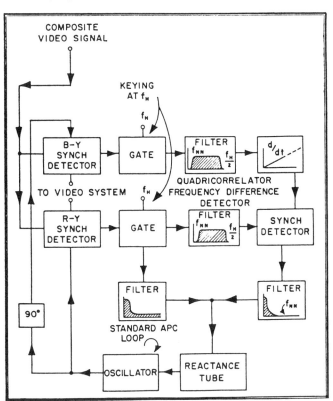

Fig. 14—Application of Fig. 12 sync system to NTSC color television receiver.

The circuit arrangement presented in block form in Fig. 14 provides one means for realizing the sync system of Fig. 12. Since this composite frequency- and phase-control system is essentially free of the previous limitations between noise bandwidth and pull-in time, the over-all system can be more readily designed for desired performance.

There appears to be a variety of frequency difference detector circuits and of linear and nonlinear interconnection arrangements which can be used to approach upper limit performance in the burst system.

The excess of performance inherent to these arrangements appears exchangeable for receiver economy and long term reliability.

The Approximate Limit of Performance Permitted by Signal Information

There are three requirements on the sync system.

(1) The static phase error shall not exceed some selected value, say 5°. It is shown in Appendexes B and C that for both passive and locked integrators this may be accomplished by use of adequately high loop gain.

(2) The rms phase error shall not exceed some selected value, say 5°, for signal-to-noise ratios at least as high as the approximate lowest level for which monochrome video picture information is acceptable; this is approximately $S_0/N_W = 1$.

The required noise bandwidth for the APC system is

$$f_{NN} = \frac{1}{2T_M} \approx \frac{1}{2}\left(3\phi_{\text{rms}}\frac{S_0}{N_W}\right)^2$$

from (2). (The effect of excess gate width is small and is neglected here for simplicity.)

(3) The stabilization time shall not be annoyingly long. For example, pull-in times shorter than 1 second are acceptable.

The minimum integration time required for frequency difference detection yielding an rms frequency error f_{rms} is shown in Appendix E to be

$$T_I = \sqrt{\frac{2}{d}\frac{1}{\pi h}\frac{N_W}{S_0}}\sqrt{\frac{f_H}{f_W}}\frac{1}{f_{\text{rms}}} \qquad (4)$$

for the signal. This is based on a pull-in range of $\pm (f_H/2)$.

It is shown in Appendexes C and D that the linear portion of the curve of Fig. 11(a) extends to a value of Δf approaching $2f_{NN}/\pi$; the control effect is strong to near f_{NN}. Then, if for frequency differences between approximately $(2/\pi)f_{NN}$ and $f_H/2$, the error in frequency difference measurement is less than $(2/\pi)f_{NN}$, pull-in will occur in time T_I. The more severe of the following two requirements then determines the frequency pull-in time T_F.

$$\begin{cases} T_I \geqq \dfrac{\pi}{2f_{NN}} \text{ Approximately} & (5) \\[2mm] f_{\text{rms}} \leqq \dfrac{1}{4}\left(\dfrac{2}{\pi}f_{NN}\right) = \dfrac{f_{NN}}{2\pi}\cdot & (6) \end{cases}$$

Combining (4) and (6),

$$T_I \geqq 2\sqrt{\frac{2}{d}\frac{N_W}{S_0}}\sqrt{\frac{f_H}{f_W}\frac{1}{f_{NN}}\frac{1}{h}}\cdot \qquad (7)$$

The same adverse tolerances used in obtaining (2) may be used here. If $d=.0352$, $h=.9$, $S_0/N_W=1$, $f_W=4.3$ mc, and $f_H=15734+$cps, then (7) becomes

$$T_I \geqq \frac{1}{f_{NN}}\cdot$$

Thus, the required frequency pull-in time is of the order of magnitude of $1/f_{NN}$ or $(\pi/2)(1/f_{NN})$. After frequency pull-in, phase pull-in occurs. (Both occur effectively simultaneously in the continuous feedback system.) The time for phase pull-in is normally less than

$$T_\phi \approx \frac{4}{f_{NN}}\cdot \qquad (8)$$

The constant in (8) depends on the shape of the passband determining f_{NN}.

Then, the stabilization time, T_S is given by

$$T_S \approx T_F + T_\phi. \qquad (9)$$

Since the required value of f_{NN} was found earlier to be 100 cps, pull-in times of the order of .05 second are possible. This is considerably shorter than is required, indicating that the information inherently contained in the signal is substantially in excess of what is required.

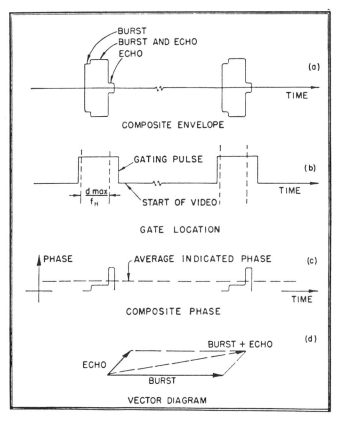

Fig. 15—Effect of echoes on the NTSC color burst.

OTHER TOPICS

Effects of Echoes

Some sketches relative to a discussion of the effect of echoes on burst sync are presented in Fig. 15. Fig. 15(a) shows one possible representation of a burst to which an echo has been added. Parameters of interest are the relative delay, the relative amplitude, and the relative phase. If the time-gate exceeds the burst width on the lagging end as indicated in Fig. 15(b) combined signals may be used to operate the burst sync system. In this case the indicated phase as a function of time is as shown in Fig. 15(c) while Fig. 15(d) is a vector diagram representing the signals of interest. Phase angles of interest are indicated for the burst phase, for the phase of the sum of the burst and echo, and for the phase of the echo. The average phase is not necessarily equal to any one of these but may often be near the phase of burst plus echo. The phase of burst plus echo is the correct reference phase for low detail large area colors. For this reason it appears possible that some extra gate width as indicated in Fig. 15(b) may give a useful and efficient exchange of noise immunity for performance in the presence of echoes. However, the existence of high order correlation between widely separated picture elements[16] may be uncommon enough to make this effect relatively unimportant.

A complete discussion of the effect of echoes in the NTSC system is beyond the scope of the present paper.

Effect of Stability of the Gate

The gate is conveniently obtained from horizontal flyback. The effect of gate stability depends on two factors: the stability of the horizontal sync system which produces the gate; and the relative widths of the gating pulse and the burst, which determines the extent to which noise jitter of the gate can be cross-modulated into the burst channel.

The fundamental physical considerations which have been presented and discussed above with regard to burst synchronization are also true of horizontal synchronization although the shape of the spectral distribution for horizontal sync introduces some additional complications. The static phase may be controlled as closely as desired, limited ultimately by transmission tolerances. The stability may be held to any desired level still permitting effectively instantaneous pull-in.

The effect of cross-modulation when it occurs is to increase the noise power for those low-frequency components to which the horizontal sync system is responsive. The horizontal sync system appears to contain more information than it needs. Stability of the gate is a design consideration but it is not a real limitation of the burst sync system.

CONCLUSIONS

The discussion above has shown that standard sync systems appear capable of completely automatic synchronization for NTSC burst sync (although without a large *excess* of performance). In the presence of strong signals the burst sync system is capable of yielding a color-carrier reference having a reliability completely determined by the gain in the receiver sync system, while noise is rejected by integrating the timing information for a suitably long period. An *effective* integration time of the order of 1/200th of a second appears appropriate. Passive integrators using controlled crystal filters, appear capable of meeting the requirements on Q, frequency stability, and rapidity of stabilization. The Standard APC System, when designed for near limit performance, appears capable of providing adequate and usable performance. This means that for reasonable operating tolerances, synchronization will always occur, and with adequate synchronization accuracy.

Improved sync systems which overcome the ultimate limitations of the standard APC sync system have been presented along with a discussion of factors leading to improvement and of the upper limit of performance permitted by the signal. These indicate that the requirement of a high order of noise immunity does not limit synchronization performance in the manner and to the degree experience with previous circuits had indicated. A large excess of attainable as compared to apparently necessary performance appears to exist.

The NTSC color-carrier reference phase synchronization signal contains adequate information for reliable performance down to levels of signal-to-noise ratio where the signals are no longer usable in picture content. A variety of circuits can provide satisfactory performance.

APPENDIX A

Phase of a Sine Wave Plus Random Noise

Derivation of the Equation

The analysis of the theoretical limits to phasing accuracy may be based on the properties of a signal composed of a sine wave plus random noise.[17] The information of each frequency component may be determined separately and then all of the information may be combined.

The problem is solved here first for a continuous (ungated) sine wave.

The probability density distribution of amplitude coefficients for a sinusoidal signal plus two-dimensional Gaussian noise is shown in Fig. 16. The signal is

$$S(t) = S \cos \omega_{SC} t. \qquad (A-1)$$

The noise may be written as

$$N(t) = a(t) \cos \omega_{SC} t + b(t) \sin \omega_{SC} t \qquad (A-2)$$

where $a(t)$ and $b(t)$ are time-varying parameters, each having a Gaussian distribution, and defined by the mean square values shown below.

$$\overline{a^2} = \overline{b^2} = \overline{N^2}. \qquad (A-3)$$

This equality results from the fact that by symmetry, $\overline{a^2} = \overline{b^2}$ while the total noise power

$$\overline{N^2} = \overline{(a \cos \omega_{SC} t + b \sin \omega_{SC} t)^2}$$

[16] E. R. Kretzmer, "Statistics of television signals," *Bell Sys. Tech. Jour.*, vol. 30, pp. 751–767; July, 1952.

[17] S. O. Rice, "Mathematical analysis of random noise," *Bell Sys. Tech. Jour.*, vol. 23, p. 282–332, July, 1944; vol. 24, pp. 46–156, Jan., 1945.

$$= \tfrac{1}{2}\overline{a^2} + \tfrac{1}{2}\overline{b^2}. \qquad (A\text{-}4)$$

For the above case it is possible to express the probability distribution of phase angles for the combination of signal and noise, relative to the phase of the signal. This, however, leads to a cumbersome expression.[18]

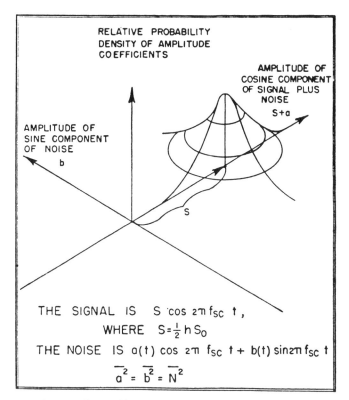

THE SIGNAL IS $S \cos 2\pi f_{SC} t$,

WHERE $S = \tfrac{1}{2} h S_0$

THE NOISE IS $a(t) \cos 2\pi f_{SC} t + b(t) \sin 2\pi f_{SC} t$

$$\overline{a^2} = \overline{b^2} = \overline{N^2}$$

Fig. 16—Probability density distribution of a sine wave and random noise.

It is more convenient to use the simplified vector diagram shown in Fig. 17.[6] Here S represents the signal, and a and b represent the cosine and sine (in-phase and quadrature) components of noise.

Then if ϕ is the phase error,

$$\phi \approx \tan \phi = \frac{b}{S + a} \approx \frac{b}{S} \qquad (A\text{-}5)$$

or, very nearly, since $b_{\text{rms}} = N$,

$$\phi_{\text{rms}} = \frac{N}{S}. \qquad (A\text{-}6)$$

This equation is a good approximation if N/S is not large; in the case where the sync measuring system is primarily responsive to the noise in quadrature with a reference signal controlled by a long time constant of integration, it is accurate enough.

Then, since

$$N = \frac{N_W}{\sqrt{f_W T_M}} = \text{noise in the noise bandwidth} \qquad (A\text{-}7)$$

$f_N = 1/T_M$, and since $S = \tfrac{1}{2} h S_0$, we obtain

[18] D. Middleton, "Some general results in the theory of noise through non-linear devices," *Quart. Appl. Math.*, vol. V, p. 471; Jan., 1948.

$$\phi_{\text{rms}} = 2\pi f_{SC} t_0' = \frac{\dfrac{N_W}{\sqrt{f_W T_M}}}{\tfrac{1}{2} h S_0}. \qquad (A\text{-}8)$$

The above equation applies for a continuous sine wave which is not gated. However, because the signal is present only a fraction d of the time, the integration is only $\sqrt{d}$ times as effective, and hence $t_0 = t_0'/\sqrt{d}$. Therefore, by substitution, the following upper limit relationship is obtained.

$$\frac{S_0}{N_W} = \frac{1}{\sqrt{d f_W T_M}} \cdot \frac{1}{f_{SC} t_0} \cdot \frac{1}{\pi h}. \qquad (A\text{-}9)$$

This is (1), presented earlier.

If the signal plus noise is passed through a limiter, the output of the limiter is approximately

$$S \cos \omega_{SC} t + b(t) \sin \omega_{SC} t$$

for signal-to-noise ratios of interest. Thus, the limiter aids in achieving the upper limit, without improving it.

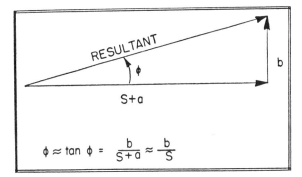

$$\phi \approx \tan \phi = \frac{b}{S+a} \approx \frac{b}{S}$$

Fig. 17—Simplified vector diagram.

When not all of the signal spectrum is used, the rms error will exceed the limiting value of t_0 computed above.

The burst may be represented by the following Fourier series

$$S(t) = \sum_{k=-k_1}^{k_2} S_k \cos \omega_k t \qquad (A\text{-}10)$$

where

$$S_k = \left(\frac{1}{2} h S_0\right) \cdot d \cdot \left(\frac{\sin dk\pi}{dk\pi}\right) = \frac{h S_0 \sin dk\pi}{2k\pi} \qquad (A\text{-}11)$$

and

$$\omega_k = \omega_{SC} + k\omega_H. \qquad (A\text{-}12)$$

Each of these carries timing information; the error associated with the measurement of any component is very nearly Gaussian. For such a case, the Principle of Least Squares[19] may be applied. Then[20]

$$\frac{1}{t_0^2} = \sum_{k=-k_1}^{k_2} \left(\frac{1}{t_{0k}^2}\right) \equiv \sum_{k=-k_1}^{k_2} \left(\frac{S_k}{N_W}\right)^2 (f_W T_M)(2\pi f_k)^2$$

[19] R. B. Lindsay and H. Margenau, "Foundations of Physics," John Wiley & Sons, Inc., New York, N. Y., chap. IV, pp. 159–187; 1936.

[20] D. Richman, "Frame synchronization for color television," *Electronics*, vol. 25, pp. 146–152; Oct., 1952.

$$= \sum_{k=-k_1}^{k_2} \left(\frac{S_0}{N_W} \right)^2 (f_W T_M)(f_{SC} + k f_H)^2 \left(\frac{h \sin dk\pi}{k} \right)^2 . \quad \text{(A-13)}$$

The factor $(1/t_{0k}^2)$ has been plotted in Fig. 5(a) as the information per component. The effective accuracy, $1/t_0$ varies as the square root of the area under the curve, for any bandwidth. Although there is an optimum weighting, the weighting is not critical in the vicinity of the correct weighting. This is a general characteristic of integration systems.

APPENDIX B

Passive Integrators

This appendix presents some equations relevant to the performance of the phase stabilized integrating filter shown in Fig. 6(a).

The basic loop parameters are as follows:

(1) The transfer characteristic of the high Q filter is $F(f)$

$$F(f) \approx \frac{1}{1 + j2 \frac{f - f_{SC}}{f_{SC}} Q} = \frac{1}{1 + j \frac{2\Delta f}{f_3}}$$

$$= \frac{1}{1 + j\pi \frac{\Delta f}{f_N}} = F(\Delta f) . \quad \text{(B-1)}$$

(2) The phase detector sensitivity, for nominal full amplitude input is $\partial E/\partial \phi$

(3) The passband characteristic of the low pass (dc) filter is $Y(f)$, where $Y(0) = 1$. Let

$$Y(f) = \frac{1}{1 + j2\pi f T} . \quad \text{(B-2)}$$

(4) The sensitivity of the phase shifter (assumed broad band) may be represented as

$$\frac{\partial \phi}{\partial E} .$$

(5) The loop gain is G

$$G = \frac{\partial \phi}{\partial E} \cdot \frac{\partial E}{\partial \phi} .$$

(6) The static phase error which would result if there were no feedback is $\Delta\phi_0$

$$\Delta\phi_0 = \arctan \left(-\frac{\pi \Delta f}{f_N} \right) \approx -\frac{\pi \Delta f}{f_N} . \quad \text{(B-3)}$$

The Static Phase Error with Feedback

The static phase error with feedback is $\Delta\phi$

$$|F(\Delta f)| \Delta\phi \cdot G \cdot Y_0 = \Delta\phi_0 - \Delta\phi = \Delta\phi_{corr}$$

$$\frac{\Delta\phi}{\Delta\phi_0} = \frac{1}{1 + |F(\Delta f)| G Y_0} \quad \text{(B-4)}$$

or, since for normal operation $F(\Delta f) \approx 1$ (very nearly) and $Y_0 \equiv Y(0) = 1$

$$\frac{\Delta\phi}{\Delta\phi_0} = \frac{1}{1 + G} . \quad \text{(B-5)}$$

Since $\Delta\phi_0 < 90°$, a loop gain of $G > 17$ makes $\Delta\phi < 5°$ always.

The Effect of the Feedback Loop Upon Noise Performance

When noise is present, the phase detector output produces a noise output, which, after filtering by $Y(f)$ produces extra phase modulation noise.

The equation written earlier can be rewritten in terms of the *phase correction*, $\Delta\phi_{corr}$, since

$$\Delta\phi_{0 \text{ effective}} = \Delta\phi_{corr} + \Delta\phi. \quad \text{(B-6)}$$

$\Delta\phi_{0 \text{effective}}$ is the equivalent phase modulation to produce the actual phase detector noise output.

Then

$$[\Delta\phi_0(p) - \Delta\phi_{corr}(p)] \cdot G \cdot Y(p) = \Delta\phi_{corr}(p)$$

or

$$\frac{\Delta\phi_{corr}(p)}{\Delta\phi_0(p)} = \frac{GY(p)}{1 + GY(p)} = \frac{G}{1 + G} \left[\frac{1}{1 + p \frac{T}{1 + G}} \right] . \quad \text{(B-7)}$$

The signals to the phase detector are

(1) The original composite signal+noise, unfiltered.

(2) The filtered signal, with a narrow band of noise having a very small rms value.

Cross beats of signal upon noise produce considerably larger output than the beatnote between noise components, which are therefore negligible.

The output noise may be expressed as a phase:

$$\frac{b(t)}{S} \approx \Delta\phi_{01}(t) \quad \text{or} \quad \frac{b(p)}{S} \approx \Delta\phi_{01}(p)$$

$$\frac{b(p) \cdot F(\Delta p)}{S} \approx \Delta\phi_{02}(p) . \quad \text{(B-8)}$$

The total phase noise is $\Delta\phi_{0 \text{effective}} = \Delta\phi^0_1 - \Delta\phi_{02}$ since, if the filter $F(f)$ were removed, the phase detector output would be identically zero.

$$\Delta\phi_0(p) = \frac{b(p)}{S} [1 - F(\Delta p)]. \quad \text{(B-9)}$$

There is little noise energy below approximately $f_N/2$ appearing at the phase detector output.

Since the transfer characteristic for this noise is

$$\frac{\Delta\phi_{corr}}{\Delta\phi_0} = \frac{G}{1 + G} \left[\frac{1}{1 + p \frac{T}{1 + G}} \right] \quad \text{(B-10)}$$

(which corresponds to a low pass filter), the following design condition may be employed to insure that the effective Q of the crystal filter will not be degraded by the feedback

$$\frac{1 + G}{2\pi T} < \frac{f_N}{2}$$

or
$$T > \frac{1+G}{\pi f_N} \cdot \qquad \text{(B-11)}$$

Transient Analysis

The response to a step in differential phase $\Delta\phi_0$, is $\Delta\phi(p)$ or $\Delta\phi(t)$

$$\frac{\Delta\phi(p)}{\Delta\phi_0} = \frac{1}{p}\left[\frac{1+pT}{1+FG+pT}\right]$$

$$= \frac{1}{pT}\left[\frac{1}{\frac{1+FG}{T}+p}\right] + \frac{1}{\frac{1+FG}{T}+p} \qquad \text{(B-12)}$$

$$\frac{\Delta\phi(t)}{\Delta\phi_0} = \int_0^t \left[\frac{1}{T}\,\epsilon^{-[1+FG/T]t}\right]dt + \epsilon^{-[1+FG/T]t}$$

$$= \frac{1}{1+FG}\left[1 - \epsilon^{-[1+FG/T)]t}\right] + \epsilon^{-[1+FG/T]t}$$

$$= \frac{1}{1+FG} + \frac{FG}{1+FG}\,\epsilon^{-[1+FG/T]t} \qquad \text{(B-13)}$$

$$= \text{steady state} + \text{transient response}.$$

For $FG \gg 1$, the transient term is negligible for $t > T$.

The time for the phase error to settle down to twice its final value may be computed, as a measure of stabilization time.

The total transient time consists effectively of an amplitude and phase transient of the high Q filter plus the transient time of the feedback loop. The transient is effectively completed in three times the time constant of the filter. Since the noise bandwidth is f_N, the time is

$$T_A \approx 3\left(\frac{1}{4f_N}\right). \qquad \text{(B-14)}$$

This overlaps with the phase loop transient time, which, neglecting amplitude effects, would be

$$T_\phi = \frac{1}{\pi f_N}\,lnG \approx \frac{1}{\pi f_N}\,ln\left[\frac{\pi\Delta f_{\max}}{f_n \Delta\phi_{\max}} - 1\right] \qquad \text{(B-15)}$$

which is based on

$$\frac{\Delta\phi}{\Delta\phi_0} = \frac{1}{1+G}\left[1 + G\epsilon^{-(1+G)(t/T)}\right]$$

$$G = \frac{\Delta\phi_0}{\Delta\phi_\infty} - 1 \approx \frac{\pi\Delta f_{\max}}{f_N \Delta\phi_{\max}} - 1$$

$$T = \frac{1+G}{\pi f_N} \cdot$$

These two pull-in times overlap.

APPENDIX C

Performance Characteristics of the Standard APC Loop

This appendix presents a description of the operating characteristics of a standard APC system. The basic parameters of the APC loop are defined. The independence of the primary parameters $\Delta\phi$ (the static phase error) and f_{NN} (the APC loop noise bandwidth) is shown; these parameters characterize the performance after the system has stabilized. The limitations of pull-in are discussed and some formulas which are derived later in Appendix D are introduced. The simple relation presented earlier for pull-in time is then obtained.

The formulas derived may be applied for designs based on any convenient set of assumed criteria.

The Basic APC Loop Parameters

(1) The output voltage ΔE of the phase detector, and the phase difference $\Delta\phi$ between the reference oscillation and the signal are related by the control characteristics. When both signals are sinusoidal,

$$\Delta E = \mu \sin \Delta\phi \qquad \text{(C-1)}$$

where ΔE is a voltage developed at the phase detector output in response to a phase difference $\Delta\phi$ between signal and reference oscillation. For operation at or very near balance,

$$\frac{\partial E}{\partial \phi} = \mu \qquad \text{volts per radian.}$$

(2) The transfer characteristic of the feedback loop filter is denoted by

$$N(\omega) = \frac{\text{output voltage}}{\text{input voltage}}$$

(3) The sensitivity of the reactance tube is denoted by

$$\beta = \frac{\partial f}{\partial E} \qquad \text{cycles per second per volt.}$$

(4) The factor $|\mu\beta| \equiv f_c$ is a characteristic parameter of the loop; the time constant $t_c \equiv 1/2\pi f_c$ is the transient time constant of the loop when $N(\omega) \equiv 1$. (This may be verified from (C-3) for $Q(\omega)$ presented later.)

(5) The static phase error, $\Delta\phi$, which results from a "free-running" frequency difference, Δf, between signal and local oscillator may be found from the preceding relations:

$$-\sin \Delta\phi = 2\pi \cdot \Delta f \cdot t_c = \frac{\Delta f}{f_c} \cdot \qquad \text{(C-2)}$$

Although (C-2) contains the appropriate signs, it is the magnitudes of the above quantities which are of interest in design work.

(6) The phase following ratio for an APC loop is

$$\frac{\text{phase variation of output phase}}{\text{phase variation of input phase}} = Q(\omega) = \frac{N(\omega)}{N(\omega) + pt_c} \cdot \qquad \text{(C-3)}$$

This is the small signal form of the differential equation which characterizes the APC loop. It is used to determine the response of the APC system to noise, after the system is synchronized.

(7) The noise bandwidth of the APC system is f_{NN}. Consistent with the usual practice, this is defined as

$$f_{NN} = \int_0^\infty |Q(\omega)|^2 \, df = \int_0^\infty Q(\omega)Q(-\omega) \, df. \quad \text{(C-4)}$$

Representative network configurations for $N(\omega)$ are shown in Fig. 18(a). For each of these networks

$$N(\omega) = \frac{1 + xpT}{1 + (1+x)pT} \quad \text{(C-5)}$$

where $T = RC$ and $p = j2\pi f = j\omega$. Then

$$Q(\omega) = \frac{1 + xpT}{1 + p(t_c + xT) + p^2(1+x)t_cT}. \quad \text{(C-6)}$$

This equation suggests one manner in which the meaning of the phase transfer ratio and noise bandwidth of an APC loop may be readily visualized. Fig. 18(b) represents a network having a voltage transfer characteristic which is identical with $Q(\omega)$ given above. If a voltage

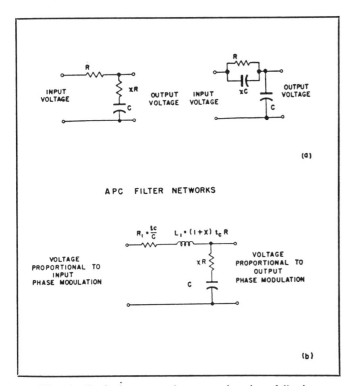

(a)

APC FILTER NETWORKS

(b)

Fig. 18—Equivalent network representing phase following ratio of an APC loop.

proportional to the phase modulation of the synchronizing signal (by noise or any other disturbance) is applied to the input of the network of Fig. 18(b), the output voltage is proportional to the phase modulation of the reference oscillator of the APC loop. The shape of the (low frequency) passband described by $Q(\omega)$ defines the small signal transient response of the loop as well as the noise bandwidth.

(8) The ratio of ac gain/dc gain through the network $N(\omega)$ is

$$m \equiv \frac{x}{1+x} \quad \text{(C-7)}$$

(from [C-5] when $pT \gg 1$). The parameter m determines the pull-in range of the APC system, when certain other parameters are specified. It is convenient therefore to express the synchronous performance in terms of m.

Also, the term xT/t_c appears often. This is written as

$$y \equiv \frac{xT}{t_c}. \quad \text{(C-8)}$$

Then, rewriting the earlier expressions in terms of these parameters,

$$N(\omega) = \frac{1 + pyt_c}{1 + p\dfrac{y}{m}t_c}. \quad \text{(C-9)}$$

$$Q(\omega) = \frac{1 + pyt_c}{1 + pt_c(1+y) + p^2\dfrac{y}{m}t_c^2}. \quad \text{(C-10)}$$

The noise bandwidth is found by integration (at the end of this Appendix C), using the definition presented earlier, to be

$$f_{NN} = \frac{1}{4t_c} \cdot \frac{1 + my}{1 + y}. \quad \text{(C-11)}$$

(9) In order to prevent resonant ringing on noise impulses, $Q(\omega)$ should have a moderately flat graph. Since the denominator of $Q(\omega)$ contains a quadratic expression, it is convenient to define a damping coefficient, K, which is defined by the following equation:

$$(1 + y)^2 = K \cdot \frac{4y}{m}. \quad \text{(C-12)}$$

Then $K = 1$ corresponds to equal roots or critical damping, $K > 1$ corresponds to overdamping and makes $Q(\omega)$ approach the shape of the single (RC) low pass filter, and $K < 1$ tends to give $Q(\omega)$ a high resonant rise.

Fig. 19 shows the shapes of $|Q(\omega)|$ and $|Q(\omega)|^2$ for several values of K, and subject to the simplifications $y \gg 1$, and $my \approx 4K$, derived below. A value of K close to 1 gives best performance.

The Synchronous Performance of the APC System

The basic equations relating to the synchronous performance of the APC system have been presented above. These are

$$-\sin \Delta\phi = 2\pi \cdot \Delta f \cdot t_c \quad \text{(C-2)}$$

$$f_{NN} = \frac{1}{4t_c} \cdot \frac{1 + my}{1 + y} \quad \text{(C-11)}$$

$$m = \frac{4Ky}{(1+y)^2}. \quad \text{(C-12)}$$

Since both tight static phase and narrow noise bandwidth are desired, it is possible to define a figure of merit for the system as $|(\sin \Delta\phi)/(\Delta f)| \cdot f_{NN}$; the smaller this product is, the better the over-all performance. However, relations above show that any arbitrarily selected figure of merit may be obtained by proper design, since, combining the above relations,

$$\left| \frac{\sin \Delta\phi}{\Delta f} \right| \cdot f_{NN} = \frac{\pi}{2} \left[\frac{1 + \dfrac{4Ky^2}{(1+y)^2}}{1+y} \right] . \quad \text{(C-13)}$$

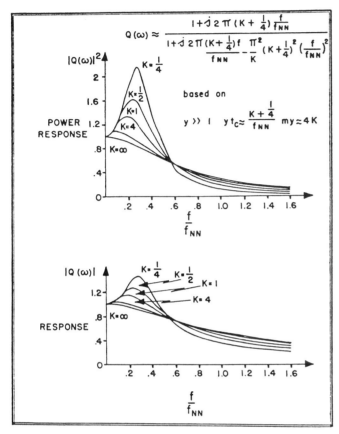

Fig. 19—APC loop small signal modulation response.

For the limiting case of a single time constant filter, $y = 0$, and then

$$\left[\left| \frac{\sin \Delta\phi}{\Delta f} \right| \cdot f_{NN} \right]_{\nu=0} = \frac{\pi}{2} . \quad \text{(C-14)}$$

Thus for the simplified filter the static phase shift and noise bandwidth are interdependent.[21],[22] However, for the filters of Fig. 18a, the parameters can be designed for whatever figure of merit is required for synchronous operation.

The above relations may usefully be written in simpler form, since, for the design ranges of interest, $m \ll 1$ and $y \gg 1$; then, very nearly $4K = my$ and hence

$$4f_{NN}t_c = m \frac{1+my}{m+my} \approx m \left(\frac{1+4K}{4K} \right) . \quad \text{(C-15)}$$

This equation will be used in expressing the pull-in performance of the system conveniently.

[21] T. S. George, "Analysis of synchronizing systems for dot-interlaced color television," Proc. I.R.E., vol. 39, pp. 124–131; Feb., 1951.
[22] K. Schlesinger, "Locked oscillator for television synchronization," Electronics, vol. 22, pp. 112–118; Jan., 1949.

The figure of merit may be written as

$$\left| \frac{\sin \Delta\phi}{\Delta f} \right| f_{NN} = \frac{f_{NN}}{f_c} \approx \frac{\pi}{2} \cdot m \left(\frac{1+4K}{4K} \right) . \quad \text{(C-16)}$$

The Transient (Pull-In) Performance

The pull-in behavior of the APC system is investigated in detail in Appendix D. The significant conclusions are as follows: The pull-in performance is expressible in terms of the relations between the parameters

$$\left(\frac{T_F}{xT} \right) \equiv \left(\frac{T_F}{yt_c} \right)$$

and

$$\left| \frac{\Delta f}{mf_c} \right| .$$

Fig. 20 shows the relation between these parameters.

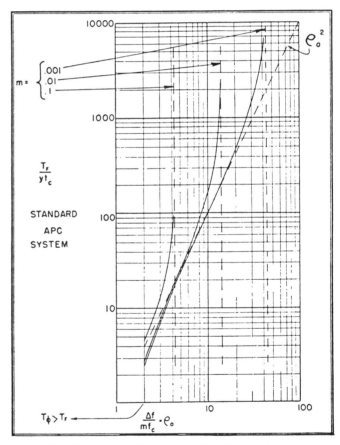

Fig. 20—Universal frequency pull-in characteristics.

The following approximation to the data represented by Fig. 20, based on (D-29), has been found useful in design work, it can also be solved for Δf:

$$T_F \approx xT \frac{\left(\dfrac{\Delta f}{mf_c} \right)^2}{1 - \dfrac{\Delta f^2}{2f_c \cdot mf_c}} .$$

If $|\Delta f/mf_c| \leq 1$ the frequency pull-in is effectively instantaneous ($T_F = 0$) but a short period is required for the phase to approach closely its stable value. If $|\Delta f/mf_c| > 1$, the system can slip cycles; often the slip is a great many cycles as this pull-in mechanism is fairly inefficient. The pull-in range is limited to the region

$$\left|\frac{f}{mf_c}\right| < \sqrt{\frac{2}{m} - 1} \qquad \begin{array}{l}\text{(C-17)}\\[1.2em]\text{(D-25)}\end{array}$$

or

$$\Delta f_{\max} = f_c \sqrt{2m - m^2} \approx \sqrt{2 f_c \cdot m f_c}$$

Then,

$$|\sin \Delta\phi| \approx |\Delta\phi| \leq \left|\frac{\Delta f_{\max}}{f_c}\right| = m \sqrt{\frac{2}{m} - 1} \approx \sqrt{2m}. \text{(C-18)}$$

If $m < (1/250)$ the phase angle after pull-in will always be less than $5°$. However, not all of the pull-in range is normally used. If

$$|\Delta\phi| < \frac{1}{2}\Delta f_{\max}, \; m < \frac{1}{62} \text{ makes } |\Delta\phi| < 5°.$$

When operation is *well within* the pull-in range the *frequency pull-in* time, T_F, which is defined as the time for the oscillator to be pulled from $|\Delta f|$ to within mf_c of the frequency of the color burst, approaches very nearly the relation

$$\frac{T_F}{yt_c} = \left(\frac{\Delta f}{mf_c}\right)^2. \qquad \begin{array}{l}\text{(C-19)}\\[1.2em]\text{(D-28)}\end{array}$$

By making m smaller and f_c larger it is possible to extend the pull-in range far enough so that the gated nature of the signal provides the only real limitation on pull-in; the range is $|\Delta f| < (f_H/2)$. The pull-in time is then expressed by the square law relation above, except near the limit of the pull-in range. Furthermore, making m smaller improves the synchronous figure of merit.

The pull-in relations may be expressed in terms of f_{NN}, since

$$yt_c = my\frac{t_c}{m} \approx 4K \cdot \frac{1}{4f_{NN}}\left(\frac{1+4K}{4K}\right) = \frac{K+1/4}{f_{NN}} \qquad \text{(C-20)}$$

and

$$mf_c = \frac{m}{2\pi t c} \approx \frac{1}{2\pi} \cdot 4f_{NN} \frac{4K}{1+4K}$$

$$= \frac{2}{\pi}\left(\frac{K}{K+1/4}\right) \cdot f_{NN} \qquad \text{(C-21)}$$

the following equation results

$$T_F f_{NN} = \lambda^2 \left(\frac{\Delta f}{f_{NN}}\right)^2 \qquad \text{(C-22)}$$

where, when f_c is large enough so that $\Delta f_{\max} \gg \Delta f$,

$$\lambda^2 \equiv \left(\frac{\pi}{2}\right)^2 \frac{(K+1/4)^3}{K^2} \geq 4.2. \qquad \text{(C-23)}$$

The approximate value 4 has been used in Figs. 8 and 9. Fig. 21 shows graphically the relation between K and λ^2. The curve has a minimum at $K = 1/2$.

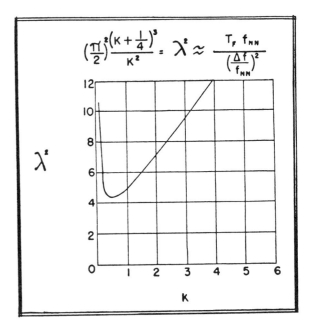

Fig. 21—Graph showing the relation between the damping coefficient, K, and the constant in the APC limit curve equation.

In view of the shape of the curve, and the normal tolerance variations of practical circuits, a value of K near 1 seems desirable. This gives good small signal transient response also. The problem of optimum design is discussed in more detail in a reference.[23]

Derivation of the Noise Bandwidth

The integration is performed as follows. Since

$$Q(p) = \left[\frac{1 + p y t_c}{\dfrac{m}{y} + p t_c(1+y)\dfrac{m}{y} + p^2 t_c^2}\right]\left(\frac{m}{y}\right), \qquad \text{(C-10)}$$

then

$$|Q^2| = Q \cdot Q^* = \left(\frac{m}{y}\right)^2 \frac{1 + y^2\theta^2}{(\theta^2 + \theta_\alpha^2)(\theta^2 + \theta_\beta^2)} \qquad \text{(C-24)}$$

where

$$\theta = \omega t_c \qquad \text{(C-25)}$$

and

$$\theta_{\alpha,\beta} = \frac{1}{2}\left\{(1+y)\frac{m}{y} \pm \sqrt{\left[(1+y)\frac{m}{y}\right]^2 - 4\frac{m}{y}}\right\} \qquad \text{(C-26)}$$

but

[23] D. Richman, "APC color sync for NTSC color television," IRE CONVENTION RECORD, part 4; presented March 23, 1953.

$$\frac{1+y^2\theta^2}{(\theta^2+\theta_\alpha{}^2)(\theta^2+\theta_\beta{}^2)} = \left[\frac{1-y^2\theta_\alpha{}^2}{\theta^2+\theta_\alpha{}^2} - \frac{1-y^2\theta_\beta{}^2}{\theta^2+\theta_\beta{}^2}\right]\cdot\frac{1}{\theta_\beta{}^2-\theta_\alpha{}^2}. \quad \text{(C-27)}$$

The above is substituted in (C-4) to give

$$\int_0^\infty Q^2(ft_c)d(ft_c) = t_c f_{NN} \quad \text{(C-28)}$$

$$= \left(\frac{m}{y}\right)^2 \frac{1}{\theta_\beta{}^2-\theta_\alpha{}^2} \int_0^\infty \left[\frac{1-y^2\theta_\alpha{}^2}{\theta_\alpha{}^2+(2\pi ft_c)^2} - \frac{1-y^2\theta_\beta{}^2}{\theta_\beta{}^2+(2\pi ft_c)^2}\right] d(ft_c).$$

Then, since d/dx arctan $x = 1/(1+x^2)$ and arctan $0 = 0$, and arctan $\infty = \pi/2$.

$$t_c f_{NN} = \left(\frac{m}{y}\right)^2 \left(\frac{\dfrac{1-y^2\theta_\alpha{}^2}{4\theta_\alpha} - \dfrac{1-y^2\theta_\beta{}^2}{4\theta_\beta}}{\theta_\beta{}^2-\theta_\alpha{}^2}\right). \quad \text{(C-29)}$$

This is simplified as follows. Since

$$\theta_\alpha + \theta_\beta = \frac{m}{y}(1+y) \quad \text{(C-30)}$$

and

$$\theta_\alpha \cdot \theta_\beta = \frac{m}{y},$$

then

$$t_c f_{NN} = \left(\frac{m}{y}\right)^2 \cdot \frac{1}{4}\left[\frac{\dfrac{\theta_\beta-\theta_\alpha}{\theta_\beta\theta_\alpha} + y^2(\theta_\beta-\theta_\alpha)}{(\theta_\beta-\theta_\alpha)(\theta_\beta+\theta_\alpha)}\right] \quad \text{(C-11)}$$

$$= \frac{1}{4}\frac{m^2}{y^2}\left[\frac{\dfrac{y}{m}+y^2}{\dfrac{m}{y}(1+y)}\right] = \frac{1}{4}\left(\frac{1+my}{1+y}\right).$$

This is the desired result.

APPENDIX D

Transient Performance of the APC Loop

This appendix provides a description and derivation of formulas relating to pull-in characteristics and pull-in time of APC loops. Exact analysis of a simplified APC loop provides useful formulas and a basis for understanding some of the phenomena relating to pull-in. This then suggests a simple approximate method for reducing the differential equation of the loop to a form which is readily solvable for the pull-in time. The results are plotted and discussed.

The Simplified Loop

The simplest form of APC network is the one for which $N(\omega) = $ a constant. See Fig. 22(a).

The basic equations are:

$$N(\omega) = m$$

$$Q(\omega) = \frac{m}{m+pt_c}$$

$$f_{NN} = \frac{\pi}{2}mf_c \quad \text{(D-1)}$$

$$|\sin\Delta\phi| = \left|\frac{\Delta f}{mf_c}\right| \le 1.$$

The differential equation of the loop is

$$m\cdot\omega_c\sin\phi = \frac{d\phi}{dt} - \Delta\omega. \quad \text{(D-2)}$$

The same equation has been shown applicable for directly synchronized oscillators.[24]

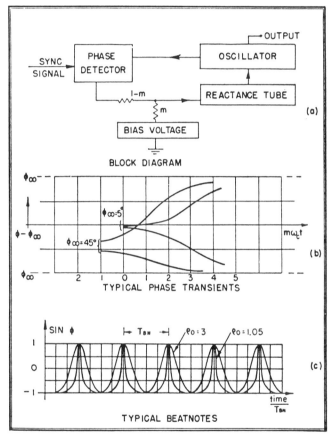

Fig. 22—Basic APC system.

This equation is equivalent to

(Filter transfer characteristic) · (Phase detector output)

= (Rate of change of phase difference)

− (Initial angular frequency difference).

The equation may be rewritten as

$$dt = \frac{d\phi}{\Delta\omega + m\omega_c\sin\phi}. \quad \text{(D-3)}$$

It has two solutions, depending on whether $\Delta\omega/m\omega_c$ is greater than or less than 1. Boundary conditions are

[24] R. Adler, "A study of locking phenomena in oscillators," Proc. I.R.E., vol. 34, pp. 351–357; June, 1946.

$$t = 0 \qquad \frac{d\phi}{dt} = \Delta\omega \quad \phi = \phi_0$$

$$t = \infty \qquad \frac{d\phi}{dt} = 0 \qquad \phi = \phi_\infty = \arcsin\left(\frac{-\Delta f}{mf_c}\right). \qquad (D\text{-}4)$$

Equation (D-3) is directly integrable.[25]

The pull-in range is $\Delta f \leq mf_c$. Within the pull-in range the phase stabilizes according to the following equation, which is the integral of (D-3) under this condition:

$$m\omega_c t \cos\phi_\infty = \ln \left| \frac{\tan\dfrac{\phi}{2} - \cot\dfrac{\phi_\infty}{2}}{\tan\dfrac{\phi}{2} - \tan\dfrac{\phi_\infty}{2}} \right|$$
$$\left| \frac{\tan\dfrac{\phi_0}{2} - \tan\dfrac{\phi_\infty}{2}}{\tan\dfrac{\phi_0}{2} - \cot\dfrac{\phi_\infty}{2}} \right| \qquad (D\text{-}5)$$

where

$$\frac{\Delta\omega}{m\omega_c} = \rho = -\sin\phi_\infty \qquad (|\rho| < 1) \qquad (D\text{-}6)$$

and

$$-\sqrt{1 - \rho^2} = \cos\phi_\infty. \qquad (D\text{-}7)$$

Typical phase transients are shown in Fig. 22(b). Phase is plotted relative to ϕ_∞ with a scale calibrated in units of $m\omega_c t$. The starting point on any curve is determined by $\phi_0 - \phi_\infty$.

An approximate time constant of stabilization is

$$\frac{-1}{m\omega_c \cos\phi_\infty} = \frac{1}{\sqrt{(m\omega_c)^2 - (\Delta\omega)^2}},$$

however, *the actual stabilization time is a function of the initial phase.*

Outside the pull-in range $\rho > 1$, and the phase as a function of time is defined by the following equation, which is the integral of (D-3) for this condition:

$$\frac{m\omega_c t \sqrt{\rho^2 - 1}}{2} = \arctan\left\{ \frac{\tan\dfrac{\phi}{2} + 1}{\sqrt{\rho^2 - 1}} \right\} \Bigg|_{\phi_0}^{\phi}. \qquad (D\text{-}8)$$

This represents a cyclic variation characterized by its wave form and its fundamental frequency, f_{BN}.

Fig. 22(c) shows examples of the cyclic relationship between $\sin\phi$ and t, $\rho_0 = \Delta f/mf_c$ being specified as 1.05 or 3. The time scale is normalized to the beatnote period $T_{BN} = 1/f_{BN}$. The period T_{BN} is such that t increases by T_{BN} when ϕ increases by 2π, and is found from the following relation:

[25] H. B. Dwight, "Tables of Integrals and Other Mathematical Data," The Macmillan Co., New York, N. Y., Integral 436.00; 1947.

$$\frac{m\omega_c T_{BN}\sqrt{\rho^2 - 1}}{2} = \pi \quad \text{when} \quad \Delta\phi = 2\pi. \qquad (D\text{-}9)$$

Then

$$T_{BN} = \frac{2\pi}{m\omega_c\sqrt{\rho^2 - 1}} = \frac{1}{\sqrt{(\Delta f)^2 - (mf_c)^2}}. \qquad (D\text{-}10)$$

This is an important relationship. It states for example, that if in the APC loop block diagram presented above the bias is adjusted so that the effective open loop frequency difference is $\Delta f (> mf_c)$, the operating beatnote frequency difference is $\sqrt{(\Delta f)^2 - (mf_c)^2}$. If the bias is a slowly varying function of time (as compared to f_{BN}), the above relationship accurately describes the variation of f_{BN} with time.

The dc bias or average dc potential developed at the reactance tube input may be determined from the above relationships. It may be expressed in terms of its effect on frequency.

Integrating the differential equation over a cycle, and dividing by the period

$$\frac{1}{T_{BN}} \oint m\omega_c \sin\phi\, dt = \frac{1}{T_{BN}} \oint \frac{d\phi}{dt} dt - \frac{1}{T_{BN}} \oint \Delta\omega\, dt \quad (D\text{-}11)$$

or

$$m\omega_c \,\overline{\sin\phi} = \frac{2\pi}{T_{BN}} - \Delta\omega \qquad (D\text{-}12)$$

and therefore, dividing by 2π,

$$mf_c \,\overline{\sin\phi} = \sqrt{(\Delta f)^2 - (mf_c)^2} - \Delta f. \qquad (D\text{-}13)$$

This is plotted in Figs. 11(a) and 23(a) which represents magnitude of the developed bias as a function of Δf.[26] In the standard loop shown later in which the bias battery is replaced by a capacitor it is proportional to the control effect which causes pull-in.

Fig. 23(a) shows that $m\omega_c \,\overline{\sin\phi}$, the average angular frequency shift, is a maximum when $\Delta\omega/m\omega_c = 1$ and decreases beyond that point, approaching zero asymptotically. When $(\Delta\omega/m\omega_c) < 1$, the phase does not shift 2π radians in a finite time. Enough bias is produced however, to shift the angular frequency by $\Delta\omega$. This bias is represented by the straight line portion, as discussed with regard to Fig. 11(a).

The Standard APC Loop

The standard APC loop is shown in Fig. 23(b). For the network shown,

$$N(p) = \frac{1 + pyt_c}{1 + p\dfrac{y}{m}t_c} = m\,\frac{1 + pyt_c}{m + pyt_c} = m + \frac{1 - m}{1 + p\dfrac{y}{m}t_c}$$

[26] In experimental work this characteristic may be measured in terms of f_{BN}. From (D-10), above, $f_{BN}^2 + (mf_c)^2 = (\Delta f)^2$.

= wideband direct transfer component

+ long time-constant integration component

= resistive component

+ capacitive component.   (D-14)

The differential equation in operational form is

$$N(p)\omega_c \sin \phi = p\phi - \Delta\omega \qquad \text{(D-15)}$$

which may be written as

$$m\omega_c \sin \phi = p\phi - \Delta\omega - \frac{1-m}{1+p\dfrac{y}{m}t_c}\omega_c \sin \phi. \qquad \text{(D-16)}$$

The term

$$\Delta\omega + \frac{1-m}{1+p\dfrac{y}{m}t_c}\omega_c \sin \phi \equiv \omega_I \qquad \text{(D-17)}$$

is the Fourier transform of a time function representing effective instantaneous impressed frequency difference.

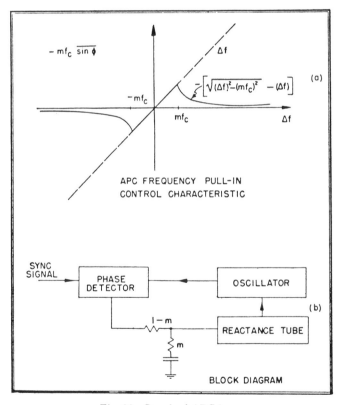

APC FREQUENCY PULL-IN
CONTROL CHARACTERISTIC

SYNC SIGNAL → PHASE DETECTOR → OSCILLATOR

1 − m

REACTANCE TUBE

m

BLOCK DIAGRAM

Fig. 23—Standard APC loop.

When this loop is turned on, or has a signal applied to it, the transient of stabilization lasts for a period of time which depends on both the initial phase and the frequency difference. However, the initial phase has only a small effect on the pull-in time and may be neglected for simplicity; the phase transient time, T_ϕ is rapid compared to the frequency pull-in time, T_F. Fig. 22(b) substantiates that for high dc loop gain ($\phi_\infty \ll 90°$) normally $m\omega_c T_\phi < 10$. Then, using (C-15),

$$10 > m\omega_c T_\phi = 4f_{NN}T_\phi \frac{K}{K+\frac{1}{4}}. \qquad \text{(D-18)}$$

If $K=\frac{1}{2}$, $T_\phi < (15/4f_{NN})$, while if $K=4$, $T_\phi < (2.7/f_{NN})$.

If the frequency difference is such that $\Delta f < mf_c$ the resistive component of loop feedback is adequate to ensure pull-in. The analysis presented above for simplified loop shows the system never slips a complete cycle.

A definition of frequency pull-in time T_F, and phase pull-in time T_ϕ is desirable; the following are useful.

If the system never slips a cycle, then the transient is defined as phase pull-in and measured in terms of the phase pull-in time, T_ϕ.

If the system slips cycles, then the period of time from the instant of switching or excitation until a definable point is reached from which the phase slip does not exceed a cycle is T_F, the frequency pull-in time.

When the initial frequency difference is such that $\Delta f > mf_c$, the long time integration component of feedback must be relied upon for pull-in.

The time constant $(y/m)t_c = y/(2\pi mf_c)$ is long compared to the loop time constant, t_c/m, since $y \gg 1$. Because of this long time constant, the average bias across the capacitor which may result from an unsymmetrical beatnote wave form from the phase detector will not change rapidly with time. It is not unreasonable therefore to integrate the differential equation for this APC loop over a cycle of beatnote.

Then

$$m\omega_c \overline{\sin \phi} = \frac{2\pi}{T_{BN}} - \overline{\omega_I} = \sqrt{\overline{(\omega_I)}^2 - (m\omega_c)^2} - \overline{\omega_I} \qquad \text{(D-19)}$$

$$\overline{\omega_I} = \Delta\omega + \overline{\frac{1-m}{1+p\dfrac{y}{m}t_c}\omega_c \sin \phi}. \qquad \text{(D-20)}$$

At this point it is necessary to recognize clearly the nature of the signal circulating in the APC loop. There are two components; there is a *cyclic* component produced as a result of the average frequency difference, and having a harmonic composition which is a function of the frequency difference and hence of time during pull-in; there is a low frequency *drift* component which represents the slow change in frequency difference which constitutes pull-in. It has been shown earlier that the generated frequency shift, $\omega_c \overline{\sin \phi}$, varies in an inverse manner with $\Delta\omega$ or $\overline{\omega_I}$; thus, frequency changes slowly except when $\overline{\omega_I}$ is very near $m\omega_c$; stated another way, almost all of the pull-in time is accrued under the condition that the rate of change of the beatnote frequency is not comparable to the beatnote frequency. Therefore, very nearly

$$\oint \frac{1-m}{1+p\dfrac{y}{m}t_c}\omega_c \sin \phi\, dt \approx \frac{1-m}{1+p\dfrac{y}{m}t_c}\oint \omega_c \sin \phi\, dt \qquad \text{(D-21)}$$

and

$$\frac{1-m}{1 + p\dfrac{y}{m}t_c}\,\omega_c \sin\phi \approx \frac{1-m}{1 + p\dfrac{y}{m}t_c}\,\omega_c\,\overline{\sin\phi}. \qquad (D\text{-}22)$$

The term $\omega_o\,\overline{\sin\phi}$ may be eliminated from the above equations, giving a first order differential equation in $\overline{\omega_I}$, the average angular frequency difference.

$$\overline{\omega_I} - \Delta\omega = \frac{1-m}{1 + p\left(\dfrac{y}{m}t_c\right)}$$

$$\cdot\frac{1}{m}\left[\sqrt{(\overline{\omega_I})^2 - (m\omega_c)^2} - \overline{\omega_I}\right]. \qquad (D\text{-}23)$$

This may be written more conveniently, dividing through by $m\omega_c$, writing $\rho = \omega_I/m\omega_c$ and $\rho_0 = \Delta\omega/m\omega_c$ and by operating on both sides with the differential operator, $1 + p(y/m)t_c$.

Then

$$\left(1 + p\,\frac{y}{m}t_c\right)(\rho - \rho_0) = \frac{1-m}{m}(\sqrt{\rho^2 - 1} - \rho)$$

or

$$\rho - \rho_0 + \frac{y}{m}t_c\frac{d\rho}{dt} = \frac{1-m}{m}(\sqrt{\rho^2 - 1} - \rho).$$

Transposing $\rho - \rho_0$ and separating the variables

$$\frac{dt}{\dfrac{y}{m}t_c} = \frac{d\rho}{\rho_0 - \rho + \dfrac{1-m}{m}(\sqrt{\rho^2 - 1} - \rho)}. \qquad (D\text{-}24)$$

This equation may be directly integrated (between the limits $\rho = \rho_0$ and $\rho = 1$) to yield T_F.

The integration is accomplished with the aid of a change of variable which permits the application of some tabulated integrals. The equations obtained are cumbersome; they are presented at the end of this appendix; they were used for the computations on which the several graphs presented are based. Fig. 20 presents the universal pull-in curves for the standard APC system. The following simplified analysis obtains the significant conclusions, in simpler form.

The limiting pull-in range may be determined as the condition which makes the required pull-in time become infinite. This occurs when the denominator of the above integrand has a real root. It will only occur when

$$\rho_0\left(\equiv \frac{\Delta f}{mf_c}\right) \geq \sqrt{\frac{2}{m} - 1}. \qquad \begin{array}{l}(D\text{-}25)\\(C\text{-}17)\end{array}$$

A simple approximate solution for the "limit-curve" may be obtained by eliminating from the equation the factor which produces the above limitation. (Specifically, the small term $(\rho_0 - \rho)$ in the denominator is omitted.)

Then, if $m \ll 1$, approximately

$$\frac{dt}{yt_c} \approx \frac{d\rho}{\sqrt{\rho^2 - 1} - \rho} = -(\sqrt{\rho^2 - 1} + \rho)d\rho \qquad (D\text{-}26)$$

and hence, integrating from $\rho = \rho_0$ to $\rho = 1$,

$$\frac{T_F}{yt_c} = \frac{\rho_0^2 - 1}{2} + \frac{\rho_0\sqrt{\rho_0^2 - 1}}{2}$$
$$- \frac{1}{2}\ln\left|\frac{\rho_0 + \sqrt{\rho_0^2 - 1}}{1}\right|. \qquad (D\text{-}27)$$

Except for ρ_0 near 1, this is closely equal to

$$\frac{T_F}{yt_c} \approx \rho_0^2 = \left(\frac{\Delta f}{mf_c}\right)^2 \qquad \begin{array}{l}(D\text{-}28)\\(C\text{-}19)\end{array}$$

which is the equation presented earlier.

The pole at $\rho_0^2 = (2/m) - 1$ can be included, writing the simplified equation as

$$\frac{T_F}{yt_c} = \frac{\rho_0^2}{1 - \dfrac{m}{2-m}\rho_0^2} \qquad 1 < \rho_0^2 < \left(\frac{2}{m} - 1\right). \qquad (D\text{-}29)$$

The exact integration of (D-24) is accomplished with the aid of the following substitution:

$$z = \sqrt{\rho^2 - 1} - \rho$$

whence,

$$-\rho = \frac{1 + z^2}{2z} \quad\text{and}\quad \frac{d\rho}{dz} = -\frac{1}{2}\left(\frac{z^2 - 1}{z^2}\right).$$

Then

$$\frac{T_F}{yt_c} = \int_{z_0}^{z_1} \frac{-\left(z - \dfrac{1}{z}\right)dz}{(2-m)z^2 + 2mf_c z + m}. \qquad (D\text{-}30)$$

The limits are

$$\begin{vmatrix}\rho = 1\\ \rho = \rho_0\end{vmatrix} \quad\text{and}\quad \begin{vmatrix}z_1 = -1\\ z_0 = \sqrt{\rho_0^2 - 1} - \rho_0.\end{vmatrix}$$

Referring to H. B. Dwight, "Tables of Integrals and Other Mathematical Data,"[25] Integrals #160.01, #160.11 and #161.11 are used.

Then

$$\frac{T_F}{yt_c} = -\Bigg\{\frac{1}{2(2-m)}\ln\left|(2-m)z^2 + 2m\rho_0 z + m\right|$$

$$- \frac{1}{2m}\ln\frac{z^2}{(2-m)z^2 + 2m\rho_0 z + m}$$

$$+ \left[\frac{2m\rho_0}{2}\left(\frac{1}{m} - \frac{1}{2-m}\right)\frac{2}{\sqrt{4(2-m)m - (2m\rho_0)^2}}\right.$$

$$\left.\cdot\arctan\frac{2(2-m)z + 2m\rho_0}{\sqrt{4(2-m)m - (2m\rho_0)^2}}\right]\Bigg\}\Bigg|_{z_0 = \sqrt{\rho_0^2 - 1} - \rho_0}^{z_1 = -1}$$

$$= \frac{-1}{2(2-m)} \ln \left| \frac{2(1-m\rho_0)}{(2-m)z_0{}^2 + 2m\rho_0 z_0 + m} \right|$$

$$+ \frac{1}{2m} \ln \left(\frac{1}{z_0{}^2} \right) \left[\frac{(2-m)z_0{}^2 + 2m\rho_0 z_0 + m}{2(1-m\rho_0)} \right]$$

$$- \left\{ 2\rho_0 \left(\frac{1-m}{2-m} \right) \frac{1}{\sqrt{m(2-m)-(m\rho_0)^2}} \right. \qquad \text{(D-31)}$$

$$\cdot \left[\arctan \frac{m\rho_0 - 2 + m}{\sqrt{m(2-m)-(m\rho_0)^2}} \right.$$

$$\left. \left. - \arctan \frac{m\rho_0 + (2-m)z_0}{\sqrt{m(2-m)-(m\rho_0)^2}} \right] \right\} .$$

APPENDIX E

Reliability of Frequency Difference Detection

This Appendix presents some mathematical derivations relating to the reliability of frequency difference detection.

The relations between rms frequency error and integration time are derived for

(a) the signal

(b) quadricorrelator frequency difference detector preceded by limiter

(c) quadricorrelator frequency difference detector alone.

Basic Signal Characteristics

The combination of signal and noise may be expressed in the following alternate forms (omitting for the moment the time gate factor)

$$S \cos \omega_{SC} + a(t) \cos \omega_{SC}t + b(t) \sin \omega_{SC}t \qquad \text{(E-1)}$$

in which the noise is related to the color subcarrier frequency, or

$$S \cos \omega_{SC}t + a_0(t) \cos \omega_0 t + b_0(t) \sin \omega_0 t \qquad \text{(E-2)}$$

in which the noise is expressed relative to the local oscillator frequency.

After limiting, the signal can be expressed as

$$S \cos (\omega_{SC}t + \phi(t)). \qquad \text{(E-3)}$$

The phase modulation due to noise is $\phi(t)$.

$$\phi(t) = \arctan \frac{b(t)}{S + a(t)} \approx \frac{b(t)}{S} \qquad \text{(E-4)}$$

as a first order approximation.

Then

$$\frac{d\phi}{dt} = \frac{S \frac{db}{dt} + \frac{d}{dt}(ab)}{(S+a)^2 + b^2} . \qquad \text{(E-5)}$$

As a second order approximation, the relationship

$$\phi(t) = \int \frac{d\phi}{dt} dt \approx \frac{b(t)}{S} + \frac{a(t)b(t)}{S^2} \qquad \text{(E-6)}$$

can be used, since $(S+a)^2 + b^2 \approx S^2$ for signals of interest.

The instantaneous frequency of the amplitude limited signal is

$$f(t) = \frac{1}{2\pi} \left[\omega_{SC} + \frac{d\phi(t)}{dt} \right] . \qquad \text{(E-7)}$$

The rms frequency error due to noise is f_{rms}.

$$f_{rms} = \frac{1}{2\pi} \left[\frac{d\phi(t)}{dt} \right]_{rms} . \qquad \text{(E-8)}$$

The signal amplitudes will also be useful in this analysis.

Then

$S = \frac{1}{2}hS_0$ = amplitude of a burst

$Sd = \frac{1}{2}hS_0d$ = average amplitude of the component at the burst frequency with gate duty cycle d.

The rms value of $b(t)$ is the square root of the noise power. If *effectively* passed through a filter of bandwidth f_H, and gated with a duty cycle d, the noise power per unit time is $d(N_W{}^2/f_W)f_H$ and hence, the first order approximation for ϕ_{rms} is

$$\phi_{rms} \approx \frac{b_{rms}}{S} = \frac{N_W}{\frac{1}{2}hS_0} \sqrt{\frac{f_H}{df_W}} . \qquad \text{(E-9)}$$

These relations are useful in evaluating the relation between integration time and reliability of the best possible frequency difference detector which might be used for the signal.

To relate reliability to time, the signal information may be averaged over a period T_I, and the rms value of the average then has improved reliability by virtue of integration. As in the case of phase information, it is convenient to use a rectangular time aperture for a standard of comparison for integrators.

Then

$$f_{rms} = \frac{1}{T_I} \left[\int_0^{T_I} [f(t) - f_{SC}]dt \right]_{rms} \qquad \text{(E-10)}$$

$$= \frac{1}{T_I} \left[\int_0^{T_I} \frac{1}{2\pi} \frac{d\phi}{dt} dt \right]_{rms}$$

$$= \frac{1}{2\pi T_I} \left[\phi(T_I) - \phi(0) \right]_{rms}$$

$$= \frac{\sqrt{2}}{2\pi T_I} \phi_{rms}$$

and therefore, using the first order approximation above,

$$f_{rms} \approx \sqrt{\frac{2}{d}} \cdot \frac{1}{h\pi T_I} \cdot \frac{N_W}{S_0} \sqrt{\frac{f_H}{f_W}} . \qquad \text{(E-11)}$$

The term $(S_0/N_W)\sqrt{f_W}$ is the signal-to-*noise-density* ratio.

The factor $(1/T_I)\sqrt{f_H}$ has the dimensions of (frequency)$^{3/2}$; such terms normally result in frequency

73

modulation noise analysis due to the triangular spectrum of the noise.[27]

The second order approximation is

$$f_{\text{rms}} \approx \frac{1}{\sqrt{2}\,\pi T_I} \left[\left(\frac{b_{\text{rms}}}{S} \right)^2 + \left(\frac{(ab)_{\text{rms}}}{S^2} \right)^2 \right]^{1/2} \qquad \text{(E-12)}$$

Note here that rms values add in quadrature.

The second term varies as

$$\left(\frac{N_W}{S_0} \sqrt{\frac{f_H}{f_W}} \right)^2$$

and for signal-to-noise ratios which at present give satisfactory monochrome video signals is small compared to the first term.

Quadricorrelator with Limiter

The quadricorrelator is shown in Figs. 13 and 14. The quadrature reference is R_Q.

$$R_Q = \sin \omega_0 t. \qquad \text{(E-13)}$$

The in-phase reference is R_I.

$$R_I = \cos \omega_0 t. \qquad \text{(E-14)}$$

The cosine beatnote is the beatnote between the input signal and R_I. This is conveniently expressed as

$$\frac{2}{S} \overline{[S \cos(\omega_{SC} t + \phi)] \cos \omega_0 t}$$

$$= \cos\left[[\omega_0 - \omega_{SC}]t - \phi(t)\right]. \qquad \text{(E-15)}$$

The sine beatnote is then

$$\frac{2}{S} \overline{[S \cos(\omega_{SC} t + \phi)] \sin \omega_0 t}$$

$$= \sin\left[[\omega_0 - \omega_{SC}]t - \phi(t)\right]. \qquad \text{(E-16)}$$

The derivative of the cosine beatnote is

$$-\left[\omega_0 - \omega_{SC} - \frac{d\phi}{dt}\right] \sin\left[[\omega_0 - \omega_{SC}]t - \phi(t)\right]. \qquad \text{(E-17)}$$

Let

$$\omega_0 - \omega_{SC} \equiv \Delta\omega \equiv 2\pi\Delta f. \qquad \text{(E-18)}$$

Then, the indicated frequency, which is the integrated output from the product of the signals expressed in (E-16) and (E-17), as multiplied in the output synchronous detector of the quadricorrelator, is, with due regard to signs,

$$f(t) = \frac{1}{\pi T_I} \int_0^{T_I} \left[\Delta\omega - \frac{d\phi}{dt}\right] \sin^2\left[\Delta\omega t - \phi(t)\right] dt. \qquad \text{(E-19)}$$

The polarity of the indicated frequency may be reversed (when so required) by transferring the differ-

[27] M. G. Crosby, "Frequency modulation noise characteristics," Proc. I.R.E., vol. 25, pp. 472–514; April, 1937.

entiating circuit from the cosine channel to the sine channel.

Then, since $\sin^2 x = \frac{1}{2} - \frac{1}{2}\cos 2x$, and since

$$\left[\Delta\omega - \frac{d\phi}{dt}\right] \cos 2[\Delta\omega t - \phi]dt$$

$$= \cos 2[\Delta\omega t - \phi] \cdot d[\Delta\omega t - \phi]$$

we obtain

$$f(t) = \Delta f + \frac{1}{2\pi T_I} \int_{t=0}^{t=T_I} d\phi$$

$$- \frac{1}{2\pi T_I} \int_{t=0}^{t=T_I} \cos 2\eta \, d\eta \qquad \text{(E-20)}$$

where

$$\eta \equiv \Delta\omega t - \phi(t).$$

Thus the output noise consists of two components: the first represents the frequency noise of the signal; it could be measured as output noise if ω_{SC} were *known*. The second represents extra noise introduced by the measurement of an unknown frequency in this circuit. Then

$$f_{\text{rms}} = \frac{1}{\sqrt{2}\,\pi T_I} \left[\phi_{\text{rms}}^2 + \text{extra noise}^2\right]^{1/2}. \qquad \text{(E-21)}$$

The extra noise is evaluated as follows:

$$\frac{1}{2\pi T_I} \int_{t=0}^{t=T_I} \cos 2\eta \, d\eta$$

$$= \frac{1}{2\pi T_I} \left[\tfrac{1}{2}\sin 2\eta(T_I) - \tfrac{1}{2}\sin 2\eta(0)\right]. \qquad \text{(E-22)}$$

Two effects are indicated:

(a) Due to the use of a rectangular time aperture, an extraneous "sampling distortion" term appears unless $\Delta\omega T_I = $ a multiple of 2π, which is therefore assumed for simplicity.

(b) The output noise has the character of random noise which is passed through a nonlinear amplifier having a gain proportional to the sine of the input. This crushes the noise peaks and reduces the rms value.

Then, since $\sin^2 x < x^2$

$$\left[\frac{1}{2\pi T_I} \int_{t=0}^{t=T_I} \cos 2\eta \, d\eta\right]_{\text{rms}} < \frac{\sqrt{2}}{2\pi T_I} \phi_{\text{rms}}. \qquad \text{(E-23)}$$

If there is substantial integration ($f_H T_I \gg 1$), the two noise components approach complete independence and add in quadrature, hence at worst,

$$f_{\text{rms}} \approx \frac{1}{\pi T_I} \phi_{\text{rms}}. \qquad \text{(E-24)}$$

Thus, the quadricorrelator, with a limiter, measures a frequency difference to within a few db of the ultimate reliability permitted by signal information. It has no "detuning" error. The stepped characteristic may be introduced to give

$$f_{\text{rms}} \approx \frac{1}{\pi T_I} \phi_{\text{rms}} \sqrt{\frac{f_H - 2f_{NN}}{f_H}} . \quad \text{(E-25)}$$

Since $f_H \gg 2f_{NN}$, the simpler equations above are adequate.

Quadricorrelator without Limiter

The input signal is

$$S \cos \omega_{SC} t + a_0(t) \cos \omega_0 t + b_0(t) \sin \omega_0 t. \quad \text{(E-2)}$$

The cosine beatnote is proportional to

$$\cos \Delta \omega t + \frac{a_0(t)}{S} . \quad \text{(E-26)}$$

The sine beatnote is

$$\sin \Delta \omega t + \frac{b_0(t)}{S} . \quad \text{(E-27)}$$

The derivative of the cosine beatnote is

$$- \Delta \omega \sin \Delta \omega t + \frac{1}{S} \frac{da_0(t)}{dt} . \quad \text{(E-28)}$$

The quadricorrelator output is

$$f(t) = \frac{1}{\pi T_I} \int_0^{T_I} \left[\Delta \omega \sin \Delta \omega t - \frac{1}{S} \frac{da_0}{dt} \right]$$

$$\cdot \left[\sin \Delta \omega t + \frac{b_0}{S} \right] dt$$

$$= \Delta f - \frac{1}{\pi T_I} \int_0^{T_I} \frac{1}{S} \frac{da_0}{dt} \sin \Delta \omega t\, dt \quad \text{(E-29)}$$

$$+ \frac{1}{\pi T_I} \int_0^{T_I} \frac{\Delta \omega}{S} b_0 \sin \Delta \omega t\, dt$$

$$- \frac{1}{\pi T_I} \int_0^{T_I} \frac{1}{S^2} b_0 \frac{da_0}{dt} dt.$$

The evaluation of the several terms is aided by integration by parts:

$$- \frac{1}{\pi T_I} \int_0^{T_I} \frac{1}{S} \frac{da_0}{dt} \sin \Delta \omega t\, dt$$

$$= - \frac{1}{\pi T_I} \sin \Delta \omega t \left. \frac{a_0}{S} \right|_0^{T_I}$$

$$+ \frac{1}{\pi T_I} \int_0^{T_I} \frac{\Delta \omega}{S} a_0 \cos \Delta \omega t\, dt. \quad \text{(E-30)}$$

Then

$$\left[- \frac{1}{\pi T_I} \sin \Delta \omega t \cdot \left. \frac{a_0}{S} \right|_0^{T_I} \right]_{\text{rms}} = \frac{1}{\pi T_I} \cdot \frac{a_{0\text{rms}}}{S} . \quad \text{(E-31)}$$

The term

$$\frac{1}{\pi T_I} \int_0^{T_I} \frac{\Delta \omega}{S} \left[b_0(t) \sin \Delta \omega t + a_0(t) \cos \Delta \omega t \right] dt \quad \text{(E-32)}$$

now appears. This is a two-dimensionally noise modulated sine wave, of the type shown in Fig. 16. The bandwidth of the noise, however, is such that it heterodynes with the carrier to produce a dc component. Then, the integral of this term has the rms value

$$\frac{\Delta \omega}{\pi S} \frac{N_{\text{rms}}}{\sqrt{f_H T_I}} = \frac{\Delta \omega}{\pi S} \frac{a_{0\text{rms}}}{\sqrt{f_H T_I}} \quad \text{(E-33)}$$

since there are $f_H T_I$ effective harmonic components. The remaining term is evaluated as follows, integrating by parts:

$$\left[\frac{1}{\pi T_I} \int_0^{T_I} \frac{a_0}{S^2} \frac{db_0}{dt} dt \right]_{\text{rms}}$$

$$= \left[\frac{1}{\pi T_I} \int_0^{T_I} \frac{b_0}{S^2} \frac{da_0}{dt} dt \right]_{\text{rms}}$$

$$= \frac{1}{\sqrt{2}} \left[\frac{1}{\pi T_I} \int_0^{T_I} \frac{1}{S^2} \frac{d}{dt} (a_0 b_0) dt \right]_{\text{rms}}$$

$$= \frac{1}{\pi T_I} \frac{(a_0 b_0)_{\text{rms}}}{S^2} . \quad \text{(E-34)}$$

Then

$$[f(t) - \Delta f]_{\text{rms}}$$

$$= f_{\text{rms}} = \frac{1}{\pi T_I} \left[\left[\left(\frac{a_{0\text{rms}}}{S} \right)^2 + \left(\frac{(a_0 b_0)_{\text{rms}}}{S^2} \right)^2 \right] \right.$$

$$\left. + \left[\frac{2\Delta f}{\sqrt{f_H T_I}} \cdot \frac{a_{0\text{rms}}}{S} \right]^2 \right]^{1/2} . \quad \text{(E-35)}$$

These terms add in quadrature as they represent independent random variables. The first bracketed term is of similar form as, but 3 db larger than, the second order signal approximation presented in (E-12), and is nearly equal to

$$\frac{1}{\pi T_I} \frac{a_{0\text{rms}}}{S} .$$

The extra noise due to amplitude modulation appears in the last term of (E-35). The ratio of the AM component of noise to the FM component of noise is near

$$\frac{2\pi \Delta f T_I}{\sqrt{f_H T_I}} . \quad \text{(E-36)}$$

When Δf is small, the quadricorrelator without a limiter approaches the limit of performance permitted by the signal. When Δf approaches $\frac{1}{2} f_H$, a poorer signal-to-noise ratio is obtained. The time T_I must be selected so that f_{rms} does not exceed some selected value, when Δf is the nominally maximum design value for pull-in range.

Equation (E-35) shows that the operation of pulling in results in a large reduction of output noise from the quadricorrelator.

Table of Symbols

Symbol	Description
$a, a(t)$	cosine component of noise at the frequency of the sync signal
$a_0, a_0(t)$	same at frequency of oscillator
$b, b(t)$	sine component of noise at the frequency of the sync signal
$b_0, b_0(t)$	same at frequency of oscillator
C	capacitor, Fig. 18
d	the duty cycle of the burst
f_c	the dc loop gain of an APC system, equal to the peak frequency holding range
f_{BN}	beatnote frequency, appearing in Appendix D
$f_H, f_{HORIZONTAL}$	the line-scanning frequency, 15,750 cps
f_N	effective noise bandwidth of a phase-detection system
$f_{N_{LIMIT}}$	the value of f_N if all of the phase information is used
f_0	frequency of the local reference oscillator
f_{NN}	the noise bandwidth of an APC loop
f_{rms}	root-mean-square frequency error of a frequency-difference detector
$f_{SC}, f_{SUBCARRIER}$	the subcarrier (color carrier) frequency
$f(t)$	indicated frequency difference
f_W	video bandwidth occupied by signal and noise
f_3	the 3 db pass-band width of a high Q filter
$F(f)$	transfer characteristic of high Q filter
$F(\Delta f)$	transfer characteristic of high Q filter measured in terms of frequency difference
G	loop gain of the phase-control system of Appendix B
h	the ratio of the peak-to-peak amplitude of the burst and the line and field sync pulses
k	index number used in Appendix A
K	a damping coefficient relating to the pass band of an APC loop
m	the resistive divider ratio of a standard APC filter, the ratio of ac gain over dc gain through the network $N(\omega)$
$N(t), N$	noise signal as a function of time
N_W	the root-mean-square noise in the entire video pass band, assumed flat over the band
$N(\omega), N(p)$	transfer characteristic of the filter of an APC loop
p	$j\omega, j2\pi f$, or d/dt as appropriate
$p(\tau)$	the relative probability density distribution function for timing data
$p(\phi)$	the relative probability density distribution function for phasing data
$Q(f), Q(\omega)$	the effective modulation pass band transfer characteristic of an APC system after synchronization
r	ratio of actual gate width to minimum burst width
R	resistor, Fig. 18
R_I	in-phase reference signal
R_Q	quadrature reference signal
S	amplitude of color burst
S_k	amplitude of kth frequency component
S_0	the amplitude of the line and field sync pulses
$S(t)$	the synchronizing signal as a function of time
T	time constant of low-pass filter $Y(f)$
T	time constant RC of Fig. 18
T_A	transient response time of high Q filter in Appendix B
T_{BN}	beatnote period
t_c	characteristic time constant of an APC loop
T_F	frequency pull-in time
T_I	integration time of a frequency-difference detector
T_M	effective integration time
$T_{M_{LIMIT}}$	the required value of T_M if all of the phase information is used
t_0	root-mean-square time error
t_0	root-mean-square error if the color synchronizing signal were present all of the time
t_{0_k}	rms timing error of the k'th frequency component
T_S	stabilization time of a synchronizing system
T_ϕ	phase pull-in time
x	constant relating to Fig. 18(a)
y	parameter defined by (C-8)
$Y(f)$	pass-band characteristic of the low-pass filter in the phase-control system of Fig. 6, Appendix B
z	variable of integration used in Appendix D
β	sensitivity of the reactance tube of an APC loop
ΔE	output voltage of the phase detector of an APC loop
Δf	frequency difference between oscillator and sync signal
Δf_{max}	maximum frequency difference from which pull-in will occur
$\Delta\phi$	the static phase error of an APC loop
$\Delta\phi$	as used in Appendix B—phase error of phase feedback system
$\Delta\phi_{corr}$	the phase correction produced by the phase feedback system of Appendix B
$\Delta\phi_{corr}(p)$	frequency spectrum of $\Delta\phi_{corr}$
$\Delta\phi_0$	static phase error of high Q filter
$\Delta\phi_{0_{effective}}$	equivalent phase modulation (re. B-6)
$\Delta\phi_0(p)$	frequency spectrum of $\Delta\phi_{0_{effective}}$
$\Delta\phi_{01}(p)$	frequency spectrum of beat between filtered signal and direct noise at the phase detector of Fig. 6(a) (re. B-8)
$\Delta\phi_{02}(p)$	frequency spectrum of beat between the direct signal and filtered noise at the phase detector of Fig. 6(a) (re. B-8)
$\Delta\omega$	angular frequency difference
η	parameter used in Appendix E
$\theta, \theta_\alpha, \theta_\beta$	parameters defined and used in Appendix C
λ^2	see (C-23) and Fig. 21
μ	transfer gain of phase detector of an APC loop
$\rho = \dfrac{\Delta f}{mf_c}$	normalized frequency difference defined in Appendix C
ρ_0	initial value of ρ
τ	time scale for a probability density
ϕ	phase angle
ϕ_1	a phase variable relating to Fig. 2(c)
ϕ_{rms}	root-mean-square phase error
ϕ_0	initial phase
ϕ_∞	static phase difference due to frequency detuning
ω_c	$2\pi f_c$
ω_H	$2\pi f_H$
ω_I	instantaneous angular frequency difference (re. D-17)
ω_k	kth angular frequency
ω_0	$2\pi f_0$
ω_{SC}	$2\pi f_{SC}$

76

Charge-Pump Phase-Lock Loops

FLOYD M. GARDNER, FELLOW, IEEE

Abstract—Phase/frequency detectors deliver output in the form of three-state, digital logic. *Charge pumps* are utilized to convert the timed logic levels into analog quantities for controlling the locked oscillators. This paper analyzes typical charge-pump circuits, identifies salient features, and provides equations and graphs for the design engineer.

I. INTRODUCTION

PHASE-LOCK loops (PLL's) incorporating sequential-logic, *phase/frequency detectors* (PFD's) have been widely used in recent years [1]–[5], [6, ch. 6]. Reasons for their popularity include extended tracking range, frequency-aided acquisition, and low cost. A *charge pump* usually accompanies the PFD, as illustrated in Fig. 1. The purpose of the charge pump is to convert the logic states of the PFD into analog signals suitable for controlling the voltage-controlled oscillator (VCO).

Good understanding of the PFD itself has been attained but very little has been published on the operation of charge pumps. In consequence, design of PLL's containing charge pumps has often proceeded as an intuitive extension of conventional PLL's. That approach obscures the special benefits and the special problems of a charge-pump PLL.

The intent of this paper is to place the design analysis of a charge-pump PLL on a sound basis so that its special features are recognized and can be either utilized or avoided, as necessary. In Section II we introduce the basic charge-pump model and derive the loop transfer function based on assumptions of small error (linearized loop) and narrow bandwidth as compared to the input frequency (continuous-time approximation).

Section III is devoted to second-order PLL's wherein it is shown that Type-II operation is obtainable even with a passive loop-filter. This behavior is contrary to that obtained in conventional PLL's and is a particular benefit associated with charge pumps.

A continuous-time approximation is not valid if the loop bandwidth approaches the input frequency. In that case, the discrete-time—or sampled—nature of the loop must be recognized. In particular, sampling introduces stability problems that do not exist in continuous time networks; the stability limit for the second-order loop is presented.

Furthermore, the control voltage (v_c in Fig. 1) has large, rectangular excursions (*ripple*) on each cycle of operation. Ripple magnitude is shown to be proportional to loop bandwidth; ripple can easily be so large as to overload the VCO. The existence of ripple places limits on the application of the simple second-order loop.

Paper approved by the Editor for Communication Electronics of the IEEE Communications Society for publication without oral presentation. Manuscript received Janaury 4, 1980; revised May 27, 1980.
The author is at 1755 University Avenue, Palo Alto, CA 94301.

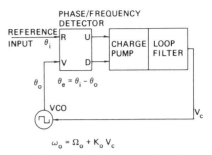

Fig. 1. Phase-lock loop with three-state phase detector and charge pump.

Filters are frequently added after the charge pump to reduce the ripple. Section IV describes the loop performance obtaining from addition of a single capacitor—the simplest-possible ripple filter. The loop is now third order (although still Type II) so analysis is more complicated. Root-locus plots are given for the continuous-time approximation. For wider bandwidths, a discrete-time, linearized analysis yields a z-plane characteristic-function from which pole locations and stability limits may be obtained. Ripple reduction factor is also set forth.

Results of a nonlinear, discrete state-variable analysis of the second-order loop are described in Part V. It turns out that transient settling times of wide-band loops obtained by discrete-time analysis are very similar to the scaled settling times of narrow loops analyzed on the ordinary continuous-time basis. Similar analysis is possible for the third-order loop, but has not been pursued.

II. MODEL

The states of a sequential-logic PFD are initiated by edges of the input waveform. In Fig. 1, if the R-input phase leads the V-input phase, then an edge of the R input sets the U (denoting "up") terminal true. The next V edge resets the U terminal false. As long as R leads V, the D (for "down") terminal remains false. Conversely, if V leads, R, a V edge sets D true and the next R edge resets D false.

Both U and D can be false simultaneously, or either one alone can be true, but both can never be true simultaneously. Therefore, a PFD has three allowable states at its two output terminals. The states will be denoted as U, D, and N, where the last connotes "null" or "neutral."

It is also possible to have *combinatorial* (or *multiplier*; see [6, ch. 6] for terminology) phase detectors with three-state logic outputs as in [7] and [8]. A combinatorial PD does not have the frequency-detector properties of the sequential PFD, but the charge-pump analyses given here apply to either type of circuit in the phase-locked condition. Matters of frequency acquisition are not treated in this paper.

Reprinted from *IEEE Trans. Comm.*, vol. COM-28, pp. 1849-1858, November 1980.

Assume that the PLL is locked and denote the frequency of the input signal as ω_i, radians/second. Let the phase error be $\theta_i - \theta_0 = \theta_e$ radians. The ON time[1] of either U or D, as appropriate, is

$$t_p = |\theta_e|/\omega_i \tag{1}$$

for each period $2\pi/\omega_i$ of the input signal. (The subscript "p" connotes "pump.")

These two features–the three-state description and the ON-time equation–completely characterize the PFD or PD for purposes of this paper.

A charge pump is nothing but a three-position, electronic switch that is controlled by the three states of the PFD. When the switch is set in the U or D position, it delivers a *pump voltage* $\pm V_p$ [Fig. 2(a) and (c)] or a *pump current* $\pm I_p$ [Fig. 2(b) and (d)] to the loop filter. In the N position, the switch is open, thereby isolating the loop filter from the charge pump and the phase detector. This open condition is not encountered in the conventional, analog PLL's and it engenders important, novel characteristics, as will be seen presently.

The loop filter can be either passive, as in Fig. 2(a) and (b), or active, as in Fig. 2(c) and (d). The significant features of the filters to be studied here are contained in the impedance $Z_F(s)$ of Fig. 2, where s is the Laplace-transform complex variable.

Most attention will be given to the current-pump, passive-filter configuration of Fig. 2(b). This choice is made partly because analysis is simplified but also because the configuration is eminently practical under many real-life conditions. It will be shown that performance of the other three configurations is readily obtained, at least approximately, from the analysis of 2(b).

Because of the switching, the charge-pump PLL is a time-varying network; an exact analysis must take account of the time variations of the circuit topology and that is a more-involved procedure than usually found in the common time-invariant networks. In particular, simple transfer-function analysis is not directly applicable to time-varying networks.

In many applications, the state of the PLL changes by only a very small amount on each cycle of the input signal. That is, the loop bandwidth is small compared to the signal frequency. In these cases we may not care about the detailed behavior within a single cycle and may be interested only in the average behavior over many cycles. By applying an averaged analysis, the time-varying operation can be bypassed and the powerful tool of transfer functions retained for our usage. The remainder of this section is devoted to the derivation of average-operation transfer functions. Be aware, though, that the per-cycle behavior can be important even for quite narrow bandwidths, as will be shown later.

Using Fig. 2(b), a pump current I_p sgn θ_e is delivered to the

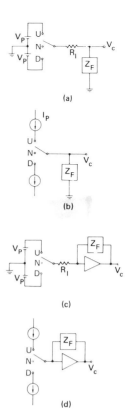

Fig. 2. Charge pumps and loop filters.

filter impedance Z_F for the time t_p on each cycle. Each cycle has a duration $2\pi/\omega_i$ seconds so, utilizing (1), the average *error current* over a cycle is

$$i_d = I_p \theta_e/2\pi \text{ amps.} \tag{2}$$

Equation (2) is also the error current averaged over many cycles, provided that both inputs are periodic–that no input cycles are missing. However, in some applications–notably, in recovery of clock from digital bit streams–edges, or pulses, of the R input will be missing at random. To avoid imposing an erroneously large error current upon the PLL, it is necessary to arrange the logic circuits to recognize the absence of R and to force the circuit into the N state upon those occasions. If the average error current on a single cycle is i_d, as in (2), and if the probability of occurrence of R (the so-called *transition density*) is denoted d, then the average error current over many cycles is $i_d d$. Where applicable, d must be factored into each of the following expressions that treats average behavior.

Oscillator control voltage is given by

$$V_c(s) = I_d(s)Z_F(s) = I_p Z_F(s)\theta_e(s)/2\pi \tag{3}$$

where $I_d(s)$ is the Laplace transform of $i_d(t)$, and similarly for the other symbols. For a locked loop (the only condition for which transfer functions are applicable, because of out-of-lock nonlinearities) the VCO phase is given by

$$\theta_0(s) = K_0 V_c(s)/s \tag{4}$$

where K_0 is the *VCO gain*[2] in radians/second/volt. These expressions, plus $\theta_e(s) = \theta_i(s) - \theta_0(s)$, lead to the loop transfer functions

$$\frac{\theta_0(s)}{\theta_i(s)} = \frac{K_0 I_p Z_F(s)}{2\pi s + K_0 I_p Z_F(s)} = H(s) \tag{5}$$

$$\frac{\theta_e(s)}{\theta_i(s)} = \frac{2\pi s}{2\pi s + K_0 I_p Z_F(s)} = 1 - H(s). \tag{6}$$

These functions apply for any Z_F.

An important property of any PLL is the *static phase error* [6, ch. 4] or *loop stress* that arises from a frequency offset $\Delta\omega$ between the input signal and the free-running frequency of the VCO. Applying the final-value theorem, as in [6], the static phase error is found to be

$$\theta_v = \frac{2\pi\Delta\omega}{K_0 I_p Z_F(0)} \text{ rad.} \tag{7}$$

The foregoing results were all obtained for the configuration of Fig. 2(b): the current switch with passive filter. Much the same expressions arise for each of the other three configurations. For an active filter, it is necessary to take the polarity reversal of the operational amplifier into account. For a voltage switch, the same equations as above occur if we let $I_p \approx V_p/R_1$. For Fig. 2(a)—voltage switch with passive filter—the resulting equations are approximate with the approximation being valid only if $|v_c| \ll V_p$.

III. SECOND-ORDER LOOP

Continuous-Time Approximation

A large preponderance of applications utilize second-order PLL's. To obtain a zero-stabilized, second-order loop, consider a loop filter function

$$Z_{F2}(s) = R_2 + 1/sC \tag{8}$$

which is produced by a series connection of a resistor and a capacitor.

To systematize the notation, define

$$\tau_2 = R_2 C s$$

$$\omega_n = (K_0 I_p / 2\pi C)^{1/2} \text{ rad/s}$$

$$\zeta = \frac{\tau_2}{2}\left[\frac{K_0 I_p}{2\pi C}\right]^{1/2}$$

$$K = \frac{K_0 I_p R_2}{2\pi} \text{ rad/s.} \tag{9}$$

[2] Notation throughout corresponds to that established in [6].

These quantities are interrelated by

$$K = 2\zeta\omega_n$$

$$K\tau_2 = 4\zeta^2$$

$$K/\tau_2 = \omega_n^2 \tag{10}$$

where K is the *loop gain*, ω_n is the *natural frequency* and ζ is the *damping factor*. Any two of the three parameters completely define the linearized, time-averaged behavior of the PLL. Substituting (8) and (9) into (5) and (6) gives the transfer functions for the second-order, charge-pump PLL. They turn out to have exactly the same form as obtained for a conventional second-order PLL [6, ch. 2]. Therefore, to the extent that the various approximations are valid, the charge-pump PLL has exactly the same small-scale behavior as conventional PLL's with the same values for the loop parameters.

To explore further, we note that $Z_F(0) = \infty$ so that the static phase error, from (7), is zero. This desirable performance is achieved with a passive filter. To approach zero static phase error in a conventional PLL requires an active filter with large dc gain. Therefore, the charge pump permits zero static phase error (Type-II response) without the need for dc amplification. This effect arises because of the input open circuit during the N state and does not necessarily depend upon use of an active current switch. The same behavior is found in any of the four configurations of Fig. 2.

Practical circuits will impose some shunt loading across the passive filter impedance. Denote the load as a resistor R_s. The actual static phase error, from (7), will be

$$\theta_v = \frac{2\pi\Delta\omega}{K_0 I_p R_s} \text{ rad.} \tag{11}$$

Shunt loading is most likely to come from input impedance of the VCO control terminal or from the switch itself. Both impedances can be made extremely large. The VCO may be varactor-tuned, which implies near-infinite resistance, and the switch is typically a reverse-biased semiconductor. Some other variety of VCO could utilize a high-impedance buffer, if necessary to isolate a small-input impedance.

When R_s is very large, then leakage current may be more significant in producing phase error. The phase error θ_b resulting from a bias current I_b injected continuously into the filter node can be calculated as

$$\theta_b = 2\pi I_b / I_p \text{ rad.} \tag{12}$$

An active filter, incorporating an ideal op amp, will, of course, obviate any static phase error from VCO control-terminal loading effects.

Although the various results above were obtained specifically for the configuration of Fig. 2(b), they also apply for the other three configurations, as noted at the end of Section II. However, the voltage switch with passive filter [Fig. 2(a)] exhibits a curious nonlinearity that may disqualify it from serious consideration in many applications. Denote the voltage on the capacitor as v_x. Pump current is $(v_p - v_x)/(R_1 + R_2)$, where $v_p = \pm V_p$ with the sign determined by the phase error direction.

For small v_x, the pump current is influenced little by the capacitor charge so v_x could be neglected in determining the approximate behavior of the circuit. If v_x should become large then it cannot be neglected. (A large v_x would be required if the VCO needed a large v_c to tune it to the proper frequency. Capacitor voltage v_x is "large" if its magnitude reaches a significant fraction of V_p; it can never exceed V_p.)

Let v_x be some positive voltage. When $v_p = +V_p$, then the pump current is $i_{p+} = (V_p - v_x)/(R_1 + R_2)$ while a negative v_p drives a pump current of $i_{p-} = -(V_p + v_x)/(R_1 + R_2)$. These currents are unequal—substantially so if v_x is large enough—so loop gain about this v_x operating point will be larger for negative phase errors than for positive. It is unlikely that any significant asymmetry can be tolerated in most applications.

Granularity Problems

All of the foregoing is based on averaged-response, time-continuous, constant-element operation of the loop. There are features arising from the actual discontinuous operation that need attention, even for narrow bandwidths. The primary features are loop stability and phase-detector ripple.

In some sense, the loop operates on a sampled basis and not as a straightforward continuous-time circuit. A sampled system almost always has more stability problems than arise in continuous-time systems. In particular, an analog, second-order PLL is unconditionally stable for any value of loop gain, but the sampled equivalent will go unstable if the gain is made too large. Prudent design requires that the stability limit be known.[3]

A linearized, sampled analysis is presented in Appendix A. The end result is the characteristic equation (denominator of the transfer function) of the sampled PLL in the z-plane, which has the form

$$D(z) = (z-1)^2 + (z-1)\frac{2\pi K'}{\omega_i \tau_2}\left[1 + \frac{2\pi}{\omega_i \tau_2}\right] + \frac{4\pi^2 K'}{\omega_i^2 \tau_2^2}$$

(13)

where $K' = K\tau_2$ may be regarded as a *normalized loop gain*, ω_i is the input frequency, and $\tau_2 = R_2 C$ is the time constant of the filter zero.

Transient response for small phase errors and loop stability are studied by examining the locations of the zeros of $D(z)$—the poles of the z-domain transfer function. The root locus shows pole locations in the z plane for varying K'; an example is sketched in Fig. 3. The shape of the locus is very similar to that of a conventional second-order loop in the s-plane [6, ch. 2].

The two poles start at $z = 1$ for $K' = 0$ and move on a

[3] Tal [9] has investigated sampled-stability of a phase-locked speed-control servo that uses a PFD and a simple lag filter. His problem differs somewhat from that considered here and his method provides an alternate approach.

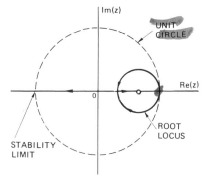

Fig. 3.   Root locus plot of second-order loop in z-plane.

circle with center at $z = (1 + 2\pi/\omega_i \tau_2)^{-1}$ for values of

$$K' < \frac{4}{(1 + 2\pi/\omega_i \tau_2)^2}.$$

For larger K', the poles lie on the real axis; one pole migrates towards the center of the locus circle and the other migrates towards $-\infty$.

The loop is stable only if the poles lie inside the unit circle. Instability results where the outbound pole crosses the unit circle at $z = -1$, as noted in Fig. 2. Normalized gain at the crossing point is

$$K' = \frac{1}{\dfrac{\pi}{\omega_i \tau_2}\left(1 + \dfrac{\pi}{\omega_i \tau_2}\right)}.$$

(14)

This value of K' is the *stability limit* and is plotted in Fig. 4.

Ripple is another granularity effect that demands attention. Upon each cycle of the PFD, the pump current I_p is driven into the filter impedance Z_F, which responds with an instantaneous voltage jump of $\Delta v_c = I_p R_2$. At the end of the charging interval $(t = t_p)$, the pump current switches off and a voltage jump of equal magnitude occurs in the opposite direction.

Frequency of the VCO follows the voltage steps so there will be frequency excursions of $\Delta\omega_0 = K_0 I_p R_2 = 2\pi K$ radians/ second for each pump pulse. The phase excursion during the pump interval t_p will be $\Delta\theta_0 = 2\pi K|\theta_e|/\omega_i$ [using (1)], so the phase jitter vanishes for $\theta_e = 0$. (A not-unexpected happening since the pump pulses are supposed to vanish for $\theta_e = 0$.)

Some applications (e.g., bit synchronizers) may be able to tolerate such frequency jitter, but others (e.g., frequency synthesizers) may require much better spectral purity.

A possibly more serious consequence of the jumps is the potential for overload of the VCO, even if the indicated ripple is allowable from a spectral-purity standpoint. Any real VCO has only a finite frequency range over which it can be tuned. If control voltages outside of this range are applied, the VCO frequency is unable to follow. (In fact, oscillations may cease or the circuit might even sustain damage.) We require that the frequency jumps remain within the allowable tuning range of the VCO under all conditions.

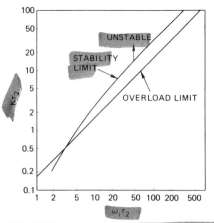

Fig. 4. Stability and overload limits for second-order loop.

As an extreme instance, the frequency jump must not exceed the input frequency. A larger jump would imply that the VCO frequency was driven negative—a meaningless status. For this extreme condition, the overload bound is $2\pi K < \omega_i$ or, in normalized form

$$K' < \omega_i \tau_2 / 2\pi. \tag{15}$$

A multivibrator, operated near the center of its tuning range, might be able to approach the excursions implied in (15). Most other oscillators will have a much smaller tuning range and therefore will be restricted to use in PLL's with much smaller values of K.

As a comparison to the stability limit, the overload limit of (15) has also been plotted in Fig. 4. It is apparent that overload is the actual restriction on loop gain; overload sets in at a lower value of gain than does instability for any practical circuit.

In all discussion of granularity effects it has been assumed tacitly that all transitions are present. If transitions can be missing at random, as in bit-clock recovery applications, then there may be a data-pattern-dependent jitter induced into the VCO phase. That problem is not treated in this paper.

IV. THIRD-ORDER LOOP

Origination

The frequency jumps inherent to the second-order loop usually cannot be accepted and additional filtering is often included within the PLL in order to mitigate the ripple. The simplest ripple filter is an additional capacitor C_3 in parallel with the earlier RC impedance, as shown in Fig. 5. Defining $b = 1 + C/C_3$, we obtain

$$Z_{F3}(s) = \left(\frac{b-1}{b}\right)\frac{s\tau_2 + 1}{sC\left(\frac{s\tau_2}{b} + 1\right)}. \tag{16}$$

Retaining the previous definition (9) for K, the closed-

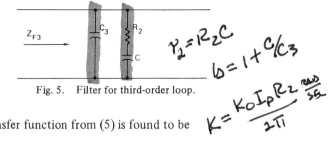

Fig. 5. Filter for third-order loop.

$\tau_2 = R_2 C$

$b = 1 + C/C_3$

$K = \dfrac{K_0 I_p R_2}{2\pi} \dfrac{rad}{sec}$

loop transfer function from (5) is found to be

$$H(s) = \frac{K\left(\dfrac{b-1}{b}\right)\left(s + \dfrac{1}{\tau_2}\right)}{\dfrac{s^3 \tau_2}{b} + s^2 + K\left(\dfrac{b-1}{b}\right)s + \dfrac{K(b-1)}{b\tau_2}} \tag{17}$$

where the continuous-operation, time-averaging assumption has been made.

Simple addition of C_3 across $R_2 + 1/sC$ ought to serve very well for the passive filter but is not likely to be satisfactory for the active filter. The operational amplifiers would be required to deliver step currents of I_p on each cycle, which is likely to be beyond the slew capabilities of most amplifiers. Rather than attempting to accommodate the current by brute force amplifiers, it is more conservative to prevent the current step from ever reaching the op amp, as in the circuits of Fig. 6. The general characteristics of these circuits ought to be much the same as those for the simple passive circuit, but some fine details will differ. This paper treats only the passive-filter, current-switch circuit.

Properties

The transfer function (17) has a denominator of third degree, so the system is a third-order PLL. In the open-loop transfer function, the additional pole is located at $s = -b/\tau_2$, which is far away from the dominant, low-frequency poles for large b. If C_3 is small compared to C ($b \gg 1$), then we should expect only high-frequency effects from the additional filtering. Low-frequency properties should be essentially the same as for the second-order loop.

In particular, the steady-state responses will be the same as for the second-order loop. The static phase error caused by a frequency offset will be zero and the phase lag caused by a frequency ramp will be $\theta_a = \omega/\omega_n^2$, [6, ch. 4] where ω is the slope of the ramp. Although the loop is third order, it is only Type II.

The s-plane root locus of (17) has been studied in [6, ch. 8] for another application. Root loci are shown in Fig. 7 for various selections of b. For large b and small-enough K' (the normalized loop gain) the dominant poles are virtually unchanged from the locations expected for the second-order loop. As K' becomes very large, the outward-bound real pole meets the extra pole coming in from $-b/\tau_2$ and the pair go complex asymptotic to a vertical line at $s = -0.5(b-1)\tau_2$. The loop could become seriously underdamped for large gain.

As b is reduced, the breakaway point for the vertical asymptote approaches closer to the low frequency portion of the locus; if $b < 9$, the locus never returns to the real axis and is underdamped for all values of K'.

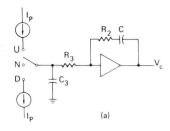

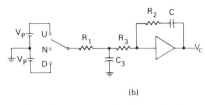

Fig. 6.   Jump suppression for active filters.

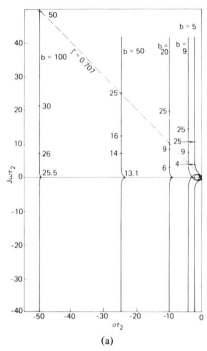

(a)

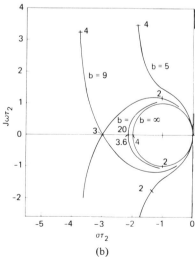

(b)

Fig. 7.   Root locus plots for third-order, type II PLL. (a) Large scale. (b) Expanded scale. Tick marks show values for $K' = K\tau_2$. (Taken from [6]. Reproduced by permission of publisher.)

If $b < 1$, then the loop is unstable for all K'. In the configuration of Fig. 5 it is impossible to have $b < 1$, but it is entirely possible in Fig. 6. (In terms of the component values, b is defined differently in Fig. 6 than in Fig. 5, but the transfer function and root locus plots have the same form for both configurations.)

This stability impairment caused by the third pole is calculated on the basis of assumed continuous-time operation. Time-discrete operation can be expected to cause even more impairment. To investigate stability of the third-order loop, an analysis similar to that given in the Appendix was performed. (The analysis is omitted here because of space constraints.)

The criterion for stability—all poles inside the unit circle—is satisfied if

$$K\tau_2 < \frac{4(1+a)}{\dfrac{2\pi(b-1)}{b\omega_i\tau_2}\left[\dfrac{2\pi(1+a)}{\omega_i\tau_2}+\dfrac{2(1-a)(b-1)}{b}\right]} \tag{18}$$

where

$$a = \exp\left(-\frac{2\pi b}{\omega_i\tau_2}\right).$$

Knowledge of the stability limit alone is not sufficient for good design; some insight into the transient response is also needed. To that end, the z-plane characteristic equation is

$$D(z) = z^3 + z^2\left[-a-2+G\left(\frac{2\pi}{\omega_i\tau_2}+\frac{(1-a)(b-1)}{b}\right)\right]$$
$$+ z\left[2a+1-G\left(\frac{2\pi a}{\omega_i\tau_2}+\frac{(1-a)(b-1)}{b}\right)\right]-a \tag{19}$$

where

$$G \triangleq \frac{2\pi K\tau_2(b-1)}{b\omega_i\tau_2}.$$

The zeros of $D(z)$ are the z-plane poles; their location defines the response to transients.

Fig. 8 shows the stability limits for several values of b. Given values for b and $\omega_i\tau_2$, any value of $K\tau_2$ below the curve yields a stable loop while any value above the curve is unstable.

Because of the extra capacitor, control voltage v_c describes a continuous, ramp-like, exponential function for each pump pulse, instead of the rectangular jump that was found for the second-order loop. The same analysis that provided the stability limit gives the ramp amplitude as

$$|\Delta\omega_0|_3 = 2\pi K\left(\frac{b-1}{b}\right)\left[\frac{b-1}{b}\left(1-e^{-\frac{b|\theta_e|}{\omega_i\tau_2}}\right)+\frac{|\theta_e|}{\omega_i\tau_2}\right] \tag{20}$$

as compared to $|\Delta\omega_0|_2 = 2\pi K$ for the second-order loop. Define $\beta = |\Delta\omega_0|_3/|\Delta\omega_0|_2$, and assume $b|\theta_e|/\omega_i\tau_2 \ll 1$ (not necessarily true, but a common condition). Then the suppression of unwanted frequency excursion provided by the extra

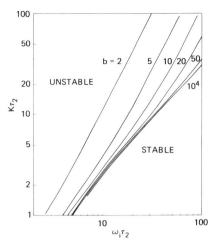

Fig. 8. Stability limits for third-order loop.

capacitor is

$$\beta \cong \frac{(b-1)\,|\,\theta_e\,|}{\omega_i \tau_2}. \qquad (21)$$

When the loop is tracking near equilibrium, $|\,\theta_e\,|$ is very small so the suppression afforded by C_3 can be substantial.

V. TRANSIENT RESPONSE

The results in all of the preceding sections—the continuous-time approximation, the z-plane characteristic function, and the stability limits—were all based upon an assumption of small phase error. That assumption fails for large phase errors such as occur during acquisition of lock. An analysis was performed for large phase errors for the second-order loop; the analysis is outlined in the Appendix.

In essence, the method is to consider the loop state variables of phase error and frequency at the instant immediately before each pump pulse. These state variables are related by difference equations which were iterated numerically on a programmable calculator. The resulting printout is a sequence of the state variables along with the times of occurrence.

Two different displays are possible: frequency- or phase-error versus time to show the familiar transient response, or frequency error versus phase error to produce a phase-plane portrait [10]. Examples of both are shown below.

The question to be addressed is the following. If bandwidth (or gain, K) is very small compared to the switching frequency ω_i, then we know that the continuous analysis provides an excellent approximation to the behavior of the charge-pump loop and we can utilize the extensive information available from the study of conventional, analog PLL's. In many applications we want to be able to use a large bandwidth. Therefore we ask, "How small can ω_i/K be made before behavior departs significantly from that predicted by the continuous-time analysis?"

Several example calculations were performed in order to explore the question. A value of $K' = 2$, corresponding to $\zeta = 0.707$ in the continuous PLL, was chosen as representative of many applications. Referring to the overload curve of Fig. 4, it can be seen that the VCO is certain to overload unless

$\omega_i \tau_2 > 12.6$, or, in other words, $\omega_i/K > 12.6/2 = 6.3$. To allow some margin on the limit, a value of $\omega_i/K = 10$ was chosen for the example calculations. For practical VCO's that is probably still too small a ratio but it will illustrate the results very well.

Transient phase error in response to a phase step of ± 6 rad and of a frequency step of $\pm 2K$ radians per second were calculated for the conditions of $\omega_i/K = 10$ and ∞. The latter corresponds to the continuous-time PLL. Transient curves are plotted in Figs. 9 and 10, respectively.

It is apparent that, even for such a low frequency as compared to bandwidth, the response of the charge-pump loop is very close to that of the classical, continuous loop.

Asymmetry between positive and negative phase errors is evident for the charge-pump loop. (The classical loop, of course, has symmetric response with respect to error polarity.) This asymmetry arises from the polarity-asymmetric dependence of pulse duration upon phase error (see the Appendix) and dwindles as ω_i/K is made larger.

The same program yields a phase-plane solution of the PLL. Example trajectories are shown in Fig. 11. Asymmetry is also apparent in this display. Each marked point represents the state of the system at the starting instant of consecutive charge pulses. The points have been connected by straight lines, to aid in following the individual trajectories, but the actual trajectory between two calculated points has not been determined and there is no reason to suppose that it would be linear.

The shape of the trajectories may seem rather peculiar; the vertical sections do not occur on the $\Delta\omega = 0$ axis as is expected from previous phase-plane plots [10]. The discrepancy arises not from the charge-pump action, but from the choice of state variables. Here the variables are θ_e and $\Delta\omega = \omega_i - \Omega_0 - K_0 v_x$, where Ω_0 is the free-running frequency of the VCO and v_x is the voltage stored on the capacitor C in the loop filter. The usual phase-plane plot uses θ_e and $\dot{\theta}_e$ as the state variables. Proportional and integral elements enter into $\dot{\theta}_e$ whereas only integral elements contribute to the frequency variable in Fig. 11. If equivalent state variables were defined, then similar, skewed trajectories would also be obtained for the classical loop.

It is clear that the loop converges towards equilibrium without difficulty, at least for the trajectories examined. Attempts were made to examine trajectories with larger initial frequency errors. An overload phenomenon intervened: a phenomenon that the program was not designed to accommodate, so an error message was produced instead of a trajectory. Inasmuch as the loop was running very close to the nominal overload as deduced from Fig. 4, the program breakdown for small ω_i/K was not pursued further, on the supposition that a practical loop would break down under even more restrictive conditions.

Trajectories were also obtained for $\omega_i/K = 100$. Much larger $\Delta\omega/K$ values could be accommodated for that condition and the program breakdown was not encountered again. With large enough initial frequency error, the loop does not converge within the phase interval $(-2\pi, 2\pi)$ but slips one or more cycles before settling. The program was not designed to accommodate phase excursions beyond $\pm 2\pi$, so no results are provided.

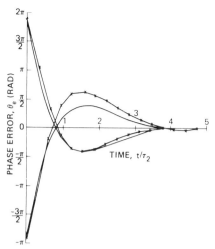

Fig. 9. Response to phase step. $\Delta\theta = \pm 6$ rad; $K\tau_2 = 2$; second-order loop. —— $\omega_i/K = \infty$; ✕—✕ $\omega_i/K = 10$.

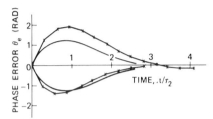

Fig. 10. Response to frequency step. $\Delta\omega/K = \pm 2$; $K\tau_2 = 2$; second-order loop. —— $\omega_i/K = \infty$, ✕—✕ $\omega_i/K = 10$.

(The phase-plane portrait for an ordinary phase detector is periodic in 2π, but the PFD portrait is more complicated. One can consider that Fig. 11 is the central region of a phase portrait but the outer regions to either side of center each extend over only -2π to 0 or 0 to $+2\pi$. The PFD portrait is not strictly periodic.)

Calculations were performed only for the second-order loop, but the third-order loop resulting from the filter of Fig. 5 is more likely to be employed in real applications. It is possible to calculate the transient response of the third-order loop in much the same way employed for second-order by taking account of three state variables. That has not yet been accomplished. Moreover, state trajectories for a third-order loop are three-dimensional and cannot be displayed readily on a two-dimensional sheet.

If the extra capacitor C_3 is effective, the third-order loop ought not suffer from VCO overload. Instead, stability limits the allowable gain for a given switch frequency (as in Fig. 8). From Fig. 7, we see that is it improbable that we would ever take b significantly less than about 10. If $K' = 2$, then the stability limit for $b = 10$ from Fig. 8 is $\omega_i/K \approx 7.5$. To obtain some stability margin, a value of ω_i/K in excess of 15 to 20 might be considered reasonable. In light of the results obtained with the second-order loop it seems fair to predict that response of the third-order charge-pump loop will be very much the same as that of the equivalent continuous-time loop.

VI. CONCLUSIONS

The conventional-wisdom rule-of-thumb has been that switching granularity effects can be neglected if the switching frequency exceeds 10 times the loop bandwidth. This paper

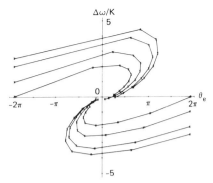

Fig. 11. Phase-plane portrait; second-order loop. $K\tau_2 = 2$; $\omega_i/K = 10$.

has shown that the rule-of-thumb is not far wrong if considered as an approximate outer limit beyond which troubles begin to appear. Somewhat more conservative design would be prudent in most circumstances.

The passive filter with current switching has been shown to have attractive properties. A fast, balanced, current-switch integrated circuit would be very helpful to the hardware designer.

The second-order loop has switching-rate frequency-excursions that are excessive for most applications. Any smoothing results in at least a third-order loop, although still Type II. Root loci for the third-order loop are presented to aid design efforts.

Transient response of practical charge-pump PLL's can be expected to be nearly the same as the response of the equivalent classical PLL.

APPENDIX

DIFFERENCE-EQUATION ANALYSIS OF CHARGE-PUMP PLL

Analysis of the charge-pump circuit is impeded by the switching of the pump current between the values $-I_p$, 0, and I_p. Moreover, the switching times are complicated functions of the relative, time-varying phases $\theta_i(t)$ of the signal input and $\theta_0(t)$ of the VCO. However, during any one switch condition, the circuit is a linear, time-invariant network and is described by linear differential equations with constant coefficients. Given the initial conditions at the start of a switching interval, it is straightforward to calculate the state variables at any time within the interval. The final state variables at the end of one interval become the initial conditions for the next interval.

If we define the phase and frequency errors at the start of a current pulse as the discrete-time state variables, then it is possible to write difference equations that describe a recursive sequence of the state. The exact difference equations were iterated to obtain the transient responses of Section V while discrete-time stability was examined for Sections III and IV by means of linearized difference equations.

This Appendix derives the linearized difference equations for the second-order PLL and shows an outline of the derivation of the exact difference equations of the second-order PLL. The process for the exact equations is shown as comments in parentheses following the corresponding portion of the linear-equations derivation.

Linearized equations were also obtained for the third-order

loop, but the derivation is not shown here because of space limitations. Exact difference equations could be found for the third-order loop, but that problem has not yet been attacked.

The notation and circuit configuration for the analysis are shown in Fig. 12. It is convenient to set the time origin to coincide with an instant of turn-on of the current switch. Observing that convention, we obtain the following equations that are valid for the entire first cycle (i.e., until the next turn-on instant of the current switch).

$$\theta_i(t) = \theta_i(0) + \omega_i t \tag{A1}$$

$$\omega_0(t) = \Omega_0 + K_0 v_c(t) \tag{A2}$$

$$\theta_0(t) = \theta_0(0) + \Omega_0 t + K_0 \int_0^t v_c(\tau)\,d\tau \tag{A3}$$

$$i_p = I_p \operatorname{sgn}\theta_e(0); \quad 0 < t < t_p \tag{A4}$$
$$= 0; \quad t_p < t < 2\pi/\omega_i$$

$$\theta_e = \theta_i - \theta_0 \tag{A5}$$

$$t_p \cong |\theta_e|/\omega_i \tag{A6}$$

$$v_c(t) = i_p R_2 + v_x \tag{A7}$$

$$v_x(t) = \frac{1}{C} \int_0^t i_p(\tau)\,d\tau. \tag{A8}$$

Suppose that the switching is initiated by an edge of the VCO waveform. Then $-I_p$ is switched on so as to retard the VCO phase; also $\theta_0(0) = 0$, and $\theta_i(0) = \theta_e(0)$, which is a negative number. The input-signal edge that shuts off the current switch occurs when $\theta_i(t) = \theta_i(t_p) = 0$; in other words, when the input phase has advanced by $\theta_e(0)$ radians at a rate of ω_i radians/second. In this case, that time is exactly $t_{p-} = |\theta_e|/\omega_i$ s.

Now suppose that the switch-on is initiated by an edge of input signal. The pump current is $+I_p$; $\theta_i(0) = 0$, $\theta_0(0) = -\theta_e(0)$, $\theta_e(0)$ is positive, and the pump current remains on until the next edge of the VCO. For the linearized analysis, that time is approximated by $t_{p+} = \theta_e(0)/\omega_i$, the same expression as for the opposite polarity of phase error.

(In actual fact, since v_c is not constant during the pump interval, the frequency of the VCO changes during $(0, t_{p+})$ so a linear equation for t_p is incorrect for positive phase error. In the nonlinear analysis it was found that the correct charging interval is a solution of a quadratic equation involving the initial conditions at $t = 0$ and the loop parameters. The quadratic solution is carried through in the numerical iteration of the nonlinear difference equations.)

Define $v_{x0} = v_x(0)$; $v_{xp} = v_x(t_p)$. Ordinary linear-network analysis methods yield

$$\theta_0(t_p) = \theta_0(0) + \Omega_0 t_p + K_0(v_{x0}t_p + i_p R_2 t_p + i_p t_p^2/2C). \tag{A9}$$

(For positive $\theta_e(0)$, setting $\theta_0(t_p) = 0$ in (A9) gives the quadratic equation for the exact value of t_{p+}, as described above.) Furthermore,

$$v_x(t_p) = v_{x0} + i_p t_p/C. \tag{A10}$$

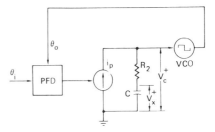

Fig. 12. Equivalent circuit of charge-pump PLL.

These equations, (A9) and (A10), are exact. (By substituting the exact solution for t_p into (A9) and (A10), the calculator program carries the state variables numerically up to t_p.)

Let t^* be the time following t_p at which the next edge—from signal or VCO, as the case may be—activates the PFD and starts a new pump pulse. Charge on the capacitor remains constant from t_p to t^*. Therefore, the VCO phase at t^* is

$$\theta_0(t^*) = \theta_0(t_p) + \Omega_0(t^* - t_p) + K_0(t^* - t_p)v_{xp}. \tag{A11}$$

Substituting (A9) and (A10) into (A11) gives

$$\theta_0(t^*) = \theta_0(0) + \Omega_0 t^*$$
$$+ K_0 \left[i_p R_2 t_p - \frac{i_p t_p^2}{2C} + v_{x0} t^* + \frac{i_p t_p t^*}{C} \right]. \tag{A12}$$

This last equation is also exact. To pursue the linearized analysis, substitute $i_p t_p \cong I_p \theta_e/\omega_i$ from (A4) and (A6), and approximate t^* by $2\pi/\omega_i$ to obtain

$$\theta_0(t^*) \cong \theta_0(0) + \frac{2\pi}{\omega_i}(\Omega_0 + K_0 v_{x0})$$
$$+ \frac{K_0 I_p \theta_e}{\omega_i} \left[R_2 + \frac{2\pi}{\omega_i C} - \frac{|\theta_e|}{2\omega_i C} \right]. \tag{A13}$$

Except for the very last term, (A13) is linear in $\theta_e(0)$. By dropping the last term—a valid approximation for small θ_e— we obtain a linear equation for the VCO phase at time t^* in terms of the initial phase and frequency and the loop parameters.

Define $\Delta\Omega = \omega_i - \Omega_0$ and recall that $\theta_e = \theta_i - \theta_0$. By the previous approximations, θ_i advances by 2π in the time interval $(0, t^*)$. With these substitutions, we obtain the linear difference equations

$$\theta_e(t^*) = \theta_e(0) + 2\pi\Delta\Omega/\omega_i$$
$$- \frac{K_0 I_p \theta_e(0)}{\omega_i}(R_2 + 2\pi/\omega_i C) - \frac{2\pi K_0 v_{x0}}{\omega_i} \tag{A14}$$

and

$$v_x(t^*) = v_{x0} + \frac{\theta_e(0)I_p}{\omega_i C}. \tag{A15}$$

(All approximations are avoided in the calculator program. The quantities $\theta_0(t_1) = 2\pi$, from (A12), and $\theta_i(t_2) = \theta_i(0) +$

$\omega_i t_2 = 2\pi$ are solved for t_1 and t_2. The smaller of these is taken as the value for t^*. Having obtained the correct values of t_p and t^*, the values for $\theta_e(t^*)$ and $\Delta\omega(t^*)$ are calculated and the process repeats with these state variables as the new initial conditions. The program starts at specified initial conditions and stops after arriving within a specified tolerance band about the zero state.)

The linearized analysis continues by taking z-transforms [11] of (A14) and (A15). Treating the initial frequency error as a frequency step gives the z-transformed equations

$$z\theta_e(z) = \theta_e(z) + \frac{2\pi\Delta\Omega z}{\omega_i(z-1)} - \frac{K_0 I_p \theta_e(z)}{\omega_i}(R_2 + 2\pi/\omega_i C)$$

$$- 2\pi K_0 V_x(z)/\omega_i \qquad (A16)$$

$$z V_x(z) = V_x(z) + \frac{\theta_e(z) I_p}{\omega_i C}. \qquad (A17)$$

Solving for $\theta_e(z)$ gives

$$\theta_e(z) = \frac{2\pi z \Delta\Omega/\omega_i}{(z-1)^2 + (z-1)\dfrac{K_0 I_p}{\omega_i^2 C}(2\pi + \omega_i C R_2) + \dfrac{2\pi K_0 I_p}{\omega_i^2 C}}.$$

$$\qquad (A18)$$

Applying notation definitions from (9), the denominator of (A18) becomes $D(z)$, as shown in (13). Analysis of pole locations follows by standard methods [11].

Linearized difference equations (three of them) were de-

rived in the same manner for the third-order loop. The volume of algebra is substantially greater than for the second-order loop so only the results are given in Section IV. Only linear approximations have been performed for the third-order loop; the exact equations have not been attempted.

REFERENCES

[1] R. C. E. Thomas, "Frequency comparator performs double duty," *EDN*, pp. 29–32, Nov. 1, 1970.

[2] J. I. Brown, "A digital phase and frequency-sensitive detector," *Proc. IEEE*, vol. 59, p. 717, Apr. 1971.

[3] *Phase-Locked Loop Data Book*, 2nd ed. Motorola, Inc., Aug. 1973.

[4] D. K. Morgan and G. Steudel, "The RCA COS/MOS phase-locked-loop," RCA, Somerville, NJ, Application Note ICAN-6101, Oct. 1972.

[5] C. A. Sharpe, "A 3-state phase detector can improve your next PLL design," *EDN*, pp. 55–59, Sept. 1976.

[6] F. M. Gardner, *Phaselock Techniques*, 2nd ed. New York: Wiley, 1979.

[7] P. Lue, "A multispeed digital regenerative repeater for digital data transmission," in *Conf. Rec., 1979 Nat. Telecommun Conf.*, paper 14.1.

[8] J. A. Afonso, A. J. Quiterio, and D. S. Arantes, "A phase-locked loop with digital frequency comparator for timing signal recovery," in *Conf. Rec., 1979 Nat. Telecommun. Conf.*, paper 14.4.

[9] J. Tal, "Speed control by phase-locked servo systems—New possibilities and limitations," *IEEE Trans. Ind. Electron. Contr. Instrum.*, vol. IECI-24, pp. 118–125, Feb. 1977.

[10] A. J. Viterbi, *Principles of Coherent Communication*. New York: McGraw-Hill, 1966, ch. 3.

[11] J. T. Tou, *Digital and Sampled-Data Control Systems*. New York: McGraw-Hill, 1959.

z-Domain Model for Discrete-Time PLL's

JERRELL P. HEIN, MEMBER, IEEE, AND JEFFREY W. SCOTT

Abstract —The well-known *s*-domain model for continuous-time phase-locked loops (PLL's) is a fundamental tool for the linearized analysis of these systems. For PLL's with digital inputs and outputs, however, a discrete-time z-domain model more accurately describes loop behavior. In this paper, a methodology is described for obtaining an accurate *z*-domain description of a discrete-time PLL. This method is an alternative approach to the analysis presented in [4]. The modeling technique transforms portions of the *s*-domain PLL model directly into the *z*-domain, requiring only straightforward algebraic manipulations even for complex loop filters. This methodology is demonstrated for a simple loop filter, and measurements from the digital signaling interface (DSI) integrated circuit are used to compare *s*-domain, *z*-domain, and time step analysis results for a more complicated loop filter. The *z*-domain model, although only incrementally more complicated than the *s*-domain model, is shown to be more accurate, especially at higher jitter frequencies.

I. INTRODUCTION

THE basic *s*-domain PLL model presented in numerous texts [1]–[3] treats a loop in the locked condition as a linear continuous-time system. The input and output waveforms are assumed to be sinusoidal and the phase detector is modeled as a linear analog multiplier with an inherent ideal low-pass filter. Although the linear continuous-time model is useful within these constraints, many PLL's operate under conditions not accurately represented by these assumptions. In particular, a large class of PLL's used most notably in data communications have digital waveforms as both inputs and outputs. For these PLL's, the phase information is contained in the digital waveform transitions and should be viewed as a discrete-time sequence. The linear, continuous-time model can approximate the operations of these loops only if the jitter frequencies of interest are much less than the incoming data transition rate.

To obtain an accurate discrete-time model of a PLL, one can write the complete set of differential equations describing the system, convert these into difference equations, linearize the equations along the way, and finally *z*-transform the result to obtain $H(z)$, the *z*-domain jitter transfer function. This procedure quickly becomes cumbersome for all but the simplest loop filters as noted by Gardner [4].

In this paper, a *z*-domain description for a PLL is presented which is only incrementally more complicated than the linear continuous-time model. The model uses the impulse invariant transformation to convert the *s*-domain description of a portion of the loop directly into the *z*-domain. In addition to the model derivation and implementation for a simple loop filter, results will be shown

comparing the continuous- and discrete-time models with a PLL timestep simulator, and actual measured devices. In Appendix A, the assumptions needed to linearize the network analysis will be discussed and it will be shown that these assumptions are identical to those required to analyze the PLL with the use of linearized difference equations. Finally, in Appendix B, the analysis methodology will be extended to PLL's with switched capacitor loop filters.

II. DISCRETE-TIME MODEL DEVELOPMENT

A. Continuous-Time PLL's and the s-Domain Model

A functional block diagram for a continuous-time PLL is shown in Fig. 1. The block diagram of the associated *s*-domain PLL model is shown in Fig. 2. This model assumes that the input waveform is a phase modulated sine wave, i.e., it has the form,

$$s_a(t) = a \cdot \sin(w_c t + \phi_i(t)).$$

The input phase modulation $\phi_i(t)$, and oscillator output $\phi_o(t)$ are continuous functions of time.

In Fig. 2, the summing node and K_p gain block represent the operation of the phase detector in the frequency domain. The phase detector is assumed to be a linear analog multiplier which multiplies the input and output waveforms. The result is a multifrequency signal which contains the phase difference information desired ($\phi_i(s) - \phi_o(s)$) in the low frequency portion of the phase detector output spectrum. The higher frequency multiplicative products are ignored in the analysis. The effect of this last assumption can be included by assuming that an ideal low-pass filter sits behind the multiplier. In practice, the loop filter following the phase detector approximates this ideal filter.

Finally, the loop filter ($F(s)$) and voltage-controlled oscillator (K_o/s) are included in the block diagram. In general, the loop filter is modeled accurately as a linear continuous-time element, especially if it is implemented with passive components. Relaxation and current ramping oscillators exhibit linear and wideband voltage-to-frequency relationships and are also accurately modeled as linear continuous-time elements.

The jitter transfer characteristic for the *s*-domain model is also shown in Fig. 2. $H(f)$ will be used to compare the accuracy of the *s*- and *z*-domain models in a later section.

The assumption inherent in the application of the *s*-domain model to a PLL are generally valid for loops operating with sinusoidal inputs and outputs (continuous-time PLL's). It will be shown, however, that for PLL's

Manuscript received October 16, 1987; revised March 18, 1988. This paper was recommended by Associate Editor C. A. T. Salama.

The authors are with AT&T Bell Laboratories, Reading, PA 19612.

IEEE Log Number 8823450.

Reprinted from *IEEE Trans. Circuits and Systems*, vol. 35, pp. 1393-1400, November 1988.

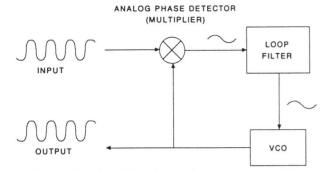

Fig. 1. Functional block diagram for continuous-time PLL.

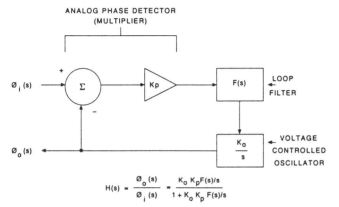

$$H(s) = \frac{\varnothing_o(s)}{\varnothing_i(s)} = \frac{K_0 K_p F(s)/s}{1 + K_0 K_p F(s)/s}$$

Fig. 2. S-domain model block diagram.

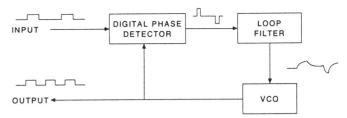

Fig. 3. Functional block diagram for discrete-time PLL.

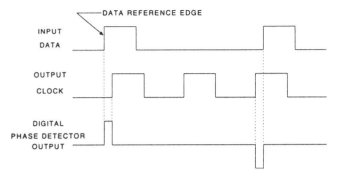

Fig. 4. Digital phase detector operation.

operating with digital inputs and outputs (discrete-time PLL's), some of the assumptions may result in significant inaccuracies in the analysis. In these cases, it is necessary to develop a discrete-time model which more accurately reflects the actual operation of the loop.

B. Discrete-Time PLL's and the z-Domain Model

The functional block diagram for a discrete-time PLL is shown in Fig. 3. The input waveform to the PLL can be described as

$$s_d(t) = a \cdot \text{sgn} \left[\sin \left(w_c t + \phi_i(t) \right) \right]$$

where sgn is the signum function. The difference between the discrete- and the continuous-time PLL structures lies in the implementation of the phase detector. The phase detector for a discrete-time PLL is a digital circuit which drives a pulse width modulated digital pulse into the loop filter. The width of the pulse is determined by the time difference between the input data reference edge and the recovered clock edge. The digital phase detector operation is depicted in Fig. 4. Phase difference information arrives at the phase detector input only when data reference edges occur. Therefore, the phase error between the data and clock is properly viewed as a discrete-time sequence with values spaced at intervals approximately equal to the time between input data reference edges. For a repetitive data pattern, the time interval between data pulses is given by: $T = 1/($recovered clock frequency · input data one's density$)$. For example, a repetitive 1,0 return-to-zero (RZ)

data pattern (as shown in Fig. 4) would have $T = 1/($recovered clock frequency $\cdot \frac{1}{2})$.

For small phase errors, the pulses driving the loop filter can be modeled accurately as weighted impulses. The accuracy of this approximation is calculated in Appendix A for a simple RC loop filter. The phase detector samples the difference between the data and clock phases at intervals of T seconds and drives the loop filter with weighted impulses. The gain factor K_p simply represents the conversion factor between input phase error and output impulse area. Therefore, the digital phase detector in a discrete-time PLL can be modeled as a summing node and gain block in the z-domain. Note that the type of blocks required to model the digital phase detector in the z-domain are the same type of blocks used to model the analog phase detector in the s-domain.

With the digital phase detector now properly modeled in the z-domain, the problem remaining is the accurate modeling of the continuous-time loop filter and VCO elements in the z-domain. For arbitrary signals driving the loop filter and VCO, it would be impossible to accurately map the elements' entire s-domain response into the z-domain. But the loop filter and VCO are not being driven by arbitrary signals, they are being driven by a series of weighted impulses from the phase detector. Therefore, it is only necessary to preserve the loop filter and VCO's impulse response in transforming from the s- to z-domain. In other words, the essential characteristics of the loop filter and VCO will be preserved if the derived discrete-time network has a unit impulse response with values equal to T spaced samples of the continuous-time impulse response. A transformation exists which guarantees exactly this type of relationship—the impulse invariant transformation.

Fig. 5 illustrates the relationship guaranteed by the impulse invariant transformation. Given a continuous-time network with impulse response $h_a(t)$, the impulse invari-

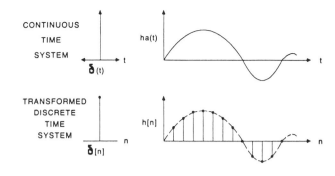

GUARANTEES :   h[n] = ha(nT)

- POLES AT s_K MAP TO $z_K = e^{s_K T}$

- ZEROS MAP DEPENDING ON POLES

Fig. 5.   Impulse invariant transformation.

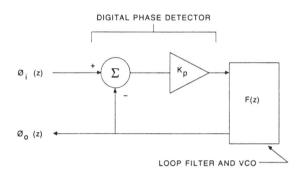

$$H(z) = \frac{\emptyset_o(z)}{\emptyset_i(z)} = \frac{K_p F(z)}{1 + K_p F(z)}$$

$$F(z) \Rightarrow \frac{K_o F(s)}{s} \quad \begin{array}{l}\text{TRANSFORMED USING IMPULSE}\\ \text{INVARIANT TRANSFORMATION}\end{array}$$

Fig. 6.   Z-domain model block diagram.

ant transformed discrete-time network will have a unit sample response $h[n]$ where the $h[n]$ values are equal to the $h_a(t)$ values sampled at intervals of T seconds [5]. The actual transformation of some function $F(s)$ to $F(z)$ is a straightforward process which will be demonstrated in a later section. Basically, the function $F(s)$ must be divided using partial fraction methods into simple $a/(s - s_k)$ and $1/s^2$ terms. Each of these terms then transforms directly into the z-domain. Poles at s_k in the s-domain map to poles at $z_k = e^{s_k T}$ in the z-domain. Zeros in the s-domain move to points in the z-plane which depend on the pole locations. Because the impulse-invariant transformation is not an algebraic mathematical mapping, the combined loop filter and VCO s-domain terms $(K_o F(s)/s)$ must be transformed together.

Once the loop filter and VCO are transformed into $F(z)$, the entire loop can be evaluated in the z-domain. The z-domain model is shown in Fig. 6 along with the jitter transfer characteristic $H(z)$. The development of the complete discrete-time PLL model involves only two steps: (1) calculation of the phase detector gain constant K_p, and

(2) transformation of the loop filter and VCO s-domain descriptions into the z-domain using the impulse invariant transformation. The model accurately represents the operation of the digital phase detector with digital data and clock inputs, and correctly transforms the continuous-time networks $F(s)$ and K_o/s into the z-domain. The application of this technique to a second-order PLL will be demonstrated in the next section.

C. Application of z-Domain Model to a Second-Order PLL

In this section, the z-domain analysis will be performed on a discrete-time PLL with a first-order loop filter. For comparison purposes, the classical s-domain model of this PLL will be described first.

The s-domain PLL model is shown in Fig. 2. A first-order transimpedance loop filter transfer function $F(s)$, (see Fig. 11) is given by

$$F(s) = R\left(\left(s + \frac{1}{RC}\right)\Big/ s\right)$$

where R and C are the series resistance and capacitance of the loop filter. Using this filter transfer function in the expression for $H(s)$ shown in Fig. 2 yields the following jitter transfer characteristic:

$$H_c(s) = RK_o K_p \frac{s + \dfrac{1}{RC}}{s^2 + RK_o K_p s + \dfrac{K_o K_p}{C}}$$

where the subscript "c" denotes the continuous-time nature of the jitter transfer expression; K_o is the voltage-to-frequency conversion gain of the VCO; and K_p is the phase detector conversion gain given by

$$K_p = dI_p/2\pi$$

where d is the one's density of the PLL input data and I_p is the magnitude of the phase detector pump current.

The loopgain expression required for root locus construction is given by

$$\text{loopgain} = K_p \cdot \frac{K_o}{s} \cdot F(s)$$

$$= K \cdot \frac{s + \dfrac{1}{RC}}{s^2}$$

where K is defined as

$$K = K_o K_p R.$$

The root locus for the PLL modeled in the s-domain is shown in Fig. 7. There are two open loop poles at the origin of the s-plane and one open loop zero at $s = 1/RC$. Note that, since the loopgain parameter K contains the one's density information, the closed-loop pole locations of the s-domain jitter transfer model depend on the input data pattern. However, it is of further interest to note that the s-domain model never predicts an unstable loop for any combination of PLL parameters.

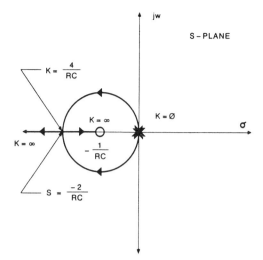

Fig. 7. S-domain root locus.

The discrete-time or z-domain analysis of the second-order PLL begins by applying the impulse invariant transformation to the term $(K_o F(s)/s)$. For the first order loop filter under investigation, this term is given by

$$\frac{K_o}{s} \cdot F(s) = RK_o \frac{s + \dfrac{1}{RC}}{s^2}$$

$$= \frac{K_o}{C}\left[\frac{1}{s^2} + \frac{RC}{s}\right].$$

The impulse response of this section of the PLL is found by taking the inverse Laplace transform, which yields

$$h_a(t) = \frac{K_o}{C}[t + RC]u(t).$$

The impulse-invariant property requires that

$$h[n] = h_a(nT)$$

$$= \frac{K_o}{C}[nT + RC]u[n]$$

where T is the effective sampling or impulse arrival rate given by $T = 1/df_{vco}$ where d is the one's density of the incoming data and f_{vco} is the VCO recovered clock frequency. The desired z-domain description of the loop filter and VCO is found by taking the z-transform of $h[n]$ (the $nT \cdot u[n]$ term is most easily transformed by invoking the differentiation property), which results in the following expression for $F(z)$:

$$F(z) = \frac{K_o}{C}\left[\frac{Tz^{-1}}{(1 - z^{-1})^2} + \frac{RC}{1 - z^{-1}}\right]$$

$$= RK_o \frac{z\left[z - \left(1 - \dfrac{T}{RC}\right)\right]}{(z - 1)^2}.$$

For simplification, let $\alpha = 1 - (T/RC)$. Then

$$F(z) = RK_o \frac{z(z - \alpha)}{(z - 1)^2}.$$

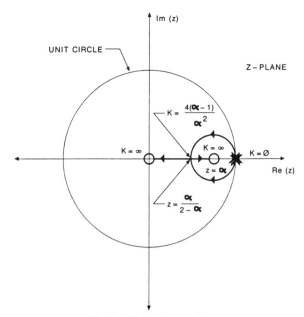

Fig. 8. Z-domain root locus.

Using this transformed filter/VCO transfer function in the expression for $H(z)$ shown in Fig. 6 yields the following jitter transfer characteristic:

$$H_d(z) = \frac{K_o K_p R}{1 + K_o K_p R} \frac{z(z - \alpha)}{z^2 - \dfrac{2 - \alpha K_o K_p R}{1 + K_o K_p R}z + \dfrac{1}{1 + K_o K_p R}}$$

where the subscript "d" denotes the discrete-time nature of the jitter transfer expression, and K_p is the phase detector conversion gain given by

$$K_p = \frac{I_p}{2\pi f_{VCO}}$$

where I_p is the magnitude of the phase detector pump current.

The loopgain expression required for z-domain root locus construction is given by

$$\text{loopgain} = K_p \cdot F(z)$$

$$= K\frac{z(z - \alpha)}{(z - 1)^2}$$

where the loopgain parameter K is again defined as $K_o K_p R$.

The root locus for the PLL modeled in the z-domain is shown in Fig. 8. There are two open loop poles at $z = 1$ and open loop zeros at $z = 0$ and $z = \alpha$. It is interesting to note that, since the open-loop zero location (α) is a function of the input data one's density, the z-domain model can predict unstable loop performance for the condition of $T > 2RC$ (in which case an open-loop zero resides on the negative real axis outside the unit circle).

For $T \ll RC$, the quantity α is close to unity. In this case, the z- and s-domain models predict similar jitter

90

transfer characteristics for jitter frequencies within the loop's bandwidth. Mathematically, the reason is that minimal aliasing occurs as the loop filter/VCO s-domain transfer characteristic is transformed to the z-domain; physically, the results predicted by the two models converge in this case because the s-domain assumption concerning the ability of the loop filter to reject phase detector high frequency output energy is valid. The two models diverge for jitter frequencies outside the loop's bandwidth where the z-domain analysis correctly models the repetitive nature of the input/output jitter spectra. For wider bandwidth loops, the z- and s-domain models will diverge even for jitter frequencies within the loop's bandwidth. The reason is that the relatively wideband loop filter does not reject high frequency phase detector components as the s-domain approximation would suggest. In the next section, it will be shown that the s- and z-domain models begin to diverge for jitter frequencies within the loop's bandwidth for practical PLL's. The two models will also be compared with experimental results taken from the DSI (Digital Signaling Interface) integrated circuit.

III. MEASURED RESULTS

In the design of the wideband PLL used for clock recovery in the DSI device, three modeling techniques were used to estimate loop performance. The jitter transfer characteristic was derived using both the linear s- and z-domain models presented in this paper. In addition, a timestep simulator was written in Fortran which performed a transient analysis on the loop in the locked condition using time domain models of the loop elements. The timestep simulation was thought to be the most accurate of the three methods because it included a number of nonlinear effects in the element models. For example, in the timestep simulator, the output of the phase detector was treated as a finite width pulse, instead of a weighted impulse and the finite pull range of the VCO was taken into account.

The nominal design parameters were entered into each of three models and the jitter response was evaluated. For the timestep simulator, a 0.1 Unit Interval (1 U.I. = 1 clock period) input jitter magnitude was used as the "small signal" input. Plots of the magnitude of $H(f)$ are shown in Fig. 9. It is seen that the timestep simulator results agree with the z-domain model results within ± 0.3 dB over all frequencies. With an input data transition rate of 193 kHz and VCO clock rate of 1.544 MHz the linear models agree within ± 0.3 dB up to about 1/20th of the input transition rate. For the z-domain model and the timestep simulator, the loop response is periodic with a period of 96.5 kHz, whereas the s-domain response continues to roll off at higher frequencies. As a result of these simulations, the z-domain model and timestep simulator were chosen as the design tools for determining loop parameters in the actual PLL.

Upon receipt of silicon, the individual loop parameters were measured for a specific device. These parameters

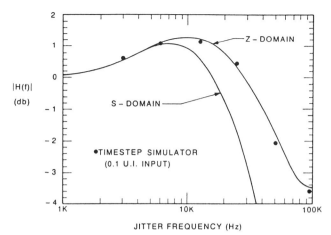

Fig. 9. $|H(f)|$ comparisons using simulators.

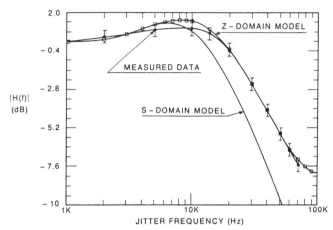

Fig. 10. $|H(f)|$ comparisons using measured data.

were then entered into both the s- and z-domain models and $H(f)$ was calculated in both cases. The results were compared with the measured $H(f)$ for the device under test. The measured $H(f)$ is defined to be the output jitter amplitude (peak-to-peak) at the input jitter frequency divided by the input jitter amplitude (pk–pk). Fig. 10 shows the three curves. It is seen that the z-domain results agree with the measured results within the ± 0.5-dB measurement error bars. Again, the s-domain $H(f)$ tracks the z-domain $H(f)$ for low jitter frequencies, but diverges at about 1/20th the input data transition rate. The agreement of the timestep simulator, fabricated PLL, and z-domain $H(f)$ verifies the accuracy of the discrete-time linear model.

IV. SUMMARY AND CONCLUSIONS

In this paper, two classes of PLL's were described: continuous-time and discrete-time PLL's. Classic s-domain analysis, although valid for continuous-time PLL's, cannot always accurately predict the behavior of discrete-time loops. The z-domain model, presented here, takes into account the sampled data nature of the digital phase detector and accurately predicts overall loop performance. The interface between discrete- and continuous-time elements is handled using the impulse invariant *s*-to-*z*-domain transformation. In this way, the entire PLL is analyzed in the discrete-time domain. Measured results from the DSI phase-locked loop support the *z*-domain model.

The modeling philosophy presented in this paper can be generalized to many systems in which discrete- and continuous-time elements coexist. One such system, a PLL with a switched capacitor loop filter driving a continuous-time VCO, is described in Appendix B. In this case, the interface waveform is staircase in nature; therefore, the step invariant transformation is appropriate. In general, if a well-defined interface waveshape exists, an appropriate transformation can be chosen to convert all *s*-domain elements into the *z*-domain. In this way, a complete linear model for the mixed continuous-time/discrete-time system can be developed.

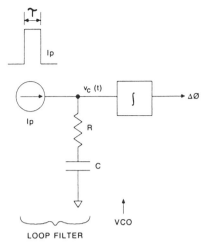

Fig. 11. PLL response to finite width pulse input.

APPENDIX A
NETWORK LINEARIZATION APPROXIMATIONS

In the derivation of the *z*-domain model for discrete-time PLL's presented in this paper, a number of approximations were made to linearize the network. In [4], Gardner uses the difference equation approach to derive a discrete-time PLL's response to an input phase ramp (frequency step). In linearizing the difference equations, a number of approximations are also required. In this appendix, each of the linearizing assumptions will be discussed. It will be shown that the assumptions required to derive the *z*-domain model through the impulse invariant transformation are the same as the assumptions required to linearize the difference equations. Therefore, both approaches lead to the same results, although the impulse invariant approach significantly reduces the amount of computation required.

Linearizing Assumptions

1) In any linear PLL model, it must be assumed that the loop filter and VCO operate in a linear fashion. As stated previously, many practical loop filters and VCO's are well modeled as linear elements.

2) In both discrete-time models, it must be assumed that phase samples occur at constant intervals and that the loop responds symmetrically to leading and lagging phase information. This implies that:

a) the phase tracking errors are small,
b) the data pattern is constant,
c) the bit-to-bit data jitter is small,
d) no significant amount of high frequency energy can propagate directly around the loop, generating asymmetric behavior for leading and lagging phase errors.

Typically, the finite high frequency response of the VCO and of ripple filters built into the loop filter remove this high frequency energy sufficiently to allow the loop to operate in a linear fashion.

3) Finally, in the model using the impulse invariant transformation, it was assumed that the pulse width modulated output of the digital phase detector could be approximated as a series of weighted impulses. The validity of this approximation can be investigated by analyzing the loop filter and VCO response to a finite width pulse for a simple RC loop filter. In Fig. 11, a simple RC charge pump loop filter is shown. The loop filter is driven with a current pulse of magnitude I_p and duration τ.

The output of the loop filter from time 0 to T (the time at which the next reference data edge arrives) is given by

$$v_c(t) = I_p R + \frac{I_p}{C} t, \qquad 0 \leqslant t \leqslant \tau$$

$$= \frac{I_p}{C} \tau, \qquad \tau \leqslant t \leqslant T.$$

The VCO integrates its control voltage to produce an output phase shift. Therefore, the total phase shift at time T caused by the finite width pulse starting at time 0 is given by

$$\Delta\phi = K_o \int_0^T v_c(t) \, dt$$

$$= K_o \left[\int_0^\tau \left(I_p R + \frac{I_p}{C} t \right) dt + \int_\tau^T \frac{I_p}{C} \tau \, dt \right]$$

$$= (I_p \tau) \frac{K_o}{2C} [2(RC + T) - \tau]. \qquad (1)$$

From this equation, one can see that the phase shift is a linear function of pulse area $(I_p \cdot \tau)$ if $2 \cdot (RC + T) \gg \tau$. For reasonable loop filter components and small tracking errors, this is a very good approximation. For the wideband loop used to verify the *z*-domain model in this paper, the RC time constant of the loop filter was ~ 40 μs, much larger than the maximum phase tracking error (< 50 ns). Therefore, approximating the finite width pulse with a weighted impulse of area $I_p \cdot \tau$ introduces negligible error in the analysis.

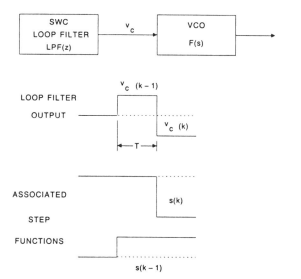

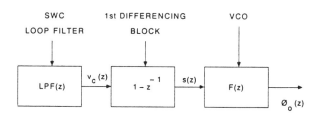

Fig. 12. Switched capacitor loop filter/continuous-time VCO interface.

F(z) ⇒ F(s) TRANSFORMED USING STEP INVARIANT METHOD

Fig. 13. Z-domain model for SWC loop filter/VCO interface using step invariant transformation.

In the difference equation analysis from [4], the $|\phi_c|/2w_iC$ term was dropped from (A13) in order to linearize the equation. This is equivalent to dropping the nonlinear τ term in (1) above. Thus, the difference equation approach also approximates the finite width pulses coming from the digital phase detector as a series of weighted impulses.

From the above analysis, it is shown that the linearizing approximations required for the impulse invariant transformation method and difference equation method are the same. Also, for most practical PLL's, the linearizing assumptions are valid and the loop operates in a linear fashion for small phase tracking errors.

APPENDIX B
EXTENSION OF MODEL TO SWITCHED-CAPACITOR LOOP FILTER/CONTINUOUS TIME VCO INTERFACE

The previous sections of this paper have dealt with a PLL architecture which consisted of a digital phase detector which was well modeled directly in the z-domain and a continuous-time loop filter and VCO which were described by s-domain transfer functions. The modeling technique presented addressed the problem of converting the continuous-time components of the PLL into the discrete-time domain when the loop filter is being driven by a weighted

series of impulses. In other PLL architectures, switched capacitor loop filters may be used to drive a continuous-time VCO. In these cases, an extension of the modeling philosophy presented in this paper can be used to develop a discrete-time linear model for the PLL.

The functional block diagram for such a switched-capacitor loop filter/continuous-time VCO interface is shown in fig. 12. The output of the loop filter is a staircase waveform which changes values at intervals of T seconds. This staircase waveform can be viewed as a superposition of step functions whose value for the kth time interval is given by the difference between the output of the SWC loop filter in the k and $(k-1)$th intervals. Quantitatively, if $s(k)$ is the output step function in the kth interval, then:

$$v_c(k) = v_c(k-1) + s(k)$$

$$= \sum_{j=0}^{k} s(j)$$

or

$$s(k) = v_c(k) - v_c(k-1).$$

Transforming this relationship into the z-domain yields

$$s(z)/v_c(z) = 1 - z^{-1}.$$

Therefore, the input to the VCO can be modeled as a summation of step functions $s(k)$ if the output of the loop filter $v_c(k)$ is passed through a first-differencing block with transfer characteristic $1 - z^{-1}$. The first-differencing block, of course, does not exist in the real system, it is only required in the model in order to represent the loop filter output as a sum of step functions.

The treatment of a continuous-time system being driven by a series of weighted step functions is analogous to the treatment of a continuous-time network being driven by a weighted sum of impulses. For a step function input to the VCO, it is necessary to use the step invariant transformation to model the VCO in the z-domain. A reconstruction filter or VCO finite frequency response can be handled by including these s-domain singularities with the basic K_o/s VCO term and step transforming the entire expression into the z-domain. The resulting z-domain model for the SWC loop filter/continuous-time VCO interface is shown in Fig. 13.

Once the continuous-time elements have been transformed into the z-domain using the step invariant method, the discrete-time analysis of the PLL can proceed completely in the discrete-time domain.

REFERENCES

[1] F. Gardner, *Phaselock Techniques*. New York: Wiley, 1979.
[2] P. Gray and R. Meyer, *Analysis and Design of Analog Integrated Circuits*. New York: Wiley, 1984.
[3] R. Best, *Phase-Locked Loops*. New York: McGraw-Hill, Inc. 1984.
[4] F. Gardner, "Charge pump phase-lock loops," *IEEE Trans. Commun.*, vol. COM-28, Nov. 1980.
[5] A. Oppenheim and R. Schafer, *Digital Signal Processing*. Englewood Cliffs, NJ: Prentice-Hall, 1975.

ANALYZE PLLs WITH DISCRETE TIME MODELING

This multi-rate, discrete time-domain model provides an accurate description of phase-lock-loop dynamics.

D ESIGN and optimization of phase-locked loops (PLLs) for frequency-synthesizer applications often require analysis of jitter and loop behavior in the time domain. Since these circuits operate on digital signals, where phase information is represented by signal edges, classical s-domain analysis is a first-degree approximation at best.[1]

Classical PLL analysis models are based on assumptions regarding the behavior of the loop and its component hardware.[2] These assumptions are that the loop is frequency locked, the loop operates in continuous time, the voltage-controlled-oscillator (VCO) signal source is an ideal integrator, and that the phase detector is a linear adder.

A PLL in the s-domain (Fig. 1) can be calculated as:

$$G(s) = K_p Z(s) \frac{K_o}{s} \qquad (1)$$

With a typical series resistance-capacitance (RC) filter configuration

JANOS KOVACS, Senior Design Engineer, Semiconductor Div., Analog Devices, Inc., 1 Technology Way, Norwood, MA 02062; (617) 937-1328.

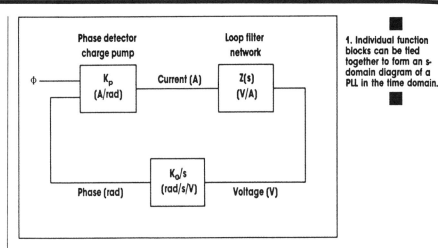

1. Individual function blocks can be tied together to form an s-domain diagram of a PLL in the time domain.

(R_1, C_1) with a smaller capacitor in parallel (C_2), the filter impedance is:

$$Z(s) = \left(R_1 + \frac{1}{sC_1}\right)$$

$$\times \frac{\dfrac{1}{sC_2}}{R_1 + \dfrac{1}{sC_1} + \dfrac{1}{sC_2}} \qquad (2)$$

assuming $C_1 >> C_2$, leads to:

$$Z(s) = \frac{1}{sC_1} \frac{1 + sR_1C_1}{1 + sR_1C_2} \qquad (3)$$

The open-loop transfer function becomes:

$$G(s) = K_p \frac{1}{sC_1} \frac{1 + sR_1C_1}{(1 + sR_1C_2)} \frac{K_o}{s} \quad (4)$$

This third-order system has two poles at $\omega=0$, one zero at $\omega_1 = 1/R_1C_1$, and another pole at $\omega_2 = R_1C_2$. The closed-loop system will be stable if the open-loop crossover frequency falls between ω_1 and ω_2, where the rolloff is 20 dB/decade. At the crossover frequency, the loop

Table 1: VCO jitter versus input frequency

Input frequency	ΔT (peak-to-peak)	Sigma
0 to 10 MHz	10 ns	3 ns
11.8 MHz	7.1 ns	1.5 ns
15.5 MHz	5 ns	0.91 ns
Note: Measurements are for the AD897 with VCO frequency set at 30 MHz.		

Reprinted with permission from *Microwaves & RF*, J. Kovacs, "Analyze PLLs with Discrete Time Modeling," pp. 224-229, May 1991.

gain is one, therefore:

$$\omega_o = (K_p K_o)R_1 = KR_1 \qquad (5)$$

The loop response can be approximated as a second-order system, Assuming $\omega_2 > > \omega_1$:

$$H(s) = \frac{G(s)}{1 + G(s)}$$

$$= \frac{K}{C_1} \frac{\left(1 + \dfrac{s}{\omega_1}\right)}{\left(s^2 + s(KR_1) + \dfrac{K}{C_1}\right)} \qquad (6)$$

The closed-loop transfer function has its own 0-dB point, ω_0. By analyzing the PLL's amplitude characteristics (Fig. 2), the natural frequency of the system can be derived:

$$\omega_n = \sqrt{\frac{K}{C_1}} \qquad (7)$$

Using the above expressions for ω_0 and ω_1, several useful formulas can be gained involving the damping factor:

$$\zeta = \frac{\omega_o}{2\omega_n} \quad \text{or} \quad \omega_1 = \frac{\omega_o}{(2\zeta)^2} \qquad (8)$$

By placing ω_1 relative to ω_0, it is possible to adjust the damping factor at will. For example, selecting critical damping would fix ω_1 at half of ω_0. A typical choice for ω_2 is about 4 to 8 times the value of ω_1. (For designers willing to experiment with the AD897 PLL, source code for a short BASIC program, "PLL filter calculations," is available from the author.)

In PLLs based on digital signals, the phase detector/charge pump presents a new current value to the loop filter only when it receives a data pulse. Between these updates, the charge pump is usually tristated (i.e., not supplying any current to the filter). PLL performance is greatly degraded if this hold mode in not efficient. Any current leakage produces a voltage change on the VCO control node. Resulting frequency drift causes pattern-dependent phase errors at the next update cycle, decreasing the error margin.

To more closely model real PLL operation, this complex frequency description must be translated into

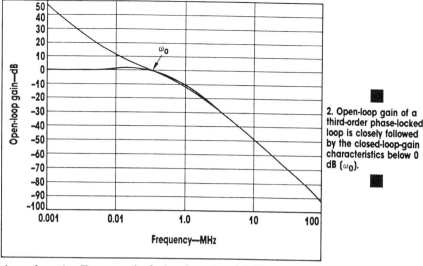

2. Open-loop gain of a third-order phase-locked loop is closely followed by the closed-loop-gain characteristics below 0 dB (ω_0).

the z-domain. For practical circuits, an impulse-invariant transformation is justified.[3] Performing this transformation for a fourth-order loop, where the finite bandwidth of the voltage to current converter found in the VCO is considered with an additional pole at ω_3, the open-loop gain is:

$$G(s) = \frac{K}{sC_1} \frac{1 + \dfrac{s}{\omega_1}}{1 + \dfrac{s}{\omega_2}} \frac{1}{1 + \dfrac{s}{\omega_3}} \frac{1}{s} \qquad (9)$$

The next step in the PLL model development is to derive the partial fraction form of the above expression:

$$G(s) = KR_1 \left(\frac{\alpha}{s} + \frac{\omega_1}{s^2}\right.$$

$$\left. - \frac{1}{\beta} \frac{1}{s + \omega_2} + \frac{\alpha\omega_2}{\beta\omega_3} \frac{1}{s + \omega_3}\right) \qquad (10)$$

where:

$$\alpha = 1 - \frac{\omega_1}{\omega_3} \quad \beta = 1 - \frac{\omega_2}{\omega_3}$$

$$\gamma = \omega_1 T_s \qquad (11)$$

The impulse-invariant z-transform

with T_s sampling period becomes:

$$G(z) = \frac{\omega_0}{f_{vco}} \left(\frac{\alpha z}{z - 1} + \frac{\gamma z}{(z - 1)^2}\right.$$

$$\left. - \frac{1}{\beta} \frac{z}{z - a} + \frac{\alpha\omega_2}{\beta\omega_3} \frac{z}{z - b}\right) \qquad (12)$$

with the new constant defined as:

$$a = \exp(-\omega_2 T_s) \qquad (13)$$

$$b = \exp(-\omega_3 T_s) \qquad (14)$$

When ω_3 approaches ω_2, the third and fourth terms in the partial-fraction form derived for $G(s)$ cancel. In this case, the original open-loop gain becomes:

$$G(s) = \frac{K}{sC_1} \frac{1 + \dfrac{s}{\omega_1}}{\left(1 + \dfrac{s}{\omega_2}\right)^2} \frac{1}{s} \qquad (15)$$

and the partial-fraction form changes to:

$$G(s) = \omega_o \left(\frac{1}{s} + \frac{\omega_1}{s^2}\right.$$

$$\left. - \frac{1}{s + \omega_2} - \frac{\omega_2}{(s + \omega_2)^2}\right) \qquad (16)$$

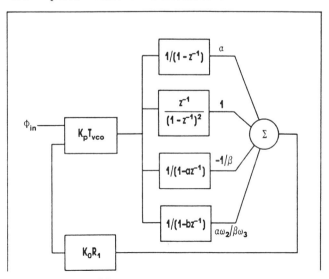

3. A PLL can be represented in the z-domain, with discrete phase-error updates.

95

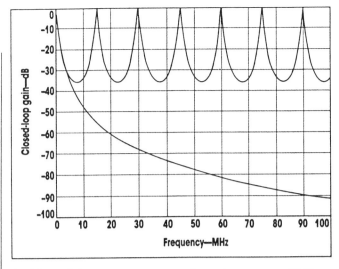

4. Closed-loop gain is plotted in the z- and s-domains, as a function of frequency, for a data rate of 15 MHz.

Assuming that $\omega_1 < \omega_2 < \omega_3$ normally holds for practical systems, the frequency-domain response can be simulated. The PLL can be represented by a weighted sum of parallel and series combinations (Fig. 3) of blocks that implement the various terms of G(z), such as:

$$g(z)_i = \frac{1}{1 - b_i z^{-1}} \qquad (17)$$

During AC analysis, the transfer functions are evaluated by setting $z = e^{j2\pi f/f_s}$. It should be noted that $f_s = f_{data}$; that is, the sampling rate is determined by the reference frequency supplied by the input data.[4] The consequences of switching to discrete time representation can be measured by comparing the amplitude transfer characteristics gained in the s- and z-domains.

The difference between the continuous and sampled time-domain results becomes even more obvious when examining a PLL's closed-loop response (Fig. 4). The act of sampling continuous signals gives rise to aliasing.[5] In the case of a lowpass transfer characteristic, spectral components of signals (including noise) at frequencies higher than one-half the sampling rate might be mirrored to the passband. The repetitive peaks of the PLL closed-loop z-domain response of $f_{data} \pm f_{3dB} \pm 2f_{data} \pm f_{3dB}, 3f_{data} \pm f_{3dB}$, and so on, will be aliased to the passband unless they are removed by filtering.

Although the input signal spectrum is limited by lowpass filtering in the read channel, the bandwidth is typically set to $f_{max} = f_{vco}/2$, which overlaps quite a few peaks at higher f_{vco}/f_{data} ratios. Aliasing should be a concern when evaluating the noise performance of these PLLs. The periodic characteristic of the transfer function also adversely effects high-frequency rolloff.

Until now, the integrator that represents the VCO in this PLL model has been included with the remainder of the loop's frequency-dependent element. However, the VCO must also be closely examined with-

in this model to ensure accurate loop simulations.

In a monolithic PLL circuit (the Analog Devices AD897) with two cross-coupled voltage-controlled ramp generators that provide 50 percent duty-cycle output voltages, the voltage ramp is generated by a voltage-controlled current source with a transconductance of g_m discharging a capacitor C. When the voltage on the capacitor becomes less than a preset threshold value, v_{th}, a flip-flop is tripped and resets the voltage on the capacitor to a maximum value v_p, thus energizing the other ramp generator. Neglecting the propagation delay in the comparator and flip-flop, the capacitor is discharged over a period of $T_0/2$, determined by the allowed voltage swing:

$$v_p - v_{th} = \frac{v_{ino} g_m}{C} \frac{T_o}{2} \qquad (18)$$

The oscillator frequency is:

$$\omega_{osc} = 2\pi \frac{g_m}{C} \frac{v_{ino}}{2(v_p - v_{th})}$$
$$= K_o v_{ino} \qquad (19)$$

where v_{ino} is the DC value of the voltage applied to the voltage-controlled current source or voltage-to-

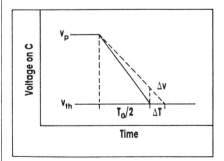

5. A VCO's period changes with control voltage; for small changes, $\Delta T \ll T_0$.

current (V/I) converter. The VCO period changes by ΔT when v_{in} changes only slightly relative to v_{ino}, due to a Δv voltage change on the capacitor (Fig. 5):

$$\Delta v = \frac{g_m}{sC} v_{in} \qquad (20)$$

with

$$\frac{\Delta v}{v_p - v_{th}} = 2\frac{\Delta T}{T_o} \qquad (21)$$

The change in phase represented in Eq. 21 can be written as:

$$\Delta\phi = \frac{2\pi g_m}{2(v_p - v_{th})} C \frac{v_{in}}{s}$$
$$= K_o \frac{v_{in}}{s} \qquad (22)$$

With a time-interval analyzer (TIA), it is possible to check how the VCO period changes with the frequency of excitation by forcing a sinusoidal voltage from a signal generator onto the VCO input and measuring the period of the oscillator output. The TIA will give a distribution with the RMS value of ΔT. Based on the above result for $v_{1n} = V_{in}\sin(\omega_{in}t)$,

$$\Delta T = \frac{1}{v_{ino}} \int v_{in} dt$$
$$= -\frac{1}{\omega_{in}} \frac{V_{in}}{v_{ino}} \cos\omega_{in}t \qquad (23)$$

that is, RMS(ΔT) should decrease with ω_{in}.

Measurements of the AD897 (Table 1) indicate flat response to $f_3 = 12$ MHz where the V/I converter's finite bandwidth affects the response. If ΔT (or $\Delta\phi$ since they differ only in a fixed multiplier at a given VCO frequency) is independent of the excitation frequency, the VCO obviously can not be considered as an integrator.

The problem lies in the crudeness

of the PLL model. Although the oscillator's operation is based on the interaction of the two ramp generators, the model does not account for the synchronous resetting action. It can be included by adding a parallel feed-forward path to the output where Δv can be measured (Fig. 6). The z^{-1} block samples the voltage on the capacitor at a rate of $f_s = f_{vco}$, which means that the voltage at the time of sampling is subtracted from the output node, resetting it to zero. Now, $\Delta\phi$ becomes:

$$\Delta\phi = K_o \frac{v_{in}}{s} (1 - z^{-1}) \qquad (24)$$

Since the term $1-z^{-1}$ adds a differentiating characteristic to the overall gain at lower frequencies, the VCO gain is flat at lower frequencies, then starts rolling off. In the AD897, the V/I converter has an exponential transfer function:

$$\omega_{vco} = \frac{2\pi}{T_o} e^{-k_{exp}v_{in}} \qquad (25)$$

and the VCO gain becomes $K_o = k_{exp}\omega_{vco}$ (i.e., it increases with frequency). The consequence is a constant fractional loop bandwidth, since $G(s)/\omega_{vco}=K_p k_{exp}R_1$. By rewriting ΔT as

$$\Delta T = \Delta\phi \frac{T_o}{2\pi}$$

$$= K_{exp}\frac{v_{in}}{s} (1 - z^{-1}) \qquad (26)$$

ΔT can be plotted as a function of frequency. Besides shifting the flat portion of the ΔT curve with the VCO frequency, the periodic nature of $1-z^{-1}$ and the fact that it peaks at multiples of one-half the sampling frequency is reflected in the oscillator's noise transfer function. This can be characterized by injecting Δv directly into the loop circuit.[6] TIA measurements will show that jitter introduced by this method changes with frequency and is periodic with the trigger or sampling frequency.

The PLL model also must include the phase detection process. Digital phase detectors apply a control signal to the charge pump, commanding it to either source or sink current. The time of pumping charge

into or out of the loop filter is equal the phase error t_ϕ measured between the input data edge and the center of the window. The window size is defined by the clock signal period present at the phase detector input; without any divider in the feedback path, it is equal to the VCO period T_{vco}.

The phase-error measurement reference center of the window is marked by one of the edges of the clock. The phase measurement oper-

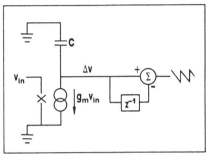

6. This VCO representation incorporates a parallel feed-forward path to measure Δv.

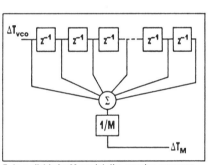

7. In a divide-by-M model, the counter averages VCO jitter over M clock periods.

ation is triggered by the arrival of a data edge. Let $t_\phi = t_{\phi i}$ be the phase error at the ith data pulse. If the data pulses are N clock cycles apart, the phase error $t_{\phi(t = 1)}$ can be calculated as:

$$t_{\phi(i + 1)} = \sum_{j=1}^{N} (T_o + \Delta T_j)$$
$$- NT_o + t_{\phi i}$$
$$= \sum_{j=1}^{N} \Delta T_j + t_{\phi i} \qquad (27)$$

In the discrete time domain, this accumulation of ΔT_j can be implemented with an adder where one input is receiving ΔT_j while the other is the value of the adder's output

delayed by one clock cycle. The block just described implements an integrator in the z-domain and its output is the position of the clock edge relative to some starting point. Phase error $t_{\phi i}$ is the difference between this reference $t_{\phi 0}$ and the accumulated clock position, and is sampled at the update rate.

The phase detection process, described in this manner, brings the integrator back into the loop. By separating the loop's elements in this way, it is possible to consider the effect of introducing dividers into the PLL's feed-forward and feed-back paths.

Classical theory suggests that adding a frequency divider between the VCO and phase detector will result in increasing VCO jitter as the division ratio M increases. Measured data contradict this prediction, however.

A divide-by-M counter can be modeled by an M-bit-long shift register, which in the z-domain is equivalent to cascading M-1 unit delay blocks (Fig. 7). Since integration of the VCO period change ΔT is performed on a cycle basis during the phase detection process, the jitter at the output of the M divider can be written as:

$$\Delta T_M = \frac{\Delta T_{vco}}{M} \sum_{m=0}^{M-1} z^{-m} \qquad (28)$$

It is clear by now that the divider implements a moving average filter, with no averaging at the lower frequencies.

The full PLL can now be built by assembling z-domain building blocks (Fig. 8). The phase detector is sampled at the data rate while other components run at the VCO frequency. The VCO frequency is actually an integer multiple of the data frequency, i.e.:

$$f_{vco} = df_{data} = [2,8]f_{data} \qquad (29)$$

The different sampling rates automatically take care of the question of how the phase detector/charge pump gain changes with frequency. In frequency synthesizer applications, there is a divider in the feed-forward path so that the VCO output frequency becomes:

$$f_{vco} = \frac{M}{N} f_{ref} \qquad (30)$$

and the data frequency becomes:

$$f_{data} = \frac{f_{ref}}{N} \qquad (31)$$

which should be the data sampling rate for the phase detector.

How does noise in the VCO effect overall jitter when the oscillator is followed by a divider stage? Since ΔT_M is given by a geometric series:

$$\Delta T_M = \frac{\Delta T_{vco}}{M} \sum_{m=0}^{M-1} z^{-m}$$

$$= \frac{\Delta T_{vco}}{M} \frac{1 - z^{-M}}{1 - z^{-1}} \qquad (32)$$

When forcing Δv in the VCO, the transfer function becomes:

$$\frac{\Delta T_M}{\Delta v} = \frac{1}{M} \frac{1 - z^{-M}}{1 - z^{-1}} (1 - z^{-1}) \quad (33)$$

or

$$\frac{\Delta T_M}{\Delta v} = \frac{1}{M} (1 - z^{-M}) \qquad (34)$$

The resulting amplitude curve is very much like the original noise transfer function in the VCO, but with z^{-1} replaced by z^{-M} and the gain reduced by a factor M. For $\Delta v=1$, the peak amplitude in dB-s is:

$$\Delta T_{M,\ dB} = 20\log\frac{2}{M} \qquad (35)$$

Varying data frequency can also affect open-loop gain. The average charge pumped into the loop filter depends on $d=f_{vco}/f_{data}$, the number of VCO cycles between data updates. In traditional s-domain models, this is considered by including d in the formula for the charge pump gain:

$$K_p = \frac{d}{d\phi}i = \frac{I_p}{2\pi d}\frac{1}{d}$$

$$= \frac{I_p}{2\pi}\frac{f_{data}}{f_{vco}} \qquad (36)$$

After replacing the filter components with their discrete time equivalent, the update rate is reflected in the sampling frequency of the z-domain network, $f_s=f_{data}$. The charge pump gain is

$$K_p = \frac{d}{d\phi}i = \frac{I_p}{2\pi}\frac{1}{f_{vco}} \qquad (37)$$

In the final multi-rate sampled data system, the same thing hap-

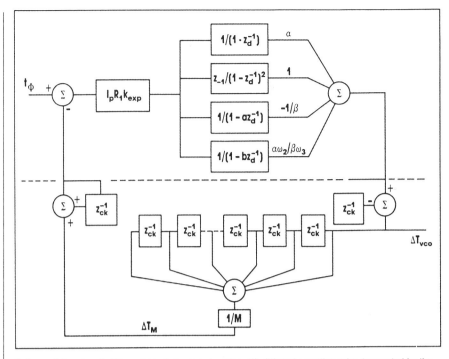

8. A complete z-domain PLL model includes two domains with different sampling rates (separated by the dotted line).

pens to f_{vco}. The charge pump gain definition is changed to reflect time-domain operation, since the phase detector that measures the phase error in seconds is:

$$K_p = \frac{d}{dt}i = I_p \qquad (38)$$

Variations in the open-loop gain with the update rate should be carefully considered as a potential source of closed-loop instability. Under normal circumstances, the open-loop crossover frequency ω_0 should fall above the loop-filter zero ω_1 since the damping factor is:

$$\zeta = \frac{1}{2}\sqrt{\frac{\omega_o}{\omega_1}} \qquad (39)$$

The loop-filter component values fix the position of ω_1 (i.e. it is independent of the data frequency, whereas ω_0 is not). Increasing the spacing between updates will decrease the open-loop gain, moving ω_0 closer to ω_1 and resulting in less and less damping. This leads to oscillatory transient behavior under closed-loop conditions.

Introducing a feedback divider with division ratio M is equivalent to adding a moving average filter between the VCO and phase detector. The transfer function of an M-point

filter is:

$$H\ (f) = \frac{\sin\left(\pi M \frac{f}{f_s}\right)}{M \sin\left(\pi \frac{f}{f_s}\right)} \qquad (40)$$

Besides the additional low-frequency rolloff in the open-loop transfer function, the filter creates an extra delay in the loop since its group delay is:

$$t_{delay} = \frac{M-1}{2}T_0 \qquad (41)$$

where T_0 is the VCO period.

High values of division factor M create the potential for unstable closed-loop behavior as the rate of change around the crossover point in the open-loop transfer function gets close to 40 dB/decade. These problems can be avoided by choosing ω_1 as low as possible, with the tradeoff being poor transient response. ••

References

1. F. Gardner, *Phase Lock Techniques, 2nd ed*, John Wiley and Sons, NY, 1981.
2. U. Rohde, *Digital PLL Frequency Synthesizers, Theory and Design*, Prentice Hall, Inc., Englewood Cliffs, NJ, 1983.
3. J. Hein and J. Scott, "z-Domain Model for Discrete Time PLLs," *IEEE Transactions on Circuits and Systems*, Vol. 35, No. 11.
4. G. Gutierrez and D. DeSimone, "An Integrated PLL Clock Generator for 275 MHz Graphic Displays," IEEE 1990 Custom Integrated Conference.
5. R. Higgins, "Digital Signal Processing in VLSI," *Analog Devices Technical Reference Books*, Prentice Hall, Englewood Cliffs, NJ, 1990.
6. "Programmable Time Delay Generation Techniques with the AD9500 Programmable Delay Generator," Analog Devices Application Notes.

Properties of Frequency Difference Detectors

FLOYD M. GARDNER, FELLOW, IEEE

Abstract—Among other applications, frequency-tracking loops are employed in digital-data receivers, either as a frequency-acquisition aid for phase-locked coherent reception, or as the sole carrier-frequency control for noncoherent reception. This article provides details of design and performance of the frequency-difference detector that lies at the heart of the loop.

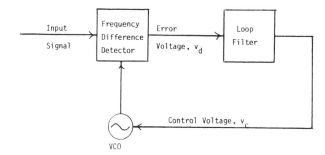

Fig. 1.   Frequency-tracking loop, simplified block diagram.

I. INTRODUCTION

RECEIVERS of modulated-carrier data signals sometimes contain frequency tracking loops. Reasons for employing such loops include the following.

• Frequency must be adjusted accurately for differentially coherent or incoherent reception.

• A Costas or other phase-locked loop for coherent reception may require frequency aiding for acquisition purposes.

A frequency-tracking loop, in its simplest form, is arranged as shown in Fig. 1. The loop consists of a frequency-difference detector (FDD), a loop filter, and a voltage-controlled oscillator (VCO).

Such a loop strongly resembles automatic frequency control (AFC) loops, which have been known for many years (e.g., [1], [2]). A conventional AFC loop employs a frequency discriminator, which relies upon a passive tuned circuit to furnish the frequency reference. By contrast, the FDD uses a local oscillator to furnish the frequency reference.

This paper examines the characteristics of a particular type of FDD in some detail. Features covered include: circuit principles; response to signals, with emphasis upon data signals; and effects of noise. Several properties important to satisfactory designs are brought to light.

Only analog implementations are treated here. Digital implementations are also of considerable interest and may be found in [3] and [4].

II. OPERATING PRINCIPLES

Simple Quadricorrelator

The best known FDD is the quadricorrelator of Fig. 2. It was first presented by Shaeffer [5] but given its name and described in some detail by Richman [6]. The quadricorrelator (and some modification thereof) is the only FDD considered in this paper.

A pair of mixers are used to convert the input passband signal into the corresponding in-phase and quadrature baseband components. For analysis purposes, the mixers are represented as ideal multipliers, but physical circuits could be switching devices, with no alteration in the results [7, p. 108]. Transfer gain of the mixer is the dimensionless factor K_m.

Outputs from the mixers consist of sum and difference frequency products between the input signal and the local oscillator. Low-pass arm filters following the mixers suppress the sum frequency and pass the difference frequency. For

Paper approved by the Editor for Communication Theory of the IEEE Communications Society for publication without oral presentation. Manuscript received April 3, 1984. This work was supported by a contract from the European Space Agency, Noordwijk, The Netherlands.

The author is a Consulting Engineer at 1755 University Avenue, Palo Alto, CA 94301.

purposes of the immediate analysis, it will be assumed that the difference-frequency component is passed without attenuation or phase shift. That assumption is removed later.

In order for the quadricorrelator to operate, the difference frequency between input signal and local oscillator must fall within the passband of the arm filters. A signal that is well outside of passband will be attenuated by the filter selectivity. Bandwidth of the arm filter therefore provides a rough estimate of the capture range of a frequency tracking loop.

Output of one arm filter (the *I*-arm in Fig. 2) is differentiated. Time constant (or "gain") of the differentiator is T_d seconds.

A perfect differentiator is assumed for ease of analysis. One can object that a perfect differentiator cannot be built and that the analysis is therefore unrealistic. In fact, perfect relative differentiation is readily obtainable and so the analysis is quite realistic. The term "relative differentiation" is explained as follows.

Let one arm filter have a low-pass filter transfer function denoted as $H_a(s)$. Let the other arm filter have a transfer function $sH_a(s)$, which is always physically realizable, provided only that $H_a(s)$ is low-pass. Then if the same signal should be applied to both filters, the output from the $sH_a(s)$ filter is exactly the derivative of the output of the $H_a(s)$ filter. Therefore, perfect relative differentiation can be achieved. Outputs of the arm filters are designated $v_I(t)$ and $v_Q(t)$.

Differentiation of the *I*-channel produces $T_d\dot{v}_I(t)$. Arm outputs are multiplied in the third multiplier to produce

$$v_d(t) = K_3 T_d v_Q(t)\dot{v}_I(t) \tag{1}$$

where K_3 is the gain of the third multiplier and has dimensions of $(\text{volts})^{-1}$.

To gain insight into the behavior of a quadricorrelator, let the input signal be a simple sinusoid

$$v_{in}(t) = V_s \cos(\omega_i t + \theta_i) \tag{2}$$

where θ_i is an arbitrary, time-invariant phase angle. Defining frequency error as $\Delta\omega = \omega_i - \omega_0$, where ω_0 is the radian frequency of the reference signals, the arm-filter outputs are calculated to be

$$v_I(t) = K_m V_s \cos(\Delta\omega t + \theta_i)$$
$$v_Q(t) = K_m V_s \sin(\Delta\omega t + \theta_i). \tag{3}$$

Reprinted from *IEEE Trans. Comm.*, vol. COM-33, pp. 131-138, February 1985.

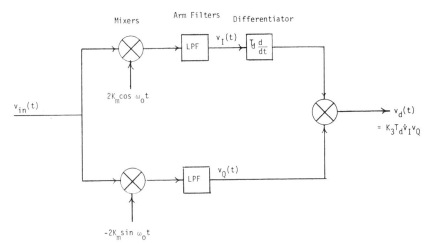

Fig. 2.   Quadricorrelator.

Differentiating and multiplying, the output of the third multiplier is found as

$$v_d(t) = -\tfrac{1}{2}\Delta\omega T_d K_m{}^2 K_3 V_s{}^2 \left[1 - \cos\left(2\Delta\omega t + 2\theta_i\right)\right]. \quad (4)$$

There are two components in this product: a dc component proportional to the frequency difference (including its sign) and a ripple component at double the difference frequency. Notice that the phase θ_i appears only in the ripple and not in the dc component.

The dc component is the useful output; it could be used for FM demodulation or as the error signal in a frequency-tracking loop. The ripple, which has the same peak amplitude as the dc component, can be a major nuisance if the circuit is used to recover modulation or if the frequency tracker is the only frequency-control element. A method of cancelling ripple will be presented below.

There are numerous variations on the basic quadricorrelator. One of them is to recognize that the differentiator need not be perfect; indeed, it is not even necessary that the box labeled "differentiator" be a high-pass network.

To see how this can be, let v_I and v_Q be as shown in (3) and let the differentiator be replaced by a selective network with transfer function $H_d(f) = A(f)\exp\left(j\phi(f)\right)$, where A is the amplitude and ϕ is the phase shift of the frequency response of the network. Denoting $\Delta f = \Delta\omega/2\pi$, the output of this network will be

$$\tilde{v}_I(t) = K_m V_s A(\Delta f)\cos\left(\Delta\omega t + \theta_i + \phi(\Delta f)\right). \quad (5)$$

(The tilde denotes a filtered signal.)

The output of the quadricorrelator is

$$\begin{aligned} v_d(t) &= K_3 v_Q \tilde{v}_I \\ &= -\tfrac{1}{2}K_3 K_m{}^2 V_s{}^2 A(\Delta f)\big[\sin\phi(\Delta f) \\ &\quad - \sin\left\{2\Delta\omega t + 2\theta_i - \phi(\Delta f)\right\}\big]. \end{aligned} \quad (6)$$

For any physically realizable transfer function $H_d(f)$, the amplitude $A(f)$ is an even function of frequency while the phase $\phi(f)$ is an odd function. Therefore, the dc component, proportional to $\sin\phi(\Delta f)$, reverses polarity as the difference frequency passes through zero, as is required for a frequency detector. This null at zero frequency difference occurs for any filter H_d whatsoever; it is a property of the quadricorrelator and not of the filter.

Frequency-difference information is provided by that portion of v_I that is rotated into phase with v_Q by the network $H_d(f)$. Since v_I and v_Q are generated $90°$ out of phase, the

most effective phase shift is $90°$. A perfect differentiator provides $90°$ phase shift at all frequencies.

Amplitude of the dc component is proportional to $\sin\phi$, whereas amplitude of the ripple is independent of ϕ. To obtain the largest possible dc component relative to ripple requires $\phi = 90°$.

There are numerous different networks that might be used for H_d; some examples include:

• a differentiator: $H_d(s) = sT_d$
• a high-pass filter: $H_d(s) = s^n H_L(s)$, where $H_L(s)$ is a low-pass filter
• a low-pass filter: $H_d(s) = H_L(s)$ (for example, [3], [8])
• a delay line: $H_d(s) = \exp\left(-sT_d\right)$ (for example, [3])
• a delay-differencing network (approximating a differentiator): $H_d(s) = 1 - \exp\left(-sT_d\right)$.

As pointed out in [8], if the phase shift of H_d becomes excessive, then $\sin\phi$ reverses sign and the FDD output has the wrong polarity for controlling frequency of the VCO. The frequency error would be increased instead of decreased by action of the tracking loop.

If phase shift exceeds $270°$, there will be one or more points of false lock where the loop comes to equilibrium at a frequency error other than zero. This phenomenon is similar to the false lock sometimes encountered in phase-locked loops [7, ch. 8].

A perfect differentiator will be assumed for H_d in the remainder of this analysis.

Balanced Quadricorrelator

Ripple may be cancelled by the balanced quadricorrelator of Fig. 3. This is a single-sideband cancellation scheme where the double-frequency component (the ripple) is cancelled and the zero-frequency components (the desired error signal) add together.

The balanced circuit, or variations thereon, have appeared previously in [8] (using low-pass filters instead of differentiators), [3], and [9], and has been mentioned but not pursued in [10].

Straightforward analysis shows that if the sinusoidal signal of (2) is applied to the balanced quadricorrelator of Fig. 3, the output voltage will be

$$v_d(t) = -\Delta\omega K_3 T_d K_m{}^2 V_s{}^2. \quad (7)$$

Ripple is gone. Moreover, the phase θ_i of the input does not appear in the output expression. Therefore, time-invariant input phases will be omitted from further consideration.

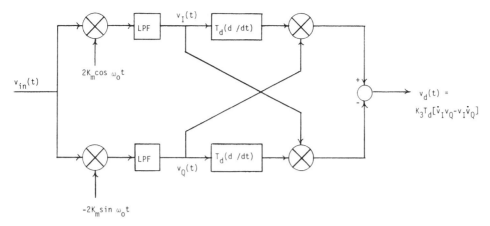

Fig. 3. Balanced quadricorrelator.

The remainder of this paper deals exclusively with balanced quadricorrelators.

III. RESPONSE TO SIGNALS AND NOISE

Bandpass Inputs

Let the input be

$$v_{\text{in}}(t) = x(t) \cos \omega_i t - y(t) \sin \omega_i t. \tag{8}$$

All bandpass signals can be reduced to this format. The coefficients $x(t)$ and $y(t)$ are real variables but no constraints are placed, yet, on their stationarity, cross correlation, or spectra.

To simplify the discussion, it will be assumed that the arm filters merely suppress sum-frequency components of the mixer outputs and do not affect the modulation components $x(t)$ and $y(t)$. Arm filter functions can be translated to a fictitious passband filter located before the mixers. Then $v_{\text{in}}(t)$ can be regarded as the output of the passband filter. This expedient permits filter effects to be included without cluttering the descriptions to follow.

If this bandpass signal is applied to the balanced quadricorrelator of Fig. 3, the output is found to be

$$v_d(t) = -K_3 T_d K_m{}^2 [\Delta\omega(x^2 + y^2) + x\dot{y} - y\dot{x}] \tag{9}$$

where the time argument has been suppressed for notational convenience. Some special examples give insight into the implications of this expression.

First, as a shorthand convenience, define

$$K_q = K_3 T_d K_m{}^2. \tag{10}$$

DSB-AM: Let $y(t) = 0$ for all t. Binary PSK is one possible example of such a signal. Then (9) reduces to

$$v_d(t) = -K_q x^2(t) \Delta\omega. \tag{11}$$

We recognize $x^2(t)$ as the squared envelope of the input signal (More generally, in (9) the squared envelope is $x^2 + y^2$.)

Mean value of the output for DSB-AM is

$$\text{Avg } [v_d] \triangleq V_d = -K_q \Delta\omega \sigma_x{}^2 \tag{12}$$

where $\sigma_x{}^2 \triangleq \text{Avg } [x^2(t)]$. (We write Avg [ ] instead of statistical expectation $E[\ ]$ to allow for time averaging as well as ensemble averaging.)

Two features are apparent from (11).

• There is no additive self-noise generated for any DSB-AM signal, no matter what the form of $x(t)$. (Self-noise is explained in [11].)

• Error voltage $v_d(t)$ is modulated by the signal envelope. When the FDD is used in a feedback frequency-tracking loop, the loop gain will fluctuate with fluctuations in the envelope.

Uncorrelated Channels: Let $E[x(t_1)y(t_2)] = 0$ for all t_1 and t_2. A band-limited signal with this property has a spectrum that is symmetric about ω_i in the passband. It can be shown that Avg $[x\dot{y}] = 0 = $ Avg $[y\dot{x}]$ and so the FDD mean output is

$$V_d = -K_q \Delta\omega(\sigma_x{}^2 + \sigma_y{}^2). \tag{13}$$

This equation has been derived previously in [10] under slightly more restrictive conditions.

PAM-QAM Signal: Let x and y be synchronous PAM data streams of the form

$$x(t) = \Sigma a_n g(t - nT)$$
$$y(t) = \Sigma b_n g(t - nT) \tag{14}$$

where $g(t)$ is a standard signaling pulse and the $\{a_n\}$ and $\{b_n\}$ are multilevel data sequences. Impose the special, but often-encountered conditions that

$$E[a_n] = 0 = E[b_n]$$
$$E[a_m a_n] = \sigma_a{}^2 \delta_{mn}$$
$$E[b_m b_n] = \sigma_b{}^2 \delta_{mn}$$
$$E[a_n b_m] = 0. \tag{15}$$

Output of the balanced quadricorrelator for such an input is

$$v_d(t) = -K_q \left[\Delta\omega \left\{ \sum_n \sum_m (a_n a_m + b_n b_m) \right. \right.$$
$$\left. \cdot g(t - nT)g(t - mT) \right\} \right]$$
$$- K_q \left[\sum_n \sum_m a_n b_m \{ g(t - nT)\dot{g}(t - mT) \right.$$
$$\left. - g(t - mT)\dot{g}(t - nT) \} \right]. \tag{16}$$

Taking statistical expectation and applying the conditions of (15) yields

$$E[v_d(t)] = -K_q \Delta\omega(\sigma_a{}^2 + \sigma_b{}^2) \sum_n g^2(t - nT). \tag{17}$$

Only the first bracketed term of (16) contributes useful average output; the second bracketed term has zero mean and only contributes pattern (or self) noise [11].

Expectation of the first term, as given by (17), is nonstationary—in fact, periodic in T. That is not surprising inasmuch as the input signal (8) and (14) is cyclostationary [12], [13]. Useful output of the circuit is the dc value, which is found by time-averaging (17) over one period, to obtain

$$V_d = -K_q \Delta\omega(\sigma_a{}^2 + \sigma_b{}^2)(1/T) \sum_n \int_0^T g^2(t - nT)\,dt$$

$$= -K_q \Delta\omega(\sigma_a{}^2 + \sigma_b{}^2)(1/T) \int_{-\infty}^{\infty} g^2(t)\,dt. \quad (18)$$

But the integral is just the energy of the pulse $g(t)$ (dissipated in a 1 Ω resistor):

$$E_g = \int_{-\infty}^{\infty} g^2(t)\,dt \quad (19)$$

so the useful dc output of the balanced quadricorrelator is

$$V_d = -K_q \Delta\omega(\sigma_a{}^2 + \sigma_b{}^2)E_g/T. \quad (20)$$

Consider pattern noise more closely: in particular, the difference of products contained in braces in the second term of (16). For $n = m$, that difference is zero; the double summation can contribute pattern noise only for $m \neq n$.

If the signaling pulse $g(t)$ is time limited to a single interval T (that is, $g(t) = 0$, $t < 0$, and $t > T$) then there is no pattern noise whatever. Pattern noise can arise only if pulses overlap.

The cancellation that is noted arises because of the balanced circuit; an unbalanced quadricorrelator (e.g., Fig. 2) not only does not afford the pattern-noise cancellation, but also contains terms that are products of pattern noise and ripple. This is just one more reason to employ a balanced circuit.

Gaussian Noise

Let the input $v_{in}(t)$ be bandpass Gaussian noise with two-sided spectral density $S_n(f)$. Bandpass Gaussian noise can be expanded in quadrature components about any arbitrary frequency ω_i to obtain

$$v_{in}(t) = n_c(t) \cos \omega_i t - n_s(t) \sin \omega_i t. \quad (21)$$

There is no implication that the spectrum is centered on ω_i or is in any way symmetric about it. Furthermore, there is no requirement that the spectrum $S_n(f)$ be symmetric about any frequency other than zero.

Adapting (8) and (9), the output of the balanced quadricorrelator, for noise input, is found to be

$$v_d(t) = -K_q[\Delta\omega(n_c{}^2 + n_s{}^2) + \dot{n}_s n_c - \dot{n}_c n_s]. \quad (22)$$

Taking statistical expectations, the average output voltage is

$$V_d = E[v_d(t)] = -K_q[\Delta\omega E(n_c{}^2 + n_s{}^2) + E(\dot{n}_s n_c - \dot{n}_c n_s)]$$

$$= -K_q[2\Delta\omega\sigma_n{}^2 + E(\dot{n}_s n_c - \dot{n}_c n_s)] \quad (22a)$$

where $\sigma_n{}^2$ is the variance of the input noise $v_{in}(t)$. To proceed further we must evaluate $E(\dot{n}_s n_c - \dot{n}_c n_s)$.

To that end we follow the approach described in [14, sect. 6-4, 8-5]. Consider a finite segment of the noise input of duration T_0. Expand $n_c(t)$ and $n_s(t)$ in Fourier series over

this interval to obtain expressions of the form

$$n_{c0}(t) = \sum_{k=1}^{\infty} [x_{ck} \cos(2\pi k/T_0 - \omega_i)t$$
$$+ x_{sk} \sin(2\pi k/T_0 - \omega_i)t]$$

$$n_{s0}(t) = \sum_{k=1}^{\infty} [x_{ck} \sin(2\pi k/T_0 - \omega_i)t$$
$$- x_{sk} \cos(2\pi k/T_0 - \omega_i)t] \quad (23)$$

where the subscripts "0" in n_{c0} and n_{s0} indicate that the series are valid only in the finite interval T_0.

These expressions are differentiated and substituted into (22). The statistical-expectation operator is next formally applied and the interval T_0 is caused to grow towards infinity while k is constrained so that $k/T_0 = f_k$ holds constant. In the limit, the cross correlations all vanish and the autocorrelations become delta functions, viz.,

$$E[x_{ck}x_{sn}] = 0$$
$$E[x_{ck}x_{cn}] = \delta_{kn}E[x_{ck}{}^2]$$
$$E[x_{sk}x_{sn}] = \delta_{kn}E[x_{sk}{}^2]. \quad (24)$$

Therefore, following [14], the expectations of the cross products are

$$E[n_c \dot{n}_s] = -E[\dot{n}_c n_s] = 2\int_0^{\infty} (\omega - \omega_i)S_n(f)\,df \quad (25)$$

wherefore the dc output is

$$V_d = -K_q\left[2\Delta\omega\sigma_n{}^2 + 4\int_0^{\infty} (\omega - \omega_i)S_n(f)\,df\right]. \quad (22b)$$

But

$$(\omega - \omega_i) = (\omega - \omega_0) + (\omega_0 - \omega_i) = (\omega - \omega_0) - \Delta\omega$$

so

$$4\int_0^{\infty} (\omega - \omega_i)S_n(f)\,df = 4\int_0^{\infty} (\omega - \omega_0)S_n(f)\,df$$
$$- 2\Delta\omega\sigma_n{}^2$$

since

$$\sigma_n{}^2 = 2\int_0^{\infty} S_n(f)\,df.$$

Therefore, the dc output of the balanced quadricorrelator is

$$V_d = -8\pi K_q \int_0^{\infty} (f - f_0)S_n(f)\,df \quad (26)$$

irrespective of the arbitrary input frequency ω_i.

In other words, the average error signal generated by the FDD is proportional to the first moment about the local reference frequency f_0 of the spectrum $S_n(f)$ of the input $v_{in}(t)$. The error signal will be zero only if the center-of-gravity (c.g.) of the input spectrum coincides with the local reference frequency. In that sense the frequency-tracking loop tracks the center-of-gravity of the input spectrum.

Conjecture: Inasmuch as the FDD is incoherent, it seems plausible that it tracks the c.g. of *any* input spectrum, not just that of bandpass Gaussian noise.

Noise Bias

This result has important consequences for circuit design. In a coherent phase-locked loop, additive noise causes phase jitter in the loop, but no bias is generated. However, (26) shows clearly that noise can generate a bias in a frequency-tracking loop, in addition to the fluctuations in tracking. An equipment designer must be able to predict the bias and often wants to be able to avoid it entirely.

Bias will be generated if the c.g. of the noise spectrum does not coincide with the local reference frequency f_0. In many receivers, the noise spectrum is shaped by passing white noise through bandpass filters in intermediate-frequency portions of the receiver. To avoid noise bias, it is sufficient to make those filters symmetric and center them on the reference frequency f_0.

There are two ways to close a tracking loop: a short-loop connection or a long loop. These options are illustrated in Fig. 4.

The bandpass filter (BPF) in the IF portion of the receiver has a fixed characteristic. If the frequency f_0 is allowed to vary, as is necessary in a short loop, then the filter center—and therefore the c.g. of the noise spectrum—cannot coincide with f_0 except by rare accident. In general, a noise bias in tracking must be anticipated whenever a short loop is employed.

By contrast, in a long loop, both f_0 and the filter center-frequency are fixed. Tracking is accomplished by controlling the frequency of an oscillator that precedes the reference source. If the filter and the reference source are properly aligned and stable, then the c.g. of the noise spectrum will always coincide with the reference frequency and no noise bias will be generated.

Signal Plus Noise

In this section, analysis of the response of the balanced quadricorrelator to signal-plus-noise is presented. The results are in the form of the spectrum of the noise output of the FDD. These results can then be applied to a frequency-tracking loop to calculate the fluctuations of tracking error.

Previous workers have analyzed signal-plus-noise from differing standpoints. Pickard [15] dealt with a simple quadricorrelator that had hard limiters in each arm. He performed his analyses entirely in the time domain and did not derive output spectra. Pawula [16] extended Pickard's work, with the same circuit; his paper has references to other predecessors.

Park [9] treated the balanced quadricorrelator and derived output spectra. He was most interested in the use of the FDD as an FM demodulator, whereas the emphasis here is on frequency tracking. A portion of Park's article deals with the balanced quadricorrelator without any limiters—the circuit of greatest interest here. He states that the FM clicks that trouble a conventional FM discriminator below threshold do not arise if there is no limiting. Also, he concludes that the limiterless circuit has better output signal-to-noise ratio if input SNR is very small. (At large input SNR's, the circuit with limiter will be superior because the limiter suppresses AM noise. It is not possible to achieve the FM noise advantage unless the AM noise is somehow removed.)

Cahn [3] performed an analysis of the extra phase fluctuation introduced by the presence of the frequency tracker in a combined phase–frequency tracking loop.

In this analysis we shall be concerned only with fluctuations in a frequency tracking loop. To that end, the input to the FDD will be assumed to be a pure sine wave plus bandpass Gaussian noise.

$$v_{\text{in}}(t) = V_s \cos \omega_i t + n(t). \tag{27}$$

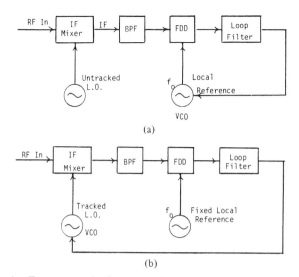

Fig. 4. Frequency-tracker loop connections. (a) Short loop. (b) Long loop.

Obviously, a data signal is not a pure sine wave so the assumed signal is a simplification of reality. The simplification affords better visibility into the operation of the circuit while, it is hoped, providing a useful approximation of performance with real signals.

The noise has a spectrum of $S_n(f)$ (two-sided), which is unrestricted other than being required to be bandpass. (Strictly speaking, the spectrum should be sufficiently band limited that foldover problems do not arise in the mixers of the quadricorrelator.)

The input may be resolved into

$$v_{\text{in}}(t) = (V_s + n_c) \cos \omega_i t - n_s \sin \omega_i t \tag{28}$$

where the time dependence of $n_c(t)$ and $n_s(t)$ has been suppressed for compactness of notation.

Assume that the arm filters remove the double-frequency mixer products, but that the filters are broad enough not to affect n_c or n_s. That is contrary to standard practice but convenient for analysis. The assumption will be removed later.

Under these conditions the filtered outputs of the mixers are

$$v_I(t) = K_m [(V_s + n_c) \cos \Delta\omega t - n_s \sin \Delta\omega t]$$
$$v_Q(t) = K_m [(V_s + n_c) \sin \Delta\omega t + n_s \cos \Delta\omega t]. \tag{29}$$

Performing the balanced-quadricorrelator operations of differentiation, multiplication, and subtraction on these arm voltages gives

$$v_d(t) = -K_q [\Delta\omega(V_s^2 + 2V_s n_c + n_c^2 + n_s^2)$$
$$+ V_s \dot{n}_s + n_c \dot{n}_s - \dot{n}_c n_s]. \tag{30}$$

(This result is the remnant after combining 30 terms of the function $\dot{v}_I v_Q - v_I \dot{v}_Q$. Output of the simple unbalanced quadricorrelator is half of the above, plus 13 distinct ripple terms.)

Average output is

$$V_d = E[v_d] = -K_q \Delta\omega V_s^2$$
$$-K_q [2\sigma_n^2 \Delta\omega + E(n_c \dot{n}_s - \dot{n}_c n_s)]. \tag{31}$$

Ordinary statistical expectations (instead of double-averaging as for a QAM input) suffice because the noise is assumed to be stationary. The terms $n_c, n_s, \dot{n}_c, \dot{n}_s$ are all zero-mean by the bandpass assumption. Variance of the noise input is σ_n^2.

The terms containing $n_c{}^2$ and $n_s{}^2$ would superficially appear to contribute to the useful dc output of the FDD. That conclusion is contrary to common sense; the terms should be regarded as an artifact of the particular selection of noise representation. The last term $-E(n_c\dot{n}_s - \dot{n}_c n_s)$—generates a component to cancel the anomaly, as has been demonstrated in (21)–(26).

As is characteristic of quadratic devices, the output in (30) is composed of $S \times S$, $S \times N$, and $N \times N$ terms [14, ch. 12]. Desired frequency-difference information is contained in the $S \times S$ term, which is independent of the noise. There is no signal suppression effect at low SNR, as would arise in a circuit containing a limiter.

Some portion of the noise in (30) is proportional to the frequency difference $\Delta\omega$. Total noise decreases as the frequency error is reduced to zero. This dependence of output noise on frequency deviation has been noted earlier [6], [9].

Frequency error in a successful frequency tracker should be very small so the error-proportional noise should also be small. In the sequel we neglect the error-proportional noise and concentrate on the last three terms of (30). Our objective is to determine their spectrum; to that end we first obtain their autocorrelation

$$E[v_d(t_1)v_d(t_2)]$$

$$= K_q{}^2(V_s{}^2 E[\dot{n}_{s1}\dot{n}_{s2}] + V_s(E[\dot{n}_{s1}\dot{n}_{s2}n_{c2}] - E[\dot{n}_{s1}\dot{n}_{c2}n_{s2}]$$

$$+ E[n_{c1}\dot{n}_{s1}\dot{n}_{s2}] - E[\dot{n}_{c1}n_{c2}\dot{n}_{s2}]) + E[n_{c1}\dot{n}_{s1}n_{c2}\dot{n}_{s2}]$$

$$- E[n_{c1}\dot{n}_{s1}\dot{n}_{c2}n_{s2}] - E[\dot{n}_{c1}n_{s1}n_{c2}\dot{n}_{s2}]$$

$$+ E[\dot{n}_{c1}n_{s1}\dot{n}_{c2}n_{s2}]).\qquad(32)$$

The subscripts 1 and 2 refer to times t_1 and t_2.

Next, the noise functions are truncated to finite time segments and expanded in Fourier series, as in (23). These functions are substituted into (32), like terms are combined, and the time segment is allowed to approach infinite extent. The details are extremely space consuming, so they are omitted. Underlying principles are identical to those of [14, sect. 6-4, 8-5].

Ultimately, it is found that the expectations of the triple products [second line of (32)] are all zero for Gaussian noise and that the autocorrelation of the remaining terms is given by

$$E[v_{d1}v_{d2}]/K_q{}^2$$

$$= 2V_s{}^2 \int_0^\infty (\omega_k - \omega_0)^2 S_n(f_k) \cos(\omega_k - \omega_0)\tau\, df_k$$

$$+ 8 \iint_0^\infty (\omega_p - \omega_0)^2 S_n(f_k)S_n(f_p)$$

$$\cdot \cos(\omega_k - \omega_p)\tau\, df_k\, df_p$$

$$+ 8 \iint_0^\infty (\omega_p - \omega_0)(\omega_k - \omega_0)S_n(f_k)S_n(f_p)$$

$$\cdot \cos(\omega_k - \omega_p)\tau\, df_k\, df_p + 4 \iint_0^\infty (\omega_m - \omega_0)$$

$$\cdot (\omega_k - \omega_0)S_n(f_m)S_n(f_k)\, df_k\, df_m\qquad(33)$$

where f_k, f_m, f_p are dummy frequency variables and $\tau = t_2 - t_1$.

The last step is to take the Fourier transform of (33) to find the spectrum of the noise on $v_d(t)$. After additional labor, the two-sided spectrum is found to be

$$S_{vd}(f)/K_q{}^2$$

$$= 2V_s{}^2(2\pi f)^2 [S_n(f_0 - f) + S_n(f_0 + f)]$$

$$+ 8 \int_0^\infty (\omega_p - \omega_0)^2 S_n(f_p)$$

$$\cdot [S_n(f_p - f) + S_n(f_p + f)]\, df_p$$

$$+ 8 \int_0^\infty (\omega_p - \omega_0)S_n(f_p)[(\omega_p - \omega_0 - 2\pi f)$$

$$\cdot S_n(f_p - f) + (\omega_p - \omega_0 + 2\pi f)S_n(f_p + f)]\, df_p$$

$$+ 8\delta(f)\left[\int_0^\infty (\omega_p - \omega_0)S_n(f_p)\, df_p\right]^2.\qquad(34)$$

The first term in the spectrum arises from $S \times N$; the other terms are all $N \times N$. The last term is a dc component that is zero only if the center of gravity of the noise spectrum coincides with f_0.

The broad-band assumption on the arm filters can now be removed. Low-pass arm filters are equivalent to a bandpass filter placed in front of the mixers; this equivalent bandpass filter is always symmetric and always centered on f_0. To take account of an arm filter, simply translate it into the equivalent bandpass IF filter. The noise spectrum $S_n(f)$ then becomes the actual input noise spectrum (often effectively white), filtered by any actual bandpass filters in the receiver and by the equivalent bandpass filter corresponding to the arm filters.

Spectrum Examples

Two particular spectra were investigated in further detail. They are given by

$$S_n(f_a) = N_0/2, \qquad |f_a - f_0| < W/2$$
$$= 0, \qquad |f_a - f_0| > W/2 \qquad(35)$$

and

$$S_n(f_b) = \frac{N_0}{2}\, \frac{\sin^2 \pi(f_b - f_0)/B}{[\pi(f_b - f_0)/B]^2}\, .\qquad(36)$$

The first is a rectangular spectrum of width W, centered at f_0. The second is the spectrum that would be imposed by an integrate-and-dump arm filter on white noise. These spectra are related in the sense that if $W = B$, then the two spectra contain equal noise powers.

Noise spectra at the FDD output are found by substituting into (34) and evaluating the integrals. The results are

$$S_{vd}(f)$$

$$= \begin{cases} 2V_s{}^2(2\pi f)^2 K_q{}^2 N_0 & (|f| < W/2) \\ 0 & (|f| > W/2) \end{cases}$$

$$+ \begin{cases} \dfrac{8}{3}\pi^2 N_0{}^2 W^3(1 - |f|/W)^3 K_q{}^2 & (|f| < W) \\ 0 & (|f| > W)\qquad(37) \end{cases}$$

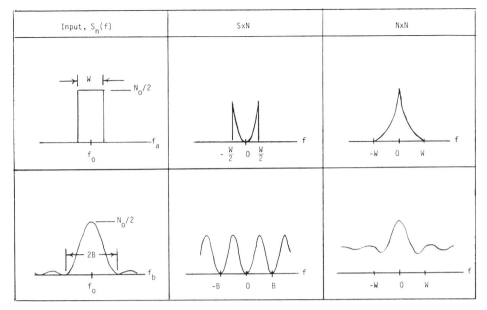

Input, $S_n(f)$	SxN	NxN

Fig. 5. Spectrum examples.

for the rectangular spectrum and

$$S_{vd}(f) = 8K_q{}^2 V_s{}^2 B^2 N_0 \sin^2 (\pi f/B)$$

$$+ 8K_q{}^2 B^3 N_0{}^2 \left(1 + \frac{\sin 2\pi f/B}{2\pi f/B}\right) \qquad (38)$$

for the integrate and dump spectrum. Since both input spectra are symmetric about f_0 there is zero dc component present.

These are two-sided spectra, valid for both positive and negative values of f. Each spectrum consists of two portions: one arises from $S \times N$ and the other from $N \times N$. The $N \times N$ portions are quite different for the two different input spectra, but the $S \times N$ portions are the same for small magnitudes of f. To see that they are the same, approximate $\sin^2 (\pi f/B)$ by $(\pi f/B)^2$, which makes the first term of (38) equal to the first term of (37).

At low frequencies at least, the $S \times N$ term is the familiar quadratic-spectrum noise that appears at the output of all FM discriminators.

The various spectra are sketched in Fig. 5. They have been drawn with $B = W$ for ready comparison between the two input shapes. Vertical scales of the output spectral components are not significant.

It is apparent that the $N \times N$ noise has its spectrum concentrated near zero frequency. A narrow-band tracking loop is likely to be more troubled by $N \times N$ noise than by $S \times N$, even for fairly large signal-to-noise ratios.

Modulated Signals

These results have all been obtained for a pure sinusoidal signal. A data signal is not a sinusoid so the noise spectra resulting from a data signal plus noise will be altered from the results shown here. The more difficult analysis needed to accommodate the modulation sidebands of the data signal will not be pursued here. Nonetheless, it is useful to speculate on the noise spectrum that would arise if a data-modulated signal replaced the sinusoid. Some possibilities include the following.

• Pattern-noise components might appear, as discussed earlier.

• The $N \times N$ output components are likely to remain the same inasmuch as they do not depend upon the signal.

• The $S \times S$ components will be those already discussed for QAM signals.

• The $S \times N$ components will be modified substantially. There are two effects that might appear.

i) The $S \times N$ components derived for the sinusoid signal are likely to be spread in frequency by convolving noise against the spread spectrum of the data signal.

ii) Additional $S \times N$ components may be generated (e.g., [9]).

In the absence of better analysis of the noisy data-signal input, an engineering approximation should provide useful guidance to the equipment designer. In two parts, the approximation is as follows.

1) Calculate useful dc error-signal output (thus obtaining the gain of the FDD) and pattern noise from the QAM analysis presented in a previous section.

2) Calculate output noise caused by the additive input noise according to the sinusoidal-signal analysis of the immediately preceding sections.

This approximation is valid only if the $N \times N$ noise is indeed dominant.

IV. CONCLUSIONS

This article has examined quadricorrelators used as frequency-difference detectors (FDD). The FDD develops an error signal that is proportional to the frequency difference between an incoming signal and a local reference oscillator.

The simplest quadricorrelator consists of a differentiator (or other, less effective network), an I–Q demodulator, and a baseband multiplier. It suffers from ripple in its output.

A balanced quadricorrelator requires an additional multiplier and differentiator; it cancels ripple and is easier to analyze.

It has been demonstrated that the loop will track the center-of-gravity of the spectrum of certain classes of inputs. It is conjectured that the loop will track the c.g. of any bandpass input spectrum.

A quadricorrelator can be used for frequency tracking of a PAM-QAM data signal. Pattern noise may arise if the data pulses overlap and the modulation is two-dimensional.

If the input noise spectrum is not symmetric on the center frequency of the data signal, the FDD (or any other frequency detector) will develop a noise bias. To avoid bias, use a long loop for tracking and a bandpass filter that is symmetric about the final demodulation frequency f_0.

Noise spectra of the FDD output have been ascertained for a sinusoidal signal plus bandpass, Gaussian noise. These output spectra depend upon filter shaping in the receiver. Two important examples of shaping are given; most practical data-filter shapes will lie between the two examples.

Both $S \times N$ and $N \times N$ noise components are generated. The $N \times N$ component is likely to dominate in a frequency tracker because the $S \times N$ spectrum goes to zero at zero frequency.

REFERENCES

[1] C. Travis, "Automatic frequency control," *Proc. IRE,* vol. 23, p. 1125, Oct. 1935.

[2] G. H. Nibbe, F. E. Towsley, and E. Durand, "AFC systems and circuits," in *Microwave Receivers,* S. N. van Voorhis, Ed. M.I.T. Radiation Lab. Series, vol. 23, ch. 3.

[3] C. R. Cahn, "Improving frequency acquisition of a Costas loop," *IEEE Trans. Commun.,* vol. COM-25, pp. 1453–1459, Dec. 1977.

[4] F. D. Natali, "AFC tracking algorithms for satellite links," in *Conf. Rec., Int. Conf. Commun.,* Boston, MA, 1983.

[5] C. F. Shaeffer, "The zero-beat method of frequency discrimination," *Proc. IRE,* vol. 30, pp. 365–367, Aug. 1942.

[6] D. Richman, "Color carrier reference phase synchronization accuracy in NTSC color television," *Proc. IRE,* vol. 42, pp. 106–133, Jan. 1954.

[7] F. M. Gardner, *Phaselock Techniques,* 2nd ed. New York: Wiley, 1979.

[8] R. W. D. Booth, "A note on the design of baseband AFC discriminators," in *Conf. Rec. Nat. Telecommun. Conf.,* Houston, TX, 1980, vol. 2, paper 24.2.

[9] J. H. Park, Jr., "An FM detector for low *S/N*," *IEEE Trans. Commun.,* vol. COM-18, pp. 110–118, Apr. 1970.

[10] D. G. Messerschmitt, "Frequency detectors for PLL acquisition in timing and carrier recovery," *IEEE Trans. Commun.,* vol. COM-27, pp. 1288–1295, Sept. 1979.

[11] F. M. Gardner, "Self-noise in synchronizers," *IEEE Trans. Commun.,* vol. COM-28, pp. 1159–1163, Aug. 1980.

[12] W. R. Bennett, "Statistics of regenerative digital transmission," *Bell Syst. Tech. J.,* vol. 37, pp. 1501–1542, Nov. 1958.

[13] W. A. Gardner and L. E. Franks, "Characterization of cyclostationary random signal processes," *IEEE Trans. Inform. Theory,* vol. IT-21, pp. 4–14, Jan. 1975.

[14] W. B. Davenport and W. L. Root, *Random Signals and Noise.* New York: McGraw-Hill, 1958.

[15] T. B. Pickard, "The effect of noise on a method of frequency measurement," *IRE Trans. Inform Theory,* vol. IT-4, pp. 83–88, June 1958.

[16] R. B. Pawula, "Analysis of an estimator of the center frequency of a power spectrum," *IEEE Trans. Inform. Theory,* vol. IT-14, pp. 669–676, Sept. 1978.

Frequency Detectors for PLL Acquisition in Timing and Carrier Recovery

DAVID G. MESSERSCHMITT, SENIOR MEMBER, IEEE

Abstract—A significant problem in phase-locked loop (PLL) timing and carrier extraction is the initial acquisition. Very narrow loop bandwidths are generally required to control phase jitter, and acquisition may depend on an extremely accurate initial VCO frequency (VCXO) or sweeping. We describe two simply implemented frequency detectors which, when added to the traditional phase detector, can effect acquisition even with very small loop bandwidths and large initial frequency offsets.

The first is the quadricorrelator, previously applied to timing recovery by Bellisio, while the second is new, and called a rotational frequency detector. The latter, while limited to lower frequencies and higher signal-to-noise ratios, is suitable for many applications and can be implemented with simpler circuitry.

1.0. INTRODUCTION

THE initial acquisition of a phase-locked loop (PLL) when used for timing or carrier extraction is a significant practical problem, since the narrow loop bandwidth generally required for jitter requirements severely restricts the pull-in range. Methods widely employed to effect acquisition include [1]

a) compromises in loop filter design,
b) highly accurate initial VCO frequency (VCXO),
c) sweeping of the VCO, and
d) in-lock detection with switching of loop filter.

In many instances, as in carrier recovery, several of these methods may be simultaneously employed.

There is a fifth method of effecting acquisition [1], which seems to have been first suggested by Richman [2], and that is to add a frequency detector (FD) to the traditional PLL phase detector (PD) in the manner of Figure 1. With a large initial VCO frequency offset, the PD output has essentially a zero d.c. output, and the FD generates a voltage proportional to the frequency difference between input and VCO, driving that difference to zero. The PD takes over when the frequency difference is small, completing the acquisition. When the PLL is in-lock, the FD output will have at the least zero mean, and optimistically will be identically zero, automatically allowing the PD and its loop filter to govern the loop dynamics. The beauty of this approach is that a crystal controlled VCO (VCXO) can often be exchanged for the additional FD circuitry in timing recovery applications, an advantageous tradeoff in this age of integrated circuitry. In carrier recover, a VCXO is

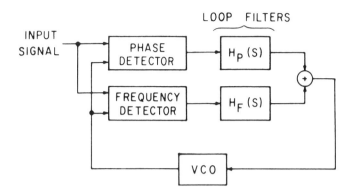

Fig. 1. PLL with Phase and Frequency Detector.

often still required because of the problem of false locking to a data sideband, but the sometimes troublesome in-lock detector and/or sweeping circuitry can be eliminated and the PD loop filter can be designed virtually independently of acquisition considerations, removing a significant burden from the designer.

This paper will discuss two specific FD's, each of which is applicable to both timing and carrier recovery. The first is the quadricorrelator described by Richman [2], which was more recently rediscovered by Pickard [3-5], Bellisio [6], and in modified form by Park [7], Cahn [8] and Citta [14]. These authors have discussed its applicability to sinusoid [2, 3] and narrowband Gaussian process [2-4] input signals, to timing recovery [6], and to Costas loop carrier recovery for biphase modulation [8]. We will show here that the quadricorrelator is more generally applicable to carrier recovery for any modulation method which has a power spectrum symmetrical about the carrier frequency. This includes most data modulation methods, with the notable exceptions of single and vestigal sideband modulation.

The second FD, called a rotational FD, is new, and unlike the quadricorrelator is implemented with predominately digital circuitry. As a consequence, its operation is limited to lower frequencies, but where applicable it is more amenable to integrated circuitry realization because of the elimination of multipliers and filtering functions. Its operation depends on detecting, with simple circuitry, the direction of rotation of the signal constellation.

For completeness we mention the papers by Oberst [9], describing an FD for two square waves (useful in frequency synthesis *), and Runge [10], describing an unrelated FD for timing and carrier recovery applications.

Paper approved by the Editor for Data Communication Systems of the IEEE Communications Society for publication without oral presentation. Manuscript received October 23, 1978; revised May 4, 1979. This research was performed for the VIDAR Division of TRW, Mountain View, CA.

The author is with the Department of Electrical Engineering and Computer Science, University of California, Berkeley, CA 94720.

* The FD's described here can be used for two sinusoids or square waves, but appear to have greater complexity than Oberst's circuits.

Reprinted from *IEEE Trans. Comm.*, vol. COM-27, pp. 1288-1295, September 1979.

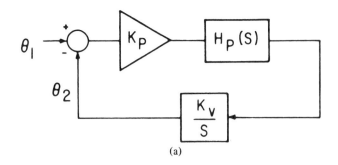

(a)

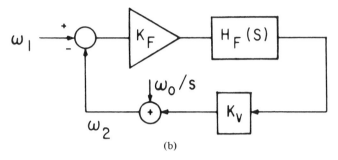

(b)

Fig. 2.   Linearized Loop Models.

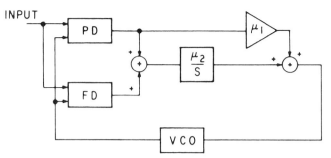

Fig. 3.   Choice of Loop Filters.

While the primary purpose of this paper is to describe and analyze the FD techniques, we first discuss in Section 2.0 the choice of loop filters. Then in Section 3.0 we focus on the quadricorrelator and rotational FD, and describe experimental results in Section 4.0.

2.0. LOOP FILTERS

Assuming the input signal to the PLL of Figure 1 is of the form $\sin(\omega_1 t + \theta_1)$, the VCO output is $\cos(\omega_2 t + \theta_2)$, and the PD and FD are both linear, the linearized models of Figure 2 result. The phase-locked loop of Figure 2(a) governs after lock has been achieved, while the frequency locked loop of Figure 2(b) governs the acquisition behavior. The design parameters are the PD and FD constant K_p and K_f, the VCO constant K_v, and the VCO free-running frequency ω_0.

The loop dynamics are governed by the standard closed loop phase transfer function

$$\frac{\theta_2(s)}{\theta_1(s)} = \frac{K_p K_v H_p(s)}{s + K_p K_v H_p(s)} \tag{2.1}$$

plus a transfer function governing acquistion

$$\omega_1(s) - \omega_2(s) = \frac{\omega_1(s) - \omega_0/s}{1 + K_f K_v H_f(s)}. \tag{2.2}$$

Bellisio [6] recommends a proportional plus integral PD loop filter,

$$H_p(s) = \mu_1 + \frac{\mu_2}{s} \tag{2.3}$$

which is a good choice since the static phase error is small [1] and the usual concern with the integrator being initially

saturated is alleviated due to the action of the FD. He also recommends that the FD use the same loop filter (that is, the summer in Figure 1 be placed in front of a single loop filter of type (2.3)). This latter choice is shown to be disadvantageous when we calculate the time response due to a step frequency change $\omega_1(s) = \omega_1/s$ from (2.2),

$$\omega_1(t) - \omega_2(t) = \frac{\omega_1 - \omega_0}{1 + \mu_1 K_f K_v} e^{-t/\tau} \tag{2.4}$$

where the time constant is

$$\tau = \frac{\mu_1}{\mu_2} + \frac{1}{\mu_2 K_f K_v}. \tag{2.5}$$

Thus, we see that, as expected, fastest acquisition occurs for K_f large, but the time constant is limited to $\tau = \mu_1/\mu_2$. Physically, this limitation on speed of acquisition is due to the proportional part of the filter, which initially reduces the frequency error and slows the charging of the integrator. The solution is to eliminate the proportional filter

$$H_f(s) = \frac{\mu_2}{s} \tag{2.6}$$

resulting in the configuration of Figure 3. The FD charges the integrator capacitor to the correct voltage to reduce the frequency error to zero (in spite of any initial saturation), and in-lock the PD maintains that charge. While (2.4) predicts that increasing K_f can result in arbitrarily fast acquisition, in practice the fact that the FD output will have a randomly fluctuating voltage on its output in-lock places a practical limit on the size of K_f.

3.0. SPECIFIC FD DESIGNS

3.1. Quadricorrelator Frequency Detector

The quadricorrelator, as shown in Figure 4, consists of two quadrature mixers, a differentiator in the in-phase channel, and a cross-correlator.** The mean value of $p(t)$ is proportional to the difference between the center frequency of the power spectrum of $r(t)$ and ω_2. While this property has been demonstrated for sinusoidal [2, 3] and Gaussian [3-5] inputs $r(t)$, it can be easily established in general. In particular, if

---

** The similarity of the quadricorrelator to the PD of a Costas loop [11] is striking.

108

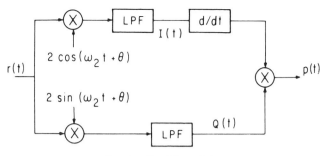

Fig. 4. Quadricorrelator.

$r(t)$ has a power spectrum symmetric about the radian frequency ω_1, it can be expanded in the form

$$r(t) = x_c(t) \cos \omega_1 t - x_s(t) \sin \omega_1 t \qquad (3.1)$$

where $x_c(t)$ and $x_s(t)$ are uncorrelated. It is shown in Appendix A, using (3.1), that

$$Ep(t) = \Delta\omega\sigma_r{}^2 \qquad (3.2)$$

where $\Delta\omega$ is the radian frequency difference,

$$\Delta\omega = \omega_1 - \omega_2 \qquad (3.3)$$

and $\sigma_r{}^2$ is the variance of $r(t)$. Thus, $p(t)$ is an unbiased estimate of $\Delta\omega$ when properly scaled by $\sigma_r{}^{-2}$.

Many data transmission modulation methods have a signal power spectrum symmetric about the carrier frequency, the most important examples being PSK and QAM [11]. The quadricorrelator is thus a suitable FD for carrier recovery with these modulation methods. For the particular case of biphase modulation, Cahn [8] has suggested a FD structure similar to the quadricorrelator, except that it includes an additional $I(dQ/dt)$ term. From the foregoing, it is evident that the simpler quadricorrelator would suffice. Bellisio [6] applied the quadricorrelator to baseband PAM timing recovery, exploiting the symmetry about the baud frequency of the pulse waveform spectrum generated by a NRZ data transition detector (differentiator followed by dead-zone quantizer).

Finally, we mention that many authors include limiters in both I and Q channels. This simplifies implementation of the correlation multiplier, which must have a very small offset to control static phase error, as well as insures a zero FD output after acquisition and eliminates the $\sigma_r{}^2$ dependence in (3.2).

3.2. Rotational Frequency Detector

The rotational FD, in contrast to the quadricorrelator, is constructed of predominately digital circuitry and includes no filtering functions. Consequently, it is particularly well suited to integrated circuit implementation, but is also inherently limited to lower frequency operation than the quadricorrelator.

The rotational FD is simplest to describe for measurement of the frequency difference between two square waves, although it offers no particular advantage for that application over circuits described by Oberst [9]. That description is given in Section 3.3, and the simple generalizations to timing and

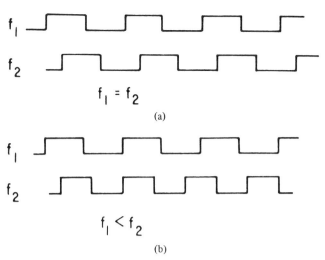

Fig. 5. Situations to be Distinguished by FD.

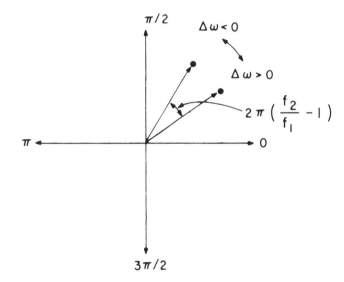

Fig. 6. Phasor Diagram of Two Successive Transitions of f_1 Relative to Phase of f_2.

carrier recovery are described in Section 3.4 and 3.5. The effect of noise and phase jitter is analyzed in Section 3.6.

3.3. Two Square Waves

Two of the three cases to be distinguished by the FD are shown in Figure 5. These cases would easily be recognized by a human observer watching the waveforms on an oscilloscope. When $f_1 = f_2$, the transitions of f_1 maintain a fixed relationship to those of f_2. When $f_1 < f_2$, the transitions of f_1 advance in phase relative to those of f_2, and vice versa when $f_1 > f_2$. An excellent way to view the situation is to draw a phasor diagram as in Figure 6. One cycle (2π radians) of f_2 is shown and the two phasors represent the relative phase of two successive transitions of f_1. The angle of rotation is readily shown to be $2\pi ((f_2/f_1) - 1)$, which is counterclockwise if $f_1 < f_2$ and clockwise if $f_1 > f_2$. Hence detecting the sign of the frequency difference is equivalent to determining the direction of rotation in Figure 6, while the magnitude of the frequency difference is related to the angle of rotation.

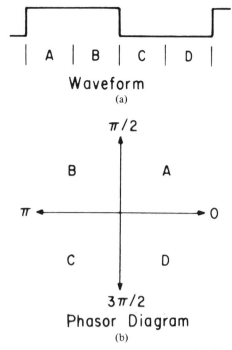

Waveform
(a)

Phasor Diagram
(b)

Fig. 7. Division of VCO Cycle into Four Quadrants.

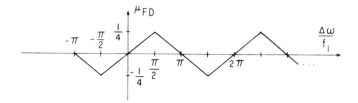

Fig. 8. FD Characteristic for Two Square Waves.

A circuit which detects the direction of rotation can be built as follows: Assume f_2 is the VCO frequency, and divide each cycle into four quadrants labeled A, B, C, and D as in Figures 7(a) and (b). This can be accomplished by actually running the VCO at four times frequency f_2, and dividing by four to obtain f_2 itself. Further assume that the PD is designed so that in-lock the PLL will maintain the positive transition of f_1 in the vicinity of the positive transitions of f_2 (other PD designs can be handled in like manner). Therefore, in-lock we would expect to observe positive transitions of f_1 predominately or exclusively in quadrants A and D. To ensure that the FD will produce an output rarely if ever in-lock, it will operate only upon the observation of positive transitions of f_1 in quadrants B and C.

Let the kth cycle of f_2 be denoted by a k-subscript. The situation $f_1 > f_2$ can be recognized by observation of C_k followed by B_{k+1}, in which case the FD generates a positive pulse. Similarly, if a B_k is followed by C_{k+1}, the FD generates a negative pulse, in recognition that $f_1 < f_2$.

The FD does not generate a pulse for every pair of f_1 transitions, since the rather special conditions of the last paragraph must be met. In particular, they will hopefully seldom be met in-lock, when no FD output is desired, since the PLL should serve to maintain the transition of f_1 in quadrants A and D.

The FD is characterized by the mean value of the pulses at its output, since that mean value serves to charge or discharge the integrating capacitor in the loop filter. That mean value is, assuming FD positive and negative output pulses have equal area,

$$\mu_{FD} = \Pr\{\text{positive pulse}\} - \Pr\{\text{negative pulse}\}. \quad (3.4)$$

If the frequency difference is $\Delta\omega$, the angle of rotation is $-\Delta\omega/f_1$ radians, and μ_{FD} can readily be calculated by assum-

ing that the phasor in cycle k is uniformly distributed from 0 to 2π radians in Figure 7(b). For example, if the angle of rotation of two successive phasors is $\phi < \pi/2$, then an FD output is generated only when the first phasor is within an angle ϕ of the π-axis, an event which has probability $\phi/2\pi$. By a simple extension of this argument, the plot of μ_{FD} of Figure 8 can be generated. The characteristic is periodic for $f_1 > f_2$, since multiple cycles of f_1 in a period of f_2 cannot be distinguished from a single cycle by the FD circuit as described. It is not periodic for $f_1 < f_2$ since, if the period of f_1 is too great, successive positive transition of f_1 will not occur within two periods of f_2 and the FD will generate no output.

As seen from Figure 8, the useful range of the FD is

$$|\Delta\omega| < \pi f_1 = \frac{\omega_1}{2}. \quad (3.5)$$

That is, a 50% offset in the initial VCO frequency f_2 can be tolerated. The range of linearity of the FD, that is, the range over which the model of Figure 2(b) is accurate, is $|\Delta\omega| \leqslant \omega_1/4$. The largest FD output is at $\Delta\omega = \pm\omega_1/4$, where the probability of an output pulse is 0.25.

3.4. Timing Recovery

In timing recovery it is standard to generate a sequence of timing pulses from the data waveform. For example, Bellisio [6] describes a circuit consisting of a differentiator and deadzone rectifier which generates data transition pulses from an NRZ data waveform. The nominal spacing between two successive pulses can be any multiple of the baud interval $T = 1/f_1$ since a pulse is only generated by a data transition. As in Section 3.3 we let f_2 be the VCO output frequency.

The FD described in Section 3.3 works for this case, where the quadrants of f_2 in which the data transition pulses occur are observed. The calculation of the FD output mean is similar, except that in addition to the requirement for two successive phasors to span the π-axis, there must be two data transitions in a row in order for an FD pulse to be generated. Thus, the FD characteristic of Figure 8 remains valid, except that μ_{FD} must be multiplied by the probability of two data transitions in a row (0.25 for equally likely independent data). The FD range is 50% of the baud rate, which is comparable to that reported by Bellisio [6] and more than adequate for the elimination of a VCXO.

3.5. Carrier Recovery

As mentioned in the introduction, the motivation for using an FD to aid acquisition in carrier recovery is somewhat dif-

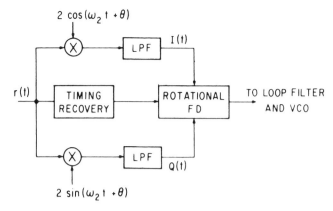

Fig. 9.   Configuration of the Rotational FD in Carrier Recovery.

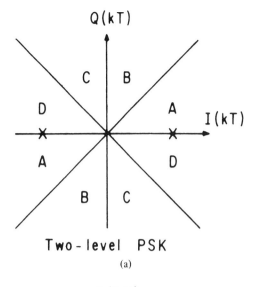

Two-level PSK
(a)

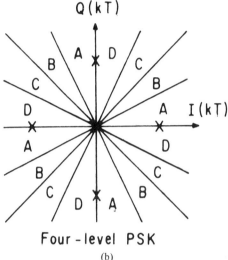

Four-level PSK
(b)

Fig. 10.   Two-Dimensional Signal Constellations and Rotational FD Thresholds (Constellation shown for $\Delta\omega = \omega = 0$).

ferent than in timing recovery, for it is usually not practical to eliminate the VCXO due to the problem of false lock to a data sideband to be described shortly. Acquisition remains a problem, however, since worst-case frequency offsets can still exceed the desired loop bandwidth by many orders of magnitude. The problem is particularly acute in microwave radio transmission, where even very accurate RF oscillator frequencies can result in absolute frequency offsets of 50 to 100 kHz, while carrier recovery loop bandwidths are more typically in the range of 0.1 to 1.0 kHz.

The signal used by the FD to extract carrier frequency offset is assumed to be of the form of (3.1), where ω_1 would typically be the carrier frequency at IF. The object of the FD is to estimate $\Delta\omega$ given by (3.3), where ω_2 is the IF local oscillator frequency generated by the VCO in our carrier recovery PLL. The configuration of the FD is shown in Figure 9. The first step is to demodulate to baseband with quadrature carriers (at frequency ω_2) just as for the Costas loop and quadricorrelator; the resulting quadrature baseband signals $I(t)$ and $Q(t)$ are given by (A.1). The FD operates on $I(t)$ and $Q(t)$, while timing recovery is performed on $r(t)$. The major requirement of the rotational FD is that timing recovery acquisition occur *before* the FD output is valid and carrier recovery acquisition is initiated. The interesting property that timing recovery can be achieved independent of carrier recovery can be seen from squaring $r(t)$ in (3.1) and eliminating the double frequency term; the result is $(x_c^2(t) + x_s^2(t))/2$, which will have a baud frequency component suitable for extraction.

The first operation of the FD is to sample I and Q at the baud interval kT; from (A.1) the result is

$$I(kT) = x_c(kT)\cos(k\Delta\omega T - \theta) - x_s(kT)\sin(k\Delta\omega T - \theta)$$

$$Q(kT) = -x_c(kT)\sin(k\Delta\omega T - \theta) - x_s(kT)\cos(k\Delta\omega T - \theta).$$

(3.6)

Consider first the case $\Delta\omega = \theta = 0$ following acquisition. The point $(I(kT), Q(kT))$ when plotted in a two-dimensional plane is, in fact, one of the data points in the two-dimensional signal constellation corresponding to the modulation method. We show two examples in Figure 10(a) and (b), two-level (biphase) and four-level (QPSK) phase-shift keying. Biphase serves as a basis of comparison to the work of Cahn [8], while

QPSK demonstrates how the technique generalizes to more complicated constellations.***

When there is a frequency difference, we recognize (3.6) as the parametric equations of a circle; that is, the signal constellation is rotated by an angle $k\Delta\omega T - \theta$. Rotation is clockwise if $\Delta\omega > 0$. The nature of the problem of false lock to a data sideband is now clearly evident; in biphase modulation rotation by $\Delta\omega T$ equal to multiples of π radians can clearly not be distinguished from $\Delta\omega T = 0$. Thus, any FD characteristic must be periodic in $\Delta\omega = \pi/T$, and the maximum useful range of any FD is

$$|\Delta\omega| \leq \pi/2T, \text{ biphase.} \quad (3.7)$$

Thus, the initial VCO frequency must not deviate from the carrier frequency in magnitude by more than one-quarter the

*** It is also possible to restrict operation of the FD to a subset of the data points, if that subset can be unambiguously identified in the face of rotation. For example, in 16-level QAM, restriction to the four inner data points results in operation identical to QPSK.

baud rate. For QPSK, (3.7) is replaced by

$$|\Delta\omega| \leqslant \pi/4T, \text{ QPSK} \qquad (3.8)$$

since false lock occurs when rotation is by $\pi/2$ radians, and the maximum VCO offset is one-eighth the baud rate.

Shown on Figure 10, in addition to the signal constellation, are the radial thresholds required for the rotational FD. Some of these thresholds can also serve as slicers for data decisions as well as for implementing a bang-bang type of PD. The radial thresholds divide the angle between each pair of adjacent data points into four quadrants A, B, C, D. In each baud interval the quadrant actually observed is independent of the data, and depends only on the angle of rotation ($k\Delta\omega T$-θ). As before, to insure infrequent FD pulses in lock when $\theta \cong 0$, operation is restricted to quadrants B and C. The actual circumstances in which an FD pulse is generated are identical to the square wave and timing recovery cases.

The FD characteristic is plotted in Figure 11. The highest probability of a pulse output is 0.25 and occurs at $|\Delta\omega T| = \alpha/2$, and the range of useful operation is $|\Delta\omega T| \leqslant \alpha$, where $\alpha = \pi/2$ for biphase and $\alpha = \pi/4$ for QPSK. Since these figures are consistent with (3.7) and (3.8) it follows that the rotational FD has as large a range of operation for carrier recovery as any FD.

3.6. Effect of Phase Jitter

The plots of μ_{FD} presented thus far have not taken into account the effects of noise and intersymbol interference. Since the rotational FD is sensitive to the angle of rotation, which in turn is influenced by these factors, there is concern that they might significantly affect FD operation.

The situation is considered in Appendix B, where it is shown that if the phase jitter is small relative to $\pi/2$, $\pi/4$, or $\pi/8$ for the timing recovery, biphase, or QPSK situations, respectively, the effect of phase jitter is virtually absent, this in spite of any statistical dependencies which may exist between successive samples of phase jitter. For phase jitter with amplitude less than twice the previously mentioned values, the effect is to change the shape of the FD characteristic (basically, round the corners), but not otherwise adversely affect its operation. Even larger phase jitter will have a significant adverse effect on FD operation, but is not likely to be encountered in practice, since the effect of this large jitter on error rate would also be substantial.

4.0. EXPERIMENTAL RESULTS

Experimental results on the use of the quadricorrelator in timing recovery were reported by Bellisio [6]. We report here on experimental results obtained in the implementation of a rotational FD in a carrier recovery application. The terrestrial microwave system to which it was applied employed a 16-point signal constellation and 10 Mbit/s data rate. References [12-13] describe timing and carrier recovery techniques which are typical for this type of system.

This particular system protection switches at an error rate of 10^{-6}, which corresponds to a baseband SNR of about 22 dB. Reliable acquisition was experimentally observed for an

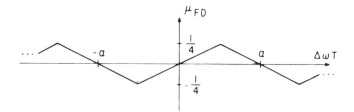

Fig. 11.   FD Characteristic for Carrier Recovery.

SNR of 15 dB or less, which is substantially lower than necessary for this type of system. Figure 12 shows the VCO control line during several acquisitions at SNR's of 22, 20, and 15 dB starting from a worst-case carrier frequency offset (133 kHz). The initial flat portion of the curves corresponds to the period of timing recovery acquisition, which must precede carrier acquisition for the rotational FD. Total acquisition time is about 15, 25, and 45 ms for the three cases.

5.0. CONCLUSIONS

The use of a FD to aid PLL timing and carrier acquisition is a very advantageous technique; the major impediment to its use appears to have been the lack of suitable FD circuits. We have described two such circuits, the quadricorrelator and rotational FD, both of which have a broad applicability. For timing recovery the rotational FD is somewhat simpler, particularly for integrated circuit implementation, since it is digital and requires no multipliers or filters; however, it is also limited to lower data rates. For carrier recovery there appears to be no great difference in the difficulty of implementation, since both circuits require quadrature mixers followed by circuitry which operates at approximately the baud rate. In some instances a substantial portion of the rotational FD circuitry can simultaneously serve other purposes (such as PD and data thresholding), in which case it becomes more attractive.

APPENDIX A

QUADRICORRELATOR OUTPUT MEAN VALUE

Let wide-sense stationary input signal $r(t)$ be written in the form of (3.1) and assume that the power spectrum of $r(t)$ is symmetrical about radian frequency ω_1 so that $x_c(t)$ and $x_s(t)$ are uncorrelated. Assuming that $\omega_1 + \omega_2$ terms are rejected by the low pass filters in Figure 4,

$$I(t) = x_c(t)\cos(\Delta\omega t - \theta) - x_s(t)\sin(\Delta\omega t - \theta)$$

$$Q(t) = -x_c(t)\sin(\Delta\omega t - \theta) - x_s(t)\cos(\Delta\omega t - \theta) \qquad (A.1)$$

where $\Delta\omega$ is given by (3.3). The crosscorrelation of $I(t)$ and $Q(t)$ is then easily shown to be, using the fact that $x_c(t)$ and $x_s(t)$ are uncorrelated,

$$R_{IQ}(\tau) = E[I(t)Q(t+\tau)]$$

$$= \frac{j}{4}(e^{j\Delta\omega\tau} - e^{-j\Delta\omega\tau})(R_c(\tau) + R_s(\tau)) \qquad (A.2)$$

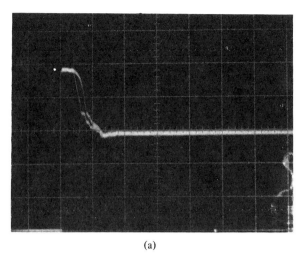

(a)

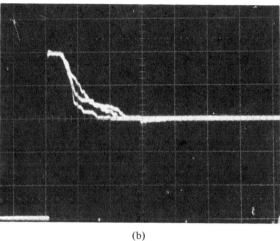

(b)

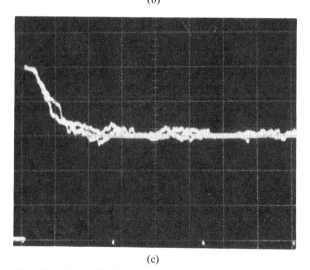

(c)

Fig. 12.   Rotational FD Carrier Recovery Acquisition. (a) SNR = 22 dB (10 ms/division). (b) SNR = 20 dB (10 ms/division). (c) SNR = 15 dB (20 ms/division).

where $R_c(\tau)$ and $R_s(\tau)$ are the autocorrelation functions of $x_c(t)$ and $x_s(t)$. The cross power spectral density, the Fourier transform of (A.2), is

$$S_{IQ}(\omega) = \frac{j}{4}\{S_c(\omega - \Delta\omega) - S_c(\omega + \Delta\omega)$$
$$+ S_s(\omega - \Delta\omega) - S_s(\omega + \Delta\omega)\}. \tag{A.3}$$

The mean of $p(t)$ is thus

$$Ep(t) = E\left[Q(t)\frac{dI(t)}{dt}\right]$$

$$= \int_{-\infty}^{\infty} (j\omega)^* S_{IQ}(\omega)\frac{d\omega}{2\pi}$$

$$= \frac{1}{4}\int_{-\infty}^{\infty} \omega\{S_c(\omega - \Delta\omega) - S_c(\omega + \Delta\omega)$$

$$+ S_s(\omega - \Delta\omega) - S_s(\omega + \Delta\omega)\}\frac{d\omega}{2\pi}. \tag{A.4}$$

Changing variables in this integral, we obtain finally

$$Ep(t) = \tfrac{1}{2}\Delta\omega \int_{-\infty}^{\infty} (S_c(\omega) + S_s(\omega))\frac{d\omega}{2\pi}. \tag{A.5}$$

Recognizing that the power in $r(t)$ is one-half the sum of the powers in $x_c(t)$ and $x_s(t)$, (3.2) follows.

APPENDIX B

We can model the effect of phase jitter by assuming the angle of rotation at the kth baud interval is $(k\Delta\omega T - \theta + \theta_k)$, where θ_k is a random phase jitter component. It is important that we *not* assume that θ_k and θ_{k+1} are independent, since intuitively dependencies should have a particularly strong influence. In order to recalculate μ_{FD} for this case, we let θ be uniformly distributed on $[0, 2\pi]$ as before. The starting angle is $(k\Delta\omega T - \theta + \theta_k)$, and the angle of rotation is $(\Delta\omega T + \theta_{k+1} - \theta_k)$. The key to simplifying the problem is to first condition on θ_k and θ_{k+1}, and take the expectation over θ, that expectation being the same as previously determined but with $\Delta\omega T$ replaced by $(\Delta\omega T + \theta_{k+1} - \theta_k)$. Thus, completing the expectation over θ_k and θ_{k+1},

$$\mu_{FD} = E[F(\Delta\omega T + \theta_{k+1} - \theta_k)] \tag{B.1}$$

where $F(\Delta\omega T)$ is the FD characteristic of Figures 8 or 11.

Equation (B.1) is exact, but for the special case where the argument of F is in the linear region with high probability, where $F(\omega) = K\omega$,

$$\mu_{FD} \cong E[K(\Delta\omega T + \theta_{k-1} - \theta_k)]$$

$$= K(\Delta\omega T + E\theta_{k+1} - E\theta_k)$$

$$= F(\Delta\omega T) \tag{B.2}$$

113

where we have made the further assumption that $E\theta_{k+1} = E\theta_k$ (not necessarily zero). The implication of (B.2) is that the phase jitter has had no effect whatsoever on μ_{FD}. Note that no assumption of independence of θ_k and θ_{k+1} has been made.

When this special case is violated, (B.1) can be used to estimate the effect. If $\theta_{k+1} - \theta_k$ has probability density $f(\cdot)$, then (B.1) becomes

$$\mu_{FD} = \int F(\Delta\omega T + \phi)f(\phi)\,d\Phi. \qquad (B.3)$$

Thus, if the argument $(\Delta\omega T + \phi)$ is not confined to the linear region of F, the effect is seen to be a smoothing of the corners of the FD characteristic. If $f(\phi)$ spans a significant portion of the period of F, then there is a significant deterioration of the FD operation.

ACKNOWLEDGMENT

The author is indebted to F. Stevens for the experimental results provided in Section 4.0.

REFERENCES

1. F. M. Gardner, *Phase-Lock Techniques,* New York: Wiley, 1966.
2. D. Richman, "Color-Carrier Reference Phase Synchronization Accuracy in NTSC Color Television," *Proc. IRE,* Vol. 42, p. 106, January 1954.
3. T. B. Pickard, "The Effect of Noise on a Method of Frequency Measurement," *IRE Trans. Information Thy,* Vol. IT-4, p. 83, June 1958.
4. R. F. Pawula, "Analysis of an Estimator of the Center Frequency of a Power Spectrum," *IEEE Trans. Information Thy,* Vol. IT-14, p. 669, September 1968.
5. S. R. J. Axelsson, "Analysis of the Quantizing Error of a Zero-Counting Frequency Estimator," *IEEE Trans. Information Thy,* Vol. IT-22, P. 596, September 1976.
6. J. A. Bellisio, "New Phase-Locked Timing Recovery Method for Digital Regenerators," *Int. Conf. Communications Record,* Philadelphia, pp. 10-17, June 1976.
7. J. H. Park, Jr., "An FM Detector for Low S/N," *IEEE Trans. Comm.,* Vol. COM-18, p. 110, April 1970.
8. C. R. Cahn, "Improving Frequency Acquisition of a Costas Loop," *IEEE Trans. on Comm.,* Vol. COM-25, p. 1453, December 1977.
9. J. F. Oberst, "Generalized Phase Comparators for Improved PLL Acquisition," *IEEE Trans. on Comm. Tech.,* Vol. COM-19, p. 1142, December 1971.
10. P. K. Runge, "Phase Locked Loops with Signal Injection for Increased Pull-In Range and Reduced Output Phase Jitter," *IEEE Trans. Comm.,* Vol. COM-24, p. 636, June 1976.
11. R. W. Lucky, J. Salz, E. J. Weldon, Jr., *Principles of Data Communication,* New York, McGraw-Hill, 1968.
12. C. W. Anderson, S. G. Barber, "Modulation Considerations for a 91 Mb/s Digital Radio," *IEEE Trans. on Communications,* Vol. COM-26, May 1978, p. 523.
13. C. R. Hogge, "Carrier and Clock Recovery for 8 PSK Synchronous Demodulation," *IEEE Trans. on Communications,* Vol. COM-26, May 1978, p. 528.
14. R. Citta, "Frequency and Phase Lock Loop," *IEEE Trans. Consumer Electronics,* Vol. CE-23, Aug. 1977, p. 358.

Analysis of Phase-Locked Timing Extraction Circuits for Pulse Code Transmission

ENGEL ROZA

Abstract—An analysis is presented of the performance of phase-locked timing extraction circuits for baseband pulse code transmission. The phase error of the extracted timing wave is influenced by the properties of three essential stages in signal processing: prefiltering, nonlinear treatment, and narrow-band filtering. The analysis enables us to calculate quantitatively the quasi-static and the dynamic part of the phase error for arbitrary but specified types of signal processing. This is more than can be done with existing theory in the case of resonant-type timing extraction circuits. Examples are given for practical cases, and conditions for optimum performance are derived.

Furthermore, the behavior of such phase-locked circuits in a chain of repeaters is investigated, and in particular, the propagation law for jitter. Byrne's model, as used for resonant-type timing extraction circuits, therefore is generalized. It is shown analytically and experimentally that by proper implementation of the timing extraction circuit, considerable improvement can be obtained as compared with resonant-type circuits.

I. INTRODUCTION

IN ORDER to regenerate pulses in a transmission system for pulse coded signals, a clock signal should be available. In self-timing repeaters, this clock signal is derived from the baseband information signal itself. This includes filtering by means of a resonant circuit or by a phase-locked loop (PLL). The properties of a single resonant circuit with respect to timing derivation have been analyzed by Sunde, Bennett, and Manley [1]–[3]. Byrne *et al.* [4] have studied the propagation of jitter of the extracted timing wave in a chain of repeaters with resonant-type clock extraction circuits. In this paper, the PLL is analyzed in these respects. In addition, a more general approach is followed, permitting the influence of undesired signal interference to be taken into account.

II. SPECTRAL DENSITY OF THE INFORMATION SIGNAL

The transmission model used consists of a signal source, sending synchronous digital impulses into an equivalent baseband transmission channel, which also includes all linear processing of the receiving and transmitting end. Generally, the bandwidth of this transmission channel will be restricted such that a well-defined transfer of information is just possible. The first Nyquist criterion states that a bandwidth, which extends from zero to a frequency between the digit frequency and its half, is sufficient.

Manuscript received November 15, 1973.
The author is with the Philips Research Laboratories, Eindhoven, The Netherlands.

The series of synchronous impulses from the signal source can be described as a superposition of a deterministic series of impulses with equal amplitudes and a stochastic series of impulses with discrete amplitudes of a random distribution. The first series is responsible for lines in the power spectrum at the digit frequency and its multiples. The second series produces the continuous part of this spectrum.

Because of the bandwidth restriction of the transmission channel, it is not possible to transfer energy at the digit frequency or its multiples. That means that nonlinear signal processing is required to derive a clock signal at the digit frequency from the received signal.

III. NONLINEAR SIGNAL PROCESSING

Two methods are commonly used which avoid undesired interference of neighboring pulses in the clock path. We shall denote this undesired interference as *interpulse interference*, which has to be distinguished from inter-symbol interference in the information path.

1) According to the first method, the input signal of the clock extractor is shaped in such a way that it fulfills Nyquist's second criterion [5]. Such a signal has the property that halfway between the centers of two successive pulses, the signal value of all other pulses is zero. This signal is then processed by a nonlinear circuit generating pulses of a short duration (shorter than one digit interval) when the signal crosses thresholds which correspond to the values of a single pulse halfway between two pulse centers. In this way, a random series of equidistant pulses originates, possessing a power spectrum with lines at the digit frequency and its multiples, so that with narrow-band filtering, a continuous clock signal can be derived. This method will be referred to as the *threshold method*.

2) According to the second method, the transmission characteristic of the clock path is changed such that an even symmetrical transfer characteristic around the Nyquist frequency is obtained. Afterwards, the signal is squared. As a result, a signal component is produced with a frequency equal to the digit frequency, modulated in amplitude, but with a constant phase. All components in the original continuous spectrum around the Nyquist frequency contribute to the power of this signal. Also, this signal has to be filtered afterwards by a narrow bandpass filter centered at the digit frequency. This process, denoted as the *symmetry method*, will be analyzed in Section VII in more detail.

The threshold method is used in systems with a digit

Reprinted from *IEEE Trans. Comm.*, vol. COM-22, pp. 1236-1249, September 1974.

115

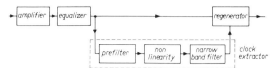

Fig. 1. Block diagram of a regenerative repeater.

frequency that is relatively low with respect to the frequency range of the applied electronic components. The symmetry method is used at higher frequencies, although its fundamental property of avoiding interpulse interference is not, or hardly, recognized. For example, the rectification of the symbols of a band-limited first-order bipolar signal has to be conceived as a nonideal application of the described principle. The bipolar coding of the first order [12] is a linear process which yields a symmetrical power spectrum around the Nyquist frequency between zero and the digit frequency. Although bandwidth limitation may somewhat deteriorate this symmetry and although rectification substitutes the squaring, the principle mentioned can readily be recognized. The resulting signal to be filtered will show phase errors caused by interpulse interference due to this imperfect execution.

The three essential parts of the clock extraction process, i.e., prefiltering, nonlinear treatment, and narrow-band filtering, are shown in the block diagram of a regenerative repeater in Fig. 1. Various possibilities for implementation will be investigated and compared in the following sections.

IV. FILTERING

At first, the description of the filter process after the nonlinear treatment of the signal will be based upon the short pulses as obtained by the threshold method. The symmetry method will be considered later. The obtained signal after the nonlinear process can be approximated by

$$s(t) = \sum_{n=0}^{N} c_n \delta(t - nT - \tau_n). \qquad (1)$$

Herein the coefficient c_n is determined by the statistical properties of the signal, and it possesses in its most simple form the binary values 0 or 1. T is the digit interval and N is related to t such that

$$t = NT + \tau, \qquad 0 \leq \tau < T. \qquad (2)$$

The quantity τ_n denotes the deviation of the nth pulse from its nominal value.

There are three possible causes for this shift:

1) additive noise in the transmission channel,

2) the input signal to the nonlinear circuit does not obey the second criterion of Nyquist,

3) the incoming symbols contain timing errors, introduced by preceding repeaters.

In a later section, the influences of more general preprocessing will be considered. It will be found that these influences can be taken into account by assigning appropriate values to c_n and τ_n. Statistical mutual dependence of those values may occur in this general outline.

It is useful to make the following remarks about c_n and τ_n a priori.

1) c_n and τ_n are assumed to be random variables in a wide sense stationary random process, i.e., their autocorrelation functions $R_c(k)$ and $R_\tau(k)$, defined by (3), are independent of n.

$$R_c(k) = \overline{c_n c_{n+k}}$$
$$R_\tau(k) = \overline{\tau_n \tau_{n+k}}. \qquad (3)$$

(The ensemble average of a random variable x will be indicated by $\bar{x}$.)

As a consequence, their discrete Fourier transforms determine their spectral densities.

2) c_n can vary randomly and considerably from digit interval to digit interval. Its spectral density, therefore, usually extends beyond the digit frequency.

3) Because of the filtering of preceding repeaters, the spectral density of τ_n can show strong components in the low-frequency part.

4) Due to the growth of the deviations τ_n along a chain of repeaters, τ_n cannot simply be considered to be small compared with one digit interval T.

A. Bandpass Filtering

Let the signal $s(t)$ of (1), represented by pulses from a current source, excite a narrow bandpass filter with center frequency as close as possible to the digit frequency. As a result, the voltage across the filter will be a harmonic function fluctuating in amplitude and phase with a nominal frequency equal to the digit frequency. The amplitude fluctuation can be eliminated by a hard limiter. The phase fluctuation deteriorates the quality of the derived clock signal.

This process has been investigated by Sunde [1] and Bennett [2] for a resonant circuit.

B. PLL Filtering

An alternative solution for obtaining a clock signal from $s(t)$ is the synchronization of the oscillator of a PLL. In order to analyze the PLL, a model for it, using digital signals, should be available. Such models have been derived in the past by Byrne [6] and Saltzberg [7]. We develop here an alternative model to show the relationship with the common linear model for sinusoidal signals so as to make possible an easy comparison with bandpass filtering. Fig. 2(a) shows the general block diagram for a large class of PLL's. It shows that the loop filter is fed by the product of the input signal $s_i(t)$ and the oscillator signal $s_o(t)$.

In the case that $s_i(t)$ and $s_o(t)$ are sinusoidal signals, the low-frequency part of the output of the multiplier is linearly proportional to the actual difference $\theta_{ea}(t)$ of the input phase $\theta_{ia}(t)$ and the oscillator phase $\theta_o(t)$. The block diagram can then be modified to Fig. 2(b). The phase of the oscillator is determined by the convolution of the phase error $\theta_{ea}(t)$ and $g(t)$. By $g(t)$ we denote the pulse response of the loop filter followed by an ideal integrator

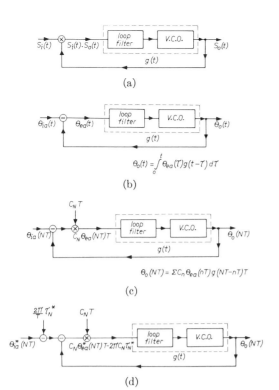

(a)

(b)

$$\theta_o(t) = \int_0^t \theta_{ea}(T)g(t-T)\,dT$$

(c)

$$\theta_o(NT) = \Sigma c_n \theta_{ea}(nT)g(NT-nT)T$$

(d)

Fig. 2. (a) General block diagram of a PLL. (b) PLL model for sinusoidal signals. (c) PLL model for digital signals. (d) PLL model for digital signals with interpulse interference.

which represents the behavior of the phase of the oscillator due to an applied signal.

So, assuming causality for $g(t)$ and $\theta_{ea}(t)$,

$$\theta_o(t) = \int_0^t \theta_{ea}(\tau)g(t-\tau)\,d\tau. \qquad (4)$$

The input signal applied to the multiplier in the case of digital signals consists of a series of equidistant pulses with random amplitudes. These pulses should have a finite width which should be, in principle, less than one digit interval T. The output of the multiplier also consists of a random series of pulses (with another shape). Because of the integrating properties of the loop, the momentary signal value of the pulses as such is not significant, but merely the integrated signal values over nonoverlapping time intervals, short with respect to the duration of $g(t)$. For this interval, the period T is a convenient choice. Consequently, the output of the multiplier can be modeled as a random series of Dirac-like impulses at discrete intervals T. The magnitudes of these impulses are proportional—linearly as a first approximation—to the product of:

1) the relative delay $\theta_{ea}(NT)T/2\pi$ of the oscillator signal with regard to the input pulse positions, and

2) the discrete pulse amplitudes c_N of the input signal.

From these considerations, the evaluation of the general block diagram of Fig. 2(a) to the PLL model for digital signals, as shown in Fig. 2(c), is evident.

For the phase of the oscillator signal, responding to an input signal $s(t)$ of (1), one can write

$$\theta_o(NT) = \sum_{n=0}^{N} c_n \theta_{ea}(nT)g(NT-nT)T. \qquad (5)$$

The correspondence with the sinusoidal model is evident if we compare (5) and (4).

$\theta_{ea}(NT)$ represents the phase difference between the *actual* phase of the input signal $\theta_{ia}(NT)$ and the phase of the oscillator $\theta_o(NT)$, so that

$$\theta_{ea}(NT) = \theta_{ia}(NT) - \theta_o(NT). \qquad (6)$$

For $\theta_{ia}(NT)$ we may write

$$\theta_{ia}(NT) = \Delta\omega_b NT - 2\pi\frac{\tau_N}{T}. \qquad (7)$$

$\Delta\omega_b$ is the detuning of the quiescent frequency of the voltage-controlled oscillator (VCO) from the angular digit frequency ω_b being

$$\omega_b = \frac{2\pi}{T}. \qquad (8)$$

τ_N has been defined by (1).

The term $\Delta\omega_b NT$ represents the *nominal* phase of the input signal $\theta_{ia}(NT)$.

Substituting (6) into (5), we find that

$$\theta_{ia}(NT) - \theta_{ea}(NT) = \sum_{n=0}^{N} c_n \theta_{ea}(nT)g(NT-nT)T. \qquad (9)$$

We are more interested in the phase error of the oscillator signal with regard to the *nominal* value $\Delta\omega_b NT$ rather than to the actual value $\theta_{ia}(NT)$ because the nominal value represents an ideal imaginary phase reference. So, substituting into (9)

$$\theta_e(nT) = \theta_{ea}(nT) + 2\pi\frac{\tau_n}{T} \qquad (10)$$

in which $\theta_e(nT)$ is the phase error with regard to the nominal input phase, we have

$$\theta_e(NT) + \sum_{n=0}^{N} c_n \theta_e(nT)g(NT-nT)T = \Delta\omega_b NT$$
$$+ \frac{2\pi}{T}\sum_{n=0}^{N} c_n \tau_n g(NT-nT)T. \qquad (11)$$

We rewrite (11) as

$$\theta_e(NT) + \sum_{n=0}^{N} \frac{c_n}{\bar{c}_n}\theta_e(nT)g_w(NT-nT) = \Delta\omega_b NT$$
$$+ \frac{2\pi}{T}\sum_{n=0}^{N} \frac{c_n \tau_n}{\bar{c}_n} g_w(NT-nT) \qquad (12)$$

with

$$g_w(NT-nT) = \bar{c}_n g(NT-nT)T \qquad (13)$$

being the normalized weighted open-loop pulse response of the PLL.

Equation (12) will be referred to as the system equation

117

of the PLL for digital signals. In Appendix I the system equation has been solved. The phase error is found to consist of a static part and a fluctuating part.

For the static part we find

$$\theta_{eo} = \theta_{eo}(1) + \theta_{eo}(2) \tag{14}$$

$$\theta_{eo}(1) = \lim_{N \to \infty} \Delta\omega_b \frac{NT}{\sum_{n=0}^{N} g_w(NT - nT)} \tag{15}$$

$$\theta_{eo}(2) = \frac{2\pi}{T} \frac{\overline{c_N \tau_N}}{\bar{c}_N} . \tag{16}$$

And for the fluctuating part we get

$$\Delta\theta_e(NT) = \sum_{n=0}^{N} b_n h_w(NT - nT) \tag{17}$$

with

$$b_n = -\frac{c_n - \bar{c}_n}{\bar{c}_n} \theta_{eo}(1) - \frac{c_n}{\bar{c}_n}\left(\frac{\overline{c_n \tau_n}}{\bar{c}_n} - \tau_n\right). \tag{18}$$

$h_w(NT)$ is the normalized pulse response of the weighted closed-loop transfer function of the PLL. Its discrete Fourier transform $H_w(l\Omega)$, defined by (94) in Appendix I, will be denoted as the phase-transfer function of the system.

We summarize the conditions which have been assumed in the derivation of the model and the solution of its system equation.

The first two conditions relate to the model. They express the extent to which the model represents the physical reality.

1) Condition of Small Bandwidth: The bandwidth can be characterized most suitably by the closed-loop noise bandwidth, that is,

$$B_L = \frac{1}{2\pi} \lim_{N \to \infty} \sum_{n=0}^{N} |H_w(l\Omega)|^2 \Omega. \tag{19}$$

The condition of small bandwidth is then mathematically expressed by

$$B_L T \ll 1. \tag{20}$$

2) Condition of Linearity: We have assumed that the loop operates in its linear range. This can only be true if the actual phase errors are small. The actual phase errors are identical with the nominal ones if the pulses of the input signal are not deviated from their nominal positions, i.e., if $\tau_n = 0$. The linearity condition is then given by

$$\theta_{eo}(1) \ll 1, \qquad (\overline{\Delta\theta_e^2})^{1/2}|_{\tau_n=0} \ll 1. \tag{21}$$

Two other conditions have been introduced during evaluation of the system equation in Appendix I. These conditions have mainly a theoretical value in order to indicate the bounds of the developed theory. Practical systems operate far from these bounds. The conditions are expressed in terms of spectral densities of the random variables, which are the discrete Fourier transforms of their autocorrelation functions.

3) Condition of Sufficient Clock Content:

$$S_{cc}(0) (2\pi B_L T)^{1/2} \ll 1. \tag{22}$$

Here S_{cc} is the spectral density of the random variable $(c_n - \bar{c}_n)/\bar{c}_n$, as defined in Appendix I by (87).

If c_n are statistical independent values of a binary sequence with mark probability Pr, (22) reduces to

$$\frac{1 - \text{Pr}}{\text{Pr}} (2\pi B_L T)^{1/2} \ll 1. \tag{23}$$

This condition is apparently strongly related to the small bandwidth condition. Furthermore, the condition of sufficient clock content requires that the spectral density S_{cc} does not vary significantly in the frequency range covered by the phase-transfer function $H_w(l\Omega)$. The name of this condition reflects the importance of the expected value $\bar{c}_n$ ($\bar{c}_n$ must be sufficiently large).

4) Condition of Restricted Interpulse Interference:

$$[R_{cc}(0)]^{1/2} S_{c\tau c}(0) (2\pi B_L T)^{1/2} \ll [R_{c\tau c\tau}(0)]^{1/2}. \tag{24}$$

$R_{cc}(0)$ and $R_{c\tau c\tau}(0)$ are the variances of the random variables $(c_n - \bar{c}_n)/\bar{c}_n$ and $c_n(\overline{c_n \tau_n}/\bar{c}_n - \tau_n)/\bar{c}_n$. $S_{c\tau c}$ is the cross spectral density of those variables, as defined by (93). It has been assumed, in addition, that $S_{c\tau c}$, like S_{cc} in the former condition, does not vary significantly in the frequency range covered by the phase-transfer function. Only strong dependence of c_n and τ_n may deteriorate this condition. As will be explained in Section VII, dependence of c_n and τ_n is caused by the mechanism of interpulse interference.

V. JITTER

Equation (17) expresses the fluctuating part of the phase error or jitter. In this form, however, it hardly produces relevant information for the circuit designer. Let us therefore discuss the formula in more detail. In Appendix II an expression for the jitter has been derived in spectral terms. For its mean-square value, which stabilizes for large N, we can write

$$\overline{\Delta\theta_e^2} = \lim_{N \to \infty} \frac{1}{N+1} \sum_{l=0}^{N} S_{bb}(l\Omega) |H_w(l\Omega)|^2. \tag{25}$$

$S_{bb}(l\Omega)$ is the spectral density of the random variable b_n or, in other terms, the discrete Fourier transform of its autocorrelation function $R_{bb}(k)$, as defined by (90). $H_w(l\Omega)$ is the discrete Fourier transform of the pulse response of the phase-transfer function of the system, as defined by (94).

Let us consider some particular cases.

1) Assume that c_n and τ_n are mutually statistically independent. This is true, for instance, if the nonlinear preprocessing is ideally performed according to the threshold method stated in Section III. We may then distinguish two contributions to the jitter: first, $\overline{\Delta\theta_{ec}^2}$ due to c_n, and second, $\overline{\Delta\theta_{e\tau}^2}$ due to τ_n:

$$\overline{\Delta\theta_e^2} = \overline{\Delta\theta_{ec}^2} + \overline{\Delta\theta_{e\tau}^2}. \qquad (26)$$

Let us first examine $\overline{\Delta\theta_{ec}^2}$. From (25) and the definition of b_n as given in (18), we obtain

$$\overline{\Delta\theta_{ec}^2} = \theta_{eo}(1)^2 \lim_{N\to\infty} \frac{1}{N+1} \sum_{l=0}^{N} S_{cc}(l\Omega) \mid H_w(l\Omega) \mid^2 \quad (27)$$

in which $S_{cc}(l\Omega)$ is the spectral density of $(c_n - \bar{c}_n)/\bar{c}_n$, which is the discrete Fourier transform of its autocorrelation function $R_{cc}(k)$, as defined by (87).

2) Equation (27) can be evaluated further if c_n are statistically independent values. Then

$$\Delta\theta_{ec}^2 = \theta_{eo}(1)^2 R_{cc}(0) \lim_{N\to\infty} \frac{1}{N+1} \sum_{l=0}^{N} \mid H_w(l\Omega) \mid^2. \quad (28)$$

Using (19), we write for (28)

$$\overline{\Delta\theta_{ec}^2} = \theta_{eo}(1)^2 R_{cc}(0) B_L T. \qquad (29)$$

By analogy to a resonant circuit, we define for a general transfer function an effective quality factor Q_{eff} as

$$Q_{\text{eff}} = \frac{\pi}{T} \frac{1}{4B_L} \qquad (30)$$

so that

$$\Delta\theta_{ec}^2 = \frac{\theta_{eo}(1)^2}{4} \frac{\pi}{Q_{\text{eff}}} R_{cc}(0). \qquad (31)$$

A similar result has been found by Bennett [2] for the resonant circuit.

3) In addition, let c_n be a symbol out of a series of marks and spaces, and let Pr denote the probability for a mark; then

$$R_{cc}(0) = \frac{1 - \text{Pr}}{\text{Pr}}. \qquad (32)$$

It is evident that the jitter vanishes if $\text{Pr} \to 1$ because the signal is then fully deterministic. For very small values of Pr, the condition of sufficient clock content no longer holds, and an exact calculation of the mean-square jitter requires numerical computation of the system equation (12).

4) Let us now consider $\Delta\theta_{e\tau}^2$.

Mutual independence of c_n and τ_n results in

$$\overline{\Delta\theta_{e\tau}^2} = \frac{4\pi^2}{T^2} \lim_{N\to\infty} \frac{1}{N+1} \sum_{l=0}^{N} S_{c\tau c\tau}(l\Omega) \mid H_w(l\Omega) \mid^2 \quad (33)$$

in which $S_{c\tau c\tau}(l\Omega)$ is the spectral density of the random variable $(c_n/\bar{c}_n)(\overline{c_n\tau_n}/\bar{c}_n - \tau_n)$.

The time domain expression can be derived directly from (17):

$$\overline{\Delta\theta_{e\tau}^2} = \frac{4\pi^2}{T^2} \frac{1}{\bar{c}_n^2} \lim_{N\to\infty} \frac{1}{N+1} \sum_{n=0}^{N} \sum_{m=0}^{N} \overline{c_n c_m (\tau_n - \bar{\tau}_n)(\tau_m - \bar{\tau}_n)}$$
$$\cdot h_w(NT - nT) h_w(NT - mT). \quad (34)$$

If, in addition, c_n are statistically independent, (34) reduces to

$$\overline{\Delta\theta_{e\tau}^2} = \frac{4\pi^2}{T^2} \frac{\overline{c_n^2}}{\bar{c}_n^2} \lim_{N\to\infty} \frac{1}{N+1} \sum_{n=0}^{N} \sum_{m=0}^{N} (\tau_n - \bar{\tau}_n)(\tau_m - \bar{\tau}_n)$$
$$\cdot h_w(NT - nT) h_w(NT - mT) \quad (35)$$

or in spectral terms

$$\overline{\Delta\theta_{e\tau}^2} = \frac{4\pi^2}{T^2} \frac{\overline{c_n^2}}{\bar{c}_n^2} \lim_{N\to\infty} \frac{1}{N+1} \sum_{l=0}^{N} S_{\tau\tau}(l\Omega) \mid H_w(l\Omega) \mid^2 \quad (36)$$

in which $S_{\tau\tau}(l\Omega)$ is the spectral density of the random variable $(\tau_n - \bar{\tau}_n)$.

In the special case of zero-mean independent values for τ_n, we find an analogous expression as found for the resonant circuit [1], [2]:

$$\overline{\Delta\theta_{e\tau}^2} = \frac{4\pi^2}{T^2} \frac{\overline{c_n^2}}{\bar{c}_n^2} \overline{\tau_n^2} \frac{\pi}{2Q_{\text{eff}}}. \qquad (37)$$

The causes for the origin of τ_n, as catalogued in Section IV, however, do not generally provide statistical independence for τ_n.

VI. SECOND-ORDER PLL

Let us consider the results obtained for a second-order PLL. The closed- and open-loop transfer functions in general notation are given by [11]

$$H_w(s) = \frac{s\omega_n(2\zeta - \omega_n/K_v) + \omega_n^2}{s^2 + 2\zeta\omega_n s + \omega_n^2} \qquad (38)$$

$$G_w(s) = \frac{s\omega_n(2\zeta - \omega_n/K_v) + \omega_n^2}{s^2 + s(\omega_n/K_v)\omega_n}. \qquad (39)$$

$H_w(l\Omega)$ and $G_w(l\Omega)$ may be regarded as sampled values of $H_w(j\omega)$ and $G_w(j\omega)$. These transfer functions are determined by three parameters: the damping factor ζ, the natural frequency ω_n, and the velocity constant K_v. As follows from (13), these parameters are weighted according to the average energy content of the signal, expressed by the expected value c_n. For K_v this weighting is linearly proportional; for ζ and ω_n it is more complicated. The effective quality factor Q_{eff} for the second-order PLL, calculated by (38), (30), and (19), amounts to

$$Q_{\text{eff}} = \frac{2\pi}{T} \frac{\zeta}{\omega_n\{1 + (2\zeta - \omega_n/K_v)^2\}}. \qquad (40)$$

This expression is for high-gain loops, i.e., loops for which

$$2\zeta \gg \frac{\omega_n}{K_v} \qquad (41)$$

independent of the velocity constant. As seen from (31) and (37), this Q_{eff} controls the jitter. Evaluation of the first contribution to the static phase error results in

$$\theta_{eo}(1) = \frac{\Delta\omega_b}{K_v}. \qquad (42)$$

The detuning $\Delta\omega_b$ is not a real constant, but is influenced by temperature, aging, and moisture, which means that

$\theta_{eo}(1)$ is a *quasi*-static error. The influence of the second part $\theta_{eo}(2)$, however, can be eliminated by an equal shift of the reference position. It may appear odd that the loop exhibits this second contribution because, as is commonly known, the response of the phase error on a phase input step should be zero in the stationary condition. The explanation comes from the fact that in our analysis, we have considered the nominal phase error instead of the actual one.

As may be seen from (42), it is apparently possible to reduce the quasi-static phase error $\theta_{eo}(1)$ to an arbitrarily small amount without influencing Q_{eff}. This is not possible for resonant-type timing extraction circuits. The corresponding expression for the quasi-static phase error, as found by Sunde and Bennett [1], [2], is

$$\theta_{eo}(1) = \beta Q \qquad (43)$$

with

$$\beta = \Delta \omega_b T / \pi \qquad (44)$$

on the assumption that

$$\beta \ll 1, \qquad \beta Q \ll 1.$$

The independence of static and dynamic phase error is, in fact, the most important advantage of the PLL system if compared with the resonant-type system.

On the other hand, faulty performance for worst case signals is possible, due to the modulation of the loop parameters by $\bar{c}_n$.

VII. INFLUENCE OF INTERPULSE INTERFERENCE

Up to now, τ_n has been regarded as an undesired time shift in an idealized model in which it is assumed that the input signal to the PLL consists of pulses short with respect to the digit interval T. In this section, however, we shall drop this assumption, and we shall show that the same model of Fig. 2(c) applies if the effects of general preprocessing, resulting in pulses of arbitrary shapes and duration, are taken into account by appropriately adapted values for τ_n and c_n.

Let $f_1(t)$ represent the waveform of a single pulse of the signal $s_1(t)$ before the nonlinear processing, so

$$s_1(t) = \sum_{n=0}^{N} a_n f(t - nT) \qquad (45)$$

in which a_n is a discrete random variable.

The signal $s_2(t)$, after a general nonlinear treatment, is given by

$$s_2(t) = k_1 s_1(t) + k_2 s_1(t)^2 + k_3 s_1(t)^3 + \cdots . \qquad (46)$$

The signal $s_2(t)$ corresponds with signal $s_i(t)$ of Fig. 2(a). Because of the integrating properties of the loop, we may approximate the output signal $s_3(t)$ of the multiplier by a series of Dirac-like impulses at interval T:

$$s_3(t) = \sum_{n=0}^{N} s_3(nT)\delta(t - nT). \qquad (47)$$

If T is short with respect to the duration of $g(t)$, which generally is true, the output signal $s_o(t)$ of the oscillator will be unaffected by this approximation. The oscillator signal consists of a sequence of elementary waveforms $f_o(t)$ with maximum duration T, which are shifted in phase by an amount $\theta_{ea}*(nT)$ from the actual phase $\theta_{ia}*(nT)$ of the fundamental signal component out of $s_2(t)$ on which $s_o(t)$ locks. Therefore, we may write for the sample values

$$s_3(nT) = \int_{-T/2}^{T/2} s_2(nT + \tau) f_o \left(\tau - \frac{T}{2\pi} \theta_{ea}*(nT) \right) d\tau. \qquad (48)$$

Later we shall argue that the actual phase of the fundamental component of $s_2(t)$ will be, in general, equal to the actual phase of the pulses of $s_1(t)$ of (45).

As long as $\theta_{ea}*(nT) \ll 1$, according to the linearity condition (21), we may approximate (48) by

$$s_3(nT) = \int_{-T/2}^{T/2} s_2(nT + \tau) f_o(\tau) \, d\tau + \frac{T}{2\pi} \theta_{ea}*(nT)$$
$$\cdot \int_{-T/2}^{T/2} s_2(nT + \tau) f_o'(\tau) \, d\tau. \qquad (49)$$

Defining now

$$c_n = \frac{T}{2\pi} \int_{-T/2}^{T/2} f_o'(\tau) s_2(nT + \tau) \, d\tau \qquad (50)$$

$$c_n \cdot \tau_n* = \frac{T}{2\pi} \int_{-T/2}^{T/2} f_o(\tau) s_2(nT + \tau) \, d\tau \qquad (51)$$

and substituting (49)–(51) into (47), we may write for the output signal $s_3(t)$ of the multiplier

$$s_3(t) = \sum_{n=0}^{N} c_n \theta_{ea}*(nT) T\delta(t - nT)$$
$$+ \sum_{n=0}^{N} \frac{2\pi}{T} c_n \tau_n* T\delta(t - nT). \qquad (52)$$

The model of the PLL for digital signals with interpulse interference can then be evaluated from Fig. 2(c) and (d).

The actual phase $\theta_{ia}*(NT)$ of the input signal $s_i(t)$ may be written as

$$\theta_{ia}*(NT) = \Delta \omega_b NT - \frac{2\pi}{T} \tau_N**. \qquad (53)$$

Here $\Delta \omega_b$, as before, is the detuning of the quiescent frequency of the VCO from the angular digit frequency $2\pi/T$ and τ_N** are the deviations of the pulses from their nominal positions.

The block diagram of Fig. 2(d) can be reduced again to that of Fig. 2(c) by defining

$$\theta_{ia}(NT) = \theta_{ia}*(NT) - \frac{2\pi}{T} \tau_N* \qquad (54)$$

$$\theta_{ea}(NT) = \theta_{ia}(NT) - \theta_o(NT). \qquad (55)$$

Combining (53) and (54), we get

$$\theta_{ia}(NT) = \Delta\omega_b NT - \frac{2\pi}{T}\tau_N \qquad (56)$$

with

$$\tau_N = \tau_N^{**} + \tau_N^{*}.$$

Apparently, we may use the same model as we had before for short pulses, provided we interpret c_n and τ_n suitably:

1) c_n is a random variable, defined by (50), which is influenced by interpulse interference;

2) τ_n consists of two parts, τ_n^{**} and τ_n^{*}. The first part represents the phase deviations of the incoming pulses. The second part originates from interpulse interference and may be calculated by (51) and (50).

About the actual phases of the input signal $s_1(t)$, defined by (45), and the fundamental signal component $s_2(t)$, it may be stated that these are equal provided

1) interpulse interference is restricted to a limited number of neighboring pulses,

2) neighboring pulses are equidistant, which is true as long as τ_n^{**} has a low-frequency character. The noise contributions to τ_n^{**} may be neglected in practical systems. Other contributions to τ_n^{**} are introduced by preceding repeaters, and will indeed have a low-frequency character because the power spectrum of τ_n^{**} is shaped by phase-transfer functions of those repeaters.

We are now able to calculate the jitter due to intersymbol interference for a single repeater, i.e., if $\tau_n^{**} = 0$. As we have seen, this jitter is fully determined by the following.

1) The properties of the transmission channel. The pulse response of this channel should be known.

2) The statistics of the digital signal. The transmission code should be known. Conditions 1) and 2) are necessary to calculate (45).

3) The nature of the essential nonlinear process. The nonlinearity should be specified to calculate (46).

4) The properties of the PLL filter. A numerical calculation can now be made by applying successively formulas (45), (46), (50), (51), and (17).

As an example, calculated results are shown in Fig. 3 for transmission channels, fulfilling Nyquist's first criterion at the input of the nonlinearity. The exhibited variables are defined as follows:

$\overline{\Delta\theta_{e\tau}^2}$ is the mean-square jitter due to $(c_n/\bar{c}_n)(\overline{c_n\tau_n}/\bar{c}_n - \tau_n)$,

$\overline{\Delta\theta_{ec}^2}$ is the mean-square jitter due to

$$(c_n - \bar{c}_n)\theta_{eo}(1)/\bar{c}_n,$$

ρ is the correlation coefficient of both contributions. The total mean-square jitter $\overline{\Delta\theta_e^2}$ can be found from

$$\overline{\Delta\theta_e^2} = \overline{\Delta\theta_{e\tau}^2} + \overline{\Delta\theta_{ec}^2} + 2\rho(\overline{\Delta\theta_{e\tau}^2}\cdot\overline{\Delta\theta_{ec}^2})^{1/2}. \qquad (57)$$

From (18) it is obvious that $(\overline{\Delta\theta_{ec}^2})^{1/2}$ can be expressed per radian quasi-static phase error.

Fig. 3(a) and (b) show the influence of the rolloff

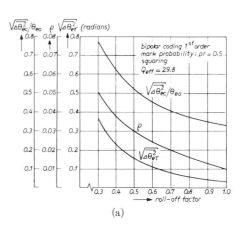

(a)

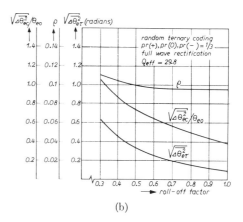

(b)

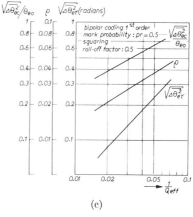

(c)

Fig. 3. (a) Jitter due to interpulse interference for Nyquist I channels and squared bipolar signals. (b) Jitter due to interpulse interference for Nyquist I channels and rectified random ternary signals. (c) Influence of effective quality factor on jitter due to interpulse interference.

factor of the channel (rolloff factor as defined in [5, ch. 5]) for two different processes. Fig. 3(c) shows the influence of the effective quality factor Q_{eff} of the PLL.

From the graphs we may conclude the following.

1) Squaring of symmetrical pulses, e.g., Nyquist I pulses, gives no guarantee of jitter-free operation, as is sometimes thought and found in the literature [8].

2) The rms value $(\overline{\Delta\theta_{e\tau}^2})^{1/2}$ is inversely proportional to the effective quality factor. Apparently, this means that the spectral density $S_{c\tau c\tau}$ of the random variable $(c_n/\bar{c}_n)(\overline{c_n\tau_n}/\bar{c}_n - \tau_n)$ increases linearly with frequency for low frequencies. This can be explained from the fact that

the average interpulse interference does not influence the symmetry of the waveforms. Consequently, the spectral density S_{crcr} is zero for zero frequency.

We may state that the effective quality factor in band-limited systems, where there is a significant influence of interpulse interference, is a more important parameter than suggested by the results of the theory for idealized systems as given by (31) and (37) and the corresponding expressions for idealized resonant-type systems.

3) The rms value $(\overline{\Delta\theta_{ec}^2})^{1/2}$ increases with the square root of the inverse effective quality factor. Apparently, the spectral density S_{cc} of the random variable $(c_n - \bar{c}_n)/\bar{c}_n$ is almost constant for low frequencies, as might be expected.

4) Even for large rolloff factors (≈ 1), there appears to be a very considerable influence of interpulse interference. Calculating $(\overline{\Delta\theta_{ec}^2})^{1/2}/\theta_{eo}(1)$ for the case of zero interpulse interference by use of (31) yields

$$(\overline{\Delta\theta_{ec}^2})^{1/2}/\theta_{eo}(1) = 0.16, \qquad \text{if } Q_{\text{eff}} = 29.8.$$

Note the difference with the results in Fig. 3(a).

5) The correlation of $\Delta\theta_{ec}$ and $\Delta\theta_{er}$ is neglectably small.

VIII. OPTIMAL PREPROCESSING

In Section III, two methods were mentioned that avoid jitter due to interpulse interference. In this section, we shall prove this statement for the symmetry method.

The equivalent transmission channel up to the nonlinearity has a symmetrical transfer function around the Nyquist frequency, yielding a signal

$$s_1(t) = \sum_{n=0}^{N} a_n f(t - nT) \cos \frac{\omega_b}{2} (t - nT). \qquad (58)$$

a_n is a discrete random variable. $f(t)$ is the pulse response of the transposed equivalent low-pass filter of the symmetrical transfer function.

$$\omega_b = \frac{2\pi}{T} \qquad (59)$$

is the digit frequency.

Squaring the signal $s_1(t)$, we find that

$$s_2(t) = \left\{ \sum_{n=0}^{N} a_n f(t - nT) \cos \frac{\omega_b}{2} (t - nT) \right\}^2$$

$$= \tfrac{1}{2}(\cos \omega_b t + 1) \Big\{ \sum\sum_{n+m=\text{odd}} a_n a_m f(t - nT) f(t - mT)$$

$$- \sum\sum_{n+m=\text{even}} a_n a_m f(t - nT) f(t - mT). \qquad (60)$$

Let the spectrum of $f(t)$ be limited in frequency:

$$F(\omega) = 0, \qquad \text{for } \omega > \omega_1. \qquad (61)$$

The signal $s_2(t)$ can then be conceived as a superposition of an amplitude-modulated signal with a nominal frequency ω_b whose sidebands reach from $\omega_b - 2\omega_1$ to $\omega_b + 2\omega_1$ and a signal with frequency components below $2\omega_1$. Let this signal be filtered by a symmetrical bandpass filter with

angular center frequency ω_b and a bandwidth limited between $\omega_b + \omega_H$ and $\omega_b - \omega_H$. As a result, the zero crossings of the filtered signal have a constant spacing T as long as

$$2\omega_1 < \omega_b - \omega_H. \qquad (62)$$

After elimination of the amplitude modulation, a jitterless clock signal is obtained. The proof for the filtering by means of a PLL can be given as follows. In the phase detector, which will be an ideally balanced multiplier, signal $s_2(t)$ of (60) is multiplied by a locked oscillator signal. Suppose, this oscillator signal does not exhibit phase fluctuations, so that it can be represented by $\sin \omega_b t$. As a result, two signal components are produced, one with a spectrum from $2\omega_b - 2\omega_1$ to $2\omega_b + 2\omega_1$, and one with a spectrum from $\omega_b - \omega_1$ to $\omega_b + \omega_1$. As long as the noise bandwidth of the loop is limited to ω_H, fulfilling condition (62), the phase of the VCO will not change.

Summarizing

In order to derive from a band-limited synchronous digital signal a clock signal which does not exhibit jitter due to interpulse interference, the following processes could be performed:

1) a linear process such that the equivalent transmission path has a symmetrical transfer function around half the digit frequency,

2) a nonlinear process such that the signal is squared after process 1),

3) a narrow-band filtering with a noise bandwidth, limited to a lower frequency value than the digit frequency minus the bandwidth of the transmission path of 1).

IX. PROPAGATION OF PHASE ERRORS

The process of jitter propagation in a chain of repeaters has been studied by several authors, including Byrne *et al.* [4], for the clock extraction procedure by a resonant circuit. We shall, in this section, use essentially the same model to derive results for more general processes.

If the phase errors in the input signal have a more or less uniform frequency spectrum in the frequency range, covered by the phase-transfer function $H(j\omega)$ of the narrow-band filter or PLL, and if fluctuations introduced by interpulse interference do not possess pronounced spectral components in this frequency range, it may be stated for the phase of the derived clock signal that

$$\Theta(j\omega) = kH(j\omega) \qquad (63)$$

in which k is a constant.

This relation can only be valid for the first repeater in the chain. The second repeater will show in its output signal not only an identical contribution of its own, but also a response on the spectral components, introduced by the first repeater. Because its own contribution is due to the structure of the pattern of digits in the input signal, the fluctuations will add coherently to the fluctuations of the preceding repeater, so that

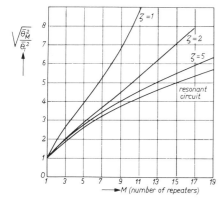

Fig. 4. Normalized absolute jitter as a function of the number of repeaters with high-gain PLL's.

$$\Theta_2(j\omega) = kH(j\omega) + kH(j\omega) \cdot H(j\omega). \quad (64)$$

The first term represents the second repeater's own contribution, and the second term the response on the spectral components originated from the first repeater. Hence, after M repeaters, we find

$$\Theta_M(j\omega) = kH(j\omega) + kH(j\omega)^2 + \cdots kH(j\omega)^M$$

$$= kH(j\omega)\frac{H(j\omega)^M - 1}{H(j\omega) - 1}. \quad (65)$$

The mean-square value $\overline{\theta_M^2}$ after M repeaters can then be calculated from

$$\overline{\theta_M^2} = \frac{1}{2\pi}\int_{-\infty}^{\infty} |\Theta_M(j\omega)|^2 \, d\omega. \quad (66)$$

This fluctuation can be denoted as absolute jitter, and has as a reference the nominal position of the digits of the original transmitted signal. This jitter is important because at too large a value, it causes foldover distortion in the decoding process of the digital signal (see [2]).

In Fig. 4 a comparison is given of the jitter propagation for clock extraction by a resonant circuit and for a second-order high-gain PLL. This result has been obtained by a numerical computation of (65) and (66). The phase-transfer function of the second-order PLL is defined by (38). The loop is supposed to be of the high-gain type as long as condition (41) is fulfilled.

It is very clear from the figure that proper performance of a chain of repeaters with high-gain PLL's requires high damping factors.

The exponential growth of the jitter for low damping factors is caused by the fact that in the Bode diagram of the phase-transfer function, the corner frequency of the zero has a lower value than the corner frequencies of the two poles, as is indicated in Fig. 5. Fluctuations with spectral components near the natural frequency are then the more increased the longer the chain is.

This is avoided for a first-order function, like a resonant circuit, a first-order loop, or a degenerated second-order loop with infinite high damping factor [Fig. 5(b)]. If the corner frequency of the zero has a value between those

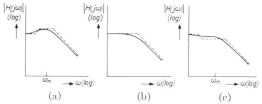

Fig. 5. Phase-transfer function for various types of timing filters. (a) High-gain PLL. (b) Resonant circuit. (c) Special low-gain PLL.

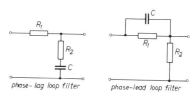

Fig. 6.

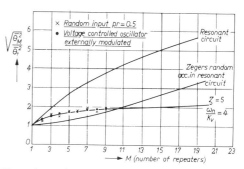

Fig. 7. Normalized absolute jitter as a function of the number of repeaters with well-designed low-gain PLL's.

of the poles, a further improvement in this respect is possible [Fig. 5(c)]. This condition is fulfilled if

$$\zeta - \zeta\left(1 - \frac{1}{\zeta^2}\right)^{1/2} < 2\zeta - \frac{\omega_n}{K_v} < \zeta + \zeta\left(1 - \frac{1}{\zeta^2}\right)^{1/2}. \quad (67)$$

It can easily be proved that this condition can be obtained by replacing the usual phase-lag loop filter by a phase-lead type (Fig. 6). Of course, the transfer function as a whole remains of the integrating type, due to the influence of the VCO, which is an ideal integrator in the mathematical model. An illustration of possible results is shown in Fig. 7. Even better results are possible than those obtained by Zegers [9], who, in a repeater with a resonant timing extraction circuit, used a scrambler in order to break the coherence of the systematic components.

An experimental chain of ten repeaters has been built to confirm the theoretical expectation. The dots in Fig. 7 represent a test of a chain operating with a fully deterministic signal, while the VCO's in succeeding repeaters are modulated coherently with a sine wave. The crosses are the results of a test with a random digital pattern without external modulation. Whether such an implementation can be used in a practical system depends on the stability and accuracy of the quiescent frequency of the VCO. Normal practice for the design of a second-order phase-

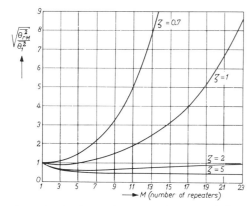

Fig. 8. Normalized relative jitter as a function of the number of repeaters with high-gain PLL's.

locked timing extraction circuit for a chain of repeaters is

1) to fix the required damping factor from propagation properties, as shown in Fig. 7,

2) to fix the natural frequency from a specification of the required noise bandwidth and acquisition time; this is possible by using the figures of Richman [10],

3) to fix the velocity constant such that the static phase error for a given stability of the quiescent VCO frequency is small enough.

An implementation according to the parameters of Fig. 7 eliminates the degree of freedom, as indicated in 3), meaning that the stability and accuracy of the VCO should be high enough to afford the desired velocity constant. Inspection of required parameters has shown that implementation with crystal-stabilized VCO's gives satisfactory results.

Apart from absolute jitter, relative jitter is also important, i.e., fluctuations between the phase of the digits of the input signal of a repeater and the phase of the derived clock signal. This type of jitter influences the error probability in the regeneration process. This jitter may be calculated, as Byrne has done, from the difference of the spectra of phase fluctuations in two succeeding repeaters, i.e.,

$$\Theta_{rM}(j\omega) = \Theta_M(j\omega) - \Theta_{M-1}(j\omega)$$
$$= kH(j\omega)^M.$$

The propagation of relative jitter for high-gain phase-locked timing extraction circuits is shown in Fig. 8.

X. CONCLUSIONS

Comparing the properties of clock extraction systems, either implemented by a resonant circuit or by a second-order PLL, we conclude the following.

1) Reduction of both quasi-static as well as dynamic phase error is by no means simply possible with resonant-type circuits: the errors are dependent; a compromise is necessary. There is no such dependence in the case of a high-gain second-order PLL.

2) In resonant-type circuits, in contrast to the signal from the PLL, the derived signal is amplitude modulated.

This modulation has to be removed by a hard limiter, which may introduce new phase errors. This is the more serious the higher the frequency is and the more the energy density of the pattern varies between wider limits.

3) The phase-transfer function of the resonant circuit is of the first order. The PLL under consideration has a second-order phase-transfer function. To avoid exponential growth of jitter in a chain of repeaters, high damping factors ($\zeta \approx 5$) are necessary for high-gain loops. If low-gain loops can be tolerated, which requires a highly stable VCO, a very small jitter accumulation can be realized by proper dimensioning.

4) The phase-transfer function of the PLL is weighted by the energy density of the signal. Faulty performance in worst case signals is possible.

A method has been given to calculate the influence of interpulse interference on phase errors for the case that the three essential parts of the clock extraction procedure, i.e., prefiltering, nonlinear process, and narrow-band filtering, are properly specified. Although this method has only been derived for the PLL system, it can be proved that it also applies to resonant-type systems. Optimum processing requires a clock path in which prefiltering yields a symmetrical transfer function around the Nyquist frequency in which nonlinearity is a squaring circuit and where narrow-band filtering fulfills certain conditions, as specified in Section VIII.

APPENDIX I

A. Solution of the System Equation

We write for the system equation (12)

$$\theta_e(NT) + \sum_{n=0}^{N} \theta_e(nT) g_w(NT - nT)$$

$$+ \sum_{n=0}^{N} \frac{c_n - \bar{c}_n}{\bar{c}_n} g_w(NT - nT)$$

$$= \Delta\omega_b NT + \frac{2\pi}{T} \sum_{n=0}^{N} g_w(NT - nT)$$

$$+ \frac{2\pi}{T} \sum_{n=0}^{N} \frac{c_n \tau_n - \overline{c_n \tau_n}}{\bar{c}_n} g_w(NT - nT). \quad (68)$$

We may split $\theta_e(nT)$ into an expected value $\overline{\theta_e(nT)}$ and a zero-mean random part $\Delta\theta_e(nT)$:

$$\theta_e(nT) = \overline{\theta_e(nT)} + \Delta\theta_e(nT). \quad (69)$$

Substitution of (69) into (68) and separation into expected values and zero-mean random contributions yields

$$\overline{\theta_e(NT)} + \sum_{n=0}^{N} \overline{\theta_e(nT)} g_w(NT - nT)$$

$$= \frac{2\pi}{T} \frac{\overline{c_n \tau_n}}{\bar{c}_n} \sum_{n=0}^{N} g_w(NT - nT) + \Delta\omega_b NT$$

$$-\sum_{n=0}^{N} \overline{\Delta\theta_e(nT) \frac{c_n - \bar{c}_n}{\bar{c}_n}} g_w(NT - nT) \qquad (70)$$

and

$$\Delta\theta_e(NT) + \sum_{n=0}^{N} \Delta\theta_e(nT)g_w(NT - nT)$$

$$= \frac{2\pi}{T} \sum_{n=0}^{N} \frac{c_n\tau_n - \overline{c_n\tau_n}}{\bar{c}_n} g_w(NT - nT) - \sum_{n=0}^{N} \overline{\theta_e(nT)}$$

$$\cdot \frac{c_n - \bar{c}_n}{\bar{c}_n} g_w(NT - nT) + \text{term} \qquad (71)$$

with

$$\text{term} \equiv -\sum_{n=0}^{N} \left\{ \Delta\theta_e(nT) \frac{c_n - \bar{c}_n}{\bar{c}_n} - \overline{\Delta\theta_e(nT) \frac{c_n - \bar{c}_n}{\bar{c}_n}} \right\}$$

$$\cdot g_w(NT - nT). \qquad (72)$$

We shall prove in Section C of this Appendix that the contribution of "term" can be neglected in most cases, and we shall then discuss the conditions for which this is allowed.

Normally, there is a pole in the origin of the open-loop transfer function of a PLL. We may then expect that in the stationary condition the expected error will be a constant, so that

$$\lim_{N\to\infty} \overline{\theta_e(NT)} = \theta_{eo} = \text{constant.} \qquad (73)$$

Substituting this in (70), we find

$$\theta_{eo} = \theta_{eo}(1) + \theta_{eo}(2) + \theta_{eo}(3) \qquad (74)$$

with

$$\theta_{eo}(1) = \Delta\omega_b NT / \left[\lim_{N\to\infty} \sum_{n=0}^{N} g_w(NT - nT) \right] \qquad (75)$$

$$\theta_{eo}(2) = \frac{2\pi}{T} \frac{\overline{c_n\tau_n}}{\bar{c}_n} \qquad (76)$$

$$\theta_{eo}(3) = -\overline{\Delta\theta_e(NT) \frac{c_N - \bar{c}_N}{\bar{c}_N}} \qquad (77)$$

provided

$$\lim_{N\to\infty} \sum_{n=0}^{N} g_w(NT - nT) \gg 1. \qquad (78)$$

Condition (78) is fulfilled because of the pole in the origin of the open-loop transfer function. After substitution of the results (73)–(77) into (71), we find for the equation for the random part $\Delta\theta_e(nT)$ neglecting "term," for large N (so that $\overline{\theta_e(nT)}$ has become independent of N), that

$$\Delta\theta_e(NT) + \sum_{n=0}^{N} \Delta\theta_e(nT)g_w(NT - nT)$$

$$= \sum_{n=0}^{N} b_n g_w(NT - nT) \qquad (79)$$

in which

$$b_n = -\frac{c_n - \bar{c}_n}{\bar{c}_n} \theta_{eo} + \frac{2\pi}{T} \frac{c_n\tau_n - \overline{c_n\tau_n}}{\bar{c}_n} \qquad (80)$$

or, using (74) and (76),

$$b_n = -\frac{c_n - \bar{c}_n}{\bar{c}_n} (\theta_{eo}(1) + \theta_{eo}(3))$$

$$+ \frac{2\pi}{T} \frac{c_n}{\bar{c}_n} \left(\frac{\overline{c_n\tau_n}}{\bar{c}_n} - \tau_n \right). \qquad (81)$$

Equation (79) can be conveniently solved by application of the z-transform, giving

$$\Delta\Theta_e(z) + \Delta\Theta_e(z)G_w(z) = \sum_{n=0}^{N} b_n G_w(z)z^{-n} \qquad (82)$$

or

$$\Delta\Theta_e(z) = \sum_{n=0}^{N} b_n \frac{G_w(z)}{1 + G_w(z)} z^{-n}. \qquad (83)$$

After the inverse transform, the result is

$$\Delta\theta_e(NT) = \sum_{n=0}^{N} b_n h_w(NT - nT), \qquad (84)$$

still under the condition of large N.

Herein $h_w(NT)$ is the inverse transform of $H_w(z)$, defined as

$$H_w(z) = \frac{G_w(z)}{1 + G_w(z)} \qquad (85)$$

which is, in fact, the weighted closed-loop transfer function.

The interrelation of θ_{eo} and $\Delta\theta_e(NT)$ via $\theta_{eo}(3)$ as expressed by (74), (77), and (81) makes a solution of these equations difficult. In Section C of this Appendix we shall prove that $\theta_{eo}(3)$ gives a negligible contribution in practical systems. The solution of the system equation is then given by (74)–(76), (84), and (81) neglecting $\theta_{eo}(3)$.

B. Definitions

For the next discussions, we shall use discrete Fourier pairs of some variables. The time interval $(N + 1)T$ will be regarded as the fundamental period. The corresponding fundamental frequency is then given by

$$\Omega = \frac{2\pi}{(N + 1)T}. \qquad (86)$$

We define discrete Fourier pairs of the following.

1) *Autocorrelation Functions:*

$$R_{cc}(kT) = \overline{\left(\frac{c_n - \bar{c}_n}{\bar{c}_n} \right) \left(\frac{c_{n+k} - \bar{c}_n}{\bar{c}_n} \right)}$$

$$S_{cc}(l\Omega) = \sum_{k=0}^{N} R_{cc}(kT) \exp{(-j\Omega Tlk)}$$

$$R_{cc}(kT) = \frac{1}{N+1} \sum_{l=0}^{N} S_{cc}(l\Omega) \exp(j\Omega Tkl). \quad (87)$$

[c_n has been defined first by (1)].

Analogously,

$$R_{\tau\tau}(kT) = \overline{(\tau_n - \bar{\tau}_n)(\tau_{n+k} - \bar{\tau}_n)}, \quad \tau_n \text{ cf. (1)}$$

$$S_{\tau\tau}(l\Omega) \Leftrightarrow R_{\tau\tau}(kT) \quad (88)$$

$$R_{c\tau c\tau}(kT) = \overline{\frac{c_n}{\bar{c}_n}\left(\frac{\overline{c_n\tau_n}}{\bar{c}_n} - \tau_n\right)\frac{c_{n+k}}{\bar{c}_n}\left(\frac{\overline{c_n\tau_n}}{\bar{c}_n} - \tau_{n+k}\right)}$$

$$S_{c\tau c\tau}(l\Omega) \Leftrightarrow R_{c\tau c\tau}(kT) \quad (89)$$

$$R_{bb}(kT) = \overline{b_n b_{n+k}}, \quad b_n \text{ cf. (81)}$$

$$S_{bb}(l\Omega) \Leftrightarrow R_{bb}(kT) \quad (90)$$

$$R_{\Delta\theta_e \Delta\theta_e}(kT) = \overline{\Delta\theta_e(nT)\Delta\theta_e(nT+kT)}, \Delta\theta_e(nT) \text{ cf. (84)}$$

$$S_{\Delta\theta_e \Delta\theta_e}(l\Omega) \Leftrightarrow R_{\Delta\theta_e \Delta\theta_e}(kT) \quad (91)$$

$$R_{(c\Delta\theta_e)^2}(kT)$$

$$= \overline{\left(\frac{c_n - \bar{c}_n}{\bar{c}_n}\right)\Delta\theta_e(nT)\left(\frac{c_{n+k} - \bar{c}_n}{\bar{c}_n}\right)\Delta\theta_e(nT+kT)}$$

$$S_{(c\Delta\theta_e)^2}(l\Omega) \Leftrightarrow R_{(c\Delta\theta_e)^2}(kT). \quad (92)$$

2) Cross Correlation Function:

$$R_{c\tau c}(kT) = \overline{\frac{c_n}{\bar{c}_n}\left(\frac{\overline{c_n\tau_n}}{\bar{c}_n} - \tau_n\right)\left(\frac{c_{n+k} - \bar{c}_n}{\bar{c}_n}\right)}$$

$$S_{c\tau c}(l\Omega) \Leftrightarrow R_{c\tau c}(kT). \quad (93)$$

3) Pulse Response:

$$H_w(l\Omega) \Leftrightarrow h_w(nT). \quad (94)$$

C. Constraints

In this section, we shall discuss the conditions for which the solution of the system equation, as derived in Section A, is true.

1) First we discuss the irrelevance of $\theta_{eo}(3)$ in (80). From (84) and (77) it follows that

$$\theta_{eo}(3) = -\overline{\frac{c_N - \bar{c}_N}{\bar{c}_N}} \sum_{n=0}^{N} b_n h_w(NT - nT). \quad (95)$$

Defining a new variable $k = N - n$, we obtain

$$\theta_{eo}(3) = -\sum_{k=0}^{N} \overline{\frac{c_N - \bar{c}_N}{\bar{c}_N}} b_{N-k} h_w(kT). \quad (96)$$

Using (81), (87), and (93), this can be written as

$$\theta_{eo}(3) = \sum_{k=0}^{N} \left[R_{cc}(-kT)\{\theta_{eo}(1) + \theta_{eo}(3)\} \right.$$

$$\left. - \frac{2\pi}{T} R_{c\tau c}(-kT) \right] h_w(kT). \quad (97)$$

By application of (87), (93), and the well-known orthogonality relationships [13]

$$\sum_{k=0}^{N} \exp(j\Omega Tlk) = N + 1, \quad \text{if } l = 0 \pmod{N+1}$$

$$= 0, \quad \text{otherwise} \quad (98)$$

we find

$$\theta_{eo}(3) = \theta_{eo}(3a) + \theta_{eo}(3b) \quad (99)$$

with

$$\theta_{eo}(3a) = \{\theta_{eo}(1) + \theta_{eo}(3)\} \frac{1}{N+1} \sum_{l=0}^{N} S_{cc}(l\Omega) H_w(l\Omega) \quad (100)$$

$$\theta_{eo}(3b) = -\frac{2\pi}{T} \frac{1}{N+1} \sum_{l=0}^{N} S_{c\tau c}(l\Omega) H_w(l\Omega). \quad (101)$$

From (100) we conclude that $\theta_{eo}(3a)$ may be neglected with regard to $\{\theta_{eo}(1) + \theta_{eo}(3b)\}$ if

$$\frac{1}{N+1} \sum_{l=0}^{N} S_{cc}(l\Omega) H_w(l\Omega) \ll 1. \quad (102)$$

$\theta_{eo}(3b)$ is neglectable in (81) if the last term of the right-hand part of (81) is much smaller than the contribution of $\theta_{eo}(3b)(c_n - \bar{c}_n)/\bar{c}_n$. Let us compare these quantities in terms of their standard deviations. $\theta_{eo}(3b)$ is neglectable provided

$$\theta_{eo}(3b)[R_{cc}(0)]^{1/2} \ll \frac{2\pi}{T}[R_{c\tau c\tau}(0)]^{1/2}. \quad (103)$$

Substituting (101), we may write

$$[R_{cc}(0)]^{1/2} \frac{1}{N+1} \sum_{l=0}^{N} S_{c\tau c}(l\Omega) H_w(l\Omega)$$

$$\ll [R_{c\tau c\tau}(0)]^{1/2}. \quad (104)$$

Let us discuss the conditions (102) and (103). In practical cases, $S_{cc}(l\Omega)$ and $S_{c\tau c}(l\Omega)$ do not change significantly in the frequency range covered by $H_w(l\Omega)$. We may write then for (102)

$$S_{cc}(0) \frac{1}{N+1} \sum_{l=0}^{N} H_w(l\Omega) \ll 1. \quad (105)$$

Also

$$\frac{1}{N+1} \sum_{l=0}^{N} H_w(l\Omega) = h_w(0). \quad (106)$$

Therefore (105) reduces to

$$S_{cc}(0) h_w(0) \ll 1. \quad (107)$$

Using result (134) of Appendix III, we may bound (107) by

$$S_{cc}(0)(2\pi B_L T)^{1/2} \ll 1. \quad (108)$$

Analogously, we may evaluate (104) to

$$[R_{cc}(0)]^{1/2} S_{c\tau c}(0)(2\pi B_L T)^{1/2} \ll [R_{c\tau c\tau}(0)]^{1/2}. \quad (109)$$

Very small values of $\bar{c}_n$ deteriorate condition (108). Therefore we denote (108) as the condition of sufficient clock content.

From the definition of $R_{crc}(kT)$ as given by (93), we may conclude that in the case of mutually statistical independence of c_n and τ_n, $R_{crc}(kT)$ is zero. In that case, (109) is always fulfilled. Because of the fact that mutual statistical dependence is caused by the mechanism of interpulse interference as described in Section VII, we denote (109) as the condition of restricted interpulse interference.

2) In this section, we shall motivate the neglect of "term" in (71). The influence of "term" can be estimated by considering it as a second-order effect. Substituting results (84) and (77) into "term" and again solving (71) for the stationary condition $(N \rightarrow \infty)$ yields a contribution $\Delta\theta_e(NT)_a$ in addition to $\Delta\theta_e(NT)$ as given by (84). Let us assume that $\Delta\theta_e(NT)_a \ll \Delta\theta_e(NT)$ and investigate whether this assumption can be verified. We may write for the total dynamic phase error $\Delta\theta_e(NT)_t$

$$\Delta\theta_e(NT)_t = \Delta\theta_e(NT) + \Delta\theta_e(NT)_a \qquad (110)$$

with

$$\Delta\theta_e(NT) = \sum_{n=0}^{N} b_n h_w(NT - nT) \qquad (111)$$

$$\Delta\theta_e(NT)_a = \sum_{n=0}^{N} \left[\frac{c_n - \bar{c}_n}{\bar{c}_n} \Delta\theta_e(nT) - \theta_{eo}(3) \right]$$
$$\cdot h_w(NT - nT). \qquad (112)$$

In the previous section, we already proved that $\theta_{eo}(3)$ is neglectable with regard to b_n, provided conditions (108) and (109) are fulfilled. For (112) there remains a relevant contribution:

$$\Delta\theta_e(NT)_b = \sum_{n=0}^{N} \frac{c_n - \bar{c}_n}{\bar{c}_n} \Delta\theta_e(nT) h_w(NT - nT). \qquad (113)$$

Its variance can be written, using Appendix II and (92), as

$$\lim_{N\to\infty} \overline{\Delta\theta_e(NT)_b{}^2} = \lim_{N\to\infty} \frac{1}{N+1} \sum_{l=0}^{N} S_{(c\Delta\theta_e)^2}(l\Omega) \mid H_w(l\Omega) \mid^2 \qquad (114)$$

in which $S_{(c\Delta\theta_e)^2}(l\Omega)$ is the spectral density of the random variable $\Delta\theta_e(nT)(c_n - \bar{c}_n)/\bar{c}_n$.

Analogously, we may write for the variance of $\Delta\theta_e(NT)$ of (111)

$$\lim_{N\to\infty} \overline{\Delta\theta_e(NT)^2} = \lim_{N\to\infty} \frac{1}{N+1} \sum_{l=0}^{N} S_{bb}(l\Omega) \mid H_w(l\Omega) \mid^2 \qquad (115)$$

in which $S_{bb}(l\Omega)$ is the spectral density of b_n, as defined by (90). The spectral density of $\Delta\theta_e(NT)$ is limited by $H_w(l\Omega)$. Supposing the condition of sufficient clock content (108) to be fulfilled, we conclude that the spectal density of $(c_N - \bar{c}_N)/\bar{c}_N$ has a very small intersection with the spectral density of $\Delta\theta_e(NT)$, so that these random variables may be regarded to be statistically independent. Then considering (92), (87), and (91),

$$R_{(c\Delta\theta_e)^2}(kT) = R_{cc}(kT) R_{\Delta\theta_e\Delta\theta_e}(kT), \qquad (116)$$

and consequently

$$S_{(c\Delta\theta_e)^2}(l\Omega) = S_{cc}(l\Omega) * S_{\Delta\theta_e\Delta\theta_e}(l\Omega) \qquad (117)$$

in which $*$ denotes a convolution.

Also, from (115)

$$S_{\Delta\theta_e\Delta\theta_e}(l\Omega) = S_{bb}(l\Omega) \mid H_w(l\Omega) \mid^2. \qquad (118)$$

After substitution of (117) and (118) into (114), we find

$$\lim_{N\to\infty} \overline{\Delta\theta_e(NT)_b{}^2} = \lim_{N\to\infty} \frac{1}{N+1}$$
$$\cdot \sum_{l=0}^{N} \{ S_{cc}(l\Omega) * S_{bb}(l\Omega) \mid H_w(l\Omega) \mid^2 \}$$
$$\cdot \mid H_w(l\Omega) \mid^2. \qquad (119)$$

Assuming that $S_{cc}(l\Omega)$ is almost constant and equal to $S_{cc}(0)$ in the frequency range covered by $H_w(l\Omega)$, we may write, using (19),

$$\lim_{N\to\infty} \overline{\Delta\theta_e(NT)_b{}^2} = 2\pi B_L T S_{cc}(0) \lim_{N\to\infty} \frac{1}{N+1}$$
$$\cdot \sum_{l=0}^{N} S_{bb}(l\Omega) \mid H_w(l\Omega) \mid^2. \qquad (120)$$

Comparing now (120) and (115), we may conclude that

$$\lim_{N\to\infty} \overline{\Delta\theta_e(NT)_b{}^2} \ll \lim_{N\to\infty} \overline{\Delta\theta_e(NT)^2} \qquad (121)$$

if the condition (108) of sufficient clock content is fulfilled. Consequently, our initial assumption $\Delta\theta_e(NT)_a \ll \Delta\theta_e(NT)$ is also allowed, and therefore "term" in (71) may be neglected.

APPENDIX II

Jitter in Spectral Terms

For the following evaluation the author is indebted to H. van den Elzen.

From (17) it follows that

$$\Delta\theta_e(NT) = \sum_{n=0}^{N} b_n h_w(NT - nT). \qquad (122)$$

Hence

$$\overline{\Delta\theta_e(NT)^2} = \sum_{n=0}^{N} \sum_{m=0}^{N} \overline{b_n b_m} h_w(NT - nT) h_w(NT - mT). \qquad (123)$$

Let

$$m = n + k. \qquad (124)$$

127

Then

$$\overline{\Delta\theta_e(NT)^2} = \sum_{n=0}^{N}\sum_{m=0}^{N} \overline{b_n b_{n+k}} h_w(NT - nT)$$

$$\cdot h_w(NT - nT - kT). \quad (125)$$

Using (90) and (94) we find

$$\overline{\Delta\theta_e(NT)^2} = \frac{1}{(N+1)^2} \sum_{n=0}^{N}\sum_{k=-n}^{N-n} R_{bb}(kT) \sum_{l_1=0}^{N} H_w(l_1\Omega)$$

$$\cdot \exp\left[\,j\Omega T(N-n)l_1\right] \sum_{l_2=0}^{N} H_w(l_2\Omega)$$

$$\cdot \exp\left[\,j\Omega T(N-n-k)l_2\right]. \quad (126)$$

After a change of the order of summation and some re-arranging, this becomes

$$\overline{\Delta\theta_e(NT)^2} = \frac{1}{(N+1)^2} \sum_{l_1=0}^{N}\sum_{l_2=0}^{N} H_w(l_1\Omega) H_w(l_2\Omega)$$

$$\cdot \sum_{n=0}^{N} \exp\left[-j\Omega Tn(l_1+l_2)\right] \sum_{k=-n}^{N-n} R_{bb}(kT)$$

$$\cdot \exp\left[-j\Omega Tkl_2\right]. \quad (127)$$

Assume for the moment that $R_{bb}(kT)$ is a periodic function with period $(N+1)T$; then

$$\sum_{k=-n}^{N-n} R_{bb}(kT) \exp\,(-j\Omega Tkl_2) = \sum_{k=0}^{N} R_{bb}(kT) \exp\,(-j\Omega Tkl_2)$$

$$= S_{bb}(l\Omega) \quad (128)$$

which is the discrete Fourier transform of the autocorrelation function $R_{bb}(kT)$.

Using (128) and the orthogonality relationship of (98), we write for (127)

$$\overline{\Delta\theta_e(NT)^2} = \frac{1}{N+1} \sum_{l=0}^{N} S_{bb}(l\Omega) H_w(l\Omega) H_w(-l\Omega) \quad (129)$$

or

$$\overline{\Delta\theta_e(NT)^2} = \frac{1}{N+1} \sum_{l=0}^{N} S_{bb}(l\Omega) \mid H_w(l\Omega)\mid^2. \quad (130)$$

By extension of the series to $N \to \infty$, the condition of periodicity of $R_{bb}(kT)$ becomes trivial. Therefore

$$\overline{\Delta\theta_e^2} = \lim_{N\to\infty} \overline{\Delta\theta_e(NT)^2}$$

$$= \lim_{N\to\infty} \frac{1}{N+1} \sum_{l=0}^{N} S_{bb}(l\Omega) \mid H_w(l\Omega)\mid^2. \quad (131)$$

APPENDIX III

Relation Between B_L and $h_w(0)$

The relation between B_L and $h_w(0)$ can be found from Cauchy's inequality, which can be written in the general form

$$\left[\sum_{l=0}^{N} a_l b_l\right]^2 \le \sum_{l=0}^{N} a_l^2 \sum_{l=0}^{N} b_l^2. \quad (132)$$

Taking $a_l = H_w(l\Omega)$ and $b_l = 1$, we find

$$\left[\sum_{l=0}^{N} H_w(l\Omega)\right]^2 \le (N+1) \sum_{l=0}^{N} \mid H_w(l\Omega)\mid^2. \quad (133)$$

Using (106) and (19), it follows that

$$h_w(0)^2 \le 2\pi B_L T. \quad (134)$$

ACKNOWLEDGMENT

The author wishes to thank N. A. M. Verhoeckx for his contributions resulting from many discussions, J. J. Martony for the experimental work, and members of T. F. S. Hargreaves' advanced development team for their stimulating questions and discussions.

REFERENCES

[1] E. D. Sunde, "Self-timing regenerative repeaters," *Bell Syst. Tech. J.*, vol. 36, pp. 891–937, July 1957.
[2] W. R. Bennett, "Statistics of regenerative digital transmission," *Bell Syst. Tech. J.*, vol. 37, pp. 1501–1542, Nov. 1958.
[3] J. M. Manley, "The generation and accumulation of timing noise in PCM systems—An experimental and theoretical study," *Bell Syst. Tech. J.*, vol. 48, pp. 541–613, Mar. 1969.
[4] C. J. Byrne, B. J. Karafin, and D. B. Robinson, "Systematic jitter in a chain of digital regenerators," *Bell Syst. Tech. J.*, vol. 42, pp. 2679–2714, Nov. 1963.
[5] W. R. Bennett and J. R. Davey, *Data Transmission*. New York: McGraw-Hill, 1965.
[6] C. J. Byrne, "Properties and design of the phase controlled oscillator with a sawtooth comparator," *Bell Syst. Tech. J.*, vol. 41, pp. 559–602, Mar. 1962.
[7] B. R. Saltzberg, "Timing recovery for synchronous binary data transmission," *Bell Syst. Tech. J.*, vol. 46, pp. 593–622, Mar. 1967.
[8] Y. Takasaki, "Timing extraction in baseband pulse transmission," *IEEE Trans. Commun.*, vol. COM-20, pp. 877–884, Oct. 1972.
[9] L. E. Zegers, "The reduction of systematic jitter in a transmission chain with digital regenerators," *IEEE Trans. Commun. Technol.*, vol. COM-15, pp. 542–551, Aug. 1967.
[10] D. Richman, "Color-carrier reference phase synchronization accuracy in NTSC color television," *Proc. IRE*, vol. 42, pp. 106–133, Jan. 1954.
[11] T. M. Gardner, *Phase Lock Techniques*. New York: Wiley, 1966.
[12] P. J. van Gerwen, "On the generation and application of pseudo-ternary codes in pulse transmission," *Philips Res. Rep.*, vol. 20, pp. 469–484.
[13] J. W. Cooley, P. A. W. Lewis, and P. D. Welch, "The finite Fourier transform," *IEEE Trans. Audio Electroacoust.*, vol. AU-17, pp. 77–85, June 1969.

Optimization of Phase-Locked Loop Performance in Data Recovery Systems

Ramon S. Co, *Member, IEEE*, and J. H. Mulligan, Jr., *Life Fellow, IEEE*

Abstract— Optimized design conditions are presented for a phase-locked loop (PLL) used as a functional block in data recovery systems with the primary function of timing recovery. A mathematical model is presented which takes into account the nonlinear and discrete-time nature of the PLL when used in data recovery applications. Performance attributes for these systems such as acquisition, tracking, and noise are considered. A systematic design procedure is presented which permits quantitative trade-offs among these performance attributes. The validation of the mathematical model and the systematic design procedure on a practical circuit implementation in CMOS technology is described.

I. INTRODUCTION

THERE is an ever-increasing need for digital data transmission and recovery. Digital signals are less susceptible than analog signals to noise and are compatible with the rapid advancements in digital technologies and digital signal processing techniques. Digitally encoded speech, for example, can be transmitted over a long distance with almost no degradation in signal quality. Central to any data transmission and recovery system is the recovery of the timing (or the clock) of the digital information. Data recovery systems have found widespread use in the digital telephone network (Bell T1 carrier system) [1]–[3], local area networks (Ethernet, Token Ring, FDDI) [3]–[5], and in disk drive data storage systems.

A simple mechanization of a digital transmission system is one wherein the digital information is transmitted in the form of rectangular pulses. The presence of a pulse, for instance, signifies binary one, and the absence of a pulse signifies binary zero. At the receiving end, the clock signal is recovered from the received pulses, and the edges of the recovered clock are used to sample the received pulses to determine the values of the binary information. Most often, the incoming pulses would have been propagated through a nonideal channel. The bandwidth limitation of the channel as well as the noise induced on the channel can impose severe restrictions on how well the clock can be recovered from the pulses, and subsequently, on how well the correct information can be detected at the receiving end.

The data and clock recovery process is typically performed through the use of a feedback control system such as a phase-locked loop [6], [7]. Current approaches [8]–[12] in the analysis of data recovery systems have assumed small

Manuscript received October 16, 1989; revised May 27, 1994. This work was conducted while R. S. Co was employed by Western Digital Corporation and a doctoral student at the University of California, Irvine, CA 92717 USA.

R. S. Co is with Pericom Semiconductor Corporation, San Jose, CA USA.

J. H. Mulligan, Jr. is with the Department of Electrical and Computer Engineering, University of California, Irvine, CA 92717 USA.

IEEE Log Number 9404027.

error signals and small loop bandwidth so that the control loop can be conveniently modeled as a linear (time-invariant) continuous-time system. Such an analysis cannot be used, in general, when the jitter on the incoming data signal is high. With high jitter on the data signal, the error signal in the PLL is large, and the dynamics of the control loop are highly nonlinear. Such an analysis cannot also be used, in general, to optimize the loop parameters (such as widening the loop bandwidth) in order to satisfy a given performance requirement. In addition, the PLL error signal is not a continuous function of time, but a train of aperiodic rectangular pulses. Moreover, by modeling the VCO as K_0/s, a sinusoidal VCO has been implicitly assumed in the phase domain. For data recovery, one is concerned in the *timing error* between the transitions of the data signal and a rectangular VCO (which produces the clock signal), not the phase error of a sinusoidal input and a sinusoidal VCO. In this paper, a mathematical model that exactly describes the dynamical behavior of PLL in data recovery systems is first developed. Given the exact model, optimization techniques are applied to find the optimal solutions according to the desired performance criteria.

In the design of a phase-locked loop for data recovery systems, one is concerned with a number of performance attributes. For instance, rapid acquisition is highly desirable. Faithful tracking of the input signal in the presence of noise or perturbation is very important for the successful recovery of data. If the recovered clock is to be used as timing for retransmission (such as in a repeater chain), the jitter impressed on the recovered clock should preferably be very small. The optimization of these performance attributes is the subject of this paper. As with any engineering design problem, the optimization of one performance attribute usually leads to the deterioration of another. This paper also provides a systematic design procedure which enables the relative importance of these performance attributes to be traded off against one another.

The results are presented in five sections. Section II contains a discussion of the basic PLL model used in a data recovery system. Section III contains a derivation of the mathematical model of the PLL when used in data recovery applications. The mathematical model is compared with the classical linear PLL model. Section IV describes the most significant performance attributes of a data recovery system, i.e., acquisition, tracking, and noise. In Section V, an objective function which can be used to optimize the performance of the data recovery system is presented. A systematic design procedure is outlined and applied to an illustrative example. In Section VI, there is

Reprinted from *IEEE Journal of Solid-State Circuits*, vol. 29, pp. 1022-1034, September 1994.

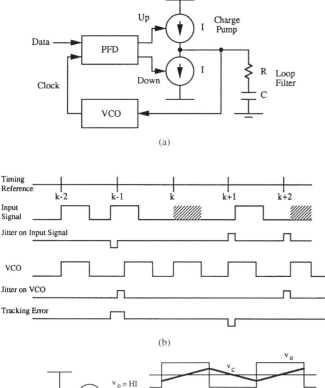

(a)

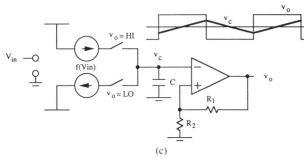

(b)

(c)

Fig. 1. (a) Phase-locked loop model of data recovery system. (b) Phase-locked loop associated waveforms. (c) Relaxation oscillator circuit diagram.

a description of the validation process for the mathematical model of the data recovery system and the systematic design procedure. The paper concludes with a summary of the principal results, including presentation of the design parameters determined to yield optimum performance.

II. DATA RECOVERY SYSTEM MODEL

The PLL model used for the data recovery system is shown in Fig. 1(a), and the associated waveforms are shown in Fig. 1(b). The circuit configuration of the PLL is also known as a *charge-pump* PLL [12]. For the example shown, data one is represented by the presence of a pulse in the first half cycle of the bit period, and data zero is represented by the absence of a pulse (shaded) in the bit period. The positioning of the pulses that represents the data depends on the line coding scheme chosen. In general, the denser the number of pulses, the easier it is to recover the clock since the timing information is imbedded on the occurrence of the pulses. It is these pulses that are tracked by the PLL.

The jitter on the data signal can be described by the time displacement of the positive transitions of the data

signal relative to the timing reference k. Each unit of the timing reference is equal to one *bit interval*. The magnitude of the jitter is typically expressed as a fraction of the bit interval. Since a pulse is not present for a data zero, the jitter is measured with respect to the positive transition of a fictitious pulse (indicating the bit boundary). Jitter magnitude can exceed one bit interval. The rate of excursion of the positive transitions of the data signal relative to the timing reference is the *jitter frequency*. The jitter on the VCO signal can be described in the same manner as the data signal jitter. The *tracking error* is the time difference between the positive transitions of the VCO signal and the data signal. It is also equal to the time difference between the VCO signal jitter and the data signal jitter.

A data sampler is generated from the VCO. It is delayed by one quarter of a bit time (for the particular example shown) relative to the VCO output, and its rising edge is used to sample the data signal. It can be seen that if the tracking error exceeds one quarter of a bit time, the data pulse would be sampled incorrectly, and a data error would result. Thus, the probability of error (bit error rate) is determined by the probability in which the tracking error exceeds one quarter of a bit interval. It is to be noted that the data sample point is a function of the line coding scheme used. In NRZ code, for instance, the data pulse is as wide as the bit interval. The optimum sample point is the middle of the bit (which is delayed by half a bit time from the VCO output), and the peak tracking error is one-half of a bit interval.

The Phase/Frequency Detector (PFD) is a *sequential* (as opposed to multiplier) type phase detector as described in [12], [13]. A sequential phase detector can be configured such that a phase comparison is made only whenever there is a data pulse (or a data transition). Whenever there is no data pulse, an error signal is not generated, and the VCO essentially free runs. The PFD in the example shown (see Fig. 1(a) and (b)) is enabled if there is a data one at its input. If the VCO signal lags the data signal, the up signal is activated (meaning the VCO is running too slowly); and if the VCO signal leads the data signal, the down signal is activated (meaning the VCO is running too fast). The duration of the up and down signals is equal to the time difference of the positive transitions of the data signal and the VCO.

The loop filter is a proportional plus integral type, i.e., series RC. The PLL is thus a second order loop. A third order loop is commonly encountered in practice in which a small capacitor is shunted from the filter node to ground. This small capacitor serves to smooth out the voltage developed across the filter resistor R whenever the charge pump (current sources) is turned on and off. The small capacitor together with the filter resistor R form a high frequency pole which is typically placed well beyond the unity gain frequency of the loop so that the loop essentially behaves as a second order loop. It is important to fully understand the behavior of a second order loop. Thus, a simpler loop filter is used for the analysis in this paper.

The VCO is modeled as a relaxation oscillator [14], [15]. A circuit diagram of the relaxation oscillator is shown in Fig. 1(c). The relaxation oscillator belongs to a class of triggered oscillators in which the output switches state when

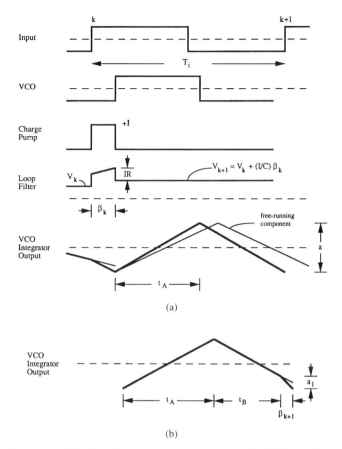

(a)

(b)

Fig. 2. (a) VCO signal lags input signal at cycle k. (b) VCO signal lags input signal at next cycle $(k+1)$, $T_i < (\beta_k + 2t_A)$.

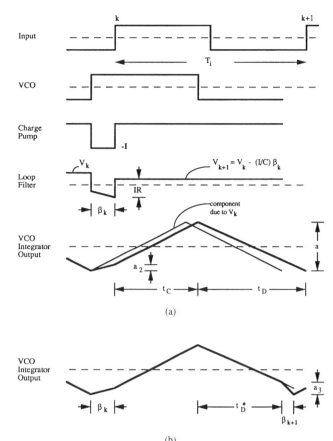

(a)

(b)

Fig. 3. (a) VCO signal leads input signal at cycle k. (b) VCO signal lags input signal at next cycle $(k+1)$, $T_i < (t_C + t_D)$.

a certain threshold is reached. The output of a relaxation oscillator is naturally a square wave which is required for the clock generation. Its time domain description is also easily formulated.

III. MATHEMATICAL MODEL

A. Formulation of Dynamical Difference Equations

The PLL is a sampled-data system, and its operation can be formulated using a difference equation description. The voltage across the capacitor in the loop filter, and the tracking (zero crossing) error between the data signal and the VCO are taken as the state variables. In the steady-state, the positive transitions of the input signal and the VCO output coincide. Since there are no zero crossing errors, the charge pump is disabled. The output voltage of the loop filter is the voltage across the capacitor C. This is the voltage necessary to hold the VCO frequency equal to the input signal frequency.

To analyze the behavior of the loop in terms of the tracking error sequence, the input to the PLL is assumed to be a train of rectangular periodic pulses. There are two possible conditions at time instant k of the input signal: either the VCO signal lags the input signal or the VCO signal leads the input signal. The waveforms when the VCO signal lags the input signal are shown in Fig. 2(a) and (b), and the waveforms when the VCO signal leads the input signal are shown in Fig. 3(a) and (b).

VCO Signal Lags Input Signal: When the VCO signal lags the input signal, the UP signal of the charge pump is

activated by the positive edge of the input signal, and is terminated by the positive edge of the VCO. The duration of the charge pump output is equal to the tracking error at time instant k, and is denoted by β_k. During this time interval, a current of magnitude $+I$ is pumped into the loop filter. If V_k denotes the voltage across the capacitor C in the loop filter prior to the turning on of the charge pump, then the voltage across the capacitor after the charge pump is turned off is equal to

$$V_{k+1} = V_k + \frac{I}{C}\beta_k. \tag{1}$$

Let "a" be the peak-to-peak threshold of the VCO integrator (see capacitor voltage V_c in Fig. 1(c)), then

$$a = (m_0 + k_i V_{k+1})t_A$$

where m_0 is the slope associated with the free-running frequency of the VCO, and k_i is the integration constant associated with the VCO gain. If we denote the VCO gain K_0 as a fractional change of the VCO free-running frequency per unit (volt) of input, then

$$k_i = K_0 m_0.$$

Also,

$$a = m_0 \frac{T_0}{2}$$

131

where T_0 is the VCO free-running period. Hence,

$$t_A = \frac{m_0 T_0/2}{m_0 + K_0 m_0 V_{k+1}} = \frac{T_0/2}{1 + K_0 V_{k+1}}. \tag{2}$$

From Fig. 2(a), it can be seen that if $(\beta_k + 2t_A)$ is equal to the period of the input signal T_i, then the positive transition of the input signal signal will coincide with the positive transition of the VCO signal at the next timing reference $k + 1$. If $T_i < (\beta_k + 2t_A)$ then the VCO lags the input at the next positive transition of the input signal (time instant $k + 1$), and if $T_i > (\beta_k + 2t_A)$ then the VCO leads the input at the next positive transition of the input signal.

a) *Calculation of β_{k+1} for $T_i < (\beta_k + 2t_A)$*: Referring to Fig. 2(b), the charge pump is reactivated at time instant $k+1$, and is held enabled (pumping a current of $+I$) until the VCO integrator reaches the negative threshold level. During the β_{k+1} time interval, the voltage drop across the loop filter consists of additional contributions from the $\mathbb{R}$ drop across the series resistor and the integrated voltage across the series capacitor. Thus,

$$a_1 = (m_0 + k_i V_{k+1})\beta_{k+1} + k_i \int_0^{\beta_{k+1}} \mathbb{R} \, dt$$

$$+ k_i \int_0^{\beta_{k+1}} (I/C)t \, dt$$
$$= m_0\beta_{k+1} + K_0 m_0 V_{k+1}\beta_{k+1} + K_0 m_0 \mathbb{R}\beta_{k+1}$$
$$+ K_0 m_0 (I/C)\beta_{k+1}^2/2.$$

Also,

$$t_B = \frac{a - a_1}{m_0 + K_0 m_0 V_{k+1}}. \tag{3}$$

Therefore,

$$T_i = \beta_k + t_A + t_B.$$

After some algebraic manipulations, there is obtained

$$(K_0/2)(I/C)[\beta_{k+1}^2] + (1 + K_0 V_{k+1} + K_0 \mathbb{R})[\beta_{k+1}]$$
$$+ (T_i - \beta_k)(1 + K_0 V_{k+1}) - T_0 = 0. \tag{4}$$

β_{k+1} is the positive real solution to the above quadratic equation.

b) *Calculation of β_{k+1} for $T_i > (\beta_k + 2t_A)$*: When the VCO signal leads the input signal at time instant $k + 1$, β_{k+1} is simply

$$\beta_{k+1} = T_i - (\beta_k + 2t_A)$$
$$= T_i - \beta_k - T_0/(1 + K_0 V_{k+1}) \tag{5}$$

since the VCO has already made a transition and is just waiting for the input signal to make its transition.

VCO Signal Leads Input Signal: When the VCO signal leads the input signal, the DOWN signal of the charge pump is activated by the positive edge of the VCO, and is terminated by the positive edge of the input signal. The duration of the charge pump output is equal to the tracking error β_k. During this time interval, a current of magnitude $-I$ is pumped into the loop filter. If V_k denotes the voltage across the capacitor in the loop filter prior to the turning on of the charge pump,

then the voltage across the capacitor after the charge pump is turned off is equal to

$$V_{k+1} = V_k - \frac{I}{C}\beta_k. \tag{6}$$

Referring to Fig. 3(a) and using the same notations as defined previously,

$$a_2 = (m_0 + k_i V_k)\beta_{k+1} - k_i \int_0^{\beta_{k+1}} , \mathbb{R} \, dt$$

$$- k_i \int_0^{\beta_{k+1}} , (I/C)t \, dt$$
$$= m_0\beta_k + K_0 m_0 V_k\beta_k - K_0 m_0 \mathbb{R}\beta_k$$
$$- K_0 m_0 (I/C)\beta_k^2/2.$$

Also,

$$t_C = \frac{a - a_2}{m_0 + k_i V_{k+1}} \tag{7}$$

$$t_D = \frac{a}{m_0 + k_i V_{k+1}}. \tag{8}$$

It can be seen that if $T_i = (t_C + t_D)$ then the positive transition of the input signal will coincide with the positive transition of the VCO signal at the next timing reference $k + 1$. If $T_i < (t_C + t_D)$ then the VCO lags the input at the next positive transition of the input (time instant $k + 1$), and if $T_i > (t_C + t_D)$ then the VCO leads the input at the next positive transition of the input.

a) *Calculation of β_{k+1} for $T_i < (t_C + t_D)$*: Referring to Fig. 3(b), the charge pump is reactivated at time instant $k+1$, and is held enabled (pumping a current of $-I$) until the VCO integrator reaches the negative threshold level. During the β_{k+1} time interval, the voltage drop across the loop filter consists of additional contributions from the $\mathbb{R}$ drop across the series resistor and the integrated voltage across the series capacitor. Thus,

$$a_3 = (m_0 + k_i V_{k+1})\beta_{k+1} + k_i \int_0^{\beta_{k+1}} \mathbb{R} \, dt$$

$$+ ki \int_0^{\beta_{k+1}} (I/C)t \, dt$$
$$= m_0\beta_{k+1} + K_0 m_0 V_{k+1}\beta_{k+1} + K_0 m_0 \mathbb{R}\beta_{k+1}$$
$$+ K_0 m_0 (I/C)\beta_{k+1}^2/2.$$

Also,

$$t_D^* = \frac{a - a_3}{m_0 + k_i V_{k+1}}. $$

Therefore,

$$T_i = t_C + t_D^*.$$

After some algebraic manipulations, there is obtained

$$(K_0/2)(I/C)[\beta_{k+1}^2] + (1 + K_0 V_{k+1} + K_0 \mathbb{R})[\beta_{k+1}]$$
$$+ T_i(1 + K_0 V_{k+1}) - T_0$$
$$+ \beta_k(1 + K_0 V_k - K_0 \mathbb{R})$$
$$- (K_0/2)(I/C)\beta_k^2 = 0. \tag{9}$$

β_{k+1} is the positive real solution to the above quadratic equation.

b) *Calculation of β_{k+1} for $T_i > (t_C + t_D)$*: When the VCO signal leads the input signal at time instant $k + 1$, β_{k+1} is simply

$$\begin{aligned}
\beta_{k+1} &= T_i - (t_C + t_D) \\
&= T_i - [T_0 - \beta_k(1 + K_0 V_k - K_0 \mathbb{IR}) \\
&\quad + (K_0/2)(I/C)\beta_k^2]/(1 + K_0 V_{k+1}) \quad (10)
\end{aligned}$$

since the VCO has already made a transition and is just waiting for the input signal to make its transition.

In summary,

i) when the VCO lags the input signal at time k

$$V_{k+1} = V_k + \frac{I}{C}\beta_k$$

and

$$t_A = (T_0/2)/(1 + K_0 V_{k+1}).$$

a) If $Ti < (\beta_k + 2t_A)$ then the VCO lags the input signal at time $k + 1$, and β_{k+1} is the positive real solution of:

$$\begin{aligned}
&(K_0/2)(I/C)[\beta_{k+1}^2] \\
&+ (1 + K_0 V_{k+1} + K_0 \mathbb{IR})[\beta_{k+1}] \\
&+ (T_i - \beta_k)(1 + K_0 V_{k+1}) - T_0 = 0.
\end{aligned}$$

b) If $T_i > (\beta_k + 2t_A)$ then the VCO leads the input signal at time $k + 1$, and β_{k+1} is computed from:

$$\beta_{k+1} = T_i - \beta_k - T_0/(1 + K_0 V_{k+1}).$$

ii) When the VCO leads the input signal at time k

$$V_{k+1} = V_k - \frac{I}{C}\beta_k$$

and

$$\begin{aligned}
(t_C + t_D) &= [T_0 - \beta_k(1 + K_0 V_k - K_0 \mathbb{IR}) \\
&\quad + (K_0/2)(I/C)\beta_k^2]/(1 + K_0 V_{k+1}).
\end{aligned}$$

a) If $T_i < (t_C + t_D)$ then the VCO lags the input signal at time $k + 1$, and β_{k+1} is the positive real solution of:

$$\begin{aligned}
&(K_0/2)(I/C)[\beta_{k+1}^2] \\
&+ (1 + K_0 V_{k+1} + K_0 \mathbb{IR})[\beta_{k+1}] \\
&+ T_i(1 + K_0 V_{k+1}) - T_0 \\
&+ \beta_k(1 + K_0 V_k - K_0 \mathbb{IR}) \\
&- (K_0/2)(I/C)\beta_k^2 = 0.
\end{aligned}$$

b) If $T_i > (t_C + t_D)$ then the VCO leads the input signal at time $k + 1$, and β_{k+1} is computed from

$$\beta_{k+1} = T_i - (t_C + t_D).$$

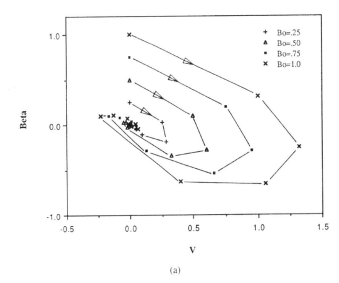

(a)

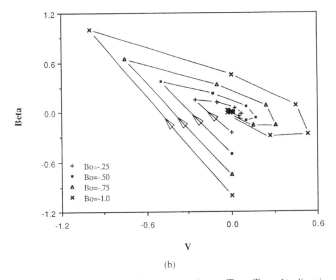

(b)

Fig. 4.   (a) State trajectories for phase step input ($T_i = T_0 = 1$); direction of trajectory is indicated by arrows. (b) State trajectories for phase step input ($T_i = T_0 = 1$); direction of trajectory is indicated by arrows.

The solutions to the set of difference equations, due to their sequential and recursive nature, can be easily generated with the aid of a computer program. Careful examination of the difference equations (by normalizing the tracking error β and the free-running period T_0 to the input signal period T_i) indicates that the unitless parameters $K_0 I T_i/C$ and $K_0 \mathbb{IR}$ are sufficient to describe the loop. The two parameters can be further reduced to $K_0 I/C$ and $K_0 \mathbb{IR}$ by setting T_i equal to unity. This is interesting because $K_0 I/C$ is related to the *natural frequency*, and $K_0 \mathbb{IR}$ is related to the *unity gain bandwidth* of the loop in linear analysis.

Fig. 4(a) and (b) show the *state trajectories* for phase step input with the tracking error β and the loop filter capacitor voltage V as state variables. Each entry in the trajectory is equivalent to an elapsed time of one clock period or one bit period. The period T_i of the input signal is equal to the free-running period T_0 of the VCO; both are normalized to one. The initial tracking error β_0 is varied (from 0.25 to 1.0 and -0.25 to -1.0) using the normalized loop parameters $K_0 I/C = 1$

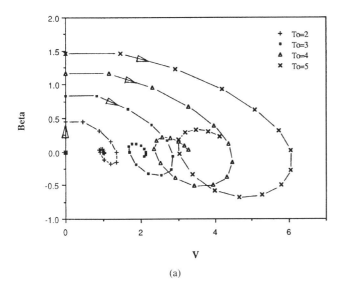

(a)

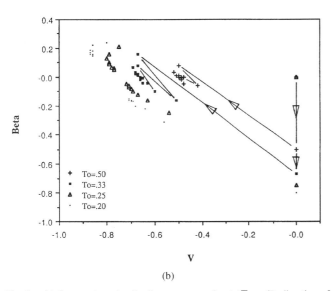

(b)

Fig. 5. (a) State trajectories for frequency step input ($T_i = 1$); direction of trajectory is indicated by arrows. (b) State trajectories for frequency step input ($T_i = 1$); direction of trajectory is indicated by arrows.

and $K_0 \mathbb{I} R = 1$. The inherent asymmetry in which the error signals are generated in the loop is evident in the asymmetry of the state trajectories for positive and negative initial tracking errors. The trajectory is the "familiar" spiral shape for positive initial tracking error; whereas, the trajectory is a "pointed" spiral for negative initial tracking errors. Recall that when the input signal leads the VCO signal, the generation of the error signal is causing the VCO transition to occur earlier from its nominal transition time. However, when the VCO signal leads the input signal, the generation of the error signal does not perturb the time of occurrence of the input signal transition (the input signal is the excitation source).

Fig. 5(a) and (b) show the state trajectories for frequency step input. The initial tracking error β_0 is equal to zero. T_i is normalized to one, and T_0 is varied (from 2 to 5 and 0.2 to 0.5) using the same loop parameters as the phase step input.

For either type of input, the steady-state solution (V_{ss}, β_{ss}) is found by setting $V_k = V_{k+1} = V_{ss}$ and $\beta_k = \beta_{k+1} = \beta_{ss}$.

Hence,

$$(V_{ss}, \beta_{ss}) = ([T_0/T_i - 1]/K_0, 0). \quad (11)$$

The system is *unstable* if the state trajectories do not converge to the steady-state solution given the initial conditions. In order to determine that the system will be *stable* over the expected range of initial conditions, the state trajectories should be computed once the loop parameters have been established for a tentative design (see Section V for an illustrative example). One can also determine from the state trajectories the *acquisition time* of the system, i.e., the number of clock cycles required for the system to reach and stay within a bounded region of the steady-state equilibrium condition. If so desired, it is also possible to plot the state trajectories for a phase and frequency step input; that is, $\beta_0 \neq 0$ and $T_i \neq T_0$.

B. Comparison With Classical Linear Model

A well-studied model used in the classical analysis of a second order loop is depicted in Fig. 6. The governing equations [13] using Gardner's notations are:

$$\omega_n = (K_0 K_d / \tau_1)^{1/2} = (K_0 (K_d/R_1)/C)^{1/2}$$
$$\zeta = \frac{\omega_n \tau}{2} = (K_0(K_d/R_1)/C)^{1/2} \frac{RC}{2}$$
$$K = 2\zeta\omega_n = K_0 \frac{K_d}{R_1} R \quad (12)$$

where ω_n is the natural frequency, ζ is the damping factor, K is the unity gain bandwidth, K_0 is the VCO gain, and K_d is the phase detector constant. Notice that the quantity $(K_d/R_1)\theta_e$ is the average error current which is driven into the RC feedback elements of the loop filter, where θ_e is the phase error. For a charge pump PLL, the average error current over a bit interval which is pumped into a similar RC loop filter is equal to $(I/2\pi)\theta_e$. Therefore, a linear continuous-time approximation can be made for the charge pump PLL by equating (K_d/R_1) equal to $(I/2\pi)$. Using this transformation, the following equivalent relations are obtained for the charge-pump PLL:

$$\omega_n = (K_0(I/2\pi)/C)^{1/2} \quad (13)$$
$$\zeta = (K_0(I/2\pi)/C)^{1/2} \frac{RC}{2}$$
$$K = 2\zeta\omega_n = K_0(I/2\pi)R.$$

These equations are in agreement with those defined by Gardner [12]. The transient response plots comparing the classical model and the mathematical model for a phase step input and a frequency step input are shown in Fig. 7(a) and (b) respectively. Note that the discrepancy in the response of the classical model from the mathematical model is particularly severe for the frequency step input.

C. Dynamical Difference Equations in the Presence of Noise

The difference equations can be modified to include the effect of noise. This is done by perturbing the period of the input data signal at each iteration cycle as shown in Fig. 8. The period of the input data at time instant k is made equal to

$$T_k = T_i + (\Delta T_{k+1} - \Delta T_k) \quad (14)$$

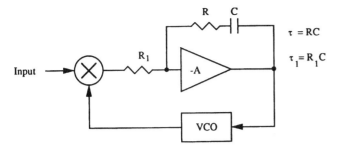

Fig. 6.   Linear model of a second order loop.

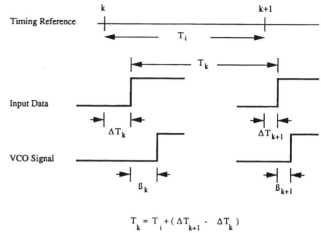

$$T_k = T_i + (\Delta T_{k+1} - \Delta T_k)$$

Fig. 8.   Signal waveforms in the presence of noise.

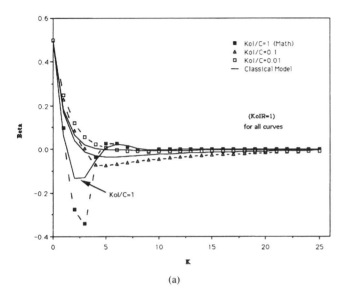

(a)

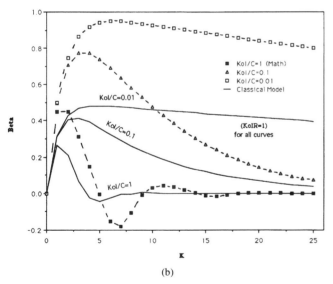

(b)

Fig. 7.   (a) Phase step input for math model (broken line) and classical model (solid line). (b) Frequency step input for math model (broken line) and classical model (solid line).

where T_i is the nominal period of the input data, and ΔT_k and ΔT_{k+1} are the jitter components of the input data at time instants k and $k+1$ respectively. The tracking error β_k is still the time difference between the positive transitions of the input data signal and the VCO signal, but the jitter on the VCO signal is now equal to $\beta_k + \Delta T_k$. Thus, the modification of the

difference equations to include the effect of noise is achieved by replacing T_i (a constant) in the original equations by T_k. The modified equations constitute a model for the PLL which is a principal result of this paper.

The statistics of the input data signal jitter ΔT_k are a function of the signal to noise ratio of the input data signal, the characteristics of the transmission medium, the particular data pattern (intersymbol interference), and the specific line code. A detailed investigation of the statistics of ΔT_k is beyond the scope of this paper. It can be shown, however, that if the data pattern is sufficiently random and the transmission channel is sufficiently narrowband, ΔT_k can be adequately modelled as a white noise process with uniform distribution.

Direct computer simulation can be performed on the difference equations if one is interested in finding an optimal design. For the purposes of the simulations, a random number generator (1000 samples) which is uniformly distributed with a peak value of 0.2 (of the normalized bit period) is used to represent the jitter component ΔT_k of the input data signal. The simulations obtained values of the rms VCO jitter (σ_n) and the rms tracking error (σ_t).

Fig. 9 shows the variation of σ_n, the rms VCO jitter (expressed as a fraction of the normalized bit period), as a function of the normalized loop parameters K_0I/C and K_0IR. It is observed that the VCO jitter is a minimum at $K_0I/C = 0.001$. No appreciable improvement can be obtained by further decreasing K_0I/C. For each K_0I/C, there is a particular value of K_0IR that corresponds to a local minimum. Fig. 10 shows the variation of σ_t, the rms tracking error (expressed as a fraction of the normalized bit period), as a function of the normalized loop parameters. The tracking error is a minimum at $K_0I/C = 0.001$. Similarly, no appreciable improvement can be obtained by further decreasing K_0I/C. Again, for each K_0I/C there is a particular value of K_0IR that corresponds to a local minimum. Further simulations were conducted for different noise levels, i.e., 0.1 peak and 0.3 peak; similar behavior and the same optimum conditions were obtained. Thus, the parameter $K_0I/C = 0.001$ yields a global optimum because it minimizes both the VCO jitter and the tracking error. There is a particular value of K_0IR that corresponds

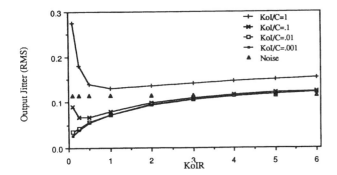

Fig. 9.   Plot of output (clock) jitter versus loop parameters (noise = 0.2 peak). (Expressed as fraction of normalized bit period).

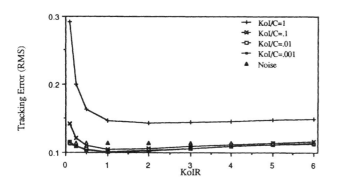

Fig. 10.   Plot of tracking error versus loop parameters (noise = 0.2 peak). (Expressed as fraction of normalized bit period).

to a global minimum for VCO jitter and another value that corresponds to a global minimum for tracking error.

D. Dynamical Equations of Different Data Patterns

So far the difference equations that are derived are valid only for a data one followed by a data one pattern, i.e., a data pulse followed by a data pulse. There are three other possible patterns. The difference equations that describe these patterns are derived as follows.

Data Zero Followed by Data Zero Pattern: Under this condition, the charge pump is inhibited and the VCO is free-running at an oscillation frequency determined by the voltage across the capacitor in the loop filter.

a) VCO lags Input at time k.

$$V_{k+1} = V_k$$
$$\beta_{k+1} = (\beta_k + 2t_A) - T_i$$
$$= \beta_k + T_0/(1 + K_0V_k) - T_i. \qquad (15)$$

β_{k+1} is positive if the VCO lags the input signal at time $k + 1$, and β_{k+1} is negative if the VCO leads the input signal at time $k + 1$.

b) VCO leads Input at time k.

$$V_{k+1} = V_k$$
$$\beta_{k+1} = (\beta_k + T_i) - 2t_A$$
$$= \beta_k - T_0/(1 + K_0V_k) - T_i. \qquad (16)$$

Data Zero Followed by Data One Pattern:

a) VCO lags Input at time k.

$$V_{k+1} = V_k. \qquad (17)$$

i) For $T_i < (\beta_k + 2t_A)$, VCO lags input at time $k+1$.

$$T_i = \beta_k + t_A + t_B \qquad \text{(see Fig. 2)}.$$

And β_{k+1} is the positive real solution of

$$(K_0/2)(I/C)[\beta_{k+1}^2]$$
$$+ (1 + K_0V_k + K_0\mathrm{I\!R})[\beta_{k+1}]$$
$$+ (T_i - \beta_k)(1 + K_0V_k) - T_0 = 0.$$

ii) For $T_i > (\beta_k + 2t_A)$, VCO leads input at time $k + 1$.

$$\beta_{k+1} = T_i - (\beta_k + 2t_A)$$
$$= T_i - \beta_k + T_0/(1 + K_0V_k).$$

b) VCO leads Input at time k.

$$V_{k+1} = V_k. \qquad (18)$$

i) For $(\beta_k + T_i) < 2t_A$, VCO lags input at time $k+1$.

$$(\beta_k + T_i) = t_A + t_D^* \qquad \text{(see Fig. 3)}$$
$$= [T_0 - \beta_{k+1} - K_0V_k\beta_{k+1}$$
$$- K_0\mathrm{I\!R}\beta_{k+1} - (K_0/2)$$
$$\times (I/C)\beta_{k+1}^2]/(1 + K_0V_k).$$

β_{k+1} is the positive real solution of:

$$(K_0/2)(I/C)[\beta_{k+1}^2]$$
$$+ (1 + K_0V_k + K_0\mathrm{I\!R})[\beta_{k+1}]$$
$$+ (1 + K_0V_k)(\beta_k + T_i) - T_0 = 0.$$

ii) For $(\beta_k + T_i) > 2t_A$, VCO leads input at time $k + 1$.

$$\beta_{k+1} = (\beta_k + T_i) - 2t_A$$
$$= \beta_k + T_i - T_0/(1 + K_0V_k).$$

Data One Followed by Data Zero Pattern:

a) VCO lags Input at time k.
Referring to Fig. 2,

$$V_{k+1} = V_k + \frac{I}{C}\beta_k$$
$$\beta_{k+1} = (\beta_k + 2t_A) - T_i$$
$$= \beta_k + T_0/(1 + K_0V_k) - T_i. \qquad (19)$$

β_{k+1} is positive if the VCO lags the input signal at time $k + 1$, and β_{k+1} is negative if the VCO leads the input signal at time $k + 1$.

b) VCO leads Input at time k.
Referring to Fig. 3,

$$V_{k+1} = V_k - \frac{I}{C}\beta_k$$
$$\beta_{k+1} = T_i - (t_C + t_D). \qquad (20)$$

β_{k+1} is positive if the VCO leads the input signal at time $k + 1$, and β_{k+1} is negative if the VCO lags the input signal at time $k + 1$.

IV. SYSTEM DESIGN CONCEPTS

There are two stages in the data recovery process. The first is the acquisition stage, and the second is the tracking and data detection stage. Acquisition is considered to be complete when the error signal has reached the steady-state (value of zero) and the VCO frequency is equal to the incoming data signal frequency. This is, however, a very stringent definition. Noise is invariably present in the system, and therefore, the error signal is never reducible to zero. An operational definition of acquisition is to consider it to have been completed when the error signal and the VCO control voltage have become equal to or less than a prescribed percentage (usually taken as 2–5%) of the bit period and the steady-state VCO control voltage, respectively. Under this condition, the PLL can start detecting (by sampling) the received signal.

A. Acquisition Performance

To assist in the acquisition of the incoming data signal, the VCO is often pretuned to an accurate local frequency reference, such as provided by a crystal oscillator. This prevents the PLL from false locking to the harmonics or the subharmonics of the incoming data signal. When the data transitions start to occur, the PLL input is switched from the local frequency reference to the incoming data signal. The voltage which is necessary to pull the VCO frequency into the incoming data signal frequency is held temporarily in the storage elements of the loop filter. With a clever design of the VCO, it can be started (so called zero phase start) so that its transitions would immediately coincide with the transitions of the incoming data signal, thereby totally eliminating the acquisition problem.

In some communication systems, information is transmitted as a group of data bits known as packets [3]. The data packet is usually preceded by a preamble. The preamble is a periodic data pattern with distinct harmonic content so that the phase-locked loop can pull into the incoming data signal even without the benefit of pretuning to a local frequency reference. For these particular systems, the acquisition time is limited to the length of the preamble.

The PLL acquisition performance is determined by examining the transient response of the loop subjected to phase step and frequency step inputs. It can be improved by having large values of $K_0 I/C$ and $K_0 \mathbb{R}$. A large value of $K_0 I/C$, however, tends to produce an undesirable damped oscillatory transient response, whereas, a large value of $K_0 \mathbb{R}$ tends to reduce the oscillatory tendency and indeed can yield a monotonic transient response.

B. Noise and Tracking Performance

When acquisition is complete, the PLL goes into the second stage of data tracking (and detection). The noise performance is measured by the jitter on the recovered clock, whereas tracking performance is measured by the tracking error. The latter determines the bit error rate performance of the system.

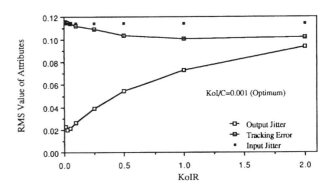

Fig. 11. Dependence of loop performance on $K_0 \mathbb{R}$ at optimum $K_0 I/C$ (expressed as fraction of normalized bit period).

It is highly desirable to have minimum values for these two attributes.

It has been found from the extensive simulations that the normalized loop parameter $K_0 I/C = 0.001$ yields a global optimum, since it minimizes both the clock (VCO) jitter and the tracking error. Given the optimum $K_0 I/C$, variation of the clock jitter and the tracking error versus the loop parameter $K_0 \mathbb{R}$ is shown in Fig. 11. The graph indicates that the clock jitter is a minimum at $K_0 \mathbb{R} = 0.025$, and the tracking error is a minimum at $K_0 \mathbb{R} = 1$. Thus, a tradeoff would have to be made in the selection of $K_0 \mathbb{R}$. A smaller value of $K_0 \mathbb{R}$ improves the clock jitter but degrades the tracking performance; whereas, a larger value of $K_0 \mathbb{R}$ improves the tracking performance but degrades the clock jitter. However, the loop bandwidth cannot be made arbitrarily small ($K_0 \mathbb{R} \to 0$) to reduce the clock jitter to zero. With an infinitely small loop bandwidth, the loop has difficulty maintaining lock to an input signal with jitter, and the clock would just drift and wander. The loop bandwidth cannot be made arbitrarily large ($K_0 \mathbb{R} \to \infty$) either to reduce the tracking error to zero. Aliasing effects become dominant as soon as the loop bandwidth exceeds half the clock frequency.

As a further note, the loop performance is not sensitive to variations in $K_0 I/C$. Actually, there is a wide range of $K_0 I/C (0.01$ to $0.0001)$ that can be considered to yield optimum performance. The choice of $K_0 \mathbb{R}$ however, is very critical.

V. PERFORMANCE OPTIMIZATION

A. Definition of Optimization Problem

An objective function which can be used in the optimization of the PLL performance is

$$\mathcal{F} = \alpha_a t_a + \alpha_t \sigma_t + \alpha_n \sigma_n \qquad (21)$$

where α_a, α_t, and α_n are weighting factors, σ_t and σ_n are the tracking and noise performance measures, and t_a is the acquisition performance measure. The optimization problem is the minimization of the objective function $\mathcal{F}$ subject to the conditions that the requirement for each of the performance attributes is met.

Two alternative definitions of the acquisition performance measure t_a have been considered. The *first method* consists of counting the number of bits (clock cycles) until the error

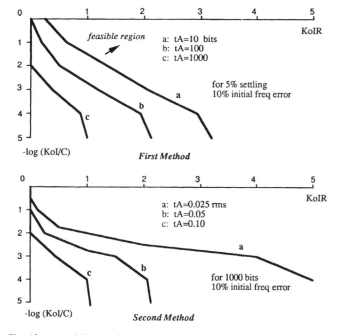

Fig. 12. Acquisition performance objective function contours.

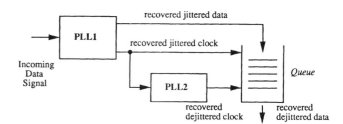

Fig. 13. A queue structure for optimum tracking and jitter.

signal and the VCO control voltage have settled to a prescribed amount. The performance measure assigned for t_a using this method has the disadvantage, however, that it has a different dimension compared to the other two attributes (σ_t and σ_n). With this approach, the acquisition time would typically be expressed as a number of bits (clock cycles), whereas the tracking error and clock jitter are expressed as rms values of fractional change of the clock period.

The *second method* consists of minimizing the rms value of the error signal samples up to the desired acquisition time and normalizing to the bit period. This measure has the advantage that the result has the same dimension as the measure of tracking and noise performance. In addition, optimization of the loop parameters with this method results in the optimization of the acquisition performance in the mean square sense. Examples of the plots of the objective function contours based on these two methods are shown in Fig. 12.

B. Existence of a Solution

For most applications, the requirements set on the performance attributes have the form:

$$\text{acquisition time} < A \text{ bits}$$
$$\text{tracking error} < T \text{ ns}$$
$$\text{clock jitter} < N \text{ ns.} \qquad (22)$$

These restrictions impose a bound on the optimization space, i.e., the loop parameters $K_0 IR$ and $K_0 I/C$, for each of the attributes. Each attribute, therefore, has its own *feasible region* [16]. A solution exists if an intersection of the feasible regions can be found.

C. The Design Process

Classes of Design Problems: Several classes of design problems are encountered in practice. These can be summarized as the following distinct possibilities:

a) The same set of loop parameters are used for acquisition, tracking, and noise.

b) A set of loop parameters is used for acquisition, and another set is used for tracking and noise. This approach permits independent optimization of the acquisition process using one set of loop parameters, and the optimization of noise and tracking (suitably weighted) with the other set.

c) The VCO is pretuned to a local frequency reference. Only optimization of the loop parameters for data tracking and clock jitter is necessary.

d) A queue is available as shown in Fig. 13 for temporary storage of jittered data. This approach permits optimization of data tracking and clock jitter independently of one another. PLL$_1$ is optimized for tracking performance for successful recovery of data, and PLL$_2$ is optimized for jitter performance to provide a jitter-free clock. PLL$_1$ is used to clock the data into the queue, and PLL$_2$ is used to clock the data out of the queue.

The first item in the list represents a general problem, and it is the most difficult one to solve. Although the design procedure to be described pertains specifically to this problem, it is also applicable to the rest since they are special cases of the first.

Design Procedure: The design problem is to choose circuit configurations and values of $K_0, I, R,$ and C so that system specifications on acquisition time, tracking error, and clock jitter are satisfied given quantities such as the data rate and the jitter on the incoming data. The procedure includes provision for a trade-off among acquisition, tracking, and jitter performance using preassigned weighting factors ($\alpha_a, \alpha_t,$ and α_n).

For convenience, the procedure is presented in several specific steps:

Step 1: Convert given requirements/constraints into the design variables $t_a, \sigma_t, \sigma_n,$ and T_i.

Step 2: Assign values for K_0 and I. K_0 is the VCO gain, defined in Section III as the fractional change in VCO center frequency ($1/T$ Hz) per volt of input voltage. The value is based on the designer's estimate of the change that will occur in frequency once a practical oscillator configuration has been selected. I, the current in the charge-pump, is a free variable. Its maximum value is determined by considerations of noise and other fluctuations which limit the accuracy of its specification and realization.

Step 3: Examine performance optimization in terms of the objective function $\mathcal{F}$ and the weights $\alpha_a, \alpha_t,$ and α_n. The feasible region is determined for each of the performance attributes from the design requirements. The intersection of the feasible regions is the set of possible solutions. A solution is not possible if an intersection cannot be found. If a solution is possible, the intersection of the feasible regions is applied as a set of boundary constraints to the system objective function $\mathcal{F}$. The preassigned weighting factors for the performance attributes are used to find the optimum normalized loop parameters ($K_0 I T_i / C = p$ and $K_0 I R = q$).

An alternate to this step is to create more contours for each of the attributes that have smaller values than the desired constraints. Creating a contour with smaller value for a particular attribute narrows the optimization space, and in effect, imposes a more stringent requirement for that particular attribute. Subsequently, an optimal solution will be found that favors the attribute whose objective function contour is reduced.

Step 4: Determine values of R and C using the equations

$$C = (K_0 I T_i)/p \quad \text{and} \quad R = q/(K_0 I).$$

The former equation results from the fact discussed in Section III upon the introduction of the normalized parameter $K_0 I T_i / C$ that use of the normalizing condition $T_i = 1$ yields $K_0 I / C$ which has been utilized throughout the ensuing development.

Step 5: Compute the state trajectories (see Fig. 4 and 5) over the expected range of initial conditions to ascertain that the loop design is stable and that the transient response is satisfactory.

Illustrative Example: Consider a data recovery system design with the following specifications:

Data Rate:	10 Mbps ($T_i = 100$ ns)
Data Jitter:	± 20 ns peak (uniform distribution)
Acquisition Time:	1000 bits
Tracking Error:	11.5 ns rms
Clock Jitter:	8.1 ns rms
Acquisition Performance Measure: (using second method)	5.0 ns rms
Performance Weightings:	$\alpha_a = \alpha_t = \alpha_n = 1$.

The tracking error is chosen such that it is equal to the rms value of the data jitter, i.e., 20 ns peak $/ \sqrt{3} = 11.5$ ns rms (for a uniform distribution, rms value = peak value$/\sqrt{3}$).

The first parameter to be chosen in the design is the VCO gain K_0. It should be large enough to pull the VCO frequency into the incoming data signal frequency. A K_0 of 50% per volt is chosen as the design value based on the choice of a ring oscillator for the VCO. It is assumed that the VCO has a center frequency with 10% accuracy. This implies that the PLL can have an initial frequency step (error) of 10%. Objective function contours for acquisition performance based on the two alternative measures discussed above are shown in Fig. 12. They were computed from the PLL model for a 10% initial error in frequency based on this assumption. The second method will be used in this example. An acquisition performance measure of 5 ns rms implies a normalized rms value of $5/100 = 0.05$. Therefore, *contour b* of Fig. 12 is selected to determine the relation between $K_0 I / C$ and $K_0 I R$ to achieve the desired acquisition performance.

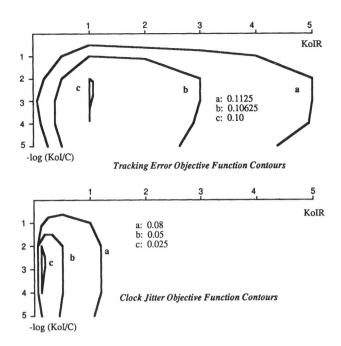

Fig. 14. Tracking and noise performance objective function contours.

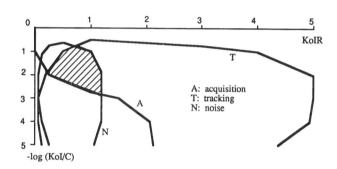

Fig. 15. Intersection of feasible regions for the illustrative example.

Plots of the objective function contours for tracking and noise performance are shown in Fig. 14. *Contour a* for tracking error provides a basis for estimating the corresponding relation between $K_0 I / C$ and $K_0 I R$ to meet the 11.5 ns requirement. *Contour a* for clock jitter also provides a comparable estimate of the relation needed to meet the 8.1 ns requirement.

The three selected contours are superimposed and are shown in Fig. 15. The shaded region which is the intersection of the three contours is the set of feasible solutions. This set of feasible solutions is applied as boundary constraints in the minimization of the system objective function $\mathcal{F}$.

The system objective function $\mathcal{F}$ for $\alpha_a = \alpha_t = \alpha_n = 1$ has a minimum at $K_0 I / C = 0.01$ and $K_0 I R = 0.5$. These values fall within the boundary constraints depicted in Fig. 15. Hence, $K_0 I T_i / C = p = 0.01$ and $K_0 I R = q = 0.5$.

With a tentative circuit implementation of the charge pump in mind, a current of 1.0 mA is selected as the value of I. The component values can now be calculated using

$$C = (0.5)(0.001)(100 \times 10^{-9})/0.01 = 0.005 \ \mu\text{F}$$

and

$$R = 0.5/(0.5)(0.001) = 1 \ \text{K}\Omega.$$

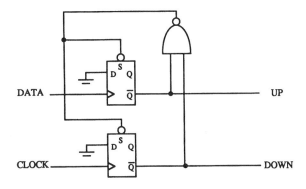

Fig. 16. Phase/frequency detector logic diagram.

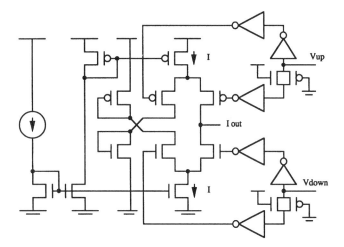

Fig. 17. Charge pump circuit schematic.

The remaining step in the design process is to compute the state trajectories to determine that the loop design is stable and that the transient response is satisfactory. The initial conditions to which the loop will be subjected to are an initial frequency error of 10% and an initial phase error which can be as much as one bit interval.

VI. VALIDATION OF RESULTS

The validity of the applicability of the mathematical model for practical system implementation was investigated using a CMOS electronic circuit having the form of Fig. 1(a). The level 7 Lattin-Jenkin-Grove MOSFET model [17] for a 1.25 μm feature size CMOS was used as a transistor model in SPICE to simulate the operation of the circuit. The logic diagram of the phase/frequency detector PFD is shown in Fig. 16, the charge-pump circuit (generator I of Fig. 1(a) schematic is shown in Fig. 17, and the VCO of Fig. 1(a) (implemented as ring oscillator circuit) is shown in Fig. 18. Note that the particular circuit configuration shown in Fig. 16 is a conventional sequential phase detector which is not suited for random data. The circuit is used in the simulation for illustrative purposes only assuming that the input is a periodic pattern.

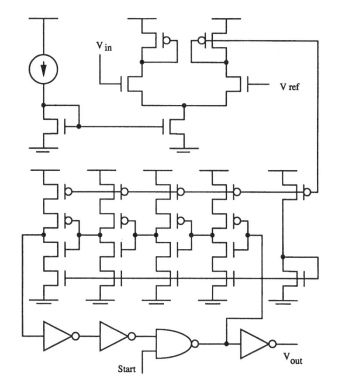

Fig. 18. Ring oscillator circuit.

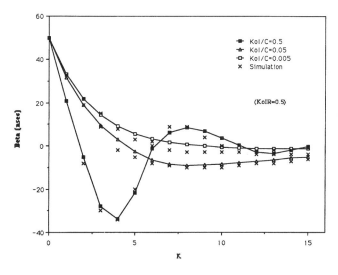

Fig. 19. Phase step input for math model and circuit simulation.

The transient response of the PLL using the CMOS circuitry was simulated for a phase step input and a frequency step input. The results are shown in Fig. 19 and 20. In the figures, the VCO frequency is 10 MHz which corresponds to a period of 100 ns. Excellent agreement between the simulated circuit and the mathematical model was obtained in all the cases.

Jitter was added into the circuit simulations by perturbing the transitions of the data signal, and the tracking error and the clock jitter were noted. These operations were performed for different noise levels and different loop parameters, including the illustrative design example presented above. Excellent agreement between the simulated circuit and the mathematical model was also obtained.

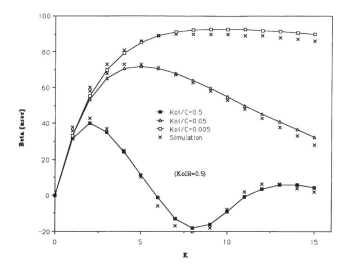

Fig. 20. Frequency step input for math model and circuit simulation.

VII. Conclusion

A mathematical model has been derived for the phase-locked loop which is commonly used in data recovery systems. The formulation of the model is exact and takes into account that a square wave voltage output of the VCO is required. The mathematical model is a difference equation which relates the tracking (zero crossing) error and the voltage across the capacitor of the loop filter at the present bit timing k to the next bit timing $k + 1$. The model includes the effect of noise (or jitter) on the input signal.

The three performance attributes of a data recovery system, namely, acquisition, tracking, and noise, are discussed in detail. The optimum loop parameters for each of these attributes are determined from the numerical solutions to the mathematical model. The solutions indicate that the normalized loop parameter $K_0 I/C = 0.001$ is optimum for both the tracking performance and the noise performance over a wide practical range of input noise levels. Wide bandwidth ($K_0 \mathbb{R} \approx 1$) is required for optimum tracking, and narrow bandwidth ($K_0 \mathbb{R} \approx 0.025$) is required for optimum jitter rejection.

Finally, a systematic design procedure for a PLL for use in data recovery systems is presented. An objective function is given which can be utilized for trade-offs among acquisition, tracking, and jitter performance. A method is also shown for determining the existence of a solution given the design conditions. The mathematical model as well as the design procedure were validated through simulations on a practical CMOS circuit implementation using the SPICE circuit simulation program.

References

[1] F. F. E. Owen, *PCM and Digital Transmission Systems*. New York: McGraw-Hill, 1982.
[2] Pub 62411, "High capacity digital service channel interface specification," *Bell System Techn. Ref.*, Sept. 1983.
[3] A. S. Tanenbaum, *Computer Networks*. Englewood Cliffs, NJ: Prentice-Hall, 1981.
[4] ANSI/IEEE Std 802.3-1985, *Carrier Sense Multiple Access with Collision Detection (CSMA/CD)*. New York: Wiley, 1985.
[5] ANSI/IEEE Std 802.5-1985, *Token Ring Access Method*. New York: Wiley, 1985.
[6] W. D. Llewellyn, M. H. Wong, G. W. Tietz, and P. A. Tucci, "A 33 Mb/s data synchronizing phase-locked loop circuit," *Int. Solid-State Circuits Conf. Tech. Dig.*, San Francisco, CA., Feb. 1988, pp. 12–13.
[7] R. H. Leonowich and J. M. Steininger, "A 45-MHz CMOS phase/frequency-locked loop timing recovery circuit," *Int. Solid-State Circuits Conf. Tech. Dig.*, San Francisco, CA, Feb. 1988, pp. 14–15.
[8] B. Saltzberg, "Timing recovery for synchronous binary data transmission," *Bell Syst. Tech. J.*, pp. 593–622, March 1967.
[9] E. Roza, "Analysis of phase-locked timing extraction circuits for pulse code transmission," *IEEE Trans. Commun.*, vol. 22, pp. 1236–1249, Sept. 1974.
[10] D. L. Duttweiler, "The jitter performance of phase-locked loops extracting timing from baseband data waveforms," *Bell Syst. Tech. J.*, pp. 37–58, Jan. 1976.
[11] J. P. Hein and J. W. Scott, "z-domain model for discrete-time PLL's," *IEEE Trans. Circuits Syst.*, vol. 35, Nov. 1988.
[12] F. M. Gardner, "Charge-pump phase-lock loops," *IEEE Trans. Commun.*, vol. 28, pp. 1849–1858, Nov. 1980.
[13] F. M. Gardner, *Phaselock Techniques*, 2nd Ed. New York: Wiley, 1979.
[14] A. A. Abidi and R. G. Meyer, "Noise in relaxation oscillators," *IEEE J. Solid-State Circuits*, vol. 18, pp. 794–802, Dec. 1983.
[15] P. R. Gray and R. G. Meyer, *Analysis and Design of Analog Integrated Circuits*, 2nd Ed. New York: Wiley, 1984.
[16] R. L. Fox, *Optimization Methods for Engineering Design*. Reading, MA: Addison-Wesley, 1971.
[17] *HSPICE Users' Manual*, Meta-Software, Inc., Campbell, CA, June 1987.

Noise Properties of PLL Systems

VENCESLAV F. KROUPA

Abstract—This is a survey paper which begins by the derivation of the general PLL noise equation and by dividing the additive noises into the passband group and the stopband group.

In the following paragraphs the behavior of all the major sources of additive noises is investigated and the practical numerical values of the respective power spectral densities are given.

In the terminating sections, guidelines for minimizing the additive noises in PLL systems and PLL frequency synthesizers are emphasized, and finally, phase-noise power spectral densities of several actual PLL frequency synthesizers are plotted in the normalized form.

The paper is accompanied by a copious bibliography.

I. INTRODUCTION

THE steadily increasing congestion in communications bands of the electromagnetic spectrum results in efforts for utilization of new ranges (e.g., microwave and optical frequencies) on one hand and better exploitation of the existing frequency allocations on the other hand. The latter task is met by adopting SSB modulation, telegraph multiplexing techniques, etc., and often by sharing the same communication channel by several services. In these cases the mutual interference is the major problem and it can be reduced only by increasing both the short-term and the long-term frequency stability of respective exciters and local oscillators.

Nowadays the long term stability is effectively solved with the assistance of frequency synthesis [1]-[4]. However, the situation with the short-term frequency stability is not so simple. Even in the ideal case when the frequency synthesizer is assumed to be a noiseless frequency transformer, we face an additive frequency noise [1] and often a too high phase-noise level after multiplication. The remedy to this difficulty may be a carefully designed phase-lock loop (PLL). However, this latter technique is more often used in frequency synthesizers only since it makes possible a substantial hardware simplification. And here the troubles start as nearly all PLL building blocks may add a sometimes substantial noise power to the useful signal. The problem is not yet generally understood and, in addition, is often underestimated. We shall therefore, in the following paragraphs, discuss theoretical backgrounds and summarize all the accessible experimental results to provide the leading lines for the design of low-noise PLL systems.

Manuscript received June 5, 1981; revised April 12, 1982. This paper was presented in part at the Third Symposium on Electromagnetic Compatibility, Rotterdam, The Netherlands, May 1979, and at the Conference on Precision Electromagnetic Measurements, Braunschweig, West Germany, June 1980.

The author is with the Institute of Radio Engineering and Electronics, Czechoslovak Academy of Sciences, 182 51 Prague 8, Czechoslovakia.

II. PLL NOISE EQUATION

In the following paragraphs we shall limit ourselves only to noises generated in the PLL building blocks and in a "low-noise" reference, leaving out the large class of cases with the reference frequency embedded in atmospheric or man-made noise which have been extensively dealt with in earlier works, from which many we shall mention only [5]-[7].

In Fig. 1 we have drawn a fairly general PLL arrangement with a phase detector (PD), a low-pass filter $F_L(s)$, and a voltage-controlled oscillator (VCO) in the forward path and a mixer $(-)$, an IF filter with the effective modulation transfer function $F_M(s)$ [1, p. 234] or [5, p. 147], and a divider $(\div N)$ in the feedback path. For completeness we have placed a divider $(\div Q)$ between the reference generator (RG) and the phase detector and a multiplier $(\times M)$ between RG and the second input to the mixer. However, we have to keep in mind that these two latter blocks, in actual PLL systems, are often replaced by more complicated frequency synthesis circuits.

Since all the noises generated or added in individual blocks in Fig. 1 are small compared with the useful signals, we have applied the rule of superposition and simply add them either at the respective inputs or outputs. Note that the subscript "n" indicates the noise signal throughout. Furthermore, the small signal theory makes it possible to use the Laplace transform approach to find the output noise of the considered PLL system or, more exactly, the respective power spectral densities.

By assuming a locked loop and by considering Fig. 1, we may write for the forward path of the loop

$$\phi_{o,n} = [(\phi_{i,n} - \phi_{o,n}')K_d + V_{\text{PD},n} + V_{F,n}]F_L(s)$$
$$\cdot \frac{K_0}{s} + \phi_{\text{osc},n} \qquad (1)$$

and for the feedback path

$$\phi_{o,n}' = (\phi_{o,n} - \phi_{m,n} + \phi_{MI,n})\frac{F_M(s)}{N} + \phi_{DN,n} \qquad (2)$$

where

$$\phi_{i,n} = \frac{\phi_{r,n}}{Q} + \phi_{DQ,n} \qquad (3)$$

and

$$\phi_{m,n} = M\phi_{r,n} + \phi_{MU,n}. \qquad (4)$$

Reprinted from *IEEE Trans. Comm.*, vol. COM-30, pp. 2244-2252, October 1982.

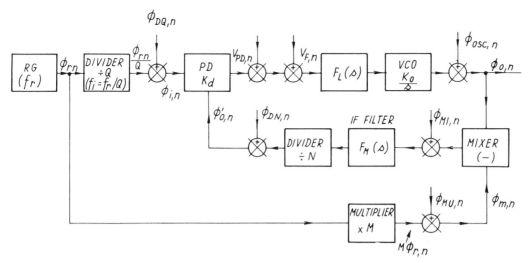

Fig. 1. Block diagram of a general PLL system with additive noise
sources.

Note that all noise components $\phi \cdots, n$ and $V \cdots, n$ are Laplace transformed quantities, i.e., $\phi \cdots, n(s)$, etc.

After introducing (2) into (1) we get for the output phase noise $\phi_{o,n}$

$$\phi_{o,n} = \left[\left(\phi_{i,n} - \phi_{DN,n} + \frac{V_{PD,n} + V_{F,n}}{K_d}\right)\frac{N}{F_M(s)} + \phi_{m,n}\right.$$

$$\left. - \phi_{MI,n}\right]\frac{F_M(s)F_L(s)K_dK_0/Ns}{1 + F_M(s)F_L(s)K_dK_0/Ns}$$

$$+ \phi_{osc,n}\frac{1}{1 + \dfrac{F_M(s)F_L(s)K_dK_0}{Ns}} \tag{5}$$

which can be simplified with the assistance of the effective loop transfer function $H'(s)$ [1, ch. 7] and (3) and (4) into

$$\phi_{o,n} = \left[\phi_{r,n}\left(M + \frac{N}{Q}\frac{1}{F_M(s)}\right) + \left(\phi_{DQ,n} - \phi_{DN,n}\right.\right.$$

$$\left.\left. + \frac{V_{PD,n} + V_{F,n}}{K_d}\right)\frac{N}{F_M(s)} + \phi_{MU,n} - \phi_{MI,n}\right]$$

$$\cdot H'(s) + \phi_{osc,n}[1 - H'(s)] \tag{6}$$

where

$$H'(s) = \frac{\dfrac{F_L(s)F_M(s)K_dK_0}{Ns}}{1 + \dfrac{F_L(s)F_M(s)K_dK}{Ns}}. \tag{7}$$

We feel that it is time to discuss the results we have arrived at until now. Let us start with the last equation.

First, we see that the loop gain $K_dK_0 = K$ is reduced in

proportion to the division factor N to a new value

$$K' = K/N = K_dK_0/N \tag{8}$$

and as a consequence we face a reduced effective natural frequency ω_n' and a prolongated settling time—the remedy might be a compensating dc amplifier incorporated into the $F_L(s)$ block.

The second difficulty arrises from the IF filter in the feedback path, respectively, from its effective modulation transfer function $F_M(s)$. In many instances it may be replaced by a mere time delay function which causes a reduction of the pull-in range $\Delta\omega_P$ [8]. However, when the IF filter is a more complicated circuit we face the danger of false locks [5, p. 151] and the degradation of the loop stability.

Finally, by investigating the behavior of the PLL transfer function $H'(s)$ in the frequency domain we easily arrive at the following conclusions: for $\omega < \omega_n'$

$$|H'(s)| \approx 1 \quad \text{and} \quad |1 - H'(s)| \approx 0 \tag{9a}$$

whereas for $\omega > \omega_n'$

$$|H'(s)| \approx 0 \quad \text{and} \quad |1 - H'(s)| \approx 1. \tag{9b}$$

The consequence is that in the PLL passband the output noise is given by

$$\phi_{o,n} \approx \phi_{r,n}\left(M + \frac{N}{Q}\right) + \left(\phi_{DQ,n} - \phi_{DN,n}\right.$$

$$\left. + \frac{V_{PD,n} + V_{F,n}}{K_d}\right)N + \phi_{MU,n} - \phi_{MI,n} \tag{10}$$

(note that we have neglected $F_M(s)$, the influence of which in the passband is small) and in the stopband by

$$\phi_{o,n} \approx \phi_{osc,n} \tag{11}$$

i.e., the PLL output noise is equal to the voltage-controlled oscillator noise. However, we shall see later that this rule of thumb is often invalidated by the improper choice of ω_n' or of the loop filter $F_L(s)$.

Now reverting to the analysis of the PLL output noise in the passband we see that the first term on the right-hand side of (10) is inevitable since it is merely a multiplied reference generator noise. The situation with the second term is not so simple; here we encounter both divider noises, the phase detector noise and the "filter noise" $V_{F,n}$, all multiplied by the division factor N. Finally, with the third term we add the multiplier and the mixer noises; however, they will be generally small compared with the second term.

As all the considered noises are random by nature and uncorrelated, we may sum the respective spectral densities.

$$
S_{\varphi o,n}(f) = S_{\varphi r,n}(f)\left[M + \frac{N}{Q}\right]^2 + \left[S_{\varphi DQ,n}(f)\right.
$$

$$
+ S_{\varphi DN,n}(f) + \frac{S_{V_{PD},n}(f) + S_{V_{F},n}(f)}{K_d{}^2}\right]
$$

$$
\cdot N^2 + S_{\varphi MU,n}(f) + S_{\varphi MI,n}(f). \qquad (12)
$$

III. NOISES GENERATED IN INDIVIDUAL PLL BUILDING BLOCKS

In the following sections we shall investigate the quantitative and qualitative share of the individual terms, in the above equation, to the total effective "reference noise."

A. Loop Filter Noise

When the loop filter $F_L(s)$ is a passive one (see Fig. 1), i.e., a simple RC lag or lag-lead network, there are two major sources of noise, namely, some types of capacitors and resistors (carbon resistors) which can generate appreciable amounts of $1/f$ noise. As a consequence the low-noise design requires their individual selection (the use of metal-film type resistors is a necessity). The second source may be the decoupling resistor R_{dc}, separating the varactor circuit from the loop filter and the phase detector. The respective noise power density is

$$
S_{V,F} = 4kT \cdot R_{dc} \approx R_{dc} \times 1.66 \times 10^{-20} \ [\text{V}^2/\text{Hz}] \quad (13)
$$

and for the typical value of the phase detector gain, $K_d = 0.3$, we get

$$
S_{\varphi,F} = R_{dc} \times 1.84 \times 10^{-19} \qquad [\text{rad}^2/\text{Hz}]. \quad (14)
$$

B. Phase Noise in dc Amplifiers

In many instances we need to introduce either an active lag-lead filter [1], [5] or merely a dc amplifier.

The design of a low-noise dc amplifier is not an easy task [9] even in instances where the tested circuit diagram is used [10]–[14]. Typical equivalent input noise voltage is only several $\text{nV}/\sqrt{\text{Hz}}$ with the corner-low-frequency between 10–100 Hz. Similar performance is also achieved with some modern IC operational amplifiers [9], [15].

C. Phase Noise in HF Amplifiers

Here, our investigations must start with the famous paper by Halford et al. [16]. They found that power spectral density of the flicker phase noise close to the carrier was approximately the same, i.e.,

$$
S_\varphi(f) \approx \frac{10^{-11.2}}{f} + S_\varphi(f)_{\text{white}} \qquad (15)
$$

for the surveyed range from 5 to 100 MHz, quite independent of the transistor type and even of the multiplication factor. Laboratory experiments proved that the intrinsic, direct phase modulation of the RF carrier by transistors was responsible for the phenomenon. The improvement—typically more than 30 dB (and up to 40 dB in some cases)—has been achieved by applying local RF negative feedback (small unbypassed resistor in the emitter—typically from 10 to 100 Ω). These findings were later supported theoretically by Healey [17] and experimentally by other authors [12], [18], [19]. Low amplifier currents and high voltages help to keep the $1/f$ noise current low. The best white noise levels $S_\varphi(f)_{\text{white}}$ reported [18] are of the order of 10^{-17}.

D. Phase Noise in Phase Detectors and Mixers

The experience with measurement of $S_{\varphi,n}(f)$ of low-noise crystal oscillators has taught that the best phase detectors are double-balanced mixers with Schottky barrier diodes in the ring configuration. A further improvement may be achieved by placing two diodes in each arm [11], [18]. Measurements performed by different authors [11], [13], [14], [18], [19] reveal

$$
S_\varphi(f) \approx \frac{10^{-14 \pm 1}}{f} + 10^{-17}. \qquad (16)
$$

On the other hand, there is not yet fully proved evidence that the logic circuit or digital phase detectors are much noisier [19], [20]. Šojdr [19] has measured noise properties of popular digital phase-frequency detectors in the range from 0.1 to 1 MHz and found

$$
S_\varphi(f) = \frac{10^{10.6 \pm 0.3}}{f}. \qquad (17)
$$

Furthermore, he verified the statement by Underhill et al. [20] that phase detectors built of ECL and CMOS logic families exhibit a better noise behavior up to about −22 dB in the flicker noise region.

E. Phase Noise in Digital Frequency Dividers

Since the frequency or phase modulation index decreases proportionally to the division factor N, the ideal noise figure is

$$
F_{\text{div}} = -20 \log N. \qquad (18)
$$

However, there is an additional noise generated in the divider itself; thus, the output phase noise is given by

$$S_{\varphi,D,n}(f) = \frac{S_{\varphi,\text{in}}(f)}{N^2} + S_{\varphi,\text{add}}(f). \qquad (19)$$

To find out the properties of the additive term we have normalized the above equation with respect to the output frequency, i.e. (cf. Section III-G),

$$\frac{S_{\varphi,D,n}}{f_{\text{out}}^2} = \sum_{k=-2}^{2} h_{k,D,n}$$

$$= \sum_{k=-2}^{2} h_k f^{k-2} + \frac{h_{1,\text{add}}}{f} + h_{2,\text{add}} \qquad (20)$$

and plotted the respective coefficients $h_{1,\text{add}}$ and $h_{2,\text{add}}$, computed from accessible experimental results [3], [19], [21]–[23] in Fig. 2. We see that the overall behavior is the same, namely, for large division factors (small output frequency) the data fit a straight line with the slope of 20 db/dec. Be referring to the lowest noise points we may write for the additive term

$$S_{\varphi,\text{add}} \approx \frac{10^{-14.7}}{f} + 10^{-16.5}. \qquad (21)$$

On the other hand, for higher output frequencies (above 1 MHz) both coefficients $h_{1,\text{add}}$ and $h_{2,\text{add}}$ become constants which indicate that the input noise predominates.

F. Phase Noise in Frequency Multipliers

We have to refer first to the above mentioned paper by Halford [16] and further to that by Baugh [24] where guidelines for the design of low-noise frequency multipliers are given (RF negative feedback—see Section III-C—and steep zero crossings). To get more information we have collected published results about noise properties of transistor frequency multipliers [12], [24]–[29] and diode frequency multipliers [12], [28], [30], [31] as well and found that both noise constants $a_{-1,\text{inp}}$ and $a_{0,\text{inp}}$ are nearly the same for properly designed transistor frequency multipliers, irrespective of the frequency, i.e.,

$$S_{\varphi MU,\text{inp}}(f) \approx \frac{10^{-14}}{f} + 10^{-16.5}. \qquad (22)$$

In diode frequency multipliers the flicker phase noise level is higher, typically

$$10^{-12.9}/f. \qquad (23)$$

We shall see later that the flicker and white phase-noise spectral densities of the best crystal oscillators are of the same order or rather worse as the respective additive terms in frequency multipliers; thus, we can conclude that the phase noise is not appreciably deteriorated by passing the signal through a properly designed frequency multiplier. (Note that this is hardly true with the diode frequency multipliers.)

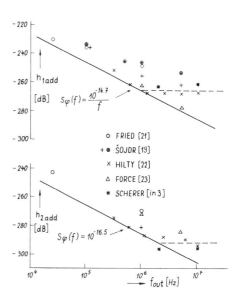

Fig. 2. Plot of the normalized additive phase-noise coefficients $h_{1,\text{add}}$ and $h_{2,\text{add}}$ in respect to the divider output frequency as measured by different authors.

G. Phase Noise in Oscillators[1]

The theory of noise in free-running oscillators has been dealt with by many authors for nearly half a century and we feel that the best we can do is to mention only a few of them [33]–[36]. However, they do not provide the information the designer of the PLL systems is looking for, at least by the first fast reading. On the other hand, when we start from the condition of the zero phase shift around the oscillating loop [37] we easily arrive at the heuristic oscillator phase-model suggested by Leeson [38], [17].

In accordance with this model any oscillator can be simplified into a loop containing a resonator and an amplifier-limiter. As a consequence its output spectral density is given by

$$S_{\varphi}(\omega)_{\text{osc},n} = S_{\varphi}(\omega)_{A,n}[1 + (\omega_0/2Q_L\omega)^2] \qquad (24)$$

where the amplifier-limiter noise is

$$S_{\varphi}(f)_{A,n} = a_{-1}/f + a_0. \qquad (25)$$

The magnitude of the flicker noise constant a_{-1} has been found experimentally [16] in the range from 5 to 100 MHz to be

$$a_{-1} = F_{-1} \cdot 10^{-11.2} \quad [\text{rad}^2]. \qquad (26)$$

The noise factor F_{-1} depends on the emitter RF feedback and may be made as small as 10^{-3}.

The white noise constant is the ratio of the noise power $\varphi_{\text{noise}}^2(t)$ to the oscillator power P_0 reduced to 1 Hz bandwidth and multiplied by a noise factor F_0:

$$a_0 = F_0 \cdot kT/P_0 \approx 4 \times 10^{-21} F_0/P_0 \quad [\text{rad}^2/\text{Hz}]. \qquad (27)$$

[1] This section is based on the paper read by the author at the EMC-79 Symposium, Rotterdam, The Netherlands [32].

145

By introducing (25) into (24) we arrive at a power law relation

$$S_{\varphi osc,n}(f) = \left(\frac{f_0}{f}\right)^2 \cdot \frac{a_{-1}}{f \cdot 4Q_L{}^2} + \left(\frac{f_0}{f}\right)^2$$

$$\cdot \frac{a_0}{4Q_L{}^2} + \frac{a_{-1}}{f} + a_0. \tag{28}$$

To compare oscillators with different output frequencies we face the difficulty that the noise sidebands close to the carrier are proportional to the square of the resonant frequency f_0. This problem is often solved by referring to the fractional-frequency power spectral density

$$S_y(f) = (f/f_0)^2 S_\varphi(f), \tag{29}$$

i.e.,

$$S_{y,osc,n}(f)$$

$$= \frac{a_{-1}}{f \cdot 4Q_L{}^2} + \frac{a_0}{4Q_L{}^2} + \frac{a_{-1}}{f_0{}^2} f + \frac{a_0}{f_0} f^2 \tag{30a}$$

$$= \frac{h_{-1}}{f} + h_0 + h_1 f + h_2 f^2 \tag{30b}$$

or to the spectral density of the phase-noise time [14]

$$S_x(f) = S_y(f)/(2\pi f)^2. \tag{31}$$

The advantage of this second normalization is the close resemblance to the actual phase noise characteristic and the ease with which $S\varphi(f)$ is calculated—the proportionality factor being $(2\pi f_0)^2$. By dropping the factor $(2\pi)^2$ we arrive at a very useful simplification:

$$S_{\varphi osc,n}(f)/f_0{}^2 = \frac{h_{-1}}{f^3} + \frac{h_0}{f^2} + \frac{h_1}{f} + h_2 \tag{32}$$

where

$$h_{-1} = a_{-1}/4Q_L{}^2; \qquad h_0 = a_0/4Q_L{}^2; \tag{33a}$$

$$h_1 = a_{-1}/f_0{}^2; \qquad h_2 = a_0/f_0{}^2. \tag{33b}$$

We have verified the validity of (28)–(32) by computing coefficients h_{-1} and h_0 for different types of oscillators from a wealth of published data and plotted them as functions of quoted Q_L. The results are shown in Fig. 3 and indicate a good agreement with the simplified oscillator noise theory. The mean value of the flicker noise constant a_{-1} is

$$a_{-1} \approx 10^{-11} \tag{34}$$

and is practically independent of the oscillator type in the whole frequency range from 5 MHz to 100 GHz; the same is true also for the white noise constant a_0, the mean value of which is

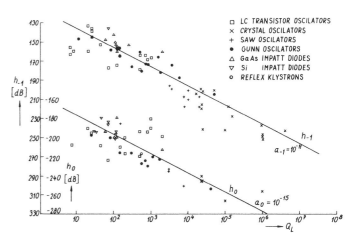

Fig. 3. Plot of the normalized phase-noise coefficients h_{-1} and h_0 in respect to the loaded Q_L of different oscillators.

$$a_0 \approx 10^{-15}. \tag{35}$$

Similar experimental verification of the dependence of h_1 and h_2 on $1/f_0{}^2$ is prevented by the lack of data for the noise power at higher Fourier frequencies. Some insight provides the investigation of noise properties of crystal oscillators discussed in the following section.

Summarizing all the above results we can write a fairly general oscillator noise equation

$$\frac{S_{\varphi osc,n}(f)}{f_0{}^2} = \frac{1}{f^3} \cdot \frac{10^{-11.6}}{Q_L{}^2} + \frac{1}{f^2} \frac{10^{-15.6}}{Q_L{}^2}$$

$$+ \frac{1}{f} \cdot \frac{10^{-11}}{f_0{}^2} + \frac{10^{-15}}{f_0{}^2} \tag{36}$$

and plot the normalized oscillator phase-noise characteristics which consist of two sets of straight lines with parameters Q_L and f_0; see Fig. 4.

H. Phase Noise in Reference Frequency Generators

The reference generator in low-noise PLL systems (frequency synthesizers) is a spectrally pure crystal oscillator. By inspecting (36) or Fig. 4 we see that a low close-to-carrier noise requires the use of resonators with the highest possible Q. It has been found earlier [39] that for the AT-cuts the intrinsic losses in the quartz crystal material are related to the resonant frequency by

$$f_0 Q \approx 1.5 \times 10^{13}. \tag{37}$$

This product is slightly lower for the advantageous SC-cuts and nearly two times larger for BT-cuts. Consequently, the low flicker and white frequency noise, i.e., small h_{-1} and h_0 in (32) and (33), requires the lowest possible frequency f_0. However, to keep dimensions of the crystal resonators in practical limits, we hardly can go below 5 MHz (cf. [39, Fig. 2]). Very often 10 MHz crystal oscillators are used as reference frequency generators.

To find practical values of h_k coefficients in (32) we have recently investigated noise characteristics of about 60 crystal oscillators in the range from 5 to 170 MHz [40] and

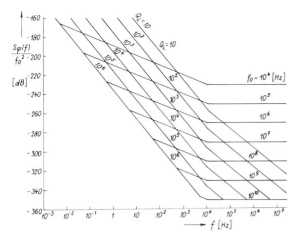

Fig. 4. Normalized phase-noise characteristics of oscillators: parameters are the loaded Q_L of the resonator and the output frequency f_0.

get for an average crystal oscillator the following noise equation.

$$\frac{S_{\varphi osc,n}(f)}{f_0^2} = \frac{1}{f^3} 10^{-37.25} f_0^2 + \frac{1}{f^2} 10^{-39.4} f_0^2$$
$$+ \frac{1}{f} \frac{10^{-12.15}}{f_0^2} + \frac{10^{-14.9}}{f_0^2}. \qquad (38)$$

By investigating the dispersion of the h_k values, particularly of the best crystal oscillators we have arrived at the conclusion that in the $1/f^3$ and $1/f^2$ regions the crystal resonator noise [41], rather than the transistor noise, is the limiting factor [42].

IV. PLL LOOP FILTERS FOR LOW NOISE BEHAVIOR

The problem has been discussed earlier by the author [43] and here only the results will be summarized.

By formulating the rule of thumb in (9) in Section II we have assumed that the transfer function $H'(s)$ has a rectangular behavior. However, this is not true in real systems and a closer investigation of the second- and third-order systems gives

$$H'(s) = \frac{s\omega_n'(2\xi - \omega_n'/AK') + \omega_n'^2}{s^2 + 2\xi\omega_n's(1 + s^2\kappa\omega_n'^2) + \omega_n'^2} \qquad (39)$$

where

$$\omega_n' = \sqrt{K'/T_1}; \quad \xi = \frac{\omega_n'}{2} T_2 + \frac{1}{AK'};$$

$$\kappa = T_3/T_2; \quad K' = K_d K_0/N \qquad (40)$$

and A is the gain of the operational amplifier used.

Since the asymptotic approximation will generally supply sufficient information for the noise behavior of the studied PLL system, we shall consider the four most important configurations.

1) Simple RC filter, i.e., $A = 1$, $\xi = \omega_n'/2K'$, and $\kappa = 0$. After introducing the normalization

$$\frac{\omega}{\omega_n} = x \qquad (41)$$

we shall find

$$H(jx) \approx -1/x^2 \qquad x \gg 1 \qquad (42a)$$

and

$$1 - H(jx) \approx 2j\xi x \qquad x \ll 1. \qquad (42b)$$

2) Passive lag-lead filter, i.e., $A = 1$, and $\kappa = 0$.

$$H(jx) \approx -j(2\xi - \omega_n'/K')/x \qquad x \gg 1 \qquad (43a)$$

$$1 - H(jx) \approx jx\omega_n'/K' \qquad x \ll 1. \qquad (43b)$$

3) Active lag-lead filter, i.e., $A \to \infty$, and $\kappa = 0$.

$$H(jx) \approx -2j\xi/x \qquad x \gg 1 \qquad (44a)$$

$$1 - H(jx) \approx -x^2 \qquad \omega_n'/AK' \ll x \ll 1. \qquad (44b)$$

4) Active lag-lead filter with an additional RC section (it has been shown [43] that for practical applications $\kappa < 0.3$).

$$H(jx) \approx -1/\kappa x^2 \qquad x \gg 1 \qquad (45a)$$

$$1 - H(jx) \approx -x^2 \qquad \omega_n'/AK' \ll x \ll 1. \qquad (45b)$$

To get a better insight we shall consider the problem of phase-locking a 100 MHz crystal oscillator and a low Q LC-oscillator to a 5 MHz reference signal. All normalized phase noise characteristics are plotted in Fig. 5. By considering first the 100-to-5 MHz PLL system we find the crossover point to be approximately 200 Hz. In instances where f_n is smaller than 200 Hz we face a large amount of additive noise, the origin of which is the attenuated 100 MHz oscillator noise; the dashed lines indicate the situation with passive filters and dot-and-dash lines indicate the improvement when the active lag-lead filter is used. Similarly, we encounter an unnecessary additive noise caused by the attenuated 5 MHz oscillator noise in cases where $f_n > 200$ Hz (dashed line). A remedy can be provided by additional filtering; however, the stability of the loop deteriorates.

In the second example of phase-locking a low Q oscillator to a crystal reference, the use of the active lag-lead filter is necessary since even for f_n equal to the crossover frequency we face a large additive noise with passive filters only (see again the dashed line in Fig. 5).

V. NOISE IN PLL FREQUENCY SYNTHESIZERS

By considering the basic PLL configuration, as shown in Fig. 1, we easily arrive, with the assistance of (12) and the condition of only a 3 dB noise increase, at the conclusion that

$$S_{\varphi,r,n}(f)\left[M + \frac{N}{Q}\right]^2 \leq N^2 \cdot S_{\varphi,\text{PLL}}(f) \qquad (46)$$

where

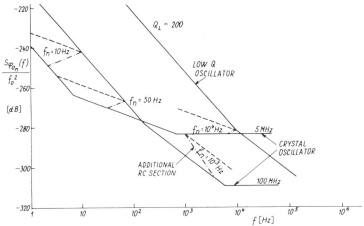

Fig. 5. Normalized phase noise of the PLL system of the 100 MHz crystal oscillator locked to a 5 MHz reference crystal oscillator and a low Q oscillator locked to the same reference oscillator.

$$S_{\varphi,\mathrm{PLL}}(f) = 2S_{\varphi,D}(f) + S_{\varphi,\mathrm{PD}}(f) + S_{\varphi,A,n}(f). \qquad (47)$$

The power spectral densities of individual noise sources in the above equation have been evaluated in Section III. In a large majority of cases where active filters are used the last term on the right-hand side of (47) will dominate. With the assistance of [9, Fig. 12] and a reasonable value for K_d, $K_d = 0.3$, we find

$$S_{\varphi,\mathrm{PLL}}(f) \approx \frac{10^{-14}}{f} + 10^{-15.5}. \qquad (48)$$

Since the flicker and white noise spectral densities of good reference oscillators in the 5 and 10 MHz frequency ranges are of the same order we arrive at the rule of thumb that the division factor N should not exceed the multiplication factor M, i.e.,

$$N \gtreqless M. \qquad (49)$$

However, this condition can hardly be met in instances where small frequency steps are desired at the synthesizer output. One solution is provided by the application of fractional frequency dividers [2, p. 74], [44], [45]. The spurious phase modulation which is often quite large but predictable [46] is suppressed by compensation.

Another solution may be provided by a subtracting PLL-system, the principle of which will be explained with the assistance of Fig. 6. In the case where N in Fig. 1 is much larger than M, the output noise of the first digital PLL in its passband is given approximately by

$$S_{\varphi,01,n}(f) \approx N^2 \cdot S_{\varphi,\mathrm{PLL},1}(f). \qquad (50)$$

With the assistance of (12) we find for the output noise (in the passband) of the subtracting loop

$$S_{\varphi,02,n}(f) \approx \frac{N^2 S_{\varphi,\mathrm{PLL},1}(f)}{P^2} + M^2 S_{\varphi r,n}(f)$$

$$+ S_{\varphi,\mathrm{PLL},2}(f). \qquad (51)$$

Generally, the last term on the right-hand side of the above equation may be neglected and the application of the 3 dB noise-increase condition reveals the second rule of thumb, i.e.,

$$\frac{N}{P} \leq M. \qquad (52)$$

By a judicious combination of both these techniques one can expect a very low additive phase noise from frequency synthesizers at frequency ranges above 100 MHz [44], [45], [47]. For generation of lower frequencies the use of dividers instead of output mixers should be preferred [21], [44], [45].

To demonstrate the state of the art we have plotted in Fig. 7 the normalized power spectral densities of the output phase noise of several commercial PLL frequency synthesizers. The advantage of the normalization is the possibility to compare the noise properties of frequency synthesizers with different output frequencies and that of the reference generator together in the same figure. The progress achieved in the last ten years is impressive.

VI. CONCLUSIONS

In this survey paper we have called the readers' attention to all major sources of additive noise in the PLL system. Furthermore, we have shown that, in the first approximation, these noises add to the input or the reference noise.

In the second we have investigated noises generated in individual PLL building blocks and tried to find numerical values for the respective power spectral densities with the assistance of the experimental findings published by different authors all over the world.

The major guideline rules for minimizing the additive noises in complicated PLL systems or PLL frequency synthesizers have been discussed in the last two sections. Finally, phase-noise power spectral densities of several commercial PLL frequency synthesizers have been plotted in normalized form in Fig. 7. On one hand, this figure demonstrates the state of the art and, on the other hand, the progress achieved in the last decade.

For readers who intend to go deeper into the problem we have collected a copious bibliography.

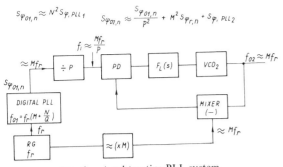

$$S_{\varphi_{01,n}} \approx N^2 S_{\varphi,PLL_1} \qquad S_{\varphi_{02,n}} \approx \frac{S_{\varphi_{01,n}}}{p^2} + M^2 S_{\varphi_{r,n}} + S_{\varphi,PLL_2}$$

Fig. 6.   A subtracting PLL system.

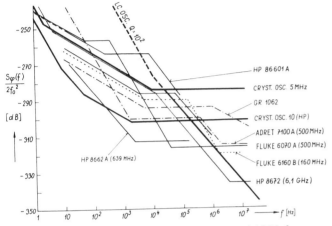

Fig. 7.   Normalized phase noise of several commercial PLL frequency synthesizers, two reference crystal oscillators, and one typical LC oscillator (VCO).

REFERENCES

[1]   V. F. Kroupa, *Frequency Synthesis, Theory, Design and Applications*.   London, England: Griffin, 1973; New York: Wiley, 1973.

[2]   J. Gorski-Popiel, *Frequency Synthesis: Techniques and Applications*.   New York: IEEE Press, 1975.

[3]   V. Manassewitsch, *Frequency Synthesizers, Theory and Design*. New York: Wiley, 1980.

[4]   W. F. Egan, *Frequency Synthesis by Phase Lock*.   New York: Wiley, 1981.

[5]   F. M. Gardner, *Phaselock Techniques*, 2nd ed.   New York: Wiley, 1979.

[6]   W. C. Lindsey, *Synchronization Systems in Communication and Control*.   Englewood Cliffs, NJ: Prentice-Hall, 1972.

[7]   W. C. Lindsey and M. K. Simon, *Telecommunication Systems Engineering*.   Englewood Cliffs, NJ: Prentice-Hall, 1973.

[8]   J. A. Develet, Jr., "The influence of time delay on second-order phase-lock loop acquisition range," in *Proc. Int. Telemetering Conf.*, vol. 1, Sept. 23–27, 1963, pp. 432–437; see also, W. C. Lindsey and M. K. Simon, Eds., *Phase-Locked Loops and Their Application*.   New York: IEEE Press, 1978.

[9]   Y. Netzer, "The design of low-noise amplifiers," *Proc. IEEE*, vol. 69, pp. 728–741, June 1981.

[10]   B. E. Blair, Ed., *Time and Frequency: Theory and Fundamentals*, Nat. Bur. Stand., Boulder, CO, Monog. 140, May 1974, p. 169.

[11]   D. G. Meyer, "A test set for the accurate measurement of phase noise on high-quality signal sources," *IEEE Trans. Instrum. Meas.*, vol. IM-19, pp. 215–227, Nov. 1970.

[12]   S. G. Andresen and J. K. Nesheim, "Phase noise of various frequency doublers," *IEEE Trans. Instrum. Meas.*, vol. IM-22, pp. 185–188, June 1973.

[13]   K. Hilty, "Messtechnik zur Bestimmung des Phasenrauschspektrums," *Technische Mitteilungen PTT*, pp. 1–15, Jan. 1980.

[14]   P. Kartaschoff, *Frequency and Time*.   London, England: Academic, 1978.

[15]   G. Erdi *et al.*, "Op amps tackle noise—And for once, noise loses," *Electron. Design*, vol. 28, pp. 65–71, Dec. 20, 1980.

[16]   D. Halford, A. E. Wainright, and J. A. Barnes, "Flicker noise of phase in RF amplifiers and frequency multipliers: Characterization, cause, and cure," in *Proc. 22nd Annu. Freq. Contr. Symp.*, Atlantic City, NJ, Apr. 1968.

[17]   D. J. Healey, "Flicker of frequency and phase and white frequency and phase fluctuations in frequency sources," in *Proc. 26th Annu. Symp. Freq. Contr.*, Atlantic City, NJ, June 1972, pp. 29–49.

[18]   F. L. Walls, S. R. Stein, J. E. Gray, and D. J. Glaze, "Design considerations in state-of-the-art signal processing and phase noise measurement systems," in *Proc. 30th Annu. Symp. Freq. Contr.*, Atlantic City, NJ, June 1976, pp. 269–274.

[19]   L. Šojdr, "Phase noise in oscillators and synthesizers" (in Czech). Dissertation, Prague, Czechoslovakia, June 1979.

[20]   M. J. Uderhill, P. A. Jordan, M. A. G. Clark, and R. I. H. Scott, "A general purpose LSI frequency synthesizer system," in *Proc. 32nd Annu. Symp. Freq. Contr.*, Atlantic City, NJ, May–June 1978, pp. 365–372.

[21]   R. Fried, "A wide-band non-heterodyned frequency synthesizer," John Fluke, Inc., Seattle, WA, 1974.

[22]   K. Hilty, private communication at the occasion of the CPEM-80 Conf., Braunschweig, West Germany, 1980.

[23]   M. Force, "Low-noise interface circuits format crystal-oscillator signals," *EDN*, vol. 25, pp. 125–129, 131, Oct. 20, 1980.

[24]   R. A. Baugh, "Low noise frequency multiplication," in *Proc. 26th Annu. Symp. Freq. Contr.*, Atlantic City, NJ, June 1972, pp. 50–54

[25]   D. J. Glaze, "Improvements in atomic cesium beam frequency standards at the National Bureau of Standards," *IEEE Trans. Instrum. Meas.*, vol. IM-19, pp. 156–160, Aug. 1970.

[26]   D. J. Healey, "$L(f)$ measurements on UHF sources comparising VHF crystal controlled oscillator followed by a frequency multiplier," in *Proc. 28th Annu. Symp. Freq. Contr.*, Atlantic City, NJ, May 1974, pp. 190–202.

[27]   F. L. Walls and A. De Marchi, "RF spectrum of a signal after frequency multiplication; Measurement and comparison with a simple calculation," *IEEE Trans. Instrum. Meas.*, vol. IM-24, pp. 210–217, Sept. 1975.

[28]   E. Bava, A. De Marchi, and A. Godone, "Spectral analysis of synthesized signals in the mm wavelength region," *IEEE Trans. Instrum. Meas.*, vol. IM-26, pp. 128–132, June 1977.

[29]   A. L. Lance, W. D. Seal, F. G. Mendoza, and N. W. Hudson, "Automating phase noise measurements in the frequency domain," in *Proc. 31st Annu. Symp. Freq. Contr.*, Atlantic City, NJ, June 1977, pp. 347–358.

[30]   E. Bava, A. De Marchi, and A. Godone, "A narrow output linewidth multiplier chain for precision frequency measurements in the 1 THz region," in *Proc. 31st Annu. Symp. Freq. Contr.*, Atlantic City, NJ, June 1977, pp. 578–582.

[31]   M. B. Bloch and A. Vulcan, "Low noise measurement techniques," in *Proc. 28th Annu. Freq. Contr. Symp.*, Atlantic City, NJ, May 1974, pp. 184–189.

[32]   V. F. Kroupa, "Electromagnetic compatibility of frequency synthesizers," in *Proc. 3rd Symp.*, Rotterdam, The Netherlands, May 1979, pp. 335–340.

[33]   I. Berstein, "On fluctuations in the neighbourhood of periodic motion of an auto-oscillating system," *Comptes Rendus de l'Academie des Sciences de l'URSS*, vol. XX, no. 1, 11–16, 1938.

[34]   W. A. Edson, "Noise in oscillators," *Proc. IRE*, vol. 48, pp. 1454–1466, Aug. 1960.

[35]   K. Kurokawa, "Noise in synchronized oscillators," *IEEE Trans. Microwave Theory Tech.*, vol. MTT-16, pp. 234–240, 1968.

[36]   K. F. Schünemann and K. Behm, "Nonlinear noise theory for synchronized oscillators," *IEEE Trans Microwave Theory Tech.*, vol. MTT-27, pp. 452–458, May 1979.

[37]   E. J. Baghdady, R. N. Lincoln, and B. D. Nelin, "Short-term frequency stability: Characterization, theory, and measurement," *Proc. IEEE*, vol. 53, pp. 704–722, July 1965.

[38]   D. B. Leeson, "A simple model of feedback oscillator noise spectrum," *Proc. IEEE*, vol. 54, pp. 329–330, Feb. 1966.

[39]   A. W. Warner, "Design and performance of ultraprecise 2.5-mc quartz crystal units," *Bell Syst. Tech. J.*, vol. 39, pp. 1193–1217, Sept. 1960.

[40] V. F. Kroupa, "Noise of crystal oscillators" (in Czech), Rep. Z-1042/1, Prague, Czechoslovakia, Nov. 1979.

[41] F. L. Walls and A. E. Wainwright, "Measurement of the short-term stability of quartz crystal resonators and the implications for crystal oscillator design and applications," *IEEE Trans. Instrum. Meas.*, vol. IM-24, pp. 15–17, Mar. 1975.

[42] V. F. Kroupa, J. Pavlovec, and L. Šojdr, "Noise in standard frequency sources," to be published.

[43] V. F. Kroupa, "Spectral properties of third order phase-locked loops," in *Proc. 5th Summer Symp. Circuit Theory*, Kladno, Sept. 1977, pp. 201–215.

[44] K. L. Astrof, "Frequency synthesis in a microwave signal generator," *Hewlett-Packard J.*, vol. 29, pp. 8–15, Nov. 1977.

[45] Special Issue on PLL Frequency Synthesizer Type 8662A, *Hewlett-Packard J.*, vol. 32, Feb. 1981.

[46] V. F. Kroupa, "Spectra of pulse rate frequency synthesizers," *Proc. IEEE*, vol. 67, pp. 1680–1682, Dec. 1979.

[47] G. Mackiw and G. W. Wild, "Microwave frequency synthesis for satellite communications ground terminals," in *Proc. 30th Annu. Symp. Freq. Contr.*, Atlantic City, NJ, June 1976, pp. 420–437.

PLL/DLL System Noise Analysis for Low Jitter Clock Synthesizer Design

Beomsup Kim*, Todd C. Weigandt, Paul R. Gray

Department of Electrical Engineering and Computer Sciences
University of California, Berkeley

*Department of Electrical Engineering
Korea Advanced Institute of Science and Technology

Abstract

This paper presents an analytical model for timing jitter accumulation in ring-oscillator based phase-locked-loops (PLL). The timing jitter of the system is shown to depend on the jitter in the ring-oscillator and an accumulation factor which is inversely proportional to the bandwidth of the phase-locked-loop. Further analysis shows that for delay-locked-loops (DLL), which use an inverter delay chain that is not configured as a ring-oscillator, there is no noise enhancement since noise jitter events do not contribute to the starting point of the next clock cycle. Finally, theoretical predictions for overall jitter are compared to behavioral simulations with good agreement.

I. Introduction

Higher clock rates in many applications such as video, audio, and data processors, requires increasingly higher performance from the clock synthesizers used to drive them. In clock recovery applications, such as data communications and disk drive read channels, as well, higher speeds require better performance from the VCOs and the overall timing recovery phase-locked-loop itself. In both types of applications clocks are generated to drive mixers or sampling circuits in which the random variation of the sampling instant, or jitter, is a critical performance parameter. The goal of this paper is to predict the timing jitter of phase-locked-loop (PLL), and delay-locked-loop (DLL) systems from the parameters of the loop and the jitter in the VCO itself. Of particular interest are ring-oscillator VCOs which are attractive from an integration and cost point of view, but suffer from larger timing jitter than traditional tuned LC-tank oscillators.

In most clock synthesis applications a VCO is locked to a low-jitter reference, often in the form of a crystal, using a phase-locked-loop (figure 1). Most of the output jitter results from noise sources in the phase detector, loop filter, and VCO. With careful PLL design, however, the jitter in the VCO is usually the dominant contributor. In clock and data recovery applications there is often a significant amount of jitter from the input source as well as the VCO. In this case, it will be shown that there is a trade-off involved in selecting the bandwidth of the PLL. A narrow bandwidth PLL rejects input jitter but does not correct VCO timing errors as quickly, leaving the total output jitter, VCO noise limited. A wide bandwidth PLL can correct VCO errors more quickly but if made too wide, leaves the system input jitter

limited.

The ring-oscillator VCO in figure 1 is popular choice in many applications. The jitter per cycle of oscillation is determined by the sum of the timing error contributions of each inverter stage in the ring [1][1]. With each cycle of oscillation the jitter variance, relative to a reference transition in the past continues to grow, unless the oscillator is configured in a PLL. In a ring-oscillator PLL, however, the total timing error is the sum of all past errors weighted by the corrective action of the loop. The total jitter is made up of the errors in the most recent cycles of oscillation, yet to be corrected by the PLL, and therefore improves for higher loop bandwidths.

Another structure popular in many applications is a DLL in which a voltage controlled delay line, (VCD) is used in place of a VCO. In this case the jitter accumulated by the end of the delay chain does not contribute to the starting point of the next cycle since the delay chain is not configured as an oscillator. The reference determines the next transition point instead. This type of system has superior jitter performance, but is only usable in some applications.

In this paper, the total output jitter for PLL and DLL systems will be determined and compared to the results of behavioral simulations. The analysis is the scope of section II and some design examples and simulations will be shown in section III.

II. PLL/DLL Jitter Analysis

Timing jitter in a ring-oscillator PLL depends on the interaction of noise in the oscillator with the dynamics of the phase-locked loop. It has been shown in [1] that the timing jitter variance at the end of a chain of inverters is given by the sum of the contributions of each stage. If each stage contributes a timing error with variance $\overline{\Delta t_n^2}$, then the total jitter at the end of N stages is $N \times \overline{\Delta t_n^2}$. In a ring-oscillator this timing error determines the starting point of the next cycle and therefore creates a permanent phase shift in the output signal. If the ring-oscillator is configured in a phase-locked-loop, however, the phase difference between the reference

---

1. The paper analyzes the output jitter of the delay cells. Each jitter source is identified and its contribution to the output jitter is calculated. It also gives guidelines for design of low jitter delay lines.

Program Supported by NSF, ARPA, and California MICRO Program

Reprinted from *Proc. of ISCAS*, June 1994.

clock and the oscillator output is detected and compensated for by the dynamics of the loop. The phase detector will sense the shift and create an error signal to change the frequency of the ring-oscillator VCO in a way which moves the phase of the output in the right direction.

Since the amount of phase adjustment is usually small, the phase error is not corrected in one clock cycle, but it is reduced gradually over the course of several cycles. The phase error may remain for up to several hundreds of cycles, depending on the bandwidth of the loop filter in the PLL.

Analysis of the accumulated phase jitter and its relation to the loop bandwidth is important for both clock synthesis and clock recovery applications. In most PLL clock synthesizer designs, the reference clock comes from a very low jitter source such as crystal oscillator. Therefore, the jitter in the ring-oscillator is the main source of the phase error in the synthesized clock. In this case the bandwidth of the loop filter determines how large the accumulated timing jitter gets. For clock recovery applications there is a trade-off involved in the choice of the loop bandwidth since the input signal that is being locked to is not ideal, but has timing jitter associated with it as well. A narrow loop-bandwidth will reduce the impact of jitter in the input signal since the loop will not try to track input fluctuations as strongly. On the other hand, this means that it will take more time to compensate for jitter events in the ring-oscillator. Previously, more attention has been paid to the first effect than the second, but both are important for high performance clock recovery applications. So for both clock synthesis and clock recovery applications, a thorough analysis of the output jitter due to the internal jitter sources is important [2].

To find the accumulated rms jitter, a PLL which uses a sequential phase detector and a charge-pumping circuit is represented by a simple discrete-time model as shown in Figure 2 [3]. The transfer function for jitter in the PLL due to the internal jitter sources is represented by (EQ 1) in z-transform domain.

$$\Theta_{on}(z) = \frac{\Theta_n(z)}{1 + K_d K_w Z_F(z) z^{-1}} \qquad \text{(EQ 1)}$$

Here the phase detector gain, $K_d = \dfrac{I_S}{2\pi}$ and VCO gain,

$K_w = \dfrac{dw}{dv}$ respectively, and I_S indicates the charge pumping current. $Z_F(z)$ is the z-transform $H(s)/s$, where $H(s)$ is the transfer function of the PLL loop filter in s-domain. In most PLL design, the second order loop filter is used and the transfer function is given by (EQ 2).

$$H(s) = \frac{a(s + n_1)}{s(s + p_1)} \qquad \text{(EQ 2)}$$

where the DC filter gain, $a = \dfrac{1}{C_P}$, a zero, $n_1 = \dfrac{1}{C_I R}$, and a

pole, $p_1 = \dfrac{C_I + C_P}{C_I C_P R}$. In most cases, the capacitor C_P does not

affect the bandwidth of a PLL and can be ignored for simplicity. Then, the loop filter is configured as a lead-lag filter composed of a resistor R, in series with a capacitor C_I. In this case,

$H(s) = \dfrac{a(s + n_1)}{s}$, $a = R$, and $n_1 = \dfrac{1}{RC_I}$. Since $n_1 T \ll 1$ in

most PLL designs, (EQ 1) can be re-written as (EQ 3),

$$\Theta_{on}(z) = \frac{(1 - z^{-1})}{1 - (1 - \varepsilon) z^{-1}} \Theta_n(z) \qquad \text{(EQ 3)}$$

where $K = K_d K_w aT$ and is actually replaced with the term ε since $K \ll 1$.

The phase jitter from the ring oscillator can be modeled as a sequence of unit step phase jumps with random magnitude. A single phase jump at time nT can be represented by (EQ 4) in the z-domain.

$$\Theta_n(z) = \frac{2\pi \Delta t_n}{T(1 - z^{-1})}, \qquad \text{(EQ 4)}$$

Here the magnitude of the error step is Δt_n. The variance of this error is shown in [1] to be proportional to the number of stages in the ring-oscillator, and the timing jitter variance contributed by each stage. Hence the output jitter in z-domain is,

$$\Theta_{on}(z) = \frac{2\pi \Delta t_n}{T(1 - (1 - \varepsilon) z^{-1})} \qquad \text{(EQ 5)}$$

For all events up to time nT, the sum of output phase shifts is represented by (EQ 6).

$$\Theta_{tot}(nT) = \sum_{k=-\infty}^{n} \frac{2\pi \Delta t_k}{T} (1 - \varepsilon)^{n-k} \qquad \text{(EQ 6)}$$

To find the rms output jitter, the expectation of the square of the sum is calculated and given by (EQ 7). Since Δt_k and Δt_l are not correlated, the $E[\Delta t_k \Delta t_l] = 0$ when $k \neq l$. When $k = l$, $E[\Delta t_k \Delta t_l]$ can be replaced by $\Delta \tau_N^2$.

$$E\left[\Theta_{tot}^2(nT)\right] = \left(\frac{2\pi}{T}\right)^2 \frac{\Delta \tau_N^2}{\varepsilon(2 - \varepsilon)} \approx \left(\frac{2\pi}{T}\right)^2 \left(\frac{\Delta \tau_N^2}{2\varepsilon}\right) \qquad \text{(EQ 7)}$$

Note that the expectation of the phase jitter is independent of nT, the time instant. Hence the r.m.s. phase jitter is,

$$\sqrt{E\left[\Theta_{tot}^2(nT)\right]} \approx \sqrt{\frac{1}{2\varepsilon}} \frac{2\pi \Delta \tau_{rms}}{T} = \alpha \frac{2\pi \Delta \tau_{rms}}{T} \qquad \text{(EQ 8)}$$

where $\Delta \tau_{rms}$ is $\sqrt{\Delta \tau_N^2}$, and $\alpha = \sqrt{\dfrac{1}{2 K_d K_w aT}}$ is defined as the

accumulation factor. The result in (EQ 8) is the r.m.s. phase jitter for a ring-oscillator PLL. From the result, the rms timing jitter in a phase-locked-loop is seen to be α times larger than the intrinsic jitter in the delay chain. The accumulation factor α is inversely proportional to the square-root of $K_d K_w aT$ and in this case shows little dependency on C_I and C_P. Therefore, as long as $n_1 T \ll 1$, $C_P \ll C_I$ and stability requirements are met [3], the jitter accumulation factor can be lowered by increasing the bandwidth of the loop filter, $w_L \approx K_d K_w a$.

An alternative scheme for clock synthesis is to use a delay-locked-loop [4]. In this case, the reference clock is fed to the input of the delay line, and the rising edge of the output of the delay line is compared to that of the reference clock. Since the rising edge of the reference clock reaches the output of the delay line after passing through all delay cells, the total delay is driven to be the same as one period of the reference clock. Also, since the output of the loop filter just changes the phase of the output of the delay line, the loop does not have any extra poles as a PLL does.

Therefore, the stability problem is relaxed and a simple capacitor loop filter can be used without any stability consideration.

In a DLL, phase jitter is not passed on from one period of the clock to the next since the output of the delay-line is not fed back to the input. Therefore we expect the jitter in a DLL to be much smaller than in a ring-oscillator based PLL. To show this quantitatively we proceed with an analysis similar to that in the previous section but with the simplified discrete time DLL model. In this case, the transfer function for output phase noise in terms of the internal jitter from the delay line is represented by (EQ 9).

$$\Theta_{on}(z) = \frac{\Theta_n(z)}{1 + K_d K_P T Z_F(z) z^{-1}} \qquad \text{(EQ 9)}$$

where K_d is phase gain and given by $\dfrac{I_S}{2\pi}$, and K_p is phase gain and given by $\dfrac{d\theta}{dv}$ when voltage controlled delay line is assumed. If the loop filter in the DLL is a single capacitor and given by $\dfrac{1}{sC} = \dfrac{a}{s}$, the transfer function (EQ 9) becomes (EQ 10).

$$\Theta_{on}(z) = \frac{(1 - z^{-1}) \Theta_n(z)}{1 + (\varepsilon - 1) z^{-1}} \qquad \text{(EQ 10)}$$

where $K_d K_P aT \ll 1$, and is replaced by the constant ε, The jitter introduced by the delay line is represented by (EQ 11) in the z-domain since in the time domain the effect of one pass down the chain is just an error impulse.

$$\Theta_n(z) = \frac{2\pi \Delta t_n}{T} \qquad \text{(EQ 11)}$$

Therefore, the variance of the total output jitter can be shown to be

$$E\left[\Theta_{tot}^2(nT)\right] = \left(\frac{2\pi}{T}\right)^2 \overline{\Delta \tau_N^2} \left(1 + \frac{\varepsilon}{2-\varepsilon}\right) \approx \left(\frac{2\pi}{T}\right)^2 \overline{\Delta \tau_N^2} \text{(EQ 12)}$$

and the rms output jitter is therefore given by (EQ 13).

$$\sqrt{E\left[\Theta_{tot}^2(nT)\right]} \approx \frac{2\pi \Delta \tau_{rms}}{T} \qquad \text{(EQ 13)}$$

This expression is very similar to the result for the PLL, given in (EQ 8), except now there is no noise enhancement factor α. Therefore a DLL provides superior timing jitter performance. How much better depends on the size of α which will be discussed in the next section.

III. Design Examples and Simulation

To verify the theoretical predictions for PLL/DLL jitter performance given in the previous sections, monte-carlo simulations were performed using a behavioral model built around the basic functional blocks pictured in Figure 3. The timing jitter generated by the noisy inverter cells in the ring-oscillator is modeled by a phase jitter noise source which adds an error phase to the ideal phase coming out of the VCO. This phase jitter is assumed white and its variance is proportional to the value of $2N(\Delta \tau_1)^2$ determined for a given ring-oscillator design using the results of [1]. The jitter is normalized to the period of the delay in order to determine the phase noise variance.

If a phase noise source is applied to a PLL with design parameters $K_d = 8.4/2\pi$ µA, $K_w = 2\pi \times 20$ MHz/volt, $R = 90\Omega$ and $T = 50$ nsec (same parameters as in [6]), then the total PLL jitter is shown in Figure 4. The total phase error wanders over a wider range than the unit variance input source. In this example the predicted accumulation factor is 25.7 and for a unit variance input noise, should yield a total PLL jitter of around 25.7. This is close to the result which was extracted from the simulation to be about 26. The simulation also indicates that the PLL shapes the free running ring oscillator phase fluctuation to a finite values with a variance ($\sigma_P^2 = 676$). Here, the input jitter variance is normalized to 1.

To reduce the jitter accumulation effect, a new design is simulated. In this design, the bandwidth of the PLL is increased by using a larger value for R ($R = 900$). In this case, the calculated jitter accumulation factor α becomes 8.1. Figure 5 shows the simulation results. Now the PLL phase error wanders over a smaller range, and changes more rapidly since the loop bandwidth is higher. The noise enhancement factor for this data is 8.3 which is very close to the predicted result of 8.1.

Figure 6 shows the jitter accumulation factor in respect to the loop bandwidth, $w_L \approx K_d K_w a$ for two input jitter values. This figure shows that the PLL output jitter decreases as the loop bandwidth increases until the external jitter becomes dominant. Therefore, the optimum bandwidth is given by the minimum point of the curve. Smaller bandwidth is preferred when the input jitter is dominant.

Using the same input noise from the delay cells, a DLL output jitter is simulated. Figure 7 shows the output jitter for the same time period. This simulation shows that DLL does not accumulate jitter and performs better for the internal jitter sources such as delay cell jitter. In this case, parameter values are taken from [2] and given as $C = 0.039$ µF, $K_d = 8.4/2p$ µA, $K_P = 2\pi$ rads/volts and $T = 50$ nsec.

For a rough experimental verification, this model for timing jitter was compared to the jitter observed in the ring-oscillator PLL described in [5]. This PLL was fabricated in a 2 µm CMOS technology and the ring-oscillator was comprised of 16 inverter delay stages running at a frequency of 30-MHz. The jitter for this case can be calculated using (EQ 14) and is a function of the jitter contribution per stage ($\Delta \tau_1$), the number of taps in the oscillator (N), and the PLL accumulation factor (α). In [1], it is shown that for the circuit parameters used in the delay cells in [5], the jitter contribution per cell was $\Delta \tau_1 = 2.09$ps. The parameters for the PLL in this design were $K_d = 20/2\pi$ µA, $K_w = 2\pi \times 5$ MHz/volt, $R = 200$ Ω, and $T = 33$ nsec, giving a jitter accumulation factor of 38.9. Therefore, the r.m.s. timing jitter for this PLL, is predicted by (EQ 14) to be 81.30 ps. This agrees well with the experimental result in [5], which was not measured exactly, but determined to be somewhere in the range of 50-100 ps.

IV. Conclusion

This analysis has shown that, including the results of [1], the jitter in a ring-oscillator is proportional to three factors; the number of stages, the jitter contribution per stage, and a PLL accumulation factor α, which is inversely proportional to the square-root of the bandwidth of the PLL. For a DLL the result is the same, except the noise enhancement factor is 1. Therefore in applications such as clock synthesis, where a DLL can be used, it is the better choice for jitter performance. To reduce the jitter enhancement in a PLL a larger loop bandwidth should be used. For applications such as clock-recovery, however, this bandwidth cannot be increased too much or it will enhance the jitter seen in the input signal.

References

[1] Todd C. Weigandt, Beomsup Kim, Paul R. Gray, "Timing Jitter Analysis for High-Frequency, Low-Power CMOS Ring-Oscillator Design", *ISCAS*, June 1994.

[2] Beomsup Kim, "High Speed Clock Recovery in VLSI Using Hybrid Analog/Digital Techniques", *UCB/ERL Memorandum*, June 1990.

[3] Floyd M. Gardner, "Charge-Pump Phase-Locked Loops", *IEEE Trans. on Communications*, vol. COM-28, no. 11, Nov. 1980.

[4] J. Sonntag, R. Leonowich, "A Monolithic CMOS 10 MHz DPLL for Burst-Mode Data Retiming", *ISSCC*, vol. 33, pp. 104-105, Feb. 1990.

[5] Beomsup Kim, David H. Helman, Paul R. Gray, "A 30 MHz High Speed Analog/Digital PLL in 2um CMOS", *ISSCC*, vol. 33, pp.104-105, Feb. 1990.

[6] National Semiconductor, *Mass Storage Handbook*, pp 2.49-2.51, 1989.

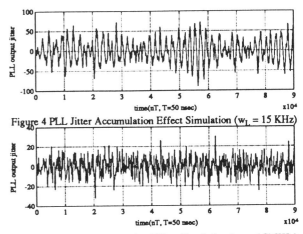

Figure 4 PLL Jitter Accumulation Effect Simulation (w_L = 15 KHz)

Figure 5 PLL Jitter Accumulation Effect Simulation (w_L = 150 KHz)

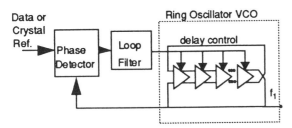

Figure 1 Ring-oscillator phase-locked-loop

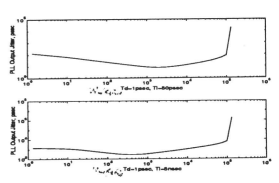

Figure 6 Jitter vs. Loop Bandwidth, VCO Noise Dominant Case (Upper Figure) and Input Noise Dominant Case (Lower Figure)

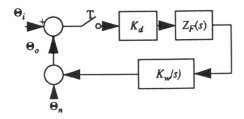

Figure 2 Simplified PLL Discrete Time Model

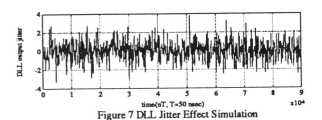

Figure 7 DLL Jitter Effect Simulation

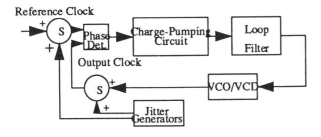

Figure 3 Jitter Simulation Setup

PRACTICAL APPROACH AUGURS PLL NOISE IN RF SYNTHESIZERS

By following a graphical analysis routine, phase noise can be predicted accurately.

NOISE and spurious signals can be analyzed for even the most complex of phase-locked loop (PLL) synthesizer architectures. Even synthesizers utilizing dual-modulus prescalers, fractional-N dividers, and translation loops can be understood.[1,2]

For example, the presence of the feedback divider makes the PLL a sampled data system.[3] However, the simpler continuous approximation is accurate if the loop bandwidth is less than 20 times the sampling rate (the comparison frequency). With this approximation, a Laplace transform analysis of the basic synthesizer yields the following results for the open-loop gain and the transfer function:

$$G(s) = \frac{K_\alpha \, K_0 \, F(s)}{s \, N} \quad (1)$$

where:

$G(s)$ = the open-loop gain, and

$$H(s) = \frac{\phi_0(s)}{\phi_i(s)} = \frac{G(s)}{1 + G(s)} =$$

MARK O'LEARY, Consultant, Comm-design, 21213B Hawthorne Blvd. Ste. 5576, Torrance, CA 90509; (213) 370-3298

$$\frac{K_\alpha \, K_0 \, F(s)}{s \, N + K_\alpha \, K_0 \, F(s)} \quad (2)$$

where:

$H(s)$ = the transfer function.

For a second-order, type-II loop:

$$F(s) = \frac{s \, \tau_2 + 1}{s \, \tau_1} \to H(s)$$

$$= \frac{2 \, s \, \xi \, w_n + w_n^2}{s^2 + 2s \, \xi \, w_n + w_n^2} \quad (3)$$

where:

$$w_n = \sqrt{\frac{K_\alpha \, K_0}{\tau_1 \, N}} \quad (4a)$$

and,

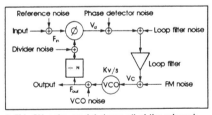

1. This PLL noise model shows all of the relevent individual noise sources.

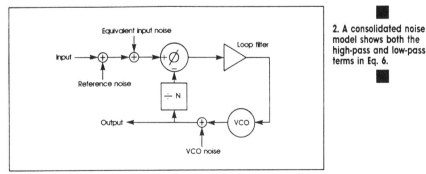

$$\xi = \frac{\tau_2}{2} \sqrt{\frac{K_\alpha \, K_0}{\tau_1 \, N}} \quad (4b)$$

A well-known PLL noise model (Fig. 1) includes the individual sources of noise within the synthesizer.[4] Each of these sources arises from a different mechanism. Loop-filter noise arises from the equivalent input noise sources of the DC amplifier, if one is used, and from logic circuit and current source noise, if a pure switching charge pump is utilized. Phase-detector noise results whether a digital or analog PLL is used. In either case, the phase detector will degrade the phase difference signal by adding white and flicker noise.

VCO noise, another oscillator noise source, is measured while the oscillator is free-running under laboratory conditions.[5,6] FM noise occurs when the VCO is operating in a synthesizer and it is subjected to several noise sources that modulate its frequency. These sources include

(continued on next page)

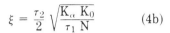

2. A consolidated noise model shows both the high-pass and low-pass terms in Eq. 6.

Reprinted with permission from *Microwaves & RF*, M. O'Leary, "Practical Approach Augurs PLL Noise in RF Synthesizers," pp. 185-194, September 1987.

control-voltage pickup (for example, capacitive coupling from nearby digital circuits and audio oscillators), noise on the VCO supply voltage (all VCO's have some frequency sensitivity to supply voltage), and vibration (through frequency sensitivity to mechanical stress). These effects are modeled by a single noise source summed with the VCO control voltage.

Divider noise occurs when a divider's output contains phase information in the position of its rising edges, which are influenced by electrical fluctuations within the divider. The divider noise is modeled as a source summed at the divider output, since contributions from higher stages are reduced by the factor, 20 log N, where N is the divide ratio. By modeling divider noise at the output, this source becomes nearly independent of N and constant for any given logic family.

All of the above noise sources can be measured and evaluated. For example, op-amp input noise data can be gleaned from data sheets; such publications contain much information on divider and phase detector noise.[7,8] However, this approach is tedious and time-consuming. The only information necessary is the synthesizer's output phase noise. Once this information is known, the PLL can be made to meet the system specification. A consolidated noise model is used to analyze the phase noise (Fig. 2).

Applying a Laplace transform analysis to this model results in the expression for output phase noise:

$$S_0(w) = N^2 \left| \frac{G(s)}{1 + G(s)} \right|^2 \left(\frac{S_r(w)}{R^2} \right.$$

$$+ S_N(w) + \frac{S_{VLF}(w)}{K_D^2} + \frac{S_{Vpd}(w)}{K_D^2} \right)$$

$$+ \left| \frac{1}{1 + G(s)} \right|^2 \left(S_{vco}(w) \right.$$

$$\left. + \frac{S_{Vfm}(w) K_0^2}{S^2} \right) \qquad (5)$$

Eq. 5 can be rewritten as,

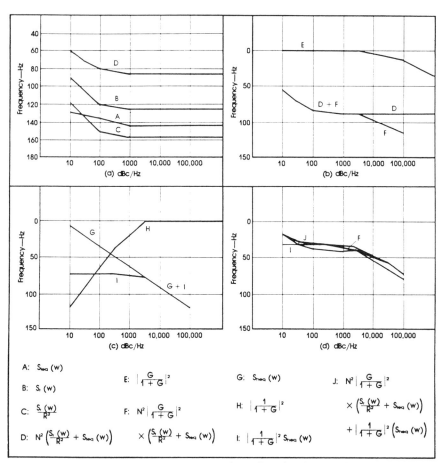

A: $S_{req}(w)$

B: $S_r(w)$

C: $\frac{S_r(w)}{R^2}$

D: $N^2 \left(\frac{S_r(w)}{R^2} + S_{req}(w) \right)$

E: $\left| \frac{G}{1+G} \right|^2$

F: $N^2 \left| \frac{G}{1+G} \right|^2$ $\times \left(\frac{S_r(w)}{R^2} + S_{req}(w) \right)$

G: $S_{heq}(w)$

H: $\left| \frac{1}{1+G} \right|^2$

I: $\left| \frac{1}{1+G} \right|^2 S_{heq}(w)$

J: $N^2 \left| \frac{G}{1+G} \right|^2$ $\times \left(\frac{S_r(w)}{R^2} + S_{req}(w) \right)$ $+ \left| \frac{1}{1+G} \right|^2 \left(S_{heq}(w) \right)$

3. A graphical solution is available, which gives phase-noise output for a PLL synthesizer.

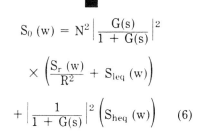

$$S_0(w) = N^2 \left| \frac{G(s)}{1 + G(s)} \right|^2$$

$$\times \left(\frac{S_r(w)}{R^2} + S_{leq}(w) \right)$$

$$+ \left| \frac{1}{1 + G(s)} \right|^2 \left(S_{heq}(w) \right) \qquad (6)$$

In Eq. 6, two kinds of noise sources have been combined. The first type of noise source, to which the loop response is low pass, is called the equivalent input noise, $S_{leq}(w)$. The second source, to which the loop response is high pass, is denoted in-circuit VCO noise, as $S_{heq}(w)$.

Note that S_{leq} is independent of the feedback divider ratio and is characteristic for a given synthesizer technology—the type of phase detector or loop filter, or the logic family of divider output stages.

The consolidated noise model is useful in that the equivalent input is easily measured, as is the VCO noise (Fig. 3). To measure equivalent input noise for a given technology, the user simply builds a synthesizer with a high division ratio and wide-

loop bandwidth, and locks it to a very clean reference frequency. Equivalent input noise is then the output phase noise reduced by 20 log N.

With this model, the importance of N in determining the output phase noise of a given synthesizer is immediately apparent. The output phase noise increases in dB as the value of 20 log N increases for offset frequencies within the loop bandwidth. Therefore, a constraint on any synthesizer design is to keep N low.

The equivalent input noise also constitutes a noise floor for the synthesizer. Regardless of how quiet the input signal is, the output phase noise will be at least 20 log N × $S_{leq}(w)$ for frequency offsets within the loop bandwidth. One effect is that output noise of PLL synthesizers is not necessarily a multiplied version of the reference, as often believed.

Another implication of Eq. 2 is the strong dependence of phase-noise performance on loop response. Both the loop bandwidth and the sharpness of loop roll-off are very important. To illustrate these concepts, Fig. 3 demonstrates a graphical so-

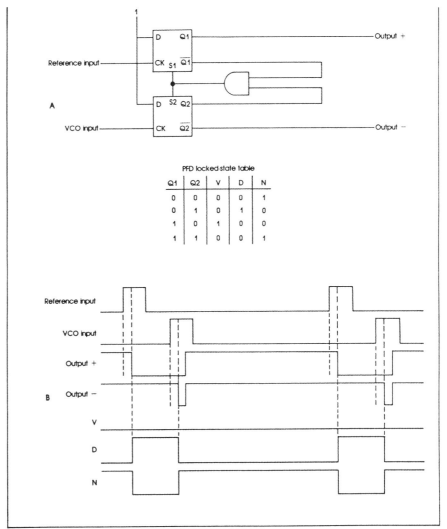

Q1	Q2	V	D	N
0	0	0	0	1
0	1	0	1	0
1	0	1	0	0
1	1	0	0	1

PFD locked state table

4. A common version of the phase/frequency detector (a) operates as illustrated by the characteristic waveforms (b).

lution to Eq. 2 for a hypothetical synthesizer.

COMPARISON FREQUENCY

Generally, the synthesizer output will have sidebands offset from the carrier frequency by the comparison frequency and its harmonics. For the simplest LO synthesizers, the comparison frequency corresponds to the step size. Therefore, comparison-frequency sidebands will limit the receiver signal-to-noise (S/N) ratio by mixing adjacent channel signals into the fixed intermediate frequency (IF) along with the desired signal.[9] If the total power in each channel slot is the same, the receiver S/N ratio is limited to B – 3 dB, where B (in dBc units) is the level of the comparison-frequency sidebands. Of course, comparison frequency does not correspond to step size for fractional-N synthesizers. Nevertheless, the sidebands must still meet certain spurious-signal specifications.

Sideband levels depend on the details of phase-detector and loop-filter implementation. For example, consider a synthesizer that is a mixed-sampled and continuous-time system. Almost always, the transition from sampled to continuous systems is done at the phase-detector output. This is because the loop filter is most easily implemented as an analog integrator, while sampled phase detectors are preferred to continuous ones. Continuous phase detectors, such as multipliers and switched mixers, are inherently low-gain, high-offset, high-drift devices. Also, their good threshold performance is not advantageous in the high S/N environment of a synthesizer. On the other hand, digital phase detectors easily interface to dividers and are less expensive.

The class of sequential digital phase detectors includes very simple circuits such as gates and flip-flops, as well as the phase/frequency detector (Fig. 4).

The phase/frequency detector (PFD) has a wide linear range (that

is, $\pm 2\pi$) and possesses the remarkable feature of functioning as a frequency comparator when it is out of phase lock.[10] Fig. 4 illustrates the locked-condition operation of a common version of the phase frequency detector. The leading edge of the signal causes its corresponding output to be set; the other input edge, after setting its output, allows both outputs to be reset after two gate delays. This creates a linear phase detector with a range of ± 1 cycle, the phase being encoded into the pulse width of the difference of the two outputs.

A device called a charge pump takes the digital output of the phase/frequency detector and produces an analog signal suitable to drive the loop filter.[11] The charge pump recognizes three independent states: pump up (U), pump down (D), and neutral (N). It sources current to the loop filter when U is true, sinks current when D is true, and is

isolated when N is true. If U, D, and N are derived from the PFD output states, as shown in Fig. 4, the charge pump will provide the following average current to the loop filter each cycle:

$$\overline{i_d(t)} = \frac{I_p\,\theta_c}{2\,\pi} \qquad (7)$$

Assuming the loop bandwidth is much less than the comparison frequency, the sampling period average describes the system:

$$I_d(s) = \frac{I_p\,\theta_e(s)}{2\pi} \qquad (8)$$

The loop filter consists of an impedance, $Z_F(s)$, so that:

$$V_c(s) = I_d(s)\,Z_F(s)$$
$$= \frac{I_p}{2\,\pi}\,\theta_e(s)\,Z_F(s) \qquad (9)$$

A charge pump may produce a pump voltage instead. When this voltage is averaged over one cycle and an active loop filter with a transfer function, F(s), is used, this results in:

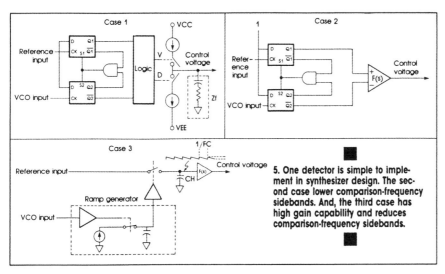

$$V_c (s) = K_d \, \theta_e (s) \, F (s) \qquad (10)$$

To summarize, the PFD accepts digital inputs, produces an analog output in conjunction with a charge pump, and causes comparison-frequency sidebands because of the pulsed nature of its output.

The other type of phase detector of practical importance is the sample-and-hold phase detector. One input clocks a ramp (if a linear characteristic is desired) or another waveform. The other input samples the waveform. The sampled voltage is held until the next cycle. This process implements a sampled phase detector and zero-order hold with a linear range of $\pm\pi$. The sample-and-hold phase detector has several advantages and disadvantages relative to phase/frequency detectors.

Advantages include reduced comparison-frequency sidebands; high gain capability resulting in noise advantages; and accurate Z-transform analysis and design. Disadvantages include the lack of built-in frequency-acquisition capability and the need for analog circuitry.

The comparison-frequency sideband levels for the three cases can now be computed (Fig. 5). The first case is the phase/frequency detector with a switching charge pump. Leakage current out of the loop-filter impedance node results in the creation of comparison-frequency

sidebands. In the phase-locked steady-state case, this current, which may include charge-pump switch varactor bias and filter-capacitor leakage currents, must be cancelled by the average charge-pump current each cycle to maintain the filter capacitor at the correct voltage. The charge-pump supplies current in pulses of peak value, I. These pulses modulate the VCO to make the sidebands. (Assume that the loop filter contains a capacitor, as is the case with the usual type-II loop.)

The level of the sidebands can be calculated as follows. In the steady state, the charge pump supplies current in pulses of peak value, I_p, and width:

$$D = \frac{I_l}{I_p} \qquad (11)$$

From Fourier analysis, the RMS level of the pulse train at the nth harmonic of the comparison frequency is as follows:

$$I_{RMS} = \frac{I_p D}{\sqrt{2}} \quad \text{for } n \ll \frac{1}{D} \qquad (12)$$

or,

$$I_{RMS} = \frac{I_l}{\sqrt{2}} \text{ for } n \ll \frac{1}{D} \qquad (13)$$

To compute the output phase modulation, it is easiest to let the phase-detector output current harmonics be the input of the loop and use the ordinary closed-loop transfer function. The following equation relates PFD output, I_d, to input Θ_e:

$$I_d = \frac{\theta_e \, I_p}{2 \, \pi} \qquad (14)$$

so,

$$\theta_e = I_l \frac{\sqrt{2} \, \pi}{I_p} \qquad (15)$$

The output response to this equivalent input (which arose due to the pulsed nature of the charge pump) is:

$$\theta_0^n = \frac{N \, H \, (nf_c) \, I_l \, \sqrt{2} \, \pi}{I_p} \qquad (16a)$$

and,

$$B = 20 \log \left[\frac{N \, H \, (f_c) \, I_l \, \sqrt{2} \, \pi}{I_p} \right] \text{dBc,}$$

$$\text{for } B < -20 \text{ dBc} \qquad (16b)$$

Assuming that N is constrained by the architecture, the sidebands can still be reduced. $H(f_c)$—where f_c equals the carrier frequency—can be decreased either by lowering the loop bandwidth or by adding a pole to the loop filter. The extra pole must be no lower than about 10 times the loop bandwidth to maintain adequate stability margin. Next, the charge pump can be improved. As evident in the above equation, I_p/I_l is a figure-of-merit for the comparison-frequency sideband performance of a charge pump. The value of f_c should be maximized in order to minimize the sidebands.

The second case, for which comparison-frequency sidebands can be computed, is a phase/frequency detector with an amplifier charge pump. The previous scheme—a PFD with a switching charge pump—has the advantages of simple implementation, rapid acquisition, and no need for a DC amplifier. However, the charge-pump switching time imposes a limitation on input zero phase-error resolution. The phase detector cannot distinguish input edges whose arrival time differences are less than the switching time of the charge pump.

Referring to Fig. 4, the transformation between the PFD's 2-bit output and the charge pump's three-state input can be accomplished using subtraction with a linear circuit instead of a digital one. The circuit for which a differential amplifier implements the charge pump and loop filter is shown in Fig. 6.

There is no limitation on the zero-phase resolution with the circuit in Fig. 6. Comparison-frequency sideband levels are calculated from the formula derived above, using the op-amp input offset current in place of I_l. The op-amp input voltage does not affect the sidebands unless a resistive buffer stage is used in the circuit before a single-ended loop filter. This is because the capacitors will hold the correct voltage to cancel the input offset voltage and obtain the necessary VCO input. If a resistive buffer stage is used, the sideband levels are:

$$B = 20 \log \left[\frac{N \, | H \, (f_c) \, | \, V_{os}}{2\sqrt{K_D}} \right] \text{dBc},$$

$$\text{for } B < -20 \text{ dBc} \qquad (17)$$

where:
$K_D = V_{cc}/4\,\pi$.

However, this is unnecessary since the charge pump, besides solving the phase-resolution problem mentioned, is also capable of a high figure-of-merit:

$$\frac{I_p}{I_l} = \frac{V_{cc}}{R \, I_{os}} \qquad (18)$$

This value can be greatly in-

creased by using a FET input amplifier. Disadvantages to the PFD/difference-amplifier approach are that it requires an op amp, op-amp noise is not negligible in some cases, and the loop bandwidth is limited by amplifier slew rate. However, these aren't normally important factors in commercial areas of the LO synthesizer performance envelope.

The third case is the sample-and-hold phase detector. In a type-II loop with a sample-and-hold phase detector, ripple due to hold-capacitor droop results in the comparison-frequency sidebands. Assuming the phase detector is well designed so that switching transients are negligible the sidebands can be calculated as follows. The phase detector output voltage, V_d, will be a sawtooth waveform (Fig. 5). From Fourier analysis, the level of V_d at the nth harmonic is:

$$V_d = \frac{A}{N\,\pi} \qquad (19)$$

And, since:

$$A = \frac{1}{F_c} \frac{d V_d}{d t} = \frac{I_l}{C_H f_c} \qquad (20)$$

Then:

$$V_d = \frac{I_l}{C_H f_c \, n} \qquad (21)$$

where:
I_l = the leakage current out of the hold-capacitor node, between sampling instants.

Using $V_d = \Theta_i K$, the ripple can be sent to the input for ease of computation. This gives a fictitious input signal at the nth harmonic of:

$$\theta_I^n = \frac{I_l}{C_H f_c \, n \, \pi \, K_D} \qquad (22)$$

Then, using Eq. 21, the output phase is:

$$\theta_0^n = \frac{N \, H \, (n f_c) \, I_l}{C_H f_c \, n \, \pi \, K_D} \qquad (23a)$$

And, the comparison-frequency sidebands are "D" dB down from the carrier, where:

$$B = 20 \log \left[\frac{N \, H \, (f_c) \, I_l}{C_H f_c \, \pi \, K_D} \right] \qquad (23b)$$

The figure-of-merit for sideband performance of the sample-and-hold phase detector is slow and unreliable. A convenient way to obtain fre-

quency-aided acquisition is to use a PFD and switching charge pump as a parallel phase detector. Outputs are then summed for both the PFD and the switching charge-pump circuits, with the PFD gain much lower so that performance is not harmed while both circuits are in phase lock. Some CMOS synthesizer parts provide both types of phase detectors for this reason. ●●

References
1. Floyd M. Gardner,"Phase-lock Techniques," *Wiley Interscience*, 2nd Ed., 1979, pp. 8-16.
2. D. Brewerton and N. Urbaneta, "Defining the elements of good design," *Microwaves & RF*, Vol. 23, Jun. 1984, pp. 79-125.
3. James A. Crawford, "Understanding the specifics of sampling in synthesis," *Microwaves & RF*, Vol. 23, Aug. 1984, pp. 120-144.
4. L.S. Cutler and C.L. Searle, "Some aspects of the theory and measurement of frequency fluctuations in frequency standards," *Proc. IEEE*, Vol. 54, Feb. 1966, pp. 136-154.
5. Gerald Sauvage, "Phase noise in oscillators: a mathematical analysis of Leeson's model," *IEEE Transactions on Instruments and Measurements*, Vol. IM-26, Dec. 1977, pp. 408-410.
6. K. Kurokawa, "Noise in synchronized oscillators," *IEEE Transactions on Microwave Theory and Techniques*, Vol. MTT-16, Apr. 1968, pp. 234-240.
7. V.F. Kroupa, "Noise properties of PLL systems," *IEEE Transactions on Communications*, Vol. COM-30, Oct. 1982, pp. 2244-2252.
8. Dieter Scherer, "Design principles and test methods for low-phase-noise RF and microwave sources," Hewlett-Packard.
9. C. John Grebenkemper, "Local oscillator phase noise and its effect on receiver performance," Watkins-Johnson Co. Technical Notes, Vol. 8, Nov./Dec. 1981, pp. 1-13.
10. J.I. Brown, "A digital phase and frequency-sensitive detector," *Proc. IEEE*, Vol. 56, Apr. 1971, pp. 717-718.
11. Floyd M Gardner, "Charge-pump phase-lock loops," *IEEE Transactions on Communications*, Vol. COM-28, Nov. 1980, pp. 1849-1858.

The Effects of Noise in Oscillators

ERICH HAFNER, MEMBER, IEEE

Abstract—An explicit expression for the output signal from an oscillator with several noise sources in the circuit is derived. This formula describes qualitatively and quantitatively the manner in which thermal and shot noise act to corrupt the performance of an ideal oscillator. The statistical properties of the signal are then evaluated, as it emerges from the oscillator stage, after passage through an output filter and after being operated on by an ideal *n*-times multiplier. Expressions are derived for the short term frequency stability, the power spectral density, and the power spectrum of the signal, as well as for the spectral density of the signal phase.

The key to the results reported is an apparently novel perturbation technique which does not require smoothing of the instantaneous nonlinearity in the basic differential equation. Discussion of the solutions shows that the instantaneous nonlinearities cause the device to act simultaneously like a linear AGC oscillator and like a high *Q* passive tuned circuit, with each aspect accorded one half the total noise excitation. Possible implications of this effect for other types of transient conditions in oscillators are indicated briefly.

INTRODUCTION

THE THEORY of noise in nearly harmonic oscillators has received considerable attention in the past, and a number of properties of the noise-perturbed signal are firmly established. However, it is very well recognized that the existing theories provide only partial descriptions of a many-sided phenomenon, and that several important questions have remained unanswered.

A common feature of the literature on the subject is the derivation of the power spectral density of the signal voltage in a noise-perturbed oscillator. This provides a convenient reference for discussing the most essential aspects of the earlier work. Some investigators [1]–[4] chose to consider the oscillator simply as a linear noise filter of very narrow bandwidth; and they arrived at results which generally agreed (a factor of two which often occurred is now known to be extraneous [5], [6]) with those derived by far more sophisticated techniques [8]–[13]. Since these latter techniques are based on the nonlinear differential equations for the noise, perturbed oscillator, a satisfactory physical explanation for this agreement could not readily be offered, primarily because the significance of the approximations involved was rather difficult to assess.

It does not appear to have been fully realized that, as is shown in Section III-C of this paper, the linear noise-filter approach contains implicit assumptions which can be met only when the device considered is indeed a linear oscillator, equipped with an external mechanism (AGC) which automatically regulates the gain of the active device, or the circuit losses. On the other hand, the most important step, consistently taken in all nonlinear analyses reported so far, is the application of the averaging principle [14] which involves smoothing of the instantaneous nonlinearity in the circuit [12]. It is shown in the Appendix that this approximation is equivalent to replacing the actual nonlinear oscillator again by a linear oscillator with AGC. The agreement just mentioned is thus to be expected, even though explicit equations describing the behavior of a linear AGC oscillator have only recently become available [15].

Whereas the power spectral density of the oscillator signal is, of course, a very useful piece of information, the derivations of the expressions for it do not contain enough parameters to provide a clear picture of what the oscillator signal is really like, and just why real oscillators do not behave as they are supposed to according to these theories. When investigating the performance of systems fed by a signal from a noise-perturbed oscillator, instead of by a pure sinusoid, it has been necessary [16] to invent working models for this signal, making numerous and varied ad hoc assumptions in the process. Also, only a limited amount of useful guidance for the development of improved devices could be extracted from the analytic investigations of the oscillator itself.

To extend the results of the earlier work, an oscillator model was chosen for the present analysis which closely resembles an actual quartz crystal oscillator. The several noise sources in the circuit are assumed at first to generate impulses of random strength at randomly spaced intervals, and the effects of white noise on the oscillator signal are obtained by appropriate summation over the disturbances caused by the individual impulses. The result is an explicit expression for the output signal from the noise perturbed oscillator. It reveals that, regardless of their location in the circuit, the effects of all noise sources are essentially equivalent. However, the source which appears directly across the output of the oscillator also contributes an additive white noise component to the output signal, which plays a very significance role in high *Q* oscillators.

Although the statistical properties of this signal are computed after the series of impulses are replaced in Section III-A by continuous random functions, the response of the oscillator to a single noise impulse is found to be a very powerful tool, peculiarly well suited for the investigation of heretofore unexplored areas of oscillator behavior. Only those aspects which pertain to noise effects are discussed in detail.

The key to the advances reported here is an appar-

Manuscript received October 29, 1965; revised November 29, 1965.

The author is with the U. S. Army Electronics Command, Ft. Monmouth, N. J.

Reprinted from *Proc. of IEEE*, vol. 54, pp. 179-198, February 1966.

ently novel technique for the solution of the perturbation equation. It is developed in the Appendix. Unlike the earlier techniques, it fully recognizes the instantaneous character of the oscillator nonlinearity, and does not employ the smoothing concept. It also provides a clear appreciation of the significance of the approximations which are made to arrive at reasonably compact expressions.

The major consequence of the presence of instantaneous nonlinearities in the circuit is that they cause the oscillator to act simultaneously like a linear AGC oscillator and like a passive tuned circuit whose effective quality factor is inversely proportional to the nonlinearity parameter, with each aspect accorded one half the total perturbing excitation. While, with white noise excitation, the former aspect is responsible for the familiar random walk phenomenon in the oscillator phase, the latter aspect contributes with each impulse a phase disturbance that decays slowly to zero, usually with a very long time constant.

In regards to white noise excitation of oscillators with time invariant circuit parameters, the existence of the tuned circuit aspect is perhaps of limited significance for practical applications. This is so primarily because the noise effects from within the oscillator loop are in many cases less important than the additive white noise from the source across the oscillator output. However, the virtual tuned circuit is also excited by sudden changes in the circuit parameters of the oscillator; and it is quite likely that, so far, the behavior of the signal phase in oscillators with time variable circuit elements has largely defied theoretical description, just because the slow decay of this excitation has not been reckoned with. It is strongly suspected that future work in this area will be considerably more successful if the presence of the virtual tuned circuit is admitted.

The impetus to the work reported here was provided by the need to determine in detail the effects of noise in lumped parameter oscillators, particularly crystal oscillators, without resort to the many intuitive concepts which are often employed. The results, however, are applicable to all major classes of oscillators, and this includes masers and lasers.

While oscillators of this latter type are governed by the laws of quantum mechanics and electrodynamics, those laws which are of consequence here can quite generally be recast in classical form [17], and the description of the device in terms of a van der Pol oscillator becomes possible. In fact, Lamb's equations [18] that pertain to lasers in single mode operation are essentially the equations for a linear AGC oscillator. The analysis there is concerned primarily with the steady state and the question of whether or not the nonlinearity is instantaneous—that is, whether or not the tuned circuit aspect exists in lasers, too—does not arise. This question could conceivably be quite important in explaining the role of frequency pulling and entrainment phenomena in multimode oscillators during build-up and transient phases of operation.[1] Evidently, it cannot yet be answered with certainty.

I. THE BASIC DIFFERENTIAL EQUATIONS

An oscillator model which closely resembles a quartz crystal oscillator and yet is still manageable analytically is shown in Fig. 1. All elements of the feedback network are considered constant unless specifically stated otherwise. The active device in the oscillator is assumed to have infinitely high input and output impedances, and to generate a current i_1, which is a nonlinear function of the output voltage of the feedback network:

$$i_1 = f(e_g). \tag{1}$$

The current generator i_s and the voltage generators v_1, v_2, and v_3 inject extraneous signals into the oscillator whose effects are to be evaluated.

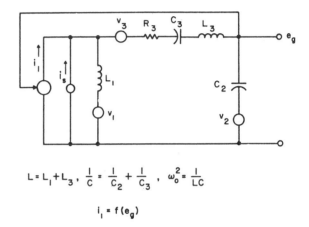

$$L = L_1 + L_3, \quad \frac{1}{C} = \frac{1}{C_2} + \frac{1}{C_3}, \quad \omega_0^2 = \frac{1}{LC}$$

$$i_1 = f(e_g)$$

Fig. 1. Oscillator Model. The feedback network is fed from an ideal current generator whose strength i_1 is a nonlinear function of the output voltage e_g. The generators i_s, v_1, v_2, and v_3 are assumed to be white noise sources. The R_3, L_3, C_3 branch approximates, by proper choice of the parameter values, the action of a quartz crystal unit.

The differential equation describing the form and behavior of the output voltage e_g is best derived by starting with the equation

$$e_g = \frac{Z_1 Z_2}{Z_1 + Z_2 + Z_3}(i_1 + i_s) + \frac{Z_2}{Z_1 + Z_2 + Z_3}(v_1 + v_3)$$
$$+ \frac{Z_1 + Z_3}{Z_1 + Z_2 + Z_3}v_2, \tag{2}$$

which follows readily from Fig. 1. Considering the Z_j as operational impedances [19]

$$Z_1 = pL_1, \quad Z_2 = 1/pC_2, \quad Z_3 = R_3 + pL_3 + 1/pC_3$$

and replacing, after some rearrangements the differentiation operator p by d/dt, one finds with (1)

$$\left[\frac{d^2}{dt^2} + \frac{L_1}{LC_2}\left(\frac{R_3 C_2}{L_1} - \frac{df(e_g)}{de_g}\right)\frac{d}{dt} + \omega_0^2\right]e_g = F(t) \tag{3}$$

[1] Multimode laser oscillators can be represented, equivalently as shown by Lamb, by an equal number of van der Pol oscillators, weakly coupled to one another.

whereby

$$L = L_1 + L_3, \quad \frac{1}{C} = \frac{1}{C_2} + \frac{1}{C_3}, \quad \omega_0{}^2 = \frac{1}{LC} = \frac{1}{L_1 C_2} \quad (4)$$

and

$$F(t) = \frac{L_1}{LC_2} \frac{di_s}{dt} + \omega_0{}^2 \frac{C}{C_2} (v_1 + v_3)$$

$$+ \left(\frac{d^2}{dt^2} + \frac{R_3}{L} \frac{d}{dt} + \frac{C}{C_3} \omega_0{}^2 \right) v_2. \quad (5)$$

Whereas the circuit in Fig. 1 contains four reactive components compared to the two customarily assumed in noise analyses of oscillators, it is important to realize that the two additional components do not cause a qualitatively more complicated behavior of the oscillator. The basic differential equation (3) is only of second order, and its essential features are not at all altered when these components are eliminated from the circuit. The ratios L_1/L and C/C_2 revert to unity for a simple LC oscillator in which $L_3 = 0$ and $C_3 \to \infty$. They carry the information required to describe the effects of the simultaneous presence of a high Q and a low Q element in the oscillator, without adding to the labor in the analysis. This information is useful when dealing with quartz crystal oscillators or with masers and lasers.

For simplicity, and to make the general applicability of the results more apparent, the following symbols will be used:

$$\frac{\omega_1 L}{R_3} \equiv Q_T \equiv \frac{1}{\gamma_T}; \quad \frac{\omega_1 L_1}{R_3} \equiv Q_N \equiv \frac{1}{\gamma_N};$$

$$\frac{L}{L_1} \equiv \overline{Q} \left(= \frac{Q_T}{Q_N} \right). \quad (4a)$$

Q_T is the effective quality factor of the entire passive feedback network; Q_N is its quality factor when the high Q element is replaced by its series resistance at ω_1. $\overline{Q}$, the ratio of these two, shall be called the reduced quality factor of the oscillator. Later on, an additional symbol Q_C, will be introduced for the quality factor of the filter following the oscillator. No Q symbol is used for the virtual tuned circuits representing certain performance aspects of the disturbed oscillator; their effective relative bandwidths will be characterized by γ, to be defined in (22).

Of major concern in this paper are the properties of the solutions of (3) when the extraneous signals are random noise. Stable, nearly harmonic solutions are assumed to exist; and we restrict ourselves to the stationary-state properties of such solutions.

If the noise sources in Fig. 1 are quiescent, (3) becomes completely deterministic, and it is well known that, with $F(t) = 0$, there are steady-state solutions which are oscillatory in nature, provided certain conditions are met [20], [21]. The presence of the noise sources is not essential for these solutions to exist; rather,

with increasing intensity, the noise causes progressively more severe random disturbances of the deterministic solutions.

The solutions of (3) with $F(t) = 0$ are perfectly periodic in the steady state and can be represented in the form

$$e_{g0} = \sum_i A_{i0} \cos (i\omega_1 t + \varphi i). \quad (6)$$

For a harmonic oscillator capable of steady-state oscillations, the A_{i0} are not all zero and ω_1 is in the neighborhood of ω_0 defined in (4). With all transients in the infinite past, the A_{i0} and φ_i are constants whose values, as well as the value of ω_1, can be determined at least in principle to any desired accuracy.[2]

The noise sources in Fig. 1 introduce transients into the system and the steady state cannot be maintained exactly; the various harmonics of the solution of (3) never have sharply defined amplitudes and phases.

Consequently we write e_g in the form

$$e_g = \sum_i (A_{i0} + a_i(t)) \cos (i\omega_1 t + \varphi_i + \phi_i(t)), \quad (7)$$

where $a_i(t)$ and $\phi_i(t)$ are stochastic variables representing the amplitude and phase disturbances caused by the noise. Equation (7), in turn, can always be written as

$$e_g = e_{g0} + u(t) \quad (8)$$

where e_{g0} is the solution (6) of the unperturbed oscillator and $u(t)$ represents all disturbances and only the disturbances.

When (8) is substituted into (3) and the fact that e_{g0} satisfies the unperturbed equation identically is considered, one finds an equation for $u(t)$ alone. If the disturbances are small, so that terms of higher order in $u(t)$ are negligible, the following perturbation equation, linear in $u(t)$ is obtained:

$$\left[\frac{d^2}{dt^2} + \frac{L_1}{LC_2} \left(\frac{R_3 C_2}{L_1} - \frac{df(e_{g0})}{de_{g0}} \right) \frac{d}{dt} \right.$$

$$+ \left. \left(\omega_0{}^2 - \frac{L_1}{LC_2} \frac{df(e_{g0})}{de_{g0}} \right) \right\} u(t) = F(t). \quad (9)$$

The time variable coefficients of (9) depend upon e_{g0}, the steady-state solution of the undisturbed oscillator, and hence are known.

Assuming higher order terms in $u(t)$ to be negligible already implies that $x_i(t)$ and $\phi_i(t)$ in (7) are small; and $u(t)$ can be written to within the same degree of approximation required to obtain (9) as

$$u(t) = \sum_i [a_i(t) \cos (i\omega_1 t + \varphi_i) - A_{i0}\phi_i(t) \sin (i\omega_1 t + \varphi_i)]. \quad (10)$$

---

[2] When (6) is substituted into (3) with $F(t) = 0$ and the principle of the harmonic balance [22] is applied, an infinite set of nonlinear algebraic equations results which can be solved by an iteration procedure. The effects of harmonic content on oscillator frequency have been studied extensively by Groszkowski [23].

The fact that this approximation is justified, in particular that

$$| \phi_i(t) | \ll 1, \qquad (11)$$

remains to be verified once $u(t)$ is computed for any given situation.

Evidently, a basic problem which must be dealt with herein is to determine the solution $u(t)$ of (9) when $F(t)$ depends upon the white noise sources in the oscillator according to (5).

The most realistic representations of the noise current i_s and the noise voltages v_1, v_2, and v_3 in $F(t)$ are series of delta functions of variable strength and occurrence. These are also the representations for which the solution of (9) is readily found and conveniently interpreted.

We therefore assume

$$i_s = \sum_k a_{sk}\delta(t - t_k)$$

$$v_r = \sum_k a_{rk}\delta(t - t_k) \qquad (r = 1, 2, 3). \qquad (12)$$

Once e_{g0} is determined and (9) has been solved for a single impulse from each one of the noise sources in the circuit, the stationary-state solution of (3) is obtained with (8) according to the linear superposition principle. We confine ourselves to an approximate solution.

II. THE DISTURBANCES OF THE FUNDAMENTAL COMPONENT

A. The Approximations

When the nonlinear terms in the current voltage characteristic (1) of the active device are small and/or the feedback network in Fig. 1 is highly selective, the second and higher harmonics in (6) are much smaller than the fundamental and can be considered negligible to a first approximation. Accordingly, (6) becomes

$$e_{g0} = A_1 \cos (\omega_1 t + \varphi). \qquad (13)$$

A first approximation to the disturbances of the fundamental frequency component of e_g can now be evaluated from (9) by letting $u(t)$ in (8) become

$$u(t) = a_1(t) \cos (\omega_1 t + \varphi) + y_1(t) \sin (\omega_1 t + \varphi) \qquad (14)$$

where

$$y_1(t) = - A_1\phi_1(t). \qquad (15)$$

Better approximations to x_1 and y_1 can still be determined from the linear perturbation equation (9) if the noise sources are weak. Successively higher harmonics must be included in e_{g0} and $u(t)$, whereby the first significant improvement should not be expected until at least the third harmonic is considered.[3] When any one

---

[3] When dealing with crystal oscillators we also note that crystal units generally have an overtone response close to the third electrical harmonic of the oscillator frequency. Under high-drive conditions, the two frequencies can coincide. The presence of this crystal response in the feedback network then becomes very important and cannot be disregarded.

or all of the extraneous sources i_s, v_1, v_2, and v_3 are strong, higher order terms in $u(t)$ are no longer negligible, even if $u(t)$ is approximated by (14) and e_{g0} by (13); (9) is then no longer adequate, and a nonlinear perturbation analysis becomes necessary. However, this latter case can be of significance in practical oscillators only when the perturbing forces are signals other than thermal or shot noise.

The two approximations of major significance for the following developments are, therefore: First, the noise sources in the oscillator are assumed weak so that the terms of higher order in the disturbances $u(t)$ are negligible. This assumption led in Section I to the linear differential equation (9) for $u(t)$. Second, the harmonic content of the output signal from the undisturbed oscillator is assumed to be very low. This assumption permits approximating e_{g0} by (13) and $u(t)$ by (14). Together the two assumptions imply that all amplitude disturbances of the fundamental component of the signal are represented by $x_1(t)$ in (14), all phase disturbances by $-y_1/A_1$.

To assure that the second approximation is reasonable, it will be assumed later that the active device in the oscillator is only weakly nonlinear and/or that the oscillator is operated at low signal levels. This assumption in turn entails that the quantity γ, to be defined later, is always very small. As shown in the Appendix, other approximations, which go beyond those stated here, are not required.

B. The Unperturbed Signal

Without imposing undue further restrictions, the current voltage characteristic (1) of the active device is assumed to be

$$i_1 = f(e_g) = g_{m0}e_g - \beta e_g{}^3. \qquad (16)$$

For (13) to be a reasonable approximation to (6), $\beta e_g{}^2 \ll g_{m0}$ is desirable, especially when the transfer impedance of the feedback network at the harmonic frequencies is not extremely low.

When (16) and (13) are inserted into (3) with $F(t)=0$, the values of A_1 and ω_1 can be determined to

$$A_1{}^2 = \frac{4}{3\beta}\left(g_{m0} - \frac{C_2 R_3}{L_1}\right) \qquad (17)$$

$$\omega_1 = \omega_0. \qquad (18)$$

The quantity

$$g_m = g_{m0} - \tfrac{3}{4}\beta A_1{}^2 \qquad (19)$$

will be recognized as the effective transconductance [24] of the active device for signals of amplitude A_1. With g_{m0} and β both positive, (19) indicates that g_m decreases monotonically with increasing A_1, as shown in Fig. 2. Because (17) requires [25]

$$\frac{C_2 R_3}{L_1} = g_m \left(= \frac{\omega_1 C_2}{Q_N}\right), \qquad (20)$$

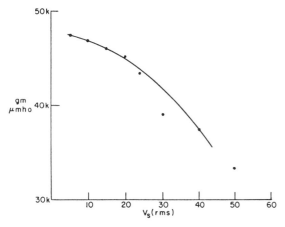

Fig. 2. Typical behavior of the effective transconductance g_m as a function of signal level. Dots are experimental points obtained on a 2N2808 transistor, rated at $i_e = 2$ mA.

the amplitude A_1 of the steady-state oscillations can be adjusted by proper choice of the circuit parameters.

In many cases g_{m0} and β can be determined from data such as plotted in Fig. 2, and the values obtained are the proper ones for use in (20). A more detailed examination of the contributing factors shows, however, that a $\beta^* \leq \beta$ should be used in the following Sections pertaining to $u(t)$, whereby β^* is the coefficient of the third-order term when the instantaneous function $f(e_g)$ in (1) is developed into a Taylor series. The coefficients of the Taylor series depend upon the bias conditions of the active device and the bias conditions change, almost invariably, with amplitude. Therefore, the g_m vs. amplitude curves, when measured under static conditions, include implicitly the effects of the even-order terms in $f(e_g)$, particularly the second-order term. The difference between β and β^* is not very significant except in oscillators with artificial level control such as AGC or lamp bridge oscillators. In AGC oscillators the change in the bias conditions is artificially magnified and utilized [25] to adjust g_{m0} and β such that (20) is satisfied for a very small value of A_1. In a lamp bridge oscillator [3] the value of R_3 or its equivalent, depending upon the actual circuit, varies with signal amplitude, while g_{m0} and β remain nominally unchanged. All amplitude dependent changes in bias conditions, or in R_3, require a finite time to become effective, governed by an RC time constant, while the first effects of a noise impulse occur instantaneously.

Therefore, for the purposes of the present analysis, the function (16) is always understood to represent the instantaneous relationship between i_1 and e_g. Any delayed action will primarily affect the envelope function of e_g and can be accounted for approximately by imposing suitable constraints on the behavior of this function after the instantaneous behavior has been established.

C. The Impulse Response

The general equation (9) for the perturbation $u(t)$ assumes, with (13), (16), and (17), the form

$$\ddot{u} + \omega_1 \gamma (1 + 2 \cos 2(\omega_1 t + \varphi)) \dot{u}$$
$$+ \omega_1{}^2 (1 - 4\gamma \sin 2(\omega_1 t + \varphi)) u = F(t) \qquad (21)$$

whereby the parameter γ is conveniently defined by either of the two equivalent expressions

$$\gamma = \frac{1}{\omega_1 C_2} \frac{1}{\overline{Q}_N} \frac{3}{4} \beta A_1{}^2 \qquad (22a)$$

$$\gamma = \frac{1}{\overline{Q}} \frac{g_{m0} - g_m}{g_m Q_N} . \qquad (22b)$$

It will become apparent later on that $2/\omega_1 \gamma$ is the time constant which controls the decay of disturbances in the oscillator. This time constant is appreciably larger than $2/\omega_1 \gamma_T$ the time constant of the passive feedback network, when the assumptions made to arrive at (21) are met: since $\beta e_{g0}{}^2$ in (16) should be much smaller than g_{m0}, it follows approximately from (22a) with (19) and (20) that

$$\gamma = \frac{1}{\overline{Q}} \frac{1}{\omega_1 C_2} \frac{3}{4} \beta A_1{}^2 \ll \frac{1}{\overline{Q}} \frac{1}{\omega_1 C_2} g_{m0} \approx \frac{1}{\overline{Q}} \frac{g_m}{\omega_1 C_2}$$
$$= \frac{R_3}{\omega_1 L} \qquad \text{(i.e., } \gamma \ll \gamma_T\text{)}. \qquad (23)$$

Equation (21) is a Mathieu equation [26]. Because it is linear, the response of the system to white noise from all sources is found by superposition of the individual impulse responses. The method used here to find these impulse responses is a cornerstone of this paper and is detailed in the Appendix.

Following the procedure used there it can be shown that the response of the system to an impulse from any one of the noise generators in Fig. 1 is, aside from differences in magnitude and phase, the same for each generator, with only one exception: the impulse from v_2 (i.e., from the source across the output) generates a system response like that caused by the other sources, and it also appears directly in $u(t)$ as an additive term. The existence and form of this additive term follows from the equations and requires no further assumptions.

For a single impulse at time t_k from each generator, i.e., when $v_1 = a_{1k}\delta(t - t_k)$, $v_2 = a_{2k}\delta(t - t_k)$, $v_3 = a_{3k}\delta(t - t_k)$ and $i_s = a_{sk}\delta(t - t_k)$, one finds as the solution of (21), except for terms with γ or γ^2 as a factor,

$$u_k(t) = \alpha_k e^{-\omega_1 \gamma (t - t_k)} \cos (\omega_1 t + \varphi) - A_1 \eta_k \sin (\omega_1 t + \varphi)$$
$$+ n_k(t) + v_2 \qquad (24)$$

whereby

$$n_k(t) = [M_k \sin \omega_1 (t - t_k) + N_k \cos \omega_1 (t - t_k)] e^{-(\omega_1 \gamma/2)(t - t_k)} \qquad (24a)$$

and the constants α_k and η_k are

$$\alpha_k = -M_k \sin (\omega_1 t_k + \varphi) + N_k \cos (\omega_1 t_k + \varphi) \qquad (24b)$$

$$\eta_k = -\frac{M_k}{A_1} \cos (\omega_1 t_k + \varphi) - \frac{N_k}{A_1} \sin (\omega_1 t_k + \varphi).$$

The parameters M_k and N_k define the effective strength of the impulses as weighted by the feedback network, and, for the circuit in Fig. 1, they are

$$M_k = \frac{\omega_1}{2\overline{Q}}\,(a_{1k} + a_{3k} - a_{2k})$$

$$N_k = \frac{\omega_1}{2\overline{Q}Q_N}\left(\frac{a_{sk}}{g_m} + a_{2k}\right). \tag{24c}$$

The significance of the solution (24) of the Mathieu equation (21) can perhaps best be appreciated when it is compared to the corresponding solution of the equation

$$\ddot{u}_p + \omega_1\gamma_T\dot{u}_p + \omega_1^2 u_p = F(t), \tag{25}$$

which describes the output of the passive feedback network in Fig. 1. The solution of (25)

$$u_{p_k} = 2[M_k \sin\omega_1(t - t_k) + N_k \cos\omega_1(t - t_k)]$$
$$\cdot e^{-\omega_1\gamma_T(t - t_k)} + v_2$$

which is easily found by standard techniques, can be written as

$$u_{p_k}(t) = \alpha_k e^{-(\omega_1\gamma_T/2)(t - t_k)}\cos(\omega_1 t + \varphi)$$
$$- A_1\eta_k e^{-(\omega_1\gamma_T/2)(t - t_k)}\sin(\omega_1 t + \varphi) + n_{p_k} + v_2, \tag{26a}$$

with

$$n_{p_k} = \tfrac{1}{2}(u_{p_k} - v_2). \tag{26b}$$

$\alpha_k\,\eta_k$, M_k and N_k are again defined by (24b) and (24c).

Evidently, the method developed in the Appendix separates (21) into two parts, with each accorded one half the strength of the original excitation. One describes the behavior due to the low-pass equivalent of the system (around zero frequency), the other due to the band-pass equivalent (around 2ω). The equations are solved in the transformed form and the results projected back again into the range around ω_1. In (24) $n_k(t)$ is the response of the band-pass part. It differs from $n_{p_k}(t)$ in (26a) only by the value of the time constant. Hence, with regard to this part of the solution, the disturbed oscillator behaves like a passive tuned circuit with quality factor $1/\gamma$.

The first two terms in (24) and (26a) represent the responses of the respective low-pass parts; here, essential differences are noted, which are brought about by the parametric pumping action apparent in (21). The low-pass part of the solution of (21) is that of a lossless tuned circuit as far as the component in-phase[4] with the pump is concerned; but the out-of-phase component is the damped response of a tuned circuit with quality factor $1/2\gamma$. In the solution of (25) the low-pass part is, of course, identical with the band-pass part.

The practice of evaluating only the low-pass response and doubling the result, correct for the passive circuit, is seen to be inapplicable in general for the system described by (21). Its applicability to other problems—

even linear problems, since (21) is linear—must evidently be carefully evaluated in every case.

The representation of the impulse response of (21) thus requires two virtual tuned circuits. To within the approximations used here, both are resonant at ω_1. Each is excited by one half the strength of the original impulse and their outputs are added linearly to the almost full strength impulse from the source across the output terminals of the network.[5] The one resulting from the low-pass equivalent shall be called the "L circuit," that from the band-pass equivalent the "B circuit." The "L circuit" performs independent operations on the two orthogonal components of its response, with the pump signal providing the reference. These can be represented by different effective quality factors and the composite shown in Fig. 3 results. The applicable Q values are indicated in the individual circuits. The circles denote the effective noise voltage generators. The distribution of the half-strength impulse to the generators in the $\parallel$ and $\perp$ circuits depends on the time t_k at which the impulse occurs in relation to the phase of the pump at $t = t_k$. This follows from (24). Because the pump is out of phase with the undisturbed oscillator signal $A_1 \cos(\omega_1 t + \varphi)$, the outputs of the $\parallel$ circuits are the phase disturbances (ϕ), those of the $\perp$ circuits the amplitude disturbances (a).

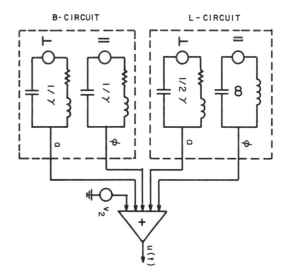

Fig. 3. Virtual circuits used to represent the components of the impulse response of the perturbed oscillator.

The $L\,\parallel$ circuit is lossless. Energy supplied to it by its share of the impulse strength is stored indefinitely; its output is the undamped sinusoid $-A_1\eta_k \sin(\omega_1 t + \varphi)$ in (24). The output of the $LV\perp$ circuit is the first term in (24), and the $B\parallel$ and $B\perp$ circuits deliver, respectively, the two orthogonal components

$$-A_1\eta_k \exp\left[-(\omega_1\gamma/2)(t - t_k)\right]\sin(\omega_1 t + \varphi)$$

---

[4] The pump phase is taken here to be that of $\sin(\omega, t + \varphi)$.

[5] Because γ is assumed very small, only a small portion of the energy in the impulse is accepted by the virtual circuits. Terms with γ or γ^2 as factor were neglected in (24)

and

$$\alpha_k \exp\left[-(\omega_1\gamma/2)(t - t_k)\right] \sin(\omega_1 t + \varphi)$$

of $n_k(t)$.

The structure of the disturbances $u(t)$ of the oscillator signal, due to a single impulse from each one of the noise sources, is thus established. It is most interesting that the properties of the particular oscillator model chosen in Fig. 1 enter (24) only by way of the parameters M_k and N_k. If the appropriate values are used for these parameters, this formula and, hence, the scheme in Fig. 3 evidently applies to any arbitrary oscillator. Those whose frequency depends strongly on fundamental amplitude already in the undisturbed case [14] will require additional terms.

D. The Perturbed Oscillator Signal

The approximation (13) requires the undisturbed oscillators to be represented as a single lossless tuned circuit with energy stored in it. According to (8) and the developments of the preceding Section, the perturbed oscillator is described by the virtual circuits in Fig. 3 if the $L\|$ circuit has been imparted a finite amount of energy initially and if the pump signal is orthogonal to the oscillations represented by this energy (i.e., e_{g0}) at all times.

The undamped oscillations in the $L\|$ circuit caused by the noise impulses at $t = t_k$ are out of phase with e_{g0}; added to it, they change its phase permanently. Since (8) with (10) is an approximation of (7), the response functions of the $L\|$ circuit represent phase disturbances only, and amplitude effects of any order do not occur in this circuit. All amplitude disturbances are represented by the outputs of the $L\perp$ and the $B\perp$ circuit, with the $B\|$ circuit contributing another phase disturbance which, significantly, decays with time.

It is important to realize that the energy supplied to the $L\|$ circuit by the impulses at t_k is necessary to effect the permanent phase changes in e_{g0}; although it is not "used up" thereby and remains stored in the circuit, it cannot be regained if no phase reference is available, and is nevertheless lost irreversibly.

Analytically, the oscillator output follows from (8), (13), and (24) as

$$e_g = [A_1 + \alpha_k e^{-\omega_1\gamma(t-t_k)}] \cos(\omega_1 t + \eta_k + \varphi) \\ + n_k(t) + v_2 \quad (27)$$

or alternatively, when $n_k(t)$ is decomposed into components parallel and normal to e_g, as

$$e_g = [A_1 + x_k(t)] \cos(\omega_1 t + \eta_k + \vartheta_k(t) + \varphi) + v_2 \quad (28)$$

whereby

$$x_k(t) = \alpha_k(e^{-\omega_1\gamma(t-t_k)} + e^{-(\omega_1\gamma/2)(t-t_k)}) \quad (28a)$$

$$\eta_k + \vartheta_k(t) = \eta_k(1 + e^{-(\omega_1\gamma/2)(t-t_k)}). \quad (28b)$$

These two forms for e_g illustrate different performance aspects of the disturbed oscillator. Equation (28) con-

forms closely to (7) and will generally be found more convenient for further work. Equation (27) emphasizes the existence of the "B circuit." It shows that the oscillator, in addition to generating e_{g0}, disturbed by the output components of the "L circuit" [i.e., by the first two terms in (24)] also acts simultaneously like a passive tuned circuit of relative bandwidth $\Delta\omega/\omega = \gamma$.

The last term in (27), $v_2 = a_{2k}\delta(t - t_k)$ is contributed to the signal by the source across the output. This term is present also as seen from (26), in the output from the passive network alone and is, thus, not peculiar to the oscillator. It still appears as an additive term in (28). Since a δ-function is infinitely large by definition, what might be its orthogonal component in reference to e_{g0} can of course not simply be taken into the argument of the cosine function. Until the oscillator signal has been acted upon by a filter in the output amplifier, v_2 must be carried as an additive term.

The solution (27) or (28) applies to oscillators whose amplitude is limited by the instantaneous nonlinearities in the circuit and whose circuit parameters are independent of time. The effects of a delayed action mechanism for amplitude control (such as changes in g_{m0} and/or the circuit losses) on this solution are complicated in the general case. An approximation is obtained rather simply, however, when the time constant T of the AGC loop and the time τ_a required for the loop to correct an amplitude disturbance obey the conditions:

$$2\pi/\omega_1 \ll T \ll \tau_a \ll 1/\omega_1\gamma.$$

Whereas the effects of the impulses at t_k on e_g are instantaneous, the AGC mechanism does not affect it immediately. The form of e_g is thus established first and can then be assumed modified by operations on the amplitude disturbances only (i.e., the concomitant effects on the signal phase are neglected). Hence, an approximation for the output of a nonlinear AGC oscillator is given by (28) if (28a) is replaced by

$$\alpha_k(t) = 2\alpha_k e^{-(t-t_k)/\tau_a}. \quad (28c)$$

The representation by virtual circuits results from Fig. 3 if the indicated Q values in both the $L\perp$ and the $B\perp$ circuits are replaced by $\omega_1\tau_a$. When τ_a goes beyond the upper limit stated above, the AGC mechanism becomes ineffectual: when T is comparable to or larger than τ_a, it becomes a disturbance. For $T\to 0$ the action becomes instantaneous. In the latter case the output signal is again given by (28) with (28a) and (28b), if the proper value for the nonlinearity is used in γ.[6]

E. The Response to White Noise

The response of the oscillator to white noise is obtained by linear superposition of the effects of all the individual impulses in (12). The undisturbed signal e_{g0}, used in deriving the response to the impulse at time t_k,

---

[6] Further details of AGC action in linear oscillators will be discussed in Section III-C.

was assumed in the Appendix to include all permanent effects of the impulses prior to t_k. Hence, the phase angle φ in (13) and, consequently, in (27) and (28) has the form

$$\varphi = \sum_{l}^{k-1} \eta_l \qquad (29)$$

and need not be carried further. The oscillator signal as disturbed by the action of the white noise source in the circuit becomes

$$e_g = (A_1 + x(t)) \cos(\omega_1 t + \bar{\eta}(t) + \vartheta(t)) + v_2 \qquad (30)$$

or alternatively, for reasonable values of γ

$$\begin{aligned}
e_g &= A_1 \cos(\omega_1 t + \bar{\eta}(t)) + x(t) \cos(\omega_1 t + \bar{\eta}(t)) \\
&\quad - A_1 \vartheta(t) \sin(\omega_1 t + \bar{\eta}(t)) + v_2 \\
&= E_C + E_A + E_\vartheta + E_N \qquad (31)
\end{aligned}$$

with

$$x(t) = \sum_k x_k(t), \quad \bar{\eta}(t) = \sum_k \eta_k, \quad \vartheta(t) = \sum_k \vartheta_k(t) \qquad (32)$$

whereby $x_k(t)$, η_k, and $\vartheta_k(t)$ are given by (28a) and (28b) together with (24b) and (24c). The summations in (32) extend over all values of k for which $t_k \leq t$. According to the developments of the preceding Section, $E_C = A_1 \cos(\omega_1 t + \bar{\eta}(t))$ in (31) is the output of the $L\|$ circuit in Fig. 3, if the latter has initially energy stored in it. E_C is thus to be considered the carrier signal. Its phase, and hence the phase of the pump signal in (21) execute a random walk due to the cumulative effect of the permanent shifts caused by the individual impulses. $E_A = x(t) \cos(\omega_1 + \bar{\eta}(t))$ is the combined output of the $L\perp$ and the $B\perp$ circuit and represents the amplitude disturbances of the carrier, whereas

$$E_\vartheta = -A_1 \vartheta(t) \sin(\omega_1 t + \bar{\eta}(t))$$

are the phase disturbances due to the $B\|$ circuit. Both E_A and E_ϑ consist of the superposition of a large number of exponentially decaying components. $E_N = v_2$ is the white noise from the source across the oscillator output.

The process indicated by (32) has considerable merit conceptually. The individual impulses in (12) are caused by the elemental phenomena involved in the transport of electrical charges and, though exceedingly numerous, are extremely weak. The addition of the effects of any one of these impulses to $x(t)$, $\bar{\eta}(t)$, and $\vartheta(t)$ in (30) or (31) (and they are to be added one by one) changes these quantities by only infinitesimal amounts. There can never be any reasonable doubt that the linear perturbation equation (21) applies and that the out-of-phase components of $u_k(t)$ can be taken into the argument of the sinusoid. It will frequently be found useful to retrace, at least mentally, the steps going from single impulses in the oscillator to the summation of their effects.

Of the various assumptions which had to be made in deriving the expressions (30) or (31) as an approxi-

mation to the solution of (3), the one regarding the absence of harmonic components, particularly of the third harmonic, in the undisturbed signal is considered to be the most serious. In general it must be expected to limit the validity of the results to oscillators operating at low signal levels. Nevertheless, the expressions do give a detailed description of the effects of noise on the oscillator signal, which becomes increasingly more accurate as the harmonic content is reduced.

III. The Statistical Properties of the Signal

A. The Continuous Noise Record

The statistical properties of e_g are more readily evaluated when the sums in (12) are replaced by continuous functions. Without discussion of the essentially philosophical questions involved thereby [27], it will be assumed from here on that the output of the noise sources in Fig. 1 is equally well described by

$$\begin{aligned}
i_s &= B_s(t) \\
v_r &= B_r(t) \qquad (r = 1, 2, 3). \qquad (33)
\end{aligned}$$

The $B_j(t)$ can be visualized as the curves that result when a series of points Δt apart are connected by a continuous line, with each point representing the integral over all noise impulses occurring during the respective time interval Δt.

While the detailed course of any $B_j(t)$ is basically unpredictable, these functions are uniquely defined if it is specified that they are independent stationary Gaussian random variables, and if their mean and variance is given. Forcing functions of the type described have been dealt with extensively in the theory of Brownian Motion and their Gaussian property is well established [28]. It is also known that they are ergodic, have zero mean, and are delta correlated; hence,

$$\begin{aligned}
\langle B_j(t) \rangle &= 0 \\
\langle B_j(t_1) B_j(t_2) \rangle &= B_j^2 \delta(t_1 - t_2) \\
\langle B_i(t_1) B_j(t_2) \rangle &= 0 \qquad i \neq j. \qquad (34)
\end{aligned}$$

The brackets here and in the following denote ensemble averages which, as generally understood, are the arithmetic mean of the quantities within the brackets, each formed with values from a particular noise record, averaged over a large number N of like records, with $N \to \infty$. The fact that the $B_j(t)$ are delta correlated [expressed by the second relation in (34)] means that the value of $B_1(t_1)$, for example, is completely independent of the value of $B_1(t)$ a moment before (i.e., at $t = t_1 - \Delta t$), and it is independent of the values of B_2, B_3, and B_s at the same or any other moment, as expressed by the third relation in (34).

With v_j now given by (33) instead of by (12), the a_{jk} in (24c) are to be replaced by

$$a_{jk} = B_j(t) \, dt \qquad (35)$$

and the sums in (32) become integrals:

167

$$x(t) = \int_0^t \alpha(\xi) \left[e^{-\omega_1\gamma(t-\xi)} + e^{-(\omega_1\gamma/2)(t-\xi)} \right] d\xi$$

$$\bar{\eta}(t) = \int_0^t \eta(\xi) d\xi$$

$$\vartheta(t) = \int_0^t \eta(\xi) e^{-(\omega_1\gamma/2)(t-\xi)} d\xi. \tag{36}$$

$\alpha(\xi)$ and $\eta(\xi)$ follow from (24b) with (24c), (29) and (35) to

$$\alpha(\xi) = - M(\xi) \sin(\omega_1\xi + \bar{\eta}(\xi))$$
$$+ N(\xi) \cos(\omega_1\xi + \bar{\eta}(\xi))$$

$$\eta(\xi) = - \frac{M(\xi)}{A_1} \cos(\omega_1\xi + \bar{\eta}(\xi))$$
$$- \frac{N(\xi)}{A_1} \sin(\omega_1\xi + \bar{\eta}(\xi)) \tag{36a}$$

with

$$M(\xi) = \frac{\omega_1}{2\bar{Q}} \left[B_1(\xi) + B_3(\xi) - B_2(\xi) \right]$$

$$N(\xi) = \frac{\omega_1}{2\bar{Q}Q_N} \left[\frac{B_s(\xi)}{g_m} + B_2(\xi) \right]. \tag{36b}$$

The difference in the definitions of $M(\xi)$, $N(\xi)$, $\alpha(\xi)$, and $\eta(\xi)$ when compared to that of M_k, N_k, $\cdots$ in (24) should cause no confusion: $M(\xi)d\xi \leftrightarrow M_k$. When $B_j(\xi) = a_{jk}\delta(\xi - t_k)$ is substituted into the above expressions, they revert to (28a) and (28b), respectively.

The properties of $M(\xi)$ and $N(\xi)$ follow directly from those of the $B_j(\xi)$:

$$\langle M(\xi) \rangle = \langle N(\xi) \rangle = 0$$
$$\langle M(\xi_1)M(\xi_2) \rangle = M^2\delta(\xi_1 - \xi_2)$$
$$\langle N(\xi_1)N(\xi_2) \rangle = N^2\delta(\xi_1 - \xi_2)$$

with

$$M^2 = \left(\frac{\omega_1}{2\bar{Q}} \right)^2 (B_1^2 + B_2^2 + B_3^2)$$

$$N^2 = \left(\frac{\omega_1}{2\bar{Q}Q_N} \right)^2 \left(\frac{B_s^2}{g_m^2} + B_2^2 \right).$$

Of special importance in the later developments is the quantity $K^2 = (M^2 + N^2)/\omega_1^2$

$$K^2 = \frac{1}{4} \frac{1}{\bar{Q}^2} \left[B_1^2 + B_2^2 + B_3^2 + \frac{1}{Q_N^2} \left(B_2^2 + \frac{B_s^2}{g_m^2} \right) \right] \tag{37}$$

which can be regarded as the weighted noise intensity of all sources in the oscillator combined.

The means of $\alpha(\xi)$ and $\eta(\xi)$ are obviously zero:

$$\langle \alpha(\xi) \rangle = \langle \eta(\xi) \rangle = 0. \tag{38a}$$

If all oscillators in the ensemble are assumed to have already been operating sufficiently long for their relative phase angles to be randomly distributed at time $t = 0$,

$\bar{\eta}(\xi)$ with respect to the ensemble is a random number between zero and 2π at any value of ξ in $0 \le \xi \le t$. Hence, ensemble averages can be employed to compute the correlation functions of $\alpha(\xi)$ and $\eta(\xi)$. Because

$$\langle \cos(\bar{\eta}(u) + \bar{\eta}(v)) \rangle = \langle \sin(\bar{\eta}(u) + \bar{\eta}(v)) \rangle = 0,$$

we find

$$\langle \alpha(u)\alpha(v) \rangle = (\omega_1^2 K^2/2)\delta(u - v)$$
$$\langle \eta(u)\eta(v) \rangle = (\omega_1^2 K^2/2)\delta(u - v)$$
$$\langle \alpha(u)\eta(v) \rangle = 0. \tag{38b}$$

The relations (38a) and (38b), together with the knowledge that $\alpha(\xi)$ and $\eta(\xi)$ are Gaussian random processes, completely characterize these functions.

The explicit expressions for the fundamental frequency component of the output from a noise perturbed oscillator have thus been derived. The more general form is given by (30), with (32), when the noise is thought of as a series of random impulses, or with (36), when the noise is represented by continuous random functions. Whereas the two descriptions of white noise (12) and (33) are equivalent, the expressions resulting from the latter are obviously more abstract and do not provide the same conceptual insight into oscillator operation afforded by the impulse response functions. Nevertheless, the expression (30) with (36), (37), (38a), and (38b) is more compact, and is more easily handled when computing the statistical properties of the oscillator signal. Because the decomposition of (30) into (31) is valid in nearly all cases of practical significance, the latter can be used whenever this is desirable.

B. The Autocorrelation Function of the Signal

The power spectral density $G_{ee}(f)$ will be computed from the autocorrelation function $\Gamma_{ee}(\tau)$ of e_g according to the well-known relations [29]

$$G_{ee}(f) = 4 \int_0^\infty \Gamma_{ee}(\tau) \cos \omega\tau \, d\tau \tag{39}$$

$$\Gamma_{ee}(\tau) = \langle e_g(t)e_g(t - \tau) \rangle. \tag{40}$$

The most convenient representation of the oscillator signal for the present purpose is given by (31), together with (36). Since all cross correlations between the terms there are zero, the autocorrelation function simply is

$$\Gamma_{ee}(\tau) = \Gamma_{CC}(\tau) + \Gamma_{AA}(\tau) + \Gamma_{\vartheta\vartheta}(\tau) + \Gamma_{NN}(\tau). \tag{41}$$

The individual terms in (41) are found with (38b) by established techniques [30], [5].

$$\Gamma_{CC}(\tau) = \frac{A_1^2}{2} e^{-(\omega_1^2 K^2/4A_1^2)\tau} \cos \omega_1\tau$$

$$\Gamma_{AA}(\tau) = \frac{\omega_1 K^2}{\gamma} \left(\frac{7}{24} e^{-\omega_1\gamma\tau} + \frac{5}{12} e^{-(\omega_1\gamma/2)\tau} \right) \cos \omega_1\tau$$

$$\Gamma_{\vartheta\vartheta}(\tau) = \frac{\omega_1 K^2}{4\gamma} e^{-(\omega_1\gamma/2)\tau} \cos \omega_1\tau$$

$$\Gamma_{NN}(\tau) = B_2^2\delta(\tau). \tag{42}$$

The relations for Γ_{AA} and $\Gamma_{\vartheta\vartheta}$ are approximations which hold whenever

$$\gamma \gg 2\omega_1 K^2/A_1^2. \qquad (43)$$

When AGC is used, $\bar{\Gamma}_{AA}(\tau)$ in (42) is to be replaced by

$$\bar{\Gamma}_{AA}(\tau) = \frac{\omega_1^2 K^2}{2}\,\tau_a e^{-\tau/\tau_a} \qquad (44)$$

where τ_a is, according to (28c), the time required by the AGC mechanism to correct an amplitude disturbance. All other terms in (42) remain the same.

The magnitude of $2\omega_1 K^2/A_1^2$ in (43) is in the order of 10^{-21}, with values of ω_1, K^2, and A_1 representative of a typical precision crystal oscillator, while the actual values of γ encountered in such devices are in the order of 10^{-8}. [See (47) in Section III-D.] The discrepancy is not always this large, but it is doubtful that the instantaneous nonlinearities are ever small enough to violate (43), even in laser oscillators [31]. Hence, in nearly all cases of practical significance, the autocorrelation function of the output signal is properly given by (41) with (42) and, where applicable, (44). It is this case that will be considered when the power spectral density function of the signal is computed in Section III-D.

The average power in the individual output components is obtained from the relations (42) with $\tau = 0$ [32]. That due to the B circuit in Fig. 3 $[\Gamma_{\vartheta\vartheta}(\tau)+\frac{1}{2}\bar{\Gamma}_{AA}(\tau)$ for AGC oscillators] is seen to be substantially smaller than that in the carrier for most practical oscillators. That is, when (43) holds, the tuned circuit aspect contributes only a very small fraction of the total output power of the oscillator. The practical significance of the B circuit is rather limited, therefore, when the disturbances due to thermal and shot noise in an oscillator with time invariant circuit parameters are considered. It is held, however, that its existence is vitally important for a proper explanation of the behavior of the signal phase in oscillators with time variable circuit elements, an area which is left for further research.

C. The Linear Oscillator

To aid the reader in relating earlier concepts to those developed here, the properties of signals from oscillators with linear active elements will now be discussed. Since the virtual circuits in Fig. 3 describe the oscillator output, these will be used thereby.

As γ decreases, the quality factors of the B circuit and of the $L\perp$ circuit increase. The responses to the individual impulses require a longer time to decay, and their cumulative effects become larger. Eventually the phase disturbances due to the $B\|$ circuit become too large to be considered as additive terms in the oscillator output (which in effect means to neglect their influence on the phase of the pump signal), and it becomes necessary to include them immediately into the phase of e_g; that is, the transition from (30) to (31) is no longer

applicable. With $\gamma \to 0$, all circuits in Fig. 3 become lossless and the oscillator output no longer shows a performance aspect identifiable with a tuned circuit of finite Q. In this transition the B circuit becomes identical with the L circuit and the two can be fused into one, which shall be called the LB circuit. Now the amplitude disturbances, as well as the phase disturbances, are undamped sinusoids; and the amplitude of the output signal also carries out a random walk, with A_1 as its mean value.

Since E_A in (30) is, with $\gamma = 0$ in (36), $E_A = 2\int_0^t \alpha(\xi)\,d\xi$ and $\vartheta(t) = \bar{\eta}(t)$, the autocorrelation function of the output signal is now

$$\bar{\Gamma}_{ee} = \bar{\Gamma}_{CC} + \bar{\bar{\Gamma}}_{AA} + \Gamma_{NN}$$

whereby

$$\bar{\Gamma}_{CC} = \frac{A_1^2}{2}\,e^{-(\omega_1^2 K^2/A_1^2)\tau}\cos\omega_1\tau$$

$$\bar{\bar{\Gamma}}_{AA} = \omega_1^2 K^2(t-\tau)e^{-(\omega_1^2 K^2/A_1^2)\tau}\cos\omega_1\tau. \qquad (45)$$

This is the situation obtained in a linear oscillator without an AGC mechanism (i.e., with $T \to \infty$, where T is the time constant of the AGC loop), provided, of course $g_m = g_{m0}$ satisfies the condition (20) for steady-state oscillations precisely.

When g_{m0} is larger than is required by this condition, the oscillations of the undisturbed oscillator build up; when it is smaller they decay, as is well known. The undisturbed oscillator can thus be represented as a tuned circuit, to be called the $\overline{LB}$ circuit, whose quality factor is $1/\gamma_e$, whereby γ_e is defined by (22b) if the difference between the quantities on the rhs and lhs of (20) is used in place of $g_{m0} - g_m$.

Amplitude noise interferes with the orderly build-up or decay of the oscillations, originally set up in the $\overline{LB}$ circuit at $t = 0$, and if its average, obtained by integrating over a time T ($\ll 2/\omega_1\gamma_e$), has the proper sign and magnitude, it can, in fact, hold the amplitude at a constant level. It is the function of a properly designed AGC mechanism to steer the value of γ_e so that this condition does occur continually. In an AGC oscillator, therefore, γ_e is a random variable of zero mean $\langle\gamma_e\rangle = 0$. The average amplitude noise energy delivered by the source during T is $\omega_1^2 K2T$; the initial gain or loss in oscillation energy during this time is $(A_1^2/2)\omega_1\gamma_e T$; since the two should balance, the root-mean-square value of γ_e is

$$\sqrt{\langle\gamma_e^2\rangle} = 2\omega_1 K^2/A_1^2.$$

It is noted that most of the oscillation energy supplied at $t = 0$ has been replaced in the process by noise energy of the proper phase after $2/\omega_1\gamma_e$ seconds, with the AGC mechanism lending or borrowing energy temporarily to smooth the random variations in the noise energy supply. The remaining amplitude variations are those occurring during the time interval $[t-\tau_a, t]$, whereby τ_a

is the time required by the AGC mechanism to remove a given amplitude disturbance. Their autocorrelation function is given by (44).

The time period τ_a depends inversely on the AGC loop gain (and directly on the reduced quality factor Q of the feedback network). When the time constant T of the loop increases beyond τ_a, the AGC action gets to be out of phase with the amplitude variations, and the slow oscillations in amplitude discussed by Golay [15] and earlier by Edson [33] will occur. To maintain minimum total disturbances requires, therefore, that the AGC loop gain be decreased as T increases. This eventually will act to limit the range of γ_e values controlled by the AGC mechanism and $\langle \gamma_e^2 \rangle \to 0$ with $T \to \infty$ results. Hence, with $T \to \infty$ the $\overline{LB}$ circuit becomes lossless and identical to the LB circuit described above. Oscillations of stable amplitude, however, are obtained only for $T \ll \tau_a \ll A_1^2/\omega_1^2 K^2$. The following discussion of the signal phase is restricted to these cases.

Assume again the $\overline{LB}$ circuit supplied at $t = 0$ with an amount of energy large when compared to the noise energy. The oscillations represented by this energy provide the reference for decomposing the noise impulse responses of the circuit into amplitude and phase disturbances and, in particular, for the AGC mechanism while removing the amplitude noise. Although the phase disturbances are time dependent with $\exp\left[-(\omega_1 \gamma_e/(2)(t - t_k)\right]$, it is the shift in the reference phase accumulating during τ_a seconds that determines their effect on the oscillations. Since the concurrent amplitude variations are eliminated after τ_a seconds, time can start anew, but with the basic oscillations at a different phase. In spite of their time dependence, the phase disturbances in a linear AGC oscillator act as though they were undamped sinusoids, which, added out of phase to a constant amplitude signal, cause its phase to execute a random walk.[7] The output of an AGC oscillator with a linear active element is thus to be represented by the output from a lossless tuned circuit, which has stored in it the energy of the basic oscillations and is excited by a noise source delivering that half of the total noise energy which produces the out-of-phase response components, and added to it the output from another tuned circuit of effective quality factor $\omega_1 \tau_a/2$ to which the second half of the noise energy is fed, producing the amplitude disturbances. That is to say, the $\overline{LB}$ circuit is equivalently represented by the LB circuit discussed above when the latter is modified to account for the AGC according to (44). The link from the scheme in Fig. 3 is thus established, even if the reverse path is not obvious. The linear model seems to give no indication that the introduction of instantaneous nonlinearities requires the LB circuit to be split into the L and B circuits shown there.

Approaching the linear oscillator from the standpoint of a passive noise filter, as was variously done in the past, leads in a most direct manner to the autocorrelation function of the oscillator output. Consider the passive feedback circuit in Fig. 1, replace R_3 by $R_3' = (\gamma_f/\gamma_T)R_3$ without affecting the strength of the noise sources, assume half the average output power removed by some appropriate mechanism, and demand that the remainder of the average output power equal $A_1^2/2$. One finds for the only adjustable parameter γ_f the value $\gamma_f = 2\omega_1 K^2/A_1^2$; that is, $\gamma_f = \sqrt{\langle \gamma_e^2 \rangle}$. With it the autocorrelation function of the output becomes

$$\overline{\Gamma}_{ee} = \overline{\Gamma}_{CC} + \overline{\Gamma}_{AA} + \Gamma_{NN}$$

with $\overline{\Gamma}_{CC}$ given in (45) and $\overline{\Gamma}_{AA}$ by (44) exactly.

Hidden in these assumptions, however, is the full description of the linear oscillator with AGC as given above. If the AGC mechanism is visualized as an amplifier following the filter, the amplifier provides the reference for decomposing the noise responses into orthogonal components. As this reference is fixed, no random walk of the output phase will occur. However, the amplitude of a noise filter output can go to zero temporarily. To maintain constant amplitude at all times, therefore, requires the AGC mechanism to be part of the feedback loop and the oscillator discussed before results. Also, it is noted that the amplitude noise is not removed by the AGC mechanism as one might be led to believe by the above assumptions. It is converted at the proper rate into the basic oscillation as mentioned before. The phase noise energy remains stored in the random phase walk and is irreversibly lost, since no absolute phase reference is available.

Whereas the passive noise filter is not the proper physical model for the linear AGC oscillator, it does lead in the most simple and direct manner to the correct autocorrelation function and, hence, power spectral density of the output from these devices. However, it is of no apparent assistance in dealing with instantaneous nonlinearities in the oscillator.

D. The Power Spectral Density of the Signal

Proceeding now to the power spectral density (PSD) of the signal, one finds $G_{ee}(f)$ from (39) with (41) and (42) as the sum of the following components:

$$G_{CC}(f) = \frac{4A_1^4/\omega_1^2 K^2}{1 + (4A_1^2/\omega_1 K^2)^2(1 - \omega/\omega_1)^2}$$

$$G_{AA}(f) = \frac{(7/12)(K^2/\gamma^2)}{1 + (1/\gamma)^2(1 - \omega/\omega_1)^2} + \frac{(5/3)(K^2/\gamma^2)}{1 + (2/\gamma)^2(1 - \omega/\omega_1)^2}$$

$$G_{\vartheta\vartheta}(f) = \frac{K^2/\gamma^2}{1 + (2/\gamma)^2(1 - \omega/\omega_1)^2}$$

$$G_{NN}(f) = 4B_2^2. \tag{46}$$

---

[7] This of course does not apply to the phase disturbances due to the $B\|$ circuit when γ is large enough to satisfy (43). The carrier provides the reference phase in that case, and the phase disturbances in response to an impulse excitation relax towards it exponentially.

As a numerical example let

$$\omega_1 L_1 = 100 \ \Omega \qquad \omega_1 = 2\pi 5 \times 10^6 \ \text{rad/sec} \qquad g_{m0} = 1/20 \ \text{mho}$$

$$R_3 = 100 \ \Omega \qquad \overline{Q} = 10^6 \qquad \beta = 10 \ \text{mho/volt}^2$$

$$\frac{1}{\omega_1 C_2} = 20 \ \Omega \qquad A_1^2/2 = 4 \times 10^{-5} \ \text{volt}^2 \qquad \gamma = 1.2 \times 10^{-8}$$

$$P_3 = (A_1^2/2)R_3(\omega_1 C_2)^2 = 10^{-5} \ \text{watts}, \qquad K^2 = 1.75 \times 10^{-31} \ \text{watts/cycle}. \qquad (47)$$

The parameters for the active device are typical for a 2N2808 transistor; the values for R_3 and $\overline{Q}$ could apply to a precision crystal unit. The power spectral density of Johnson noise is $4kTR$, that of shot noise in a transistor $2kTg_m$; hence, because of (42) and (39),

$$B_j^2 = kTR_j$$
$$B_s^2 = \tfrac{1}{2}kTg_m. \qquad (48)$$

The resistive components R_1 and R_2 have not been considered so far because, other than affecting the values of Q_T and Q_N, their effect upon the results does not extend significantly beyond their action as noise generators. Rather pessimistic assumptions [34] about the effective values of R_1 and R_2 in a transistor oscillator lead to 10 ohms each. With the above values and $kT = 5 \times 10^{-21}$ W/s, one finds, for the peak densities (in watts per cycle) and normalized half widths at midheight (in cycles/cycle) of the individual terms in (46), respectively

$$G_{cc}[1.4 \times 10^8, 3.5 \times 10^{-20}], \qquad G_{AA1}[7.1 \times 10^{-16}, \ 3.5 \times 10^{-8}],$$

$$G_{AA2}[2 \times 10^{-15}, 1.75 \times 10^{-8}], \qquad G_{\vartheta\vartheta}[1.2 \times 10^{-15}, 1.75 \times 10^{-8}];$$

and $G_{NN} = 2 \times 10^{-20}$ watts/cycle.

The power spectral density function cannot be observed directly. The power spectrum $P(\bar{f})$ which is observable can be defined as

$$P(\bar{f}) = \int_0^\infty |H(\bar{f}, f)|^2 G_{ee}(f) \, df \qquad (49)$$

whereby $H(\bar{f}, f)$ is the transfer function of the filter being used in the observation of the spectrum and $\bar{f}$ its mean frequency. The output power from the filter, when it is tuned to the carrier frequency, i.e., when $\bar{f} = f_1$, must then be considered the signal power and the definition of the noise-to-signal power ratio at the frequency $\bar{f}$ becomes

$$N/S = P(\bar{f})/P(f_1). \qquad (50)$$

It is obvious that the noise-to-signal power ratio depends upon the bandwidth of the filter. This parameter is implicit in $H(\bar{f}, f)$. The form factor of the filter determines whether $P(\bar{f})$ (i.e., the output power from the filter when it is tuned to $\bar{f}$), is indeed due to $G_{ee}(f)$ in the neighborhood of $f = \bar{f}$, or instead is due to the carrier in the tail of the filter characteristic.

When evaluating the ratio (50) analytically, it is frequently possible to approximate the PSD curves given in (46) by their limiting curves, truncated at their peak

height, as indicated in Fig. 4. The limiting curves are

$$G_{cc}(f) \to (K^2/4)(1 - \omega/\omega_1)^{-2}$$

$$G_{AA}(f) \to K^2(1 - \omega/\omega_1)^{-2}$$

$$G_{\vartheta\vartheta}(f) \to (K^2/4)(1 - \omega/\omega_1)^{-2}$$

$$G_{NN}(f) \to 4B_2^2. \qquad (51)$$

They are, in conformance with a general property of Lorentzian curves, independent of A_1^2 and γ^2, respectively.

Because G_{NN} does not decrease with increasing frequency separation from the carrier, it crosses the other curves at a certain value of $F = f - f_1$, and from there on dominates the PSD of the oscillator output. The crossover point depends, according to (37), on the strength of the noise sources and on the reduced quality factor $\overline{Q}$ of the oscillator. In an LC oscillator $\overline{Q} = 1$, and the crossover occurs too far out from the carrier frequency to be of practical significance. The white noise from the source across the output has no noticeable effect on the PSD of the LC oscillators. When $\overline{Q}$ is large, however, such as in quartz crystal oscillators, the crossover is very close to the carrier (about 3 parts in 10^6 in the above numerical example), and it is the white noise component that determines the N/S ratio of the oscillator in many applications. A very significant improvement in this ratio can often be achieved in this case by using an output filter of sufficiently narrow bandwidth [35]. The relations pertaining to this case are given in the following Section. In all cases, it is obvious that the most effective way to improve the N/S ratio is to operate the oscillator at as high a power level as is possible, subject to limitations by higher order effects.

E. The Effects of the Output Filter

Up to now it was assumed implicitly that the signal at the output of the oscillator stage is available directly for observation or actual use. This however, is rarely the case. Normally the oscillator signal is fed to an isolation amplifier or to some other devices which contain tuned elements. It is the output of these devices that is actually observed, and the modifications of the signal in passage through them must be considered. Of primary interest is the effect of the tuned elements.

Although it is obvious that more effective filters can be used, consider as a simple example the circuit in Fig. 5. When the filter is tuned to ω_1 (i.e., $L_c C_c = 1/\omega_1^2$), and it is assumed that, because of (23),

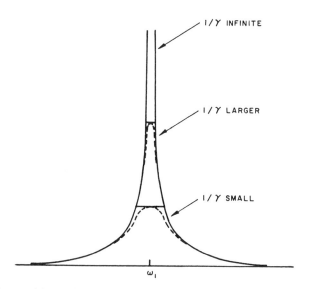

Fig. 4. Illustration of general property of PSD curves. Truncated limiting curves are useful in approximate calculations involving PSD functions.

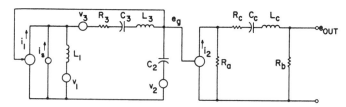

Fig. 5. Model of oscillator with output amplifier. The filter in the output amplifier serves primarily to shape the white noise component in e_g, which stems from v_2. The effect of the filter on the carrier and on the noise response of the oscillator loop is negligible in a high Q oscillator.

$$Q_c \equiv 1/\gamma_c \equiv \frac{\omega_1 L_c}{R_a + R_b + R_c} \ll 1/\gamma, \qquad (52)$$

the equivalent to (30) becomes to a very good approximation

$$e_{\text{out}} = (A_1 + \bar{x}(t)) \cos(\omega_1 t + \phi(t)) \qquad (53)$$

with

$$\bar{x}(t) = x(t) + x_n(t)$$
$$\phi(t) = \bar{\eta}(t) + \vartheta(t) + \vartheta_n(t), \qquad (53a)$$

and

$$x_n(t) = \frac{\omega_1}{Q_c} \int_0^t B_2(\xi) e^{-(\omega_1 \gamma_c/2)(t-\xi)} \cos[\omega_1 \xi + \bar{\eta}(\xi)]\, d\xi$$

$$\vartheta_n(t) = \frac{\omega_1}{Q_c} \int_0^t B_2(\xi) e^{-(\omega_1 \gamma_c/2)(t-\xi)} \sin[\omega_1 \xi + \bar{\eta}(\xi)]\, d\xi \quad (53b)$$

where $x(t)$, $\bar{\eta}(t)$ and $\vartheta(t)$ are still defined by (36). The equivalent to (21) becomes

$$e_{\text{out}} = E_C + E_A + E_\theta + E_N \qquad (54)$$

whereby only the definition of E_N is changed from $E_N = B_2(t)$ to

$$E_N = \frac{\omega_1}{Q_c} \int_0^t B_2(\xi) e^{-(\omega_1 \gamma_c/2)(t-\xi)} \cos \omega_1(t-\xi)\, d\xi. \quad (54a)$$

The expressions (53) and (54) then are explicit forms of the oscillator signal after it has passed through the tuned output stage.

All relations in Section III-B for e_g apply equally for e_{out}, if only $\Gamma_{NN}(\tau)$ and $G_{NN}(f)$ in (42) and (46), respectively, are replaced by

$$\Gamma_{NN}(\tau) = \frac{\omega_1}{Q_c} \frac{B_2{}^2}{2} (1 + Q_c/Q_T) e^{-(\omega_1 \gamma_c/2)\tau} \cos \omega_1 \tau \quad (55a)$$

$$G_{NN}(f) = \frac{2 B_2{}^2 (1 + Q_c/Q_T)}{1 + (2Q_c)^2 (1 - \omega/\omega_1)^2}. \qquad (55b)$$

The limiting curves for $G_{NN}(f)$ are given by

$$G_{NN}(f) \to (B_2{}^2/2Q_c{}^2)(1 + Q_c/Q_T)(1 - \omega/\omega_1)^{-2}. \quad (56)$$

F. The Properties of the Signal Phase

When evaluating the properties of the signal phase it is important that the effects of the output filter be included in the considerations. This eliminates the difficulties with the additive white noise component in (30) which, according to the comments in Section II-D, cannot be decomposed into orthogonal components until acted upon by the filter. Hence, the signal representation (53) must be used. Because of $\bar{\eta}(t)$ in (53a), the phase of the signal is only a weakly stationary random process; however, it does have stationary independent increments, and the autocorrelation function of $\phi(t)$ can be readily computed to

$$\langle \phi(t)\phi(t-\tau)\rangle$$
$$= (\omega_1{}^2 K^2/2 A_1{}^2)\left(t - \tau + \frac{2}{\omega_1 \gamma} + \frac{3}{\omega_1 \gamma} e^{-(\omega_1 \gamma/2)\tau}\right)$$
$$+ (\omega_1 B_2{}^2/2 A_1{}^2 Q_c)[Q_c/2Q_T + (1 + Q_c/Q_T)e^{-(\omega_1 \gamma_c/2)\tau}]. \quad (57)$$

With some care in dealing with the first term, (57) can be used in (39) to obtain the spectral density of the signal phase. Except for some δ functions at the origin, it is

$$G_{\phi\phi}(f) = (2K^2/A_1{}^2)(\omega_1/\omega)^2$$
$$+ (12K^2/A_1{}^2\gamma^2)[1 + (2/\gamma)^2(\omega/\omega_1)^2]^{-1}$$
$$+ (4B_2{}^2/A_1{}^2)(1 + Q_c/Q_T)[1 + (2/\gamma_c)^2(\omega/\omega_1)^2]^{-1}. \quad (58)$$

A sketch of the three terms in (58) is shown in Fig. 6. The respective limiting curves for the second and third term are

$$(3K^2/A_1{}^2)(\omega_1/\omega)^2; \quad (B_2{}^2/A_1{}^2 Q_c{}^2)(1 + Q_c/Q_T)(\omega_1/\omega)^2 \quad (58a)$$

and, as before, it may be adequate for practical purposes to approximate these terms by their truncated limiting curves as indicated by the dotted lines in Fig. 6.

The expression (58) for the spectral density of the signal phase can be compared to the expressions for G_{ee} or for $G_{CC} + G_{\vartheta\vartheta} + \frac{1}{2} G_{NN}$ [using the relations (46) with (55b)]. The latter is the PSD of the output signal with all amplitude disturbances removed. It will be noted that, in spite of the similarities of the corresponding terms, there is no simple relationship between these expressions. In the most general case the seven parameters entering them must be evaluated from independent

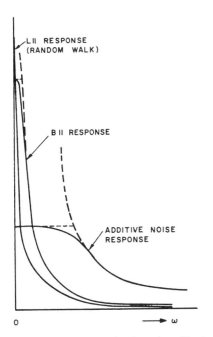

LI RESPONSE
(RANDOM WALK)

B II RESPONSE

ADDITIVE NOISE
RESPONSE

o $\longrightarrow \omega$

Fig. 6.   Sketch of spectral density of oscillator phase
disturbance components.

measurements to specify the different aspects of the signal. These parameters are A_1, ω_1, Q_c, Q_T, γ, K^2, and B_2^2. Only when the additive noise component due to the source across the output of the oscillator stage predominates over all other terms, can the properties of the power spectrum of the output signal be related directly to those of the spectrum of the phase and vice versa. In all cases it is important to recognize the distinction between these two spectra.

IV. THE SHORT TERM FREQUENCY STABILITY

The frequency of a periodic or quasi-periodic signal can be obtained by integrating the phase θ over a time τ and dividing the result by τ:

$$\omega_\tau = \frac{1}{\tau} \int_{t-\tau}^{t} d\theta = \frac{\theta(t) - \theta(t-\tau)}{\tau}. \quad (59)$$

In general, ω_τ is a function of the time t at which the integration is carried out. When $\theta(t)$ can be assumed to have the form

$$\theta(t) = \omega_1 t + \phi(t), \quad (60)$$

whereby ω_1 is a constant and $\phi(t)$ a random variable of zero mean, the short term frequency stability of the signal for integration times of τ seconds can be defined as

$$S(\tau) = \sqrt{\left\langle 4 \frac{(\omega_\tau - \omega_1)^2}{\omega_1^2} \right\rangle}. \quad (61a)$$

With (59) and (60) this becomes

$$S(\tau) = \frac{2}{\omega_1 \tau} \left\{ \langle (\phi(t))^2 \rangle + \langle (\phi(t-\tau))^2 \rangle \right.$$
$$\left. - 2\langle \phi(t)\phi(t-\tau) \rangle \right\}^{1/2}. \quad (61b)$$

When $\phi(t)$ is known, ω_τ and $S(\tau)$ are readily evaluated; however, the reverse process, that is the determination of the characteristics of the signal phase from the properties of ω_τ, can be carried out only in a very restricted sense.

The form of the output signal of the oscillator appropriate for use in this Section is given by (53), since the effects of the output filter must be included in the analysis. The autocorrelation function of $\phi(t)$ required in (61b) was derived in Section III-F. With (57) one finds the short term frequency stability of the signal as

$$S(\tau) = \left\{ (2K^2/A_1^2\tau^2)[\tau + (6/\omega_1\gamma)(1 - e^{-(\omega_1\gamma/2)\tau})] \right.$$
$$\left. + (4B_2^2/\omega_1 Q_c A_1^2\tau^2)(1 + Q_c/Q_T)(1 - e^{-(\omega_1\gamma_c/2)\tau}) \right\}^{1/2}. \quad (62)$$

K^2 in (62) is given by (37), with (4a) and (48), γ by (22); and $Q_c = 1/\gamma_c$ is the quality factor of the output filter in in Fig. 5. The conditions (23) and (52) are assumed satisfied.

The first term in (62), $(2K^2/A_1^2\tau)^{1/2}$, corresponds qualitatively to the familiar result obtained when only the random walk in the signal phase is considered, while the second term in the first bracket, with

$$(1 - \exp[-(\omega_1\gamma/2)\tau])$$

as factor, is due to the output from the $B\|$ circuit in Fig. 3. With $\gamma \to 0$ the first component of (62) becomes $(8K^2/A_1^2\tau)^{1/2}$, in full agreement with the corresponding expressions found in the literature.

The second component of (62) is due to the additive white noise component in e_g and depends strongly on the properties of the output filter.

With the numerical values for the various parameters as chosen in (47) and a $Q_c = 10$, (62) becomes

$$S^2(\tau) = \frac{44 \times 10^{-28}}{\tau^2} \left[\tau + \frac{3}{0.2}(1 - e^{-0.2\tau}) \right]$$
$$+ \frac{8 \times 10^{-24}}{\tau^2}(1 - e^{-1.6 \times 10^6 \tau}). \quad (63)$$

The two components of $S(\tau)$ as given by (63) are drawn in Fig. 7. The overall character of the first component shows a $\tau^{-1/2}$ dependence on integration time. The transition of the curve from a higher to a lower level occurs at a value of τ, which depends upon γ and, hence, upon the nonlinearity of the active device. A large value of γ is desirable to push this transition to shorter integration times. The overall level of this component of $S(\tau)$, in an oscillator whose noise generators have a given strength, depends primarily upon the reduced quality factor $\overline{Q}$ and upon the signal amplitude.

The contribution to $S(\tau)$ of the second component in (63) varies with τ^{-1} for integration times larger than the time constant of the output filter. In the above example the latter is less than one microsecond; and at $\tau = 1$ sec this contribution is still greater than that of the first component by more than one order of magnitude. It can be reduced, as apparent from (62) and as indicated

173

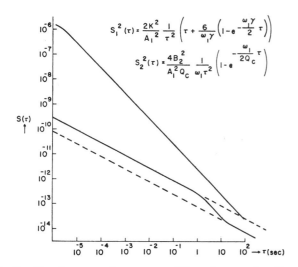

$$S_1^2(\tau) = \frac{2K^2}{A_1^2} \frac{1}{\tau^2} \left(\tau + \frac{6}{\omega_1\gamma} \left(1 - e^{-\frac{\omega_1\gamma}{2}\tau} \right) \right)$$

$$S_2^2(\tau) = \frac{4B_2^2}{A_1^2 Q_c} \frac{1}{\omega_1\tau^2} \left(1 - e^{-\frac{\omega_1}{2Q_c}\tau} \right)$$

Fig. 7. The short term frequency stability of a crystal oscillator as a function of integration time is the sum of the two solid curves. The lower curve, given by $(S_1^2)^{1/2}$, is due to effects inside the oscillator loop. It includes the effects of the random walk phenomenon in the carrier phase and of the noise response of the oscillator. The upper curve, given by $(S_2^2)^{1/2}$, is due to the additive white noise from the source in the oscillator's output (v_2 in Fig. 5). Its significance can be reduced by the use of high Q filters in the output amplifier. The numerical values of (47) apply.

previously, by increasing the quality factor of the output filter. When Q_c is larger than 10^4, the short term stability of the signal frequency at $\tau = 1$ sec is in this example essentially given by the first term in (63).

An important conclusion can be drawn from the fact that the last term in (62) is for $Q_c \ll Q_T$ independent from the properties of the oscillator feedback network and the active device characteristics. If this term is much larger than the first, the short term frequency stability of the oscillator can be expressed approximately by

$$S(\tau) \doteq \frac{1}{\tau} \sqrt{4kTR_3/\omega_1 Q_c A_1^2} = (2/\omega_1\tau)\sqrt{V_N^2/V_S^2} \quad (64a)$$

whereby

$$V_N^2 = \omega_1 B_2^2/2Q_c \quad (64b)$$

is the mean square noise voltage measured at the amplifier output when the crystal unit is disconnected, and V_S^2 the mean square output voltage when the oscillator is operating normally. The advantage of (64) for practical design work is obvious.

The relationship between the power spectrum of the signal or the spectrum of the phase and the short term frequency stability is rather simple only when $S(\tau)$ in (62) is dominated by the last component, i.e., when the signal disturbances of interest are due to the additive white noise from the oscillator output. In the general case such a relationship is again best established, as in Section III-F, via the basic parameter A_1, ω_1, Q_c, Q_T, γ, K^2, and B_2^2.

V. The Signal After Frequency Multiplication

For many applications it is necessary to multiply the frequency of the oscillator signal, and the properties of

the signal at the output of the multiplier have to be known. The short term frequency stability of the multiplied signal is still properly represented by the expressions derived in Section IV, provided the multiplier itself does not contribute significant noise components. The spectral characteristics of the signal, however, are modified by the multiplication process, and the nature of these modifications under idealized conditions will now be determined.

An ideal n times multiplier removes all amplitude disturbances and multiplies the phase of the signal by n, hence, with the signal as given by (53) as an input, the multiplier output is

$$e_m = A_1 \cos n[\omega_1 t + \phi(t)]. \quad (65)$$

It can be approximated for small $\vartheta(t)$ by

$$e_m = A_1 \cos n[\omega_1 t + \bar{\eta}(t)]$$
$$\quad - A_1[\vartheta(t) + \vartheta_n(t)] \sin n(\omega_1 t + \bar{\eta}(t))$$
$$= E_{mC} + E_{m\theta}, \quad (66)$$

and the power spectral density of e_m can be computed following the procedure used in Section III-D. One finds

$$G_{e_m e_m} = \frac{4A_1^4/n^2\omega_1^2 K^2}{1 + (4A_1^2/n\omega_1 K^2)^2(1 - \omega/n\omega_1)^2}$$
$$\quad + \frac{n^2 K^2/\gamma^2}{1 + (2n/\gamma)^2(1 - \omega/n\omega_1)^2}$$
$$\quad + \frac{n^2 B_2^2(1 + Q_c/Q_T)}{1 + (2n/\gamma_c)^2(1 - \omega/n\omega_1)^2}. \quad (67)$$

The character of the change in the PSD brought about in the multiplication process is best appreciated by first considering the limiting curves of $G_{e_m e_m}$ and of

$$G_{\bar{e}_{out}\bar{e}_{out}} = G_{CC} + G_{\theta\theta} + \tfrac{1}{2}G_{NN},$$

respectively, which are, term by term

$$G_{e_m e_m} \rightarrow (K^2/4)(1 - \omega/n\omega_1)^{-2} + (K^2/4)(1 - \omega/n\omega_1)^{-2}$$
$$\quad + (B_2^2/4Q_c^2)(1 + Q_c/Q_T)(1 - \omega/n\omega_1)^{-2}, \quad (68a)$$
$$G_{\bar{e}_{out}\bar{e}_{out}} \rightarrow (K^2/4)(1 - \omega/\omega_1)^{-2} + (K^2/4)(1 - \omega/\omega_1)^{-2}$$
$$\quad + (B_2^2/4Q_c^2)(1 + Q_c/Q_T)(1 - \omega/\omega_1)^{-2}. \quad (68b)$$

Whereas these curves are identical on a relative frequency scale, the peak densities of the noise terms increase with n^2 while the peak density of the carrier decreases with $1/n^2$ due to the multiplication. Nevertheless, the PSD of the carrier of the multiplied signal is equal to that of a signal derived from an oscillator operating at $\omega_1' = n\omega_1$ in every detail, provided A_1 and K^2 are equal.

For many practical applications it is of interest to know which conditions are potentially capable of providing the highest signal-to-noise ratio at a given frequency. The above equations are quite useful to answer a number of questions in this area. A few examples will be discussed below by comparing the N/S ratio of

174

two signals, one derived with the aid of an ideal n times multiplier from Oscillator I, which operates at ω_1, the other derived from Oscillator II which operates directly at $\omega_1' = n\omega_1$. Amplitude disturbances are assumed removed from both signals. Each oscillator is considered under two conditions I-1, I-2, and II-1, II-2, respectively, whereby 1 and 2 refer to

1) no additive noise $K^2 \gg B_2^2/2Q_c^2$
2) white additive noise $K^2 \ll B_2^2/2Q_c^2$.

Condition 1) occurs either when the oscillator is free running, or when an output filter of sufficiently high quality factor is used to suppress the additive noise to insignificant levels. Condition 2) can occur only in oscillators with large reduced quality factor $\overline{Q}$, and only over that range of frequencies away from the carrier where G_{NN} in (55b) is effectively constant (i.e., where the additive noise is essentially white).

The N/S is computed according to (50), whereby under Condition 1) only the limiting curves are considered. $H(F, f)$ is assumed to represent a rectangular window whose pass band extends from f_a to f_b and, with $F = 0$, is wide enough for P_S to be given by $A_1^2/2$ to a good approximation. F is the mean frequency separation of the filter pass band from the carrier at $nf_1 = f_1'$, defined by

$$F^2 = (f_1' - f_a)(f_1' - f_b).$$

One finds for the N/S ratios of the high frequency signals under Condition 1)

I-1 $\qquad N_{\mathrm{I}}/S_{\mathrm{I}} = (K_{\mathrm{I}}^2/A_{1\mathrm{I}}^2)(n^2 f_1^2/F^2)(f_b - f_a)$

II-1 $\qquad N_{\mathrm{II}}/S_{\mathrm{II}} = (K_{\mathrm{II}}^2/A_{1\mathrm{II}}^2)(f_1'^2/F^2)(f_b - f_a)$ $\qquad$ (69)

and under Condition 2)

I-2 $\qquad N_{\mathrm{I}}/S_{\mathrm{I}} = (2n^2 B_{2\mathrm{I}}^2/A_{1\mathrm{I}}^2)(f_b - f_a)$

II-2 $\qquad N_{\mathrm{II}}/S_{\mathrm{II}} = (2B_{2\mathrm{II}}^2/A_{1\mathrm{II}}^2)(f_b - f_a)$ $\qquad$ (70)

whereby the parameters of the two oscillators are identified by the subscripts I and II, respectively.

Evaluating the ratios $(N_{\mathrm{I}}/S_{\mathrm{I}})/(N_{\mathrm{II}}/S_{\mathrm{II}})$ permits estimation of the relative merits of deriving a desired signal from a lower frequency oscillator by multiplication or directly from an oscillator operating at the required frequency.

In many cases, the noise intensities in the two oscillators can be assumed to have about the same magnitude. The ratio $K_{\mathrm{I}}^2/K_{\mathrm{II}}^2$ is then determined primarily by the reduced quality factors $\overline{Q}_{\mathrm{I}}$ and $\overline{Q}_{\mathrm{II}}$ of the oscillator, i.e.,

$$k_{\mathrm{I}}^2/k_{\mathrm{II}}^2 \approx \overline{Q}_{\mathrm{II}}^2/\overline{Q}_{\mathrm{I}}^2)$$

while the signal amplitudes are determined by the dissipated power, i.e.,

$$A_{1\mathrm{II}}^2/A_{1\mathrm{III}}^2 \approx P_{\mathrm{I}}^2/P_{\mathrm{II}}^2.$$

If, furthermore, B_2^2 is assumed to be about an order of magnitude smaller than the sum of all sources entering K^2, the relative merits can be estimated from the following table.

TABLE I

	I 1	I 2
II 1	$\overline{Q}_{\mathrm{II}}^2 P_{\mathrm{II}}/\overline{Q}_{\mathrm{I}}^2 P_{\mathrm{I}}$	$\dfrac{n^2 F^2}{f_1'^2}\dfrac{P_{\mathrm{II}}\overline{Q}_{\mathrm{II}}^2}{P_{\mathrm{I}}}$
II 2	$\dfrac{f_1'^2}{F^2}\dfrac{P_{\mathrm{II}}}{P_{\mathrm{I}}\overline{Q}_{\mathrm{I}}^2}$	$n^2 P_{\mathrm{II}}/P_{\mathrm{I}}$

When the ratios shown in Table I are smaller than one, the Oscillator I-multiplier combination is to be preferred; when they are larger, the high frequency oscillator has the lower N/S ratio under the conditions stated. The upper left ratio shows that, if the additive noise is suppressed, the oscillator with the larger $\overline{Q}^2 P_S$ product will be superior, regardless of the multiplication ratio. If additive noise is present in both oscillators, however, the lower right ratio shows that regardless of its reduced quality factor, the lower frequency oscillator is almost invariably an inferior source. The upper right ratio indicates that a free running microwave oscillator can give a lower N/S ratio than a quartz crystal oscillator multiplier combination, if no high Q output filter is used in the crystal oscillator and n is large. The lower left ratio might be of interest when the signal derived from a quartz crystal oscillator under optimum conditions is compared to the output of a maser, where the additive noise is not readily suppressed.

Whereas these relations are valid only under a number of restrictive conditions, they do illustrate the very serious degradation of the signal properties caused by the additive noise from the source across the output of the oscillator stage. If this noise is eliminated, however, only the properties of the feedback network and the final frequency determine the N/S ratio of the signal, but not the multiplication factor.

Conclusions

A major result of this paper is the derivation of the explicit form of the output signal from a noise perturbed oscillator. It permits the complete evaluation of any desired performance aspect, such as the spectral properties of the signal and of the signal phase, as well as the short term frequency stability, and it clarifies the interrelations between them. Although a specific oscillator model was chosen to formulate the basic differential equation, the results apply to all oscillators whose frequency does not depend on signal amplitude in first order. They are expressed in terms of a few general parameters all of which, except one, can be determined from the impulse response function and the noise output of the passive feedback network alone. Only the parameter γ, which depends upon the magnitude of the instantaneous nonlinearity in the circuit, requires a knowledge of the active device characteristics.

The expression for the output signal was derived by means of a novel first-order perturbation technique which, in contrast to earlier work, does not require smoothing of the instantaneous nonlinearity. Consideration of the instantaneous nonlinearities reveals a parametric pumping action to occur within the oscillator loop which, however, affects only the low-pass equivalent of the disturbed system, but not the band-pass equivalent. Two dissimilar performance aspects are thus created and the oscillator appears to act simultaneously like a linear AGC oscillator and like a high Q passive tuned circuit. The investigation of the significance of this effect in oscillators with time variable circuit parameters and in frequency pulling, pushing, and entrainment phenomena is believed to be a very fruitful area for further research.

APPENDIX

SOLUTION OF THE PERTURBATION EQUATION

A. The General Solution

The basic equation (3) for the noise perturbed oscillator contains no approximations or restrictive conditions, other than that the circuit elements are independent of time. The perturbation equation (9)

$$\ddot{u} + \frac{L_1}{LC_2}\left(\frac{R_3 C_2}{L_1} - \frac{df(e_{g0})}{de_{g0}}\right)\dot{u}$$

$$+ \left(\omega_0^2 - \frac{L_1}{LC_2}\frac{d}{dt}\frac{df(e_{g0})}{de_{g0}}\right)u = F(t) \quad (71)$$

results from it when $u(t)$ in (8) is assumed small, so that higher order terms are negligible. A small $u(t)$ can, regardless of its time dependence, cause only small variations in e_g. Hence, (9) already contains the basic premise of the method of the slowly varying amplitude and phase [26], and this method cannot be applied again to impose additional restrictions on $u(t)$.

In the general case e_{g0}, the steady-state signal in the unperturbed oscillator can be represented as

$$e_{g0} = \sum_n [a_n \cos n(\omega_1 t + \varphi) + b_n \sin n(\omega_1 t + \varphi)], \quad (72)$$

and with it we find

$$\frac{df(e_{g0})}{de_{g0}} = \sum_n [\alpha_n \cos n(\omega_1 t + \varphi) + \beta_n \sin n(\omega_1 t + \varphi)]$$

where the α_n and β_n are (nonlinear) aggregates of the constants a_n and b_n in (72) and the coefficients of the various powers of e_{g0} when $f(e_{g0})$ in (1) is developed into a Taylor series. When $u(t)$ is then assumed to have the form

$$u(t) = \sum_n [x_n(t) \cos n(\omega_1 t + \varphi) + y_n(t) \sin n(\omega_1 t + \varphi)], \quad (73)$$

(71) can be written as

$$\sum_{n=1}^{\infty} [\ddot{x}_n + g_n(t)] \cos n(\omega_1 t + \varphi)$$

$$+ \sum_{n=1}^{\infty} [\ddot{y}_n + h_n(t)] \sin n(\omega_1 t + \varphi) = F(t) \quad (74)$$

whereby the

$$g_n(t) = g_n(\dot{x}_1, x_1, \dot{x}_2, x_2, \cdots; \dot{y}_1, y_1, \dot{y}_2, y_2, \cdots)$$

$$h_n(t) = h_n(\dot{x}_1, x_1, \dot{x}_2, x_2, \cdots; \dot{y}_1, y_1, \dot{y}_2, y_2, \cdots) \quad (75)$$

are linear functions in the $\dot{x}_q, x_r, \dot{y}_s, y_t$ with constant coefficients.

It is noted that only identities have been used in transforming (9) into (74). We can furthermore introduce the identity

$$F(t) = F(t)\cdot 1 = F_c(t)\cos(\omega_1 t + \varphi) + F_s\sin(\omega_1 t + \varphi)$$

$$F_c(t) = F(t)\cos(\omega_1 t + \varphi)$$

$$F_s(t) = F(t)\sin(\omega_1 t + \varphi) \quad (76)$$

into (74) and we find (71) to be identical with

$$[\ddot{x}_1 + g_1(t) - F_c]\cos(\omega_1 t + \varphi)$$

$$+ [\ddot{y}_1 + h_1(t) - F_s]\sin(\omega_1 t + \varphi)$$

$$+ \sum_{n=2}^{\infty} [\ddot{x}_n + g_n(t)]\cos n(\omega_1 t + \varphi)$$

$$+ \sum_{n=2}^{\infty} [\ddot{y}_n + h_n(t)]\sin n(\omega_1 t + \varphi) = 0. \quad (77)$$

Equation (77) is of the form

$$\sum_{n=0}^{\infty} G_n(t)\psi_n(t) = 0 \quad [G_0(t) = 0] \quad (78)$$

whereby $\{\psi_n(t)\}$ is a complete set of orthogonal functions in the interval $[t, t+2\pi/\omega_1]$. Solving (71) is thus reduced to determining the functions $G_n(t)$ such that (78) is satisfied for all t.

The trivial solution of (78) is

$$G_n(t) = 0 \quad n = 1, 2, 3, \cdots. \quad (79)$$

Hence, if a set of functions $\{^{(0)}x_n\}$ and $\{^{(0)}y_n\}$ is found which satisfies the simultaneous equations

$$\ddot{x}_1 + g_1(t) = F_c(t) \qquad \ddot{y}_1 + h_1(t) = F_s(t)$$

$$\ddot{x}_2 + g_2(t) = 0 \qquad \ddot{y}_2 + h_2(t) = 0$$

$$\ddot{x}_3 + g_3(t) = 0 \qquad \ddot{y}_3 + h_3(t) = 0 \quad (80)$$

$$\vdots \qquad\qquad\qquad \vdots$$

it will, inserted into (73), define a $u(t)$ which satisfies (9). The condition (79) is thus clearly sufficient to define a solution $u(t)$ of (9).

Aside from (79), however, there exists a countably infinite number of other conditions which satisfy (78). These are obtained by demanding that the terms in the sum (78) cancel in pairs, triplets, quadruplets and so on. For example, (78) is satisfied if

$$G_1(t)\psi_1(t) + G_2(t)\psi_2(t) = 0, \quad G_n(t) = 0, \quad n = 3, 4, \cdots. \quad (81)$$

This condition is met, of course, when (79) holds; but it is also met when $G_1(t) = \lambda_1\psi_2(t)$, $G_2(t) = -\lambda_1\psi_1(t)$ with λ_1

176

an arbitrary constant. This leads to

$$\ddot{x}_1 + g_1(t) = F_c + \lambda_1 \sin(\omega_1 t + \varphi),$$
$$\ddot{y}_1 + h_1(t) = F_s - \lambda_1 \cos(\omega_1 t + \varphi)$$
$$\ddot{x}_2 + g_2(t) = 0$$
$$\ddot{y}_2 + h_2(t) = 0. \qquad (82)$$
$$\vdots$$
$$\vdots$$

The set of functions which satisfies these equations can be written as $\{^{(0)}x_n + \lambda_1{}^{(1)}x_n\}\{^{(0)}y_n + \lambda_1{}^{(1)}y_n\}$ whereby the $^{(0)}x_n$, $^{(0)}y_n$ are the solutions of (80) and the $\lambda_1{}^{(1)}x_n$, $\lambda_1{}^{(1)}y_n$ are the particular integrals for the sinusoidal forcing terms in (82). Proceeding in this manner, it can be shown that any possible choice of the $G_n(t)$, other than (79), which satisfies (78), leads to solutions $\{x_n\}$ and $\{y_n\}$, which consist of the integrals of (80) plus the particular integrals in response to some (in every case well defined) combination of the ψ_n as forcing terms, multiplied by an arbitrary constant λ_j. In all cases, it is necessary, therefore, that the equations (80) be satisfied.

Considering all possible choices of the $G_n(t)$, the solution of (77) can be written as

$$\{x_n\} = \left\{ {}^{(0)}x_n + \sum_{j=1}^{\infty} \lambda_j{}^{(j)}x_n \right\}$$

$$\{y_n\} = \left\{ {}^{(0)}y_n + \sum_{j=1}^{\infty} \lambda_j{}^{(j)}y_n \right\}. \qquad (83)$$

When the development leading to (82) is traced backwards and (82) is written in the form (74), it is found that the left-hand side of (74) remains unaltered, while the right-hand side becomes

$$F(t) + \lambda_1[\sin(\omega_1 t + \varphi)\cos(\omega_1 t + \varphi) - \cos(\omega_1 t + \varphi)\sin(\omega_1 t + \varphi)]$$
$$= F(t) + \lambda_1 0.$$

This result is true for all possible choices of the $G_n(t)$ other than (79). Hence, when the $^{(j)}x_n$, $^{(j)}y_n$ are inserted into (73), the resulting $^{(j)}u(t)$ is but the complementary function of (9), which is, of course, contained already in $^{(0)}u(t)$. All solutions of (77) which are not based on (79) are redundant. Therefore, all λ_j in (83) can be set equal to zero without affecting the solutions in any way.

It follows that the equations (80) are necessary and sufficient to determine, with (73), a solution $u(t)$ of (71), and that this solution is the general solution of that equation.

B. An Approximate Solution

The equation (77) was obtained by neglecting higher order terms in $u(t)$. Its solution will thus only give the first-order approximation to the disturbances of the oscillator signal. When the perturbing forces are thermal and shot noise, knowledge of this first-order approximation is adequate, beyond any reasonable doubt, for all practical purposes. However, to determine it fully requires solving the infinite set of simultaneous linear equations (80), a task which clearly calls for an iteration procedure.

The zero-order solution is found by setting all $\dot{x}_n$, x_n; $\dot{y}_n$, y_n for $n \geq 2$ equal to zero; that is, by assuming $u(t)$ to be given by (14):

$$u(t) = x_1 \cos(\omega_1 t + \varphi) + y_1 \sin(\omega_1 t + \varphi). \qquad (84)$$

Then x_1 and y_1 are to be determined from the first two equations in (80); that is, from the equations which result when the coefficients of the fundamental frequency terms in (77) are required to be zero for all t. These equations are of the form

$$\ddot{x}_1 + c_1\dot{x}_1 + c_2 x_1 + c_3\dot{y}_1 + c_4 y_1 = F_c$$
$$\ddot{y}_1 + d_1\dot{x}_1 + d_2 x_2 + d_3\dot{y}_1 + d_4 y_1 = F_s \qquad (85)$$

whereby the c_k and d_k are constants which depend upon the parameters of the nonlinear current voltage characteristic of the active device and upon the harmonic amplitudes of e_{g0}. The latter, of course, must be determined from the homogeneous part of (3).

The evaluation of these constants becomes very simple if it is justified to approximate e_{g0} by its fundamental component, as in (13). It then is also adequate to approximate the Taylor series $f(e_{g0})$ by (16), since the second power in e_{g0} contributes no fundamental component to (77), and the fourth and higher order terms are usually small enough to be neglected in nearly harmonic oscillators.

With these approximations the equations (85) become

$$\ddot{x}_1 + 2\omega_1\dot{y}_1 + 2\omega_1\gamma\dot{x}_1 = F_c$$
$$\ddot{y}_1 - 2\omega_1\dot{x}_1 - 2\omega_1{}^2\gamma x_1 = F_s \qquad (86)$$

which are most easily verified when (84) and (76) are inserted into the Mathieu equation (21). And, to determine $u(t)$ to within these approximations, it remains to solve (86) when $F(t)$ is given by (5) with (12).

The impulse response functions of the system (86) can be found in the following manner. With

$$i_s = 0, \quad v_1 = v_2 = 0, \quad v_3 = a_{3k}\delta(t - t_k) \qquad (87)$$

F_c and F_s in (76) become

$$F_c = (\omega_1{}^2/\overline{Q})a_{3k}\delta(t - t_k)\cos(\omega_1 t + \varphi) \qquad (88)$$
$$F_s = (\omega_1{}^2/\overline{Q})a_{3k}\delta(t - t_k)\sin(\omega_1 t + \varphi).$$

Since the initial conditions for x_1 and y_1 and their first derivatives are zero when, at $t = t_k$ the oscillator is assumed to be in that state which would result had all previous impulses been in the infinite past, the Laplace transform of (86) is

$$(p^2 + 2\omega_1\gamma p)X_{13} + 2\omega_1 p Y_{13} = L_c e^{-pt_k}$$
$$-2\omega_1(p + \omega_1\gamma)X_{13} - p^2 Y_{13} = L_s e^{-pt_k} \qquad (89a)$$

whereby

$$L_c = (\omega_1{}^2/\overline{Q})a_{3k}\cos(\omega_1 t_k + \varphi)$$
$$L_s = (\omega_1{}^2/\overline{Q})a_{3k}\sin(\omega_1 t_k + \varphi). \qquad (89b)$$

The characteristic equation of (89a)

$$p[p^2(p + 2\omega_1\gamma) + 4\omega_1{}^2(p + \omega_1\gamma)] = 0 \qquad (90a)$$

can be approximated by

$$p(p + \omega_1\gamma)(p^2 + \omega_1\gamma p + 4\omega_1^2) = 0 \qquad (90b)$$

when $\gamma^2 \ll 4$.

The inverse transforms of the solutions X_{13} and Y_{13} of (89a) are then, except for terms with γ^2 as a factor

$$x_{13k} = -\frac{L_s}{2\omega_1} e^{-\omega_1\gamma(t-t_k)}$$

$$+ \frac{1}{2\omega_1} e^{-(\omega_1\gamma/2)(t-t_k)} \left[L_c \sin 2\omega_1(t - t_k) + L_s \cos 2\omega_1(t - t_k) \right]$$

$$+ \frac{\gamma}{2\omega_1} e^{-(\omega_1\gamma/2)(t-t_k)} \left[L_c \cos 2\omega_1(t - t_k) - \frac{L_s}{2} \sin 2\omega_1(t - t_k) \right].$$

$$\qquad (91)$$

$$Y_{13k} = \frac{L_c}{2\omega_1}$$

$$+ \frac{1}{2\omega_1} e^{-\omega_1\gamma/2(t-t_k)} \left[L_s \sin 2\omega_1(t - t_k) \right.$$

$$\left. - L_c \cos 2\omega_1(t - t_k) \right]$$

$$- \frac{\gamma}{4\omega_1} e^{-\omega_1\gamma/2(t-t_k)} \left[L_s \cos 2\omega_1(t - t_k) \right.$$

$$\left. + \frac{L_s}{2} \sin 2\omega_1(t - t_k) \right]$$

It is noted that approximating e_{g0} by (13) and $f(e_{g0})$ by (16) is a matter of convenience only and is not necessary to solve (85). Likewise, approximating (90a) by (90b), that is, assuming $\gamma^2 \ll 4$, results in a simplification of the expressions (91), but (86) can, of course, be solved for arbitrary values of γ. Since the parameter γ determines the effective bandwidth of the oscillator, it is realized that the narrow band approximation need not be introduced at any point to solve (71). It is also realized, however, that (13) will be a rather poor approximation of e_{g0} if γ is not very small. Hence, even assuming $\gamma \ll 1$ is thoroughly consistent with the approximation (13), and the terms with γ as a factor in (91) need not be considered further.

When (91), without the γ terms, is inserted into (84), one finds as the approximate solution of (21) for a single impulse from v_3,

$$u(t) = (\omega_1/2\overline{Q})a_{3k} \cos(\omega_1 t_k + \varphi) \sin(\omega_1 t + \varphi)$$

$$- (\omega_1/2\overline{Q})a_{3k} \sin(\omega_1 t_k + \varphi) e^{-\omega_1\gamma(t-t_k)} \cos(\omega_1 t + \varphi)$$

$$+ (\omega_1/2\overline{Q})a_{3k} e^{-(\omega_1\gamma/2)(t-t_k)} \sin \omega_1(t - t_k). \qquad (92)$$

When the expressions corresponding to (92) are derived for a single impulse from each of the other noise generators contributing in $F(t)$ and the results are added, the relations (24) shown in the text are obtained. The Laplace transforms of derivatives of delta functions required thereby are readily found according to established techniques [36].

C. Comparison with Prior Art Results

Previous investigations of the effects of noise in oscillators have led to results the essence of which is equivalent to the statement that the perturbation $u(t)$ is given by

$$u(t) = (\omega_1/\overline{Q})a_{3k} \cos(\omega_1 t_k + \varphi) \sin(\omega_1 t + \varphi)$$

$$- (\omega_1/\overline{Q})a_{3k} \sin(\omega_1 t_k + \varphi) e^{-\omega_1\gamma(t-t_k)} \cos(\omega_1 t + \varphi) \quad (93)$$

instead of by (92). The most conspicuous difference is the absence in (93) of the third term in (92), which is recognized as the response function of a tuned circuit of relative bandwidth γ, excited by one half of the strength of the original impulse. The impulse response (93) corresponding to the prior art agrees in form with the first two terms in (92), but is twice as strong.

Considering the discussions of the oscillator perturbations in the Sections II-D through III-C of the text it is noted that (92) will assume the form (93) when the instantaneous nonlinearity in the oscillator is zero (i.e., $\gamma = 0$), and when the amplitude disturbances are corrected by a delayed action mechanism (i.e., AGC). In that case, however, $\omega_1\gamma$ in (93) must be interpreted to mean the reciprocal of τ_a, the time required for the AGC mechanism to correct a given amplitude disturbance. (When the time constant T of the AGC loop is short, τ_a is almost directly proportional to the AGC loop gain and is nearly independent of T, a fact that follows from Golay's equations [15].)

The techniques used in the literature for the analytic treatment of noise perturbed nonlinear oscillators differ substantially [8]–[13]. But upon closer study one can find that, indeed, all include, at various stages of the development, the assumption that there are no instantaneous nonlinearities in the circuit. The present results, thus, are not in conflict with those derived earlier; they extend them to the more realistic case where the presence of instantaneous nonlinearities in the circuit is admitted.

The most commonly used procedure employs the averaging principle [14], which involves smoothing of the instantaneous nonlinearity [12]. Its application in effect reduces a differential equation of the type

$$\ddot{e} + \omega_0\left(\frac{1}{Q} - \alpha + \beta_1 e^2\right)\dot{e} + \omega_0^2 e = \omega_0^2 E(t), \qquad (94)$$

which describes rather generally the voltage e in a van der Pol [20] oscillator under the influence of a disturbing noise voltage $E(t)$ to [7]

$$\ddot{e} + \omega_0\left(\frac{1}{Q} - \alpha + \beta A^2\right)\dot{e} + \omega_0^2 e = \omega_0^2 E(t). \qquad (95)$$

A, thereby, is the amplitude of the oscillator voltage which is apparently always approximated at the very beginning by $e = AV \cos \omega t$; (95), however, no longer describes the behavior of a truly nonlinear oscillator. It refers to an oscillator with a linear active element whose effective transconductance $(\alpha - \beta A^2)$ is controlled by an amplitude sensitive mechanism with a small, but finite, time constant. The smoothing process cannot physically be accomplished within the nonlinear active element; rather, the signal from a linear oscillator must be rectified externally and the information obtained used to steer the gain of the device.

The perturbation equation corresponding to (95) with $e = e_0 + u$ is not (21), but

$$\ddot{u} + \omega_0^2 u = \omega_0^2 E(t), \qquad (96)$$

plus an additional equation, similar to that used by

Golay [15], which describes the behavior of the amplitude disturbance under the influence of the delayed action mechanism.

The Mathieu equation (21) and, hence, ultimately the solution (92) is obtained when the parametric pumping action [13] that occurs in the oscillator because the $\beta e^2 = \frac{1}{2}\beta A^2(1+\cos 2\omega t)$ term in (94) is properly represented in the perturbation equation. Conversely, when the perturbations are assumed to be given by (93), the existence of an instantaneous nonlinearity in the oscillator is ignored.

The difference between the present results and those of the earlier work can also be illustrated by tracing the development of the equations (86) back to the form (74), that is, to

$$(\ddot{x}_1 + 2\omega_1 \dot{y}_1 + 2\omega_1 \gamma \dot{x}_1)\cos(\omega_1 t + \varphi)$$
$$+ (\ddot{y}_1 - 2\omega_1 \dot{x}_1 - 2\omega_1^2 \gamma x_1)\sin(\omega_1 t + \varphi) = F(t) \quad (97)$$

which, with

$$u_a = x_1(t)\cos(\omega_1 t + \varphi)$$
$$u_\vartheta = y_1(t)\sin(\omega_1 t + \varphi) \quad (98)$$

can be rewritten as

$$\ddot{u}_a + 2\omega_1 \gamma \dot{u}_a + \omega_1^2 u_a + \ddot{u}_\vartheta + \omega_1^2 u_\vartheta = F(t). \quad (99)$$

The earlier results (93) are obtained when (99) is assumed to be identical [11] with

$$\ddot{u}_a + 2\omega_1 \gamma \dot{u}_a + \omega_1^2 u_a = F_c \cos(\omega_1 t + \varphi)$$
$$\ddot{u}_\vartheta + \omega_1^2 u_\vartheta = F_s \sin(\omega_1 t + \varphi). \quad (100)$$

While any pair of solutions u_a and u_ϑ of (100) satisfies (99), the converse is true only when γ in (99) is zero, that is, when the oscillator contains no instantaneous nonlinearity. Replacing (99) by (100) thus involves smoothing of the instantaneous nonlinearity. The first equation in (100) is then equivalent to the additional equation mentioned above in connection with (96). It represents the action of the AGC mechanism, that is, $\omega_1 \gamma$ in (100) is to be considered the reciprocal of τ_a as explained before.

According to the discussions in Section III-C, the solutions (92) and (93) are stochastically equivalent when $\gamma \leq 2\omega_1 K^2/A_1^2$. Whereas the response of the L circuit satisfies the equations (100) for any value of γ, the fact that the solution of (99) representing the response of the B circuit does not satisfy (100) becomes increasingly more important as γ exceeds $2\omega_1 K^2/A_1^2$, and the solutions (92) must be used to describe the oscillator behavior.

References

[1] A. Spälti, "Der Einfluss des thermischen Widerstandsrauschens and des Schrotteffektes auf die Störmodulation von Oscillatoren," *Bulletin des Schweizerischen Electrotechnischen Vereins*, vol. 39, pp. 419–427, June 1948.

[2] R. M. Lerner, "The effects of noise on the frequency stability of a linear oscillator," *Proc. Nat. Electr. Conf.*, vol. 7, pp. 275–280, 1951.

[3] W. A. Edson, *Vacuum Tube Oscillators*. New York: Wiley, 1953, p. 367 ff.

[4] J. P. Gordon, H. J. Zeiger, and C. H. Townes, "The maser, a new type of microwave amplifier, frequency standard and spectrometer," *Phys. Rev.*, vol. 99, pp. 1264–1274, 1955.

[5] W. A. Edson, "Noise in oscillators," *Proc. IRE*, vol. 48, pp. 1454–1466, August 1960.

[6] A. Blaquiére and P. Grivet, "Comments on normalized equations of the regenerative oscillator—noise, phase-locking and pulling,'" *Proc. IEEE (Correspondence)*, vol. 53, pp. 518–519, May 1965.

[7] P. Grivet and A. Blaquiére, "Masers and classical oscillators," *Proc. of the Symposium on Optical Masers*, p. 72, 1963.

[8] I. L. Berstein, "On fluctuations in the neighborhood of periodic motion of an auto-oscillating system," *Doklady Akad. Nauk.*, vol. 20, p. 11, 1938.

[9] P. I. Kuznetsov, R. L. Stratonovich, and V. I. Tikhonov, "The effects of electrical fluctuations on a vacuum tube oscillator," *J. Exp. Theor. Phys. USSR*, vol. 1, pp. 510–519, 1955; in Russian, vol. 28, pp. 509–523, 1955.

[10] S. M. Rytov, "Fluctuations in oscillating systems of the Thomson type," *Soviet Physics*, vol. 2, pp. 217–235, 1956; in Russian, *J. Exp. Theor. Phys. USSR*, vol. 29, pp. 304–333, 1955.

[11] A. Blaquiére, "Effet du Bruit de Fond sur la Fréquence des Auto-Oscillateurs à Lampes," *Ann. Radio Elect.*, vol. 8, pp. 36–80, January 1953. "Spectre de Puissance d'un Oscillateur Non-Linéaire Perturbé par le Bruit," *Ann. Radio Elect.*, vol. 8, pp. 153–179, August 1953.

[12] P. Grivet and A. Blaquiére, "Nonlinear effects of noise in electronic clocks," *Proc. IEEE*, vol. 51, pp. 1606–1614, November 1963.

[13] J. A. Mullen, "Background noise in nonlinear oscillators," *Proc. IRE*, vol. 48, pp. 1467–1473, August 1960.

[14] N. Kryloff and N. Bogoliuboff, *Introduction to Nonlinear Mechanics*. Princeton, N. J.: Princeton University Press, 1949, pp. 12, 14, 27, 28.

[15] M. J. E. Golay, "Normalized equations of the regenerative oscillator—noise, phase-locking, and pulling," *Proc. IEEE*, vol. 52, pp. 1311–1330, November 1964.

[16] E. J. Baghdady, R. N. Lincoln, and B. D. Nelin, "Short term frequency stability: characterization, theory, and measurement," *Proc. IEEE*, vol. 53, pp. 704–722, July 1965.

[17] E. T. Jaynes and F. W. Cummings, "Comparison of quantum and semiclassical radiation theories with application to the beam maser," *Proc. IEEE*, vol. 51, pp. 89–109, January 1963.

[18] W. E. Lamb, Jr., "Theory of an optical maser," *Phys. Rev.*, vol. 134, pp. A1429–A1450, 1966.

[19] Y. P. Yu, "Application of network theorems to transient analysis," *J. Franklin Inst.*, vol. 248, pp. 381–398, 1949.

[20] B. Van der Pol, "The non-linear theory of electrical oscillations," *Proc. IRE*, vol. 22, pp. 1051–1086, September 1936.

[21] P. Le Corbeiller, "Two-stroke oscillators," *IRE Trans. on Circuit Theory*, vol. CT-7, pp. 387–398, December 1960.

[22] W. J. Cunningham, *Introduction to Non-Linear Analysis*. New York: McGraw-Hill, 1958.

[23] J. Groszkovski, *Frequency of Self-Oscillations*. New York: Macmillan, 1964, pp. 186, 293.

[24] N. W. McLachlan, *Ordinary Non-Linear Differential Equations*. Oxford: Clarendon Press, 1950, p. 110.

[25] H. J. Reich, *Functional Circuits and Oscillators*. Princeton, N. J.: van Nostrand, 1961, pp. 327, 353.

[26] N. W. McLachlan, *Theory and Application of Mathieu Functions*. Oxford: Clarendon Press, 1947.

[27] S. Chandrasekhar, "Stochastic problems in physics and astronomy," in *Selected Papers on Noise and Stochastic Processes*, N. Wax, Ed. New York: Dover, 1954, p. 22.

[28] Ming Chen Wang and G. E. Uhlenbeck, "Theory of the Brownian motion," in *Selected Papers on Noise and Stochastic Processes*, N. Wax, Ed. New York: Dover, 1956, p. 122.

[29] S. O. Rice, "Mathematical analysis of random noise," *Bell Sys. Tech. J.*, vol. 23, pp. 282–332, 1944, and vol. 24, pp. 46–156, 1945, eq. (2.1–5).

[30] J. S. Bendat, *Principles and Applications of Random Noise Theory*. New York: Wiley, 1958, p. 79.

[31] N. Bloembergen, "Wave propagation in nonlinear electromagnetic media," *Proc. IEEE*, vol. 51, pp. 124–131, January 1963.

[32] W. R. Bennett, "Methods of solving noise problems," *Proc. IRE*, vol. 44, pp. 609–638, May 1956.

[33] W. A. Edson, "Intermittent behavior in oscillators," *Bell Sys. Tech. J.*, vol. 24, pp. 1–22, 1945.

[34] E. R. Chenette, "Low noise transistor amplifiers," *Solid State Design*, vol. 5, pp. 27–30, February 1964.

[35] The importance of auxiliary filtering in improving the spectrum was already pointed out by Edson in [5] above.

[36] W. B. Davenport, Jr. and W. L. Root, *An Introduction to the Theory of Random Signals and Noise*. New York: McGraw-Hill, 1958.

A Simple Model of Feedback Oscillator Noise Spectrum

INTRODUCTION

This letter contains brief thoughts on the following points.

1) The relationships among four commonly used spectral descriptions of oscillator short-term stability or noise behavior.
2) A heuristic derivation, presented without formal proof, of the expected spectrum of a feedback oscillator in terms of known oscillator parameters.
3) Some experimental results which illustrate the validity of the simple model.
4) Comments on the effect of nonlinearity, specific spectral requirements for several applications, choice of resonator frequency and active element, and expected spectrum characteristics of several oscillator types.

SPECTRAL MODELS OF PHASE VARIATIONS

Consider a stable oscillator whose measurable output can be expressed as

$$v(t) = A \cos \left[\omega_0 t + \phi(t) \right].$$

It is common to treat $\phi(t)$ as a zero-mean stationary random process describing deviations of the phase from the ideal. The frequency domain information about phase or frequency variations is contained in the "power" spectral density $S_{\dot\phi}(\omega_m)$ of the phase $\phi(t)$ or, alternatively, in the "power" spectral density $S_\phi(\omega_m)$ of the frequency $\dot\phi$. By analogy to modulation theory, we use ω_m to mean the modulation, video, baseband, or offset frequency associated with the noise-like variations in $\phi(t)$. The units of $S_\phi(\omega_m)$ are radians²/cps bandwidth or dB relative to 1 radians²/cps BW; $S_{\dot\phi}(\omega_m)$ is expressed in (radians/sec)² per c/s BW [1] [2]. The two are related by $S_{\dot\phi}(\omega_m) = \omega_m{}^2 S_\phi(\omega_m)$.

$S_{\dot\phi}(\omega_m)$ can also be expressed in terms of the equivalent rms frequency deviation Δf_{rms} in a given video bandwidth. Further, subject to the limitations that $\overline{\phi^2} \ll 1$ (small total modulation index) and that AM $\ll$ FM components, the normalized RF power spectrum $G(\omega - \omega_0)$ is identical to the two-sided spectrum of the phase $S_\phi(\omega_m)$; i.e., RF sidebands relative to the carrier are down by $S_\phi(\omega_m)$ expressed in decibels relative to 1 radian²/BW.

RELATION TO OSCILLATOR INTERNAL NOISE

A basic requirement on an oscillator noise model is that it show clearly the relationship of the spectrum of the phase $S_\phi(\omega_m)$ to the known or expected noise and signal levels and resonator characteristics of the oscillator. A simple picture can be constructed using a model of a linear feedback oscillator. Minor corrections to the results are necessary to account for nonlinear effects which must be present in a physical oscillator. Assume a single resonator feedback network of fractional bandwidth $2B/\omega_0 = 1/Q$, where Q is the operating, or loaded, quality factor. For small phase deviations at video rates which fall within the feedback half-bandwidth $\omega_0/2Q$, a phase error at the oscillator input due to noise or parameter variations results in a frequency error determined by the phase-frequency relationship of the feedback network, $\Delta\theta = 2Q\dot\phi/\omega_0$. Thus, for modulation rates *less* than the half-bandwidth of the feedback loop, the spectrum of the *frequency* $S_{\dot\phi}(\omega)$ is identical (with a scale factor) to the spectrum of the uncertainty of the oscillator input phase due to noise and parameter variations. This uncertainty will be denoted $\Delta\theta(t)$, and its two-sided power spectral density $S_{\Delta\theta}(\omega_m)$.

For modulation rates *large* compared to the feedback bandwidth, a series feedback loop is out of the circuit. At these modulation rates, the power spectral density of the output *phase* $S_\phi(\omega)$ is identical to the spectrum of the oscillator input phase uncertainty $S_{\Delta\theta}(\omega_m)$.

For a physical oscillator the spectrum $S_{\Delta\theta}(\omega_m)$ of the input phase uncertainty $\Delta\theta(t)$ is expected to have two principal components. One component is due to phase uncertainties resulting from additive white noise at frequencies around the oscillator frequency, as well as noise at other frequencies mixed into the pass band of interest by

Manuscript received December 10, 1965; revised December 29, 1965

nonlinearities. The second component is due to parameter variations at video frequencies which affect the phase (such as variations in the phase shift of a transistor due to carrier density fluctuations in the base resistance). The additive noise component of $S_{\Delta\theta}(\omega_m)$ is identical to the spectral density of the noise voltage squared relative to the mean square signal voltage. For white additive noise, this component is flat with frequency. For a feedback oscillator with an effective noise figure F, the two-sided $S_{\Delta\theta}(\omega) = 2FKT/P_S$; P_S is the signal level at oscillator active element input.

The video spectrum of parameter variations is found typically to have a power spectral density varying inversely with frequency (a $1/\omega_m$ or $1/f$ spectrum). The total power spectral density of oscillator input phase errors is of the form $S_{\Delta\theta}(\omega_m) = \alpha/\omega_m + \beta$, where α is a constant determined by the level of $1/f$ variations and β is $= 2FKT/P_S$ for two-sided spectra.

To find $S_\phi(\omega_m)$ or $S_{\dot\phi}(\omega_m)$, we use the fact that

$$\text{for} \quad \omega_m < \frac{\omega_0}{2Q} \qquad S_\phi(\omega) = \left[\frac{\omega_0}{2Q} \right]^2 S_{\Delta\theta}(\omega_m)$$

$$\omega_m > \frac{\omega_0}{2Q} \qquad S_{\dot\phi}(\omega) = S_{\Delta\theta}(\omega_m).$$

A suitable composite expression is

$$S_\phi(\omega_m) = S_{\Delta\theta} \left[1 + \left(\frac{\omega_0}{2Q\omega_m} \right)^2 \right].$$

This yields an asymptotic model for $S_\phi(\omega)$ shown on log-log scales in Fig. 1.

The model can be summarized as follows.

$S_\phi(\omega_m)$ decreases with ω_m
  at 9 dB/octave up to the point where $1/f$ effects no longer predominate.
  at 6 dB/octave from that point up to the feedback loop half-bandwidth.
  at 0 dB/octave above that frequency up to a limit imposed by subsequent filtering.
$S_{\dot\phi}(\omega)$ decreases at 3 dB/octave up to the first breakpoint, is flat with frequency up to the feedback baseband bandwidth, and increases at 6 dB octave above that point.

The case where $1/f$ effects predominate only for frequencies small compared with the feedback loop bandwidth is shown here as an

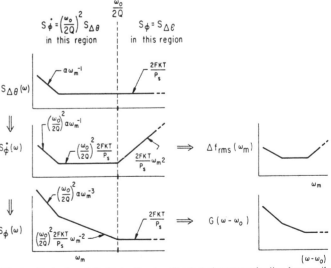

Fig. 1. Derivation of Oscillator Spectra. The logical sequence leading from oscillator parameters to spectrum characteristics is presented here. The power spectra of output phase or frequency are derived from the spectrum of input phase uncertainties and from the oscillator feedback bandwidth. The calculable constants of the oscillator are FKT, P_s, and $\omega_0/2Q$; the $1/f$ constant α is not accurately predictable but can be inferred from data. The amplitude spectrum of frequency deviation and the RF spectrum can be derived as shown, subject to limitations discussed in the text.

Reprinted from *Proc. of IEEE*, pp. 329-330, February 1966.

example. For a high-Q oscillator, $1/f$ effects in $S_{\Delta\theta}$ can predominate out to a modulation rate exceeding $\omega_0/2Q$; in this case there is no 6 dB per octave region in $S_\phi(\omega)$. A similar spectrum results where large additive noise in following amplifier stages or measuring equipment obscures the oscillator internal noise, except at very low modulation frequencies.

Note that there is a portion of the curve $S_\phi(\omega_m)$ which is proportional to $1/\omega_m^2$, leading to a $1/\omega_m$ or $1/f$ variation for rms phase deviation. This is often confused with the true $1/f$ effects associated with parameter variations leading to the $1/f$ portion of the curve for $S_{\Delta\theta}(\omega_m)$ and $S_{\dot\phi}(\omega_m)$. These two are not the same thing; "$1/f$" refers to a power spectral density rather than an amplitude spectrum.

In practice, the measurable $S_\phi(\omega_m)$ is always modified by subsequent bandlimiting filtering and by additive noise contributed by following amplifiers. It is conceivable that, for a two-terminal oscillator, the filtering action of the resonator eliminates the additive phase noise component for $\omega_m > (\omega_0/2Q)$.

Experimental Verification

Measurements were taken on a stable microwave signal source[1] designed to have a spectral purity limited only by the oscillator, which was a 100 M/s crystal oscillator. This unit employs two large-jump step recovery diode multipliers with amplification between them. The data are presented in Fig. 2 in comparison with a model derived from the following constants:

Feedback bandwidth = 16 kc/s
P_s = −4 dBm
F = 9 dB
KT = −174 dBm in 1 c/s BW
Multiplication ratio = 100 = 40 dB
$N^2 2FKT/P_s$ = +40 + 3 + 9 − 174 + 4 = −118 dB.

This leads to an asymptotic value for $S_\phi(\omega_m)$ of −118 dB relative to 1 radian²/BW in 1 c/s bandwidth, i.e., a carrier-to-sideband ratio of 118 dB. The "$1/f$" region (9 dB/octave) constant α is estimated for best data fit.

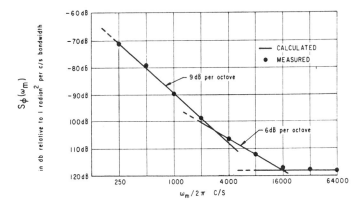

Fig. 2. $S_\phi(\omega_m)$ for Stable Microwave Signal Source. The data presented here is the average of two independent measurements which were in excellent agreement. These measurements were made at X Band on the multiplied output of a 100 Mc/s voltage controlled crystal oscillator having a 16 kc/s feedback half-bandwidth. Since this bandwidth can be reduced by a considerable factor without exceeding the present state of the art, the data is not intended to represent ultimate attainable levels, but rather serves as an illustrative example. The $1/f$ constant is chosen for best data fit. Slopes and other calculated parameters are derived from known oscillator characteristics.

Nonlinear Effects

The data was based on an estimated transistor noise figure of 9 dB. This was taken high to account for nonlinear mixing of noise at third harmonic and higher frequencies which is mixed into the pass band of interest by second harmonic periodic parameter variations

[1] 9.5 Gc/s Solid State Local Oscillator PN 31-007191, manufactured by Applied Technology, Inc., Palo Alto, Calif. Measurements are average of values measured by the author and D. J. Healey, III, Westinghouse Corp., Baltimore, Md., using Spectra Electronics SE-200 and Westinghouse proprietary noise test sets.

caused by the nonlinearity. The excellent fit of the data implies that this degradation of effective noise figure may well be an adequate description of the effect of nonlinearity.

Video Frequency Range of Interest

A number of applications which have been dealt with in this issue of the Proceedings may be summarized in terms of the video frequency range of interest. Space systems and Doppler radar applications are of particular interest to the author. For these two, interest lies in the range of a few c/s up to 100 kc/s. Space applications typically concentrate on the range where, for a crystal oscillator, $S_{\dot\phi}(\omega_m)$ is proportional to $S_{\Delta\theta}(\omega_m)$ [3], while Doppler radar applications place additional emphasis on the region above the oscillator feedback loop bandwidth [4]. Both applications typically require microwave systems which employ multiplication from the oscillator frequency.

Choice of Oscillator Frequency for Crystal Oscillator-Multiplier

It is of interest to inspect the effect of oscillator frequency upon the output spectrum of an oscillator-multiplier system having a fixed output frequency. Two assumptions which aid the calculation are a) constant oscillator input signal-to-noise ratio, and b) resonator Q varying inversely with the oscillator frequency ω_0. Under these assumptions a comparison of two oscillator frequencies yields the following results.
1) For $\omega_m < (\omega_0/2Q)$, of the lower frequency oscillator, the multiplied output $S_\phi(\omega_m)$ is identical for either choice.
2) For $\omega_m \gg (\omega_0/2Q)$, the output $S_\phi(\omega_m)$ varies as the square of the multiplication ratio (i.e., inversely as the square of the oscillator frequency).
This can be verified by a simple graphical construction.

Choice of Active Element in a Transistor Oscillator

It is apparent that $1/f$ variations and nonlinearity can have significant deleterious effects on the attainable low levels of $S_\phi(\omega_m)$. In the light of suggestions by O. Mueller that microthermal effects [6] contribute to $1/f$ noise in transistors, it is suggested that AGC oscillators using large area transistors having high power capabilities may provide simultaneous improvements in $1/f$ level and in nonlinear effects.

Spectrum Characteristics of Microwave Solid State Sources

The spectrum model given here allows simple prediction of spectrum shape and level for microwave sources of the types discussed by Johnson et al [5]. Comparison with their data shows good agreement —their measurements for crystal oscillator units extend to $\omega_m \gg (\omega_0/2Q)$, while microwave oscillators are characterized by Q factors such that, for the measurements cited, $\omega_m < (\omega_0/2Q)$.

Acknowledgment

The author is pleased to acknowledge helpful discussions with members of IEEE Subcommittee 14.7, of which this letter may be considered a brief summary. Prepublication access to all of the papers contained in this issue is also freely acknowledged. The influence of L. S. Cutler, J. A. Mullen, and W. L. O. Smith has been of special value in the preparation of this correspondence.

D. B. Leeson
Applied Technology, Inc.
Palo Alto, Calif.

References

[1] L. S. Cutler and C. L. Searle, "Some aspects of the theory and measurement of frequency fluctuations in frequency standards," this issue, page 136.
[2] E. J. Baghdady, R. N. Lincoln, and B. D. Nelin, "Short-term frequency stability: characterization, theory, and measurement" Proc. IEEE, vol. 53, pp. 704-722, July 1965.
[3] R. L. Sydnor, J. J. Caldwell, and B. E. Rose, "Frequency stability requirements for space communications and tracking systems," this issue, page 231.
[4] D. B. Leeson and G. F. Johnson, "Short-term stability for a Doppler radar: requirements, measurements, and techniques," this issue, page 244.
[5] S. L. Johnson, B. H. Smith, and D. A. Calder, "Noise spectrum characteristics of low-noise microwave tubes and solid-state devices," this issue, page 258.
[6] O. Mueller, "Thermal feedback and 1/f-flicker noise in semiconductor devices," 1965 Internat'l Solid State Circuits Conference, Digest, p. 68 ff.

Noise in Relaxation Oscillators

ASAD A. ABIDI AND ROBERT G. MEYER, FELLOW, IEEE

Abstract —The timing jitter in relaxation oscillators is analyzed. This jitter is described by a single normalized equation whose solution allows prediction of noise in practical oscillators. The theory is confirmed by measurements on practical oscillators and is used to develop a prototype low noise oscillator with a measured jitter of 1.5 ppm rms.

I. INTRODUCTION

LOW noise in the output of an oscillator is important in many applications. The noise produced in the active and passive components of the oscillator circuit adds random perturbations to the amplitude and phase of the oscillatory waveform at its output. These perturbations then set a limit on the sensitivity of such systems as receivers, detectors, and data transmission links whose performance relies on the precise periodicity of an oscillation.

A number of papers [1]–[6] were published over the past several decades in which theories were developed for the prediction of noise in high-Q LC and crystal oscillators which are widely used in high-frequency receivers. Noise in these circuits is filtered into a narrow bandwidth by the high-Q frequency-selective elements. This fact allows a relatively simple analysis of the noise in the oscillation, the results of which show that the signal to noise ratio of the oscillation varies inversely with Q.

Relaxation oscillators are an important class of oscillators used in applications such as voltage-controllable frequency and waveform generation. In contrast to LC oscillators, they require only one energy storage element, and rely on the nonlinear characteristics of the circuit rather than on a frequency-selective element to define an oscillatory waveform. These circuits have recently become common because they are easy to fabricate as monolithic integrated circuits.

Due to their broad-band nature, these oscillators often suffer from large random fluctuations in the period of their output waveforms, termed the *timing jitter*, or simply, *jitter* in the oscillator. In an application such as an FM demodulator, the relaxation oscillator in a phase-locked loop will be limited in its dynamic range, and hence sensitivity, due to this jitter. There are no systematic studies in the litera-

ture, however, on either measurements of jitter in relaxation oscillators, or on an analysis of how noise voltages and currents in the components of the oscillator randomly modulate the periodic waveform. Such an analysis was perhaps discouraged by the nonlinear fashion in which the oscillator operates, as shall become evident in Section III below.

Despite this state of affairs, circuits designers have successfully used methods based on qualitative reasoning to reduce the jitter in relaxation oscillators. The purpose of this study has been to develop an explicit background for these methods. By analyzing the switching of such oscillators in the presence of noise, circuit methods are developed to reduce the timing jitter.

The results of this study have been verified experimentally, and a prototype low jitter oscillator was built with jitter less than 2 parts per million, nearly an order of magnitude better than most commonly available circuits.

II. OSCILLATOR TOPOLOGIES AND DEFINITION OF JITTER

One of the most popular relaxation oscillator circuits is the emitter-coupled multivibrator [7] with a floating timing capacitor shown in Fig. 1, which uses bipolar transistors as the active devices. Transistors $Q1$ and $Q2$ alternately switch on and off, and the timing capacitor C is charged and discharged via current sources I. Transistors $Q3$ and $Q4$ are level shifting emitter followers, and diodes $D1$ and $D2$ define the voltage swings at the collectors of $Q1$ and $Q2$. A triangle wave is obtained across the capacitor and square waves at the collectors of $Q1$ and $Q2$.

This circuit is sometimes modified for greater stability against temperature drifts, and other types of active devices are used, but in essence it is always equivalent to Fig. 1. The oscillator operates by sensing the capacitor voltage and reversing the current through it when this voltage exceeds a predetermined threshold.

Another common relaxation oscillator is shown in Fig. 2(a), which uses a grounded timing capacitor. The charging current is reversed by the Schmitt trigger output, whose two input thresholds determine the peak-to-peak amplitude of the triangle wave across the capacitor. The block diagram of this circuit is shown in Fig. 2(b). As the ground in an oscillator is defined only with respect to a load, the circuit of Fig. 1 is also represented by the block diagram of Fig. 2(b).

Manuscript received November 16, 1982; revised April 14, 1983. This research work was supported by the U.S. Army Research Office under Grant DAAG29-80-K-0067.
A. A. Abidi is with the Bell Laboratories, Murray Hill, NJ 07974.
R. G. Meyer is with the Electronics Research Laboratory, University of California, Berkeley, CA 94720.

Reprinted from *IEEE Journal of Solid-State Circuits*, vol. SC-18, pp. 794-802, December 1983.

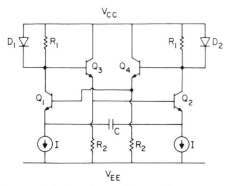

Fig. 1. Cross-coupled relaxation oscillator with floating timing capacitor.

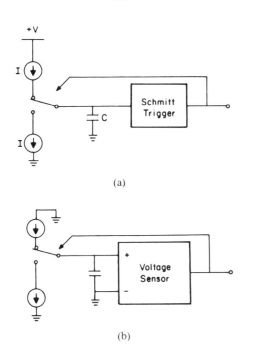

(a)

(b)

Fig. 2. (a) Relaxation oscillator with grounded timing capacitor. (b) Topological equivalent of oscillator.

The voltage sensor in such oscillators is a bistable circuit. When the capacitor voltage crosses one of the two trigger points at its input, the sensor changes state. The sensor has a vanishingly small gain, while the capacitor voltage is between these trigger points; but as a trigger point is approached, the operating points of the active devices in the sensor change in such a way that it becomes an amplifier of varying gain. The small-signal gain of the circuit is determined by an internal positive feedback loop, and becomes unboundedly large at the trigger point, causing the sensor to switch regeneratively and change the direction of the capacitor current.

In contrast to the linear voltage waveform on the capacitor (Fig. 3), the currents in the active devices of the sensor circuit are quite nonlinear because of this regeneration. The slope of these currents increases as the trigger point is approached, as shown in (Fig. 4), so that noise in the circuit is amplified and randomly modulates the time at which the circuit switches. Thus, the noisy current of Fig. 5 produces the randomly pulsewidth modulated waveform of

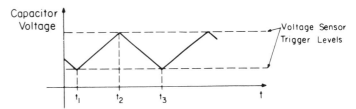

Fig. 4. Typical waveform of the current in a switching device in a relaxation oscillator.

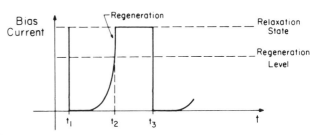

Fig. 4. Typical waveform of the current in a switching device in a relaxation oscillator.

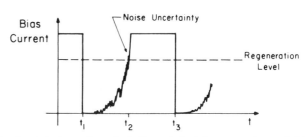

Fig. 5. Actual waveform (including noise) of the current in a switching device in a relaxation oscillator.

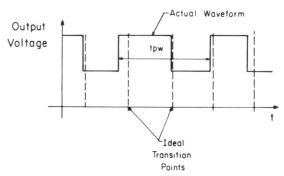

Fig. 6. Output voltage waveform of a relaxation oscillator showing the effect of noise.

Fig. 6. The timing jitter may be defined in terms of the mean μ_t and standard deviation σ_t of the pulsewidth as

$$\text{jitter} = \sigma_t / \mu_t \qquad (1)$$

which is usually expressed in parts per million.

To determine the noise in FM demodulation, it is more desirable to know the frequency spectrum of an oscillation with jitter. However, the problem is more clearly stated, and solved, in the time domain, and the spectrum of the jitter should, in principle, be obtainable from a Fourier transformation.

III. The Process of Jitter Production

The switching of a floating capacitor oscillator in the presence of noise is now analyzed by examining the circuit of Fig. 7 as it approaches regeneration. The device and parasitic capacitance are assumed to be negligibly small. Suppose $Q2$ conducts a small current I_1, while $Q1$ carries the larger current $2I_0 - I_1$. The current flowing through the timing capacitor produces a negative-going ramp at V_4, causing an increase in $V_{BE}(Q_2)$ and thus in the current through Q_2. A single stationary noise source I_n is assumed to be present in the circuit, as shown in Fig. 7. The equations describing the circuit are

$$V_{BE_1} = V_T \log_e \frac{2I_0 - I_1}{I_S} \tag{2}$$

$$V_{BE_2} = V_T \log_e \frac{I_1}{I_S} \tag{3}$$

$$V_C = (2I_0 - I_1)R + V_{BE_2} - (I_1 - I_n)R - V_{BE_1} \tag{4}$$

$$\frac{dV_C}{dt} = \frac{I_0 - I_1}{C} \tag{5}$$

where $V_T = kT/q$ and I_S is the reverse leakage current at the base of the transistor.

Substituting (2) and (3) into (4), and applying (5), the differential equation describing the circuit is

$$\left(\frac{V_T}{2I_0 - I_1} + \frac{V_T}{I_1} - 2R \right) \frac{dI_1}{dt} + \frac{R \, dI_n}{dt} = \frac{I_0 - I_1}{C}. \tag{6}$$

This may be rewritten as

$$\frac{dI_1}{dt} = \left(\frac{1}{\dfrac{V_T}{2I_0 - I_1} + \dfrac{V_T}{I_1} - 2R} \right) \left(\frac{I_0 - I_1}{C} - \frac{R \, dI_n}{dt} \right) \tag{7}$$

where the right-hand side consists of two terms, one due to the autonomous dynamics of the circuit and one due to noise. As I_1 increases, the denominator of the right-hand side diminishes until it becomes zero when

$$I_1 = I_R \approx \frac{V_T}{2R} \tag{8}$$

where $I_1 \ll I_0$; I_1 must satisfy this inequality for the circuit to oscillate. I_R is defined as the *regeneration threshold* of the circuit: upon exceeding it, and in the absence of any device or stray capacitance, I_1 would change at an infinite rate until one of the circuit voltages limits. Accordingly, the circuit is said to be in the *relaxation mode* while $0 < I_1 < I_R$, and in *regeneration* when $I_R < I_1 < 2I_0$. (I_1 may be the current through either $Q1$ or $Q2$, depending on the particular half cycle of oscillation.)

Suppose that at some time $t = t_A$ the circuit is in relaxation so that the current $I_1 = I_A$, and that it builds up the threshold of regeneration $I_1 = I_R$ at time $t = t_2$. Equation

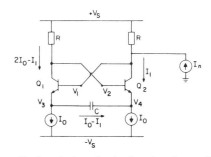

Fig. 7. Generalized equivalent circuit of a relaxation oscillator for noise analysis.

(6) can then be integrated from t_A to t_2 as follows:

$$\int_{I_A}^{I_R} \left\{ \frac{V_T}{2I_0 - I_1} + \frac{V_T}{I_1} - 2R \right\} dI_1$$
$$= \int_{t_A}^{t_2} \left\{ \frac{I_0 - I_1}{C} \right\} dt - R \{ I_n(t_2) - I_n(t_A) \} \tag{9}$$

so that

$$V_K = \frac{I_0}{C}(t_2 - t_A) - \int_{t_A}^{t_2} \frac{I_1}{C} dt - R \{ I_n(t_2) - I_n(t_A) \} \tag{10}$$

where V_K, the left-hand side of (9), is a constant which depends only on the choice of initial current I_A, I_0, and, from (8), on V_T and R.

The influence of the noise current on the instant of switching is now evident from (10): random fluctuations in the value of $I_n(t_2)$ must induce corresponding fluctuations in t_2 so that the sum of the terms on the right-hand side of (10) remains equal to the constant V_K. More precisely, if $t_A = 0$ and $t_2 = T$ (the half-period of the oscillation), then

$$F(I_n(T), T) \overset{\text{def}}{=} \frac{I_0}{C} T - \int_0^T \frac{I_1(t)}{C} dt - R \{ I_n(T) - I_n(0) \}$$
$$= \text{constant} \tag{11}$$

and thus

$$\delta F = \frac{\partial F}{\partial I_n} \bigg|_T \delta I_n(T) + \frac{\partial F(I_n, T)}{\partial T} \delta T = 0 \tag{12}$$

which implies

$$-R \delta I_n(T) + \left\{ \frac{I_0}{C} - \frac{I_1(T)}{C} \right\} \delta T = 0 \tag{13}$$

so that

$$\delta T = \frac{R}{\left\{ \dfrac{I_0 - I_1(T)}{C} \right\}} \delta I_n(T). \tag{14}$$

By definition, $I_1(T) = I_R$ so

$$\delta T = \frac{R}{\left\{ \dfrac{I_0 - I_R}{C} \right\}} \delta I_n(T). \tag{15}$$

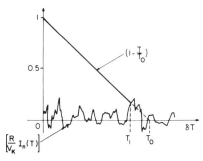

Fig. 8. Graphical interpretation of (19).

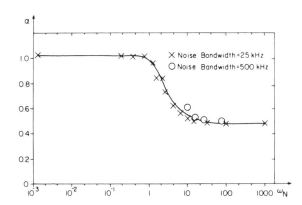

Fig. 9. Measured values of parameter α versus normalized noise bandwidth ω_N.

The main result of this paper lies in interpreting this equation as follows: the *variation* in switching times δT is equal to the variation in the times of first intersection of a $ramp(I_0 - I_R/C)T$ with the noise waveform $RI_n(T)$, as shown in Fig. 8. Note that the x axis in this figure is T, and *not* "real" time t. Thus, while the current waveform is nonlinear, as in Fig. 5, and while the time T, at which it starts to regenerate is given by the integral equation (10), *variations* in T appear as if they were produced by the first crossing of a noisy threshold by a linear ramp.

The statistics of T are contained in (10) in terms of the distribution of $I_n(t)$. This is the well-known equation for the first-crossings' distribution of a deterministic waveform with noise [8] and does not have a convenient closed form solution.

Nevertheless, a useful empirical result has been obtained for the distribution of T which allows the jitter in relaxation oscillators to be predicted quite accurately. This result is based on the following observations.

1) The linearized equation (15) describes completely the *deviations* in T; therefore, it contains information on the statistics of T.

2) It is plausible that the jitter will be directly proportional to the rms noise current for a fixed slope of the timing waveform.

3) If the dominant noise power in the oscillator lies at high frequencies, it acts to reduce the mean period of oscillation. This is evident from Fig. 8, where the ramp will almost always cross a positive-going peak of the noise.

4) If low-frequency noise is dominant, the resulting jitter will be greater than it would be if the same noise power were concentrated at higher frequencies. Again, Fig. 8 shows this, because if noise varies slowly compared to the ramp rate, the first-crossing will occur both above and below the x axis.

These observations can be quantitatively summarized as

$$\sigma(\delta T) = \sigma(T) = \alpha \frac{RC}{I_0 - I_R}\sigma(I_n) \qquad (16)$$

where σ denotes the standard deviation of its argument and α is a constant of proportionality which by 1) and 2) above depends on whether the noise power is contained at high or at low frequencies. Equation (15) changes to (16) in going from the variation in T in one switching event to the standard deviation of an ensemble of such events.

The constant α varies with the relative slope of the ramp to the rms slope of the noise. For white noise which has been low-pass filtered, the rms slope can be defined relative to the ramp rate by ω_N, where

$$\omega_N \overset{def}{=} 2\pi \times \frac{\text{rms noise voltage} \times \text{noise bandwidth}}{\text{voltage ramp rate}} \qquad (17)$$

and where the rms noise voltage is responsible for modulating the first-crossing instant of the voltage ramp of Fig. 8.

The dependence of α on ω_N for low-pass filtered white noise is shown in Fig. 9. It was obtained from measurements on different oscillator circuits, and at varying noise levels, as described in Section V of this paper. This dependence should be the same for white noise in any relaxation oscillator. For $\omega_N \ll 1$, α approaches unity, as 4) above suggests, because the deviation in oscillator period faithfully follows the meanderings of the noise waveform. For $\omega_N \gg 1$, α is asymptotic to about 0.5 because only the positive peaks of noise determine the first-crossing, in accordance with 3).

If the effect of all the noise sources in an oscillator is represented by the single source I_n, then using the appropriate α, the jitter in its output can be predicted by (16). For example, in Fig. 7 devices $Q1$ and $Q2$, while forward-biased, act as voltage followers for the various noise voltage sources in the circuit, so that I_n is the rms sum of these noise voltages divided by the node resistance at the collector. This is true for all noise sources except the noise current flowing through the timing capacitor which is integrated into a voltage by the capacitor. As shown in Appendix I, its contribution to the jitter is usually negligible.

IV. INTERPRETATION AND GENERALIZATION OF RESULTS

We emphasize that the linear result of (16), which describes the variations in the nonlinear waveform of Fig. 5, is not merely an outcome of an incremental analysis of the problem, which would go as follows. As the loop approaches regeneration, the dc incremental gain increases and the small signal bandwidth due to the device and

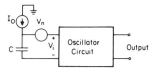

Fig. 10. Representation of noise in the relaxation oscillator by an equivalent input generator V_n.

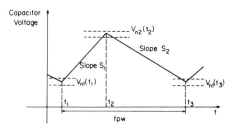

Fig. 11. Capacitor voltage waveform in a relaxation oscillator.

parasitic capacitance decreases in inverse proportion, so that the incremental gain is infinite and the bandwidth zero at the onset of regeneration. Having entered the regeneration regime, the effect of the capacitance is to limit the maximum rate of change of the circuit waveforms. Thinking in terms of signal-to-noise ratio, the incremental device current is the signal which must compete with the amplified noise in the circuit to determine the time of switching. From the considerations of gain and bandwidth above, the rms value of white noise would become infinite at the regeneration threshold, so reducing the "signal"-to-noise ratio to zero. This implies infinite jitter, which is obviously not the case in reality. Such an approach demonstrates the inadequacy of thinking of this problem in terms of small signals.

A complete, large scale analysis shows that jitter production is better understood by thinking of the oscillator as the simplified threshold circuit of Fig. 10. The equivalent noise at the input of the circuit adds to the linear voltage waveform on the timing capacitor to produce an uncertainty in the time of regeneration. It is important to realize that Fig. 10 represents the oscillator only for an incrementally small time before the onset of regeneration.

This result is independent of the type of the oscillator circuit, and of the nature of the active devices used in it.

The analysis of the grounded capacitor oscillator in Appendix II, and further generalizations given elsewhere [10] show that the jitter in any relaxation oscillator is given by the following expression:

jitter = $\dfrac{\text{rms noise voltage in series with timing waveform}}{\text{slope of waveform at triggering point}}$

$\times$ a constant. $\qquad$ (18)

V. Experimental Results

To verify the formulas for phase jitter developed above, and also to develop low noise oscillator circuits, it is necessary to be able to measure jitter with a resolution of about 1 ppm. This entails obtaining the distribution of pulsewidths from an accurate pulsewidth meter while the oscillator runs at a low frequency; the jitter is then the standard deviation of this distribution. However, the short-term thermal drift of the oscillation frequency can overwhelm the variations in cycle-to-cycle jitter, making

the measurement of the latter very difficult. By placing the oscillator in a phase-locked loop with a crystal-derived reference frequency, and by designing the loop filter with a cutoff well below the oscillation frequency, the thermal drift can be compensated, while the cycle-to-cycle jitter produced by noise frequencies above this cutoff remains unaffected. Appropriate precautions must be taken against fluctuations in the power supply, and against ground loop noise.

The effect of a noise voltage V_n over a complete cycle of oscillation is shown in Fig. 11, where the timing capacitor waveform is assumed to be asymmetrical for generality. As the device current I_1 in Fig. 7 always equals I_R at switching, the limits of the capacitor voltage have to fluctuate in response to the noise. Therefore, as the capacitor voltage is a continuous waveform, the fluctuations at times t_1, t_2, and t_3 all contribute to δT. Thus,

$$\delta T = \frac{V_n(t_1)}{S_1} + \frac{V_n(t_2)}{S_1} + \frac{V_n(t_2)}{S_2} + \frac{V_n(t_3)}{S_2}. \qquad (19)$$

The standard deviation of the random variable δT is

$$\sigma(\delta T) = \alpha \times \left\{ \text{mean value of} \left\langle \frac{V_n^2(t_1)}{S_1^2} + V_n^2(t_2)\left(\frac{1}{S_1} + \frac{1}{S_2}\right)^2 + \frac{V_n^2(t_3)}{S_2^2} \right\rangle \right\}^{1/2} \qquad (20)$$

where $V_n(t_1)$, $V_n(t_2)$, and $V_n(t_3)$ are statistically independent values of the noise voltage $V_n(t)$, and α is the constant of proportionality defined in (16). When $S_1 = S_2 = S$ in magnitude,

$$\sigma(\delta T) = \alpha\sqrt{6}\,\frac{V_n(\text{rms})}{S} \qquad (21)$$

where $\sigma(\delta T)$ is the jitter.

Measurements were made by dominating the oscillator's internal noise sources by an externally injected low-pass filtered white noise current. Two different oscillator circuits, described below, were used to obtain the experimental data. Histograms for the distributions of pulsewidths for injected noise with $\omega_N = 0.2$ and 1000 are shown in Fig. 12, where this range of ω_N was obtained by adjusting both the power and bandwidth of the injected noise. These histograms are unimodal and approximately symmetric, and they fit the normalized Gaussian function to within experimental error. Such experiments were also used to obtain the curve of α versus ω_N of Fig. 9, with the jitter defined as the standard deviation of these histograms.

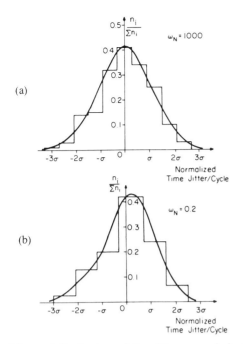

(a)

(b)

Fig. 12. Measured distributions of time jitter per cycle in the AD537 oscillator with (a) $\omega_N = 1000$ and (b) $\omega_N = 0.2$; fractional number of samples of single pulses versus their pulse widths.

To verify the general result of (18), the following experiments were performed:

1) For a fixed injected rms noise, the jitter was measured as a function of the period of oscillation; the results are shown in Fig. 13.

2) For a fixed period of oscillation, the jitter was measured for a varying rms input noise; the results are plotted in Fig. 14.

The straight line fit of the data confirmed (18). In both 1) and 2), the range of the independent variable was restricted such that $\omega_N \gg 1$, thus ensuring that α was at its lower asymptotic value of 0.48, so that variations in α did not confound the measurements. The data of Fig. 13 were obtained from the AD 537 [9] floating capacitor oscillator, with a noise voltage applied in series with the voltage reference; Fig. 14 was obtained from measurements on a discrete component grounded capacitor oscillator [10], with a noise current injected into the Schmitt trigger. The reduction in mean period of oscillation predicted by 3) in Section III was also observed.

The most important application of this theory is in determining the jitter due to the inherent noise sources in an oscillator circuit. However, it is essential to know the bandwidth that applies to these noise sources in determining their total noise power. The small-signal bandwidth changes with the approach to the regeneration threshold when the circuit starts to behave like an integrator, as discussed in Section IV. While a detailed analysis of this is beyond the scope of this paper, the situation can be examined qualitatively. Consider the circuit described by (7) in the relaxation regime, and with all devices in the active region, when, say, $I_1 = 0.1 \times I_R$. If the spectrum of the superposition of all the noise sources on I_1 is considered as the "output" noise variable, the noise bandwidth of

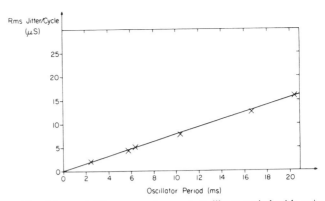

Fig. 13. Measured jitter per cycle versus oscillator period with noise injected into the AD537.

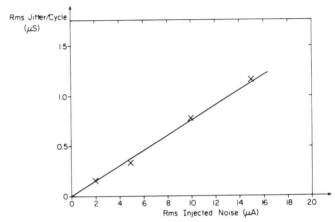

Fig. 14. Measured jitter per cycle versus injected noise amplitude in a grounded capacitor oscillator.

the circuit can be defined as the 3 dB down frequency of this spectrum. The equivalent input noise source of Fig. 10 is then this superposed noise in I_1 referred to a voltage source in series with the capacitor.

As I_1 approaches I_R, the circuit acts more like an integrator, so that fluctuations in the switching instant T are proportional to the fluctuations in the initial conditions of the integrator. The latter are simply produced by the noise in the circuit when the approach towards regeneration is started, which may roughly be defined as the time when $I_1 = 0.1 \times I_R$.

The spectral density of some of the noise sources in the circuit will depend on I_1, but usually they make a small contribution to the total noise, and their value at $0.1 \times I_R$ is a good approximation.

The noise bandwidth of the oscillator can be measured experimentally in two ways. The first relies on the availability of a white noise source with an adjustable output filter whose cutoff extends beyond the bandwidth to be measured. In response to a constant injected noise power, with the filter cutoff being progressively increased, the jitter will drop by 3 dB at the noise bandwidth of the oscillator. Alternatively, if the noise source has a fixed cutoff frequency, it can be used to modulate the amplitude of a carrier frequency, which is then injected into the oscillator. The random modulation will produce jitter, and as the carrier frequency is increased, the jitter will reduce by 3 dB at the noise bandwidth. This relies on the fact that while the jitter is determined by the amplitude fluctuations (Fig. 8), the frequency components of the injected noise are concentrated around the carrier frequency, and increase with it.

As an example, the inherent jitter of the AD 537 VCO can be predicted. The noise bandwidth was measured to be 16 MHz using the second method described above. This 16 MHz value gives a fair agreement with SPICE noise simulations of the circuit biased at $I_1 = 0.1 \times I_R$, despite the fact that the exact values of the device capacitances were not available. The noise spectral density was simulated to be 1.3×10^{-8} V/$\sqrt{\text{Hz}}$ using SPICE, and was primarily due to the current sources and the base resistances of the level shift and switching transistors. Thus,

$$V_n = 1.3 \times 10^{-8} \times \sqrt{16 \times 10^6} = 52 \ \mu\text{V rms}$$

At an oscillation frequency of 1 kHz, the slope of the timing capacitor ramp was 2.6×10^3 V/s, giving $\omega_N = 2.01$ and thus $\alpha = 0.8$ from Fig. 9. The predicted jitter is

$$\sigma(T) = \sqrt{6} \times 0.8 \times \frac{52 \times 10^{-6}}{2.6 \times 10^3}$$
$$= 39 \text{ ns}$$
$$= 39 \text{ ppm}.$$

The measured jitter from several samples had a mean value of 35 ppm, an excellent agreement in view of the approximate values of the active device noise models.

VI. A Low Jitter Oscillator Circuit

Many applications require even lower values of jitter than that of the AD 537 which is one of the lowest jitter monolithic VCO's widely available. The theory developed in this paper allows oscillators to be designed to a specified noise performance; as an example, a circuit with 1 ppm jitter was designed.

In the block diagram of Fig. 10, suppose that the timing voltage on the capacitor is a triangle wave with a peak-to-peak value V and period T. The slope of the ramp is

$$S = 2V/T \qquad (22)$$

so the fractional jitter is

$$J = \frac{\alpha\sqrt{6}}{2} \frac{V_n}{V} \qquad (23)$$

where V_n is the rms noise voltage. Thus, to obtain a small jitter, it is necessary to reduce V_n and increase V within the constraints of the circuit. V is limited by the power supply of the circuit, and V_n depends on both the characteristics of the active devices and the circuit topology. In the AD 537 topology (Fig. 15), for example, many devices additively contribute to the total noise in series with the timing capacitor because the functions of regeneration and threshold voltage detection are combined into one circuit.

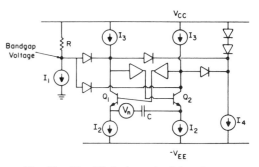

Fig. 15. Simplified schematic of the AD537.

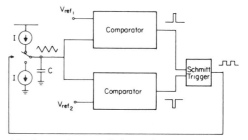

Fig. 16. Low-noise grounded capacitor oscillator topology.

In the grounded capacitor circuit [Fig. 2(a)], however, the current switch is separate from the regenerative Schmitt trigger; their noise contributions can thus be minimized independently.

Fig. 16 shows a variation of this topology where the Schmitt trigger is driven by fast pulses produced by high gain comparators following the timing capacitor. The result (18) shows that the contribution of the noise in the Schmitt trigger to the total jitter is inversely proportional to the slope of the waveform which drives it, and so is very small. Instead, the input noise of the comparators, which are driven by a slow ramp, determines the jitter. This scheme was used because very low input noise comparators are easily available, although there is no fundamental reason why a Schmitt trigger of comparable noise could not be designed. The complete schematic of the discrete component oscillator circuit is shown in Fig. 17. The differential comparators have a gain of 100 and an equivalent input noise of 6.3 μV rms over a noise bandwidth of 75 kHz. Using ± 6 V power supplies, the timing capacitor waveform was 8.8 V peak-to-peak, so (23) predicted the jitter to be 0.9 ppm rms. At an oscillation frequency of 1 kHz, the cycle-to-cycle jitter was measured to be 1.5 ppm rms; the additional jitter probably came from inadequate decoupling in the circuit.

An oscillator working from a single 5 V power supply with a jitter of about 1 ppm would be desirable in many systems applications. Such a circuit has been developed [12] on the basis of the results above, and the jitter measured to be less than 1 ppm at a 1 kHz oscillation frequency.

VII. Conclusions

Noise in relaxation oscillators can be described by a single normalized equation, which allows the jitter in any

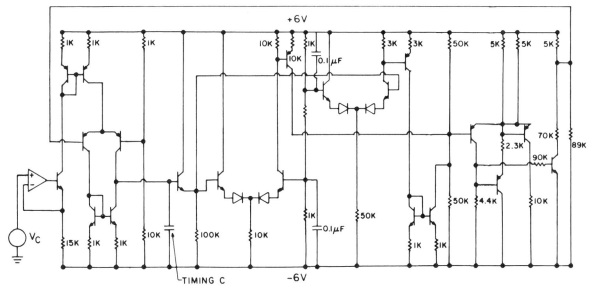

Fig. 17. Complete schematic for the low noise grounded-capacitor oscillator.

such oscillator to be predicted. The importance of this equation is that it linearizes the nonlinear regenerative waveforms in the oscillator. Further, it suggests those circuit topologies which promise low jitter, as demonstrated by an experimental prototype, whose jitter was measured to be 1.5 ppm at 1 kHz.

APPENDIX I

Effect of Noise Current Through the Timing Capacitor

In the schematics of Fig. 1 and Fig. 2 it can be seen that noise in the current sources appears directly in series the timing capacitor. In practice, these current sources are almost always active sources whose noise can be represented by a band-limited output noise current generator $I_{nc}(t)$. The capacitor then acts as a low-pass filter and the contribution V_{nc} from I_{nc} to the equivalent input noise voltage V_n of the oscillator (see Fig. 10) is

$$V_{nc}(t_r) = \frac{1}{C}\int_0^{t_r} I_{nc}(t)\, dt \qquad (A1)$$

where t_r is the time during which I_{nc} charges C. If I_{nc} is subject to a single-pole frequency roll off with bandwidth B and flat spectral density $S_{nc}(f)$, then it can be shown that [11]

$$(V_{nc})(\text{rms}) = \frac{1}{C}\sqrt{S_{nc}(f)} \times \sqrt{t_r} \qquad (A2)$$

where it is assumed that $t_r \gg 1/B$.

The rms noise contribution in V_n due to I_{nc} as given in (A2) can be compared with the contribution V_{nn} from I_n at the collector

$$(V_{nn})(\text{rms}) = (I_n)(\text{rms})R.$$

The relative importance of these two terms is now ex-

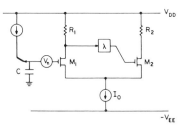

Fig. 18. Grounded capacitor oscillator's Schmitt trigger using MOS transistors.

amined. Taking some typical values, if 1 mA current sources produce the shot noise,

$$S_{nc}(f) = 3.2 \times 10^{-22}\ \text{A}^2/\text{Hz}$$

and if $t_r = 1$ ms and $C = 0.5\ \mu\text{F}$, then from (A2)

$$(V_{nc})(\text{rms}) = 1.1 \times 10^{-6}\ \text{V}.$$

In comparison, when I_n is the shot noise in 1 mA of dc current, and it is band-limited to 16 MHz by the circuit capacitances, then $R = 500\ \Omega$ gives

$$(V_{nn})(\text{rms}) = \sqrt{3.2 \times 10^{-22} \times 16 \times 10^6} \times 500$$
$$= 36 \times 10^{-6}\ \text{V}.$$

Thus $V_{nc} \ll V_{nn}$, and the difference is even larger when additional contributions to V_{nn} are considered.

APPENDIX II

Noise in an MOS Grounded Capacitor Oscillator

A simplified circuit of a grounded capacitor oscillator consisting of only the timing capacitor and the Schmitt trigger is shown in Fig. 18, where the latter uses MOS active devices with square law characteristics. The current I through device $M1$ is driven by the capacitor voltage and makes the Schmitt trigger approach its regeneration threshold. The noise in the circuit is represented by the equivalent source V_n, and adds to the timing capacitor ramp in the manner of Fig. 10; it could equally well have been represented as a noise current source within the Schmitt circuit.

The circuit equations are as follows:

$$\frac{I_0 t}{C} + V_n = V_{GS_1} + V_S = V_T + \left(\frac{I}{\beta}\right)^{1/2} + V_S \quad \text{(B1)}$$

$$\lambda(V_{DD} - IR_1) = V_{GS_2} + V_S = V_T + \left(\frac{I_0 - I}{\beta}\right)^{1/2} + V_S. \quad \text{(B2)}$$

Subtracting (B2) from (B1)

$$\frac{I_0 t}{C} + V_n - \lambda V_{DD} + \lambda IR_1 = \left(\frac{I}{\beta}\right)^{1/2} - \left(\frac{I_0 - I}{\beta}\right)^{1/2} \quad \text{(B3)}$$

and differentiating (B3),

$$\frac{I_0}{C} + \frac{dV_n}{dT} + \lambda R_1 \frac{dI}{dt} = \frac{1}{2\beta}\left\{\left(\frac{\beta}{I}\right)^{1/2} + \left(\frac{\beta}{I_0 - I}\right)^{1/2}\right\} \frac{dI}{dt} \quad \text{(B4)}$$

rewriting which

$$\frac{dI}{dt} = \frac{\dfrac{I_0}{C} + \dfrac{dV_n}{dt}}{\dfrac{1}{2\beta}\left\{\left(\dfrac{\beta}{I}\right)^{1/2} + \left(\dfrac{\beta}{I_0 - I}\right)^{1/2}\right\} - \lambda R_1}. \quad \text{(B5)}$$

The regeneration threshold $I_R = 1/4\beta\lambda^2 R_1^2$ is the current at which the denominator of the right-hand side becomes zero. If this happens at $t = T$ relative to some time $t = 0$, then integrating (B5) from 0 to T gives

$$V_K = \frac{I_0 T}{C} + \{V_n(T) - V_n(0)\} \quad \text{(B6)}$$

where V_K is a constant voltage, depending only on the circuit parameters. This is exactly the same equation as (10), and gives the same result for jitter as (16).

REFERENCES

[1] J. L. Stewart, "Frequency modulation noise in oscillators," *Proc. IRE*, vol. 44, pp. 372–376, Mar. 1956.
[2] W. A. Edson, "Noise in oscillators," *Proc. IRE*, vol. 48, pp. 1454–1466, Aug. 1960.
[3] J. A. Mullen, "Background noise in nonlinear oscillators," *Proc. IRE*, vol. 48, pp. 1467–1473, Aug. 1960.
[4] M. G. E. Golay, "Monochromaticity and noise in a regenerative electrical oscillator," *Proc. IRE*, vol. 48, pp. 1473–1477, Aug 1960.
[5] P. Grivet and A. Blaquiere, "Nonlinear effects of noise in electronic clocks," *Proc. IEEE*, vol. 51, pp. 1606–1614, Nov. 1963.
[6] J. Rutman, "Characterization of phase and frequency instabilities in precision frequency sources: Fifteen years of progress," *Proc. IEEE*, vol. 66, pp. 1048–1075, Sept. 1978.
[7] A. B. Grebene, "The monolithic phase-locked loop–A versatile building block," *IEEE Spectrum*, vol. 8, pp. 38–49, Mar. 1971.
[8] D. Middleton, *Statistical Communication Theory*. New York: McGraw-Hill, 1960.
[9] B. Gilbert, "A versatile monolithic voltage-to-frequency converter," *IEEE J. Solid-State Circuits*, vol. SC-11, pp. 852–864, Dec. 1976.
[10] A. A. Abidi, "Effects of random and periodic excitations on relaxation oscillators," Univ. California, Berkeley, Memo UCB/ERL M81/80, 1981.
[11] E. Parzen, *Stochastic Processes*. San Francisco: Holden-Day, 1962.
[12] T. P. Liu and R. G. Meyer, private communication.

Analysis of Timing Jitter in CMOS Ring Oscillators

Todd C. Weigandt, Beomsup Kim*, Paul R. Gray

Department of Electrical Engineering and Computer Science
University of California, Berkeley

* Department of Electrical Engineering
Korea Advanced Institute of Science and Technology

Abstract

In this paper the effects of thermal noise in transistors on timing jitter in CMOS ring-oscillators composed of source-coupled differential resistively-loaded delay cells is investigated. The relationship between delay element design parameters and the inherent thermal noise-induced jitter of the generated waveform are analyzed. These results are compared with simulated results from a Monte-carlo analysis with good agreement. The analysis shows that timing jitter is inversely proportional to the square root of the total capacitance at the output of each inverter, and inversely proportional to the gate-source bias voltage above threshold of the source-coupled devices in the balanced state. Furthermore, these dependencies imply an inverse relationship between jitter and power consumption for an oscillator with fixed output period. Phase noise and timing jitter performance are predicted to improve at a rate of 10 dB per decade increase in power consumption.[1]

I . Introduction

Ring oscillators are widely used in phase-locked-loops (PLL) for clock and data recovery, frequency synthesis, clock synchronization in microprocessors, and many applications which require multi-phase sampling [1] [2]. In many such applications, clock signals are generated to drive mixers or sampling circuits in which the random variation of the sampling instant, or jitter, is a critical performance parameter. In some applications the frequency domain equivalent of jitter, called phase noise, is important. A block diagram of a typical PLL using a ring-oscillator for multi-phase clock generation is shown in figure 1. Jitter requirements in typical applications range from on the order of 100 picoseconds r.m.s. down to less than 5 picoseconds in very high-speed communications receivers, for example.

Jitter can arise from many sources, including inadvertent injection of signals from other parts of the circuit through the power supply. However, interfering sources like these can often be minimized by the use of circuit techniques such as differential implementations. In a fully optimized design the main source of timing jitter is the inherent thermal and/or shot noise of the active and passive devices that make up the inverter cell. 1/f noise is usually not of practical importance since it is rejected by the PLL loop filter, and does not effect the stage-to-stage delay in a DLL. Therefore minimizing the impacts of thermal and shot noise in the basic inverter cells becomes the key to attaining low timing jitter.

1. Research supported by NSF, ARPA, and the California MICRO Program

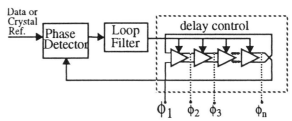

Figure 1. Ring-oscillator phase-locked-loop with multi-phase sampling

This paper attempts to determine analytically and through simulation the relationship between the design parameters of the inverter cell used in the ring-oscillator and the resulting noise-induced jitter. The class of circuits analyzed is source-coupled differential delay cells with resistive loads, implemented in CMOS technology, where the loads are realized by PMOS transistors in the triode region (figure 2). This particular implementation has proven useful in practical applications because of its high speed and rejection of supply noise [1]. In this paper we will first consider jitter for the individual delay stages in a ring-oscillator, and then look at the implications for design of the overall ring-oscillator phase-locked-loop.

II . First Order Timing Jitter Analysis

The period of a ring-oscillator is determined by the number of stages in the ring and the delay for each stage. Accompanying each cycle of oscillation is a random timing error due to noise. The goal of this section is to determine the contribution of thermal noise sources in an ECL type inverter circuit, like that shown in figure 2, to the timing jitter of the ring-oscillator.

In this analysis, each inverter stage in a ring-oscillator is assumed to contribute a nominal time delay, t_d, and a timing error, $\Delta\tau$, to each cycle of oscillation. The timing error has a mean of zero and a variance denoted by $\Delta\tau_1^2$. To first order, the delay per stage is measured from the time when the outputs begin switching to the time when the differential output reaches zero, as illustrated in figure 3. With this assumption the nominal delay per stage is given by

$$t_d \cong V_{PP} \left(\frac{C_L}{I_{SS}} \right) \qquad (1)$$

where I_{SS} / C_L is the output slew rate and V_{PP} is one half the full differential output swing. The load capacitance, C_L, is the total capacitance at the output of each inverter.

The random component of the timing delay is estimated using the first crossing approximation ([3]), illustrated in figure 4. Here, the simplifying assumption is made that the next stage begins switch-

Reprinted from *Proc. of ISCAS*, June 1994.

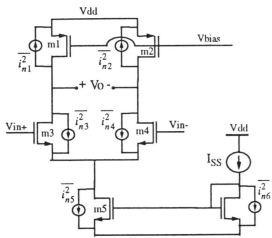

Figure 2. Differential delay cell with noise sources

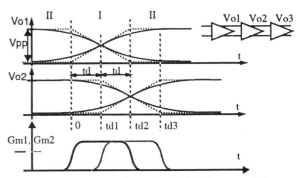

Figure 3. Output waveforms for CMOS inverter chain

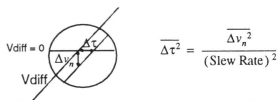

$$\overline{\Delta \tau^2} = \frac{\overline{\Delta v_n^2}}{(\text{Slew Rate})^2}$$

Figure 4. First crossing approximation for timing jitter

ing when the differential output voltage crosses zero, and the error in the actual time of crossing is the timing error passed on to future stages in the delay chain. Figure 4 shows that an error voltage at the nominal time of crossing shifts the actual time by an amount proportional to the voltage error divided by the slew rate of the output. Using this approximation, the timing error variance is given by :

$$\overline{\Delta \tau_1^2} \cong \overline{\Delta v_n^2} \times (\frac{C_L}{I_{SS}})^2 \qquad (2)$$

The voltage noise variance in equation (2) is the sum of contributions of each of the thermal noise sources in figure 2. The contribution of these noise sources to the differential output voltage is actually time varying in nature since they change as the circuit switches. In this section the simplifying assumption is made that the voltage noise variance will be the same as if the circuit where in equilibrium. In this case traditional noise analysis techniques [4] apply and the output referred voltage noise can be determined by integrating the noise spectral density over the bandwidth of the low-pass filter formed by the load resistor and the gate capacitance of the next stage. If this result is combined with (2) and (1), the r.m.s. timing jitter error for one stage normalized to the time delay per stage can be shown to be :

$$\frac{\Delta \tau_{1rms}}{t_d} \cong \frac{\Delta v_{rms}}{V_{PP}} = \sqrt{\frac{2kT}{C_L}} \bullet (\sqrt{1 + \frac{2}{3} a_v}) \bullet \frac{1}{V_{PP}} \qquad (3)$$

Interestingly, the ratio of the timing error to the time delay per stage is just given by the ratio of the r.m.s. voltage noise to the voltage swing, V_{PP}. The voltage noise has the familiar kT/C dependence, and is proportional to another term called the noise contribution factor, ξ. In this case $\xi = \sqrt{1 + (2/3) a_v}$ where a_v is the small-signal gain of the inverter. The NMOS noise contribution is given by the second term in this expression and is proportional to the gain since, for a fixed output bandwidth, higher gain implies higher transconductance and hence a larger noise contribution. The PMOS contribution is the first term which in this case is just one.

III. Second Order Analysis

The first order analysis neglects many important contributions to noise. A more thorough analysis must consider the time varying nature of the noise sources, the effects of the tail current noise sources, and interactions between stages.

Time varying noise sources

The assumption that the voltage noise variance is the same as its equilibrium value is not valid for the NMOS differential pair transistors since each side switches from fully on to fully off, during which the transconductance, and hence the noise contribution changes dramatically. Furthermore the tail current noise, although rejected by the circuit when balanced, contributes to the output voltage noise during other parts of the switching transient.

To simplify the analysis we break up the noise contributions into two piecewise constant regions of operation, as shown in figure 3. The tail current noise seen at the output is assumed zero while in balanced mode (I), and fully on during the unbalanced mode (II). The NMOS differential pair noise source contribution is approximately zero for the unbalanced mode[1], and is approximated as constant for the balanced region of operation. The contribution of the triode-region PMOS noise sources is nearly constant for both. To find the voltage noise at the output as a function the current noise sources, analysis is carried out in the time domain using autocorrelation functions and convolution. The result is a new noise contribution factor which captures the time dependence of the output voltage noise.

$$\xi = \sqrt{1 + \frac{2}{3} a_v (1 - e^{-t/\tau}) + \frac{2\sqrt{2}}{3} a_v e^{-t/\tau}} \qquad (4)$$

This equation shows that as the circuit begins switching (t=0), the diff. pair noise contribution (second term) rises exponentially from

---

1. In the unbalanced mode, one side of the differential pair is off, and the other side's contribution is reduced by emitter degeneration.

zero to its equilibrium value considered previously. The third component of the noise contribution factor is due to the tail current noise source and decays exponentially from its equilibrium value in the unbalanced mode (II) towards zero when switching begins. It can be shown that the time constant τ is approximately equal to the time delay of the stage, in which case the exponentials in this expression reduce to constants at the time of interest, t_d. This means that the noise contribution factor is relatively insensitive to most design parameters except gain.

Inter-stage Interaction

Figure 4 shows that for a typical CMOS inverter chain the switching times of adjacent stages overlap and there are times when more than one stage is in the active region of amplification. In this case it is not sufficient to consider the noise contribution of a single inverter alone since noise from one inverter may be amplified and filtered by the next stage, contributing to the jitter in the subsequent stages in that manner. A better model is to consider two successive stages, and determine the voltage noise at the output of the second stage directly from the thermal noise current sources in the first stage.

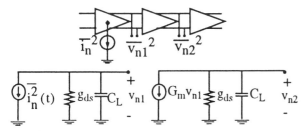

Figure 5. Extended circuit model for inter-stage interaction

Analysis for this case yields a slightly different noise contribution factor ξ than before, and an increase in the voltage noise variance by a factor of $1/2\,(a_v)^2$. With some re-arrangement, the new normalized timing jitter expression can be shown to be :

$$\frac{\Delta\tau_{1\,rms}}{t_d} = \sqrt{\frac{kT}{C_L}} \bullet \frac{1}{(V_{GS} - V_T)} \bullet \xi \qquad (5)$$

This means the normalized r.m.s. timing jitter is actually given by the ratio of the *kT/C* noise level to the gate bias voltage above threshold for the balanced state, $(V_{GS} - V_T)$. If the second term is brought under the radical, then it is apparent that this fundamental timing error can also be expressed as the ratio of the thermal noise energy level to the electrical energy stored on the gate capacitance of the next stage.

IV. Simulations

A monte-carlo approach to transient noise analysis was taken to simulate the jitter performance of ring-oscillators in SPICE. This approach includes the effects of time-varying transconductances and inter-stage interaction. Figure 6 shows that, as expected, the normalized timing jitter improves with the square root of C_L. In the graph, C_L is scaled by changing the gate width. The gate width and current are scaled proportionally so as to keep $(V_{GS}-V_T)$ constant and fix the delay per stage. Since the static power consumption is proportional to I_{SS}, jitter improves with the square root of power consump-

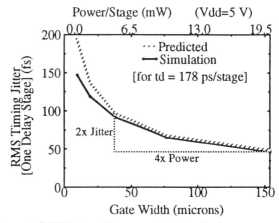

Figure 6. RMS timing jitter versus inverter size / power per stage

V. Design Implications

When designing a ring-oscillator, the parameter of interest is the jitter per cycle of oscillation. The analysis to this point has investigated the intrinsic jitter per delay stage, and we now extend these results to consider the jitter of the overall ring-oscillator. The jitter per cycle of oscillation can be used to determine the total PLL jitter for a ring-oscillator configured in a phase-locked-loop, and can also be used to predict the oscillator's phase noise spectrum.

Cycle-to-cycle jitter

Suppose the goal is to design a ring-oscillator with a fixed period, T_0, and minimal jitter. For an N-stage configuration the period of the oscillator is given by $2N \times t_d$, and the total jitter variance for once cycle of oscillation is given by $2N \times \overline{\Delta\tau_1^2}$, provided noise sources in successive stages are independent. Using the results of the last section, the jitter per cycle of oscillation, or cycle-to-cycle jitter, can be shown to be

$$\overline{\Delta\tau_N^2} = \overline{\Delta\tau_1}^2 \times \frac{T_0}{t_d} = \frac{kT}{I_{SS}} \frac{a_v\xi^2}{(V_{GS} - V_T)} \times T_0 \qquad (6)$$

where the substitution $2N = T_0 / t_d$ is used so that the jitter can be expresses as function of T_0, rather than N.

To design for low jitter, $(V_{GS}-V_{TN})$ should chosen as large as possible. The inverter gain term, a_v, is the result of inter-stage amplification consideration. For designs where this is a factor (more true of CMOS than bipolar), this implies that a for a fixed delay and fixed current, the jitter improves with lower gain per stage. Inverter gain must be kept greater than one, however, for oscillation to occur. The noise contribution factor, ξ, is a weak function of most design parameters except gain. For many CMOS designs, a_v is kept in the range of 1.5-3, and ξ ranges from 1.3 to 1.9.

The main result of equation (7) is that with everything else fixed, the timing jitter variance improves linearly with an increase in supply current. Since power consumption depends on the quiescent current level, this implies, at least for the class of circuits considered here, a direct trade-off between power consumption and timing jitter.

Interestingly, the implications of equation (7) to *first order* do not change with changes in supply voltage, technology scaling, and configuration. If $(V_{GS}-V_{TN})$ is proportional to the supply voltage, then for a constant jitter, decreasing the supply voltage requires increasing the supply current by the same amount. This means that the power consumption stays the same. Scaling of the gate length gives access to higher speeds, but equation (7) shows that for a fixed T_0, the jitter is proportional to the current itself, and does not depend directly on the gate length. Velocity saturation effects have not been neglected, to first order, either, since no form for the current equation has been assumed. Another interesting result, is that the jitter variance does not depend on the exact configuration of the oscillator itself. Each of the configurations in figure 7, for instance, have the same period if inverters with the same $(V_{GS} - V_T)$ and I_{SS} are used; but by equation (7), they also have the same jitter variance. Power is minimized, in this case, by using the configuration with as few delay stages as necessary.

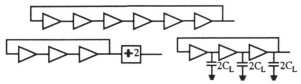

Figure 7. Multiple oscillator configurations with same period

Equation (7) shows that cycle-to-cycle jitter variance is proportional to the period, T_0, itself. A better figure of merit, however, is the jitter normalized to the period of oscillation. The r.m.s. jitter as a percentage of the output period $(\Delta\tau_{Nrms} / T_0)$ actually varies as $1 / \sqrt{T_0}$. This implies that higher frequency oscillators will have poorer jitter for the same power consumption.

Overall ring-oscillator PLL Jitter

In a ring-oscillator the variance of the timing error relative to a fixed reference transition, grows with each successive period of oscillation unless the oscillator is configured in a PLL. Analysis in [5] [6] shows that the total r.m.s. jitter when locked in a PLL will be α times the cycle-to-cycle jitter, where α is a multiplying factor which is inversely proportional to the bandwidth of the PLL. A wider bandwidth PLL corrects timing errors more quickly, resulting in a smaller overall jitter and an earlier roll-off point in figure 8. The minimum practical value of α is limited by clock feed-through, and other PLL design issues. α is typically in the range of 10-100. For a delay-locked-loop [4], jitter is not accumulated between periods, and α is effectively equal to one.

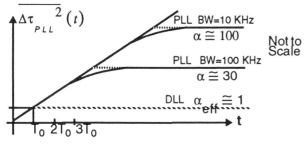

Figure 8. Jitter variance vs. time for reference transition at t=0

Ring-oscillator phase noise

Phase noise is an important figure of merit for oscillators used in RF applications. An ideal oscillator spectrum is an impulse in the frequency domain at the carrier frequency f_0. A real oscillator has its energy spread over a narrow bandwidth around f_0, and phase noise is a measure of the noise power in a 1-Hz bandwidth at a frequency f_m, offset from the carrier, relative to the total power of the oscillator. Phase noise can be determined from ring-oscillator timing jitter in a few ways. One way is to note that the accumulated phase error is a Wiener process, in which case its power spectrum can be shown to be a lorentzian. The other is to relate the spectral density of the normalized frequency fluctuations to the instantaneous error in the period of oscillation, $(\Delta\tau / T_0)$, and then use the relationships in [4] to arrive at the spectral density of phase fluctuations. With some re-arrangement of terms, the phase noise spectrum can be shown to have the following form :

$$S_\Phi(f_m) \cong \frac{f_0}{f_m^2}\left(\frac{\Delta\tau_{rms}}{T_0}\right)^2 = \left(\frac{f_0}{f_m}\right)^2 \bullet \frac{kTa_v\xi^2}{I_{SS}(V_{GS}-V_T)} \quad (7)$$

The phase noise is related to the ratio of the offset frequency to the oscillator frequency, and falls off at a rate of *20 dB / decade* for higher offsets. For a given frequency offset from carrier, the phase noise improves with higher inverter cell supply currents. Therefore phase noise is expected to improve with power consumption at a rate of 10 *dB* / decade.

VI . Conclusions

This paper has analyzed the relationship between design parameters of ECL type inverter cells and the resulting thermal-noise-induced jitter. The jitter per stage was shown to depend on the ratio of the *kT/C* noise level to the $(V_{GS}-V_T)$ bias point. The cycle-to-cycle jitter of a ring oscillator was shown to improve with larger bias currents and the normalized jitter was proportional to $1/\sqrt{T_0}$, indicating inherently higher jitter for higher speeds. Finally, the overall jitter in a PLL and the phase-noise of a ring oscillator were determined from the cycle to cycle jitter, and phase noise was predicted to improve at a rate of 10 dB / decade increase in power consumption.

VII . References

[1] B. Kim, D.N. Helman, P.R. Gray, "A 30-MHz Hybrid Analog/ Digital Clock Recovery Circuit in 2 mm CMOS," *IEEE Journal of Solid-State Circuits*, Vol. 25, No. 6, pp. 1385-1394, December 1990.

[2] M. Johnson, E.L. Hudson, "A Variable Delay Line PLL for CPU-coprocessor Synchronization," *IEEE Journal of Solid-State Circuits*, Vol. 23, No. 5, pp. 1218-1223, October 1988.

[3] A. A. Abidi, R. G. Meyer, "Noise in Relaxation Oscillators," *IEEE Journal of Solid-State Circuits*, Vol. 18, No. 6, pp. 794-802, December 1983.

[4] P. R. Gray and R. G. Meyer, "Analysis and Design of Analog Integrated Circuits", John Wiley & Sons, NY, 1984

[5] B. Kim, T. C. Weigandt, P. R. Gray, "PLL/DLL System Noise Analysis for Low Jitter Clock Synthesizer Design," *ISCAS '94 Proceedings*, June 1994.

[6] J.A. McNeill, "A 200 mW, 155 MHz, Phase-Locked-Loop with Low Jitter VCO", *ISCAS '94 Proceedings*, June 1994.

Analysis, Modeling, and Simulation of
Phase Noise in Monolithic Voltage-Controlled Oscillators

Behzad Razavi

AT&T Bell Laboratories, Holmdel, NJ07733

Abstract

In this paper, the phase noise of monolithic voltage-controlled oscillators is formulated with the aid of a linearized model. A new definition of Q is introduced and three mechanisms leading to phase noise are identified. A simulation technique using sinusoidal noise components is also described.

I. Introduction

Low-noise voltage-controlled oscillators (VCOs) are an integral part of high-performance phase-locked systems such as frequency synthesizers used in wireless tranceivers. While most present implementations of RF VCOs employ external inductors to achieve a low phase noise, the trend towards large-scale integration and low cost mandates monolithic solutions. For example, ring oscillators have been proposed as a suitable candidate [1], but their phase noise is generally known to be "high."

This paper describes an approach to analyzing, modeling, and simulating the phase noise of monolithic VCOs, with particular attention to CMOS ring oscillators. Following an analysis of a general oscillatory system and a new definition of the quality factor, Q, we employ a linearized model of ring oscillators to predict the phase noise with reasonable accuracy.

II. General Oscillatory System

Consider the linear feedback system depicted in Fig. 1. The system oscillates at $\omega = \omega_0$ if the transfer function

$$\frac{Y}{X}(j\omega) = \frac{H(j\omega)}{1 + H(j\omega)} \qquad (1)$$

goes to infinity at this frequency, e.g., $H(j\omega_0) = -1$. (We call ω_0 the "carrier frequency.") The phase noise

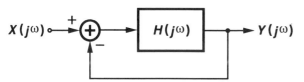

Fig. 1. General oscillatory system.

observed at the output is a function of: 1) sources of noise in the circuit, and 2) how much the feedback system rejects (or amplifies) various noise components. Modeling each source of noise as an input, $X(j\omega)$, to the system, we first quantify the latter effect for frequencies close to ω_0. If $\omega = \omega_0 + \Delta\omega$, then $H(j\omega) \approx H(j\omega_0) + \Delta\omega dH/d\omega$ and the noise transfer function is

$$\frac{Y}{X}[j(\omega_0 + \Delta\omega)] = \frac{H(j\omega_0) + \Delta\omega \dfrac{dH}{d\omega}}{1 + H(j\omega_0) + \Delta\omega \dfrac{dH}{d\omega}}. \qquad (2)$$

Since $H(j\omega_0) = -1$ and for most practical cases $|\Delta\omega dH/d\omega| \ll 1$, (2) reduces to

$$\frac{Y}{X}[j(\omega_0 + \Delta\omega)] \approx \frac{-1}{\Delta\omega \dfrac{dH}{d\omega}}. \qquad (3)$$

This equation indicates that a noise component at $\omega = \omega_0 + \Delta\omega$ is multiplied by $-(\Delta\omega dH/d\omega)^{-1}$ when it appears at the output of the oscillator. In other words, the noise power spectral density is shaped by

$$\left|\frac{Y}{X}[j(\omega_0 + \Delta\omega)]\right|^2 = \frac{1}{(\Delta\omega)^2 \left|\dfrac{dH}{d\omega}\right|^2}. \qquad (4)$$

This is illustrated in Fig. 2. As we will see later, (4) assumes a simple form for ring oscillators.

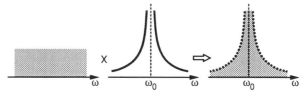

Fig. 2. Noise shaping in oscillators.

To gain more insight, let $H(j\omega) = A(\omega)\exp[j\Phi(\omega)]$. Thus, (4) can be written as

$$\left|\frac{Y}{X}[j(\omega_0 + \Delta\omega)]\right|^2 = \frac{1}{(\Delta\omega)^2[(\dfrac{dA}{d\omega})^2 + (\dfrac{d\Phi}{d\omega})^2]}. \qquad (5)$$

Reprinted from *Proc. CICC*, pp. 323-326, May 1995.

We define the open-loop Q as

$$Q = \frac{\omega_0}{2} \sqrt{(\frac{dA}{d\omega})^2 + (\frac{d\Phi}{d\omega})^2}. \quad (6)$$

The *open-loop Q* is a measure of how much the *closed-loop* system opposes variations in the frequency, as is better seen when (5) and (6) are combined:

$$\left| \frac{Y}{X}[j(\omega_0 + \Delta\omega)] \right|^2 = \frac{1}{4Q^2}(\frac{\omega_0}{\Delta\omega})^2, \quad (7)$$

a familiar form previously derived for simple LC oscillators [2]. It is interesting to note that in an LC tank at resonance, $dA/d\omega = 0$ and (6) reduces to the conventional definition of Q: $\omega_0(d\Phi/d\omega)/2$. As will be seen later, in ring oscillators $dA/d\omega$ and $d\Phi/d\omega$ are of the same order and only the more general definition proposed in (6) can be used.

III. LINEARIZED MODEL OF CMOS VCOs

Submicron CMOS technologies have demonstrated potential for RF phase-locked systems [3]. Fig. 3 shows a fully differential 3-stage ring oscillator suitable for such applications. To calculate the phase noise, we model the signal path in the VCO with a linearized (single-ended) circuit as in Fig. 4. Here, R and C represent the output resistance and the load capacitance of each stage, respectively, ($R \approx 1/g_{m3} = 1/g_{m4}$), and $G_m R$ is the gain required for steady oscillations. The noise of each differential pair and its load devices is modeled as current sources I_{n1}-I_{n3}, injected onto nodes 1-3, respectively.

Before calculating the noise transfer function, we note that the circuit of Fig. 4 oscillates if, at ω_0, each stage has unity voltage gain and 120° of phase shift. Thus, $\omega_0 = \sqrt{3}/(RC)$, and $G_m R = 2$, and the open-loop transfer function is given by

$$H(j\omega) = \frac{-8}{(1 + j\sqrt{3}\frac{\omega}{\omega_0})^3}. \quad (8)$$

Therefore, $|dA/d\omega| = 9/(4\omega_0)$ and $|d\Phi/d\omega| = 3\sqrt{3}/(4\omega_0)$. It follows from (5) that if a noise current I_{n1} is injected onto node 1 in the oscillator of Fig. 4, then its power spectrum is shaped by

$$\left| \frac{V_1}{I_{n1}}[j(\omega_0 + \Delta\omega)] \right|^2 = \frac{R^2}{27}(\frac{\omega_0}{\Delta\omega})^2. \quad (9)$$

This equation is the key to predicting various phase noise components in the ring oscillator.

IV. ADDITIVE AND MULTIPLICATIVE NOISE

Modeling the ring oscillator of Fig. 3 with the linearized circuit of Fig. 4 entails a number of issues. While

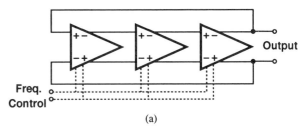

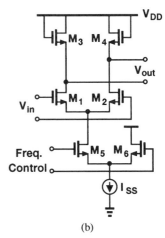

Fig. 3. CMOS VCO. (a) Block diagram, (b) implementation of one stage.

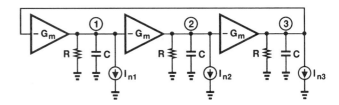

Fig. 4. Linearized model of the VCO.

the stages in Fig. 3 turn off for part of the period, the linearized model exhibits no such behavior. Furthermore, the dependence of the delay upon the tail current I_{SS} is not reflected in Fig. 4. In order to incorporate these effects, we identify three types of phase noise.

A. Additive Noise

Additive noise consists of components that are directly added to the output as shown in Fig. 2 and formulated by (4) and (9).

To calculate the additive phase noise in Fig. 4 with the aid of (9), we note that for $\omega \approx \omega_0$ the voltage gain in each stage is close to unity, and the total output phase noise power density due to I_{n1}-I_{n3} is

$$|V_{1tot}[j(\omega_0 + \Delta\omega)]|^2 = \frac{R^2}{9}(\frac{\omega_0}{\Delta\omega})^2 \overline{I_n^2}, \quad (10)$$

where it is assumed $\overline{I_{n1}^2} = \overline{I_{n2}^2} = \overline{I_{n3}^2} = \overline{I_n^2}$. For the differential stage of Fig. 3(b), the noise current per unit bandwidth is equal to $\overline{I_n^2} = 8kT(g_{m1} + g_{m3})/3 \approx 8kT/R$. Thus,

$$|V_{1tot}[j(\omega_0 + \Delta\omega)]|^2 = 8kT\frac{R}{9}(\frac{\omega_0}{\Delta\omega})^2. \quad (11)$$

Additive phase noise is predicted by the linearized model with high accuracy if the stages in the ring oscillator turn off for only a small portion of the period. In a 3-stage CMOS oscillator designed for the 900-MHz range, the differential pairs turn off for less than 10% of the period. Furthermore, at zero-crossing points —where most of the phase noise is generated— all stages are on. Therefore, the linearized model emulates the CMOS oscillator with reasonable accuracy. A simple measure of this accuracy is the error in the oscillation frequency of the model with respect to that of the actual circuit. This error remains below 10% for a 3-stage ring and 20% for a 4-stage ring.

Since additive noise is shaped according to (11), its effect is significant only for components close to the carrier frequency.

B. High-Frequency Multiplicative Noise

The nonlinearity in the differential stages of Fig. 3, especially as they turn off, causes noise components to be multiplied by the carrier (and by each other). If the input/output characteristic of each stage is expressed as $V_{out} = \alpha_1 V_{in} + \alpha_2 V_{in}^2 + \alpha_3 V_{in}^3$, then for an input consisting of the carrier and a noise component, e.g., $V_{in}(t) = A_0 \cos\omega_0 t + A_n \cos\omega_n t$, the output exhibits the following important components:

$V_{out1}(t) \propto \alpha_2 A_0 A_n \cos(\omega_0 \pm \omega_n)t,$
$V_{out2}(t) \propto \alpha_3 A_0 A_n^2 \cos(\omega_0 - 2\omega_n)t,$
$V_{out3}(t) \propto \alpha_3 A_0^2 A_n \cos(2\omega_0 - \omega_n)t.$

Note that $V_{out1}(t)$ appears in band if ω_n is small, i.e., if it is a *low-frequency* component, but in a fully differential configuration, $V_{out1}(t) = 0$ because $\alpha_2 = 0$. Also, $V_{out2}(t)$ is negligible because $A_n \ll A_0$, leaving $V_{out3}(t)$ as the only significant cross-product.

Simulations indicate that the feedback in the oscillator yields approximately equal magnitudes for $V_{out3}(t)$ and the original component at ω_n. Thus, the nonlinearity folds all the noise components below ω_0 to the region above and vice versa, effectively doubling the noise power predicted by (11). Such components are significant if they are close to ω_0 and are herein called high-frequency

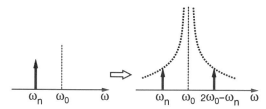

Fig. 5. High-frequency multiplicative noise.

multiplicative noise. This phenomenon is illustrated in Fig. 5.

C. Low-Frequency Multiplicative Noise

Since the frequency of oscillation in Fig. 3 is a function of the tail current in each differential pair, noise components in this current modulate the frequency, thereby contributing phase noise. Depicted in Fig. 6, this effect can be significant because, in CMOS oscillators, ω_0 must be adjustable by approximately $\pm20\%$ to compensate for process variations, thus making the frequency quite sensitive to noise in the tail current.

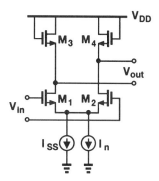

Fig. 6. Carrier modulation by tail noise current.

To quantify this phenomenon, we note that since variations in the tail current modulate the impedance of M_3 and M_4, the resistor R in the linearized circuit can be modeled as the sum of a constant term and a small current-dependent term. It can be proved that if the noise current $I_n = I_{n0}\cos\omega_n t$, then two current components described by $\frac{\sqrt{3}}{4}I_{n0}\cos(\omega_0 \pm \omega_n)t$ appear in the signal path and hence are multiplied by the transfer function in (9). Thus,

$$|V_n|^2 = \frac{R^2}{48}(\frac{\omega_0}{\Delta\omega})^2 |I_n|^2, \quad (12)$$

where $\Delta\omega = \omega_n$. This mechanism is illustrated in Fig. 7.

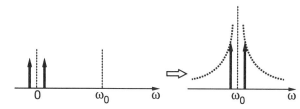

Fig. 7. Low-frequency multiplicative noise.

It is seen that modulation of the carrier brings the low frequency noise components of the tail current to the band around ω_0. Thus, flicker noise in I_n becomes particularly important.

In the differential stage of Fig. 3(b), two sources of low-frequency multiplicative noise can be identified: noise in

I_{SS} and noise in M_5 and M_6. For comparable device size, these two sources are of the same order and must be both taken into account.

V. SIMULATION

To simulate the phase noise in the oscillator and verify the accuracy of the above derivations, we use a small sinusoidal "noise" current that is injected onto different nodes of the circuit. This approach is justified by the fact that random noise can be expressed as a Fourier series of sinusoids with random phase [4].

Designed to operate at 970 MHz, each oscillator is simulated in the time domain for 2 μsec with 30-psec steps and the resulting output waveform is processed by Matlab to obtain the spectrum. Shown in Fig. 8 is the output spectrum of the linearized model in response to a sinusoidal current with 2-nA amplitude at 980 MHz. The vertical axis represents $10 \log V_{rms}^2$. The observed magnitude of the 980-MHz component is in exact agreement with (10).

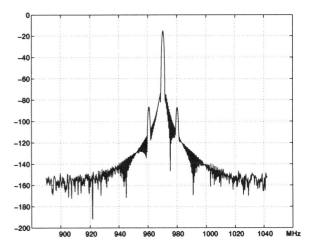

Fig. 9. Simulated spectrum of CMOS VCO.

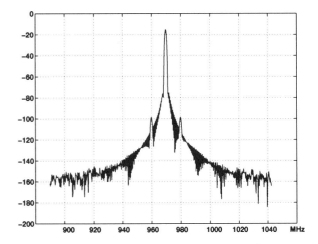

Fig. 10. Simulated spectrum of CMOS VCO with tail current noise.

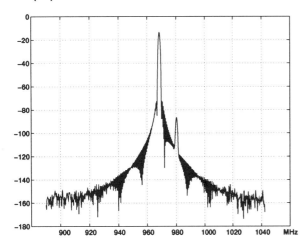

Fig. 8. Simulated spectrum of linearized model.

A similar test on the CMOS oscillator of Fig. 3 yields the spectrum in Fig. 9. Note that the component at 980 MHz has approximately the same magnitude as that in Fig. 8, indicating that the linearized model is indeed an accurate representation. As explained in Section IV, the 960-MHz component originates from third-order mixing of the carrier and the 980-MHz component and essentially doubles the phase noise.

Low-frequency multiplicative noise is simulated by modulating the tail current of one stage in the CMOS oscillator with a 2-nA 10-MHz sinusoid. The resulting sideband magnitudes, shown in Fig. 10, closely agree with (12).

VI. CONCLUSION

Analysis of a general oscillatory system leads to a linearized model of ring oscillators that predicts the phase noise with reasonable accuracy. Quantified in this paper are three mechanisms, namely, additive noise, high-frequency multiplicative noise, and low-frequency multiplicative noise, that contribute to the phase noise of oscillators. Simulations confirm the accuracy of the model and the derivations.

REFERENCES

[1] A. A. Abidi, "Radio-Frequency Integrated Circuits for Portable Communications," *Proc. CICC*, pp. 151-158, May 1994.

[2] D. B. Leeson, "A Simple Model of Feedback Oscillator Noise Spectrum," *Proc. IEEE*, pp. 329-330, Feb. 1966.

[3] B. Razavi, "A 3-GHz 25-mW CMOS Phase-Locked Loop," *VLSI Circuits Symp. Dig. Tech. Papers*, pp. 131-132, June 1994.

[4] S. O. Rice, "Mathematical Analysis of Random Noise," *Bell System Tech. J.*, pp. 282-332, July 1944, and pp. 46-156, Jan. 1945.

Part 2
Building Blocks

THE design of high-performance phase-locked systems requires a solid understanding of the limitations of the circuit building blocks. In this part, a number of papers on the design of oscillators and phase and frequency detectors are collected. The first paper describes criteria for reliable operation of high-frequency oscillators, while the next two present techniques achieving several decades of variation in the oscillation frequency.

The next three papers introduce phase/frequency detectors for periodic data, useful components for increasing the capture range of PLLs. The four papers following deal with the problem of phase and frequency detection when the data is random, an important topic in the context of clock and data recovery. The last paper in this part describes a phase detector that performs data alignment as well.

Additional References

S. K. Enam et al., "A 300-MHz CMOS voltage-controlled ring oscillator," *IEEE J. Solid-State Circuits,* vol. 25, pp. 312–315, February 1990.

K. Kato et al., "A low-power 128-MHz VCO for monolithic PLL IC's," *IEEE J. Solid-State Circuits,* vol. 23, pp. 474–479, April 1988.

J. Maneatis and M. Horowitz, "Precise delay generation using coupled oscillators," *IEEE J. Solid-State Circuits ,* vol. 28, pp. 1273–1282, December 1993.

M. Soyuer and R. G. Meyer, "Frequency limitations of a conventional phase-frequency detector," *IEEE J. Solid-State Circuits,* vol. 25, pp. 1019–1022, August 1990.

Start-up and Frequency Stability in High-Frequency Oscillators

Nhat M. Nguyen, *Student Member, IEEE*, and Robert G. Meyer, *Fellow, IEEE*

Abstract—Start-up criteria in harmonic oscillators are explored. Conventional criteria for start-up prediction are shown to be necessary but not always sufficient for high-frequency oscillators, due to the effects of parasitic elements. Design methods are derived to allow design of reliable oscillators with a well-defined frequency of oscillation. The theory is confirmed with the design, fabrication, and characterization of a 2-GHz monolithic *LC* oscillator.

I. INTRODUCTION

OSCILLATOR circuits are widely used in communication systems and instrumentation applications. Oscillators can be broadly classified into two groups: relaxation oscillators and harmonic oscillators. A *relaxation* oscillator tends to have poor phase noise characteristic and high harmonic content as the circuit switches back and forth between two astable equilibrium states [1]. A *harmonic* oscillator is capable of producing a near-sinusoidal signal with good phase noise and high spectral purity. Harmonic oscillators usually use *LC* resonant circuits [2], crystals [3], or dielectric resonators [4] for defining the oscillation frequency.

This paper is concerned with the analysis and design of reliable harmonic oscillators for use in the microwave frequency range (>1 GHz). At high frequencies, parasitic elements introduce significant excess phase shift and modify the oscillation frequency. More importantly, the parasitic elements increase the order of the system and can result in the creation of phenomena not predicted by first-order theory. Conventional low-frequency design techniques are thus not generally applicable in the design of microwave oscillators. In particular, widely used methods of predicting the existence of oscillatory behavior are not always valid and can provide misleading results. In Section II, we briefly review the two fundamental oscillator models: the feedback model and the negative-resistance model. The start-up conditions for use in these

models are reexamined in Section III and additional circuit techniques are proposed. An undesirable oscillation behavior, namely, multi-oscillation, is described in Section IV. A design procedure for reliable microwave oscillators is proposed in Section V and is confirmed with an experimental implementation.

II. OSCILLATOR MODELS

The analysis of harmonic oscillators can be based on two fundamental models: the *feedback* model or the *negative-resistance* model. Depending on the oscillator configuration and characteristic, one model may be preferred over the other. The feedback model is shown in Fig. 1 where an oscillator circuit is decomposed into a forward network and a feedback network, both of which are typically multiport networks. If the circuit is unstable about its operating point (poles in the right half of the *S* plane), it can produce an expanding transient when subject to an initial excitation. As the signal becomes large, the active devices in the circuit behave nonlinearly and limit the growth of the signal. Since an oscillator is an autonomous circuit, electronic noise in the circuit or the power-supply turn-on transient can provide the excitation that initiates the oscillation buildup. The linear behavior of a feedback circuit is typically studied using the loop-gain quantity, defined as the product of the forward and feedback transfer functions $a(s)$ and $f(s)$:

$$T(s) = a(s) f(s). \qquad (1)$$

The expression $1 - T(s) = 0$ gives the characteristic equation of the circuit from which the poles are found. For ease of reference, we use the term feedback oscillator to denote an oscillator circuit that has a well-defined feedback loop and can be analyzed using the feedback model.

The negative-resistance model is shown in Fig. 2, where an oscillator circuit is separated into a one-port active circuit and a one-port frequency-determining circuit. The function of the active circuit is to produce a small-signal negative resistance about the operating point of the oscillator and to couple with the frequency-determining circuit in defining the oscillation frequency. The frequency-determining circuit is usually a linear time-invariant circuit and is signal independent. In Fig. 2, the active and frequency-determining circuits are assumed characterized by impedance quantities $Z_a(s)$ and $Z_f(s)$, respectively. They can also be characterized in terms of an ac-

Manuscript received July 24, 1991; revised November 24, 1991. This work was supported by the U.S. Army Research Office under Grant DAAL03-87-K0079 and by a grant from IBM.

N. M. Nguyen was with the Department of Electrical Engineering and Computer Sciences and the Electronics Research Laboratory, University of California, Berkeley, CA 94720. He is now with Avantek Inc., Newark, CA 94560.

R. G. Meyer is with the Department of Electrical Engineering and Computer Sciences and the Electronics Research Laboratory, University of California, Berkeley, CA 94720.

IEEE Log Number 9106624.

Reprinted from *IEEE Journal of Solid-State Circuits*, vol. SC-27, pp. 810-820, May 1992.

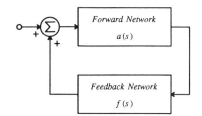

Fig. 1. Feedback model.

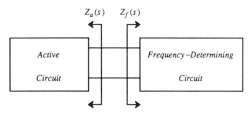

Fig. 2. Negative-resistance model.

tive admittance $Y_a(s)$ and a frequency-determining admittance $Y_f(s)$. The characteristic equation of a negative-resistance oscillator can be derived from either the expression $Z_a(s) + Z_f(s) = 0$ or $Y_a(s) + Y_f(s) = 0$. The negative-resistance model is widely used in the design of microwave oscillators due to its simplicity and the ease of characterizing the one-port negative resistance. We use the term negative-resistance oscillator to denote an oscillator circuit where a negative resistance can be identified and the circuit can be analyzed using the negative-resistance model.

As a basic requirement for producing a self-sustained near-sinusoidal oscillation, an oscillator must have a pair of complex-conjugate poles in the right-half plane (RHP)

$$p_{1,2} = \alpha \pm j\beta. \qquad (2)$$

While this requirement does not guarantee an oscillation with a well-defined steady state, it is nevertheless a necessary requirement for any oscillator. When excited by an arbitrary input, the RHP poles in (2) give rise to a sinusoidal signal with an exponentially growing envelope

$$x(t) = Ke^{\alpha t} \cos (\beta t) \qquad (3)$$

where K is set by initial conditions. The growth of this signal is eventually limited by nonlinearities in the circuit.

The application of either the feedback model or the negative-resistance model is sufficient for analyzing the linear behavior of an oscillator circuit. This analysis is essentially a study of the poles of the circuit. An oscillator must be unstable about its bias point or, equivalently, have poles in the RHP if an oscillation buildup is to take place.

III. START-UP CONDITIONS

In this section we investigate oscillator start-up criteria and propose circuit techniques for analyzing the linear oscillatory behavior. Both feedback and negative-resistance oscillators are considered.

A. Feedback Oscillators

For a feedback oscillator circuit, the fulfillment of the following conditions is often used as an indication that the circuit is unstable [5]:

$$\text{Ph } \{T(\omega_z)\} = 0 \qquad \text{Mag } \{T(\omega_z)\} > 1. \qquad (4)$$

In (4), T denotes the loop gain expression and ω_z denotes the zero-phase frequency at which the total phase shift through both the forward and feedback networks is zero. It is important, however, to note that there are oscillator circuits that meet the condition (4) but are, nevertheless, stable circuits and hence cannot produce an oscillation. Equivalently stated, the fulfillment of condition (4) does not necessarily imply that a circuit is unstable.

The above point is illustrated through use of the Pierce oscillator, whose ac schematic is shown in Fig. 3 [6]. For simplicity, circuit elements r_b, r_o, and C_μ of the bipolar transistor are neglected. After lumping the remaining parasitic elements of the transistor in with the appropriate passive elements, we can show that the loop gain of the circuit is [7]

$$T(s) = -T_o \frac{s^2LC + s\dfrac{L}{R} + 1}{a_3 s^3 + a_2 s^2 + a_1 s_1 + a_0} \qquad (5)$$

where

$$T_o = g_m R_1$$

$$a_3 = LC_2 R_1 C_1 \left(1 + \frac{C}{C_1} + \frac{C}{C_2}\right)$$

$$a_2 = L(C + C_2) + \frac{L}{R} R_1 (C_1 + C_2)$$

$$a_1 = \frac{L}{R} + R_1 (C_1 + C_2)$$

$$a_0 = 1.$$

The poles of the circuit can be determined from the roots of the expression $1 - T(s) = 0$. Fig. 4 shows the root locus as a function of the transconductance $g_m (= I_C/V_T)$. Quantities g_{m1} and g_{m2} denote the bias current range in which the circuit has a pair of RHP poles. From this plot we observe that with adequate dc loop gain ($g_{m1}R_1 \leq T_o \leq g_{m2}R_1$), the complex-conjugate poles enter the RHP from the left-half plane (LHP) of the S plane. In this situation the circuit is unstable and will produce a growing sinusoidal signal in response to an excitation. If the loop gain is too large ($g_m > g_{m2}$), however, it is interesting to note that the complex poles reenter the LHP. Under this condition the circuit is stable and would produce a decaying signal in response to an excitation. The Bode plot for the case $g_m > g_{m2}$ is shown in Fig. 5. We observe that there are two frequencies where (4) is clearly satisfied but the circuit is nevertheless stable from the locus of Fig. 4. To further confirm this observation, we generate the Nyquist plots in Fig. 6(a) and (b) for $g_{m1} < g_m < g_{m2}$ and

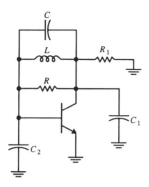

Fig. 3. Pierce oscillator ac equivalent.

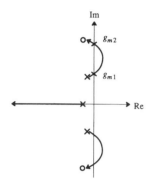

Fig. 4. Root locus for the circuit of Fig. 3.

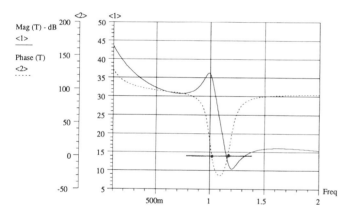

Fig. 5. Bode plot for the circuit of Fig. 3.

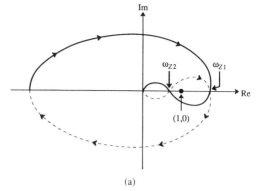

(a)

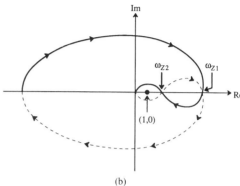

(b)

Fig. 6. Nyquist diagrams corresponding to (5). (a) $g_{m1} < g_m < g_{m2}$. (b) $g_m > g_{m2}$.

$g_m > g_{m2}$, respectively. It is helpful to recall the Nyquist criterion which can be stated in simplified form as:

If the polar plot of $T(\omega)$ plus its mirror image encircles the point $(1, 0)$ in a clockwise direction as ω varies from zero to infinity, the circuit is unstable [8].

In Fig. 6(a), there are two clockwise encirclements of the point $(1, 0)$, which indicates the existence of two RHP poles. In Fig. 6(b), the net clockwise encirclement of the point $(1, 0)$ is zero, confirming that for $g_m > g_{m2}$, the circuit is indeed stable.

The above analysis affirms that the start-up condition (4) is not always sufficient for predicting the circuit instability and should be used with the full knowledge of its limitation. As a rule of thumb, condition (4) is valid if it holds at only one frequency ω_z.

B. Negative-Resistance Oscillators

The negative-resistance model of Fig. 2 is often used in the design of microwave oscillators. Assume that the active and frequency-determining circuits are modeled by two impedances $Z_a = R_a + jX_a$ and $Z_f = R_f + jX_f$. The following start-up condition has been widely used as an indication for circuit instability [9]–[11]:

$$R_a(\omega_x) + R_f(\omega_x) < 0 \qquad (6a)$$

$$X_a(\omega_x) + X_f(\omega_x) = 0. \qquad (6b)$$

In the above equations, frequency ω_x denotes a frequency at which the total reactive component $X_a + X_f$ equals zero. It is important to distinguish the frequency ω_x from the frequency ω_z since, for any oscillator circuit, the frequency ω_x that fulfills the negative-resistance condition (6) is not necessarily the same frequency that fulfills the feedback condition (4). This point is further illustrated in the Appendix. It is also important to emphasize that an underlying assumption in (6) is that the current entering the active circuit in the steady state must be near-sinusoidal. If, instead, the voltage across the active circuit is near-sinusoidal, the active and frequency-determining circuits should be modeled in terms of parallel admittances $Y_a = G_a + jB_a$ and $Y_f = G_f + jB_f$, where G and B denote the conductance and susceptance. The dual start-up condition is then

$$G_a(\omega_x) + G_f(\omega_x) < 0 \qquad (7a)$$

$$B_a(\omega_x) + B_f(\omega_x) = 0. \qquad (7b)$$

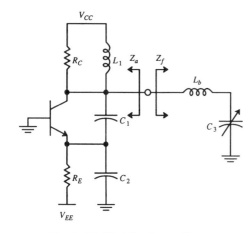

Fig. 7. Modified Colpitts oscillator.

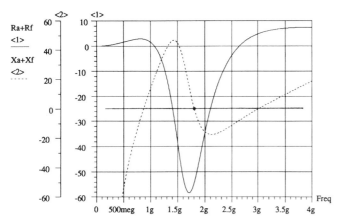

Fig. 8. Impedance plot for the circuit of Fig. 7 (C_1 = 3 pF, C_2 = 2 pF, C_3 = 4 pF, R_C = 200, L = 6 nH, L_b = 2 nH, I_C = 10 mA, 5× device).

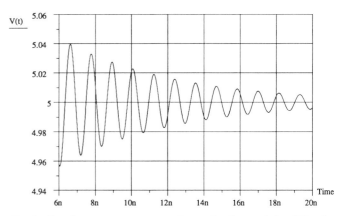

Fig. 9. Transient response corresponding to the characteristic of Fig. 8.

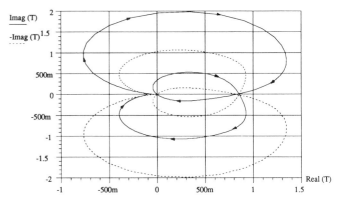

Fig. 10. Nyquist diagram corresponding to the characteristic of Fig. 8.

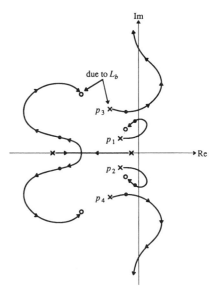

Fig. 11. Root locus (not to scale) for the circuit of Fig. 7. The poles corresponding to the characteristic of Fig. 8 are denoted by the ● symbols. This locus is a function of bias current and is based on the characteristic equation derived from the expression $Z_a(s) + Z_r(s) = 0$ of the circuit.

We now show that conditions (6) and (7) are not always valid for predicting circuit instability in negative-resistance oscillators, particularly those operating in the microwave region where the effects of parasitic elements become more significant. We consider the widely used modified Colpitts oscillator shown in Fig. 7. In this circuit, C_3 is a varactor used for tuning the oscillation frequency and L_b is parasitic series inductance. With the active and frequency-determining circuits defined as shown, the total resistance $R_a + R_f$ and reactance $X_a + X_f$ are plotted in Fig. 8 for the given set of circuit parameters. At a frequency of 1.8 GHz, condition (6) is satisfied. Without understanding the limitations of this start-up condition, one would be inclined to assert that the circuit is unstable and is capable of producing an expanding transient. However, the simulated transient response corresponding to the characteristic of Fig. 8 shows a decaying sinusoidal waveform instead of a growing waveform (Fig. 9), indicating that the circuit is stable. As a check we generate the Nyquist plot for the circuit in Fig. 10. We observe that since there is no clockwise encirclement of the point (1, 0), the circuit is indeed stable. This point is further confirmed by the root locus shown in Fig. 11 where all the poles are in the LHP. An indication that condition (6) may give misleading results in this circuit is that the

reactive plot shown in Fig. 8 crosses zero at multiple frequencies.

In summary, conventional methods of predicting oscillator start-up are not always valid and can provide misleading results. Special attention should be given when the conventional feedback condition (4) is met at multiple frequencies ω_z, or when the imaginary part of the negative-resistance condition (6) or (7) crosses zero at multiple frequencies ω_x. The Nyquist and root-locus analyses are fundamental powerful techniques for studying the linear behavior of oscillator circuits.

IV. MULTI-OSCILLATION

In the microwave region, the effect of parasitic elements in an oscillator circuit can give rise to a *multi-oscillation* phenomenon, in which two or more oscillations exist simultaneously in steady state. In other words, there are parasitic or unwanted oscillations existing together with a main oscillation. Due to the multiple oscillations, the net steady-state signal is severely distorted and can cause unwanted spurious effects in many applications. It is worth mentioning that the multi-oscillation phenomenon is quite different from that of a multimode oscillator [12]. A multimode oscillator uses an arbitrary number of tuned circuits to specify a well-defined set of oscillation frequencies, and when subject to an injected instruction signal, the circuit oscillates at only one of these well-defined frequencies.

The circuit configuration shown in Fig. 7 can also be used for studying multi-oscillation behavior. In Fig. 12, impedance $Z_a + Z_f$ as a function of frequency is generated for a new set of circuit parameters. We see that the start-up condition (6) is now satisfied at three frequencies, indicating a potential problem in the oscillator. In order to obtain further insight into the circuit operation, we consider the root locus as a function of bias current shown in Fig. 13. We see that there exists a bias range in which the circuit can possess two pairs of RHP poles

$$p_{1,2} = \alpha_1 \pm j\beta_1$$

and

$$p_{3,4} = \alpha_2 \pm j\beta_2. \qquad (8)$$

Given an initial impulse, the circuit can thus produce a signal of two concurrently growing sinusoids

$$x(t) = K_1 e^{\alpha_1 t} \cos (\beta_1 t) + K_2 e^{\alpha_2 t} \cos (\beta_2 t) \qquad (9)$$

where K_1 and K_2 are set by initial conditions. Simulated time-domain waveforms during transient buildup and at steady state are shown in Fig. 14. Fig. 14(a) agrees with the linear behavior predicted in (9). Fig. 14(b) displays the steady-state waveform which is quite distorted, having two distinct frequency components. This plot shows the simultaneous presence of two near-sinusoidal oscillations.

The advantages of using the root-locus analysis in oscillator design is apparent from the above analysis. This technique allows us to determine the exact location of all the poles in the circuit and to identify the circuit elements that cause undesirable behavior. For the circuit under consideration, one way to eliminate the multi-oscillation phenomenon is to minimize the effect of parasitic series inductance. This could be done with multiple parallel bond wires and board or substate layout optimization.

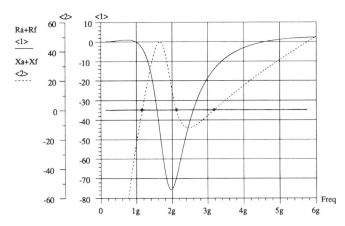

Fig. 12. Impedance plot for the circuit of Fig. 7 ($C_1 = 3$ pF, $C_2 = 2$ pF, $C_3 = 2$ pF, $R_C = 500$, $L = 6$ nH, $L_b = 2$ nH, $I_C = 10$ mA, $7\times$ device).

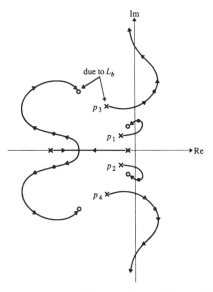

Fig. 13. Root locus (not to scale) for the circuit of Fig. 7. The poles corresponding to the characteristic of Fig. 12 are denoted by the • symbols.

V. OSCILLATOR DESIGN METHODOLOGY

In this section we study the interaction between oscillator circuit components and then propose a design methodology based on the negative-resistance model for realizing reliable oscillators. The design methodology should achieve the following objective: given an active circuit, a frequency-determining circuit is systematically chosen such that the resultant oscillator circuit achieves a reliable oscillation buildup and has a single predictable frequency of oscillation in the steady state.

It is useful to distinguish two types of active circuits. The first group of active circuits consists of one-port devices or circuits whose I–V characteristics contain a negative-resistance region. Examples of such active circuits are the tunnel diodes, avalanche diodes [13], regenerative circuits, negative resistors constructed from op-amp circuits [14], and inductive transformer-coupled configurations [15]. We use the abbreviation VCNR to denote a voltage-controlled negative resistor and ICNR to denote a current-controlled negative resistor. An active circuit is

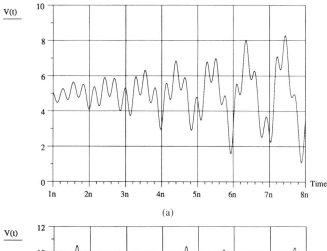

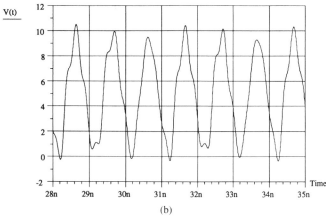

Fig. 14. Transient response corresponding to the characteristic of Fig. 12:
(a) start-up response and (b) steady-state response.

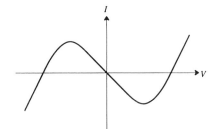

Fig. 15. *I–V* characteristic of a VCNR.

a VCNR if the port current of the circuit is a single-valued function of the port voltage (Fig. 15). It has been shown that the most appropriate frequency-determining circuit for a VCNR that results in a well-behaved harmonic oscillator is a parallel *RLC* resonant circuit, and for an ICNR it is a series *RLC* resonant circuit [16], [17].

The second group of active circuits does not possess a negative-resistance region in its *I–V* characteristics but instead depends on *LC* elements (often connected in feedback configurations) in order to produce a small-signal negative resistance about the operating point of the circuit. Examples of such circuits are the Colpitts, Pierce, and Hartley oscillators [3]. It is for this group of active circuits that a systematic design methodology is proposed here. Since an active circuit of this kind depends on energy-storage *LC* elements in order to produce a small-signal negative resistance, the negative resistance is frequency dependent. That is, if the active circuit is represented by an equivalent impedance, the resistive component is only negative in a finite frequency band, referred to as the negative-resistance band. In addition to the negative resistive component, the active circuit also has a reactive component, either inductive or capacitive. The design methodology given as follows is based on the negative-resistance model and involves three steps.

1) Active-Circuit Characterization: This step determines the negative-resistance band. Because no oscilla-

tion buildup is possible outside this band, the negative-resistance band must be designed to cover the frequency range of interest. Often this involves the selection of the appropriate *LC* elements and active devices.

2) Frequency-Determining Circuit Selection and Verification: This step selects a frequency-determining circuit that together with the active circuit forms a well-defined oscillator. The selection is based on the start-up condition for negative-resistance oscillators and is confirmed with either the Nyquist or root-locus analysis for validity. The frequency-determining circuit is assumed limited to either a parallel *LC* or a series *LC* resonant circuit.

3) Large-Signal Analysis: This step is concerned with the steady-state amplitude and frequency of oscillation.

The above design methodology is now illustrated with a design example. The circuit under consideration is a widely used microwave oscillator configuration (Fig. 16). One advantage of this configuration is that if the device capacitance C_μ of the transistor can be neglected (the only significant parasitic element across the base–collector junction), an output signal can be taken across the resistor R_C without introducing a loading effect to the circuit. This is true since there is insignificant linkage between the collector and the base, and the load R_C is in series with the current source representing the collector–emitter junction. The function of the inductor L_1 is to produce a negative resistance seen at the emitter port.

For simplicity, we neglect the effects of R_1, R_C, C_μ (internal base–collector capacitance), and C_{bx} (external base–collector capacitance). The active impedance can be approximated by

$$Z_a(s) = \frac{1 + sC_\pi r_b + s^2 L_1 C_\pi}{g_m\left(1 + s\dfrac{C_\pi}{g_m}\right)} \quad (10)$$

where g_m, r_b, and C_π are, respectively, the transconductance, base resistance, and base–emitter capacitance of the transistor. By separating the active impedance into a real component and an imaginary component, we obtain

$$Z_a(j\omega) \approx \frac{1}{g_m}(1 - \omega^2 L_1 C_\pi) + j\omega\,\frac{r_b C_\pi}{g_m}. \quad (11)$$

We note that the real component of (11) is frequency de-

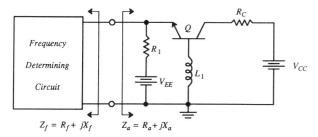

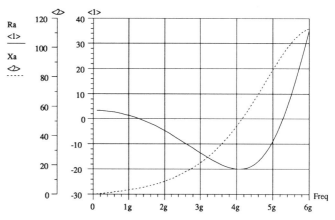

Fig. 16. Microwave oscillator.

pendent and is negative if

$$\omega > \left(\frac{1}{L_1 C_\pi} \right)^{1/2}. \qquad (12)$$

Since capacitor C_π is typically several picofarads and inductor L_1 several nanohenries, this equation suggests that this circuit can perform well into the microwave range. In practice, this circuit can be used to produce oscillation at frequencies close to the device f_{max}. As frequency increases, it is important to take into account the effects of C_μ and C_{bx}. The upper limit of this circuit's negative-resistance band can be shown to be constrained by the smaller of the self-resonant frequency of the integrated inductor L_1 or the frequency $[L_1(C_{bx} + C_\mu)]^{-(1/2)}$. In Fig. 17, the simulated active impedance shows a negative-resistance band of about 4 GHz, extending from 1 to 5 GHz. In addition to the negative resistance component, the active impedance also has an inductive component.

With the active circuit characterized, the second step is to determine a frequency-determining circuit that together with the active circuit forms a reliable oscillator. The objective is to select a frequency-determining circuit such that the resultant circuit has only one pair of RHP poles. The selection is based on the start-up condition for negative-resistance oscillators. Even though this condition may provide misleading results regarding the linear behavior of the circuit, it nevertheless provides us with an initial direction in selecting a frequency-determining circuit. It is important to emphasize that this selection process should be confirmed with either the Nyquist or root-locus analysis.

Since the active reactance of Fig. 16 is inductive, the frequency-determining circuit must be capacitive so that the total reactive component can be tuned out at the frequency of interest. One frequency-determining circuit could simply be a capacitor. If we want to minimize electronic noise and distortion content in the output signal of this circuit, we should use a high-Q LC resonant circuit. While it is well understood that a voltage-controlled (current-controlled) negative resistor requires a parallel (series) resonant circuit for reliable oscillation, it is not apparent as to which type of resonant circuits should be used for an LC-dependent negative resistor. The Clapp and Seiler oscillators, for example, differ from the Colpitts oscillator by virtue of the frequency-determining circuit. Whereas the Colpitts uses an inductor, the Clapp uses a

Fig. 17. Active-circuit characterization ($L_1 = 2$ nH, $I_C = 10$ mA, $R_C = 50$ Ω, 7× device).

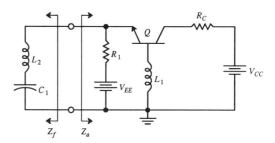

Fig. 18. Complete schematic of the microwave oscillator.

series resonant circuit and the Seiler uses a parallel resonant circuit [18]. For the circuit of Fig. 16, a series LC resonant circuit is used as shown in Fig. 18 even though a parallel resonant circuit could also be used. As multiple LC elements are used to form the frequency-determining circuit, extreme care must be taken to ensure that a multioscillation phenomenon does not take place. There should be, therefore, only one frequency ω_x at which the start-up condition holds. This requirement can be satisfied if the total reactance plot is monotonic. Fig. 19 shows the simulated impedance of the circuit. As a check for the circuit of Fig. 18, the root-locus and Nyquist plots are given in Fig. 20. These plots confirm the existence of a well-defined pair of RHP poles. Due to inductor L_2, we observe that the circuit has a pair of high-Q poles. This is desirable since it is observed in practice that any well-behaved harmonic oscillator possesses a pair of RHP poles near the $j\omega$ axis.

Upon the completion of the first two steps, the linear behavior of the circuit of Fig. 18 is fully understood. While these steps are used for achieving a reliable oscillation buildup and for predicting the oscillation frequency, they are not valid for predicting the circuit behavior in steady state. Nonlinear analysis must be used in order to predict the amplitude of oscillation and the output power level of the oscillator. Usually a nonlinear oscillator analysis starts with the assumption that the steady-state signal is near-sinusoidal and then proceeds to predict the amplitude of oscillation. The circuit of Fig. 18 is studied with a general near-sinusoidal analysis based on the negative-resistance model [19]. This method implicitly

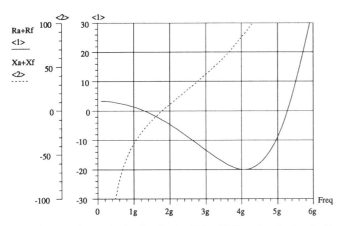

Fig. 19. Impedance plot for the circuit of Fig. 18 ($L_2 = 2$ nH, $C_1 = 3$ pF, $L_1 = 2$ nH, $I_C = 10$ mA, $R_C = 50$ Ω, 7× device).

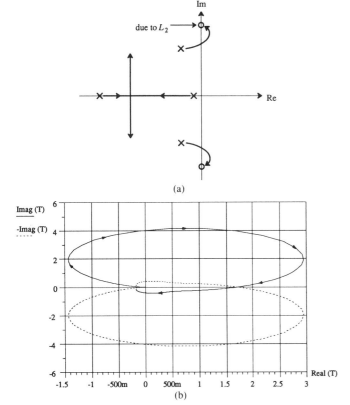

(a)

(b)

Fig. 20. (a) Root locus and (b) Nyquist diagram for the circuit of Fig. 18.

uses the concept of the *describing function* [20] to represent the nonlinearity of the active circuit by an approximate linear transfer function. For a multiport nonlinear circuit, the describing function is usually defined as the ratio of the phasor representing the fundamental component of the output to the phasor representing the sinusoidal input. For a one-port nonlinear circuit, the describing function is similarly defined with the input and the output being either the port current or port voltage. Note that even though the input is sinusoidal, the output may not be because of the nonlinearity. For this analysis [19] to be valid, either the current through or the voltage across the active circuit must be sinusoidal or near-sinusoidal in the steady state. If there is a well-defined near-sinusoidal

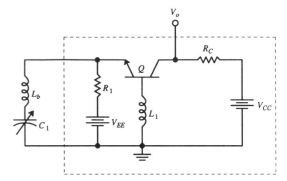

Fig. 21. Complete schematic of the 2-GHz oscillator.

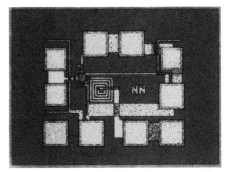

Fig. 22. Die photograph of the microwave oscillator.

oscillation in the steady state and, furthermore, the current is near-sinusoidal, the following oscillation condition has been derived [19]:

$$R_a^{(1)}(\omega_o, A_o) + R_f(\omega_o) = 0 \qquad (13a)$$

$$X_a^{(1)}(\omega_o, A_o) + X_f(\omega_o) = 0. \qquad (13b)$$

In (13), the frequency-determining impedance is assumed independent of signal amplitude while the active impedance is dependent on both the signal amplitude and frequency. Quantities ω_o and A_o are the frequency and amplitude of oscillation, respectively. The superscript (1) is used to emphasize that the active impedance is evaluated at the oscillation frequency (fundamental frequency). It can be shown that (13a) and (13b) correspond to the conservation of real and complex energy, respectively [21]. The dual-oscillation condition of (13) is

$$G_a^{(1)}(\omega_o, A_o) + G_f(\omega_o) = 0 \qquad (14a)$$

$$B_a^{(1)}(\omega_o, A_o) + B_f(\omega_o) = 0. \qquad (14b)$$

To confirm the theoretical analyses, the techniques described above were used in the design of the Si monolithic microwave oscillator shown in Fig. 21. The circuit was fabricated and characterized. Capacitor C_1 is an off-chip varactor for frequency tuning. Bond-wire inductance L_b together with C_1 form a series LC resonant circuit. L_1 is a monolithic inductor of 1.5 nH [22]. Resistor R_C is matched to the system impedance of 50 Ω. The bias current I_C is 10 mA. The circuit was fabricated in an oxide-isolated Si bipolar IC process with peak $f_T = 9$ GHz.

A die photograph of the oscillator is shown in Fig. 22. The circuit achieves a measured negative-resistance band

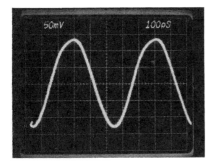

Fig. 23. Measured output waveform.

of 2.5 GHz extending from 1.5 to 4 GHz. With $C_1 = 3.9$ pF and $L_b = 0.7$ nH, the measured output waveform taken from the collector node is shown in Fig. 23. The measured oscillation frequency is approximately 2 GHz compared to the simulated oscillation frequency of 2.1 GHz. The output power is -6.5 dBm and was measured across a 50-Ω off-chip load. The simulated power is -4.0 dBm. The power loss is attributed to the package loss and the ac-bypass circuits that were used in the RF testing.

VI. CONCLUSION

Both theoretical and experimental investigations of microwave oscillators have been presented. Conventional start-up conditions have been shown to be necessary but not sufficient conditions for reliable oscillation. By combining these conditions with Nyquist and root-locus analyses, and the steady-state oscillation condition, we can design predictable and reliable harmonic oscillators. The phenomena explored in this paper can provide explanations to many examples of unpredictable and unreliable oscillation behavior observed in practice.

APPENDIX
RELATING OSCILLATION FREQUENCY TO START-UP CONDITIONS

The zero-phase frequency as defined in the start-up condition (4) is often used as an estimate for the steady-state oscillation frequency. Because of the nonlinearities in the active devices, the zero-phase frequency is not exactly equal to the steady-state frequency of oscillation. Nevertheless, it is a good approximation in oscillator circuits that have a pair of "high-Q" RHP poles lying close to the $j\omega$ axis. In this appendix we study whether the frequency ω_x of the start-up condition for negative-resistance oscillators can also be utilized for predicting the oscillation frequency.

The circuit used for illustration is the transformer-coupled oscillator shown in Fig. 24(a). Fig. 24(b) shows the simplified ac equivalent circuit. The loop gain of the circuit can be shown to be

$$T(s) = \frac{1}{n}\left(g_m - \frac{1}{nZ_\pi}\right)\frac{sL}{s^2LC + s\dfrac{L}{R} + 1} \quad (A1)$$

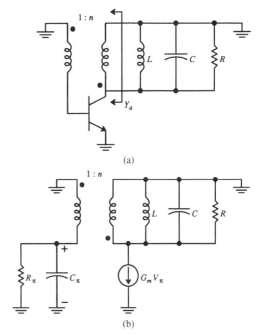

Fig. 24. (a) Transformer-coupled oscillator (bias circuit not shown) and (b) simplified ac equivalent circuit.

where

$$Z_\pi(s) = \frac{r_\pi}{1 + sr_\pi C_\pi} \quad (A2)$$

$$r_\pi = \frac{\beta_o}{g_m} \quad (A3)$$

$$C_\pi \approx g_m\tau_F. \quad (A4)$$

Substituting (A3) and (A4) into (A2) yields

$$Z_\pi(s) = \frac{\beta_o}{g_m}\frac{1}{1 + s\beta_o\tau_F}. \quad (A5)$$

Using (A5) in (A1) and solving for the zero-phase frequency gives

$$\omega_z = \left[L\left(C + \frac{\tau_F}{nR\eta}\right)\right]^{-(1/2)} \quad (A6)$$

where $\eta \equiv 1 - (1/n\beta_o)$. The admittances $Y_a(s)$ and $Y_f(s)$ of Fig. 24(a) are

$$Y_a(s) = -\frac{1}{n}\left(g_m - \frac{1}{nZ_\pi}\right) \quad (A7)$$

$$Y_f(s) = \frac{1}{R} + \frac{1}{sL} + sC. \quad (A8)$$

Substituting $s = j\omega$ into (A7) and (A8) yields

$$Y_a(j\omega) = -\frac{g_m\eta}{n} + j\omega\frac{g_m\tau_F}{n^2} \quad (A9)$$

$$Y_f(j\omega) = \frac{1}{R} + \frac{1}{j\omega L} + j\omega C. \quad (A10)$$

We now assume that the start-up condition (7) is valid.

From (A9) and (A10) the circuit is unstable if

$$\text{Real:} \quad -\frac{g_m \eta}{n} + \frac{1}{R} \le 0 \Leftrightarrow g_m \ge \frac{n}{R}\frac{1}{\eta} \quad \text{(A11)}$$

$$\text{Imag:} \quad -\frac{1}{\omega_x L} + \omega_x C + \omega_x \frac{g_m \tau_F}{n^2} = 0$$

$$\Leftrightarrow \omega_x = \left[L\left(C + \frac{g_m \tau_F}{n^2} \right) \right]^{-(1/2)}. \quad \text{(A12)}$$

Using (A11) in (A12) gives

$$\omega_x \le \left[L\left(C + \frac{\tau_F}{nR\eta} \right) \right]^{-(1/2)}. \quad \text{(A13)}$$

By comparing (A13) to (A6), we see that $\omega_x \le \omega_z$ for this circuit.

We now determine the poles of the circuit from the following characteristic equation, derived from either the expression $1 - T(s) = 0$ or $Y_a(s) + Y_f(s) = 0$:

$$s^2 L\left(C + \frac{g_m \tau_F}{n^2} \right) + sL\left(\frac{1}{R} - \frac{g_m \eta}{n} \right) + 1 = 0. \quad \text{(A14)}$$

Assume that the poles are complex conjugate, i.e., $p_{1,2} = \alpha \pm j\beta$. We can show that

$$\alpha = \frac{1}{2}\left(\frac{g_m \eta}{n} - \frac{1}{R} \right)\left(C + \frac{g_m \tau_F}{n^2} \right)^{-1}$$

$$\beta = \left\{ \left[L\left(C + \frac{g_m \tau_F}{n^2} \right) \right]^{-1} - \left[\frac{1}{2}\left(\frac{g_m \eta}{n} - \frac{1}{R} \right) \right. \right.$$

$$\left. \left. \cdot \left(C + \frac{g_m \tau_F}{n^2} \right)^{-1} \right]^2 \right\}^{1/2}.$$

The poles are in the RHP if

$$\alpha \ge 0 \Leftrightarrow g_m \ge \frac{n}{R}\frac{1}{\eta}. \quad \text{(A15)}$$

Note that (A11) and (A15) give the same requirement for quantity g_m and thus confirm the validity of the start-up condition (7) in this circuit. If $g_m \approx (n/R)(1/\eta)$, quantity α is approximately equal to zero. Under this condition the poles lie close to the $j\omega$ axis and

$$\omega_x \approx \left[L\left(C + \frac{\tau_F}{nR\eta} \right) \right]^{-(1/2)} = \omega_z.$$

The above analysis on the transformer-coupled oscillator suggests the following:

a) The frequency ω_z that fulfills the feedback condition (4) is, in general, not equal to the frequency ω_x that fulfills the negative-resistance condition (6) or (7).

b) If the RHP poles of the oscillator are high-Q, the frequency ω_x is almost the same as the frequency ω_z. Under this condition, frequency ω_x can be used for estimating the oscillation frequency.

The conclusions drawn above can also be obtained from similar analysis of other oscillator circuits.

ACKNOWLEDGMENT

The authors wish to acknowledge the contributions of Prof. D. Pederson and P. Kennedy during many simulating discussions. The authors are also grateful to Signetics Corp. for HS3 fabrication of the prototype oscillator and of Avantek Inc. for the use of its microwave measurement laboratory.

REFERENCES

[1] D. O. Pederson and K. Mayaram, *Analog Integrated Circuits for Communication—Principles, Simulation and Design.* Boston: Kluwer Academic, 1991.
[2] B. Parzen, *Design of Crystal and Other Harmonic Oscillators.* New York: Wiley, 1983.
[3] M. E. Frerking, *Crystal Oscillator Design and Temperature Compensation.* New York: Van Nostrand Reinhold, 1978.
[4] I. Bahl and P. Bhartia, *Microwave Solid State Circuit Design.* New York: Wiley Interscience, 1988.
[5] A. B. Grebene, *Bipolar and MOS Analog Integrated Circuit Design.* New York: Wiley Interscience, 1984, ch. 11.
[6] M. A. Unkrich and R. G. Meyer, "Conditions for start-up in crystal oscillators," *IEEE J. Solid-State Circuits*, vol. SC-17, no. 1, pp. 87–90, Feb. 1982.
[7] R. D. Middlebrook, "Measurement of loop gain in feedback system," *Int. J. Electron.*, pp. 485–512, Apr. 1975.
[8] E. M. Cherry and D. E. Hooper, *Amplifying Devices and Low-Pass Amplifier Design.* New York: Wiley, 1968.
[9] G. Gonzalez, *Microwave Transistor Amplifiers—Analysis and Design.* Englewood Cliffs, NJ: Prentice-Hall, 1984, ch. 5.
[10] W. El-Kamali, J. Grimm, R. Meierer, and C. Tsironis, "New design approach for wide-band FET voltage-controlled oscillators," *IEEE Trans. Microwave Theory Tech.*, vol. MTT-34, no. 10, pp. 1059–1063, Oct. 1986.
[11] S. Maas, *Nonlinear Microwave Circuits.* Norwood, MA: Artech House, 1988, ch. 12.
[12] W. A. Edson, "Frequency memory in multi-mode oscillators," *IRE Trans. Circuit Theory*, pp. 58–66, Mar. 1955.
[13] G. Gibbons, *Avalanche-Diode Microwave Oscillators.* Oxford: Clarendon, 1973.
[14] L. O. Chua, C. A. Desoer, and E. S. Kuh, *Linear and Nonlinear Circuits.* New York: McGraw-Hill, 1987.
[15] R. G. Meyer, "Advanced integrated circuits for communication," EECS-242 Class Notes, Univ. California, Berkeley, 1989.
[16] B. van der Pol, "The nonlinear theory of electric oscillations," *Proc. IRE*, vol. 22, pp. 1051–1086, 1934.
[17] N. M. Nguyen, "Monolithic microwave oscillators and amplifiers," Ph.D. dissertation, UCB/ERL M91/36, Univ. California, Berkeley, May 1991.
[18] J. K. Clapp, "Frequency stable LC oscillators," *Proc. IRE*, vol. 42, pp. 1295–1300, Aug. 1954.
[19] K. Kurokawa, "Some basic characteristics of broadband negative resistance oscillator circuit," *Bell Syst. Tech. J.*, pp. 1937–1955, July–Aug. 1969.
[20] J. D'Azzo and C. Houpis, *Feedback Control System Analysis and Synthesis.* New York: McGraw-Hill, 1966, ch. 18.
[21] G. Hachtel, "Semiconductor integrated oscillators," Ph.D. dissertation, Univ. California, Berkeley, 1965.
[22] N. M. Nguyen and R. G. Meyer, "Si IC-compatible inductors and LC passive filters," *IEEE J. Solid-State Circuits*, vol. 25, no. 4, pp. 1028–1031, Aug. 1990.

MOS Oscillators with Multi-Decade Tuning Range and Gigahertz Maximum Speed

MIHAI BANU, MEMBER, IEEE

Abstract —This paper studies the realization of fully integrated MOS oscillators with multi-decade tuning range and top speed exceeding 1 GHz. This fast operation requires the use of submicrometer fabrication technology. In order to overcome the analog circuit limitations of the latter, the design of the oscillator is simplified and, in turn, highly redundant digital and analog control capability is provided. This strategy allows the implementation of certain oscillator features at the system level. A brief review of possible oscillator structures is given and it is found that the best approach for high-speed, wide-range specifications is the relaxation network of the constant current charge type. Circuit techniques are presented to increase the speed of the latter, based on active use of parasitics and simplified feedback networks. NMOS and CMOS implementations are discussed and compared. The design and performance of an experimental submicrometer NMOS oscillator is presented. This device covers the 100-kHz to 1-GHz frequency range and has a robust structure.

I. INTRODUCTION

MODERN monolithic-circuit applications in communications, signal processing, etc. have challenging requirements in terms of ever-increasing complexity, operating frequency, precision, and reliability. This warrants the extensive use of digital techniques, whenever possible. The necessary capability for digital high-speed VLSI is provided by the new submicrometer MOS integrated circuit technologies.[1] However, there are basic functions such as signal amplification or signal generation whose implementation relies fundamentally on analog circuitry. The analog capability of the MOS processes mentioned above is limited by small available voltage levels and degraded device characteristics. In this context and in connection with the on-chip signal generation problem, it will be seen that the concept of a high-speed, wide-range digital- and voltage-controlled-oscillator (DVCO) [1] is beneficial, effectively simplifying the analog circuit design problem. The following discussions assume the goal of accomplishing total system integration.

Variable-frequency oscillators (VFO's) are basic signal-generating blocks frequently needed in systems. For example, analog and digital phase-locked loops (PLL's) use

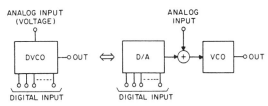

Fig. 1. Definition of DVCO.

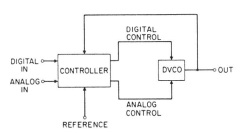

Fig. 2. General scheme for on-chip signal generation using a DVCO.

voltage-controlled oscillators (VCO's) and digital-controlled oscillators (DCO's), respectively. VCO's are realized normally with tuned or relaxation analog networks [2], while standard DCO's are implemented digitally using programmable counters that divide the frequency of a fixed signal [3]. The DVCO is a generalization of the previous types of oscillators and its functional equivalent is shown in Fig. 1. The use of conventional D/A converters is not necessary; e.g., a voltage-controlled ring oscillator with digital-controlled number of stages constitutes a valid DVCO implementation.

The availability of both analog and digital controls allows system configurations better than otherwise possible. By using a digitally assisted analog design approach, the sensitivity of the system performance to circuitry can be minimized. This is essential for submicrometer MOS circuits to overcome their analog limitations, and is similar to the idea of improving the performance of an inherently limited circuit through correction at the system level [4], [5]. Therefore, the general scheme of Fig. 2 is suggested as specifically suitable for on-chip signal generation. The DVCO is required to provide only the minimum perfor-

Manuscript received May 11, 1988; revised August 12, 1988.
The author is with AT&T Bell Laboratories, Murray Hill, NJ 07974.
IEEE Log Number 8824188.
[1] Other high-speed technologies are available, but MOS technology remains the preferred choice for VLSI.

Reprinted from *IEEE Journal of Solid-State Circuits*, vol. SC-23, pp. 474-479, April 1988.

mance that is fundamentally related to circuitry: speed and range. This simplifies its design considerably. The burden of other features such as voltage/digital-to-frequency linearity or temperature stability is carried by the controller at the system level. Feedback and external references may be used in general and the controller itself may have analog and digital inputs. Depending on the controller, the system in Fig. 2 could be a complex wide-band frequency synthesizer or just a stand-alone precision VCO. An example that shows the flexibility of the structure in Fig. 2 is an analog PLL with digital acquisition aid [6]. This system could be designed such that the slow-varying DVCO temperature drifts are automatically corrected via its digital input; the effect of changing one least significant bit is assumed small enough not to disturb the analog PLL operation.

The approach of Fig. 2 is limited by the oscillator circuit-sensitive features, whose improvement is, therefore, of fundamental importance. In this paper, techniques will be given for the design of fully integrated DVCO's with multi-decade tuning range and gigahertz maximum speed, appropriate for submicrometer MOS fabrication. As a consequence of the simplified design objectives, the resulting circuits will be robust and digital-process compatible.

II. INADEQUACY OF RESONANCE TECHNIQUES FOR HIGH-SPEED, WIDE-RANGE APPLICATIONS

According to a conventional view, high-frequency operation and wide-band coverage are incompatible oscillator specifications. Traditionally, it has been accepted that these features are realizable only separately with tuned and relaxation networks, respectively [7]. The improvement of either the range of tuned oscillators or the speed of relaxation oscillators will be studied.

The design approach based on tuned networks gives excellent results for high-speed, narrow-band oscillators [7], [8]. Unfortunately, the frequency coverage limitation is difficult to remove for fundamental reasons [9]. The resonance phenomena taking place in tuned networks are described mathematically as functions giving large phase variations for small frequency variations. This property drastically limits the operating frequency of such circuits. Since the usual resonance elements, such as LC tanks or crystals, cannot be integrated, it is totally impractical to attempt wide frequency coverage by use of many components external to the chip. In fact, even the use of one such component is not desirable.

A class of tuned networks contains circuits synthesized only with resistors, capacitors, and amplifiers [7]. These can be fully integrated either by MOSFET-C [10] or switched-capacitor [11] methods. Operation over wide frequency ranges is possible by including programmable resistor or capacitor arrays in their structures or by employing simulated resonance [12]. However, these approaches have not been proven feasible for high-speed operation. In fact, it is unlikely that such a development will be possible

in the future due to the complicated topologies needed and the necessary use of high-gain amplifiers.

We conclude that tuned oscillator methods are not adequate for high-speed, wide-tuning-range implementations.

III. PRINCIPLES OF HIGH-SPEED, WIDE-RANGE RELAXATION OSCILLATORS

Unlike the previous case, there are no fundamental reasons to prevent the realization of high-speed, wide-range multivibrators. In general, these relaxation networks can be fully monolithic because they do not make use necessarily of nonintegrable components. Since the multivibrator capability for very large frequency coverage is well established [7], [13], it remains to be shown that their traditional slow speed can be increased. This task is partly accomplished at the fabrication level by use of submicrometer MOS technology [14]. In addition, at the circuit level, new techniques will be applied. These are discussed next.

In order to ensure fundamentally a maximum speed of operation, all circuitry will be developed from the high-speed design philosophy that one should have: a) topologies with minimum parasitics, b) complexities as small as possible, and c) no reactive elements other than the unavoidable parasitic capacitances. Since the synthesis of oscillators requires reactive elements, the parasitics themselves will be employed as useful components. Rules a) and c) maximize the analog bandwidth and b) minimizes the number of switching delays, always present in relaxation networks. A systematic search for the optimum oscillator structure follows, using the previous principles as qualifying criteria. It is profitable to start with a study of the DVCO frequency control mechanism, a fundamental and critical part of its structure.

There are several possible ways to accomplish the DVCO frequency-range programmability: topological changes in ring oscillators, capacitor or resistor switching in RC charge-type relaxation networks, and current-source switching in constant-current-charge-type relaxation networks. The first approach is suitable for high-speed operation but is not practical for large numbers of digital control bits because it is not possible to program the ring-oscillator topology in a binary style. Switching properly ratioed resistors or capacitors to vary an RC time constant accomplishes easy programmability but this scheme suffers from a large number of necessary interconnections. The associated parasitics prohibit operation at high frequencies. Most circuits following the first and the second approaches also have the disadvantage that they must use different analog and digital control mechanisms. When the two types of controls are operated at the same time, undesirable long transients can be generated.

The current-source switching method and the associated constant-current-charge (discharge) oscillator configuration fit better the high-speed, wide-range requirements. The MOS transistors in saturation are natural voltage-controlled current sources that can be easily ratioed and

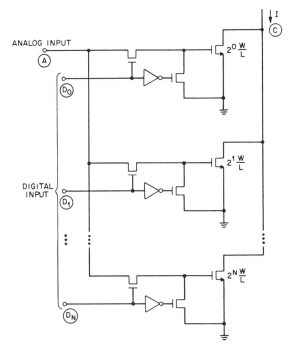

Fig. 3.   Digital and analog current control block.

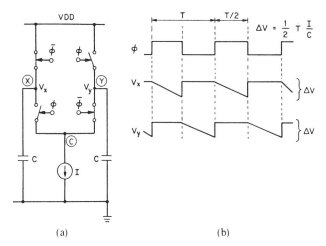

Fig. 5.   (a) Charging scheme capable of using parasitic timing capacitors. (b) Waveforms for the circuit of (a).

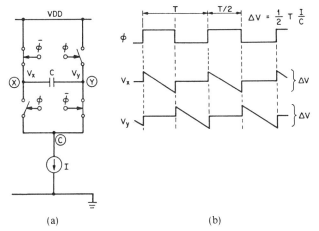

Fig. 4.   (a) Conventional charging scheme incapable of using parasitic timing capacitors. (b) Waveforms for the circuit of (a).

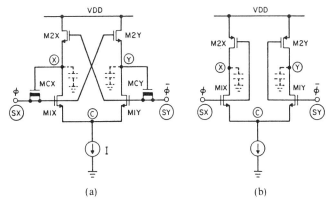

Fig. 6.   (a) NMOS charging scheme. (b) CMOS charging scheme.

interconnected. Thus, digital and voltage controls are provided in a unified way through a single current value, as illustrated in Fig. 3. This current can be varied over a wide range but when it is set, the only parasitic capacitances interfering with the operation of the oscillator are the ones at node C.

A constant-current-charge (discharge) oscillator consists of two distinctive parts: a capacitor charging (discharging) block which will be discussed first, and a feedback block to be described in Section V. The standard method of emitter-coupled multivibrators, illustrated in Fig. 4, is not optimum for operation at high frequencies because the floating timing capacitor is connected in series with the grounded parasitics, always present at the terminals of the switches. The value of this timing element cannot be reduced arbitrarily without essentially affecting the operation of the scheme. In fact, it must be large compared to

the parasitics. Therefore, the maximum switching speed of the structure is limited by the timing element and not by the intrinsic time constants of the transistors and their interconnections. Furthermore, the circuit is sensitive to the practical finite ON resistance of the top switches. Notice that the voltage drop on these switches, which is a direct function of the current I, modifies the potentials at nodes X and Y in a detrimental way.

Improved performance is obtained with the grounded-capacitor method of Fig. 5. The timing capacitors are in parallel with the parasitics and can be reduced to zero value, their function being performed by the latter. The maximum switching speed is the intrinsic one of the structure. In addition, since the timing parasitic capacitors are switched at only one terminal, the circuit is insensitive to all switches' nonideal ON resistance. A potential problem with this scheme is the practical mismatch between the two capacitors. It is relatively simple to avoid such significant mismatches by employing a symmetric and tight layout [1]. Topologies with grounded timing capacitors have been recognized as superior in the past [15], but without advantageous use of parasitics.

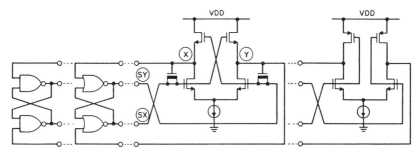

Fig. 7. Synthesis attempt of DVCO's with one-latch feedback.

IV. MOS REALIZATION OF THE HIGH-FREQUENCY CURRENT SWITCHING SCHEME

Fig. 6 shows the previous scheme applied to NMOS and CMOS. Dual circuits also exist with reversed transistor types. Ignoring temporarily the two depletion devices in Fig. 6(a), it is seen that each switch has been implemented with a transistor. Cross-coupling of the left and the right current paths is necessary in NMOS due to the use of only one type of MOSFET.

The operation of these circuits is modified from the ideal case by nonidealities. When the transistor $M1X$ or $M1Y$ is turned on, the charges at the respective nodes X or Y quickly redistribute with the ones at node C, independently of the current I. As a result, the voltages V_X or V_Y jump downward by a substantial amount because all capacitances involved are on the same order of magnitude. This is not desirable because for proper operation, V_X and V_Y must ramp down under the control of the current source. Also, when the switching waveform ϕ changes states, charges are injected into nodes X and Y via the gate-to-source and gate-to-drain parasitic capacitors of the transistors. In the case of the CMOS circuit the resulting effect opposes the previous voltage jump, but in the case of the NMOS circuit it adds to it. The difference comes from the cross-coupling of devices. The latter situation is worsened further by the fact that V_X and V_Y in the NMOS scheme are charged only to V_{DD} minus one transistor threshold voltage.

The practical implication of the previous considerations is that the CMOS circuit can be used directly but the NMOS version, with excessive nonideal effects, must be improved. A simple way of doing this is by connecting two compensating charge-pumping depletion devices as shown in Fig. 6(a). Current-controlled voltage ramping is thus reinstalled but at the expense of a slight decrease of top operating speed due to larger total capacitance. This drawback is minimized by connecting the depletion transistors so as to make use of their nonlinear gate-to-channel capacitance: from Fig. 6(a) notice that the pumping is accomplished mostly in the upward ϕ transitions; otherwise the devices are off part of the time. A different technique for decreasing the effect of nonidealities without speed penalties, valid for both NMOS and CMOS, is restricting the high-voltage level of ϕ and forcing the transistors $M1X$ and $M1Y$ into saturation. Then they act as cascodes for the current source and partially "mask" the parasitics at

node C from the discharging nodes. This technique is applied easily to CMOS implementations [6].

V. FEEDBACK CIRCUITRY AND DVCO SYNTHESIS

In order to produce oscillations, the previous schemes need to be complemented by signal-reversing feedback from nodes X, Y to SX, SY. The additional circuitry generates the switching waveforms ϕ and $\bar{\phi}$ by reversing the output signals of the current discharging mechanisms. It effectively contains a voltage comparator and a latch. A change in the state of the latch is to take place only when the voltage comparator detects that V_X or V_Y has fallen below a predetermined value. In this way the circuit will periodically "relax" between two unstable states. Assuming that the response of the reversal block is much faster than the discharge time of the capacitors, the maximum excursion ΔV of V_X and V_Y is practically kept constant. Since ΔV is directly proportional to the switching waveform period and to the current I, the frequency of oscillation depends linearly on I. Increasing the latter, a high-speed limitation is eventually approached when the delay through the reversal block becomes significant compared to the oscillation period. The minimum speed attainable depends on the leakage currents at the discharging nodes.

Many signal reversal blocks are possible, but the ones with minimum complexity are of utmost importance for high-speed applications. For example, the reversal block of the typical CMOS multivibrator described in [7] is unnecessarily large and slow. Considerably faster designs are possible. The simplest conceivable reversal block is just a latch with NOR or NAND gates, as shown in Fig. 7. Synthesis of four DVCO versions could be attempted because the conditions for generation of instability appear to be met. However, a closer analysis of all possibilities shows that the resulting circuits contain positive dc feedback that freezes them into stable states. For example, consider the NMOS NOR version that can be visualized readily in the center of Fig. 7. If nodes X and Y are high at the same time, even temporarily, the outputs of the NOR-gate latch are both driven low. Then, the common-source transistors turn off and there is no path to discharge the parasitic capacitors at nodes X and Y. The circuit will remain latched in this state. Such a situation results not only due to an unfavorable initial condition but also due to switching transients during the operation of the scheme. ADVICE simulations [16] clearly show this to be the case.

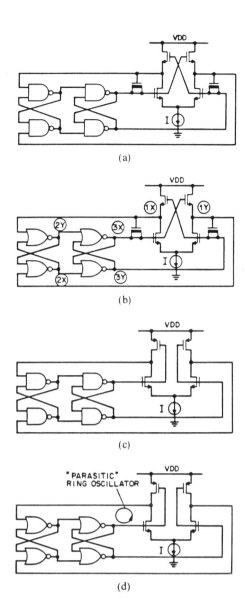

Fig. 8. (a) NMOS DVCO with two-NAND-latch feedback. (b) NMOS DVCO with two-NOR-latch feedback. (c) CMOS DVCO with two-NAND-latch feedback. (d) Nonoperational CMOS DVCO with two-NOR-latch feedback.

The next level of complexity in which positive dc feedback is no longer present, is adding a second latch. All circuits illustrated in Fig. 8 oscillate as expected except for the CMOS NOR-gate version shown in Fig. 8(d). This failure is due to multiple oscillatory modes. The switching sequence of the two NOR-gate latches is such that two cascaded effective inverters are active at a time. Adding to this the CMOS inverters in the discharging block, two "parasitic" ring oscillators with three stages are created within the DVCO. The same situation does not happen in the NMOS NOR-gate circuit of Fig. 8(b) because the cross-coupling of transistors breaks these undesired loops.

The three valid DVCO's developed and their reverse transistor-type duals have common properties. They are analog monolithic transistor-only circuits, similar in this respect to the result of [17]. Their implementation requires only a standard digital process and their structure is robust, digital-like. A noncritical capacitor (dis)charging is the only analog process performed internally. The topological and electrical symmetry which is realizable in practice by correct layout and matching ensures that the two output signals are identical in shape and opposite in phase. The duty cycle at the input of the charging block is approximately 50 percent, the transitions being separated by only one gate delay. This is a useful feature when driving circuits requiring nonoverlapping clocks such as dynamic shift registers. The top speed is approximately equal to that of a ring oscillator with five stages (fan-in and fan-out of 2); for maximum control current the discharging block switches as a digital gate. Thus, if implemented with present fine-line technologies providing 100 ps or less gate delays the proposed circuits will work at speeds in excess to 1 GHz. The current control is responsible for a tuning range of several decades. Layout area and power consumption are small due to the simple structure.

In terms of the highest frequency attainable, the various DVCO versions are not equivalent mainly because of the different logic speeds. The best NMOS circuit is that of Fig. 8(b) since the NOR gates are faster than the NAND gates. In CMOS, the dual of the circuit of Fig. 8(c) using pseudo-NMOS NOR gates is optimum [6].

VI. SUBMICROMETER NMOS DESIGN

An experimental NMOS NOR DVCO as in Fig. 8(b) was designed for an available NMOS process [14]. One-micrometer minimum design rules were used consistently except for the 0.75-μm coded transistor channel lengths. Eleven digital bits were provided to obtain a high degree of control redundancy. In order to keep the total parasitic capacitance of the programmable current source within manageable values, only the respective first five most-significant-bit transistors were ratioed in terms of their widths. They had minimum channel lengths. The other transistors of the current source were ratioed in lengths. This created an unavoidable systematic mismatch, practically unessential due to the high degree of control redundancy. Since the circuit contains a relatively small number of devices there were no special constraints on maximum power dissipation. As a result, the NOR gates could be designed powerful enough to ease the driving of the off-chip loads. This conservative approach also minimized the effect of interconnections. Thus, all enhancement transistors were selected to have channel lengths of 60 μm. The proper ratioing of the depletion-load transistors required a choice of 20 μm for their channel lengths. In order to minimize the parasitics, the transistors critical for high speed were split in two identical parts that shared a common drain in the case of the enhancement devices or a common source in the case of the depletion devices. The necessary gate area of the bootstrapping transistors was found through computer simulation to be about 130 μm^2.

The nonoverlapping DVCO output signals at nodes $3X$ and $3Y$ could drive the super-buffer shown in Fig. 9. In turn, the latter was designed to drive an off-chip load as

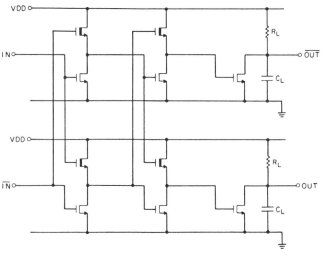

Fig. 9. NMOS super-buffer.

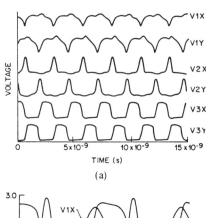

(a)

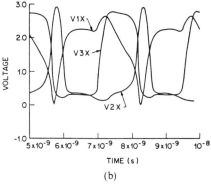

(b)

Fig. 10. (a) Typical ADVICE simulation for the NMOS DVCO with two-NOR-latch feedback. (b) Details of simulation of (a).

large as 25 Ω and 10 pF. This was accomplished by having open-drain output devices with channel width of 200 μm (the channel length is, of course, 0.75 μm coded). Due to the powerful NOR gates used, the output super-buffer contained only three stages and yet did not substantially load the DVCO.

A typical ADVICE simulation is shown in Fig. 10. For the control setting chosen, the circuit oscillates at approximately 400 MHz. Fig. 10(a) illustrates all DVCO node voltages in relation to each other. Notice how the first latch behaves like a voltage comparator, generating short pulses when the signals at nodes $1X$ and $1Y$ drop below a certain value. These pulses flip the state of the second

latch. The nonoverlapping property of the signals at nodes $3X$ and $3Y$ is apparent. More details can be seen in Fig. 10(b). Only one of the two symmetric aspects of the operation is shown. Node $1X$ is initially charged up to V_{DD} minus one transistor threshold voltage. When the signal at node $3X$ goes high, the bootstrapping mechanism injects charges into node $1X$ that compensates for parasitic leaks (see Section IV). As seen, the voltage of this node is in fact increased due to an intentionally conservative overcompensation. Then, the constant current source discharges the node in a ramp-like fashion. The node continues to be discharged even after the first latch is triggered because finite time is required for the signal to travel around the oscillator loop. Notice that the charging time of node $1X$ is not critical as long as it is shorter than four gate delays. This condition is automatically met since the charging takes place in approximately one gate delay.

The layout was done according to the design assumptions mentioned previously. No special attempt was made to accurately match the transistors of the analog/digital control block; some trivial matching rules were, however, observed as a matter of routine. The immediate possible application of this DVCO, a high-frequency, wide-band clock recovery circuit, has no requirements on the digital-to-frequency linearity. For other cases, correction at the system level can be applied, as discussed in Section I.

VII. EXPERIMENTAL RESULTS

The fine-line DVCO, designed and laid out as in the previous section, was fabricated, tested, and found fully operational. The chip photomicrograph is shown in Fig. 11. The switching block and the latches are in the center, the digital/analog-controlled current source is situated at right (11 digital inputs and one analog input going to the second pad from right can be seen), and the three-stage output buffer is at left. All results presented here were obtained probing at wafer level and with 3-V power supply. In order to bring the oscillator output signals off the chip reasonably uncorrupted by reflections, transmission lines with 50-Ω characteristic impedance were used. This was possible because the probe card itself contained such transmission lines up to the contacting needles. For the same reason, the biasing of the open-drain devices had to be applied through standard "bias T" networks, properly terminated by the measuring instruments.

Devices from several lots were tested. Their performance was essentially the same except for their maximum speed which varied approximately ±25 percent around 1.1 GHz. As expected, there was a strong correlation between the top speed observed and the effective transistor channel lengths in the wafers tested. Typically, the devices that had estimated effective channel lengths of 0.5 μm or smaller operated at maximum speeds in excess to 1 GHz.

The high-frequency capability and very wide tuning range of the DVCO are illustrated in Fig. 12. By changing the analog and digital control inputs, the output frequency

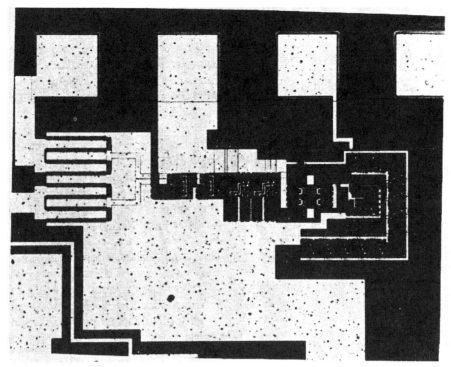

Fig. 11. Chip photograph of experimental DVCO.

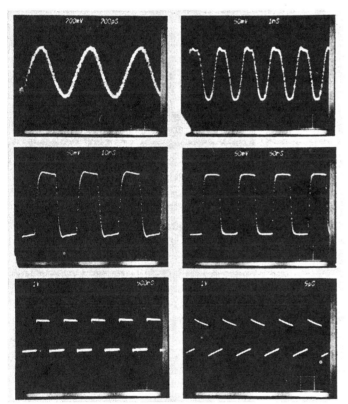

Fig. 12. Typical experimental DVCO output waveforms for various analog and digital control inputs; on the bottom right trace the frequency is slightly outside the passband of the bias T network used for testing. In the 1, 10, and 50 ns per division traces the signal was attenuated by 10.

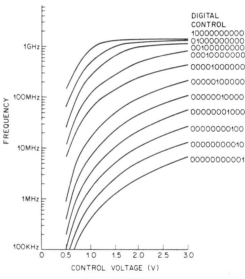

Fig. 13. Typical experimental DVCO frequency characteristics.

of this chip could be varied from a maximum of approximately 1.4 GHz down to about 80 kHz, the tuning range exceeding four decades in frequency. A more quantitative characterization of the DVCO is given in Fig. 13. Each curve represents the output frequency as a function of the analog control voltage for a given digital input; only 11 out of 2047 such curves are shown. All frequencies in the range are covered continuously with a high degree of redundancy. The gain of the voltage-controlled part depends on the bias point of the analog input and is increasingly larger as the input voltage decreases. The nonuniformity in the spacing of the characteristics, more apparent for the top five curves, can be explained by the transistor mismatches in the controlled current source and of course by the saturation phenomenon discussed in Section II. In the photograph of Fig. 8, the five most-significant-bit control lines, exiting the DVCO on the bottom, are seen to

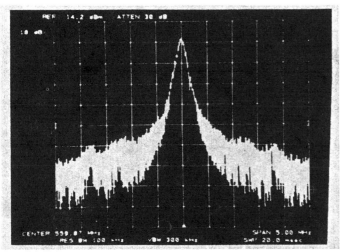

Fig. 14. Typical experimental DVCO output spectrum; 10 dB/div vertically, 559.87-MHz center frequency, 5-MHz span, 100-kHz resolution bandwidth.

come from transistors laid out differently from the rest. The shapes of the top five curves of Fig. 13 are also influenced by the respective transistor short-channel effects. As discussed in Section I, the digital–analog highly redundant control feature of the device can be used to effectively correct the oscillator characteristics within system environments.

A picture of the oscillator output spectrum at approximately 560 MHz is shown in Fig. 14. This corresponds to a worst-case phase jitter measurement since the trace contains residual amplitude modulation and was generated at wafer level with practically no noise isolation from the environment.

The active area including the output driver was 0.075 mm^2 and the power dissipation was 50 mW. The circuit proved to be very robust: a version with 1-μm coded channel lengths and no other design changes operated as well as the original one except, of course, for lower maximum speed.

VIII. Conclusions

Fully integrated variable-frequency oscillators with gigahertz top speed and multi-decade frequency coverage can be realized in submicrometer MOS technology. The proposed circuits are relaxation networks of the constant-current (dis)charge type, use only parasitics as timing capaci-

tors, have symmetrical topology and simplified feedback blocks, and can be controlled with high degree of redundancy by analog and digital means. The latter property is essential in providing system capability for performance improvement. In this way full advantage is taken of the high-speed performance of submicrometer MOS technologies while circumventing some of their analog circuit limitations. As a result of the simplified design, these transistor-only oscillator structures have small complexity, can be implemented with a standard digital process, and are robust. The theory was verified with an experimental submicrometer NMOS oscillator that was fully operational.

References

[1] M. Banu, "A 100 KHz–1 GHz NMOS variable-frequency oscillator with analog and digital control," in *ISSCC Dig. Tech. Papers*, 1988, pp. 20–21.
[2] F. M. Gardner, *Phaselock Techniques*. New York: Wiley, 1979, ch. 6.
[3] R. E. Best, *Phase-Locked Loops Theory, Design, and Applications*. New York: McGraw-Hill, 1984, ch. 4.
[4] M. H. Wakayama and A. A. Abidi, "A 30-MHz low-jitter high-linearity CMOS voltage-controlled oscillator," *IEEE J. Solid-State Circuits*, vol. SC-22, no. 6, pp. 1074–1081, Dec. 1987.
[5] T. Liu and R. G. Meyer, "A 250 MHz monolithic voltage-controlled oscillator," in *ISSCC Dig. Tech. Papers*, 1988, pp. 22–23.
[6] M. Banu, "Design of high-speed, wide-band MOS oscillators for monolithic phase-locked loop applications," in *Proc. IEEE Int. Symp. Circuits Syst.*, 1988, pp. 1673–1677.
[7] A. B. Grebene, *Bipolar and MOS Analog Integrated Circuit Design*. New York: Wiley, 1984, ch. 11.
[8] K. Matsumoto *et al.*, "A 700 MHz monolithic phase-locked demodulator," in *ISSCC Dig. Tech. Papers*, 1985, pp. 22–23.
[9] J. Millman and C. C. Halkias, *Integrated Electronics: Analog and Digital Circuits and Systems*. New York: McGraw-Hill, 1972, ch. 14.
[10] M. Banu and Y. Tsividis, "An elliptic continuous-time CMOS filter with on-chip automatic tuning," *IEEE J. Solid-State Circuits*, vol. SC-20, pp. 1114–1121, Dec. 1985.
[11] B. Hosticka, R. W. Brodersen, and P. R. Gray, "MOS sampled data recursive filters using switched capacitor integrators," *IEEE J. Solid-State Circuits*, vol. SC-12, pp. 600–608, Dec. 1977.
[12] NASA Tech Briefs, "Oscillator or amplifier with wide frequency range," June 1987, p. 26.
[13] F. V. J. Sleeckx and W. M. C. Sansen, "A wide-band current-controlled oscillator using bipolar-JFET technology," *IEEE J. Solid-State Circuits*, vol. SC-15, no. 5, pp. 875–881, Oct. 1980.
[14] K. J. Orlowsky, D. V. Speeney, E. L. Hu, J. V. Dalton, and A. K. Sinha, "Fabrication demonstration of 1-1.5 μm NMOS circuits using optical trilevel processing technology," in *IEDM Tech. Dig.*, 1983, p. 538.
[15] J. F. Kukielka, and R. G. Meyer, "A high-frequency temperature-stable monolithic VCO," *IEEE J. Solid-State Circuits*, vol. SC-16, no. 6, pp. 639–647, Dec. 1981.
[16] L. W. Nagel, "ADVICE for circuit simulation," in *Proc. 1980 IEEE Int. Symp. Circuits Syst.* (Houston, TX), Apr. 28, 1980.
[17] Y. Tsividis, "Minimal transistor-only micropower integrated VHF active filter," *Electron. Lett.*, vol. 23, no. 15, pp. 777–778, July 16, 1987.

A Bipolar 1 GHz Multi-Decade Monolithic
Variable-Frequency Oscillator

Jieh-Tsorng Wu

Variable-frequency oscillators (VFOs), or voltage-controlled oscillators (VCOs), are the crucial elements limiting the performance of many phase-locked loop, modulator, and demodulator applications. In the past, most high-speed bipolar VFO integrated circuits were based upon use of an emitter-coupled multivibrator, because of its suitability for monolithic implementation, high-speed, and symmetrical output waveforms[1,2]. However, for emitter-coupled multivibrator architecture, there is an inherent design conflict between maximum and minimum oscillation frequencies, making it difficult to achieve high-speed and wide-band operation simultaneously[2]. This paper describes the design of a VFO that exhibits no such constraint. The circuit can be optimized for maximum-speed oscillation without compromising low-frequency performance. Implemented in a bipolar technology with 10GHz peak ft, the VFO covers a three-decade frequency range with a maximum oscillation frequency exceeding 1GHz.

The VFO is based on the architecture of the constant-current charge and discharge relaxation multivibrator[1]. Although similar approaches are found in previous design, this maximizes both frequency range and speed in bipolar technologies[1,3]. As shown in Figure 1, the VFO consists of an oscillation frequency control (OFC), a Schmitt-trigger flip-flop (STFF), and an output buffer. The OFC and STFF together constitute a relaxation multivibrator. The OFC alternately charges and discharges node VC1 and node VC2, depending on the state of the complementary signals, VFF1 and VFF2. The STFF then senses the voltages at VC1 and VC2, and switches VFF1 and VFF2 accordingly. Also shown in Figure 1 is a set of representative waveforms for the VCs and VFFs. When VFF1 is high and VFF2 low, VC1 remains constant at Vtp, while VC2 is discharged from Vtp to a threshold voltage, Vbm, set by the STFF. When VC2 reaches Vbm, the STFF initiates a state transition regeneration, changing VFF1 to low and VFF2 to high, which in turn switches the state of the OFC and repeats the relaxation process, but reversing the role of VC1 and VC2.

The oscillation frequency control (OFC) is shown in Figure 2. The complementary signals, VFF1 and VFF2, control the current switch Q5 - Q6. When VFF1 changes from low to high and VFF2 high to low, the VC1's discharging path through Q1 is turned off, and VC1 is charged through Q3. At the same time, the VC2 charging path through Q4 is cut off, and VC2 is discharged through Q2. The discharging rate is set by I1, which is controlled by VF. While VC2 is discharged through Q2, Q4 must remain off under normal operation. Thus, the voltages at VC1 and VC2 have a lower bound, and it can be shown that the voltage swing at these two nodes, ΔVC, must satisfy:

$$\Delta VC \leq R1 \times I2 \qquad (1)$$

for the OFC to function properly.

The Schmitt-trigger flip-flop (STFF) is shown in Figure 3. The STFF combines the functions of two Schmitt trigger comparators and a flip-flop. The two Schmitt triggers are (Q11, Q12, Q17) and (Q13, Q14, Q18). Note that $Q11 \sim Q14$ are all Darlington pairs. Due to the positive feedback, the current switch Q15 - Q16 has two stable states, i.e., the current I11 flows through either one of the transistors in steady state. When VFF1 is high and VFF2 low, Q15 is one, and the Schmitt trig-

ger comparator (Q11, Q12, Q17) is activated. The comparator compares VC2 with a reference voltage at the base of Q12. When VC2 drops near the reference voltage, a regeneration process in the STFF is triggered, which switches VFF1 low, VFF2 high, and activates the (Q13, Q14, Q18) comparator. The voltage swing at node VC1 and VC2, ΔVC can be approximated by:

$$\Delta VC \approx R3 \times I11 \qquad (2)$$

By using a current mirror, an external current, IF, is duplicated as the discharging current of the OFC, I1, and therefore, controls the oscillation frequency of the VFO, f_o. f_o can be expressed as:

$$\frac{1}{f_o} = 2 \left[\frac{qc}{IF + \Delta I} + tst \right] \qquad (3)$$

where tst is the VFO equivalent state transistion time, qc is the total amount of charge that has to be removed for VC1 or VC2 to drop from Vtp to Vbm, and ΔI is the offset current between the external frequency control current, IF, and the actual discharging current for qc. The offset current ΔI is mainly contributed by the base currents of the transistors associated with nodes VC1 and VC2, and the IF signal path. The charge qc is provided by the parasitic capacitors at nodes VC1 and VC2; thus, the operation of the VFO requires no special on-chip or external capacitor.

The VFO is implemented in a self-aligned polysilicon emitter bipolar process with peak ft of 10GHz. A micrograph of the die is hown in Figure 4. The area occupied by the OFC and STFF is 492x286μm. The entire circuit is designed to operate from a single +5V power supply; the total power dissipation of the OFC and STFF is 65mW. The chip is mounted on an RF hybrid package for evaluation. Figure 5 shows the VFO output waveforms at various IF values. The duty cycles of these waveforms are all close to 50%, indicating good symmetrical match among the circuit elements. The VFO oscillation frequency, f_o, is measured with IF varying from 0.1μA to 1mA, at ambient temperature of 23°C and 110°C. The averaged f_o temperature coefficients are -1.1, -0.21, -0.12%/°C at IF = 0.1μA, 10μA, and 1mA respectively. The solid lines in Figure 6 are calculated by using Equation 3 with constants qc, dI, and tst extracted from measurement data. Figure 7 shows the effects of temperature variation on qc, ΔI, and tst. The averaged temperature coefficients for these parameters are respectively 0.21, -1.2, and 0.065%/°C.

Acknowledgement

The author thanks R. Dugan, G. Flower, B. Lai, and D. Lee for technical discussions, also P. Petruno and J. Aukland for support.

[1] Grebene, A.B., "Bipolar and MOS Analog Integrated Circuit Design", John Wiley & Sons, 1st ed., 1984.
[2] Kato, K., et al., "A Low-Power 128MHz VCO for Monolithic PLL IC's", IEEE Journal of Solid-State Circuits, Vol. SC-23, p474-479, April, 1988.
[3] Banu, M., "MOS Oscillators with Multi-Decade Tuning Range and Gigahertz Maximum Speed", IEEE Journal of Solid-State Circuits, Vol. SC-23, p1386-1393, Dec. 1988.

Reprinted from *ISSCC Dig. Tech. Papers*, pp. 106-107, February 1991.

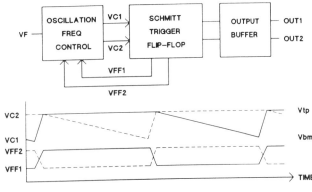

FIGURE 1 — VFO block diagram

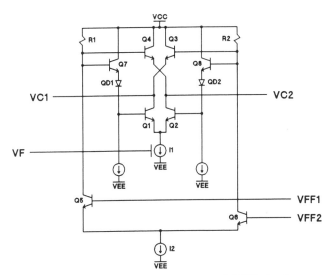

FIGURE 2 — Oscillation frequency control (OFC) circuit
schematic

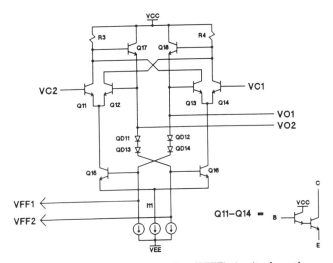

FIGURE 3 — Schmitt-trigger flip-flop (STFF) circuit schematic

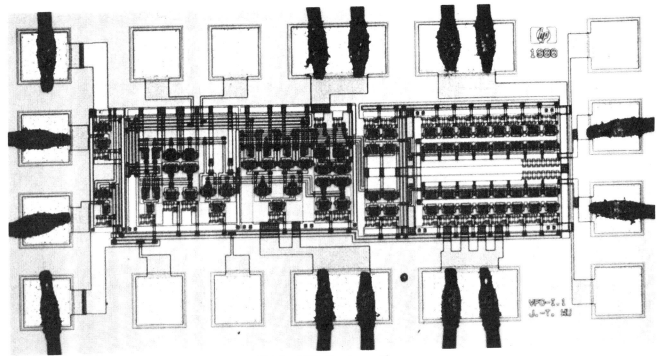

FIGURE 4 — VFO die micrograph

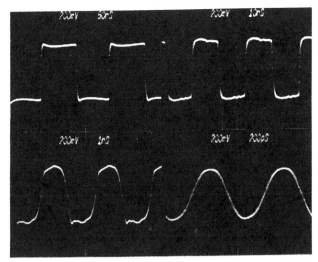

FIGURE 5 — VFO output waveforms (with 6dB attenuation), IF = 1μA, 10μA, 100μA, and 1mA, with corresponding f_0 = 4.3MHz, 27MHz, 270MHz, and 1.02GHz

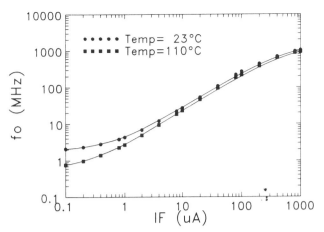

FIGURE 6 — VFO oscillation frequency, f_0, vs. frequency control current, IF

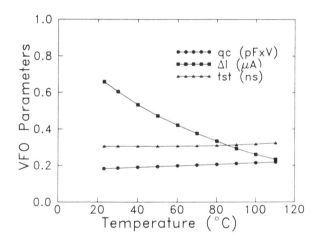

FIGURE 7 — VFO constants vs. ambient temperature

A Digital Phase and Frequency-Sensitive Detector

J. I. Brown

Abstract—A description is given of a novel digital phase and fre-quency-sensitive detector suitable for use in phase-locked loops operating over several octaves in frequency. When the loop is in balance neither the carrier nor its harmonics appear at the output of the detector. The error signal is large and independent of frequency difference when out of phase lock. The performance of the detector in a phase-locked loop operating over two decades of frequency is illustrated.

Phase-locked loops (PLL) are widely used in communication systems and instrumentation. A common constraint is the limited capture range. Baldwin and Howard [1] have proposed a method of achieving maximum capture range by reducing the harmonic content of the output of the phase-sensitive detector and eliminating the loop filter. In instrumentation a PLL is often used in situations where the phase error is always small,

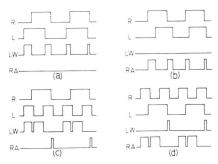

Fig. 1. Various combinations of input phase and frequency differences and the resultant desired outputs from which the flow table of the logic is designed. (a) Local oscillator in phase lead. (b) Local oscillator in phase lag. (c) Local oscillator frequency high. (d) Reference frequency high.

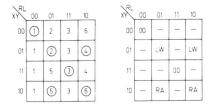

Fig. 2. Condensed flow table and output map derived from Fig. 1. Stable states circled.

i.e., the loop is in a static rather than a dynamic state. This suggests the possibility of designing a phase-sensitive detector (PSD) with zero harmonic output when in balance. The combination of such a PSD in a loop with zero steady-state error (proportional plus integral loop filter) offers a system with potentially infinite capture range and negligible phase ripple

at balance. To take advantage of the large capture range the PSD should provide a correction signal as large as possible when out of phase lock. Conventional PSDs such as analogue multipliers and exclusive OR gates [2] have large harmonic outputs at balance and provide an error signal which decreases inversely with difference frequency when out of lock [3]. Nield [4] has described a PSD with zero output at balance but its perfor-mance with gross frequency differences is ambiguous. Other phase and frequency detectors use two-mode operation to achieve fast capture [5], [6].

The new PSD is a simple gated sequential logic circuit which has been designed using classical synthesis techniques [7]. It has two separate out-puts, a raise and a lower, and it accepts digital reference and local oscillator signals which should be of unity mark space ratio when in lock. (This can always be achieved by passing the inputs through binary dividers.) When either input leads or lags the other the appropriate output (raise or lower) is energized for a time proportional to the phase error; both positive and negative going transitions of each input can generate an output, and only when transitions on both inputs occur simultaneously is there zero output.

The four fundamental combinations of inputs and the desired outputs are shown in Fig. 1. From these a primitive flow table is developed which can be condensed to the flow table and output map in Fig. 2.

The Boolean relations derived from Fig. 2 are

$$X = R \cdot L + R \cdot \overline{Y} + L \cdot X$$
$$Y = R \cdot L + L \cdot \overline{X} + R \cdot Y$$
$$RA = X \cdot \overline{Y}$$
$$LW = \overline{X} \cdot Y$$

where X and Y are the secondary variables, R is the reference input, L is the local oscillator input, RA is the raise frequency output, and LW is the lower frequency output. Fig. 3 shows the resultant realization in transistor-transistor logic (TTL) gates.

The PSD produces an average output which is proportional to phase error over the range $-\pi$ to $+\pi$. When a frequency error exists, the ap-propriate output has an average value which approaches half full scale as the frequency difference increases, while the opposite output approaches zero. It should be noted that the PSD is only suitable for use in systems with

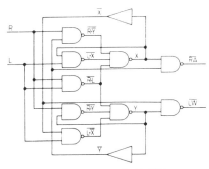

Fig. 3. The new PSD realized in TTL gates.

Reprinted from *Proc. of IEEE*, pp. 717-718, April 1971.

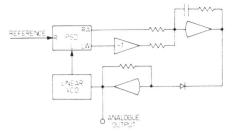

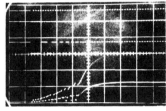

Fig. 4. A wide range frequency to voltage converter using a phase-locked loop incorporating the new phase-sensitive detector.

Fig. 5. Response of the frequency to voltage converter to a frequency step from 60–1200 Hz. Top: reference input; middle: loop local oscillator; bottom: control voltage to VCO 5 ms/div.

low noise level since otherwise the detector will attempt to lock the loop to the number of zero crossings rather than the fundamental input frequency.

This PSD has been incorporated into a loop which converts a frequency in the range 50–5000 Hz to an analogue voltage with an accuracy of 0.1 percent (Fig. 4). The capture time is of the order of 30 ms. The output from the loop filter is passed through an exponential transfer function before being applied to the linear VCO. This provides rapid locking at both high and low frequencies and maximizes the range of stable operation. The analogue output is taken from the input of the VCO. Fig. 5 shows the response of the loop to a step change in frequency from 60–1300 Hz. Relocking takes place in 15 ms.

REFERENCES

[1] G. L. Baldwin and W. G. Howard, "A wide-band phase-locked loop using harmonic cancellation," *Proc. IEEE* (Lett.), vol. 57, Aug. 1969, pp. 1464–1465.
[2] G. Pasternack and R. L. Whalin "Analysis and synthesis of a digital phase-locked loop for FM demodulation," *Bell Syst. Tech. J.*, vol. 47, Dec. 1968, pp. 2207–2238.
[3] G. S. Moschytz, "Miniaturized RC filters using phase-locked loop," *Bell Syst. Tech. J.*, vol. 44, May–June 1965, pp. 823–870.
[4] P. N. Nield, "Zero crossing phasemeter with sense indication," *Electron. Eng.* (London), vol. 40, May 1968, pp. 282–284.
[5] D. Richman, "The DC quadricorrelator: a two-mode synchronization system," *Proc. IRE*, vol. 42, Jan. 1954, pp. 288–299.
[6] G. G. Gassmann, "Neue Phasen- und Frequenzvergleichschaltungen," *Arch. Elek. Übertragung*, vol. 15, Aug. 1961, pp. 359–376.
[7] M. P. Marcus, *Switching Circuits for Engineers*. Englewood Cliffs, N. J.: Prentice-Hall, 1962.

A 3-state phase detector can improve your next PLL design

Amazingly simple, this technique boasts wider operating ranges and needs less output filtering. Do you know the principles involved?

C. Andrew Sharpe, The Bendix Corp.

The growing army of digital and analog users of phase-locked loops (PLL's) already know that PLL's are superior to standard detection methods. These control devices can readily restore phase (and frequency) of a signal drowned in noise.

A key element in a PLL is the phase detector, which typically is of the cross-coupled latch, set-reset flip-flop, or exclusive-OR variety. Unfortunately, these phase-detector circuits, although simple with respect to hardware, suffer from several major drawbacks:

- The output information for phase detection is contained in the width of the output pulse, which requires substantial filtering to remove the reference frequency content.
- This class of detectors often responds to harmonics of the input signals, restricting the operating range of the input frequency.
- If some type of memory does not detect a frequency difference, the capture range (maximum frequency difference between the two inputs before the system can start to lock) is one of the design variables that

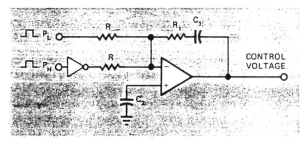

Fig. 1—**A positive input** on P_L lowers the control voltage, while a positive input on P_H causes it to rise. The very small pulse width required to compensate for op-amp leakage and finite gain has an unusually low reference component.

Fig. 2—**Since the machine has two output lines,** two states are immediately evident: states A and B, one for $f_A > f_B$, the other for $f_B > f_A$. When the machine is in state A, repetitive positive transitions on its A input will not change the state.

directly oppose the other loop constraints, such as filtering and external noise.

The 3-state phase detector, however, eliminates these objections.

Control is the goal

This new approach to phase detection can best be understood by considering how the op-amp arrangement of **Fig. 1** generates a control voltage to drive the variable phase/frequency element. Observe that a positive input on P_L will lower the control voltage, while a positive input on P_H will cause it to rise. If $P_L = P_H$, no change in the control voltage will occur.

With this method of control-voltage generation, then, the problem of frequency/phase detection reduces to activating the proper input line that will drive the circuit into lock. Once phase lock is established, the operation requires no further change in control voltage, and neither line will be active. (Of course, the latter is an idealized condition, since leakage currents and finite op-amp gain always require some average voltage out of the phase detector. Nevertheless, the very small pulse width typically required has a

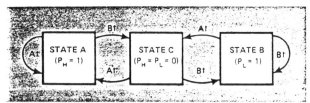

Fig. 3—A single positive transition on input B (or A) after repeated transitions on A (or B) sends the machine to state C, where both outputs are low—even if both frequencies are unequal. As the frequencies begin to track, the width of the pumping pulse will be directly proportional to the phase difference.

reference component substantially smaller than that produced by the previously mentioned techniques.)

"State-ing" the problem

Before we can develop a suitable digital control circuit, we must establish some fundamental laws of frequency comparison:

- For two pulse trains, if $f_A > f_B$, then at some point in time two or more reference transitions of pulse train A will occur between two reference transitions of pulse train B.
- If $f_A = f_B$, at <u>no</u> time will two reference transitions of pulse train A occur between two reference transitions of B.

By applying these postulates, you can build a 3-state phase detector in a manner that avoids the traditional and lengthy asynchronous-machine design problem. The goal is to combine logic devices that will:

- provide positive pulses on output line one if $f_A > f_B$, a condition which may be detected by looking for two consecutive pulses on the A input line
- provide pulses on output line two if $f_B > f_A$
- hold HIGH output line one for the duration of the phase difference <u>if</u> $f_A = f_B$ and the positive transition of A leads that·of B
- hold HIGH output line two for the duration of the phase difference <u>if</u> $f_A = f_B$ and the positive transition of B leads that of A.

By defining two output lines for the machine, then, two states are immediately evident as shown in **Fig. 2.**

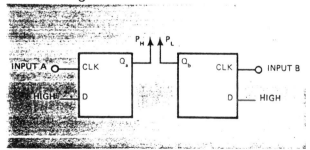

Fig. 4—A leading-edge triggered "D" type flip-flop with its "D" input tied high satisfies the requirements of states A and B.

Three states are better than two

If the machine is in state A (or B), repetitive positive transitions on the A (or B) input will obviously have no effect on that state. But if a transition should occur on B (or A) after consecutive transitions on A (or B), you do not want to go directly to state B (or A). Why? Because if state A does transition directly to state B, one output will always be HIGH, thus violating the criteria that $P_L = P_H = $ ZERO when $f_A = f_B$ and the phase angle equals zero. Getting around this problem requires adding an intermediate third state, C, to the machine (**Fig. 3**).

Now, a positive transition on the B (or A) input occurring after repetitive transitions on the A (or B) input will send the machine to state C. At state C outputs P_H and P_L both will be LOW (no pumping action). Notice here that regardless of the initial state, two such transitions on the A (or B) input will put the machine into state A (or B) signifying that the A (or B) input has the higher repetition rate.

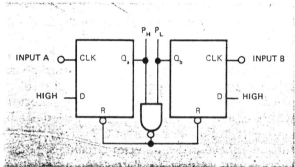

Fig. 5—State C is easily achieved by adding a NAND gate that resets the state A and B flip-flops. This, then, is the basic dual-D phase/frequency detector.

Phasing out your differences

As the rate of A (or B) decreases to equal B (or A), the inputs to the machine will have the phase of A (or B) leading the phase of B (or A). This results in the same output being held HIGH for both higher frequency and leading phase, satisfying all of the goals previously stated.

When the machine is initially turned on, it can be in any of the three states. But as soon as two consecutive transitions appear on one of the inputs, the machine will go to the state with the higher input frequency and then alternate between that state and state C. Note that only the state corresponding to the input line of higher frequency will be pulsed HIGH.

As the frequencies become identical, the width of the pumping pulse will be directly proportional to the phase differences. And as the phase difference approaches zero, the output pulse also will become zero.

This completes the state diagram for a 3-state

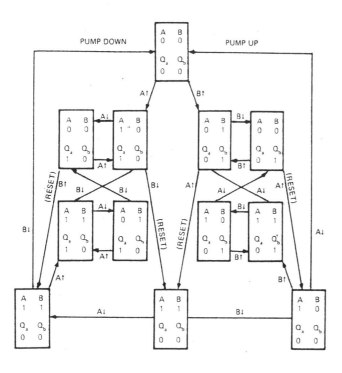

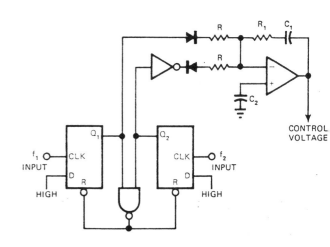

Fig. 6—This state-transition diagram explores all possible input/output conditions of the dual-D flip-flop 3-state circuit of **Fig. 5.**

phase/frequency detector. Since the dynamic characteristics of this particular state map are based on positive transitions, implementation must be accomplished by using a storage element that changes states only on positive transitions. Rather than attempt the classical exercise in asynchronous machines, the following method simplifies the design procedure considerably.

Okay, bring on the hardware

Since states A and B respond identically to their A and B inputs, the same circuit can be used for both, taking advantage of symmetry. Recall that once either state has been reached by a positive input transition, the circuit's output must remain HIGH for consecutive positive transitions. A leading-edge triggered type-D flip-flop with the input tied HIGH meets this requirement (**Fig. 4**).

In a state C network, if the machine is in either state A or B, a transition on the opposite input must reset both output lines to ZERO. But when the machine is in state A and the leading edge of a pulse appears on the B input, both outputs, Q_a and Q_b, will be HIGH. If this condition can be

detected and used to set both devices to ZERO immediately, then state C has been realized. A NAND gate driving the reset inputs of the state A and B networks achieves this operation (**Fig. 5**).

Comparing the workings of this circuit to the final version of the state diagram (**Fig. 6**), you can see that the model has been successfully constructed for all possible input combinations. This state-transition table also indicates all possible input/output combinations, thus assuring that no "locked-up" conditions exist.

The lines labelled "reset" correspond to the NAND gate's resetting flip-flops to ZERO. Out of a possible 16 states (two inputs plus two outputs = 2^4), 12 stable conditions exist. Note that the four remaining states (Q_a and Q_b—both HIGH) generate an immediate reset back to a stable state. **Fig. 7** shows the complete frequency/phase comparator and the 3-state charge pump (control-voltage generator).

The digital network used in this realization is only one example of an infinite number of logic-circuit combinations that can be employed. Virtually every major IC manufacturer offers at least one device that will perform the digital function; and many of the phase-locked loop packages available today will use a network satisfying the state-diagram requirements. However, the dual-D version just described is the easiest to understand, since by choosing an edge-triggered type-D flip-flop with negative-going reset, we have eliminated many of the difficulties encountered in determining a circuit by its schematic.

Applying the concept

A practical example should put the concept of a 3-state phase/frequency detector in proper per-

Fig. 7—The completed 3-state phase/frequency detector combines the frequency/phase comparator of **Fig. 5** with **Fig. 1's** control-voltage generator (tri-state charge pump).

spective. So let's consider the design of a digital frequency synthesizer in which channel change from the low end to the high end of the frequency band occurs by changing the modulus of a programmable divider, **(Fig. 8).**

When the channel change occurs at t = 0, the VCO frequency cannot instantaneously change to the new desired channel, due to the time constant of the system. However, <u>N changes instantaneously</u>, providing the variable input of the phase detector with a sometimes radically different frequency. In fact, it is not unusual for the variable-channel phase-detector input frequency to equal five or 10 times the reference input frequency at t = 0+.

The frequency/phase detector will perform the desired detection to bring the VCO frequency to the new value ($N \times f_{ref}$) by applying pulses on the proper phase-detector output. This will, in effect, incrementally slew the control line to the new voltage. The rate at which the "slew" occurs—hence the time for wide-channel change—depends on the average value of the pulse. Therefore, attention must be given to the average pump voltage (available as a function of frequen-

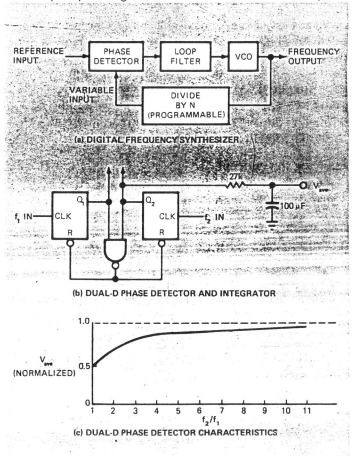

(a) DIGITAL FREQUENCY SYNTHESIZER

(b) DUAL-D PHASE DETECTOR AND INTEGRATOR

(c) DUAL-D PHASE DETECTOR CHARACTERISTICS

Fig. 8—The digital frequency synthesizer (a), one application of 3-state phase detectors, ranges across the frequency band as the modulus of the divider changes. The integrator output **(b)** determines the normalized dual-D phase-detector characteristic **(c).**

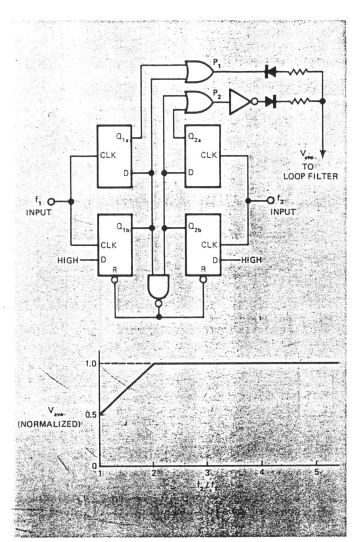

Fig. 9—Quad-D circuit shortens lock time when f_2 is significantly greater than f_1.

cy) to determine the lock time of the system. **Fig. 8b** shows the normalized characteristic of the dual-D phase detector and the method of measurement (**Fig. 8c**). In the illustrated circuit, V_{ave} (normalized) = $(1 - f_1/2f_2)$—providing the pulses are not harmonically synchronous.

Note that when f_2 occurs at a much higher rate than f_1, the output will be HIGH nearly all the time. When f_2 is very slightly greater than f_1, however, the output will be HIGH approximately 50% of the time (over a long interval).

In typical wide-channel change cases where f_2 = $5f_1$ or $10f_1$, the average voltage will be above 90% of the pulse height. So, for this region, V_{ave} = $V_{pulse\ height}$ is a reasonable approximation. In fact, whenever $f_2 > f_1$ (or vice versa), it is generally useful to assume that V_{ave} = $V_{pulse\ height}$ as a first approximation—and realize that this performance can never be achieved. The actual error will depend on the initial frequency ratio, but magnitudes of 10 or 20% are typical.

Why not save some "time"?

Fig. 9 shows a configuration that speeds up lock time by clamping the proper output line HIGH when $f_2 = 2f_1 + \epsilon$. This quad-D contains two additional storage elements. In combination with the action of the dual-D configuration, these added elements maintain the proper output HIGH as soon as three consecutive pulses appear on one input between two pulses on the other input.

The quad-D system performs as a normal dual-D for phase detection when alternating pulses are present, and it transitions into the expanded region as f_2 approaches $2f_1$. When $f_2 > 2f_1$, the normalized output voltage is the full pump-pulse height. This results in a much better approximation for calculating lock time when large frequency steps are involved.

Quad-D implementation is such that it is a direct replacement for the dual-D version. Since the phase-detector gain for a phase-locked loop (V/rad) is calculated only when the two input frequencies are identical (differing only in phase), the gain remains the same. For a small maximum channel change where $f_2(\text{max.}) = 1.5f_1$, no noticeable operating difference exists. In systems where $f_2 = 5f_1$ or $10f_1$, however, the quad-D significantly improves lock times, with the magnitude of the improvement depending upon the initial frequency separation. □

GaAs Monolithic Digital Phase/Frequency Discriminator

I. Shahriary, G. Des Brisay, S. Avery, P. Gibson

Hughes Aircraft Company, Space and Communications Group
Los Angeles, CA 90009

ABSTRACT

A GaAs monolithic digital phase/frequency dis-
cirminator with a linear characteristic at the
zero phase crossing has been developed at
Hughes SCG. The maximum operating frequency of
480 MHz is about ten times faster than that of
existing silicon devices. These devices are
currently used in low noise phase-locked loops.
We discuss the design, operation, and use of
this device.

INTRODUCTION

Digital phase/frequency discriminators are an
essential part of modern wideband frequency synthesizers. For good performance, the digital
phase/frequency discriminators need to have
linear characterisitcs and a predictable scale
factor, especially near the zero phase point.
The Motorola digital phase/frequency discriminator has been available as an IC for over a decade. Silicon versions of the discriminator include the CMOS Motorola MC4044 and ECL Fairchild
11C44 which have operating frequencies of under
50 MHz. The crossover distortion of these discriminators near zero phase is well known. For
example, in its ECL data book, Fairchild presents
curves comparing crossover distortion of the
MC4044 with that of the 11C44 which, while improved, is still nonlinear (see Figure 1).
We have developed a GaAs monolithic digital phase/
frequency discriminator that has no detectable
crossover distortion and operates up to 480 MHz.
The design and performance of this discriminator
is presented in the following sections.

DESIGN

The logic diagram of the GaAs monolithic digital
phase/frequency discriminator is shown in Figure
2. It consists of four cross-coupled gate configurations known as RS latches number I, II, III,
and IV. Two additional gates are labeled reset
gates. In order to describe the circuit's operation, let us assume for initial conditions that
latches I and II are in their reset condition (Q's
low, $\bar{Q}$'s high) and that latches III and IV are in
their set condition (Q's high, $\bar{Q}$'s low). Also
assume that both input signals A and B are in a
low logic state. When one of the input signals
(assume A) goes high, latch I becomes set and discriminator output F_A goes high. At this time, the
other latch outputs remain unchanged. Any further
action by A at this time produces no latch changes.
Next, if input B goes high, it sets latch II producing a logic one at output F_B. At this moment
both of the input signals of the reset gate have
become high forcing its output low and resetting
both latches III and IV. This action terminates
the output pulses at the Q outputs of latches I and
II by forcing them low. This pulse termination
occurs at both outputs whether or not either or
both input signals have returned to a low logic
level. If either input had returned to its low
logic level prior to termination of the output
pulses, the affected latch (I or II) would complete its reset at this time forcing its Q high.
This in turn would set the associated latch (III
or IV). If either or both of the input signals

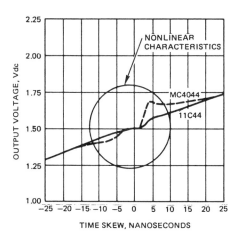

Fig. 1 Performance of Silicon Phase Detectors
(Fairchild ECL Data Book)

Reprinted from *IEEE GaAs IC Symp. Dig. of Tech. Papers*, pp. 183-186, October 1985.

delay going low until after the output pulse terminations occur, the affected Q outputs of latches I and II will remain low until the latch input goes low. It will then go high setting its associated latch (III or IV).

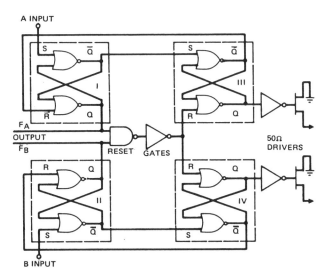

Fig. 2 Phase/Frequency Discriminator Logic Diagram

The discriminator is an edge-triggered device. The first input to go high causes its associated output to go high. The last of the two input signals to go high causes its associated output to go high and initiates the termination of both output pulses as described above. This action produces output pulses that can be differenced by an amplifier to produce a linear voltage versus phase discriminant (approximately ±2 radians) of the phase relationship between the appropriate edges of the two input pulse streams. Figure 3 is a timing diagram showing input pulses and discriminator outputs for a condition where the phase of input B lags the phase of input A.

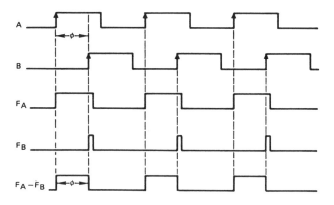

Fig. 3 Discriminator Timing Diagram

Another characteristic of the discriminator is that it will produce an output of the correct sense in response to a frequency difference of the two input pulse streams. This occurs very naturally since the higher frequency input will consistently be the first to set its associated latch (after a common reset), and the lower frequency input will consistently be the one to produce the common reset. This will result in minimum width pulses occurring at the output of the discriminator associated with the low frequency input and varying width pulses occurring at the other output. The monolithic GaAs discriminator was designed and simulated using standard buffered field effect transistor (FET) logic. The output latches are buffered to drive 50Ω and can be easily interfaced to ECL logic. Power consumption is near 0.5 watt with 60 percent of the power consumed in the output buffers. The integrated circuit was processed at Avantek, Inc., Santa Clara, California. A photomicrograph of the chip is shown in Figure 4. This chip measures 40 x 50 mils.

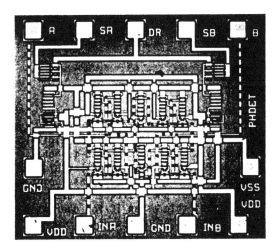

Fig. 4 Photograph of GaAs Monolithic Phase Detector

PERFORMANCE

To determine the maximum frequency of operation, assume that input A and output F_A are both high (logic 1) and input B and output F_B are both low (logic 0). Now if at time t=0, input B goes high, two gate delays ($2 t_g$) later, output F_B will go high. From the logic diagram it can be seen that from the time that output B goes high, given that output A is already at logic one, it takes five gate delays ($5 t_g$) for latches I and II to be reset forcing outputs F_A and F_B to go low. Thus seven gate delays after input B goes high, both outputs are reset. During this period of seven gate delays, input A may go low but may not return to logic 1. The limiting case is when signal A and B are 180° out of phase with a half period of seven gate delays (see Figure 5). Therefore, the maximum operating frequency is given by

$$F_{max} = \frac{1}{14t_g}$$

Assuming an average gate delay of 150 ps, the maximum frequency of operation will be about 500 MHz. This frequency is in agreement with the measured maximum frequency of 480 MHz.

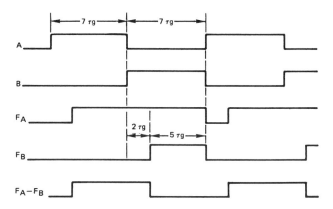

Fig. 5. Timing Diagram at Maximum Frequency

The sensitivity of the discriminator can be derived from the timing diagram of Figure 5. Note that the average value of the output signal is

$$V\ out\ av = \frac{\phi}{360} \times Amplitude$$

For GaAs buffered FET Logic, the amplitude is 2 volts. Therefore, the discriminator sensitivity is $\frac{2}{360}$ = 5.5 mv/deg.

Linearity of the discriminator was determined using the test configuration of Figure 6. Two phase-aligned signals were applied to the phase discriminator inputs, to the A input directly, and to the B input through a variable delay line. The amount of delay inserted into the B signal was measured by a calibrated oscilloscope. The discriminator outputs were sent to a differential integrating amplifier with a gain of 7.5, forming the error voltage measured by a programmable digital voltmeter. The input frequency was 400 MHz. Plots of error voltage versus phase difference for three discriminators selected at random are shown in Figures 7 and 8. From the slopes of the resulting curves, taking into account the amplifier gain, the discriminator sensitivity is 4 to 4.5 mv/deg.

Phase Detector Application

The digital phase/frequency discriminator has been incorporated into a phase-locked loop. A

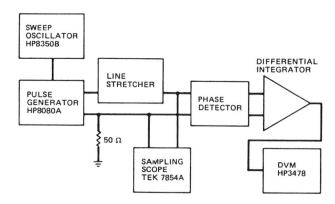

Fig. 6 Test Setup for Measuring the Discriminator Characteristic

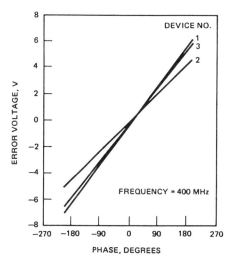

Fig. 7 Measured Discriminator Characteristic (±180°)

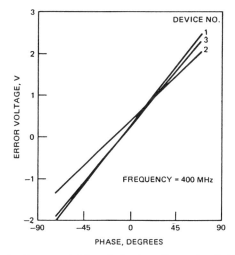

Fig. 8 Measured Discriminator Characteristic Near Lock

block diagram of the loop is shown in Figure 9. This application emphasizes the low noise characteristics of the monolithic GaAs phase detector when operated in a closed loop. Voltage controlled oscillators with bandwidths from 500 MHz to 1 GHz and centered at 10 GHz were used. Divider ratios were 2 to 100. The reference frequency was varied from 10 to 450 MHz with no apparent degradation in performance. Figure 10 is the spectrum analyzer display of the phase-locked loop output with a reference frequency of 450 MHz. The noise level shown is more than 100 dB down from the signal. (Note the resolution bandwidth is 10 kHz.)

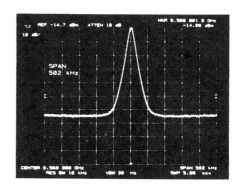

Fig. 10   Phase-lock Loop Performance

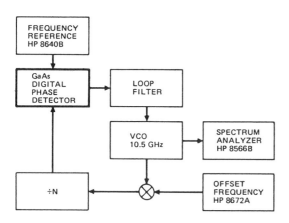

Fig. 9   Phase-lock Loop With Discriminator

CONCLUSIONS

A GaAs phase/frequency discriminator with excellent linearity and low noise performance has been developed. It operates at signal speeds to 500 MHz. It has been used in a phase-locked loop with a noise level more than 100 dB below the carrier. This indicates promising performance in future frequency synthesizers.

A NEW PHASE-LOCKED TIMING RECOVERY
METHOD FOR DIGITAL REGENERATORS

By

J. A. Bellisio
Bell Telephone Laboratories, Incorporated
Holmdel, New Jersey   07733

ABSTRACT

Timing can be extracted at low cost from a random baseband digital data signal by locking a local oscillator
to the reshaped bit stream. To avoid use of a very
stable oscillator, a narrowband phase-locked loop is
forced to synchronize to the scrambled but otherwise
unrestricted data by means of an auxiliary frequency
sensitive loop. The novel comparator which can measure
the phase and frequency difference between a local
oscillator and a random data signal is described. The
method is simple to implement even at high frequency,
synchronizes rapidly, and has excellent acquisition and
jitter control performance when operated in a cascade
of regenerators. A loop with an acquisition range of
greater than ±20 percent of the data rate and 0.1 percent noise bandwidth has been demonstrated.

INTRODUCTION

Whenever it is necessary to regenerate or process a
digital signal, the correctly phased companion timing
wave must be made available. Because this periodic
clock signal contains virtually no information, it
should be inexpensive to obtain. Older timing methods[1]
using code restrictions, extra signaling levels, added
timing waves or separate transmission paths pay a great
deal for the clock. It has been demonstrated that a
statistically random data stream contains sufficient
information to synchronize a properly constructed phase
locked loop (PLL) retiming circuit.[2]

The PLL has been one of the major factors in reducing
the cost and raising the quality of digital timing.
Before describing a further improvement in this technique we must be aware of a number of considerations
which are unique to digital systems. Figure 1 is a
block diagram of a basic digital regenerator and a
typical timing recovery circuit is shown in Figure 2.
The equalized baseband data signal is prefiltered and
applied to a nonlinear circuit in the timing information extractor. Prefilter shapes to optimize performance are described in the literature.[3,4] In one
realization the data is differentiated, full wave
rectified and clipped.[2,5] The overall effect is to
produce a standardized, unipolar pulse coincident with
every transition of the input data. In a binary, non-
return-to-zero random data stream, not all transitions
are present. Data scrambling, that is, modulation by
a pseudo-random sequence, can insure a long term
average transition density of 50 percent but there
still can be long intervals without timing information.
The pulses occur with the probability of a transition
in the random data stream (usually 1/2). The mean
position of each pulse is at the data transition time
but there is a statistical time deviation resulting
from intersymbol interference and noise.[6] The stream
of pulses forming the input to the PLL is thus a
cyclostationary random process.[5] It is the function
of the PLL to generate a stable timing wave locked to
the best estimate of the transition times. The PLL
must be capable of extrapolating over long periods when
no input pulses are available. This filtering requirement dictates the use of a very narrow band loop

resulting in a small pull-in range. Instead of using
a very stable voltage controlled oscillator (VCO) to
insure pull-in, we describe an acquisition range
enhancing circuit with the unique properties required
for digital clock recovery.

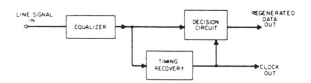

Fig. 1: Basic Regenerator For Baseband Data

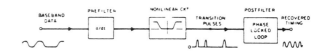

Fig. 2: Typical Timing Recovery Circuit

THE TIMING PLL

In designing the PLL to follow the timing information
extractor, a major consideration is that the phase
detector must be able to work with the random transition pulses as input. There should be no large
excursions in the phase detector output during intervals
containing no transitions. Many phase detection
schemes employing logic elements (counters or flip-
flops) work with sinusoidal signals but fail with data
inputs by producing spurious outputs when transitions
are missing.[7] A detector with a quasi-linear multiplier can be arranged to produce no output when either
input is absent and can thus ignore missing transitions.

We would expect the PLL to have high loop gain near
d-c to minimize static phase offset resulting from
oscillator mistuning. As previously mentioned, the
loop should have a narrow bandwidth to provide good
averaging over the statistics of the input pulse
stream. Closed loop bandwidths on the order of 10^{-4}
of the bit rate are not uncommon in timing circuits.

When in lock, a PLL can be considered as a linear
feedback control system coupling small variations in
the phase of the input to the phase of the VCO output.
The magnitude of the transfer function relating input
to output timing jitter (as phase noise is called in
digital systems) can exceed unity over some range of
jitter frequencies. That is, for some frequencies,
jitter out of the PLL can be greater than the input

Reprinted from *IEEE Int. Comm. Conf. Rec.*, vol. 1, pp. 10-17 to 10-20, June 1976.

jitter. The jitter amplification will be cumulative for a chain of identical regenerators and must be well controlled.[6] The shape of the jitter transfer function, usually not a major consideration in the design of isolated PLL systems can thus be of controlling importance in a digital system.

Before a phase locked oscillator can produce a usable timing signal, the loop must lock to the input signal. The pull-in range of the PLL must necessarily be sufficient to achieve lock under all conditions of signal and circuit variation. In a digital transmission system consisting of a chain of regenerators of the type shown in Figure 1, the time required between application of signal to the line and lock of all timing circuits can become crucial. If the individual regenerator timing circuits are such that no loop can lock until the loop in the previous regenerator has locked, then the pull-in time for a long line increases linearly and usually excessively with the number of regenerators. Individual timing circuits should be designed so as to cause all regenerators to pull-in on a nearly simultaneous basis. This is an important criterion in the consideration of pull-in enhancement schemes and usually makes sweep and switched bandwidth methods unsuitable.

The basic dilemma of elementary PLL design is that the pull-in range, static phase offset and jitter smoothing characteristics are strongly coupled. Since the pull-in range of a simple PLL is of the same order as the bandwidth, the designer has been forced to use a very stable, expensive, voltage controlled oscillator in digital applications.

A PHASE-FREQUENCY LOCKED TIMING METHOD

The key to solving the basic PLL problem is a separation of pull-in and oscillator stability considerations from the performance requirements after phase lock has been achieved. If acquisition is neglected we can design a narrow band PLL timing system which satisfies all transient and jitter filtering requirements using conventional linear PLL techniques. A low frequency integrator (high dc gain) should be included in the loop to reduce to zero the static phase offset from oscillator mistuning. If this PLL is built with a low stability VCO and initial conditions are such that the loop is locked, then system performance will be as desired. If the VCO is initially unlocked, then the PLL loop bandwidth (constrained to be narrow by jitter considerations) will be much too small to insure pull-in. This is true a fortiori with the above design since the second loop integration will quickly accumulate offset voltages and drive the VCO to an extreme. If a frequency comparator is added to the PLL as in Figure 3, the unlocked VCO is forced toward the narrow pull-in range of the PLL and lock will be achieved for a wide range of conditions. The loop integrator then insures that the VCO remains locked permanently. The frequency comparator produces an output proportional to the sign and magnitude of the frequency difference between VCO and data stream. Whenever the VCO is unlocked, voltage is applied to the loop integrator such as to cause the VCO to slew toward the data frequency. The VCO eventually will fall within the pull-in range of the PLL, the loop will lock, and the frequency difference will be zero. The output of the frequency comparator can then also be zero and this added portion of the system will have no impact on PLL dynamics. If there is a residual offset voltage from the frequency comparator, phase lock will still occur provided that the PLL can overcome the offset. Note that if for some reason pull-in is not achieved, the sign of the frequency difference will change and frequency slewing will reverse. Indeed, with the PLL

portion of the system disabled, the system behaves as a frequency locked loop (FLL) which tends to reduce the frequency offset to zero.

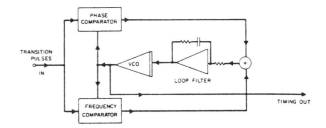

Fig. 3: Phase/Frequency Locked Loop

The acquisition enhancing technique described above is commonly used in PLL systems with deterministic inputs.[7] It will now be demonstrated that a frequency comparator workable with a random data stream can be constructed easily and inexpensively.

THE PHASE/FREQUENCY COMPARATOR

The spectral component near the bit rate contained in the data transition pulses of Fig. 2 can be represented as

$$S(t) = D(t)\cos\left(\omega_d t + \phi_d(t)\right) \qquad (1)$$

The frequency and phase of the periodic process underlying the pulses are ω_d and ϕ_d.

The data transition pulses are applied to two separate multipliers as in Figure 4. The multipliers are driven by in-phase and quadrature components of the local oscillator at ω_0. The two phase difference signals extracted by low-pass filtering then represent the in-phase and quadrature phase detection components. The outputs of the low-pass filters are approximately

$$V_I(t) = \frac{D(t)}{2}\cos\left((\omega_0-\omega_D)t - \phi_d(t)\right) \qquad (2)$$

$$V_Q(t) = \frac{D(t)}{2}\sin\left((\omega_0-\omega_D)t - \phi_d(t)\right) \qquad (3)$$

The assumptions here are that (1) the low-pass filter rejects all multiplication products with frequencies around $(\omega_0+\omega_D)$ or higher and (2) the variations in $D(t)$ or $\phi_d(t)$ do not cause significant a.m. and p.m. components of the $(\omega_0+\omega_D)$ term to appear at low frequency. These are modest constraints for ω_0 close to ω_d.

The two phase detector signals contain the information which indicates whether the local oscillator frequency is above or below the data frequency. Signal $V_I(t)$ leads $V_Q(t)$ by 90 degrees when $\omega_0 > \omega_d$. This phase relation reverses when $(\omega_0-\omega_d)$ changes sign. Fig. 4 includes a circuit to exploit this property. The phase detector signals after low pass filtering are applied to quantizers. The quantizer outputs have the signs of $V_Q(t)$ and $V_I(t)$. The output of the differentiator following the V_I quantizer consists of positive and negative pulses occurring near the zero crossings of $V_I(t)$. Note that all of the circuits following the phase detectors operate in the frequency range characterized by the difference frequency $(\omega_0 - \omega_d)$ and are thus relatively slow circuits. The square wave and pulses derived above are then multiplied together in a circuit which can be realized with logic gates. Because of the phase relationship between V_I and V_Q, the multiplier produces all positive pulses for $\omega_0 > \omega_d$, and all negative pulses for $\omega_0 < \omega_d$. The sign of the pulses show the direction of the frequency offset. The rate of pulses is proportional to the slip rate. Very importantly, when $\omega_0 = \omega_d$ there are no pulses and the output is identically zero except for the small offset voltage of the multiplier itself. This offset can be made quite small by designing the multiplier to take advantage of the fact that its only allowed outputs will be positive pulses, negative pulses, or zero.

slip pulses is then held off during gaps in the input but there is little change in acquisition performance from a "cold" start.

EXPERIMENTAL RESULTS

A timing recovery circuit has been constructed for a clock rate of 44.7 Mb/sec. It was instructive to disconnect the feedback paths and investigate the performance of just the frequency comparator.

The dc component of the slip pulses is shown in Fig. 5. The important features of the method are clearly demonstrated. The circuit built with conventional components works with pseudo-random data to generate a frequency difference signal. No resonant elements or precise phase comparators are required. The circuit produces almost no output when there is no frequency difference. The sign of the frequency correction voltage is always correct for use in the phase/frequency locked loop over the entire offset frequency range measured.

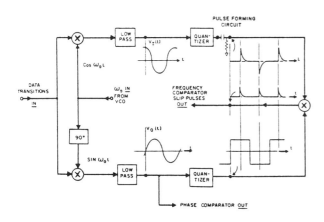

Fig. 4: Phase/Frequency Comparator

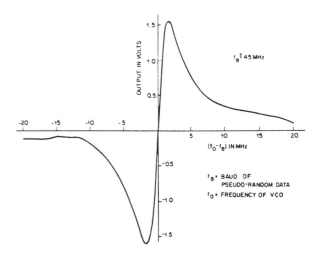

Fig. 5: Frequency Comparator Characteristic

THE PHASE/FREQUENCY LOCKED LOOP (P/FLL)

The pulses from the frequency comparator, Fig. 4, are summed with the phase comparator output as in Fig. 3 and are applied to the PLL loop filter. Note that signal V_Q contained within the frequency comparator can be conveniently used as the phase comparison. When lock is obtained, V_Q is then forced toward zero and V_I to its maximum value holding the V_I quantizer in a fixed state. Pulses are only produced for complete half-cycle slips between clock and data. After lock there is no differentiator output. The frequency comparator then remains quiet for small phase deviations and has no effect on PLL dynamics.

When the input signal is small for a long period of time, the outputs of both low pass filters drop toward zero. To guard against the production of spurious pulses, the low pass filters of Fig. 4 can be shaped to provide higher gain near d-c.* The production of

When the phase and frequency locking loops were combined, a pull-in range of greater than ±20 percent of the bit rate was obtained. The parameters of the proportional plus integral PLL were such that a noise bandwidth of 0.1 percent with .04 dB of jitter peaking was realized.

CONCLUSION

The P/FLL timing extraction technique described should provide a general solution to an old problem. Timing can be extracted from a random data stream without resort to extra signaling levels or bits, high stability crystal oscillators, special phase detectors or other high accuracy circuit elements. The system exhibits very low static phase offset, good jitter filtering and transfer characteristics, fast acquisition, and the possibility of construction with wide circuit margins and no initial adjustment. This P/FLL method could find use in other applications involving inputs which are usually strong but have the probability of occasionally vanishing.

---

* Suggested by D. H. Wolaver.

References

1. M. R. Aaron - "PCM Transmission in the Exchange Plant," BSTJ 41, January, 1962.

2. F. D. Waldhauer - "A 274 MB/S Regenerative Repeater for T4M," ICC 1975, IEEE CatA, 1975 CHO 971-2 CSCB.

3. L. E. Franks, J. P. Bubrouski, "Statistical Properties of Timing Jitter in a PAM Timing Recovery Scheme," IEEE Trans on Comm, COM-22, No. 7, July, 1974.

4. E. Roza, "Analysis of Phase-Locked Timing Extraction Circuits for Pulse Code Transmission," IEEE Trans on Comm, COM-22, No. 9, September, 1974.

5. W. R. Bennett, "Statistics of Regenerative Digital Transmission," BSTJ, 37, November, 1958.

6. D. L. Duttweiler, "The Jitter Performance of Phase Locked Loops Extracting Timing from Baseband Data Waveforms," BSTJ, 55, January, 1976.

7. J. F. Oberst, "General Phase Comparators for Improved Phase Locked Loop Acquisition, IEEE Trans. Com. Tech., COM-19, No. 6, December, 1971.

"A PHASE-LOCKED LOOP WITH DIGITAL FREQUENCY COMPARATOR FOR TIMING SIGNAL RECOVERY"*

J.A.AFONSO, A.J.QUITÉRIO and D.S.ARANTES

DEPTº ENGENHARIA ELÉTRICA
UNIVERSIDADE ESTADUAL DE CAMPINAS
C.P.1170 - 13100 - Campinas - SP - BRAZIL

ABSTRACT

A Phase-Locked-Loop with a Digital Frequency Comparator for clock extraction in a PCM regenerative repeater is presented. With such a Frequency and Phase-Locked-Loop (FPLL) it is possible to extract the timing signal in a PCM repeater without resort to stable and expensive Voltage Controlled Oscillators. In addition, a small noise bandwidth and a wide pull-in range can be easily obtained, thus improving performance and increasing reliability. Unlike previous solutions which use analog frequency comparators, here we present a completely digital frequency comparator which is simple to implement and which presents a wide margin against spurious outputs produced by input jitter.

1. INTRODUCTION

The performance of a timing signal extraction circuit is in general characterized by the total RMS jitter of the recovered clock signal. The proper control of this jitter is extremely important in a long chain of PCM regenerative repeaters. In fact, the jitter accumulated in a chain of repeaters is the limiting factor on the maximum number of repeaters that can be used.

The total mean-square value of the jitter, $\overline{\Delta\theta_e^2}$, at the output of a PLL clock extraction circuit, as found by Roza {1}, is given by

$$\overline{\Delta\theta_e^2} = \alpha_1 \frac{\theta_{e0}^2}{Q_{eff}} + \alpha_2 \frac{1}{Q_{eff}} \qquad (1)$$

where α_1 and α_2 are constants which depend only on the statistics of the incoming signal, θ_{e0} is the quasi-static phase error and Q_{eff} is the effective Quality Factor of the circuit, which is given by

$$Q_{eff} = \frac{\pi}{4TB_L} \qquad (2)$$

where B_L is the closed-loop noise bandwidth.

An expression similar to (1) has also been found by Bennett {2} for a resonant circuit.

From (1) we see that, in principle, the jitter can be made arbitrarily small by helding θ_{e0} fixed and by making Q_{eff} arbitrarily large.

For a second order PLL the quasi-static phase error is given by

$$\theta_{e0} = \frac{\Delta\omega}{K} \qquad (3)$$

where $\Delta\omega$ is the frequency detuning and K is the open loop gain. Moreover, for large values of K the Q_{eff} is given approximately by

$$Q_{eff} = \frac{2\pi}{T} \cdot \frac{\xi}{\omega_n(1+4\xi^2)} \qquad (4)$$

On the other hand, for a resonant circuit it has been shown by Bennett {2} and Sunde {3} that

$$\theta_{e0} = \beta Q \qquad (5)$$

where

$$\beta = \frac{T\Delta\omega}{\pi} \qquad (6)$$

on the assumption that $\beta \ll 1$ and $Q \ll 1$.

From (1), (3) and (4) we see that the mean-square jitter can be made negligible in a PLL circuit by making θ_{e0} small and Q_{eff} large. On the other hand, from (5) it is clear that in a resonant circuit the static phase error and quality factor are strongly coupled and they cannot be both optimized to reduce the jitter to a negligible value.

The above discussion shows us that, due to the independence of θ_{e0} and Q_{eff}, a PLL with high open-loop gain is superior to a resonant clock extraction circuit. This result has been originally

---

* This work was supported by Telecomunicações do Brasil S/A, under contract TELEBRÁS/UNICAMP 139/76.

demonstrated by Roza {1}.

In designing a standard PLL extraction circuit, however, one is faced with many problems. For example, in order to achieve a satisfactory jitter performance, the closed-loop bandwidth should be sufficiently narrow compared with the bit rate of the incoming signal (high Q_{eff}). But a narrow bandwidth implies a correspondingly narrow pull-in range {4}; and this, in turn, forces the designer to use a very stable and expensive Voltage Controlled Oscillator.

The conflict of narrow closed-loop bandwidth and wide pull-in range in a PLL can be resolved with the use of a Frequency Comparator in order to help the PLL to acquire lock. This idea was first introduced by Richman {5}, for Color TV applications, and then recently by Bellisio {6} for clock recovery in PCM repeaters. Different approaches have also been proposed by Ghosh {7} and by Ghosh and Foster {8}.

As pointed out by Ghosh {7}, the solution given Bellisio {6}, which is based mainly on analog multipliers, may be difficult to implement {9}.Here we present a simpler solution which uses a completely digital frequency comparator.

The frequency comparator presented by Bellisio has a built-in phase detector, but the one presented here can be used with virtually any kind of phase detector.
By the time we were finishing this work we came across the work of Ghosh and Foster {8},which presents an alternative solution to this problem.

2. THE FREQUENCY AND PHASE-LOCKED LOOP

The block diagram of an FPLL is shown in Fig. 1 below.
The inner feedback loop is a standard PLL consisting of a phase-detector, a loop filter and a VCO. The frequency comparator in the outer loop should produce a useful output only as long as the inner loop is out of lock. When phase lock is achieved the frequency comparator output should be zero.

With the phase-detector output disconnected

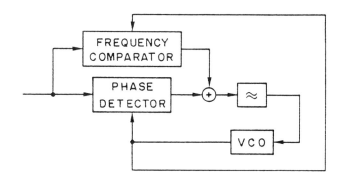

Fig. 1 - Frequency and Phase-Locked Loop

the circuit works as a Frequency-Locked Loop. In this case the VCO will be forced to oscillate with a frequency close to the incoming signal frequency. Here the error voltage across the VCO input will correspond to a Static Frequency Error. Now, roughly speaking, if this static frequency error is such that the VCO frequency is within the pull-in range of the Phase-Locked Loop itself, then phase lock will be achieved when the phase-detector is reconnected.

3. THE DIGITAL FREQUENCY COMPARATOR

The block diagram of the Digital Frequency Comparator that we are proposing is shown in Fig. 2 below.

We will assume that the DATA INPUT is a random stream of rectangular pulses of width equal to T/2, which are obtained from the equalized incoming signal by non-linear processing.

The VCO output and its 90° delayed version are sampled (copied) by the positive-going edges (occurring at t_i) of the DATA INPUT. We will also assume that the VCO output is a square wave. We can then identify four distinct regions corresponding to the possible values of the logic variables Q_1 and Q_2, as follows:

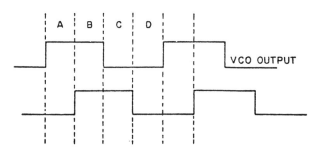

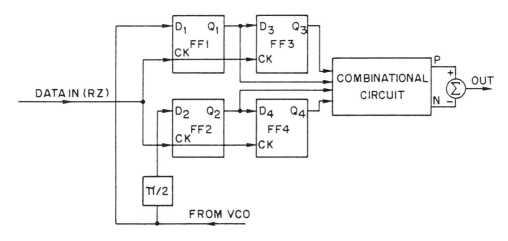

Fig. 2 - Digital Frequency Comparator

If transition t_i occurs in region A, then Q_1=1 and Q_2=0; if in region B, then Q_1=1 and Q_2=1; in C Q_1=0 and Q_2=1 and in D Q_1=0 and Q_2=0. We then have four states represented by 10, 11, 01 and 00. The outputs of Filp-Flops 1 and 2 store the state corresponding to the present transition (at t_i) and the outputs of Flip-Flops 3 and 4 store the state at the previous transition (at t_{i-1}).

Since the transitions are obtained from a random data signal and since we are assuming a frequency difference between VCO and incoming signal, it then follows that the state copied at time t_i can be any of the states A, B, C or D. Therefore, the following state diagram describes all possible state transitions:

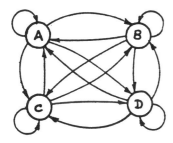

were each path identifies the states copied by two adjacent transitions of the incoming data signal. It is then clear that the inputs to the Combinational Circuit in Fig. 2 are these two "adjacent states". In other words, each single path in the state diagram can be thought of as the input to the Combinational Circuit.

Since the input data signal is random, the state diagram represents a random process with a given statistics (not necessarily Markovian) which depends on the statistics of the data signal and on the frequency difference.

The Combinational Circuit is designed in such a way that a "positive" pulse is produced when a given single path (or a set of single paths) occurs and a "negative" pulse is produced when another single path (or another set of single paths) occurs.

One simple and reasonable choice for the Combinational Circuit works as follows:
    1. A "positive" pulse is produced when transition A to B occurs.
    2. A "negative" pulse is produced when transition B to A occurs.

The average value of the "negative" pulses are then subtracted from the average value of the "positive" pulses in order to produce the final frequency error voltage at the output of the frequency comparator. A simple realization for this particular logic is given in Fig. 3 below.

The frequency comparator characteristic (output voltage versus frequency difference) is plotted in Fig. 4 for the combinational circuit as chosen above and for an input frequency f_i=2.048 Mbits/sec. The frequency difference was obtained by varying the other input, f_0, corresponding to the VCO input. It is clear from this figure that the frequency comparator produces a useful output over a wide frequency range about the input frequency. In this case the "positive" and "negative" pulses were stretched by means of a monostable circuit (1 μs duration). Note that the plot of Fig. 4 was obtained under the idealized situation where the input signal is a nice binary RZ pseudo-random data.

One possible configuration for the whole system, including the phase detector, the VCO and interfaces, is presented in Fig. 5, as suggested in {8}.

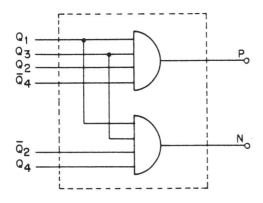

Fig. 3 - A simple choice for the combinational circuit

in {8}.

4. CONCLUSIONS

A Frequency and Phase-Locked Loop (FPLL) has been presented for clock extraction. The circuit uses a Digital Frequency Comparator which is simple to implement, provides a wide pull-in range, a small noise bandwidth and requires no stable and expensive Voltage Controlled Oscillator.

One important and critical point in an FPLL circuit is that the frequency comparator should produce no spurious output when the system is phase locked. Since an equalized data stream at the receiving end of a dispersive channel is likely to present some amount of jitter, due to noise and intersymbol interference, it is then imperative that no spurious pulses be produced by the jittering transition times. This can be taken care of by a proper choice of the phase-detector and of the non-linear signal processing preceding the FPLL. For example, for the phase-detector given in Fig. 5 the FPLL will phase lock in such a way that the positive-going clock transitions sample the DATA INPUT signal at the center of the pulses. In this situation the

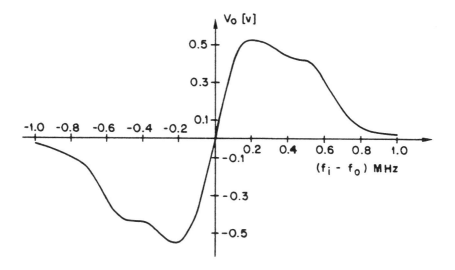

Fig. 4 - Transfer Characteristic of the Digital Frequency Comparator for f_i=2.048 MHz and 1μs pulse duration.

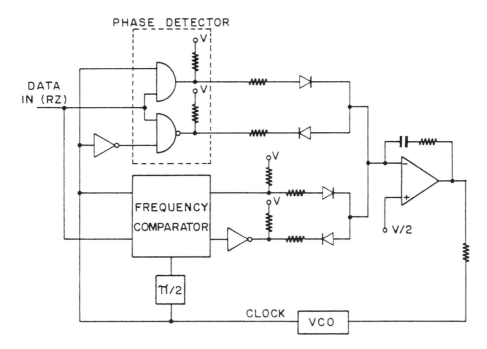

Fig. 5 - Implementation of the Frequency and
Phase-Locked Loop with an active loop
filter.

frequency comparator produces no output as long as
the positive-going edges of the DATA INPUT are
confined to an interval of $\pm 180^{\circ}$ around their
average position (boundary between regions C and
D). This is certainly a wide margin against input
jitter.

REFERENCES

1. E. ROZA, "Analysis of Phase-Locked-Loop Timing
Extraction Circuits for Pulse Code Transmission"
IEEE Trans. on Comm., COM-22, n⁰ 9, September,
1974.

2. W.R.BENNETT, "Statistics of Regenerative Digital
Transmission", BSTJ, v.37, November, 1958.

3. E.D.SUNDE, "Self-Timing Regenerative Repeaters",
BSTJ, v.36, July, 1957.

4. F.M.GARDNER, "Phase-Lock- Techniques", John
Wiley, N.Y., 1966.

5 D.RICHMAN, "Color-carrier Reference Phase
Synchronization Accuracy in NTSC Color Television", Proc. IRE, v.42, January, 1954.

6. J.A.BELLISIO, "A New Phase-Locked Timing
Recovery Method for Digital Regenerators", IEEE
ICC Records, 1976.

7. S.GHOSH, "The Bandwidth of Phase-Lock Loops",
Electronic Design, March, 1978.

8. S. GHOSH and C.FOSTER, "A Phase-Frequency
Locked Loop for Random Input Data", to appear.

9. T.L.MAIONE; D.D.SELL and D.H.WOLAVER,"Practical
45-Mb/s Regenerators for Lightwave Transmission'
BSTJ, v.57, July-August, 1978.

241

Clock Recovery From Random Binary Signals

J. D. H. Alexander

Indexing terms: Clocks, Pulse circuits, Synchronisation

A circuit for detecting timing errors between a binary signal and a local clock pulse generator is described. Three binary samples are compared and logical control signals for the clock are derived.

Clock recovery is often essential for the regeneration of distorted binary signals. The paucity of published work in this field has been noted in a recent paper.[1] Some techniques are outlined in Bennett and Davey.[2] A simple scheme is described in this letter in which zero or datum crossings of a distorted binary signal are measured as early or late events when compared with the transitions of a local clock wave. The circuit is simple to build in t.t.l. and may be used as the detector for an analogue or digital phase-locked loop to achieve clock synchronisation.

An idealised eye diagram is shown in Fig. 1. Two samples are taken in each nominal bit interval. Samples are taken close to midbit and changeover times. Midbit samples taken at times A, C, and E, may be called a, c and e, the changeover samples then being b and d. These samples are transformed into binary variables and processed digitally.

A suitable circuit is shown in Fig. 2. Signal S is the output of a limiter. This signal is a binary function given by the zero, or datum, crossing of the distorted analogue waveform. Samples of S are taken by clocked D-type monostables D_1 and D_3. Clock pulse trains CKM and CKC are at the nominal data rate, and CKM is arranged to be near to the midbit instant whilst CKC occurs at the changeover time. Complementary squarewave clock waveforms and edge triggered monostables ensure exactly interleaved sampling.

Sample a is transferred to D_2 at time C when D_1 stores sample c. At this time, sample b is transferred to D_4. This enables the variables a, b and c to be examined simultaneously at the outputs of D_1, D_2, and D_4. If the midbit clock is early, $a = b$, and $b = c$ randomly. Similarly, if the midbit clock is late, $b = c$ and $a = b$ randomly.

The binary variables a, b and c are related to the early–late situations by the following four rules:

(a) If $a = b$ and $b \neq c$, the clock is late.

(b) If $a \neq b$ and $b = c$, the clock is early.

(c) If $a = b = c$, no decision is possible.

(d) If $a = c \neq b$, no decision is possible.

Let E represent early, L represent late and X represent indecision. The eight possible combinations of abc and the conclusion drawn from the four rules are shown in Table 1.

Table 1

a	b	c	Conclusions
0	0	0	X
0	0	1	L
0	1	0	X
0	1	1	E
1	0	0	E
1	0	1	X
1	1	0	L
1	1	1	X

A set of control functions may be deduced by noting that the conclusion entry is reflected. If the pure binary sequence abc is expressed as a Grey sequence rst, the XLE functions are independent of r, as shown in Table 2. This shows that $\bar{X}$ is

Table 2

r	s	t	Conclusions
0	0	0	X
0	0	1	L
0	1	1	X
0	1	0	E
1	1	0	E
1	1	1	X
1	0	1	L
1	0	0	X

the modulo-2 sum of s and t, and L is given by t:

$$\bar{X} = s \oplus t$$

But, since

$$s \oplus t = a \oplus b \oplus b \oplus c$$

$$\bar{X} = a \oplus c$$

and

$$X = \overline{a \oplus c}$$

also

$$L = b \oplus c$$

An oscillator may be designed to operate at either of three frequencies: f_0, $f_0 + f_x$ and $f_0 - f_x$. An example of this class of oscillator is the logical phase-controlled oscillator. Frequency control requires a 2 bit word, and the relationships

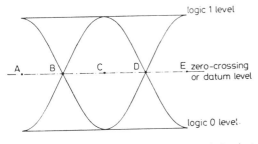

Fig. 1 *Eye diagram showing synchronous timing instants*

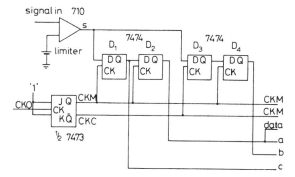

Fig. 2 *Sampling circuit and clock pulse generator*

are shown in Table 3. Any appropriate 3-frequency oscillator

Table 3

$$P = \overline{a \oplus c}$$
$$Q = b \oplus c$$

Control word $P\ Q$	Frequency
0 0	f_0
0 1	f_0
1 0	$f_0 + f_x$
1 1	$f_0 - f_x$

may be used in conjunction with the early–late detector to realise a phase-locked clock recovery system.

An alternative technique involves the generation of a 3-level signal for frequency control of a v.c.o. A 3-valued variable A may be generated:

$$A = (a \oplus b) \text{ minus } (b \oplus c)$$

The relationships between the binary word abc and the 3-valued variable A are shown in Table 4. Circuits for the generation of the logical control signals, P, Q and the 3-valued variable A for analogue loops are shown in Fig. 3.

When established as part of a synchronised phase-locked loop, the sample a may be taken as a true midbit sample of the distorted binary signal. An alternative use for the synchronised changeover clock CKC is control of an integrate-and-dump detector operating directly on the unsliced binary signal.

Table 4

a	b	c	$a \oplus b$	$b \oplus c$	A	early–late
0	0	0	0	0	0	X
0	0	1	0	1	-1	L
0	1	0	1	1	0	X
0	1	1	1	0	$+1$	E
1	0	0	1	0	$+1$	E
1	0	1	1	1	0	X
1	1	0	0	1	-1	L
1	1	1	0	0	0	X

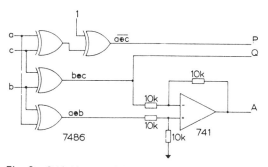

Fig. 3 *2 bit binary and 3-level analogue early–late detectors*

References

1 BETTS, J. A., BROOM, R. S., COOK, S. J., and CLARK, J. G.: 'Use of pilot tones for real-time channel estimation of h.f. data circuits', *Proc. IEE*, 1975, **122**, (9), pp. 887–896
2 BENNETT, W. R., and DAVEY, J. R.: 'Data transmission' (McGraw-Hill, 1965), chap. 14

A Si Bipolar Phase and Frequency Detector IC for Clock Extraction up to 8 Gb/s

Ansgar Pottbäcker, Ulrich Langmann, *Member, IEEE*, and Hans-Ulrich Schreiber

Abstract—A phase and frequency detector IC is presented that operates up to an NRZ bit rate of 8 Gb/s. The IC comprises a phase detector (PD), a quadrature phase detector (QPD), and frequency detector (FD). In the PD and QPD the VCO signal and the quadrature VCO signal are sampled by the NRZ input signal. The two beat notes provided by this operation are subsequently processed in the FD. The superposition of the FD output and the PD output signals are then fed into a passive loop filter (lag/lead filter). The loop filter and the VCO are external components. The measured pull-in range is $> \pm 100$ MHz at 8 Gb/s. The measured rms time jitter of the extracted clock is less than 1.9 ps for a pseudorandom bit sequence (PRBS) length of 2^{23}–1. A 0.9-µm 12-GHz f_T silicon bipolar process was used to fabricate the chip with a total power consumption of 1.4 W.

I. Introduction

AT THE receiving end of an optical-fiber transmission link, a clock signal has to be provided for the processing of the data signal, such as signal regeneration and/or demultiplexing. The clock has to be extracted from the arriving binary data themselves, for which, normally, an NRZ format is chosen. Regarding the factors of cost of implementation, flexibility, and reliability, an approach employing IC's for clock recovery is superior to other methods like, for instance, filtering by a discrete narrow-band filter. So far, clock recovery based on IC's without narrow-band filtering has been demonstrated up to only 4 Gb/s [1]. The implementation of a flexible and reliable clock recovery IC for the bit rate range of beyond 4 Gb/s has not been demonstrated yet.

There is a twofold advantage of using a phase-locked loop (PLL) for the clock-recovery process: first, it can be implemented as a monolithic IC and, second, its dynamic properties can easily be adjusted by a proper choice of the filter parameters. However, a well-known limitation of conventional PLL's is the fact that the pull-in range is not much larger than the noise bandwidth. In other words, if the noise bandwidth is reduced by a proper filter design, the pull-in range decreases accordingly. The targets of small noise bandwidth and a large pull-in range (capture range) cannot be met at the same time by a mere PLL approach.

Manuscript received May 12, 1992; revised July 10, 1992. This work was supported by the Forschungsinstitut der Deutschen Bundespost TE-LEKOM.

The authors are with Mikroelektronik-Zentrum, Ruhr-Universität Bochum, D-4630 Bochum, Germany.

IEEE Log Number 9203628.

Therefore, an acquisition aid is indispensable for clock recovery since a narrow-band active filtering of the (preprocessed) data signal is desirable. This leads to a clock recovery approach based on a phase and frequency detector (PFD) as part of a phase- and frequency-locked loop (PFLL).

So far, three different schemes have been proposed for combined phase and frequency detection: 1) the quadricorrelator [1], 2) the rotational frequency detector [2], and 3) a PFLL employing a special training fill word at the input for supporting the pull-in process [3]. The PFD concept presented below (see also [4]) works without signal preprocessing, internal filtering, and phase shifting, required for the quadricorrelator approach. In contrast to the rotational frequency detector, our PFD here needs fewer circuit components, is based on a simple circuit scheme, and—this should be stressed here—can operate at higher bit rates, since relatively slow beat notes are processed for frequency detection.

II. Operation Principle and Circuit Diagrams

Fig. 1 shows the block diagram of the PFD in a PFLL. The first part of the detector is composed of two identical sample-and-hold cells which serve as phase detector (PD) and quadrature phase detector (QPD). At every transition of the input data stream the VCO signal and the delayed VCO signal are sampled. This operation provides beat notes at the outputs $Q1$ and $Q2$, if the frequency of the VCO and the bit rate frequency are unequal. When sample-and-hold cells as phase detectors are used, the NRZ input data signal can be processed directly; that is, there is no need for providing and adjusting an additional time constant by, for example, an external delay line or an on-chip half-bit delay generator (as in [1]) for a preprocessing unit. This simplification is a major advantage.

The frequency detector (FD) now processes the beat notes of the PD and the QPD and delivers a frequency difference signal at the output $Q3$. The open-collector outputs of the PD and the FD are connected at the external low-pass filter providing a superposition of both signals. The filter output delivers a clear dc component that drives the loop towards lock.

An external transistor is used in the loop filter for dc coupling the VCO tuning input and the IC outputs. The on-chip outputs could not be connected directly to the fil-

Reprinted from *IEEE Journal of Solid-State Circuits*, vol. SC-27, pp. 1747-1751, December 1992.

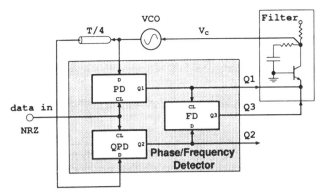

Fig. 1. Block diagram of the PFD IC.

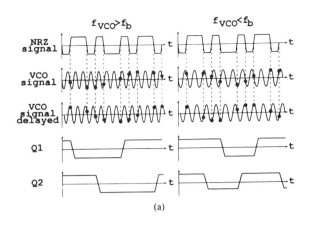

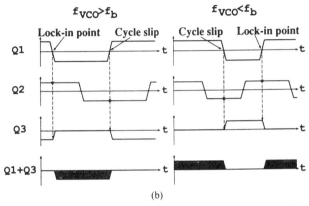

(a)

(b)

Fig. 2. (a) Schematic timing diagram of the PD/QPD. (b) Schematic timing diagram of the PD, QPD, and FD.

ter because of the relatively low collector–emitter breakdown voltage of the bipolar process employed.

If not provided by the VCO, the quadrature component needed for the detector IC can be generated using a delay line. Simulations and measurements showed that the exact value of the phase shift is not critical and values within a range of 45° up to 135° are tolerable.

In order to illustrate the operation principle of the phase and frequency detector, Fig. 2(a) shows schematic timing diagrams of the input and output signals for a VCO frequency f_{VCO} higher and lower than the bit rate frequency f_b, respectively. At every transition of the input data stream the VCO signal and the delayed VCO signal are sampled and the analog voltage levels, as indicated in this figure by dots, are stored internally. Then the sampled signals are fed to a limiting output buffer yielding the presented beat notes $Q1$ and $Q2$.

Fig. 2(b) illustrates how the signals $Q1$ and $Q2$ are processed by the FD. Whenever a transition of $Q1$ occurs while $Q2$ is low, the sign of the transition corresponds to the sign of the frequency difference between the VCO and the bit rate. Therefore, at a rising edge of $Q1$ the FD output $Q3$ is set to minus one and at a falling edge to plus one, thus reducing the frequency difference. Any transition of $Q1$, while $Q2$ is high, resets $Q3$. Note that, once lock is achieved, $Q3$ remains zero since around a stable lock-in point the signal $Q2$ is always high. The operation scheme of the frequency detector can be described by the logic diagram of Table I. It will be shown below that this ternary logic scheme can easily be realized as a differential emitter-coupled logic. As indicated in Fig. 2(b), the superposition of the signals $Q1$ and $Q3$ in the loop filter provides a clear dc component driving the loop towards lock. The ensuing acquisition process is strictly monotonic and predictable.

The pull-in range is limited by two conditions: 1) it cannot be larger than half the hold-in range (due to the 50% duty cycle of the frequency detector output), and 2) at least one transition of the NRZ signal should (statistically) occur within one quarter of the beat note period. The latter condition ensures that the phase position of the signals $Q1$ and $Q2$ can be analyzed correctly by the FD.

The circuit diagram of the PD and the QPD is depicted in Fig. 3. The circuits are designed with a differential two-

TABLE I
LOGIC TABLE OF THE FREQUENCY DETECTOR

$Q1$	$Q2$	$Q3$
X	1	0
Rising	0	−1
Falling	0	+1

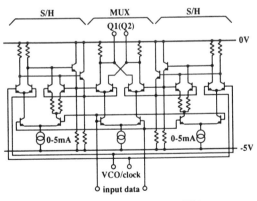

Fig. 3. Circuit diagram of the PD's.

level current-mode logic in order to optimize the operating speed and to avoid problems caused by internal crosstalk.

The PD/QPD is composed of two sample-and-hold cells and a multiplexer. Depending on the sign of the data signal, one sample-and-hold cell tracks the input signal, for

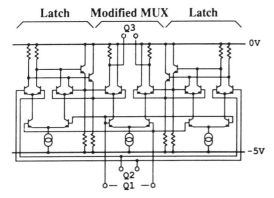

Latch　Modified MUX　Latch

Fig. 4. Circuit diagram of the FD.

instance the left one in Fig. 3, whereas the other holds the previous sampled voltage level. Note that the multiplexer selects only the sample-and-hold cell that is in the hold mode, thus preventing a direct connection between input and output.

If the data signal changes sign, the cell that was in the tracking mode switches to the hold mode and stores the current voltage level while the other cell starts tracking the input signal. In the hold mode, the sample-and-hold cell has an internal closed-loop gain near unity, thus sustaining the charge stored in the parasitic transistor capacitors. Therefore, no separate hold capacitors are required for an operation above 100 Mb/s.

The value of the closed-loop gain can be adjusted externally by the transconductance of the current switches, which depends on the static current impressed by the current sources. The current sources can be set to a value between 0 and 5 mA. An adjustment of the loop gain is only necessary for applications below 1 Gb/s; a current of 1 mA (3 mA) ensures a hold time of about 2 ns (20 ns). If the closed-loop gain is greater than one, corresponding to a current higher than about 3 mA, the PD/QPD operates in a digital mode since the sampled voltage levels are regenerated up to the full internal voltage swing. In the digital mode, however, the clock jitter of the locked loop increases if no other contribution to the overall jitter is dominant (like VCO jitter or input data jitter or jitter by intersymbol interference in the PFD). The binary phase signal that passes the high-frequency part of the lag/lead filter toggles the VCO around the bit-rate frequency. This bang-bang operation leads to a phase error depending on the digital voltage swing, the high-frequency loop gain, and the time between two transitions in the NRZ signal.

The circuit scheme of the FD is displayed in Fig. 4. The required logic function is implemented with standard series gating techniques. The only unusual part of the circuit scheme is the modified multiplexer, which generates the desired ternary logic levels. In the two normal states of the current switches, the full current is switched to one of the load resistors and no current to the other. In a third state, the current is fed to neither of the resistors. A conventional open-collector output buffer, which is not shown in Fig. 4, transforms the internal levels into three external current levels.

The chip was fabricated at the Microelectronic Center of the Ruhr-University Bochum. A self-aligned polysilicon technology without both polysilicon emitter and trench-isolated structures was used [5]. The rather conservative 1.5-μm lithography results in an effective emitter stripe width of about 0.9 μm. The measured transit frequency f_T is about 12 GHz at $V_{CE} = 1$ V. The fact that the PFD operates up to 8 Gb/s demonstrates that our concept fully utilizes the bandwidth given by the bipolar process.

Fig. 5 shows a chip die photograph of the PFD IC. The overall size of the chip is about 1.6×2.3 mm^2. The three main processing blocks are: the PD, the identical QPD, and the FD. The remaining blocks are the input stage (at the left side) and a buffer stage. These stages perform buffering and level shifting as required. In order to avoid transit-time differences, the layout symmetry is preserved as far as possible.

First, the chip was measured under open-loop conditions, which means that the VCO frequency was directly adjusted by a steady voltage source. The frequency difference chosen between f_{VCO} and f_b was about 100 MHz. Fig. 6(a) gives the results if $f_{VCO} > f_b$. The upper trace shows the input data stream and the lower traces the beat notes at $Q1$ and $Q2$ and the output $Q3$ of the FD. The average voltage of $Q3$ is about half the voltage swing according to the mark–space ratio of 50%. This dc component would drive the closed loop towards lock. Jitter of the input data signal and intersymbol interference led to the observed "oscillations" of the beat note $Q1$ close to the transition region. This affects the effective pull-in range, because the average dc voltage of $Q3$ is slightly reduced by spikes in the $Q3$ signal occurring during these oscillations.

The results of the same measurement configuration, but with reversed frequency difference, are shown in Fig. 6(b). Compared to the previous figure, the signal $Q3$ now has a positive dc offset and the phase positions of $Q1$ and $Q2$ are inverted.

The acquisition process measured at a bit rate of 5.2 Gb/s is displayed in Fig. 7. In order to get a periodic representation of the acquisition process, the internal currents of the sample-and-hold cells are switched from 0 to 3 mA and vice-versa using the above-mentioned control port provided for the PD and QPD subcircuits.

The upper trace shows the voltage at the tuning port of the external VCO. The lower traces show the inverted phase signal, the quadrature phase signal, and the inverted output signal of the FD. The external loop filter was built of discrete elements and has a low-pass time constant of about 150 μs.

The loop starts in an unlocked state. In this state the internal currents of the sample-and-hold cells are switched off. (The outputs show some noise caused by internal and external crosstalk.)

In the following state, the acquisition process, the cur-

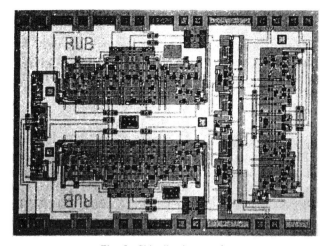

Fig. 5. Chip die photograph.

$f_{VCO} > f_b$

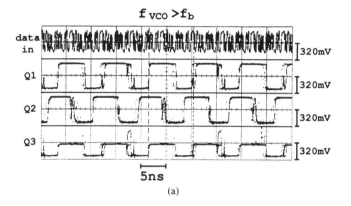

(a)

$f_{VCO} < f_b$

(b)

Fig. 6. Measured signal waveforms at 8 Gb/s: (a) PFLL open with $f_{VCO} > f_b$, and (b) PFLL open with $f_{VCO} < f_b$.

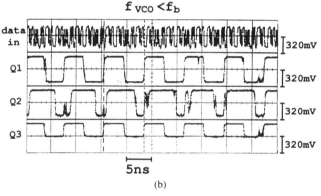

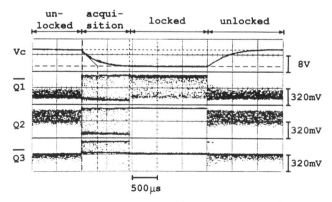

Fig. 7. Measured acquisition of the PFLL at 5.2 Gb/s.

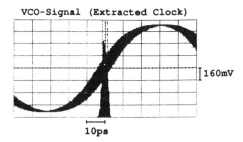

VCO-Signal (Extracted Clock)

Fig. 8. Measured jitter histogram of the locked VCO at 8 Gb/s.

TABLE II
PERFORMANCE LIST OF THE PFD CHIP

Measured max. bit rate	8 Gb/s
Exp. pull-in range	> ±200 MHz at 5.2 Gb/s
(with lag/lead LPF)	> ±100 MHz at 8 Gb/s
rms phase jitter	‹ 1.5 ps (2^{15}–1 PRBS)
(external VCO)	1.9 ps (2^{23}–1 PRBS)
Power dissipation	1.4 W
Transistor count	150
Chip size	1.63×2.30 mm^2

rent is switched to the normal value of 3 mA. The signals $Q1$ and $Q2$ now toggle with the beat frequency between the low and high level with no clear dc component in contrast to the signal $Q3$. This signal is now responsible for driving the loop towards lock. The time constant of the acquisition process is about two times the low-pass filter time constant.

Once lock is achieved, the output of the FD remains zero. The output of the QPD is constantly high, which indicates the locked state. The dc component needed to sustain lock is now supplied by the output of the PD.

Finally, the transition back to the unlocked state is shown on the right end of Fig. 7. Here the internal currents of the sample-and-hold cells are switched off again. The overall voltage swing at the tuning port of the external VCO is about 8 V, which corresponds to a frequency swing of about 400 MHz.

Fig. 8 shows the jitter histogram of the VCO output at 8 GHz if locked to 8 Gb/s. The measured rms time jitter, observed at the oscilloscope, is about 1.5 ps if a PRBS length of 2^{15}–1 is applied. The measured systematic rms time jitter of the measuring equipment alone is about 1.2 ps.

Table II summarizes the most important specification of the PFD. The maximum bit rate is about 8 Gb/s. The measured pull-in range with a passive loop filter is greater than ±200 MHz at 5.2 Gb/s and greater than ±100 MHz at 8 Gb/s.

The pull-in range of ±200 MHz at 5.2 Gb/s roughly corresponds to half the hold-in range, which is limited (in this setup) by the passive loop filter. A larger pull-in range can be expected if an active filter is used (larger dc loop gain). The frequency limitation of the sample-and-hold operation of the PD's leads to the observed reduction of the pull-in range at 8 Gb/s.

The measured rms phase jitter is 1.5 ps with a PRBS length of $2^{15}-1$ and 1.9 ps with a PRBS length of $2^{23}-1$. The power dissipation of the chip is about 1.4 W, with more than half of the value being consumed by the buffer stages. The transistor count is about 150.

IV. CONCLUSIONS

The PFD presented here needs no signal preprocessing and internal filtering. Therefore, it is applicable for a wide frequency range. As pointed out above, the simple circuit scheme permits high-speed operation. So far, an operating speed of up to 8 Gb/s has been achieved, based on a low-cost silicon production technology with conservative lithography. The operation speed can be extended to even higher bit rates by using a more advanced silicon bipolar process. Very low jitter operation of less than 1.9 ps rms was demonstrated. The noise bandwidth could easily be lowered furthermore without affecting the pull-in range. The pull-in range of ± 100 MHz at 8 Gb/s is considered as sufficiently large for practical requirements. If desired, it could be extended by applying an integrator in the loop filter. Eventually, a self-timing regenerator IC with a wide locking range is feasible by adding an on-chip VCO and a D-flip-flop.

REFERENCES

[1] H. Ransijn and P. O'Connor, "A PLL-based 2.5-Gb/s clock and data regenerator IC," *IEEE J. Solid-State Circuits*, vol. 26, no. 10, pp. 1345–1353, Oct. 1991.
[2] L. DeVito, J. Newton, R. Croughwell, J. Bulzacchelli, and F. Benkley, "A 52 MHz and 55 MHz clock-recovery PLL," in *ISSCC Dig. Tech. Papers*, 1991, pp. 142–143.
[3] R. C. Walker, T. Hornak, C.-S. Yen, J. Doernberg, and K. H. Springer, "A 1.5 Gb/s link interface chipset for computer data transmission," *IEEE J. Selected Areas Commun.*, vol. 9, no. 5, pp. 698–703, June 1991.
[4] A. Pottbäcker, U. Langmann, and H.-U. Schreiber, "An 8 Gb/s Si bipolar phase and frequency detector IC for clock extraction," in *ISSCC Dig. Tech. Papers*, Feb. 1992, pp. 162–163.
[5] H.-U. Schreiber, "A simple self-aligned Si bipolar transistor for high-speed integrated circuits," *IEEE Trans. Electron Devices*, vol. 36, pp. 1212–1213, June 1989.

A Self Correcting Clock Recovery Circuit

CHARLES R. HOGGE, Jr., MEMBER, IEEE

Abstract—Conventional approaches to the problem of extracting clock from NRZ data do not automatically hold the clock in the center of the data eye. Other means must be used to keep the clock properly centered in the eye at the decision flip-flop. A new approach to the problem is described. The circuit is both simple and self correcting.

I. INTRODUCTION

THE RECOVERY of synchronous clock from NRZ data requires a nonlinear process since the spectrum of the signal is a continuous $(\sin x)/x$ function with no predominate clock related component. The conventional approach to the problem is to pass the signal through a limiter, then to differentiate the square pulses, then to rectify the differentiated signal to produce a reference signal having a strong component at clock frequency. That reference signal is then filtered with either a narrow-band phase locked loop or a SAW filter. In either case, clock extraction occurs in a spur circuit off the data path and the alignment of the clock in the center of the data "eye" is not self correcting. A new approach (patent pending) to the problem is described. The same flip-flop that synchronously detects the incoming data, making a jitter free logic decision each clock period, is a part of a phase detector. That phase detector is in a closed loop which includes the clock VCXO. The closed loop forces the clock to be positioned at the center of the eye and is inherently self correcting.

II. DESCRIPTION OF THE CIRCUIT

Fig. 1 shows the circuit in its simplest form. The NRZ data signal is applied to both the Type D flip-flop U1, and the exclusive OR U2. U1 is both the retiming decision circuit for the receiver and a part of the clock recovery circuit. The retimed data from U1 is applied to U2, U3, and U4. U1 is clocked with true CLK and U4 is clocked with inverted CLK. The X-OR output from U3 is a fixed width square pulse the width of half a clock period for each transition of the retimed data. The X-OR output from U2 is a variable width pulse for each transition of the data signal—the width of which depends on the position of the clock within the eye opening. When the leading edge of the clock is centered properly, the width of the variable width pulse is identical to that of the fixed width pulse, plus and minus edge jitter. The two pulses are filtered lightly and applied to the active loop filter to produce the error voltage for controlling the clock VCXO.

Manuscript received May 20, 1985; revised July 22, 1985.
The author is with Rockwell International, Dallas, TX 75801.

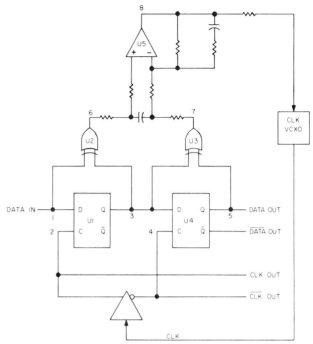

Fig. 1. Self correcting clock recovery circuit.

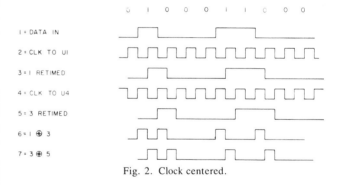

Fig. 2. Clock centered.

Fig. 2 is the timing diagram for an arbitrary data signal (containing three logic ones and seven logic zeros) with the clock properly centered within the data bit interval. The waveform numbers correspond with the node numbers in Fig. 1. Note that the pulse patterns from the outputs at 6 and 7 have identical average values—resulting in zero error voltage from the active loop filter at node 8 of Fig. 1. The active loop filter consists of differential amplifier U5 and its feedback network. In conjunction with the integration contribution by the VCXO, a second order loop with a low-pass filter and phase-lead correction is realized. The effect of combining the differencing and integrating functions into one circuit is the same as if we

Reprinted from *IEEE Journal of Lightwave Technology*, vol. LT-3, pp. 1312-1314, December 1985.

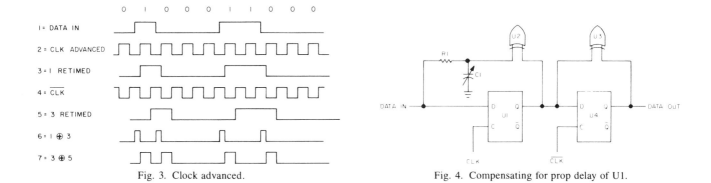

Fig. 3. Clock advanced.

Fig. 4. Compensating for prop delay of U1.

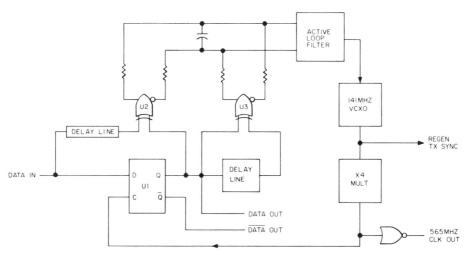

Fig. 5. 565-MHz clock recovery circuit.

were to separately integrate the two pulse waveforms (nodes 6 and 7) and then subtract the result to produce the correction voltage at node 8.

Fig. 3 is a similar timing diagram but with the clock signal advanced relative to the center of the data bit interval. That causes the logic one pulses at node 6 to become more narrow while those at node 7 remain the same width as in Fig. 2. The result is that the average value of the pulse pattern at node 6 shifts more negative than that at node 7—producing a negative correction voltage at node 8. Similarly, when the clock is retarded, the logic one pulses at node 6 become wider while those at node 7 again remain the same width and a positive correction voltage results.

It is not necessary for the data to have an equal number of logic ones and zeros over any time interval—only that a sufficient number of transitions in either direction occur to keep the loop stable.

The phase detector gain factor varies with the data activity. For square pulses at nodes 6 and 7, the gain factor can be easily calculated for any assumed transition density. As an example, assume a 1-V logic swing and one data transition within five bit intervals. A shift in the clock position of $\pi/8$ will shift the dc level at node 6 by 25 mV. Thus, the gain factor for that transition density is 63.66 mV/rad. As the transition density decreases, the open loop gain decreases but the closed loop bandwidth also de-

creases so the effective Q of the circuit increases. The circuit is thus able to accommodate occasional long strings of ones or zeros which can occur even with scrambled data.

The circuit of Fig. 1 works well as shown as long as the propagation delay through U1 is negligibly small relative to one bit interval. In cases where it is not negligibly small, as with the receiver for the LTS-135 at 135 Mbit/s, [2], and LTS-3139 at 139 Mb/s, provision is made to compensate for the delay through U1 by adding comparable delay between the originating node and U2, as shown in Fig. 4. The same result can be obtained with a delay line or with appropriate gate delay.

The clock recovery circuit for the LTS-1565 lightwave system (which operates at 565 Mbit/s) uses a variation of the above, as shown in Fig. 5. The propagation delay of U1 is compensated for with a test select plug-in delay line. The delay line is a 50-Ω microstrip circuit on a ceramic substrate. The fixed width pulse is obtained by using a similar delay line rather than a second retiming flip-flop to provide the half bit interval delay at the second input to U3. The exclusive OR/NOR gates need not be fast enough to deliver square pulses since their outputs are integrated. It is important, however, that they are reasonably well matched in terms of the dc components resulting from integrating their respective outputs when their inputs are identical. Fig. 5 shows resistive summing to take advantage of all four available outputs from U2 and U3. This provides balanced inputs to the loop filter centered midway between logic levels regardless of the data transition density.

One precautionary note is in order. This circuit is not a phase-frequency detector. It, therefore, requires that the clock source be either crystal stabilized or that a sweep acquisition circuit be included.

III. ADVANTAGES OF THIS CIRCUIT

By the nature of this circuit, with the loop closed around the decision flip-flop, the clock position is self correcting. Other approaches to clock recovery which do not include this feature have to address the problem by other means. When the clock reference signal is extracted from the incoming data in a spur circuit, the resulting jitter free clock applied to the decision flip-flop will have gone through several stages of active and one or more passive (high Q L C or SAW filter) circuits before it reaches its destination. Careful timing alignment between that clock and the data is required and that alignment must be maintained for the life of the product. The effects of temperature and power supply variations as well as component aging must be taken into account. When all of these factors are considered, it is clear that a circuit, such as the one described here, that is both simple and self correcting has both performance and cost advantages.

Incidentally, rather than using the error voltage derived from this circuit to control a VCXO, it could be used to adjust a phase shifter to center the clock from a SAW filter if there is some compelling reason to want to use a SAW filter. But then the opportunity to filter the noise in the much more narrow loop filter of Figs. 1 and 5 would be lost.

IV. RESULTS AND CONCLUSIONS

The circuit of Fig. 3 has a measured sensitivity of between 65 mV/rad and 82 mV/rad over the temperature range of 0–+75°C using the F100107 at 565 Mb/s with a pseudorandom bit sequence of 2^{23}-1. The overall clock recovery loop has a 45-KHz 3 dB bandwidth and is heavily damped so that the noise transfer function has a maximum peak of 0.003 dB based on a worst case analysis of all elements of the circuit.

In conclusion, a new approach to extracting clock from NRZ data is presented. The approach has inherent performance and economic advantages over commonly used approaches. While it has been applied in a wide range of lightwave products (spanning two decades of data rates) its use is by no means restricted to lightwave systems. It has been implemented with off-the-shelf ECL and TTL flip-flops and exclusive OR/NOR gates. The next logical step would be to implement it on a single chip.

ACKNOWLEDGMENT

The author would like to thank J. A. King for developing the microstrip delay lines used at 575 Mbit/s and for performing all of the laboratory measurements on this circuit at 90, 135,139, and 565 Mbit/s.

REFERENCES

[1] F. M. Gardner, "Rapid synchronization: Carrier and clock recovery for high-speed digital communication," *Microwave Syst. News,* vol. 6, no. 1, p. 57, 1976.
[2] K. Y. Maxham *et al.,* "Rockwell 135-Mbit/s lightwave system," *J. Lightwave Technol.,* vol. LT-2, no. 4, Aug. 1984.

Part 3
Modeling and Simulation

PHASE-LOCKED systems are difficult to simulate, primarily because they involve vastly different time scales. As an extreme, yet realistic, example, consider a 2-GHz frequency synthesizer with a reference input frequency of 200 kHz. Simulation of such a circuit requires a time step on the order of 50 ps to provide adequate resolution in the output, and also a total length of at least 10 ms to allow the loop to lock. Thus, the circuit must be simulated for roughly 200 million points.

In this part, six papers on the modeling and simulation of PLLs are presented. The first two papers employ macromodeling in SPICE to greatly simplify the circuit, while the third paper develops nonlinear mathematical models for each building block of a PLL. The next three papers utilize high-level behavioral representation to simulate phenomena such as tracking, acquisition, and phase noise in circuits as nonlinear as charge-pump PLLs.

An Integrated PLL Clock Generator for 275 MHz Graphic Displays

German Gutierrez

Dan DeSimone

Brooktree Corporation
9950 Barnes Canyon Rd.
San Diego, CA 92121

Quadic Systems Inc.
29B Hutcherson Dr.
Gorham, Maine 04038

Abstract

A monolithic phase-locked loop (PLL) and clock generator/driver has been developed to provide clock frequencies up to 275 MHz from low cost crystals for use with high resolution RAMDACs (random-access-memory and digital-to-analog converter). Unlike previous integrated PLL's, this PLL clock generator simplifies video clock generation design since it requires no coils or varicaps. The internal capacitor for an emitter coupled multivibrator achieves high quality, with little coupling to the substrate. This integrated circuit incorporates all of the building blocks of a PLL, including a crystal amplifier which will work with a wide range of crystal frequencies. The only external components required are two resistors and two capacitors which operate at low loop filter frequencies. An innovative method of efficiently simulating PLL's was also developed.

Introduction

As color display resolution increases the need for inexpensive sources of master clock becomes important. For a 2k by 1.5k pixel display, for example, typical "pixel rate" is 263 MHz. The required jitter specification for this clock is as low as 0.25 nS, and there can be no low frequency periodic disturbances, because they will be visible in the display. It becomes mandatory, unless low price, quality, resonators become available, to use a phase-lock technique to multiply up the frequency of an inexpensive low frequency crystal. The purpose of this project was to develop a monolithic, easy to use, PLL and pair it with all the clocking functions required by today's popular RAMDACs. These clocks are provided by a dedicated combination of ECL logic, for the pixel frequencies, and TTL logic for the control of loading and multiplexing. The design challenge consisted of integrating inherently noisy TTL clock drivers with a sensitive PLL, and required several layout and isolation tricks. The ease of use is related to the absence of external high frequency components; only low frequency loop filter and biasing components are required.

PLL's are traditionally difficult to use because they either require an external VCO, or because the onboard VCO requires external RF components. Controlling the layout of a circuit that oscillates at 275 MHz is far from trivial. And, in general, because of the narrow tuning range of some VCO's, several external components need to be changed to vary the output frequency range. Integrating all of the VCO, at these frequencies, not only makes the PLL less difficult to use, but provides a "black box" solution with

ready to use signals. Furthermore, the integrated VCO can have much wider tuning range than the externally tuned types.

Monolithic PLL's have been recently receiving considerable attention. Several papers report work both in bipolar and CMOS processes, and show that this technique is maturing in monolithic form. We believe that the overall performance of our PLL clock driver requires bipolar design. The main limitation of CMOS is in realizing good ECL and TTL output drivers. It also seems that current mode logic of bipolar design is superior in power supply noise rejection.

Functional Description

The integrated phase-locked loop/clock generator consists of a crystal amplifier, a PLL, a voltage reference, and the clock generation functions, as shown in Fig. 1. The crystal amplifier can be used to build a Pierce-type crystal oscillator or it can be driven by an external TTL clock. The PLL consists of a multivibrator voltage controlled oscillator (VCO), a digital phase-frequency detector, a current-out charge pump and an externally programmable divider. The VCO in turn consists of a current controlled oscillator and an operational amplifier connected as a voltage to current converter. The voltage reference of 1.2 volts is available for setting the range of the RAMDAC's converters.

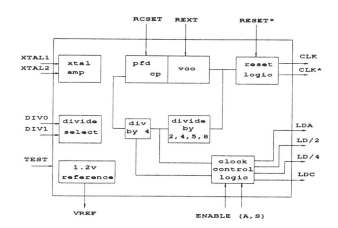

Fig 1. Block diagram of PLL clock driver

Reprinted from *Proc. CICC*, paper 15.1, May 1990.

The chip provides all the clock generation functions of similar clock drivers for RAMDAC's. Two of the TTL load clocks are enabled either synchronously or asynchronously and can be used to stop reading the RAM during retrace. A reset pin stops the pixel clock for the number of cycles required to clear a RAMDAC. There is also ld/2 and ld/4 functions of the load clocks which are used for muxing of banks of external video RAM's.

In addition, this clock generator has multiple test modes. For instance the loop can be opened at several places to allow full testability of all the building blocks. Test signals are routed to the blocks under test and their individual responses can be observed. An external voltage source can directly drive the VCO to measure its linearity, power supply rejection, temperature coefficient, or any other function. However, the most comprehensive test for a PLL is to make it lock, since this requires the simultaneous operation of several blocks.

Main Loop Components

The VCO is a standard emitter coupled multivibrator with some enhancements to improve temperature coefficient [1], [2]. The capacitor which determines the frequency range is built with a three layer metal process, with the bottom metal used as a substrate shield. The two top plates make up two capacitors with their electrodes cross connected to further cancel common mode injection [3]. The size of this capacitor in relation to parasitics, and the high linearity voltage to current converter provide a VCO of linear characteristics. An external resistor in the voltage to current converter allows selection of the VCO range. Extended VCO range increases the sensitivity to stray noise, but this is offset by the lower time constant set by the lower external resistor. The observed upper range of the VCO exceeds 450 MHz and scaling down the timing capacitor predicts operation up to 700 MHz. The measured phase noise close to the fundamental is low, which is attributed to the quality and internal location of the timing capacitor.This type of VCO has a wide tuning range and does not require external components like coils or varactors.

The charge pump (CP) is an I-2I design. Only NPN transistors switch sinking currents off and on. Slow PNP transistors make up the fixed sourcing current. The switching transients are controlled to provide matched rising and falling times in the output pulse. As shown in Fig. 3, the current sources are mirrored from the PNP to the NPN side to achieve matching. When the PLL is locked, only one NPN source is on and no net current flows out of the chip. The bandgap voltage reference for these is designed to offset the dynamic temperature coefficient in the charge pump. A mismatch of the current sources results in leakage into the external loop filter, requiring periodic corrective pumping. Worse than just creating a small phase offset, the periodic pumping may cause the VCO to produce a frequency modulated output, with phase error drifting out of specification between corrections. The instant the charge pumps, a jitter is also produced in the pixel clock, due to the "zero" in the loop filter. Because of the loop divider, the jitter is observed every N cycles of the pixel clock, where N is the divide ratio. All these effects are reduced by current matching and by the selection of large loop integration time constant.

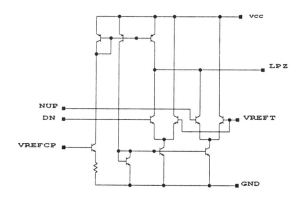

Fig 3. Charge Pump schematic

The crystal amplifier was selected to give optimum performance in terms of frequency offset, bandwidth, biasing and ease of use [4]. It is basically a linear amplifier of fixed gain and known input and output impedance. The amplifier has been made to oscillate with crystal frequencies from 3 MHz to 27MHz, which shows it has a good gain-bandwidth product.

Transient Simulation

We used HSPICE to simulate the transient operation of the PLL. The loop building blocks are all converted to operate on phase, which prevents having to have oscillators. As shown in Fig. 5, a subtractor, a divider, and an integrator were required, and were built up from combinations of controlled sources [5]. The integrator, for instance, is made up of a VCVS (voltage controlled voltage source) driving a 1 ohm resistor in series with a 1 farad capacitor. We refined our simulation by adding an approximation to a digital phase frequency detector. Since under lock the charge pump feeds very narrow impulses, we can approximate it as a narrow amplitude modulated pulse [6],[7]. The voltage

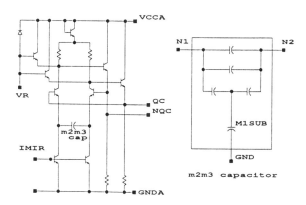

Fig 2. VCO schematic

control resistor (vcr) is switched by a narrow pulse at the expected lock frequency. It transfers, through the VCCS, an amount of charge proportional to the phase difference, and this goes to the loop filter. We verified the predicted stability for a time discontinuous loop in this way. Note that one can easily add parasitics and non-linearities to the simulator.

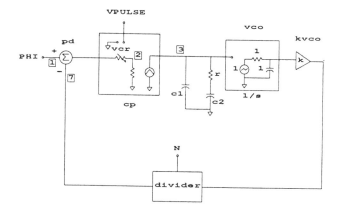

Fig 4. Transient simulation model

The simulator helps develop a "feel" for the stability of the PLL. We used the option "sweep" to verify operation over several ranges of loop components, divide ratios, VCO gains and lock frequency. Figure 5 shows a family of curves where we are varying the N, the divide number, at values of 16, 24 and 32. The next figure shows, form top to bottom, the input and feedback phase waveforms, at the input to the subtractor (pd), the impulse-like phase error, at the output of sampler (cp), and the loop filter voltage.

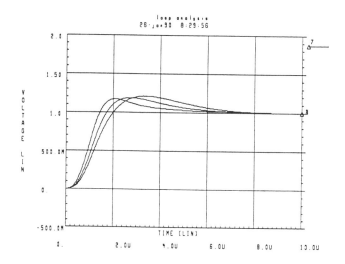

Fig 5. Family of curves for varying N

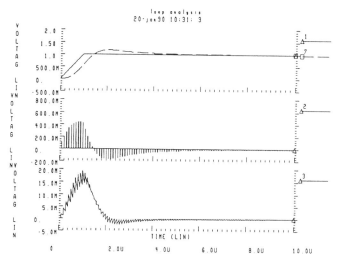

Fig 6. Details of transient simulation

Layout

The VCO and the output buffers were laid-out on opposite sides of the chip and run from separate power supplies. Substrate contacts surround the VCO and isolate it further. The digital side is divided into the noisy output buffers and the core logic. The output buffers are also shielded by a substrate contact barrier.

The chip is shown in the photograph in figure 7. The upper right corner holds the on board timing capacitor and VCO. The left side contains the output buffers. The bipolar process which produced this part has an ft of 6 GHz, and uses lateral pnp structures, schottky diodes, and a three layer metal system. The third layer of metal is very useful both for making the die smaller and for improving analog performance.

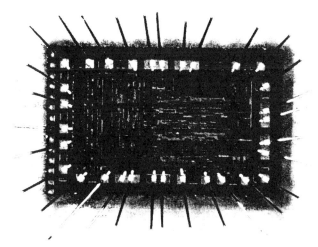

Fig 7. Photograph of PLL clock driver

Measured Results

A typical range of frequency over which the PLL will lock, without changing the external resistor, is 80 MHz to 275 MHz. This is a wide range and it allows the users a lot of flexibility in clock frequency selection without component changes. We have measured good phase noise without isolating the analog ground, using a single 5 V supply with local decoupling networks. Supply noise rejection was measured at 1 Vpp up to 10 KHz. Low power consumption of 800 mW maximum permits using a 28 pin PLCC package for smaller board space. This part has been demonstrated driving a 1280 by 1024 display with no observable pixel jitter or low frequency wandering of the screen. The photograph shows the pixel clock locked at 256 MHz and one of the TTL load clocks. The input crystal was 8.0 MHz and the divide ratio 32. For the middle range of graphic display frequencies we find crystals of around 10 MHz to give most flexibility and higher performance. This is related to the choice programmable divide ratios and the size of external capacitors required to resonate the lower frequency crystals.

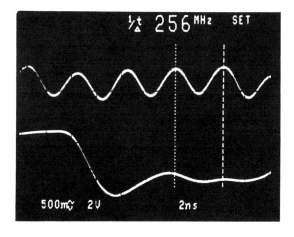

Fig 8. Pixel clock and one load clock

Conclusion

We have shown a single chip PLL clock driver for high resolution graphic displays, which operates in excess of 275 MHz and is easy to use. One direction to optimize the design will be to trade speed for power consumption. We can predict similar performance at half the power, or twice the present speed, after circuit optimization. Another area of circuit improvement is temperature coefficient and voltage sensitivities, in particular to apply the PLL technology to more general uses like clock recovery and signal demodulation. Edge control of fast output drivers will reduce radiation, which, in some environments, can result in no-lock conditions; an improvement here would make the PLL less sensitive to PCB layout.

Several variations of this part are under development. More intelligence in the control of timing, synchronizing and deskewing are being added. This PLL technology breakthrough is broad enough that it will also be incorporated in future ATE and Imaging products.

Acknowledgements

This project was jointly developed by Quadic Systems, which provided PLL experience, and Brooktree, which carried out resimulation and chip integration. Credit is due to Bryan Peter and Benny Chang, of Quadic Systems, who designed the VCO and Crystal Amplifier, respectively, and developed simulations and outstanding documentation.

References

[1] Gray, Paul R. and Meyer, Robert G., "Analysis and Design of Analog Integrated Circuits, Second Edition", John Wiley and Sons, 1984.

[2] Gilbert, Barrie, "A Versatile Monolithic Voltage-to-Frequency Converter", IEEE J. Solid-State Circuits, Vol.SC-11, No. 6, December 1976.

[3] Wakayama, M.H. and Abidi A. A., "A 30-MHz Low-Jitter High-Linearity CMOS Voltage-Controlled Oscillator", IEEE J. Solid-State Circuits, Vol. SC-22, pp. 1074-1081, Dec. 1987.

[4] Matthys, Robert J., "Crystal Oscillator Circuits", John Wiley and Sons, 1983.

[5] Epler, Bert, "SPICE2 Application Notes for Dependent Sources", IEEE Circuits and Devices, Vol 3, No. 5, September 1987.

[6] Gardner, Floyd M., "Charge-Pump Phase-Locked Loops", IEEE T. on Communications, Vol. Com-28, No. 11, November 1980.

[7] Hein, P. H. and Scott, J. W., "z-Domain Model for Discrete-Time PLL's", IEEE T. on Circuits and Systems, Vol. 35, No. 11, pp.1393-1400, Nov. 1988.

The Macro Modeling of Phase Locked Loops for the Spice Simulator

Mark Sitkowski

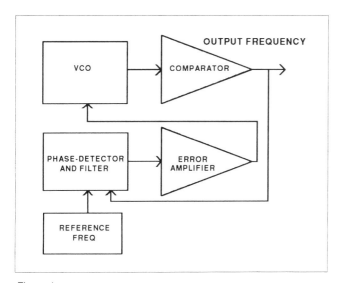

Figure 1

The phase-locked loop has always presented several problems in terms of macro-modeling. Four main elements comprise a basic phase-locked loop: a voltage-controlled oscillator producing a square-wave output; a phase-detector; a filter; and error amplifier (Fig. 1). Each of these, in itself, presents several problems in terms of circuit design, but the principal difficulty is the vast disparity between the frequencies involved.

The local oscillator, for example, runs at frequencies which are an order of magnitude higher than the frequencies appearing at the output of the phase-detector. This leads to the

Figure 2

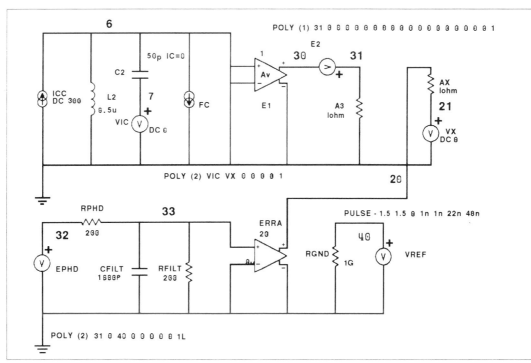

Reprinted from *IEEE Circuits and Devices Magazine*, pp. 11-15, March 1991.

calculation of many data points, since the simulator sampling rate must be referred to the highest frequency.

Most conventional approaches to phase-locked loop modelling produce macro-models which are unusable, because of their complexity. This complexity is due, in part, to the inclusion of semiconductor junctions and, also, to the use of voltage-controlled oscillators designed by conventional electronic circuit design techniques. Simulation over reasonable time intervals takes an unacceptably long time.

The approach described in this article produces a macro-model which uses no semiconductor junctions, and no regenerative feedback. The voltage-controlled oscillator is implemented by means of a circuit producing 'infinite ringing,' and the comparator characteristics are obtained via an odd-order polynomial transfer function.

Performance of the complete macro-model (Fig. 2) is such that 8000 point simulations, corresponding to some 150 cycles of local oscillator, may be performed in about 15 minutes on a platform such as the Sun 3/60, or Apollo DN4500.

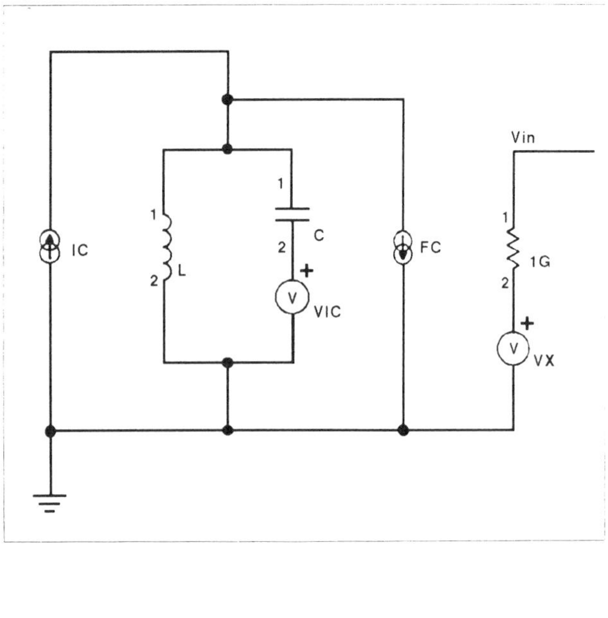

Oscillators

Oscillators are, in themselves, problematic, due to the large number of iterations necessary to evaluate regenerative circuits. Any attempt at modelling a controlled oscillator using traditional oscillator circuit design techniques will lead to problems. SPICE slows down appreciably when simulating any regenerative circuit, partly due to the iteration count, and partly due to timestep control. Also, if the loop gain is made high to reduce the number of iterations, there is a risk that SPICE will fail to converge. If you're very unlucky, SPICE will issue the error message which strikes fear into the most stalwart: "Internal timestep too small." Additionally, any use of semiconductor devices will cause the simulation speed to be unacceptably slow when simulating over practical time intervals. Thus, it would seem that the correct approach is to use no semiconductors, and no positive feedback.

In fact, it is best not to use an oscillator at all, but to employ the technique of 'infinite ringing.' If a tuned circuit formed by an inductor and capacitor is set into oscillation, it will 'ring' at its resonant frequency, without the need for positive feedback of any kind. This response is realized because the gain comes from the 'Q' of the circuit, and the feedback mechanism, necessary to sustain oscillations, is replaced by energy storage and transfer between the reactive elements (which require very little effort on the part of the SPICE simulator). The choice of series or parallel circuits is arbitrary, depending on whether a maximum voltage or maximum current resonance is desired or, perhaps, whether voltage or current frequency control is desired. The control of frequency is simply achieved by varying the instantaneous value of either the inductor or the capacitor. The control process has infinite resolution, such that the frequency

Figure 3

Figure 4

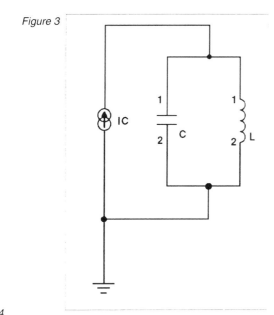

can be changed with zero delay, within a half-cycle or less.

Since the circuit elements used within SPICE are all ideal, it is a very simple matter to design a lossless tuned circuit, set to resonate at the center frequency of the controlled oscillator. For a parallel tuned circuit, the drive is provided by a direct current source connected across the circuit, as shown in Figure 3. If "IC = 0," SPICE will charge the capacitor when the simulation starts, thus causing the tuned circuit to oscillate. Because there is no damping, the circuit will oscillate almost forever. Connecting any kind of component (like a voltage-controlled source) to the circuit will cause a small damping effect, presumably because its input impedance is not infinite, but equal to the reciprocal of the minimum conductance parameter, G_{MIN}, set by default to 1E-12, and resettable with the .OPTIONS card. Phase-locked loops require a square-wave output, and so the sinusoidal waveform must drive a comparator. Thus, if the sinusoid is made large enough, any damping effects can be ignored. Accordingly, the source current is set to 300 amps, to ensure a very high output voltage at resonance. However, if a constant amplitude is really important, a loose feedback loop can be incorporated by making the bias current dependent on the output voltage.

Voltage Controlled Oscillator

The next step is to make the frequency of the oscillator voltage-variable. To control the oscillation frequency, it is necessary to vary the value of either the inductor or the capacitor, in proportion to an external voltage. Although SPICE contains voltage-variable capacitors, they are either part of a semiconductor device, or their characteristics

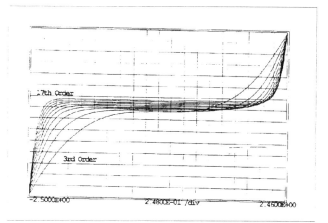

Figure 6

are controlled by a polynomial, defined in terms of the terminal voltage. In addition, most voltage-variable capacitors described in the literature also use the terminal voltage as the control and, accordingly, are not appropriate for this application.

What is needed is a capacitor whose value is proportional to an external voltage that does not appear across the tuned circuit. The actual implementation of a variable capacitor is accomplished by the usual technique of multiplying the instantaneous capacitive current by a factor, and adding it to the original current. This method owes it origins to the hardware method of obtaining apparently large values of capacitance, by placing small capacitors between the base and collector of a transistor. The apparent capacitance between collector and emitter is then approximately equal to the product of h_{FE} and the base-collector capacitance.

To achieve this result with SPICE, a zero-value voltage source is placed in series with the capacitor in order to measure its current. Next, a current controlled current source is placed in parallel with it. Since we need a control input, the current controlled current source is configured as a multiplier by using the appropriate polynomial to define its transfer function. Its output is now proportional to the product of two currents, so the control voltage has to be converted to a current analogue. This may be achieved either with yet another controlled source or, more simply, with a 1-ohm resistor.

Figure 4 shows the configuration described above, and shows the parallel tuned circuit with a voltage-variable capacitor incorporated. The capacitive current is measured by the zero-

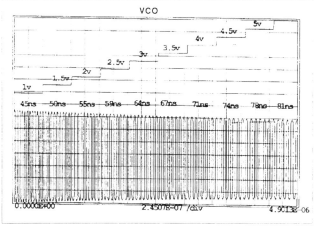

Figure 5

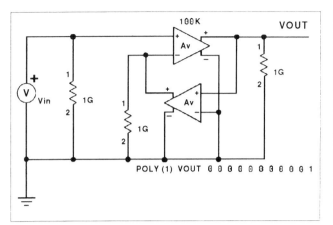

Figure 7

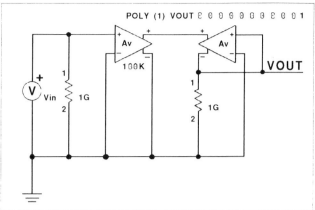

Figure 8

value voltage source VIC, while VX measures the current proportional to the input voltage. Current source IC provides bias drive for the tuned circuit, while controlled current source FC supplies the voltage-proportional capacitive current.

The variation of the frequency with applied voltage is shown in Figure 5, where the input waveform is a series of voltage steps, varying between 1 and 5.5 volts. The period of the output waveform varies between 45 ns and 81 ns.

Comparator

There are several approaches to the design of a minimal comparator. One which can be dismissed immediately, is that which attempts to accurately model an integrated circuit comparator; the cost in primitives is too great.

A simplified version of the integrated circuit comparator appears in the traditional macro-models. In such a case, the function is approximated by a high-gain amplifier, followed by amplitude-limiting, via a series resistor, clamping diodes and voltage sources. Although such an approach produces a good result, it still uses an excessive number of primitives which, accordingly, slows the simulation.

If we approach the problem from basic principles, it may be seen that the ideal comparator performs the mathematical signum function with an offset and scaling factor. An examination of the odd-order polynomial functions (Fig. 6), shows that they are symmetrical about the y-axis. The slope of the function after the knee (which occurs for an x-axis value of unity) depends on the actual order of the polynomial.

This function may be used to model a comparator, by one of several methods, each of which has certain merits. One method uses a dependent voltage source having such a transfer characteristic in a feedback loop around an amplifier (Fig. 7). When the input is below a volt, the feedback is minimal, and the amplifier delivers its full output. However, as the input crosses unity, the feedback becomes several orders of magnitude higher, and the output is limited to approximately 1 volt. The gain of the amplifier determines the slope of the dc transfer function, which affects the trigger level of the comparator. Amplifier gain may, to a certain extent, be traded off against the order of the polynomial, which has a similar effect on the comparator.

The above design produces good results, limiting the output at around 1.06 volts for the values shown. However, it does require one ground return resistor more than the next method.

A second approach is shown in Figure 8. A buffer, whose gain determines the slope of the dc transfer characteristic of the comparator, drives a load resistor, to which it is coupled via a dependent source. The source output is a function of the voltage across the load resistor, and is connected so as to oppose the input voltage. Its transfer function is a 9th order polynomial.

When the load voltage is less than unity, the input waveform appears virtually unchanged but, the nearer the load voltage gets to unity, the higher the voltage subtracted from the input waveform, so the load voltage stays at about 1.44 volts, given the values shown. This design simulates marginally faster than the previous one, and uses one primitive less.

Phase Detector

The phase detector used in many phase-locked loops is a four-quadrant analogue

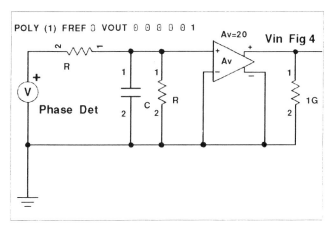

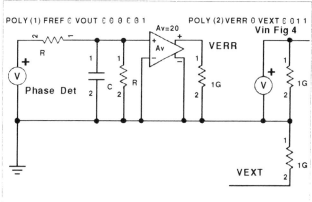

Figure 9

Figure 10

multiplier. The design of the detector with SPICE primitives is trivial, because the multiplier is itself a SPICE primitive. A nonlinear voltage-controlled voltage source, with a transfer function defined as:

POLY(2) FREF 0 VOUT 0 0 0 0 0 1

performs the necessary function, where *FREF* is the reference frequency or the input to the phase-locked loop, and V*OUT* is the comparator output (Fig. 9).

Low-pass Filter and Error Amplifier
The low-pass filter is taken from a standard design. The amplifier is merely a voltage-con-

trolled voltage source, which drives the 1 ohm current sensing resistor in Figure 4, at the point V$_{in}$. Some phase-locked loops have the facility of applying external voltages to the error amplifier, in order to perform FM demodulation. If this is necessary then, instead of the direct connection shown, the error amplifier should drive an adder (Fig. 10).

Complete SPICE Netlist
The SPICE netlist for the complete phase-locked loop is fairly compact (see Table). The simulator control cards have been left intact so that it may be tested directly. For the sake of ease of cross-reference, the names of the circuit elements correspond in most cases to those used in the text.

The performance of this PLL has not been optimized in any way, nor is it intended to resemble any actual device. The filter's time constant is intentionally set to provide an underdamped response, just to examine the characteristics of the loop. The linearity of the VCO has not been adjusted, and the damping (negative, in this case) of the tuned circuit has been left untouched.

However, as a macro-model, it performs extremely well, permitting the simulation, within a realistic time, of one or two hundred cycles of local oscillator. In addressing each component of a PLL individually, the macro-model provides the user a large measure of flexibility in applying the technique to real-world devices.

—*Mark Sitkowski*
Mentor Graphics Inc.
6/670 Canterbury Rd.
Box Hill
Victoria 3128
Australia

Table: SPICE Netlist for the PLL
```
PLL
.TRAN 1N 8U UIC
.PRINT TRAN V(6) V(20) V(31) V(40)
V(32) V(33)
ICC 0 6 300
C2 6 7 50P IC=0
VIC 7 0 DC 0
L2 6 0 0.5U
FC 6 0 POLY(2) VIC VX 0 0 0 0 1
E1 30 0 6 0 1
E2 30 31 POLY(1) 31 0 0 0 0 0 0 0 0
0 0 0 0 0 0 0 0 0 1
R3 31 0 1
VREF 40 0 PULSE -1.5 1.5 0 1N 1N
22N 48N
RGND 40 0 1G
EPHD 32 0 POLY(2) 31 0 40 0 0 0 0 0 1
RPHD 32 33 200
CFILT 33 0 1000P
RFILT 33 0 200
ERRA 20 0 33 0 20
RX 20 21 1
VX 21 0 DC 0
.END
```

263

Modeling and simulation of an Analog Charge-Pump Phase Locked Loop

Sumer Can
Signetics Corporation
Linear LSI Division
Sunnyvale, California 94086

and

Yilmaz E. Sahinkaya
Lockheed
Palo Alto Research Laboratory
Palo Alto, California 94304

SUMER CAN was born in Koyulhisar, Turkey, on December 24, 1947. He received his Y.Muh. (M.Sc.) degree in electrical engineering from Istanbul Technical University (Turkey) and his M.A.Sc. degree in electrical engineering from the University of Toronto (Canada) in 1972 and 1977 respectively. He worked at the Istanbul Technical University (1972-1974) as an instructor and at the University of Toronto (1974-1978) as a graduate assistant. In 1978, he joined the industry and worked at Litton Systems (Canada) Ltd., National Semiconductor Corp., Sperry-Univac, and Magnetic Peripherals Inc. Since 1984, he has been employed by Signetics Corporation, Sunnyvale, California in Linear LSI Division as a member of technical staff. His interests are bipolar/MOS integrated circuit design, circuits and system simulation, nonlinear dynamics, and chaotic circuits. Mr. Can is the author of several technical publications and patent disclosures in the area of circuits and systems.

YILMAZ E. SAHINKAYA received his B.Tech and M.S. degrees in mechanical engineering from the Loughborough College of Technology (England) and the University of Michigan in 1961 and 1962 respectively. He received his Ph.D. in electrical engineering from CalTech in 1969. His industrial experience includes Allis Chalmers Research Labs (1962-1964), Jet Propulsion Labs (1968-1970), Control Data Corporation (1970-1974), Bimsa (1976-1977), Control Data Corporation (1977-1985), and Lockheed Palo Alto Research Laboratory (1985-present).

His specialty is the design and development of large and complex control systems using computer simulation methodology. His current interest is the design and development of intelligent control systems utilizing microprocessors. He has published many papers in national conferences and currently is preparing a textbook entitled, "Applied Computer Simulation for Control Engineers."

ABSTRACT

We describe a nonlinear computer simulation model of an Analog Charge-Pump Phase Locked Loop (ACP-PLL). Offsets of the Phase Detector and Analog Charge-Pump are modeled as disturbances to simulate their effects on the steady-state phase error of the loop.

An extensive computer simulation study is carried out for the design of an example Analog CP-PLL using the proposed nonlinear model and comparing it to the conventional linear model. Results demonstrate the linear model is not sufficient to fully analyze and predict the behaviour of the Analog CP-PLL.

INTRODUCTION

The Analog CP-PLL is an electronic control system whose simplified block diagram is in Figure 1. The s_R and s_O are applied to the inputs of the Phase Detector which produces an output voltage s_D corresponding to the phase difference of these two input signals. The Phase Detector is followed by an Analog Charge-Pump. It consists of a Transconductance Amplifier with a capacitor at its output, forming an integrator. Note that this analog charge-pump is different from the Charge-Pump described by Gardner (1980) which is basically made of a current source/sink driven by a sequential-logic phase/frequency detector. The output of the loop filter is amplified and applied to the input of the Voltage Controlled Oscillator (VCO). This in turn produces an output signal at a frequency corresponding to the change of control voltage v_B such that it reduces the phase difference between s_R and s_O.

The operational characteristics of the Analog CP-PLL (Figure 1) are

(1) When the reference frequency f_R equals VCO frequency f_O, there is a 90° phase angle between s_R and s_O ($\phi_E = \phi_D - 90° = 0$).

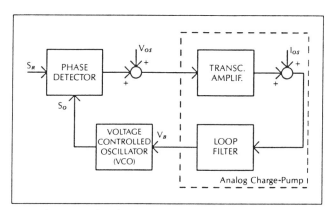

Figure 1. Analog Charge-Pump Phase Locked Loop (ACP-PLL).

Reprinted with permission from *Simulation*, S. Can and Y. E. Sahinkaya, "Modeling and Simulation of an Analog Charge Pump Phase-Locked Loop," vol. 50, pp. 155-160, April 1988.

(2) When the reference frequency changes to a new value, the VCO frequency takes up the same value within a required time interval, called "pull-in time" or "lock-in time." During this control action, the phase angle error ϕ_E, taken with respect to the 90° phase angle mentioned above, changes from zero to a positive or negative value and finally becomes zero when $f_R = f_o$ at the end of the pull-in time.

(3) When the reference frequency remains unchanged and a disturbance signal (i.e., V_{os} and/or I_{os}) enters into the system, the phase angle error changes from zero to a positive or negative value. The value of ϕ_E is proportional to the level of the offset signal.

The control action during the pull-in time is also known as the acquisition mode of operation. The control action, after frequency lock following a disturbance, is known as the tracking mode of operation. As in the following sections, all three major components of the Analog CP-PLL have nonlinearities.

Hence, the Analog CP-PLL is a nonlinear feedback control system whose system parameters and stability conditions which satisfy the prescribed performance requirements cannot be determined by well-known techniques such as "Root-Locus" and "Bode Plot" from the classical control theory. An efficient technique (Mitchell 1978; Dost and Liu 1985) to accomplish the task is use the power and flexibility of system simulation languages such as ACSL, CSSL-IV, or DSL/VS.

In this article, the usage of CSSL-IV is demonstrated in the determination of time-domain response characteristics of an example Analog CP-PLL.

The performance criteria of the Analog CP-PLL are set as:

(1) Settling time < 2 μs

(2) Overshoot < 20%

(3) Steady-state phase error < 3°

in response to the application of a 10% reference frequency step at $f_R = 10$ MHz; and Phase Detector offset $V_{os} = 10$mV; and Charge-Pump offset $I_{os} = 75$ μA in the form of step inputs.

MATHEMATICAL MODELING

Analog CP-PLL (see Figure 1) contains three main subcircuits: Phase Detector, Charge-Pump and Loop Filter, and Voltage Controlled Oscillator (VCO). We will briefly describe the operation of these subcircuits and present their mathematical models.

The main differences of this model from the classical models are that:

(1) The offsets of the Phase Detector and Charge-Pump are introduced into the model as disturbances.

(2) The Loop Filter is characterized with its state-space equations and used as a sub-block in the simulation.

Phase Detector: For a simplified circuit schematic of a typical analog phase detector see Figure 2. Apply the output of the VCO denoted by s_o to the upper transistor pairs Q3,Q4 and Q5,Q6. Apply the PLL's reference signal denoted by s_R to the lower transistor pair Q1,Q2. The input signals are given by:

$$s_R = A_R \sin(2\pi f_R t + \phi_R) \quad (1)$$

$$s_o = A_o \sin(2\pi f_o t + \phi_o) \quad (2)$$

where

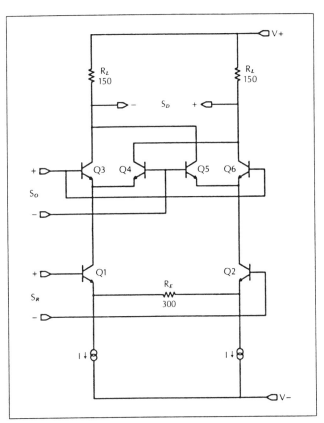

Figure 2. Simplified phase detector schematic.

A_R, A_o are the amplitudes; f_R, f_o are the frequencies and ϕ_R, ϕ_o are the phase angles of ACP-PLL's reference input and the VCO output, respectively.

An expression for the phase detector output signal S_D can be derived by observing its operational characteristics as follows:

• When s_R and s_o are in phase and have positive polarities, assume A_o is large enough so that transistors Q3, Q6 are fully turned "on" and transistors Q4, Q5 are fully turned "off." The differential pair Q1, Q2 and cascode transistors Q3, Q6 configuration yield a voltage of $s_D = (2R_L/R_E)s_R$ across load resistors R_L.

• When the input signals are in phase and have negative polarities, assume A_o is large enough so that transistors Q4, Q5 are fully turned "on" and transistors Q3, Q6 are fully turned "off." Since s_R is of negative polarity, differential pair Q1, Q2 and cascade transistors Q4, Q5 configuration yield a voltage of the same magnitude and polarity as in the previous case across load resistors R_L. Hence, the input reference signal is amplified. Also, its negative swinging half is inverted, producing an output signal s_D as a periodic function at twice the frequency of s_R with a maximum positive average dc level.

• When s_o lags s_R by 180°, using the same argument given above, s_D again is a similar periodic function with a maximum negative average dc level.

• When s_o lags s_R by 90°, s_D is a similar periodic function with a zero average dc level.

• When s_o lags s_R by any phase angle less than 90°, s_D is a periodic function with a positive average dc level. When s_o lags s_R by any phase angle more than 90°, s_D is a periodic function with a negative average dc level. In Figure 3 is the

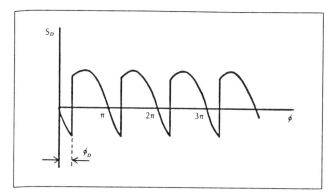

Figure 3. Phase detector output for a phase angle difference ϕ_D.

phase detector output signal s_D for a phase angle difference of ϕ_D between s_O and s_R.

Based on the above discussion an expression for s_D is given by:

$$s_D = (2R_L/R_E) s_R. \text{ Sign }\{s_O\} \qquad (3)$$

in which $(2R_L/R_E)$ is the dc gain of the phase detector, and Sign $\{s_O\}$ is defined by:

$$\text{Sign }\{s_O\} = +1 \qquad \text{for } s_O > 0 \qquad (4a)$$

$$\text{Sign }\{s_O\} = 0 \qquad \text{for } s_O = 0 \qquad (4b)$$

$$\text{Sign }\{s_O\} = -1 \qquad \text{for } s_O < 0 \qquad (4c)$$

From Figure 3, an average value of s_D is given by:

$$s_D = (2R_L/\pi R_E)*(\int_{t_o}^{\pi} A_R \text{Sin}\varphi \, d\varphi \; - \int_0^{t_o} A_R \text{Sin}\varphi \, d\varphi) \qquad (5)$$

carrying out the integration in Eq. 5 yields:

$$s_D = (4R_L A_R/\pi R_E) \text{Cos}\phi_D \qquad (6)$$

Since $\phi_E = \phi_D -90°$, Eq. 6 can be rewritten as:

$$s_D = (4R_L A_R/\pi R_E) \text{Sin}\phi_E \qquad (7)$$

Charge-Pump and Loop Filter: A simplified circuit schematic of an Analog Charge-Pump is in Figure 4. The differential output of the phase detector is applied to the input of differential pair Q1, Q2. The output current I_{CP} of this transconductance amplifier charges and discharges the capacitor C_{CP}. For $R_{CP} = 0$, the output current of the charge-pump is given by:

$$I_{CP} = I_{CS} \text{ Tanh} (s_D/V_T) \qquad (8)$$

where

I_{CS} is the constant current source; V_T is given by kT/q in which k is the Boltzmann constant, T is the absolute temperature, and q is the electronic charge. When $R_{CP} \neq 0$, the expression given in Eq. 8 can be simplified to a linear relationship as:

$$I_{CP} = (1/R_{CP}) s_D \qquad (9)$$

The Loop Filter forms by series connection of a resistor R_X and

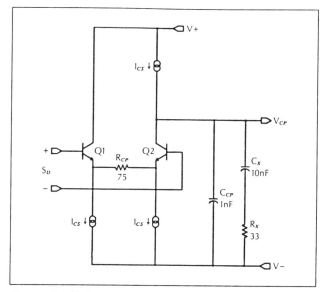

Figure 4. Simplified schematic of the Analog Charge-Pump and Loop Filter.

a capacitor C_X. The state equations for the configuration in Figure 4 are given by:

$$\overset{\circ}{v}_{CP} = -[(b-1)/\tau] v_{CP} + [(b-1/\tau] v_{CX} + [(b-1)/C_X] I_{CP} \qquad (10)$$

$$\overset{\circ}{v}_{CX} = (1/\tau) v_{CP} - (1/\tau) v_{CX} \qquad (11)$$

where

$\tau = R_X C_X;\; b = 1 + C_X/C_{CP}$.

Voltage Controlled Oscillator: A negative resistance LC type oscillator is considered as a VCO in this article. A simplified schematic of such oscillator is in Figure 5. The differential pair Q1, Q2 which is connected in a positive feedback configuration forms a negative resistor. The combination of an inductor L, a capacitor C_V and the negative resistor produces the oscillations at the collector of Q1. The amplitude of the oscillations is controlled by a subcircuit which varies the negative resistor by changing the tail current of differential pair Q1, Q2. The amplifier A drives a frequency divider circuit. The frequency of oscillation is given by:

$$f_O = K_D/2\pi [L \; C_V(v_B)]^{1/2} \qquad (12)$$

where

K_D is coefficient introduced by the frequency divider ($K_D = 1/2$) circuit which follows the VCO and v_B is the output voltage of the buffer/amplifier following the charge-pump and is given by:

$$v_B = K_B v_{CP} \qquad (13)$$

A voltage controlled capacitor (varactor) is used to realize frequency control by voltage. The capacitance C_V versus the buffer voltage v_B characteristic is nonlinear. However, the f_O versus v_B characteristic for a particular varactor used in our example is almost linear. Hence, Eq. 12 can be linearized and re-written as:

$$f_O = K_O v_B + f_{OC} \qquad (14)$$

266

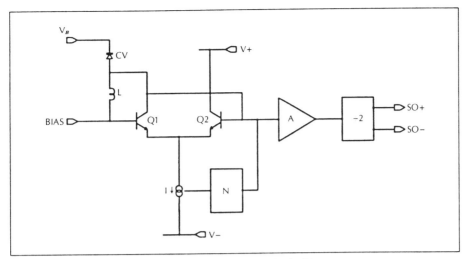

Figure 5. Simplified schematic of VCO.

where f_{oc} is the VCO's frequency of oscillation at $v_B = 0$, which is called center frequency (or free running frequency). Equation 14 is used in the linear analysis part of this study. The output of the VCO is given by:

$$s_O = A_O \, Sin\,[2\pi f_O(v_B)\,t + \phi_O(v_B)] \qquad (15)$$

where

$$\overset{\circ}{\phi}_O = f_O(v_B) \qquad (16)$$

In Figure 6 is the complete block diagram corresponding to the nonlinear mathematical model developed for the Analog CP-PLL. Note that the offset voltage of the phase detector v_{os} and the offset current of the charge-pump i_{os} are introduced as disturbance signals in Figure 6.

COMPUTER SIMULATION

A computer simulation of the proposed Analog CP-PLL model is carried out in the following steps:

Step1. *Linear Analysis.* For small changes of ϕ_E, $Sin\,\phi_E$ is approximately equal to ϕ_E. Therefore, Eq. 7 can be linearized as:

$$s_D = (4R_L \, A_R / \pi R_E)\,\phi_E \qquad (17)$$

Using linearized models of phase detector, transconductance amplifier and VCO given by Eqs. 17, 9, and 14 respectively, we obtain the block diagram of the linearized ACP-PLL as in Figure 7.

When the phase detector and charge-pump offsets are introduced into the system as $v_{os} = v_{os}\,u(t)$ and, $i_{os} = I_{os}\,u(t)$ where $u(t)$ being a unit step function, from the block diagram in Fig-

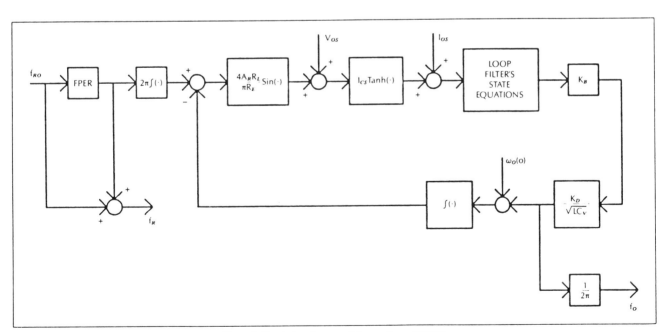

Figure 6. Simulation diagram of the nonlinear ACP-PLL.

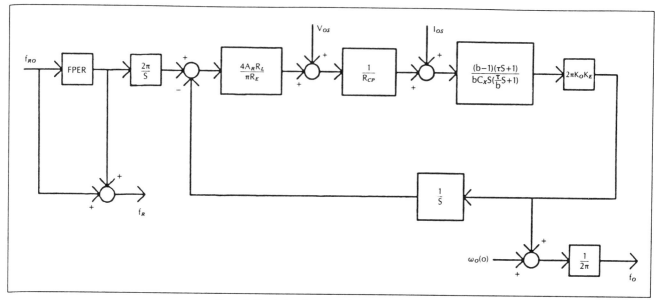

Figure 7. Simulation diagram of the linearized ACP-PLL.

ure 7, it can be shown that the steady-state phase error (Gardner 1979) due to offsets is:

$$\phi_e = (\pi R_E / 4 A_R R_L) [V_{OS} + R_{CP} I_{OS}] \quad (18)$$

From Figure 7, the closed-loop transfer function (when offsets are zero) is obtained as:

$$\phi_O / \phi_R = F(s) / [1 + F(s)] \quad (19)$$

where

$$F(s) = K(s + 1/\tau) / [s^2(s + b/\tau)] \quad (20)$$

in which

$$K = [8 R_L A_R (b-1) K_B K_O] / [R_E R_{CP} C_X] \quad (21)$$

$$\tau = R_X C_X \quad (22)$$

Substituting Eq. 20 into Eq. 19 we obtain:

$$\phi_O(s) / \phi_R(s) = K(s + \tau) / [s^3 + (b/\tau)s^2 + Ks + (K/\tau)] \quad (23)$$

where the characteristic equation is:

$$s^3 + (b/\tau)s^3 + Ks + (K/\tau) = 0 \quad (24)$$

The roots of Eq. 24 determine the response characteristics of the linear system as in Figure 7. Here, the system parameters used are $A_R = 0.2V$; $K_B = 2$; $K_O = 3.67 *10^6$ Hz/V; $C_X = 10nF$; $R_X = 33$ ohm; $b = 11$; $R_L = 150$ ohm; $R_E = 300$ ohm. From the Root Locus diagram (not shown here) of Eq. 24, the roots of the characteristic equation and the gain which yield an acceptable system performance are determined. As a result, an initial value of 150 ohm is selected for the resistor, R_{CP}, of the linearized charge-pump.

A computer simulation program using CSSL-IV was developed (Can and Sahinkaya 1986) for the linear ACP-PLL and simulated for a 10% frequency step (i.e., FPER = 0.1) and zero offsets. The ACP-PLL response is in Figure 8a.

Step2. *Nonlinear Analysis.* A CSSL-IV program is also developed (Can and Sahinkaya 1986) for the nonlinear ACP-PLL model of Figure 6. The response of the nonlinear ACP-PLL for a same input frequency step as in linear ACP-PLL is in Figure 8b. From Figure 8, the following conclusions are drawn:

(1) The linear model *is not adequate* for predicting the response characteristics of the ACP-PLL.

(2) The "pull-in time" as predicted by the nonlinear model, with previously assumed circuit parameters, is about 14 µs which is about twice the value obtained from the linear model, and seven times larger than the maximum allowable value of 2 µs. Note that here, $R_{CP} = 150$ ohms. To reduce the "pull-in time" to an acceptable value, the loop gain must be increased without exceeding the stability limit. At this point, for the design method used in choosing the parameter values one may refer to a recent article by Gardner (1980).

Several computer runs have been made to optimize the design. Only the most interesting results are given here. The VCO frequency (see Figure 8b) loses its synchronism with the reference frequency for about 4 cycles before a "lock-in" is achieved. This clearly demonstrates the importance of the nonlinear analysis since the linear model will not show such characteristics. In Figure 9 is the response for final selection of parameter R_{CP} as 75 ohms for a frequency step of 10%.

In Figure 10 is the step response of the ACP-PLL when the linearized charge-pump model is replaced with the nonlinear model and other parameters are kept the same as in Figure 9. With the nonlinear charge-pump, the "lock-in" time is further reduced from approximately 5 µs (see Figure 9) to less than 2 µs.

CONCLUSIONS

A nonlinear mathematical model is developed for the simulation of Analog Charge-Pump Phase Locked Loops. Using CSSL-

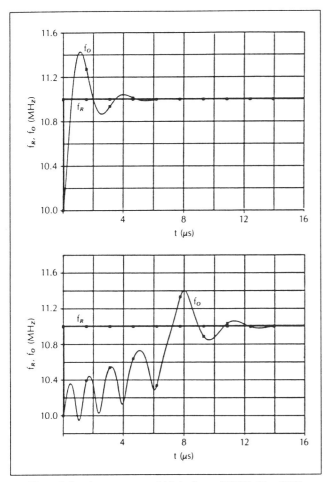

Figure 8. Step input response of (a) the linear ACP-PLL (R_{CP}=150Ω) (b) the nonlinear ACP-PLL with linear charge-pump (R_{CP}=150Ω).

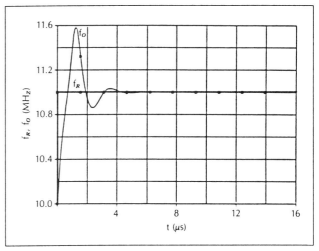

Figure 9. Step input response on nonlinear ACP-PLL with linear charge-pump (R_{CP}=75 ohms).

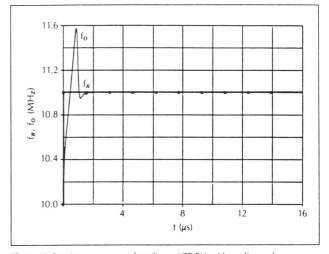

Figure 10. Step input response of nonlinear ACP-PLL with nonlinear charge-pump (R_{CP}=0).

IV, it was once again clearly shown that the linearized model is not sufficient to fully analyze and predict the response characteristics of the loop. It is almost mandatory to use computer based nonlinear analysis methodology for the design of Analog Charge-Pump Phase Locked Loops. The Phase Detector and Charge-Pump offsets are introduced as disturbances into the model and their influence on the ACP-PLL response characteristics is investigated.

REFERENCES

Can, S. and Y. E. Sahinkaya. 1986. "A Computer Simulation Model For An Analog Charge-Pump Phase Locked Loop." In *Proceedings of the 1986 Summer Computer Simulation Conference* (Reno, Nevada, July 28-30). SCS, San Diego, Calif., 886-892.

Dost, M. H. and C. C. Liu. 1985. "Variable Frequency Clock Study Using DSL/VS." In *Proceedings of the 5th International Conference on Mathematical Modelling* (U.C. Berkeley, Calif., July 29-31.)

Gardner, F. M. 1979. *Phaselock Techniques.* John Wiley & Sons Publishing Co., New York, N.Y.

Gardner, F. M. 1980. "Charge-Pump Phase Locked Loops." *IEEE Transactions on Communications* COM-28, no. 11 (Nov.): 1849-1858.

Mitchell, E. L. 1978. "Phase Locked Loop Techniques: Interactive Simulation With the Advanced Continuous Simulation Language (ACSL)." In *Proceeding of the 1978 Summer Computer Simulation Conference* (Newport Beach, Calif.). SCS, San Diego, Calif., 444-451.

MIXED-MODE SIMULATION OF PHASE-LOCKED LOOPS

Brian A.A. Antao,[1] Fatehy M. El-Turky[2] and Robert H. Leonowich[2]

[1] Department of Electrical Engineering,
Vanderbilt University,
Nashville, TN 37235

[2] AT&T Bell Laboratories,
Allentown, PA 18103

Abstract

This paper describes behavioral models for mixed-mode and multi-level simulation of phase-locked loops(PLLs). PLLs are a difficult class of systems to evaluate using conventional circuit simulators because of the mixed analog digital signals involved, and extensively long runtimes required to capture the performance. Behavioral modeling techniques and a mixed-signal simulator, the AT&T Bell Laboratories ADAMS simulator, are used to overcome these limitations. An all analog PLL is simulated at the behavioral level to measure the lock-in characteristics as well as the tracking range. A high-speed digital PLL is simulated at the behavioral level, as well as at a multi-level using device-level models for the phase detector, to measure the lock-in time and detect false locking.

1   Introduction

Phase locked loops are a class of systems that find use in many applications ranging from data recovery in communication systems to clock synthesizer circuits in digital systems[2]. PLLs have varying implementations: all analog, mixed analog-digital, or all digital. The most commonly used PLL configurations have digital inputs and outputs with the intermediate signals being analog in nature. The behavior of a PLL prior to locking is highly non-linear and stochastic and therefore hard to evaluate. The system parameters are usually designed based on a linearized model which is valid once locking occurs. Simulation of a PLL design is essential to verify the correct operation, ie locking to occur. Simulation of a PLL is plagued with two major bottlenecks: 1) The PLL system and the signals are mixed analog-digital in nature 2) A large number of clock cycles have to be simulated to obtain meaningful results. The mixed-signal nature of the PLL can be dealt with by the use of a conventional circuit simulator such as SPICE. However device level simulation results in impractical run-times to obtain the performance of the PLL. To overcome this drawback, efforts in the past have focused on simulating a PLL at a behavioral level[3, 5]. These attempts at the behavioral simulation of a PLL are based on

developing a customized stand-alone simulator for specific PLLs at a very abstract level. The PLL is seldom used as a stand-alone system, and one would like to simulate the PLL functionality within the context of a larger system. Attempts at macromodeling PLLs in SPICE like simulators does not address the mixed-signal problem or alleviate the runtime bottleneck altogether, and are at best coarse approximations for sophisticated PLL components such as the sequential phase/frequency detector[8].

Behavioral modeling and simulation is vital in order to expedite the simulation of full chip or board-level designs. Behavioral level simulators that can handle both analog and digital sections, allow a designer to make quick feasibility analysis of the entire system, verifying the functionality and experimenting with different configurations. The AT&T Bell Laboratories mixed-signal behavioral simulator, ADAMS, being used as the general simulator provides the user with the flexibility of defining custom behavioral models at varying levels of abstraction. The behavioral models can be defined in terms of 1) State-space models; 2) algebraic expressions; and 3) mixed algebraic and differential equations[1]. These models are written in ABCDL (Analog Behavior Circuit Description Language). The individual simulation models that represent the subsystems are linked together to simulate the functional behavior of an entire system. In this paper we describe the mixed-mode simulation of PLLs, where we utilize behavioral models and a new generation mixed-signal simulator to overcome the traditional bottlenecks of efficiently simulating PLLs.

2   Modeling Methodology

The key to expediting simulation runtimes is to use higher level behavioral models that represent the essential characteristics of the PLL. The granularity of these models can be varied to tradeoff accuracy versus simulation speed. The considerable speed obtained through the use of behavioral models, allows one to undertake simulation for many clock cycles to determine characteristics such as the tracking range. These models are simulated in the ADAMS simulation framework, which allows mixed signal (analog-digital) and multi-level (behavioral-device level) simulation. Two

Reprinted from *Proc. CICC*, paper 8.4, May 1993.

approaches can be undertaken for behavioral modeling, 1) A top-down hierarchical decomposition methodology and 2) A bottom-up hierarchical abstraction methodology. The top-down approach is followed in the design of new systems. Here at the system conceptualization and design phase, the overall system characteristics have to be designed and functionality has to be verified. Behavioral models are used to get quick estimates of the system parameters. Use of higher-level behavioral models speeds up simulation and provides leverage for iteratively optimizing the system parameters.

The bottom-up methodology is used in the simulation and verification of existing designs. The bottom-up approach is also useful for extracting and parameterizing behavioral models for use in future design processes. These behavioral models can then be used to simulate a system such as the PLL within a larger system. Since it is too expensive to simulate a complete PLL at the device level, extracted behavioral models are used for simulating the entire PLL. We cite experiences and simulation results of two PLL configurations that illustrate both the behavioral modeling paradigms.

The PLLs are modeled by decomposition into major functional blocks – phase detectors, loop filters, voltage controlled oscillators, charge pumps, etc. These behavioral models are parameterized for use in different PLL applications. Behavioral models for each of the functional blocks are written in ABCDL(Analog Behavioral Circuit Description Language) and are simulated by the ADAMS simulator.

3 Behavioral modeling an Analog PLL

An all analog PLL is simulated to evaluate the system-level characteristics, and verify that these parameters yield a functional PLL configuration. Using the mixed-mode simulator, these models were used to determine PLL characteristics such as Lock-in time, step response and the tracking range in very reasonable computer run-times. The PLL configuration is shown in figure 1, and consists of an analog multiplier phase detector, a loop filter and an analog voltage controlled oscillator. Each of the PLL components are modeled at the behavioral level. The behavioral models are as follows:
Analog multiplier phase detector:

$$V_{out} = K_d * V_{in1} * V_{in2} \tag{1}$$

Loop filter:

$$F(j\omega) = \frac{1}{1 + j\omega\tau_1} \tag{2}$$

and $\tau_1 = R_1 C$
Analog VCO:
The output frequency of the VCO is a linear function of the control voltage, the linear VCO characteristics is expressed in terms of the center frequency as

$$\omega_o = \omega_c + K_o(v_c(t) - V_{co}) \tag{3}$$

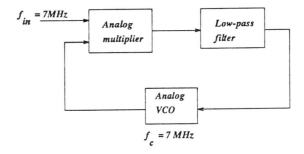

Figure 1: Analog Phase Locked Loop

V_{co} is the VCO control voltage at center frequency. The phase angle of the ouput is computed as

$$
\begin{aligned}
\phi_2(t) &= \int \omega_o(t)dt \\
&= \int [\omega_c + K_o(v_c(t) - V_{co})]\,dt \\
&= \omega_c t + K_o \int (v_c(t) - V_{co})dt \tag{4}
\end{aligned}
$$

The above analog VCO model is implemented using a state equation.

The models were parameterized, and a specific PLL design, adapted from [6]with the following characteristics was simulated. The key parameter values used in this simulation are:
K_d = Analog Multiplier gain = 3.72.
τ_1 = Loop filter pole = 7×10^5 rad/sec.
f_c = VCO center frequency = $7MHz$.
K_o = VCO gain = $30KHz/volt$.
V_{co} = VCO control voltage at center frequency = $0V$.
This PLL configuration was simulated to determine the acquisition characteristics and the tracking range. A fixed reference frequency signal was applied to the input and the time taken by the PLL to settle at this frequency was measured to determine the lock-in time, which was approximately $12\mu s$. Figure 2 shows shows the step response. To measure the response to a frequency step using transient simulation, an identical VCO used in the PLL was connected to the reference input, and a step input signal was applied to the input VCO, the output of the loop filter was measured to obtain the step response. A similar configuration was used to measure the tracking range of the PLL. A staircase input signal was applied to the reference VCO, and the PLL output characterisitcs show the tracking capabilities of the PLL. Figure 3 shows the tracking performance of the PLL.

4 Behavioral modeling a high-speed PLL

Figure 4 shows the high-speed digital PLL. The behavioral

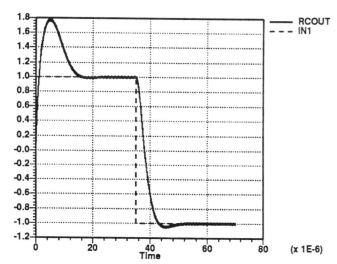

Figure 2: Step response of analog PLL

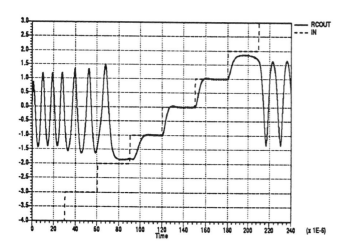

Figure 3: Tracking performance of analog PLL

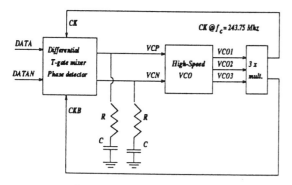

Figure 4: High speed PLL design

models for a high-speed PLL(HSPLL) were developed using a bottom-up approach where a novel CMOS integrated circuit PLL design was used as the basis [4]. The PLL has a differential configuration with a center frequency of 243.75MHz, and includes a differential transmission gate mixer phase detector, complementary first order loop filters, a high speed digital VCO and 3X multiplier. The behavioral models for each of the components were extracted from the CMOS transistor and gate level implementation, and the PLL was simulated at the behavioral level using these models.

The output of the phase detector is a continuous analog signal, whose high frequency components are filtered by the loop filter. The static control voltage at the center frequency is 2.5 volts. The behavior of the phase detector is similar to that of a complementary XOR gate phase detector. The loop filter is modeled by using discrete RC elements. The VCO output is a three phase clock which is combinatorally multiplied by the 3X multiplier to generate the high frequency complementary VCO clock outputs. The VCO operates in two regions, a cutoff region where the output frequency is 10MHz for control voltage $v_c \leq 1V$, and in the linear region the VCO gain is $47.5Mhz/V$. The VCO is simulated by using behavioral equations that model the VCO characteristics, and generate the 3-phase digital clock output.

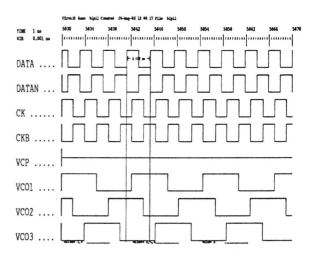

Figure 5: Simulation response of the HSPLL

Figure 5 shows the locked signals of this PLL when simulated with an input signal at a frequency equal to the VCO center frequency. The PLL has a lock-in time of $4\mu s$ A multi-level simulation of this PLL was carried out by replacing the behavioral-level model of the transmission gate phase detector, with a full transistor-level ADVICE [7] subcircuit, using CSIM models for the MOS transistors. Figure 6 shows the lock-in process of the multilevel simulation compared with the all behavioral simulation of this PLL. VCP1 is the response with an all behavioral model and VCP is the response with the mixed behavioral and device-level models. The CPU time for the multi-level models versus an all behavioral model was 5min 17.00sec versus 3min 15.21sec for simulating $10\mu s$ on a SUN SparcStation1+. The slight deviation in the lock-in characteristics can be attributed to two factors, 1) The device-level parasitics feature in the multi-level simulation, 2) In the behavioral model, the transmission gate impedance was approximated to a fixed value.

False-Locking is a phenomenon that can occur in many PLL designs. When the input reference frequency is varied about the center frequency, typically at a higher frequency, the PLL may falsely lock on to a harmonic frequency. This PLL configuration was simulated at an input reference frequency of 1.5 times the center frequency, ie at 365MHz, and it was observed that the PLL falsely locked at 219 MHz. This represents a harmonic lock with a period of $13.7ns$ or 5 input cycles to every 3 VCO cycles.

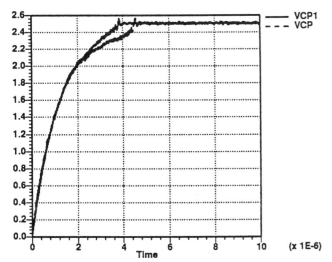

Figure 6: Multi-level simulation of HSPLL

5   Conclusions

By using behavioral modeling techniques, and a mixed signal simulator, PLLs can be efficiently simulated in very reasonable runtimes. The behavioral model finds use in the high-level system design process, as well as in speeding up the simulation of the PLLs. These behavioral models can be parameterized for use across various system designs. The multi-level simulation capability represents a significant advancement in the state-of-the-art, which allows a designer to approach the design process in a systematic manner. A designer would thus start the design process by using an all behavioral model of the PLL and focusing on getting the system parameters correct. Once satisfactory performance is obtained with a set of system parameters, circuit-level implementations can be realized for the various subsystems. Multi-level simulation provides an inexpensive way of evaluating how the circuit-level implementation along with all the device-level parasitics would impact the performance of

the overall system by selectively replacing behavioral models with their full device models. Multi-level simulation would also serve as an aid to fine tune the behavioral models for more accurate future simulations.

References

[1] B.A.A. Antao and F.M. El-Turky, "Automatic analog model generation for behavioral simulation", IEEE Custom Integrated Circuits Conference, May, 1992.

[2] R.E. Best, *Phase-Locked Loops: Theory, Design and Applications*, McGraw-Hill book co., NY, 1984.

[3] S. Can, and Y.E. Sahinkaya, "Modeling and simulation of an analog charge-pump phase locked loop", *Simulation*, vol.50, pp 155-160, April 1988 .

[4] R.H. Leonowich, "A High speed, wide tuning range, monolithic CMOS voltage contr olled oscillator utilizing coupled ring oscillators", AT&T Bell Laboratories internal design document, in preparation.

[5] E.Liu and A.L. Sangiovanni-Vincentelli, "Behavioral representations for VCO and detectors in Phase-Lock systems", IEEE Custom Integrated Circuits Conference, May 1992.

[6] V. Manassewitsch, *Frequency synthesizers: Theory and design*, 3rd edition, John Wiley & sons inc., NY, 1987.

[7] L.W. Nagel, "ADVICE for circuit simulation", Proceedings International symposium on circuits and systems, 1980.

[8] M. Sitkowski, "The macro modeling of phase-locked loops for the SPICE simulator", *IEEE Circuits and Devices*, pp 11-15, March 1991.

Behavioral Representations for VCO and Detectors in Phase-Lock Systems

Edward Liu and Alberto L. Sangiovanni-Vincentelli
Department of Electrical Engineering & Computer Sciences
University of California
Berkeley, California 94720

Abstract

This paper presents behavioral representations for detectors and voltage-controlled oscillators that are independent of circuit architectures. Parameter extraction techniques are described. Finally, parameter extraction for a VCO is demonstrated, and an example PLL constructed using the models is simulated and verified against actual chip measurements.

1 Introduction

Behavioral simulation is used in our constraint-driven, top-down hierarchical design methodology[1] to verify the design early. In digital domain, the behavior of a circuit is defined to be the function to be implemented independent of the architecture and/or schematics. Behavioral simulation in digital circuits can reduce substantially the design cycle. In the analog domain, behavioral simulation is often confused with macromodeling where an approximate model is built with components that are typical circuit components such as capacitors, controlled sources, and resistors. The notion of behavior as given for digital circuits could be extended to the analog domain. However, since the high-level analog functions are very simple, the power of behavioral simulation to uncover potential problems is greatly reduced. In analog circuits, malfunctioning is mostly due to the non-ideal behavior of the components. Thus, an effective behavioral simulation strategy has to be based on high-level models of *second order effects*. We have developed models for A/D converters that are based on these principles[2][3][4]. To have a useful behavioral simulation environment, we need to develop models for most of the high-level analog functions. In this paper, we propose behavioral representations for voltage-controlled oscillators (VCO) and detectors that are essential circuit components in any phase-lock system. The representations include sufficient second order effects for realism, are *independent* of the circuit component architectures, and are general for use in many diverse phase-lock system modeling applications. The paper is organized as follows: Behavioral models for sinusoidal and square wave VCO's are presented in Section 2 and 3, respectively. In Section 2.3, results of parameter extraction for a VCO are compared with SPICE and macromodeling results. A behavioral model for detectors and parameter extraction techniques are presented in Section 4. Finally, in Section 5, a phase-locked loop application is presented with simulation results compared with actual measurements.

2 Sinusoidal VCO

A sinusoidal voltage-controlled oscillator (VCO) ideally accepts as input a voltage, v, and outputs a sinusoidal signal whose frequency is linearly proportional to the input voltage. In a practical VCO, the frequency is nonlinearly related to the input voltage, the output contains harmonic components (distortion), and the phase has random variations (phase noise). Previous work[5][6] includes macromodels for SPICE which do not model distortion or noise effects. In our model, non-ideal behaviors can be captured by a two stage model shown in Figure 1, where stage

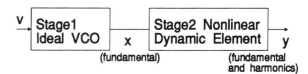

Figure 1: Two stage VCO model

1 is an ideal VCO to generate a signal, x, followed by a nonlinear dynamic stage 2 whose output, y, contains fundamental and harmonic frequencies. Given the input v, the output of stage 1, x, is given by

$$f = g(v) \tag{1}$$

$$\Phi(t) = 2\pi \int_0^t f(\tau)d\tau + \Psi(t) \tag{2}$$

$$x(t) = cos(\Phi(t)) \tag{3}$$

where g is the nonlinear relationship between the frequency f and v, Φ is the instantaneous phase, and Ψ is any stationary random process to represent phase noise (usually Gaussian with mean ψ_0 and variance σ^2). Since f and Ψ are in general time-varying, x is not sinusoidal. When such a signal is fed into the nonlinear dynamic stage 2, the output y is given by

$$y = \sum_{n=1}^{\infty} \int \cdots \int h_n(\tau_1, \ldots, \tau_n)x(t-\tau_1)\ldots x(t-\tau_n)d\tau_1 \ldots d\tau_n \tag{4}$$

where h_n are the Volterra kernels[7] of stage 2. To solve (4), we make the **quasi-static** approximation.

2.1 Quasi-static approximation

In a practical sinusoidal VCO, the phase noise Ψ is small, the control range is small, and the control voltage varies much slowlier than the frequency f. Therefore, we propose to determine VCO model parameters for a range of *constant* values of v using (4). The quasi-static approximation means that *during simulation for time-varying $v(t)$, we use the model parameters corresponding to instantaneous values of v*. If v is a constant, then x is a pure sinusoidal, and (4) has solution,

$$y = \sum_{n=0}^{N} |A_n(\omega)|cos(n\Phi(t) + \angle A_n(\omega)) \tag{5}$$

Reprinted from *Proc. CICC*, paper 12.3, May 1992.

where N is the order of approximation, $\omega = \frac{d\Phi}{dt}$ is the constant frequency, and A_n are complex coefficients given by

$$A_0(\omega) = \frac{1}{2}H_2(\omega, -\omega)e^{j\Psi}, A_1(\omega) = (H_1(\omega) + \frac{3}{4}H_3(\omega, \omega, -\omega))e^{j\Psi},$$

$$A_2(\omega) = \frac{1}{2}H_2(\omega, \omega)e^{j\Psi}, A_3(\omega) = \frac{1}{4}H_3(\omega, \omega, \omega)e^{j\Psi}, etc$$

Summarizing, the VCO model consists of stage 1 described by (1), (2), and (3), and stage 2 described by (5). The model parameters are the function g, the phase noise Ψ, and the complex coefficients A_n.

2.2 VCO parameter extraction

Parameters g, Ψ, and coefficients A_n can be estimated from VCO outputs $y(t, v)$ for different fixed values of v obtained from SPICE simulations or laboratory measurements. The following steps are followed: (a) For a constant input v, estimate from output y the fundamental frequency, f, using the sinusoidal minimum error method[8] or the complex demodulation method[9]. In the latter case, the idea is to multiply $y(t, v)$ with $e^{j2\pi \hat{f}t}$, followed by low pass filtering to get $z(t)$, where $\hat{f}$ is our guess of f. If $\hat{f} = f$, the phase of $z(t)$ is constant; otherwise, it is slowly time-varying for $\hat{f} \approx f$. Using Newton-Raphson[10], solve for f such that the phase of $z(t)$ is constant. (b) With f known, the magnitude and phase of the fundamental and harmonics can be similarly calculated using demodulation. Specifically, magnitude is the average of $2|z(t)|$ and phase is the average of $-\angle z(t)$. (c) Repeat (a) and (b) for different v to estimate the nonlinear function g for equation (1), as well as for more data points on $A_n(\omega)$ in (5). (d) Estimate σ for (2) from laboratory measurements or by hand analysis[11]. (Notice that SPICE does not simulate noise in the time domain.) (e) Assign value for ψ_0, which is the only initial condition.

2.3 Example VCO parameter extraction

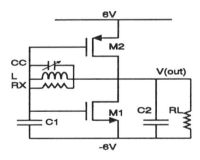

Figure 2: Pierce oscillator

Shown in Figure 2 is a schematic of a Pierce oscillator with varactor tuning that consists of circuit elements $M1 = 15\mu m/1\mu m$, $M2 = 35\mu m/1\mu m$, $L = 129.75nH$, $C1 = C2 = 20pF$, $RL = 100\Omega$, $RX = 1M\Omega$, and voltage-controlled capacitor $CC = 2pF/V$. Its behavioral model parameters (Figure 3) were extracted from three SPICE transient simulations for three different control voltage v. Figure 4 shows the SPICE output for $v = 0$ and the residual error in the 5th order (N=5) behavioral approximation used in this case. In contrast, the SPICE

v	0	1	2
n	(gain,phase)	(gain,phase)	(gain,phase)
1	(2.88V, -3.72)	(2.88V, -4.10)	(2.88V, -4.47)
2	(1.63mV, -0.89)	(2.00mV, -1.29)	(2.02mV, -1.77)
3	(78.2mV, -3.02)	(78.0mV, -4.17)	(78.2mV, -5.27)
4	(30.2μ V, -1.90)	(45.6μ V, -0.92)	(36.3μ V, -1.08)
5	(6.65mV, -1.37)	(6.63mV, -3.29)	(6.69mV, -5.12)

Figure 3: Behavioral parameters for Pierce oscillator

macromodels in [5] do not model harmonics, so the lower error bound is given by the first order residual error shown in Figure 5. Thus, the behavioral model is *more accurate* than macromodels, as well as *faster* since evaluating the behavioral model only involves evaluating the function in (5).

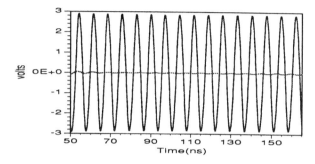

Figure 4: SPICE simulations and residual error in behavioral approximation

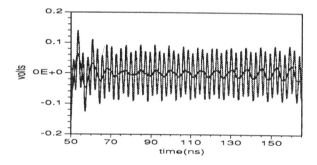

Figure 5: Residual errors in first and fifth order approximation

2.4 Error analysis

To estimate the error of the quasi-static approximation when the control input is not constant, we let the frequency, f, be modulated by a sinusoidal signal with small amplitude (narrow band frequency modulation). For instance,

$$f = g(v) = f_c + \beta f_m cos(2\pi f_m t)$$

where β is the modulation index, and f_m is the modulation frequency. Assuming the phase noise is negligible, it can be shown that the error of the quasi-static approximation is bounded by an error term proportional to β. As a result, when the modulation index β is sufficiently small, the error becomes negligible.

3 Square wave VCO

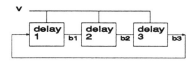

Figure 6: Square wave VCO with $m = 3$ delays

In addition to oscillators with sinusoidal outputs, there are oscillators with square wave outputs. In contrast to the sinusoidal VCO where the waveform is represented by (5), the waveform of a square wave is represented by the zero crossings (low-to-high and high-to-low transition times). In general, a VCO consists of m delay elements connected in a ring structure (Figure 6). Each of the m outputs are phased shifted by $\frac{1}{2mf}$ seconds, where f is the frequency of the VCO. We represent the jth transition time of output i with $y_i(j)$ as shown in Figure 7. Similar to

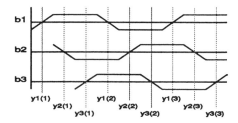

Figure 7: VCO square waveforms

the sinusoidal VCO, the frequency is controlled by the input v using equation

$$f_i = g_i(v) \tag{6}$$

where g_i represents control relationship for element i, and f_i represents the effective frequency of delay element i. Unfortunately, v varies considerably in a square wave VCO; therefore, the quasi-static approximation is not appropriate in this case. The transition times are determined by the delay between adjacent delay elements. For example, $y_i(j)$ for $i \neq 1$ is equal to $y_{i-1}(j)$ plus the delay of element i. Because of this relationship, it is convenient to use a recursive definition for $y_i(j)$,

$$\pi = 2\pi m \int_{y_{i-1}(j)}^{y_i(j)} f_i(\tau)d\tau + \Psi_i(y_i(j)) - \Psi_i(y_{i-1}(j)), i \neq 1 \tag{7}$$

$$\pi = 2\pi m \int_{y_m(j-1)}^{y_i(j)} f_i(\tau)d\tau + \Psi_i(y_i(j)) - \Psi_i(y_m(j-1)), i = 1 \tag{8}$$

where Ψ_i is the random phase noise of element i. Summarizing, the model consists of $(6),(7)$, and (8). Model parameters are the number of delay elements m, the voltage-control relationship g_i, the phase noise Ψ_i, and the initial condition $y_1(1)$.

4 Detectors

A detector compares two signals (input waveform and VCO waveform). Traditionally, circuit designers represent a detector by its phase characteristic which plots the average product of the input waveform and the VCO waveform as a function of the phase difference between the two waveforms. One problem with this modeling approach is that the phase characteristic is

defined only when the input waveform and the VCO waveform have the same frequency. As a result, this approach cannot be used during phase-locked loop acquisition. Another problem is that the phase characteristic depends on the shape of the waveforms, so the characteristic is not defined until input signals are applied. As a result, the phase characteristic is not appropriate as a behavioral representation.

Detectors are classified into two broad categories[12]: multipliers (zero memory circuits) and sequential circuits (with memory). In this paper, sequential circuits are represented by state machines with analog inputs, a and b, analog outputs $g(a,b)$, and clock triggered by zero crossings of a and b. Moreover, the same representation can be used for multipliers if they are considered as state machines with a single state and no clock. In general, all detectors can be conveniently represented by a state transition table such as the one shown in Figure 8 for a phase-frequency detector. Signals a, b, and c are the input, VCO, and output waveforms, respectively. The first column is the state number; the second column is the next state if a crosses zero; the third column is the next state if b crosses zero; t_d are gate delays of next state logic; the last column is the output $g(a,b)$ for each state. State transition tables for common detectors are shown in Figure 9. Since all known detectors can be written in this representation, this representation is general and independent of the circuit architecture.

state	a crosses (next, delay)	b crosses (next, delay)	c g(a,b)
0	$(1, t_d)$	$(2, t_d)$	0.0
1	$(1, t_d)$	$(0, t_d)$	Iup
2	$(0, t_d)$	$(2, t_d)$	Idown

Figure 8: Behavioral model for phase frequency detector

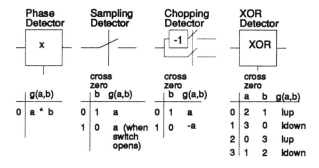

Figure 9: Common detectors

4.1 Detector parameter extraction

For a given detector type, the next state logic is fixed. Non-idealities are the transition delays, t_d, and deviations of the output function. t_d is computed by measuring the total gate delays in the next state logic. Typically, $g(a,b)$ is very simple; for instance, $g(a,b) = c_0 + c_1 a + c_2 b + c_3 ab$. Its non-idealities include constant offset error in c_0 and multiplier gain error in c_3. All four coefficients $c_0 \ldots c_3$ can be extracted using four DC SPICE simulations with different $a's$ and $b's$, and solving for the coefficients in a system of four equations.

5 Phase-lock loop simulation

Using the models presented, we constructed a high-level model of a real phase-lock loop[13]. The loop filters are modeled with differential equations, the VCO is modeled with a square wave VCO with one delay element ($m = 1$, output fed directly back to input), and the phase-frequency detector is modeled as in Figure 8 with $t_d = 0$. Relevant system parameters, such as acquisition time and jitter gain function, are computed from transient simulations. In our prototype, the VCO output, b, is modeled as a switching function, which is completely characterized by the transition points, $y_1(j)$. Traditional integration algorithm for differential equations cannot handle this type of system because b is non-differentiable. However, the problem is much easier because the differential equation solver needs only determine the transition time points, $y_1(j)$, which are defined by (8). For example, in (8), we need to find the upper integration limit, $y_1(j)$, to satisfy the equation. As a result, *we solve for $y_1(j)$ by using an iterative solver on top of an integration algorithm.* For prototyping, we implemented the iterative false position method[10] on top of the trapezoidal integration algorithm to search for the points. Using this algorithm, the simulation speed is dramatically increased, while preserving numerical stability.

A prototype simulator has been implemented using the C++ language on a DECstation 5000. Figure 10 shows the output frequency as the input frequency is stepped from 120MHz to 138.5MHz, compared with the output frequency measured from an experimental chip. From Figure 11, measured overshoot (34%) and settling ($8\mu s$) agree with simulation (33%, $7\mu s$). Next, the input frequency is varied to probe the system for the jitter gain function. From the data, -3db cross-over frequency is estimated to be 470KHz which agrees with the theoretical value of 410KHz. For both examples, the circuit was simulated from time zero to $40\mu s$ (using DEC5000 CPU time 17 minutes), corresponding to 5165 VCO cycles. If the same analysis is done with SPICE, the time step size should be about $100ps$ since the VCO period is $8.3ns$. So, assuming SPICE takes $0.5s$ to compute a time point, SPICE would take at least 2 days for each of the examples.

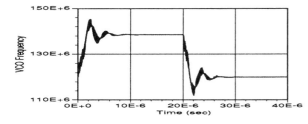

Figure 10: Simulated PLL acquisition

6 Summary

We have presented behavioral models for VCO and detectors in phase-lock systems. Parameter extraction techniques to identify the models have been described, and the techniques were used to extract parameters for an actual VCO. Furthermore, we used these models to model and simulate an actual PLL. The simulation time is significantly lower than an estimated

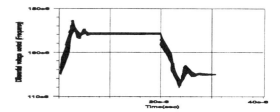

Figure 11: Measured PLL acquisition

SPICE simulation time, and the simulation results agree with measured chip data. Finally, because our models are independent of circuit architecture, we can model and simulate a wide range of PLL circuits. We have presented a way to model and simulate PLL's in a reasonable amount of time with minimal loss of accuracy.

Acknowledgements

The authors acknowledge the many helpful discussions by the members of CAD and IC groups at U. C. Berkeley. We especially thank Prof. P. R. Gray for his helpful advice on the square wave VCO model. This project is supported by the Semiconductor Research Corporation. Its support is gratefully acknowledged.

References

[1] H. Chang, et al. "A top-down, Constraint-Driven Design Methodology for Analog Integrated Circuits", *Proc. IEEE CICC*, May 1992

[2] E. Liu, et al. "A Behavioral Representation for Nyquist Rate A/D Converters", *Proc. IEEE ICCAD*, Nov 1991

[3] E. Liu, et al. "Behavioral Modeling and Simulation of Data Converters", *Proc. IEEE ISCAS*, May 1992

[4] G. Gielen, et al. "Analog Behavioral Models for Simulation and Synthesis of Mixed Signal Systems", *Proc. EDAC*, March 1992

[5] Mark Sitkowski, "The Macro Modeling of Phase Locked Loops for the Spice Simulator", *IEEE Circuits and Devices Magazine*, v7 n2 March 1991 p. 11-15

[6] E. Tan, *Phase-Locked Loop Macromodels*, Master's Thesis, (Call no. T7.49 1990 T283) U.C. Berkeley, August 1990.

[7] D. Weiner, J. Spina *Sinusoidal Analysis and Modeling of Weakly Nonlinear Circuits*, pages 274-278, Van Nostrand Reinhold Company, New York, 1980

[8] B. Boser, et al. "Simulating and testing oversampled analog-to-digital converters", *IEEE Trans. Computer-Aided Design*, vol. 7, pp. 668-674, June 1988

[9] D. Brillinger *Time Series Data Analysis and Theory, Expanded Edition* McGraw-Hill, New York, 1981

[10] W. Press, et al. *Numerical Recipes The Art of Scientific Computing*, pages 263-266, Cambridge University Press, 1988.

[11] A. Abidi, R. Meyer, "Noise in Relaxation Oscillators", *IEEE Jornal of Solid-State Circuits*, vol. SC-18, No. 6, December 1983

[12] F. M. Gardner, *Phase-lock Techniques* John Wiley & Sons, Inc., New York, 1979

[13] K. M. Ware "A High-Frequency Integrated CMOS Phase-Locked Loop", *IEEE Journal of Solid-State Circuits*, vol. 24, No. 6, December 1989

Behavioral Simulation Techniques for Phase/Delay-Locked Systems

Alper Demir   Edward Liu   Alberto L. Sangiovanni-Vincentelli   and   Iasson Vassiliou

Department of Electrical Engineering & Computer Sciences
University of California, Berkeley, CA 94720

Abstract

This paper presents behavioral simulation techniques for phase/delay-locked systems. Numerical simulation algorithms are compared and the issue of numerical noise is discussed. Behavioral phase noise simulation for phase/delay-locked systems is described. The role of behavioral simulation for phase/delay-locked systems in our top-down constraint-driven design methodology, and in bottom-up verification of designs, is explained with examples. Accuracy and efficiency comparisons with other methods are made. Simulation techniques are described in the framework of phase/delay-locked systems, but simulation methodology and the results attained in this work are applicable to the behavioral simulation of mixed-mode nonlinear dynamic systems.

1. Introduction

Phase-locked loops (PLLs) and delay-locked loops (DLLs) are an important part of the class of mixed-mode nonlinear dynamic systems which includes PLLs, DLLs, oversampling data converters, switching power supplies. PLLs and DLLs have many applications; they are used in receivers, clock generators [3], clock recovery [4], data synchronization [5], etc.

In general, mixed-mode nonlinear dynamic circuits are stiff systems. Traditional simulators such as SPICE spend large amounts of CPU time on computing the accurate waveforms of the digital circuits, yet the waveforms of the much slower analog circuits determine the system behavior. Consequently, traditional simulators take too much time to compute the system behavior by simulating the whole system over the interval required by the analog circuits.

When a system is originally conceived in our top-down constraint-driven hierarchical design methodology [1], the realization of the digital circuits, as well as the details of the analog circuits, have not been decided yet. So, an accurate circuit simulation is neither possible nor desired. Therefore, designers desire to efficiently obtain the analog waveforms and the *timing* of the digital waveforms rather than the actual digital waveforms.

Key specifications for PLLs are acquisition time, capture range, lock range and phase noise/jitter. Phase noise is a major and very important PLL specification. For instance, this noise can cause data errors in communication systems, create inaccurate readings in instrumentation systems such as spectrum analyzers. Prediction of phase noise through simulations is crucial at the early stages of the top-down design. Phase noise simulation imposes tight restrictions on the numerical simulation algorithms to be used. *Numerical noise* created by the simulation algorithm should be negligible when compared with the phase noise to be computed for accurate results.

In this paper, in Section 2 below, we compare several algorithms for their numerical noise behavior and accuracy-efficiency trade-offs. In Section 3, behavioral phase noise simulation is discussed. In Section 4, use of behavioral simulation in bottom-up verification of designs is explained. The role of behavioral simulation for phase/delay-locked systems in our top-down constraint-driven design methodology is elucidated in Section 5.

2. Numerical Algorithms

Behavioral models of circuit components such as voltage-controlled oscillators (VCOs), phase detectors (PDs), and phase/frequency detectors (PFDs) which operate only on timing information have already been developed [2]. These are high-level mathematical representations of the components. They include differential equations, difference equations, transfer functions, statistical distributions, tables or arithmetic expressions [1]. Behavioral models which consist of these representations capture the input-output behavior of circuits, and they are independent of circuit architecture. They describe the first-order behavior of the circuit as well as the second-order analog effects in order to get a realistic idea of the performance of the overall system. With these models, a wide variety of phase/delay-locked systems can be simulated.

Simulation of systems, which are composed of the interconnection of such behavioral models, can be done by using a modified integration method. Digital circuits are simulated with an event-driven simulator, while the analog circuits are simulated with a differential equation solver. By interfacing the analog simulator with the digital, using a protocol, the two simulators are synchronized. At the interface, integration must pause at breakpoints which correspond to the exact *times* when the digital signals switch. Because switching times are generally not known ahead of time, they must be solved for during the integration. For instance, the time-derivative of instantaneous phase in a square wave VCO model is expressed as,

$$\dot{\phi}(t) = 2\pi f(v(t)) \qquad (1)$$

where $v(t)$ is the VCO control voltage, and $f(v)$ is, in general, a nonlinear function relating the effective frequency of VCO to the control voltage [2]. The waveform at the output of the VCO is represented by zero crossings, which occur when $\phi(t)$ takes values which are integer multiples of π. At a time point during the integration of (1), if a zero crossing occurs in between the previous time point and current time point, the integration algorithm should track back and hit the point where the zero crossing occurs, as illustrated in Fig. 1. There are two schemes to solve for the timing of switch-

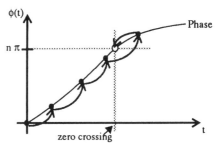

Fig. 1. Integration algorithm with zero crossings

ing events:

• *Interpolation method.* Switching times are estimated based on a

Reprinted from *Proc. CICC*, paper 21.3, May 1994.

polynomial interpolation using the information from current and previous time points. For instance, with a first-order interpolation method, let a switching event be detected between the time points t and $t + \Delta t$, where Δt is the current time step size. Then, the time for the switching event is estimated by

$$t_s = t + \frac{\phi(t_s) - \phi(t)}{\phi(t + \Delta t) - \phi(t)} \qquad (2)$$

$$\phi(t_s) = n\pi \qquad n = 1, 2, 3...$$

Iterative method. Switching times are calculated using a bisection search (Fig. 2). If at time $t + \Delta t$, the phase $\phi(t)$ is larger

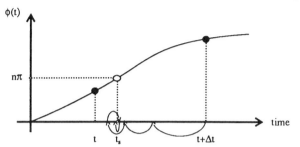

Fig. 2. Iterative search for switching times

than $n\pi$ while at t it was smaller, then there has been a switching event in between. To calculate the switching time, we divide the present time step by 2. Then we evaluate $\phi(t + \Delta t/2)$. If this is still to the right of the switching time, we repeat the bisection procedure. If it is to the left, then we return the control to the integration algorithm. To prevent the algorithm from taking too much of a time step, we limit the maximum step size to $1/4$ of the last time step.

We have implemented an integration algorithm, including these schemes as options, to simulate phase/delay-locked systems. The integration algorithm is a variable time-step, variable order one based on the implicit backward differentiation formula [6][7]. First-order (linear) and second-order interpolation methods were implemented for estimation of the timing of switching events. The iterative method explained above, which can compute the timing of switching events to a desired accuracy, was also implemented. A behavioral model for a charge-pump PLL (Fig.3) [8], which consists of a ring-oscillator VCO, a PFD, a charge-pump and a second-order loop filter, was built to analyze the behavior of the numerical algorithms.

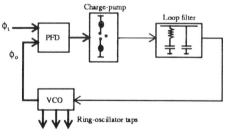

Fig. 3. Charge-pump PLL

The algorithms were compared by doing phase noise simulations of the above PLL. (Please refer to Section 3 for a discussion of behavioral phase noise simulation). Accuracy, in this context, is used to mean the accuracy of the solutions for the timing of switching events.

First, a comparison of the first-order and second-order interpolation methods was made, which was in favor of the first-order method with a very slight accuracy difference. Then, the first-order interpolation method was compared with the iterative method. There are basic differences in the behavior of these methods. The accuracy of the iterative method can be set to any desired value. However, there is *no* control on the accuracy of the interpolation method. During an unlocked state, while PLL is acquiring the input, the accuracy of the interpolation method decreases considerably. When PLL is "in lock", accuracy is inversely proportional to the noise level in the circuit. This is, in fact, a desired property, because more accuracy is needed for smaller noise levels. On the other hand, the accuracy of the iterative method stays at the desired/set level regardless of the state or noise level of the circuit. Fig. 4 shows the response of the above PLL to a frequency step at its input in the presence of noise. Even after acquisition, deviations around the mean value occur because of the noise in the circuit. Fig. 4 also shows the error in the solution of the timing of switching events for the interpolation method and the iterative method. The error for the iterative method in Fig. 4 is not visible on the plot that uses the same scale as the plot for the interpolation method, because the error for the iterative method was several orders of magnitude smaller. The details can be seen in the blow-up plot.

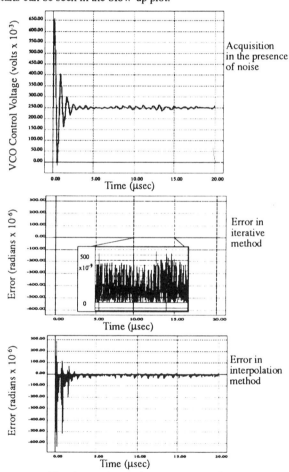

Fig. 4. Acquisition in the presence of noise

For the PLL considered above, the accuracy of the interpolation method was found to be sufficient for doing phase noise simulations when the circuit is "in lock". The results obtained by the interpolation method were verified by using the iterative one.

Although this was the case for the PLL considered, there is no guarantee on the accuracy of the interpolation method in general. On the other hand, the accuracy of the iterative method does not depend on the type, state or noise level of the circuit. The accuracy can also be set to a desired level where numerical noise will not interfere with the results.

3. Phase Noise Simulation

Phase noise/jitter is a very important specification for phase/delay-locked systems. Design for low phase noise is probably the most challenging part of the design of such systems. The difficulty of design for low phase noise is amplified by the lack of efficient simulation tools for its prediction. Prediction of phase noise through simulations is crucial at the early stages of the top-down design. Any modeling or design methodology, and simulation tools, which do not address this problem are not suitable for high performance phase/delay-locked system design.

The models we use in behavioral simulation include statistical second-order effects [2]. Analysis of a phase/delay-locked system in the presence of noise, when the system is *not* "in lock", is a hard problem. Fortunately, almost all of the time, a detailed observation of the system, when it is in the unlocked state, is not required as long as the loop is *guaranteed* to lock on to the input in the specified amount of time. This can be assured by simulations. At this point, it is assumed that the system is "in lock" for phase noise simulation. The expression, "in lock", is used in a time-averaged sense here, not in the instantaneous sense. For instance, for a PLL (say with a ring-oscillator VCO, charge-pump and PFD), instantaneous frequency of the VCO will have variations around a mean value because of the noise in the circuit. The mean value of the VCO frequency will be equal to the input frequency. It is assumed that phase noise performance of the circuit is above a level where it does not lose lock (in a time-averaged sense). This is a reasonable assumption, because if it does not hold, that means that we are confronted with a bad design which has deterministic performance problems. It is important to understand that "in lock" assumption does not mean we are using a linear model for the circuit. Linearity assumption might be validated for very low noise levels, but higher noise levels will excite the nonlinearities. For behavioral phase noise simulation, we use the full nonlinear models of the components, which are described in [2].

The method that is used for phase noise simulations is Monte Carlo in time domain. The CPU time disadvantage of the Monte Carlo method disappears because of the high efficiency of behavioral simulation. As discussed in Section 2, having a numerical algorithm with *controllable* numerical noise makes it possible to predict phase noise accurately. In behavioral phase noise simulation, circuit, which is "in lock", is simulated over a period of time while the delay cells add jitter. Then, the characteristics of jitter at the output is calculated by a statistical analysis over that time period. An example about phase noise simulation will be given in Section 5.

4. Behavioral Simulation in Bottom-Up Verification

The traditional approach to the bottom-up verification of phase/delay-locked system designs is to use a transistor-level simulator such as SPICE. Transistor-level simulation of a phase/delay-locked system takes too much time to be practical, because the system is stiff. Macromodeling these systems in a circuit simulator was proposed to circumvent the efficiency problems of transistor-level simulation [9][10]. This approach also results in impractical simulation times, because many circuit elements are needed for accurate models. Still, the accuracy attained by the macromodeling approach

is not as high as needed for verification of designs. On the other hand, behavioral simulation achieves the goal of verifying complex system behavior efficiently. This is made possible by accurate statistical behavioral models which include analog second order effects. Evaluations of behavioral models are fast which results in an efficient system simulation.

Bottom-up verification using behavioral simulation is done in two steps as illustrated in Fig. 5:

• *Set up the behavioral models for the components.* The model parameters are extracted using SPICE from the transistor level description of components.

• *Simulate the system in time-domain using the behavioral simulator to calculate the performance measures.* Behavioral simulation is much faster than circuit simulation.

Fig. 5. Two-step verification with behavioral simulation

To illustrate this procedure, component behavioral models for a bipolar PLL [11] were set up, and behavioral simulation was used to analyze the acquisition characteristics. Fig. 6 shows SPICE domain extraction of the relation between the effective frequency of VCO and the control voltage ($f(v)$ in (1)). Other model parameters are

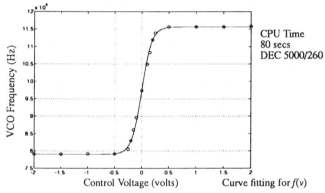

Fig. 6. Extraction with SPICE

extracted in a similar way from the transistor level description. Model extraction is done only once for a circuit. Then created models are used in many behavioral simulations to analyze the acquisition characteristics, stability, timing jitter and others. Fig. 7 shows the response of the modeled PLL to a frequency step at its input (both SPICE and behavioral simulation).

5. Behavioral Simulation in Top-Down Design

Behavioral simulation and modeling reduces the design time considerably in a top-down approach. Designers can verify system design early, with behavioral models of system components, before investing time in detailed circuit implementation. Design space exploration can be done efficiently. In our top-down, constraint-driven design methodology [1], given a set of circuit specifications (circuit characteristics, the design rules, the technology, and user options), a *mapping* is made to schematics or layout. Behavioral simulation and optimization are used for architecture selection. Once a particular architecture is chosen, behavioral simulation and optimization are used to map the architecture and circuit specifications to the detailed specifications of components or subblocks.

The role of behavioral simulation in top-down design of a

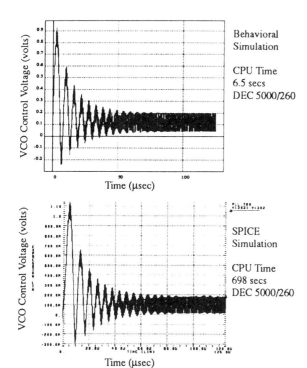

Behavioral
Simulation

CPU Time
6.5 secs
DEC 5000/260

SPICE
Simulation

CPU Time
698 secs
DEC 5000/260

Fig. 7. Behavioral and SPICE simulation of acquisition

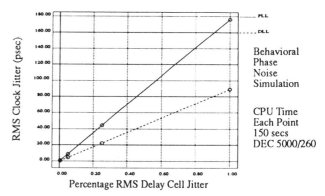

Behavioral
Phase
Noise
Simulation

CPU Time
Each Point
150 secs
DEC 5000/260

Fig. 9. Clock jitter for PLL and DLL clock generator tures. Ring-oscillator VCO for PLL, as well as the delay line of DLL, has 5 delay cells. Reference clock frequency is 50 Mhz.

From Fig. 9, we conclude that DLL has better phase noise performance, when compared with a PLL, for fixed percentage delay cell jitter. Then, the relationship between clock jitter and percentage delay cell jitter is used to predict amount of jitter allowable an a delay cell, given a clock jitter allowance. In this way, the clock jitter constraint is mapped on to the delay cell jitter constraint.

phase/delay-locked system will be illustrated with a multi-phase clock generator. A PLL with a ring-oscillator VCO or a DLL can be used to generate multi-phase clocks (Fig. 8). These two architectures were compared for their phase noise performance using behavioral simulation. Fig. 9 shows the relationship between clock jitter and percentage delay cell jitter for a given design of both architec-

6. Conclusions and Future Work

We have presented behavioral simulation techniques for phase/delay-locked systems. Numerical algorithms and behavioral phase noise simulation have been described. The role of behavioral simulation for phase/delay-locked systems in our top-down constraint-driven design methodology, and in bottom-up verification of designs, has been explained with examples. We are in the process of top-down, constraint-driven design of a clock generator circuit for a RAMDAC using the simulation tools we have presented. As we move down the hierarchy of design, we will create new tools and/or modify the existing ones as need arises.

Acknowledgements

This project is supported by DARPA and MICRO. Their support is gratefully acknowledged.

PLL Multi-Phase Clock Generator

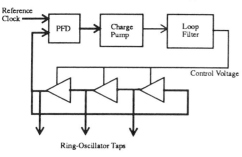

Ring-Oscillator Taps

DLL Multi-Phase Clock Generator

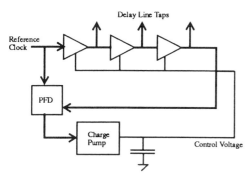

Fig. 8. PLL and DLL multi-phase clock generator

References

[1] H. Chang, A. Sangiovanni-Vincentelli, F. Balarin, E. Charbon, U. Choudry, G. Jusuf, E. Liu, E. Malavasi, R. Neff and P. Gray, "A top-down constraint-driven design methodology for analog integrated circuits", *IEEE CICC*, May 1992.

[2] E. Liu and A. Sangiovanni-Vincentelli, "Behavioral representations for VCO and detectors in phase-lock systems", *IEEE CICC*, May 1992.

[3] I.A. Young, J.K. Greason and K.L. Wong, "A PLL clock generator with 5 to 110 MHz of lock range for microprocessors", *IEEE Journal of Solid-State Circuits*, vol. 27, No. 11, November 1992.

[4] T.H. Lee and J.F. Bulzacchelli, "A 155-MHz clock recovery delay- and phase-locked loop", *IEEE Journal of Solid-State Circuits*, vol. 27, No. 12, December 1992.

[5] K.M. Ware, H-S. Lee and C.G. Sodini, "A 200-MHz CMOS phase-locked loop with dual phase detectors", *IEEE Journal of Solid-State Circuits*, vol. 24, No. 6, December 1989.

[6] T.L. Quarles. *Analysis of Performance and Convergence Issues for Circuit Simulation.* Ph.D. Thesis. (Call No: T7.6 1989 Q82 ENGI) U.C. Berkeley, April 1989.

[7] EECS 219. *Circuit Theory and Computer-Aided Analysis.* Lecture Notes. U.C. Berkeley. Fall 1991.

[8] F.M. Gardner, "Charge-pump phase-lock loops,", *IEEE Transactions on Communications,* vol. COM-28, No. 11, November 1980.

[9] M. Sitkowski, "The macro modeling of phase-locked loops for the SPICE simulator", *IEEE Circuits and Devices Magazine,* vol. 7, No. 2, March 1991.

[10] E. Tan. *Phase-Locked Loop Macromodels.* Masters' Thesis. (Call No: T7.49.1990 T283 ENGI) U.C. Berkeley, August 1990.

[11] P.R. Gray and R.G. Meyer. *Analysis and Design of Analog Integrated Circuits* p.624. Second Edition. John Wiley & Sons. 1984.

Part 4
Phase-Locked Loops

IN this part, 13 papers dealing with monolithic realization of PLLs are collected. Aiming at communication applications, the first five papers describe PLLs with combinational phase detectors and are ordered according to their speed.

The remaining papers, all on CMOS implementations, are concerned with clock generation and deskewing in digital and mixed-signal systems. They all employ phase/frequency detectors together with charge pumps to achieve a wide capture range and a small static phase error. The speed of these circuits has scaled along with that of their corresponding systems, going from approximately 20 MHz to 250 MHz over a period of 8 years.

Additional References

J. McNeill and R. Croughwell, "A 150-mW 155-MHz phase-locked loop with low jitter VCO," *Proc. ISCAS*, London, UK, June 1994.

W. O. Keese, "Dual PLL IC achieves fastest lock time with minimal reference spurs," *RF Design*, pp. 30–38, August 1995.

G. Baldwin, "A wideband phase-locked loop using harmonic cancellation," *Proc. IEEE*, vol. 57, pp. 1464–1465, August 1969.

M. Mizuno et al., "A 0.18-μm CMOS hot-standby phase-locked loop using a noise-immune adaptive-gain voltage-controlled osillator," *ISSCC Dig. Tech. Papers*, San Francisco, CA, pp. 268–269, February 1995.

A Monolithic Phase-Locked Loop with Post Detection Processor

ENJETI N. MURTHI, MEMBER, IEEE

Abstract—This paper details the design and fabrication of a high-frequency (50-MHz) phase-locked loop with a post detection processor which allows the detection of FSK signals with few external components. The circuit operates with a single 5-V supply and has TTL compatible inputs and outputs.

Manuscript received August 14, 1978; revised September 28, 1978.
The author is with Signetics Corporation, Sunnyvale, CA 94086.

I. INTRODUCTION

MONOLITHIC phase-locked loops (PLL's), ever since their introduction, have found wide use in a number of applications. However, they have suffered from some limitations, particularly in terms of power supplies necessary for operation, the frequency range over which they may be used, and the difficulties present in terms of their system interfacing. Furthermore, extensive external circuitry has generally been necessary for the demodulation of frequency shift keyed (FSK) signals. This paper details the design of a monolithic PLL with a unique output stage which can be used for both FSK and linear demodulation. The frequency of operation is more than twice that possible with previously available circuits and has logic compatible inputs and outputs, and only a single 5-V supply is needed for its operation.

II. CIRCUIT DESCRIPTION

Before considering the details of the circuit design, it is convenient to consider the block diagram of the NE564, as shown in Fig. 1. The PLL part of the circuit consists of the standard phase comparator (with filter) and voltage controlled oscillator (VCO) with an additional limiter. The post detection processor consisting of the dc retriever and Schmitt trigger is necessary for the demodulation of FSK signals, which essentially involves conversion of the demodulated signal at the PLL output into a logic compatible digital signal. The complete schematic of the 564 appears in Fig. 2. To understand the operation of the circuit, it is convenient to consider the various parts separately.

III. LIMITER

The limiter ahead of the phase comparator is necessary to improve amplitude modulation (AM) rejection, which is conveniently achieved by eliminating the amplitude variations in the input FM signal. It should be pointed out that this scheme does to some extent impair the adjacent channel interference rejection characteristics when the PLL is used for demodulation of standard FM broadcast signals. However, since the present PLL is more oriented towards demodulation of FSK-type signals, the benefits of a limiter more than offset the lack

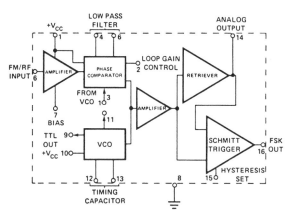

Fig. 1. Block diagram of the PLL with post detection processor.

of it. Signal limiting is conveniently accomplished with a differential amplifier whose output voltage is clipped by diodes D_1 and D_2 (see Fig. 3). Schottky diodes are used because their limiting occurs between 0.3 and 0.4 V instead of the 0.6-0.7 V for regular IC diodes. This lower limiting level is helpful in biasing, especially for 5-V operation. When limiting, the dc voltage across R_2 and R_3 remains at the Schottky-diode voltage. Good high-frequency performance for Q_2 and Q_3 is achieved with current levels in the low milliampere range.

When demodulating FSK signals, the input is a TTL signal of 0-5-V amplitude referenced to ground. If this signal were applied to the base of Q_2 in Fig. 3, it would appear as a base-emitter voltage which leads to saturation problems. For this reason, it is preferable to use vertical p-n-p's at the input and modify the circuit to that shown in Fig. 4. The logic signal now appears as a collector-base voltage for the vertical p-n-p and is easily handled. It is conceivable that there may be saturation problems even in this case, when the input signal swings towards ground, which is solved by making Q_1 a Schottky transistor. Transistor Q_5 is in turn diode biased by D_3 and D_4 (see Fig. 2) which places the base voltages of Q_1 and Q_5 at approximately 1.0 V. This same biasing network establishes a 1.3-V bias at the base of Q_{13} for biasing the phase comparator section. A differential output signal from the input limiter is applied to one input of the phase comparator (Q_9-Q_{12}) after buffering the level shifting through the Q_7-Q_8 emitter-followers.

IV. PHASE COMPARATOR

The phase comparator section of the 564 is shown in Fig. 5. It is basically the conventional double-balanced mixer commonly used in PLL circuits with a few exceptions. The transconductance g_m for the Q_{13}-Q_{14} differential amplifier is

Reprinted from *IEEE Journal of Solid-State Circuits*, vol. SC-14, pp. 155-161, February 1979.

285

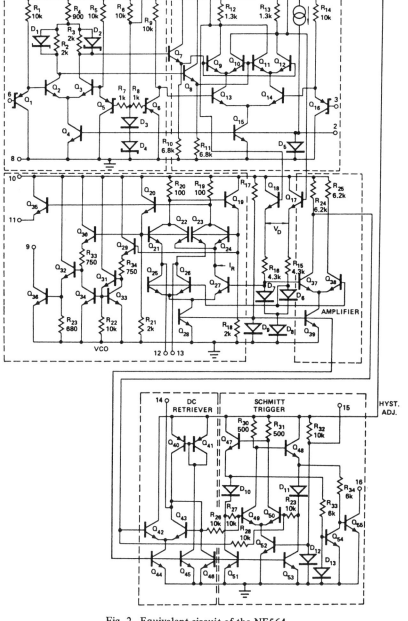

Fig. 2. Equivalent circuit of the NE564.

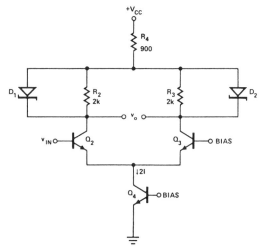

Fig. 3. Basic limiter stage.

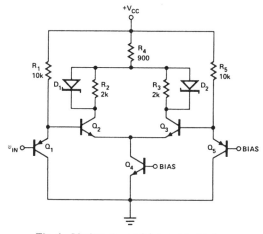

Fig. 4. Limiter stage with input buffering.

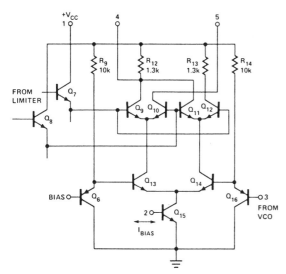

Fig. 5. Phase comparator section.

directly proportional to the mirror current in Q_{15}. Thus, by externally sinking or sourcing the current at pin 2, g_m can be changed to alter the phase comparator's conversion gain, K_d. The nominal current injected into this node by the internal current source is 0.75 mA for 5-V operation. If this current is externally removed by gating, the phase comparator can be disabled, and the VCO will operate at its free-running frequency.

The current level established in Q_{15} of Fig. 5 determines all other quiescent currents in the phase comparator (Q_9–Q_{14}). Currents through R_{12} and R_{13} set the common-mode output voltage from the phase comparator (pins 4 and 5). Since this common-mode voltage is applied to the VCO to establish its quiescent currents, the VCO conversion gain (K_0) also depends upon the bias current at pin 2.

V. VCO

The VCO is of the basic emitter-coupled astable type with several modifications included to achieve the high-frequency TTL compatible operation, while maintaining low-frequency drift with temperature changes. The basic oscillator in Fig. 6 consists of Q_{19}, Q_{20}, Q_{21}, and Q_{23} with current sinks of Q_{25} and Q_{26}. The VCO action is obtained by altering the ratio of currents in Q_{25}–Q_{26} and the dummy current sink Q_{27}. The total current is maintained constant by Q_{28} so that the drop across R_{19}–R_{20} remains unchanged when the frequency is changed. This kind of achieving VCO action facilitates a differential design to be maintained with relevant advantages. The master current sink of Q_{28} keeps the total current constant; by altering the ratio of currents in Q_{25}–Q_{26} and the dummy current sink of Q_{27}, VCO action is obtained.

The input drive voltage for the VCO is made up of common- and difference-mode components from the phase comparator. After buffering the level shifting through Q_{17}–Q_{18} and R_{15}–R_{16}, the VCO control voltage is applied differentially to the base of Q_{27} and to the common bases of Q_{25} and Q_{26}.

The VCO control voltages from the phase comparator are the pin 4 and pin 5 voltages, or

$$V_4 = V_{C9} = V_{B18} = V_{CM} + \tfrac{1}{2} V_{DM} \tag{1}$$

$$V_5 = V_{C12} = V_{B17} = V_{CM} - \tfrac{1}{2} V_{DM} \tag{2}$$

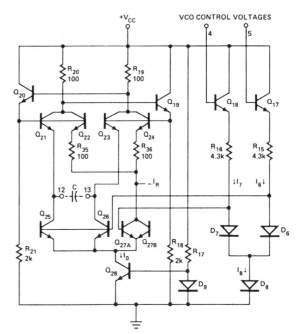

Fig. 6. VCO section of the PLL.

where V_{CM} and V_{DM} are the respective common- and difference-mode voltages.

Emitter-followers Q_{17} and Q_{18} convert these control voltages into control currents through D_6 and D_7 which can be expressed as

$$I_6 = \frac{1}{R_{15}} \left[V_{CM} - \frac{1}{2} V_{DM} - 3V_{BE} \right] \tag{3}$$

$$I_7 = \frac{1}{R_{16}} \left[V_{CM} + \frac{1}{2} V_{DM} - 3V_{BE} \right]. \tag{4}$$

These individual currents are summed in D_8 and become, with $R_{15} = R_{16} = R$,

$$I_8 = I = I_6 + I_7 = \frac{2}{R} (V_{CM} - 3V_{BE}). \tag{5}$$

Writing I_6 and I_7 as functions of the total I current gives

$$I_6 = \frac{I}{2} \left(1 - \frac{V_{DM}}{RI} \right) \tag{6}$$

$$I_7 = \frac{I}{2} \left(1 + \frac{V_{DM}}{RI} \right). \tag{7}$$

Now consider variations in I_6 and I_7 while I remains constant. Let x indicate the current imbalance such that

$$I_6 = (1 - x) I = \frac{I}{2} \left(1 - \frac{V_{DM}}{RI} \right) \tag{8}$$

$$I_7 = xI = \frac{I}{2} \left(1 + \frac{V_{DM}}{RI} \right) \tag{9}$$

where $0 \leqslant x \leqslant 1$. Thus x is defined to be

$$x = \frac{1}{2} \left(1 + \frac{V_{DM}}{RI} \right). \tag{10}$$

Currents I_6 and I_7 establish proportional currents in Q_{25}, Q_{26}, and Q_{27} in a manner similar to the analysis above since the current in Q_{28} is a constant, or

$$I_0 = I_{C28} = I_{E25} + I_{E26} + I_{E27A} + I_{E27B}.$$

Gilbert [1] has shown that the D_6–D_7 diode pair will cause identical differential currents to be reflected in both the Q_{25}–Q_{26} and the Q_{27A}–Q_{27B} differential amplifier pairs. Consequently, the constant current of I_0 jointly shared by the differential amplifier pairs will divide in each pair with the same x factor imbalance as in (10).

$$I_{E25} + I_{E26} = xI_0 \tag{11}$$

$$I_{E25} = I_{E26} = \frac{x}{2} I_0 \tag{12}$$

$$I_{E27A} + I_{E27B} = (1 - x)I_0 \tag{13}$$

$$I_{E27A} = I_{E27B} = \left(\frac{1-x}{2}\right) I_0. \tag{14}$$

Now consider placing a capacitor between the collectors of Q_{25} and Q_{26} (pins 12 and 13). Oscillation will occur with the capacitor alternately being charged by Q_{21} and Q_{23} and constantly discharged by Q_{25} and Q_{26}. When the Q_{21} and Q_{22} pair conducts, Q_{23} and Q_{24} will be off causing a negative ramp voltage to appear at pin 13 and a constant voltage at pin 12, as shown in Fig. 7. During the next half-cycle, the transistor roles and voltages are reversed. Capacitor discharge is via Q_{25} and Q_{26} which act as constant-current sinks with current amplitudes as in (12).

During each half-cycle, the capacitor voltage changes linearly by $2\Delta V$ in ΔT s where

$$\Delta V = 2R_{20} I_0 \left(\frac{x}{2} + \frac{1-x}{2}\right) = R_{20} I_0 \tag{15}$$

and

$$\Delta T = \frac{C2\Delta V}{I_{E25}}. \tag{16}$$

Combining these two equations with (12) gives a half-period of

$$\Delta T = \frac{4CR_{20}}{x}. \tag{17}$$

Utilizing (10) with the ΔT expression gives the desired VCO frequency expression of

$$f_0 = f_0' \left(1 + \frac{V_{DM}}{RI}\right) = f_0' \left[1 + \frac{V_{DM}}{2(V_{CM} - 3V_{BE})}\right] \tag{18}$$

where f_0' is the VCO's free-running frequency given by

$$f_0' = \frac{1}{16 R_{20} C}. \tag{19}$$

Equation (18) shows that the oscillator frequency is a linear function of the differential voltage from the phase comparator. Resistors R_{35} and R_{36} function to insure that an initial current imbalance exists between the Q_{25}–Q_{26} transistor pair and the dummy Q_{27}. This imbalance insures that the oscillator is self-starting when power is first applied to the circuit.

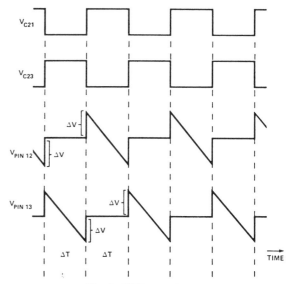

Fig. 7. VCO waveshapes.

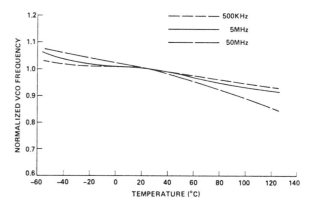

Fig. 8. Temperature behavior of the free-running frequency of the VCO.

The temperature behavior of the VCO frequency can be determined by an examination of (19). Assuming that the capacitor C, which is external, is a constant, the frequency f_0' is inversely proportional to the resistor R_{20}. Since monolithic resistors have a well-defined positive temperature coefficient, the temperature behavior of the frequency is exactly the same as that of the resistor but of the opposite sign. If no temperature compensation were applied, the temperature coefficient of frequency would be about -1200 ppm/°C. However, this number is greatly improved by introducing a component I_R into the VCO as shown in Fig. 6. The current I_R is made to have a temperature coefficient exactly equal to that of the resistor. This scheme of improving temperature stability is different from methods used in the past [2], the advantage of this scheme being that the temperature coefficient is well defined and easy to compensate. As shown in Fig. 8, the frequency variation with temperature is excellent at the lower frequencies and quite acceptable at 50 MHz. The VCO conversion gain is determined as

$$K_0 = \frac{\partial f_0}{\partial V_{DM}} = \frac{f_0'}{RI} \quad \text{Hz/V} \tag{20}$$

which is valid as long as the transistors' V_{BE} changes are small with respect to the common-mode voltage.

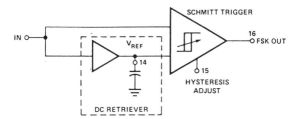

Fig. 9. Post detection processor for FSK.

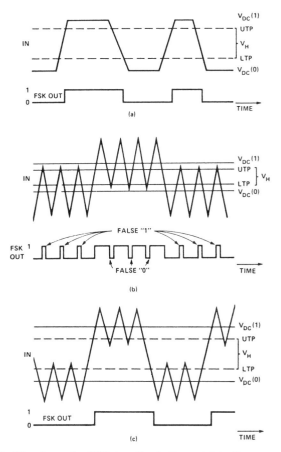

Fig. 10. Waveshapes for FSK decoding in the post detection processor. (a) Low data rates with negligible carrier feedthrough. (b) False FSK outputs due to feedthrough and low hysterisis. (c) Increased hysterisis restores FSK output in the presence of feedthrough.

VI. AMPLIFIER

The difference-mode voltage from the phase comparator is extracted and amplified by the amplifier in Fig. 2. The single-ended output from this amplifier serves as an input signal for both the Schmitt trigger and a second-differential amplifier. Low-pass filtering with a large capacitance at pin 14 produces a stable dc reference level as the second input to the Schmitt trigger. When the PLL is locked, the voltage at pin 14 is directly proportional to the difference between the input frequency and f_0'. Thus pin 14 provides the demodulated output for an FM input signal.

VII. SCHMITT TRIGGER

In FSK applications the pin 14 voltage will assume two different voltage levels corresponding to the mark and space input frequencies. A voltage comparator could be used to sense and convert these two voltage levels to logic compatible levels. However at high data rates, V_{DM} will contain a considerable amount of the carrier signal which can be removed by extensive filtering. Normally this complex filtering requires quite a few components, most all of which are external to the monolithic PLL. Also, since the control voltage for the comparator depends upon K_0 and the deviations of the mark and space frequencies from f_0', the filtering has to be optimized for each different system utilized. However, the necessary dc reference level for the comparator is present in the PLL but buried in carrier frequency feedthrough which appears as noise in the system. A Schmitt trigger with variable hysterisis can be used successfully to decode the FSK data without the need for extensive filtering.

Consider the system shown in Fig. 9 where the input signal is the single-ended output derived from the amplifier section of the 564. The dc retriever functions to establish a dc reference voltage for the Schmitt trigger. The upper and lower trigger points are adjustable externally around the reference voltage giving the variable hysterisis. For very low data rates, carrier feedthrough will be negligible, and the ideal situation depicted in Fig. 10 results. An increased data rate produces the carrier feedthrough shown in Fig. 10(b) where false FSK outputs result because the feedthrough amplitude exceeds the hysteresis voltage. Having the capability to increase the hysteresis as in Fig. 10(c) produces the desired FSK output in the presence of the carrier feedthrough.

Another important factor to be considered is the temperature drift of the f_0' in the VCO. Small changes in f_0' will change the dc level of the input voltage to the Schmitt trigger. This dc voltage shift would produce errors in the FSK output in narrow-band systems where the mark and space deviations in

f_{in} are less than the f_0' change with temperature. However, this effect can be eliminated if the dc or average value of the amplifier signal is retrieved and used as the reference voltage for the Schmitt trigger. In this manner, variations in the f_0' with temperature do not affect the FSK output.

VIII. LAYOUT AND PERFORMANCE

A chip photograph appears in Fig. 11 which shows the careful attention paid to layout so that the high-frequency performance of the circuit is not impaired in any manner. This is particularly true in terms of the ground placement, which is extra wide. The chip was processed using a thin-epi dual-layer metal Schottky process.

The correction diagram for 5-V operation as a FSK demodulator is shown in Fig. 12. A capacitor will be necessary at pin 14 depending upon the demodulated frequency. It is to be

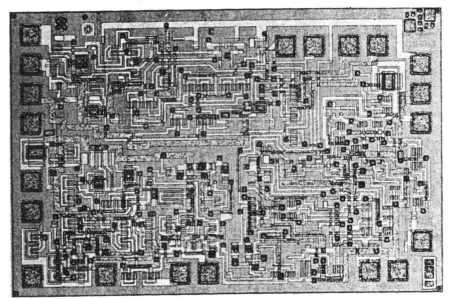

Fig. 11. Photomicrograph of the chip.

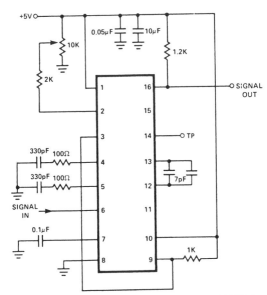

Fig. 12. Connection diagram for demodulation of FSK signals.

TABLE I
ELECTRICAL CHARACTERISTICS—$V^+ = 5$ V, $T_A = 25°$C

	Parameter	Test Conditions	Typ. Perf.		Unit
			50 MHz	5 MHz	
f_o	Lock range	$I_2 = 400$ μA	40	50	%
	Frequency change with supply voltage	V+ = 4.5V to 5.5V	4	3	%/V
	Demodulated output voltage	±1% input deviation	10	14	mVrms
		±10% input deviation,	100	140	
	Output voltage linearity		4	3	%
	Signal to noise ratio		35	40	dB
	AM rejection		30	35	dB
I_{CC}	Supply current	5V	30	30	mA
I_{LC}	Leakage current	Pin 9	1	1	μA

noted that no external filter or comparator are necessary. The loop gain of the PLL can be controlled by varying the current at pin 2, so that large gains for fast capture and small gains for low noise are attainable. The characteristics, as shown in Table I, clearly indicate the excellent performance achieved by the final circuit.

290

IX. Conclusion

A new high-performance phase-locked loop capable of operating at high frequencies has been built and is in production. Its characteristics make it aptly suited for FSK demodulation as well as clock recovery systems in telecommunications.

Acknowledgment

The author would like to express his appreciation to G. Kelson for useful and helpful discussions and to F. Adamic for his excellent work in fabricating the devices. He would also like to thank Dr. A. Connelly and J. Keith for their help.

References

[1] B. Gilbert, "A new wide-band amplifier technique," *IEEE J. Solid-State Circuits*, vol. SC-3, pp. 353–365, Aug. 1968.
[2] R. R. Cordell and W. G. Garrett, "A highly stable VCO for application in monolithic phase-locked loops," *IEEE J. Solid-State Circuits*, vol. SC-10, pp. 480–485, Dec. 1975.

A 200-MHz CMOS Phase-Locked Loop with Dual Phase Detectors

KURT M. WARE, MEMBER, IEEE, HAE-SEUNG LEE, MEMBER, IEEE,
AND CHARLES G. SODINI, MEMBER, IEEE

Abstract —A high-frequency integrated CMOS phase-locked loop (PLL) including two phase detectors is presented. The design integrates a voltage-controlled oscillator, a multiplying phase detector, a phase-frequency detector, and associated circuitry on a single die. The loop filter is external for flexibility and can be a simple passive circuit. A 2-μm CMOS p-well process was used to fabricate the circuit. The loop can lock on input frequencies in excess of 200 MHz with either or both detectors and consumes 500 mW from a single 5-V supply. The oscillator is a ring of three inverting amplifiers, and it draws power from an internal supply voltage regulated by an on-chip bandgap reference. This combination serves to reduce the supply and temperature sensitivities of the oscillator. The measured oscillator supply sensitivity is less than 5 percent/V; oscillator temperature variation is 2.2 percent in the range of 25 to 80°C. The typical oscillator tuning range is 112 to 209 MHz. The multiplying phase detector and phase-frequency detector (PFD) exhibit input-referred phase offsets of < 4° and −24°, respectively. A numerical system simulation program was written to explore the time-domain behavior of an idealized model based on the phase-locked loop design.

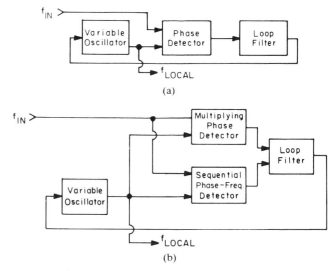

Fig. 1. (a) Single-loop PLL. (b) Dual-loop PLL.

I. INTRODUCTION

THE COMMERCIAL success of local area networks and the demand for higher data rates have recently increased the need for inexpensive high-frequency phase-locked loops [1]. Data storage and RF data communications applications have also added to this need [2]. Silicon CMOS is a natural technology for these circuits because the high production volume of digital CMOS circuits has significantly reduced the unit cost. In addition, because many phase-locked loops are part of a larger CMOS digital or analog system, the potential for a higher level of integration exists in such applications [3].

Dual-loop phase-locked loops are an active topic of research, mostly due to the desire to improve the trade-off between acquisition behavior and locked behavior [4], [5]. In this paper, a 200-MHz CMOS phase-locked loop (PLL) incorporating a multiplying phase detector and phase-

frequency detector is described. The loop filter is external for flexibility, and it may be a simple passive circuit. Either detector may be used by itself, or the two may be used together in certain applications.

Fig. 1(a) shows the block diagram of a conventional single-loop PLL, which contains a variable oscillator, phase detector, and loop filter. Fig. 1(b) shows the block diagram of a dual-loop PLL which makes use of a multiplying phase detector and a sequential phase-frequency detector.

There are two reasons to include both a phase detector and a frequency detector on the PLL integrated circuit. First, the appropriate type of detector is quite application dependent. For example, a PLL employing a multiplying phase detector is well suited to locking on a data pulse stream, whereas a PLL employing a phase-frequency detector (PFD) is well suited to frequency synthesis since the input signal does not have missing transitions [6], [7]. Second, frequency detection can be used to improve PLL signal acquisition. Consider a data synchronizer application where the PLL input is a pulse stream with packets of data separated by recognizable periods of 100 percent pulse density. The dual-loop PLL of Fig. 1(b) can be used as a fast acquisition data synchronizer. The PFD aids frequency acquisition because the multiplying phase detector provides insufficient frequency feedback when the loop

Manuscript received May 11, 1989; revised August 10, 1989. This work was supported by Analog Devices, AT&T, DEC, General Electric, IBM, and Texas Instruments Incorporated, and in part by DARPA under Contract N00014-87-K-0825.
K. M. Ware was with the Department of Electrical Engineering and Computer Science, Massachusetts Institute of Technology, Cambridge, MA 02139. He is now with the Jet Propulsion Laboratory, Pasadena, CA 91109.
H.-S. Lee and C. G. Sodini are with the Department of Electrical Engineering and Computer Science, Massachusetts Institute of Technology, Cambridge, MA 02139.
IEEE Log Number 8931197.

Reprinted from *IEEE Journal of Solid-State Circuits*, vol. SC-24, pp. 1560-1568, December 1989.

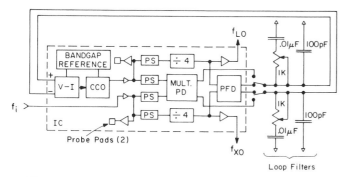

Fig. 2.   Integrated circuit block diagram with example loop filter.

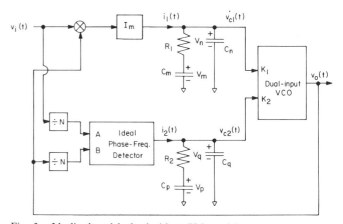

Fig. 3.   Idealized model of a dual-loop PLL used for system simulator.

bandwidth is smaller than the input frequency difference [7]. Frequency detection must be enabled during the periods of 100-percent pulse density and disabled during the data packets.

Fig. 2 shows a block diagram of the integrated circuit and a simple passive loop filter connected to the PFD output. The variable oscillator consists of three parts: a voltage-to-current converter, a current-controlled oscillator, and a bandgap reference. The reference is used to reduce the supply and temperature sensitivity of the oscillator. Each block marked PS is a phase splitter, which generates a differential output to drive the multiplying phase detector and the frequency dividers. Frequency division by a factor of 4 allows the PFD to operate properly at the maximum oscillator frequency. Outputs f_{LO} and f_{XO} are taken from the dividers to drive external pins. Both the multiplying phase detector and the PFD generate differential output currents.

In this paper, Section II gives an overview of a simulation program used to explore the behavior of an idealized dual-loop phase-lock system based on the integrated circuit design. In Section III, the PLL circuits are described. Experimental results are presented in Section IV.

II. NUMERICAL SIMULATION OF THE PLL

The response of a PLL to limited phase variations can be found by examining a linearized system model and using frequency-domain methods [7]. Due to its nonlinear behavior, the response to large phase variations cannot be found in a similar fashion. Some loops can be described by ordinary differential equations, which in nearly all cases must be integrated by computer for a particular input [8]. A loop employing a PFD cannot be described by a differential equation unless simplifying approximations are made [9].

For a particular input, the time-domain response of a PLL can be found with a circuit simulator such as SPICE [10]. However, simulating a PLL with many transistors is expensive, especially because the system is often characterized by widely spaced natural frequencies. For example, circuit simulation of a high-frequency narrow-band phase-lock system requires very small time steps to model the oscillator output, but the time scale of interest, such as the acquisition time, could be orders of magnitude larger than

the time-step size. As a result, conventional circuit simulation of a complex PLL is impractical.

Simulation at the system level requires less computer time than circuit simulation because idealized models are used. Although some nonideal behavior is not modeled, system simulation retains the major characteristics of the actual system and allows one to explore design trade-offs [11]. A computer program was written to simulate the time-domain response of a restricted class of PLL's using idealized models for the loop components. The program can simulate the response of a loop employing the multiplying phase detector, the PFD, or both detectors. In the last case, numerical simulation is essential to assessing acquisition behavior.

Fig. 3 shows the system modeled by the simulator. The input signal model is $v_i(t) = \sin(\theta_i(t))$, and the output signal model is $v_o(t) = \sin(\theta_o(t))$. A constant I_m sets the magnitude of the multiplier output so $i_1(t) = I_m \cdot \sin(\theta_i(t)) \cdot \sin(\theta_o(t))$. The ideal PFD changes state only when the A and B inputs cross zero with positive slope, and its output current is $+I_p$, 0, or $-I_p$, depending on the state [9]. The voltage-controlled oscillator (VCO) output frequency is $\omega_o(t) = 2\pi f_c + K_1 v_{c1}(t) + K_2 v_{c2}(t)$, where f_c is the center frequency.

The oscillator output frequency $\omega_o(t)$ is integrated numerically using the first-order forward Euler method to find the output phase $\theta_o(t)$. The forward Euler method is used because other methods require an estimate of the control voltages at time t in order to compute their value at t. The nature of the PFD logic makes it difficult to use other integration methods because a small change in the estimate of the control voltages could cause the PFD model to change state and lead to a divergent solution. The time-step size is conservatively set to ensure that the natural frequencies of the discretized system fall in the stable region.

The time steps are chosen as follows. When the PFD logic enters a new state at time t_0, the current VCO and input frequencies are used to estimate the end time t_e of that state. The time-step size is set to $\Delta t = (t_e - t_0)/M$, where M corresponds to the target number of time steps per single-logic-state interval. The estimate for t_e is revised

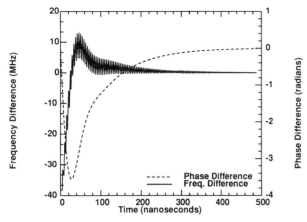

Fig. 4. Simulated step response of a PLL employing a PFD.

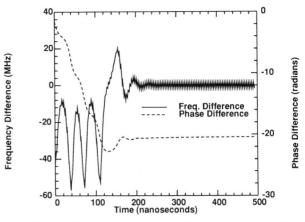

Fig. 5. Simulated step response of a PLL employing a multiplying phase detector.

for each time step. The last time step in the single-logic-state interval is sized to terminate on the most recent estimate of t_e. To assure reasonable accuracy, the value of M was adjusted until the actual number of time steps per interval was equal to the target number for a wide-band loop. Several wide-band loop simulations were carefully examined to assess the correctness of the method.

Capacitor voltages v_m, v_n, v_p, and v_q are treated as state variables. Because $i_2(t)$ is constant during the time step $(t_0, t_0 + \Delta t)$, it is possible to exactly solve for $v_p(t_0 + \Delta t)$ and $v_q(t_0 + \Delta t)$. This calculation yields $v_{c2}(t_0 + \Delta t)$.

During the time step, one can approximately solve for $v_{c1}(t_0 + \Delta t)$. The current $i_1(t)$ is $I_m \cdot \sin(\theta_i(t)) \cdot \sin(\theta_o(t))$. If one assumes that the time step is small enough to assure that $(\omega_i - \omega_o)\Delta t \ll 1$ and $(\omega_i + \omega_o)\Delta t \ll 1$, one can then substitute small-angle approximations for their sine and cosine. Under the approximation, we may solve for $v_m(t_0 + \Delta t)$ and $v_n(t_0 + \Delta t)$ in terms of their value at the beginning of the interval. This calculation yields $v_{c1}(t_0 + \Delta t)$.

Loops employing only one of the phase detectors can be simulated by setting K_1 or K_2 to zero. In addition, because the loop filters are linear, the response of a system with two phase detectors connected to a single loop filter can be derived by superposition. If $R_1 = R_2$, $C_m = C_p$, and $C_n = C_q$, the capacitor voltages in the single filter will be equal to the sum of the voltages on their counterparts in the system with two loop filters.

A sample of the simulator output is shown in Fig. 4. For this simulation, K_1 was set to zero to simulate the response of a loop employing the PFD alone. The input frequency was stepped at $t = 0$ from 191 to 229 MHz. The frequency and phase difference plotted correspond to the difference between the input signal and the simulated oscillator output.

Loops employing only a multiplying phase detector were also simulated. If the initial frequency difference is large, the desired part of the multiplying phase detector output can lie outside the loop bandwidth, and it is attenuated. Acquisition in this case can be very slow or, if an offset is present in the loop, fail to occur entirely [7]. Fig. 5 shows the response of a loop employing the multiplying phase detector given an input frequency step at $t = 0$ from 191 to 235 MHz. The cumulative phase difference decreases several cycles during acquisition and comes to a new equilibrium at a phase corresponding to -3.25 cycles.

The simulator duplicates the behavior predicted by linearized models where applicable and shows the expected nonlinear behavior. The simulator was used to efficiently explore the acquisition behavior of various single-loop and dual-loop systems.

III. CIRCUIT DESCRIPTIONS

In the following sections, the PLL subsystems will be described in more detail. In particular, the voltage-controlled oscillator, multiplying phase detector, and the phase-frequency detector circuits are presented.

A. Voltage-Controlled Oscillator

The voltage-controlled oscillator includes a voltage-to-current converter, a current-controlled ring oscillator, and a bandgap reference. The ring consists of three inverting amplifiers that combine an inverter in parallel with a current-controlled inverter. The inverter is a fast circuit that would make a high-frequency ring oscillator, whereas the current-controlled inverter permits the oscillator to be tuned over a suitable range. The supply and temperature sensitivities of this design are improved by using an internal reference.

Fig. 6 shows the current-controlled oscillator circuit. The node labeled V_{RR} is an internal supply voltage whose nominal value is 0.5 V. The ring oscillator includes $M1-M25$; $M26-M33$ constitute a voltage follower to set V_{RR} to the same voltage as the V_{REF} input. Devices $M1-M3$ control the oscillation frequency by setting the bias of the current-controlled inverters. The three inverting amplifiers in the ring are made up of devices $M4-M21$ whereas $M22-M24$ buffer the output signal. The first inverting amplifier in the ring consists of $M7$ and $M13$ (the inverter) and $M4$, $M8$, $M14$, and $M19$ (the current-controlled inverter).

Devices $M26-M33$ constitute a simple two-stage op-amp connected in a voltage-follower configuration. If vari-

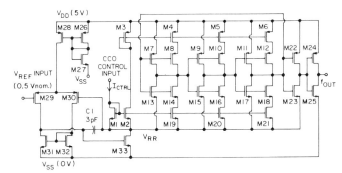

Fig. 6.   High-frequency ring-oscillator schematic.

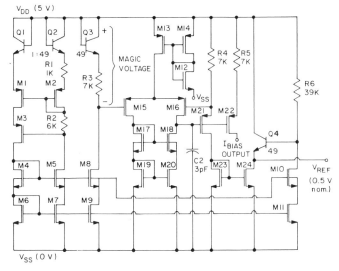

Fig. 7.   Bandgap current source and voltage reference schematic.

ations in V_{DD} are duplicated exactly by V_{REF}, $V_{DD} - V_{RR}$ is constant, and the ring-oscillator supply sensitivity to V_{DD} will be greatly reduced. This design was fabricated using a p-well process; by tying the wells of $M13-M21$ to their respective source terminals, the oscillator is nearly isolated from substrate voltage variations.

The op amp uses $M33$ for the second stage; its width is large to assure the device remains saturated when its drain-to-source voltage is 0.5 V. Pole-splitting compensation is performed by $C1$, which is a sandwich of metal-2, metal-1, and p-well material. Simulations indicated that $C1$ improves the op-amp open-loop phase margin to 79°.

The center frequency temperature sensitivity of the uncompensated oscillator was measured at -2000 ppm/°C. The voltage-follower input V_{REF} has a nominal value of -4.5 V, referred to V_{DD}, on which an intentional temperature coefficient (TC) of $+2400$ ppm/°C is induced. Through the oscillator center frequency sensitivity to V_{RR}, the ring-oscillator temperature sensitivity is greatly reduced.

The reference voltage V_{REF} is generated on-chip by the circuit in Fig. 7. This circuit is an adaptation of a basic CMOS bandgap current source [12], and it also generates a bias current I_{BIAS} used in the voltage-to-current converter. Devices $Q1$, $Q2$, $R1$, $R2$, and $M1-M11$ serve to generate bias voltages across $R3$ and $R6$, which are proportional to

absolute temperature (PTAT). Devices $M12-M20$ form an op amp that sets the voltages across $R4$ and $R5$ equal to the magic voltage across $Q3$ and $R3$, and $M21$ and $M22$ act as source followers. The op amp is compensated by $C2$, which improves the op-amp open-loop phase margin to 72° according to simulation. A temperature-independent current is mirrored by $M23$ and $M24$ through the emitter of $Q4$, which, with $R6$, creates the V_{REF} output.

The n-p-n transistors here are substrate–well–n$^+$ devices that are available in any p-well CMOS process. The collector terminals of all such devices are connected through the substrate to V_{DD}. Transistor $Q1$ is constructed with a single 10-μm square emitter; $Q2-Q4$ have 49 such emitters in parallel.

Ordinarily, the place of $R2$ would be taken by a diode-connected p-channel transistor, and the circuit would require a supply voltage greater than 5 V. With $R2$, $|V_{DS}|$ of $M1$ is reduced, and the circuit can operate from 5 V with standard threshold transistors.

All resistors shown are made from polysilicon material. The magic voltage across $Q3$ and $R3$ is designed to exhibit a TC equal to that of resistors $R4$ and $R5$ to make the I_{BIAS} output current (250-μA nominal) supply and temperature independent. The voltage $V_{DD} - V_{REF}$ has a nominal value of 4.5 V along with a temperature coefficient of

$$\mathrm{TC}_{V_{REF}} = \frac{3.9\ \mathrm{V}}{4.5\ \mathrm{V}} \cdot \frac{1}{T} + \frac{-2.3\ \mathrm{mV}}{4.5\ \mathrm{V}} \qquad (1)$$

or $+2400$ ppm/°C at room temperature.

Since the loop filter output is a control voltage, and the circuit of Fig. 6 is tuned by a current, a voltage-to-current converter is required. The differential control voltage is converted to a single-ended current output by a differential stage whose transconductance is largely determined by a polysilicon resistor. When the input differential voltage is zero, the output current is set by I_{BIAS}. The output current varies in a nearly linear fashion from zero to full scale for differential inputs from approximately -1.25 to $+1.25$ V.

SPICE simulation of the voltage-controlled oscillator using early device models indicated that the oscillator tuning range would be 170 to 270 MHz and that the transfer coefficient of the oscillator would be 15 MHz/V near the center frequency.

B. Multiplying Phase Detector

Practical multiplying phase detectors fall into three groups: linear [13], chopping [6], [14], [15], and EXCLUSIVE-OR [16]. The chopping multiplier and EXCLU-SIVE-OR circuit generate a signal whose low-frequency components approximate those of the linear multiplier, but their high-frequency components will differ. In phase-lock applications, the loop filter attenuation tends to minimize the effect of these differences. Linear multipliers are often based on the double-balanced Gilbert cell [17]. The chopping multiplier can be based on a single-balanced Gilbert

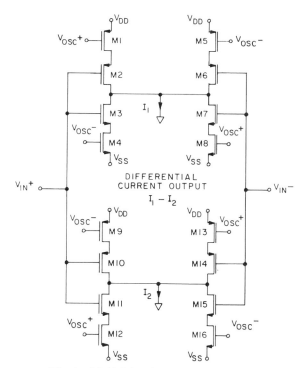

Fig. 8. Multiplying phase detector schematic.

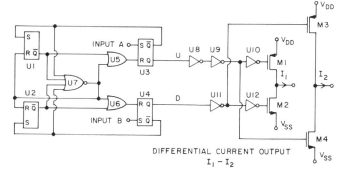

DIFFERENTIAL CURRENT OUTPUT
$I_1 - I_2$

Fig. 9. PFD block diagram.

cell [6], [14] or switching between $+1$ and -1 amplifiers [15]. In the latter case, one may replace the amplifiers with a phase-splitter circuit, which generates differential outputs from a single-ended input.

In this work, a differential-output chopping multiplier was implemented. Fig. 8 shows the multiplying phase detector design. The differential signals V_{OSC+} and V_{OSC-} are generated from the VCO output by a phase splitter. Another phase splitter generates V_{IN+} and V_{IN-} from the PLL input signal. Transistors $M1-M8$ form one chopping multiplier and use tri-state gates to alternately switch the V_{IN+} and V_{IN-} signals to the output. When V_{OSC} is low, the tri-state inverter consisting of $M1-M4$ is enabled, whereas the inverter consisting of $M5-M8$ goes into high-impedance mode. When V_{OSC} is high, their roles are interchanged. If the output were unloaded, the resulting output voltage would correspond to the EXCLUSIVE-OR function of its inputs. Tying the output to incremental ground creates a current-mode output representing the product of the local oscillator output and the PLL input signal. Devices $M9-M16$ are the same as $M1-M8$ but are switched on the opposite phase of V_{OSC}.

The average output of an ideal multiplier is zero when the inputs are sinusoids of the same frequency and $\pm 90°$ out of phase. If we consider phase differences near $-90°$, an actual multiplying phase detector will have zero average output at a phase that is offset slightly. This offset between the ideal model and the actual circuit is the input-referred phase offset. The differential configuration in Fig. 8 is intended to reduce the input-referred phase offset.

Let us concentrate on one output I_1. If the input phase difference is $-90°$ and the load is purely capacitive, the output voltage will not settle at the midpoint between V_{DD} and V_{SS}; it will be at some voltage V_{DC} due to the difference in average current output between the n- and p-channel devices. Because the matching circuit, including devices $M9-M16$, is switched on the opposite phase, its output voltage should also be V_{DC}. The differential output voltage should be zero in this case and therefore be free of offset.

Supply voltage variations can cause some change in the magnitude of the phase characteristic but should not increase the input-referred phase offset because such variations will not affect the switching duty cycle.

If the V_{IN} and V_{OSC} signals do not have 50-percent duty cycle, a phase detector offset will result. The phase-splitter output signals are designed to have nearly 50-percent duty cycle at high frequencies so the phase detector offset is relatively small.

C. Phase-Frequency Detector

The phase-frequency detector in this work is based on the conventional design, with an edge-sensitive state machine controlling an output charge pump [9]. For compatibility with the multiplying phase detector design, a differential output circuit was designed. A block diagram of the phase-frequency detector design is shown in Fig. 9.

If the leading edge of input A precedes the leading edge of input B, $U1-U7$ will generate a pulse on U whose width corresponds to the time difference between the two leading edges. This pulse turns on $M1$ and $M4$, and the output differential current is then positive. Conversely, if the leading edge of B precedes A, a pulse appears on D, which turns on $M2$ and $M3$, generating a negative differential output current.

The maximum input frequency of the phase-frequency detector is significantly lower than the maximum oscillator output frequency. If the PLL input signal does not have missing or extra transitions, one may precede the PFD inputs with two identical frequency dividers. Since the PFD itself is intolerant of missing or extra transitions, this was not considered to be a restriction.

In this work, the frequency dividers preceding the PFD each consist of a cascade of two high-speed toggle flip-flops. The toggle flip-flop design is shown in Fig. 10. This is an adaptation of a GaAs design whose reported maximum toggle rate corresponds to $1/(2\tau_p)$, where τ_p is the gate propagation delay [18]. The input clocks CLK+ and

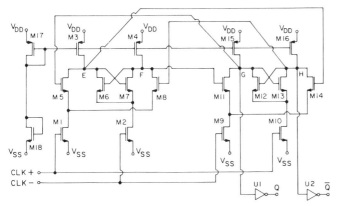

Fig. 10. Toggle flip-flop schematic.

CLK − are driven by a phase splitter or a previous divider stage, so they are complementary and have a nominal duty cycle of 50 percent. This is a differential master–slave configuration. The gates of $M5$ and $M8$ constitute the differential D and $\bar{D}$ inputs to the master flip-flop stage. When CLK + is high, $M5$ and $M8$ transfer the input state to nodes E and F. When CLK + goes low and CLK − goes high, the master state is held by a regenerative pair $M6$ and $M7$. Nodes E and F drive the slave stage, which is identical but switched on the opposite clock phases. The ON times of $M1$ and $M2$ overlap, as do those of $M9$ and $M10$, so the state is smoothly transferred from master to slave. Nodes G and H are the slave stage outputs. Node G is connected to the D input, and node H is connected to the $\bar{D}$ input; a logical inversion occurs between the slave stage output and the master stage input. This makes the circuit act as a toggle flip-flop. Inverters $U1$ and $U2$ serve to buffer the output and restore it to full voltage swing. The maximum input frequency of the divider exceeds the maximum frequency of the oscillator.

IV. EXPERIMENTAL RESULTS

The prototype circuit was fabricated by MOSIS using their scalable CMOS 2-μm p-well process [19]. The process has two levels of metal interconnect and a nominal oxide thickness of 400 Å. A labeled photograph of the die is shown in Fig. 11. Each prototype circuit includes two complete PLL's, a VCO test circuit, and a bandgap reference test circuit. Major subsystems are labeled on the left-hand PLL in Fig. 11. The active area of the PLL with both detectors is 1.5×3.7 mm^2. The measured data presented in this section were taken from devices fabricated on two separate MOSIS runs sent to the same foundry.

A minor error was found in the bandgap reference design. To test the circuits as received, the oscillator temperature and supply sensitivities were measured with $V_{DD} = 6.0$ V. In this case, $V_{DD} - V_{REF}$ was close to the nominal value of 4.5 V. Because the oscillator frequency is primarily dependent on $V_{DD} - V_{REF}$ and much less dependent on $V_{REF} - V_{SS}$, these measurements should reflect the true circuit performance. All other measurements presented here were conducted with the nominal supply $V_{DD} = 5.0$ V.

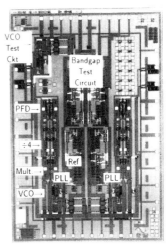

Fig. 11. Die photograph of the prototype PLL and test circuits.

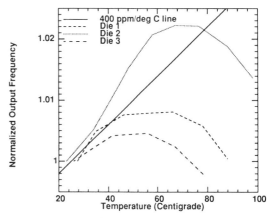

Fig. 12. Normalized oscillator temperature measurements.

The mean oscillator center frequency was measured at 190 MHz. Chip-to-chip center frequency variations of 148 to 234 MHz (± 23 percent) were observed. The tuning range of a typical device was 112 to 209 MHz, which represents a reach of ± 30 percent about the midpoint of the range. Although the center frequency varied widely from chip to chip, the oscillator range as a percentage of the center frequency was quite uniform. Out of 19 prototypes, 85 percent of the devices could reach frequencies from 128 to 187 MHz. All of the devices could reach 140 to 162 MHz, so the tuning range was deemed adequate to allow for chip-to-chip variation.

The worst-case supply sensitivity of the oscillator with the on-chip reference was measured at 4.7 percent/V, which is a fourfold improvement over the measured sensitivity when no on-chip reference was used. Fig. 12 shows the measured temperature dependence for three oscillators, along with a 400 ppm/°C line for comparison. The worst-case oscillator temperature variation was restricted to a 2.2-percent range for temperatures between 25 and 80°C, equivalent to an average temperature coefficient of 400 ppm/°C. This is a considerable improvement over the −2000 ppm/°C temperature coefficient of the uncompensated oscillator.

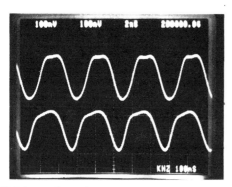

Fig. 13. PLL internal waveforms, locked on 200-MHz input, using the PFD. Top trace: buffered input signal. Bottom trace: buffered oscillator output. Vertical scale is 2 V/div; horizontal scale is 2 ns/div.

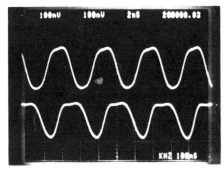

Fig. 14. PLL internal waveforms, locked on 200-MHz input, using the multiplying phase detector. Top trace: buffered input signal. Bottom trace: buffered oscillator output. Vertical scale is 2 V/div; horizontal scale is 2 ns/div.

When employing the phase-frequency detector, the PLL can lock on a 200-MHz sine wave, as shown in Fig. 13. The PLL input signal was amplified and limited by the input buffer circuit. These waveforms were measured on the chip at buffered probe pads, representing the PLL input signal and the oscillator output (see Fig. 2). A low-capacitance active probe was used to measure the signals, attenuating them by a factor of 20. The loop bandwidth was estimated to be 500 kHz. The loop can lock on input frequencies corresponding to most of the oscillator range. Pull-in occurs spontaneously over the entire lock range. This photograph shows a steady-state phase difference near 0°, as expected. Steady-state phase error measurements indicated that the input-referred phase offset of the PFD was −6° at its input. This corresponds to a −24° offset at the input of the frequency dividers, which contribute negligible phase offset.

When employing the multiplying phase detector, the PLL can lock on a 200-MHz sine wave, as shown in Fig. 14. The PLL input signal was amplified and limited by the input buffer circuit. These waveforms were measured on the chip at buffered probe points, representing the PLL input signal and the oscillator output. The loop bandwidth was estimated to be 1 MHz. The loop can lock on input frequencies corresponding to most of the oscillator range. The pull-in range varies, depending on the loop filter, but can correspond to most of the lock range. This photograph shows a steady-state phase difference near −90°, as expected. Steady-state phase error measurements indicated

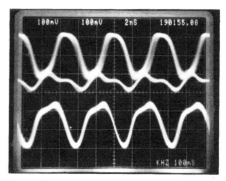

Fig. 15. PLL eye diagram, locked on 190-Mbit/s RZ data pulse stream input, using the multiplying phase detector. Top trace: buffered input signal. Bottom trace: buffered oscillator output. Vertical scale is 2 V/div; horizontal scale is 2 ns/div.

that the input-referred phase offset of the multiplying phase detector was less than 4°.

When employing the multiplying phase detector, the PLL can also lock on a data pulse stream. Fig. 15 shows an eye diagram, where the PLL input was a 190-Mbit/s repeating 64-bit return-to-zero (RZ) data pulse stream, and the oscilloscope was triggered on the trailing edge of the oscillator output. Due to the sizable eye opening, a decision circuit strobed by the oscillator could reliably regenerate the incoming data. An external delay was adjusted to place the trailing edge of the oscillator output in the center of the eye. The loop bandwidth was estimated to be 600 kHz.

The detector outputs are compatible and can be connected together directly. We note that the nominal steady-state input phase difference for the PLL employing the multiplying phase detector alone is −90° compared with 0° when the PFD alone is used. When the two are used in combination, the phase characteristic is a composite, and the steady-state input phase difference lies between −90 and 0°. For this integrated circuit, the nominal steady-state phase difference is −81°; the frequency dividers preceding the PFD serve to reduce the effect of the PFD on the input phase difference. Measurements of the closed-loop behavior in this case were consistent with the composite phase characteristic model.

To demonstrate the applicability of the PLL with dual detectors, a prototype fast acquisition data synchronizer test circuit was designed and fabricated. A block diagram of the chip is shown in Fig. 16. The chip was designed to regenerate non-return-to-zero (NRZ) data, which is converted on chip to a RZ data pulse stream suitable for synchronization by the PLL. A half-bit delay and an EXCLUSIVE-OR circuit were used to implement the NRZ-to-RZ converter.

The carrier detector controls the PFD charge pump so that frequency detection is enabled during periods when the PLL input has 100-percent pulse density and disabled when the PLL input pulse density is less than 100 percent. The carrier detector employs a leading edge detector, a retriggerable one shot whose time constant determines the expected data rate, and a state register that keeps a record of the previous nine ZERO-TO-ONE transitions. A logic cir-

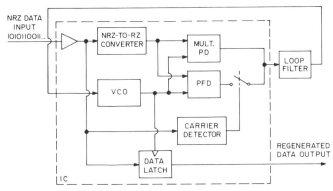

Fig. 16. Fast acquisition data synchronizer block diagram.

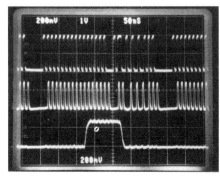

Fig. 17. Carrier detector measured waveforms. Top trace: NRZ data stream input, 1 V/div. Middle trace: edge detector output (internal), 4 V/div. Bottom trace: carrier detector output (internal), 4 V/div. Horizontal scale is 50 ns/div.

cuit generates the CD output. When an 18-bit 101010 ⋯ pattern is detected, the CD output goes high and stays high until the first ZERO-to-ONE transition after the pattern is broken and data are present.

The system shown in Fig. 16 is superior to one in which the CD output is used to independently multiplex the phase detectors. Because each phase detector results in a different steady-state input phase difference, a phase difference transient will result when the CD output changes. In the first case, the multiplying phase detector is always enabled, and the input steady-state phase difference is $-81°$ when the PFD is enabled and $-90°$ when the PFD is disabled. In the second case, the input steady-state phase difference is $0°$ when the PFD is used alone and $-90°$ when the multiplying phase detector is used alone. Thus, the shift in the phase difference required in the first case is much smaller than the second case—the system shown in Fig. 16 is superior. The details of the transient and its effect on the system performance will depend on the loop filter parameters and the application requirements.

The complete data synchronizer test circuit could not be tested due to a layout error. However, subsystem tests indicated the feasibility of the basic concept. Fig. 17 shows measured carrier detector waveforms with an NRZ input data rate of 180 Mbits/s. The CD output rising edge follows an 18-bit 101010 ⋯ pattern; the falling edge follows the first ZERO-to-ONE transition that does not match the minimum spacing.

TABLE I
SUMMARY OF MEASUREMENTS ON THE PHASE-LOCKED LOOP

VCO Mean Center Frequency	190 MHz
VCO Center Frequency Variation (Worst Case)	148 to 234 MHz (± 23%)
VCO Tuning Range (Typical)	112 to 209 MHz (± 30%)
VCO Transfer Coefficient	40 MHz/V
VCO Supply Sensitivity	4.7 %/V (max.)
VCO Temperature Variation	2.2% (max.) 25°C and 80°C 400 ppm/°C avg.
PFD Transfer Coefficient	370 µA/radian
PFD Input Referred Phase Offset	−24°
Mult. PD Transfer Coefficient	640 µA/radian
Mult. PD Input Referred Phase Offset	< 4°
PLL Supply Voltage	5 V
PLL Power Consumption	500 mW
PLL Active Area	1.5 mm × 3.7 mm

V. SUMMARY

A 200-MHz phase-locked loop with a multiplying phase detector and a phase-frequency detector has been constructed using a 2-µm CMOS process. An on-chip bandgap reference has been used to significantly improve the VCO supply and temperature sensitivities. The presence of dual phase detectors increases the application flexibility of the circuit, and some applications may benefit from the use of both phase and frequency detection. The entire PLL circuit operates from a single 5-V supply and consumes 500 mW when locked on an input at 200 MHz while driving two 50-Ω outputs at 50 MHz. A summary of the measured data is given in Table I.

ACKNOWLEDGMENT

The authors wish to thank J. K. Roberge for his suggestions regarding chip testing and P. Ferguson at Analog Devices, Inc. for providing access to apparatus for modifying test devices. They also wish to express their gratitude to MOSIS for fabricating the devices.

REFERENCES

[1] R. H. Leonowich and J. M. Steininger, "A 45-MHz CMOS phase/frequency-locked loop timing recovery circuit," in *ISSCC Dig. Tech. Papers*, 1988, pp. 14–15.
[2] W. D. Llewellyn, M. M. H. Wong, G. W. Tietz, and P. A. Tucci, "A 33 Mb/s data synchronizing phase-locked loop circuit," in *ISSCC Dig. Tech. Papers*, 1988, pp. 12–13.
[3] D. A. Hodges, P. R. Gray, and R. W. Broderson, "Potential of MOS technology for analog integrated circuits," *IEEE J. Solid-State Circuits*, vol. SC-13, pp. 285–294, June 1978.
[4] J. Scott *et al.*, "A 16 Mb/s data detector and timing recovery circuit for token ring LAN," in *ISSCC Dig. Tech. Papers*, 1989, pp. 150–151.
[5] J. D. Blair *et al.*, "A 16 MBPS adapter chip for the token-ring local area network," in *ISSCC Dig. Tech. Papers*, 1989, pp. 154–155.
[6] R. R. Cordell, J. B. Forney, C. N. Dunn, and W. G. Garrett, "A 50 MHz phase- and frequency-locked loop," *IEEE J. Solid-State Circuits*, vol. SC-14, pp. 1003–1009, Dec. 1979.
[7] F. M. Gardner, *Phaselock Techniques*, 2nd ed. New York: Wiley, 1979.
[8] A. J. Viterbi, *Principles of Coherent Communication*. New York: McGraw-Hill, 1966, ch. 3.
[9] F. M. Gardner, "Charge-pump phase-lock loops," *IEEE Trans. Commun.*, vol. COM-28, pp. 1849–1858, Nov. 1980.
[10] T. Quarles, A. R. Newton, D. O. Pederson, and A. Sangiovanni-Vincentelli, *SPICE 3B1 User's Guide*, Univ. of California, Berkeley, Apr. 1987.

[11] S. Can and Y. E. Sahinkaya, "Modeling and simulation of an analog charge-pump phase locked loop," *Simulation*, vol. 50, pp. 155–160, Apr. 1988.

[12] P. R. Gray and R. G. Meyer, *Analysis and Design of Analog Integrated Circuits*, 2nd ed. New York: Wiley, 1984, pp. 736–737.

[13] E. N. Murthi, "A monolithic phase-locked loop with post detection processor," *IEEE J. Solid-State Circuits*, vol. SC-14, pp. 155–161, Feb. 1979.

[14] M. Soyuer and R. G. Meyer, "A 350 MHz bipolar monolithic phase-locked loop," in *Proc. CICC*, 1988, pp. 9.6.1–4.

[15] B. J. Hosticka, W. Brockherde, U. Kleine, and R. Schweer, "Non-linear analog switched-capacitor circuits," in *IEEE Int. Symp. Circuits Syst.*, 1982, pp. 729–732.

[16] C. S. Park and R. Schaumann, "Design of a 4-MHz analog integrated CMOS transconductance-C bandpass filter," *IEEE J. Solid-State Circuits*, vol. 23, pp. 987–996, Aug. 1988.

[17] B. Gilbert, "A precise four-quadrant multiplier with subnanosecond response," *IEEE J. Solid-State Circuits*, vol. SC-3, pp. 365–373, Dec. 1968.

[18] C. A. Liechti *et al.*, "A GaAs MSI word generator operating at 5 Gbit/s data rate," *IEEE Trans. Microwave Theory Tech.*, vol. MTT-30, pp. 998–1006, July 1982.

[19] C. Tomovich, "MOSIS—A gateway to silicon," *IEEE Circuits Dev. Mag.*, vol. 4, pp. 22–23, Mar. 1988.

High-Frequency Phase-Locked Loops in Monolithic Bipolar Technology

MEHMET SOYUER, MEMBER, IEEE, AND ROBERT G. MEYER, FELLOW, IEEE

Abstract —Circuit design techniques for realizing high-frequency, low-power phase-locked loops (PLL's) in monolithic silicon bipolar technology are discussed. A varactor-tuned voltage-controlled oscillator (VCO), an analog phase detector, and a bandgap reference have been utilized as building blocks. A test circuit fabricated in a 2-μm bipolar process exhibited a maximum center frequency of 350 MHz, and the PLL pull-in range was larger than ± 1 percent. The circuit operates from a 5-V supply and dissipates 270 mW.

I. INTRODUCTION

PHASE-LOCKED LOOP (PLL) circuits are widely used in various communications and control systems applications, including frequency synthesis, clock signal (timing) recovery, and modulation and demodulation of signals. Monolithic realizations of PLL's have the potential for very high frequency operation (due to small parasitics) and low cost. Silicon bipolar technology is attractive for the realization of high-frequency monolithic PLL's because of its frequency capability, precision, and matching. However, the maximum operating range of most currently available low-power monolithic PLL circuits is limited to frequencies below 150 MHz [1]–[3]. Higher speed means more power dissipation in general [4]–[6]. Therefore, the realization of high-frequency performance with minimum power dissipation is an important goal.

A PLL circuit is essentially a negative feedback control system. The frequency of the voltage-controlled oscillator (VCO) is controlled by the phase detector output through the feedback loop and is made equal to the frequency of the input signal with a phase difference. There are several outputs available from a PLL which can be useful depending on the application. In demodulation of the FM signals, the loop filter output provides the modulating signal. In timing recovery applications, the VCO output waveform is used as the recovered clock signal, and the ultimate goal is to minimize the phase error (static and dynamic) between

the input signal and the VCO and at the same time to filter out the noise components [7], [8]. To minimize the jitter (dynamic phase error) caused by the input noise, the loop bandwidth must be very small compared to the input frequency. This, in turn, reduces the pull-in range available from the loop. The initial VCO center-frequency offset is one of the main contributors to the phase error together with the dc offsets in the phase detector and loop amplifier. Therefore, it should be minimized against temperature and supply variations. When used as a narrow-band filter, as in timing recovery, a PLL with a supply and temperature-compensated VCO requires a relatively narrower pull-in range and this makes the loop-filter design easier. Also, a high-frequency crystal is not required if the VCO can be compensated for temperature and supply variations. Although this design approach involves a trade-off between different PLL parameters, most significantly between pull-in range and noise bandwidth, it is relatively easy to implement such a PLL without any extra circuitry for frequency acquisition. Aided frequency acquisition techniques require extra circuitry and may degrade the PLL frequency performance. They also require more power consumption and larger chip area [3], [6], [9]. Therefore, the best strategy for the high frequencies seems to be that one should keep the circuits simple enough so that they can be easily implemented.

Using the general approach outlined above, several types of PLL building blocks will be discussed in the following sections. Experimental results are presented for a test circuit fabricated in a 2-μm oxide-isolated bipolar technology.

II. VOLTAGE-CONTROLLED OSCILLATOR DESIGN

It is possible to minimize the noise contribution of the PLL by employing a VCO with a small noise bandwith. Harmonic VCO's employ an inductor and a capacitor for frequency selection. Therefore, it is possible to design a VCO with a small noise bandwith if a high-Q tank circuit is available. Varactor-controlled VCO's are attractive since an on-chip junction capacitance with a known temperature dependence can be used as a varactor. The control voltage can then be derived from an on-chip bandgap reference to cancel the temperature dependence of the VCO center

Manuscript received September 7, 1988; revised November 15, 1988. This work was supported by the U.S. Army Research Office under Grant DAAL03-87-K0079. M. Soyuer was supported by the Scientific and Technical Research Council of Turkey.

M. Soyuer was with the Department of Electrical Engineering and Computer Sciences, University of California, Berkeley, CA 94720. He is now with IBM Research Division, T. J. Watson Research Center, Yorktown Heights, NY 10598.
R. G. Meyer is with the Department of Electrical Engineering and Computer Sciences, University of California, Berkeley, CA 94720.
IEEE Log Number 8927707.

Reprinted from *IEEE Journal of Solid-State Circuits*, vol. SC-24, pp. 787-795, June 1989.

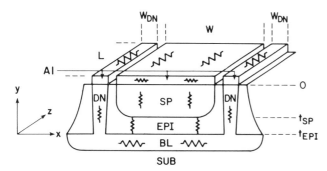

Fig. 1. Unit cell structure for varactor.

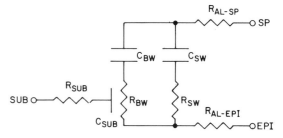

Fig. 2. Unit cell model for varactor.

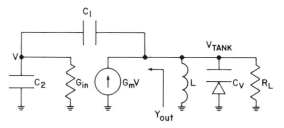

Fig. 3. VCO configuration with varactor.

frequency. Assuming a low-loss external inductor, the noise bandwith of the VCO then depends on the quality factor of the varactor. Consequently, a varactor-controlled harmonic VCO with low jitter and low TC is a good choice for high-frequency narrow-band PLL implementations in monolithic technology but requires a high-quality junction capacitor and an external inductor.

The quality factor, Q_v, of a varactor is a function of bias and frequency [10]. At low frequencies, the shunt conductance that represents the leakage current limits Q_v, whereas at high frequencies, the series resistance is more significant. Therefore, for frequencies greater than a few megahertz Q_v can be written as

$$Q_v = \frac{1}{\omega R_v C_v} \qquad (1)$$

where R_v is the series resistance of variable capacitance, C_v. In other words, the R_v–C_v product must be minimized to achieve high Q_v at high frequencies. Although Q_v increases with reverse bias, this increase is limited by the breakdown voltage and the minimum design value of C_v. Usually, a high-frequency bipolar process has a deep-p (DP) implant for the extrinsic base and a shallow-p (SP) implant for the intrinsic base. Therefore, in conjunction with the n-type epi layer, these can be used to form a varactor diode without any extra processing steps. The SP-EPI structure shown in Fig. 1 will be considered as a possible unit cell for Q_v optimization.

As shown in Fig. 1, a pair of deep-n (DN) plugs connect the buried layer (BL) to the collector contacts to reduce the collector resistance. Base resistance is reduced by covering the whole SP area with metal (Al). If the R_v–C_v product of this unit structure can be minimized with respect to the planar dimensions W and L, then any number of these cells can be connected in parallel without degrading the optimum Q_v. This problem is obviously three-dimensional, but some simplifying assumptions can still be made. As shown in Fig. 2, one can consider the bottom-wall and sidewall capacitances separately and model the metal losses in the z direction by R_{AL}. Assuming that both branches have a quality factor larger than 10, the quality factor of their parallel combination is given by

$$Q_{BW\|SW} = \frac{1}{\omega} \frac{C_{BW} + C_{SW}}{R_{BW}C_{BW}^2 + R_{SW}C_{SW}^2}. \qquad (2)$$

It can easily be shown that (2) is a function of W but not of L [11]. Therefore, by taking the derivative of (2) with respect to W, one can find W_{opt} which maximizes $Q_{BW\|SW}$. Then, for a given number of unit cells n, R_{AL} can be found as

$$R_{AL} = R_{sAL} \frac{L}{n} \left(\frac{1}{W_{opt}} + \frac{1}{W_{DN}} \right) \qquad (3)$$

where R_{sAL} is the sheet resistivity of metal lines and W_{DN} is the width of the DN region. For the process parameters under consideration, the method outlined above yields a W_{opt} of 40 μm. Then for a zero-bias value of 10 pF, and choosing $n = 5$, the required unit cell length is 100 μm. Neglecting the contact resistance and the skin effect, the estimated Q_v at 1 GHz is around 20.

Because of the various assumptions made in modeling this three-dimensional distributed structure, the estimated values above are optimistic and should be used with caution. However, they are useful in estimating an upper limit for Q_v. Note that the substrate capacitance, C_{SUB}, can have a very low quality factor. An ac short circuit between the epi and the substrate terminals will minimize the effect of the substrate junction.

It will be assumed that the external inductor used in the VCO has a quality factor at least equal to that of the varactor diode so that a tank circuit with a quality factor greater than 10 can be designed. An attractive VCO configuration including a gain stage, a capacitive feedback network, and a tank circuit is shown in Fig. 3. Extensive SPICE simulations have been performed with common-base (Colpitts oscillator) and emitter-coupled pair (ECP) amplifiers with an equivalent loading of 250 Ω to 1 kΩ assumed from the tank circuit. Although the Colpitts oscillator can operate up to several gigahertz with a 10-GHz bipolar process, the ECP oscillator functions better below 500 MHz, providing larger negative conductance. There-

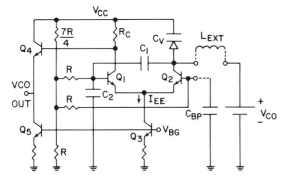

Fig. 4.   Schematic of the varactor-tuned VCO.

fore, the ECP oscillator is a better choice for frequencies from 100 MHz to near 1 GHz.

The complete VCO schematic is shown in Fig. 4. The ECP topology has the additional advantage of an isolated output from the tank circuit besides its high input impedance. The total harmonic distortion is also smaller compared to a single-transistor oscillator since the tanh-nonlinearity of a differential pair has only odd harmonics. The effect of the varactor substrate junction is minimized by connecting the epi side to the supply potential. Positive feedback is provided by the MOS capacitors C_1 and C_2. The worst-case series resistance of the MOS capacitors can be estimated to be less than 5 Ω using an approach similar to that for the varactor. External inductance and capacitance can be used to adjust the center frequency, and an external resistor can be used to obtain the optimum amplitude of oscillation. A clock amplifier follows the VCO to provide a differential drive to the phase detector. The base voltages of the ECP transistors are chosen to be slightly larger than $2V_{BE}$ to keep the current source in the forward-active region. Assuming that the largest control voltage, V_{CON}, available is $V_{CC} - V_{BE}$ and the peak oscillation amplitude is about one V_{BE}, the dynamic range of V_{CON} is limited to $V_{CC} - 4V_{BE}$ to keep the common-base transistor in the forward-active region.

The ac schematic of the VCO including the chip and package parasitics is shown in Fig. 5. The center frequency of the VCO can be approximated as

$$f_o = \frac{1}{2\pi\sqrt{(L_{EXT} + L_{bw} + L_{pl})(C_V + C_x)}} \quad (4)$$

where

$$C_x = \frac{C_1(C_2 + C_p)}{C_1 + C_2 + C_p} + C_{pad} + C_{pintognd} + C_{pintopin} \quad (5)$$

and

$$C_V = \frac{C_0}{\left(1 + \frac{V_R}{V_{bi}}\right)^m}. \quad (6)$$

In the equations above, C_p is the total parasitic capaci-

tance across C_2 including the input capacitance of the gain stage and the substrate capacitance of C_1, and C_0 is the zero-bias value of the varactor capacitance, C_v. The epi side of the capacitor C_1 is connected to C_2 in order not to load the tank circuit with its substrate capacitance. The epi side of the capacitor C_2 is connected to the ground potential. At high frequencies, bonding wire and package lead inductance together with pad and pin capacitance will affect the frequency of operation, and therefore they are also included in (4) and (5).

A built-in potential, V_{bi}, of 0.4 V and a grading coefficient, m, of 0.2 are specified for the process used to fabricate the test circuit. If V_R is between 0.8 and 2.8 V, then C_v varies between $0.66C_0$ and $0.80C_0$. For a minimum value of $C_x = 1.5$ pF, C_0 must be larger than 11 pF so that the tuning range reduction due to C_x is less than 20 percent. The tuning range that can be obtained with these values is -5 and $+3$ percent. The lower limit on C_x depends on the package used and the upper limit on C_0 depends on the practical values of external inductance, hence again on the package, if the chip area is not a crucial factor. For a center frequency of 500 MHz, a total inductance of 11 nH is required if C_0 is 11 pF and V_R is 1.8 V. If the package contributes a minimum pin inductance of 2 nH then the external inductance should be 9 nH. The varactor also has a series inductance of 2 nH due to the supply pin; therefore, its self-resonance frequency is 1.3 GHz. Consequently, it will be useful to use multiple pins for the supply to reduce the parasitic inductance for frequencies above 500 MHz. This will also help to reduce the effect of the substrate junction on the varactor impedance. In the final test circuit design, a C_0 value of 13.5 pF is used. The VCO will have a gain of $-0.05\omega_o$ and $+0.03\omega_o$ rad/(s·V) with this varactor. The tuning range can be increased by 30 percent by shifting the V_R range to between 0.4 and 2.4 V and limiting the oscillation amplitude to less than 400 mV.

Finally, we consider the temperature and supply compensation of the VCO center frequency in conjunction with the loop-amplifier design. Fig. 6 shows the schematic of the loop amplifier and the VCO. A two-stage amplifier with a gain of 50 is designed to achieve a high-gain second-order PLL. Assuming a worst-case VCO gain of $0.03\omega_o$ rad/(s·V) and a worst-case phase-detector gain of 0.1 V/rad, a PLL with this loop amplifier introduces a phase error of less than 4° for a pull-in range of ± 1 percent [7]. For an equal contribution from the offset voltages, the total offset voltage from the loop amplifier and the phase detector must be less than 7 mV. In order to minimize the contribution of the level-shifting diodes and the second stage to the input offset, the first-stage gain is designed to be 20. Large-area devices are used to improve the matching throughout the loop amplifier. The bias current of the first-stage ECP should be kept low to minimize the input offset current and to maximize the input impedance. The bias currents of the second-stage ECP and the emitter followers are higher to achieve a low output impedance. The bandwidth of the loop amplifier is

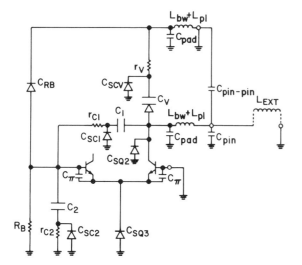

Fig. 5. Alternating-current schematic of the VCO with parasitics.

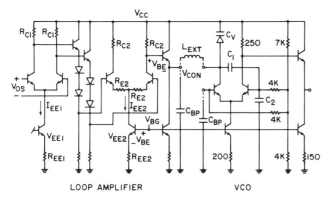

Fig. 6. Loop amplifier and VCO.

not very critical as long as it passes the largest frequency difference between the input and the VCO without much attenuation during capture.

The low-frequency control voltage at the amplifier output can be written as

$$V_{CON} = V_{CC} - V_{BE} - 0.5 R_{C2} I_{EE2}$$
$$- 0.5 \frac{0.5 I_{EE1} R_{C1}}{V_T} \frac{0.5 I_{EE2} R_{C2}}{V_T + 0.5 I_{EE2} R_{E2}} V_{OS} \quad (7)$$

where V_{os} is the sum of the offset voltages from the phase detector output and amplifier input. The reverse bias across the varactor diode is

$$V_R = V_{CC} - V_{CON} \quad (8)$$

and this voltage appears in series with the built-in potential, V_{bi}, of the varactor diode. Therefore,

$$V_{bi} + V_R = V_{bi} + V_{BE} + 0.5 R_{C2} I_{EE2}$$
$$+ 0.125 \frac{I_{EE1} R_{C1}}{V_T} \frac{I_{EE2} R_{C2}}{V_T + 0.5 I_{EE2} R_{E2}} V_{OS}. \quad (9)$$

Let us assume that

$$V_{OS} = \pm 0.2 V_T \quad (10)$$

$$I_{EE1} = \frac{V_{EE1}}{R_{EE1}} \quad (11)$$

and

$$I_{EE2} = \frac{V_{EE2}}{R_{EE2}} \quad (12)$$

where V_{EE1} and V_{EE2} are the required reference voltages. If $0.5 I_{EE2} R_{E2} \gg V_T$, then

$$V_{bi} + V_R \simeq V_{bi} + V_{BE} + 0.5 V_{EE2} \frac{R_{C2}}{R_{EE2}}$$
$$\pm 0.05 V_{EE1} \frac{R_{C1}}{R_{EE1}} \frac{R_{C2}}{R_{E2}}. \quad (13)$$

One possible approach to minimizing the temperature dependence of the expression above is to set

$$0.5 \frac{R_{C2}}{R_{EE2}} \frac{\partial V_{EE2}}{\partial T} = - \frac{\partial (V_{bi} + V_{BE})}{\partial T} \quad (14)$$

and

$$\frac{\partial V_{EE1}}{\partial T} = 0. \quad (15)$$

If

$$V_{EE2} = V_{BG} - V_{BE} \quad (16)$$

and

$$C_v = \left[\frac{K \epsilon_S^\alpha}{V_{bi} + V_R} \right]^m \quad (17)$$

where K, α, and m are constants with temperature, then from (14)

$$\frac{R_{C2}}{R_{EE2}} = 2 \left[\frac{\dfrac{\partial V_{BE}}{\partial T} + \dfrac{\partial V_{bi}}{\partial T} - (V_{bi} + V_R) \alpha \dfrac{\partial \epsilon_S}{\epsilon_S \partial T}}{\dfrac{\partial V_{BE}}{\partial T}} \right]. \quad (18)$$

Consequently, V_{EE1} can be set equal to the bandgap reference voltage V_{BG}, and V_{EE2} can be set equal to the voltage difference between V_{BG} and V_{BE}. In this way, the varactor capacitance, hence the VCO center frequency, can be made independent of supply and temperature to a first-order approximation.

From SPICE simulations, $\partial V_{BE}/\partial T$ and $\partial V_{bi}/\partial T$ are found to be -1.5 and -2.9 mV/°C, respectively. Furthermore, α and $\partial \epsilon_S/\epsilon_S \partial T$ are assumed to be 2 and 200 ppm/°C, respectively [12]. Then from (18), $R_{C2} \simeq 7 R_{EE2}$

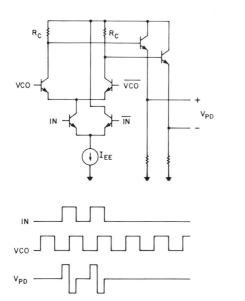

Fig. 7.   Single-balanced modulator (tristate phase detector).

for $V_{bi} = 0.4$ V and $V_R = 2.0$ V. The bandgap reference used in this design is a simple two-cell reference circuit without any second-order correction [13], [14].

III.   PHASE-DETECTOR DESIGN

Analog PLL's may employ either analog or digital phase detectors. In this paper, we will only consider the analog phase detectors which can be implemented either with modulator (mixer) type circuits or combinatorial logic gates. Most common analog phase detectors are of the modulator type. A single-balance modulator-type circuit is shown in Fig. 7. The differential output waveform can have three levels; hence this circuit is also called a tristate phase detector [9]. Whenever there is no input signal, the phase detector output is ideally zero. Therefore, the VCO asymmetry has almost no effect on the tristate phase detector output for the case where there is no input signal.

There are two parameters which define the static performance of a phase detector: its gain and dc offset. First, let us assume that all the input transitions are present; i.e., the PLL input is a deterministic signal. It can easily be shown that the phase-detector gain is

$$K_{PD} = \frac{R_C I_{EE}}{\pi} \tag{19}$$

for a single-balanced modulator assuming large input signals. In a practical circuit, V_{BE} mismatch between the output emitter followers and R_C mismatch between the load resistors contribute to the dc offset. Therefore, large-area devices and wide resistors should be used to minimize these contributions. Bias currents must be kept low enough to minimize the effect of the emitter resistor mismatch in these devices. For large input signals, the dc offset is given

by

$$V_{OS} = \Delta V_{BE} + \frac{I_{EE}\Delta R_C}{4} \tag{20}$$

for a single-balanced modulator.

Now, let us consider the random data input case and find the static phase error introduced by the tristate phase detector using the following equation:

$$\Theta_e = \frac{V_{OS}}{K_{PD}}. \tag{21}$$

If there are no input transitions, the second term in (20) becomes zero. Since the first term is present all the time and can only be corrected when there are transitions in the data, the effective phase-detector gain to be used in (21) is reduced by a factor of 2 (assuming equally likely data) for the first term in (20). Therefore, the static phase error for a single-balanced modulator is

$$\Theta_e = \pi \left[\frac{2\Delta V_{BE}}{I_{EE}R_C} + \frac{\Delta R_C}{4R_C} \right]. \tag{22}$$

The dynamic performance of a phase detector is of interest during capture. There is a minimum pull-in range requirement from a PLL, which is given by

$$\Delta\omega_p \geq K_{AMP}K_{VCO}(V_{OS} + V_{AOS}) \tag{23}$$

where K_{AMP} is the loop-amplifier gain, K_{VCO} is the VCO gain, and V_{AOS} is the acquisition offset voltage, which is a function of the duty cycles of the input and VCO [15]. As the frequency difference goes to zero, the phase-detector output waveform approaches the phase-detector characteristics. It can easily be shown that when one of the inputs has a duty cycle different from 50 percent, an analog phase detector will have a trapezoidal characteristic instead of a triangular one for large input amplitudes. Whether the average of the phase-detector characteristics, V_{AOS}, over a period will still be zero depends on the type of circuit and the duty cycle of the other input. For a tristate phase detector, the average output as the frequency difference goes to zero can be shown to be

$$V_{AOS} = R_C I_{EE} D_{IN}(1 - 2D_{VCO}) \tag{24}$$

where D_{IN} and D_{VCO} are the duty cycles of the input and the VCO, respectively. Hence, it is possible to reduce this acquisition offset voltage by employing narrow input pulses. For comparison, it can be shown that an input signal with a duty cycle less than $1/3$ yields a smaller acquisition offset voltage when a tristate phase detector is used instead of a double-balanced modulator [11].

For random data, D_{IN} is zero when there is no input transition at the PLL input, that is, when a zero bit is received, and V_{AOS} becomes zero for a tristate phase detector from (24). Using (20), (23), and (24) for equally likely random data, one can show that

$$\Delta\omega_P \geq K_{AMP}K_{VCO}\left[\Delta V_{BE} + \frac{\Delta R_C I_{EE}}{8}\right.$$

$$\left. + 0.5 R_C I_{EE} D_{IN}(1 - 2D_{VCO})\right] \quad (25)$$

for a single-balanced modulator. A similar analysis for a double-balanced modulator shows that the tristate phase detector is a better choice if the duty cycle of the equally likely data waveform is less than 2/3 [11]. If the VCO duty cycle is 50 percent, the tristate phase detector always has a better dynamic performance for both deterministic and random input signals.

At low frequencies, it is usually assumed that there is a 90° phase difference between the two phase detector inputs when the PLL is in lock. However, this phase difference is also duty cycle dependent. In other words, the phase difference which produces a zero average output is a function of the input duty cycles. The zero output point can be found by plotting the phase-detector characteristics for arbitrary duty cycles, D_{IN} and D_{VCO}. The final result is given by

$$\Phi_o = \pi D_{IN} \quad (26)$$

for a tristate phase detector. As it can be seen from (26), random variations in the VCO duty cycle do not create random variations in the stable operating phase point of the PLL.

At high frequencies, Φ_o may be different from 90° even for inputs with 50 percent duty cycle, since there is a phase shift introduced by the phase-detector circuit itself. This phase shift depends on the frequency, process, temperature, and supply and can degrade the circuit performance above a few hundred megahertz. The phase-detector bias current, I_{EE}, can be derived from a bandgap reference to reduce the temperature and supply dependence. The bandwidth of the phase detector must be large enough to pass the largest frequency difference during capture and must lie well above the loop bandwidth so that no undesirable phase shift is introduced.

For small-signal PLL inputs, the tristate behaves like a sinusoidal phase detector. Following an analysis similar to that in [16], we find the tristate phase-detector gain for input signal amplitudes less than $2V_T$ to be

$$K_{PD} = \frac{R_C I_{EE}}{\pi} \frac{V_{in}}{2V_T} \quad (27)$$

where V_{in} is the input signal amplitude to the lower differential pair of the tristate phase detector. Under small-signal operating conditions, the circuit can be considered as a cascode-connected amplifier for every half cycle of the VCO waveform. In this case, the amplifier bandwidth is equal to the bandwidth of the phase-detector

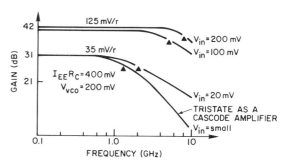

Fig. 8. Tristate phase-detector gain versus frequency.

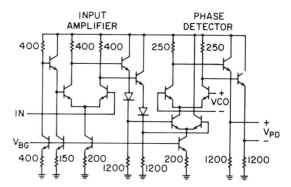

Fig. 9. Input amplifier and tristate phase detector.

gain. Computer simulations show a small-signal bandwidth of 1.3 GHz for typical circuit values, as shown in Fig. 8. Increasing the input amplitude increases the bandwidth of the phase-detector gain, as expected. Fig. 8 also shows three curves corresponding to a tristate bandwidth of 2, 5, and 8 GHz which are obtained from the large-signal simulations with SPICE. Therefore, the phase-detector bandwidth is not a limiting factor if the operating frequency range is below 1 GHz.

The complete circuit including the input amplifier is shown in Fig. 9. The maximum (low-frequency large-signal) phase-detector gain is designed to be 160 mV/rad. For 1-mV V_{BE} mismatch and 1-percent resistor mismatch, V_{OS} is 2.25 mV from (20). When there are no input transitions, V_{OS} reduces to 1 mV. Then, the static phase error for equally likely data is 1.17° from (22). The total offset that has to be overcome during capture is 4.125 mV from (25) assuming D_{IN} is 0.50 and D_{VCO} is 0.49. The input amplifier has a small-signal gain of 12 and a bandwidth of 650 MHz. It helps to improve the phase-detector gain for low-level signals. When considered with large inputs, it behaves like an ECL-NOR gate with complementary outputs to provide sufficient balanced drive to the phase detector.

IV. MEASURED RESULTS

The complete schematic of the PLL circuit is shown in Fig. 10. Although the complete circuit can be housed in a 14-pin small-outline (SO) package with low parasitic values, the test chip was housed in a 40-pin dual-in-line (DIL)

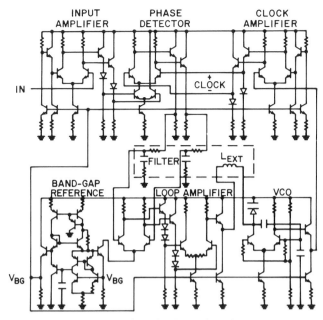

Fig. 10. Complete PLL schematic.

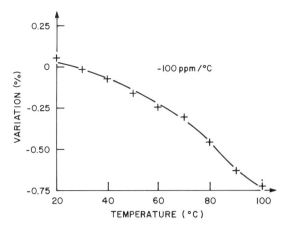

Fig. 11. VCO temperature stability at 193 MHz.

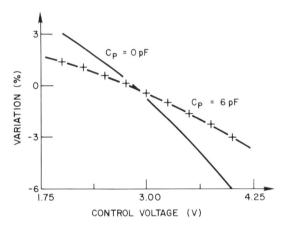

Fig. 12. VCO tuning range at 188 MHz.

package due to the large number of test pads. This degraded the high-frequency performance of the circuit considerably, as expected.

The varactor measurements showed that the devices have reasonably high quality factors but the 40-pin DIL package causes significant Q loss at high frequencies. The measurements also showed that the reverse bias should have a TC of $+5$ mV/°C for the varactor capacitance to stay constant with temperature. This is somewhat larger than the estimated values, suggesting that the parameters m, K, and α are also functions of bias and temperature, especially for complicated structures which cannot be modeled by an abrupt or a linearly graded junction. In general, a design iteration is necessary when an accurate varactor model is not available.

Open-loop measurements on each PLL block were made, beginning with the phase detector whose gain was measured to be 140 mV/rad. This is close to the design value of 160 mV/rad. The measured output offset was 1 mV when both outputs were high. The major contributor to this offset is believed to be the emitter-resistance mismatch of the devices used as the emitter followers. A mismatch of 0.3 Ω at 3.3-mA bias current could produce this offset. The input amplifier preceding the phase detector has a measured gain of 13 at low frequencies.

The loop-amplifier gain was measured as 50 with an input offset voltage of 0.25 mV. The first-stage gain was measured to be 17. This gain is large enough to reduce the input-referred offset coming from the level shifting diodes to 0.2 mV. The nominal value of the output voltage is 2.9 V with a dynamic range of 2.3 V. The control voltage has a measured TC of -3.6 mV/°C. In other words, the reverse bias across the varactor diode will have a TC of $+3.6$ mV/°C. This is 28 percent less than the desired value of $+5$ mV/°C. Therefore, the varactor still has a positive TC which is not compensated.

Fig. 11 shows the temperature stability of the VCO center frequency with the compensation circuitry. As the center frequency was increased to 250 MHz and beyond, the oscillations stopped at temperatures higher than 70°C due to the increased losses in the VCO tank circuit. The actual TC without any fixed parasitic capacitance in the tank circuit is estimated to be around -150 ppm/°C. This is mainly a result of the uncompensated TC of the varactor and the TC of the inductances in the circuit. The measured voltage coefficient values for all samples were better than 0.3 percent/V up to 250 MHz for a supply variation of ± 5 percent at 5 V.

The tuning range of the VCO is a function of the fixed capacitance introduced by the package and the printed-circuit board including the socket. Fig. 12 shows the tuning range at 188 MHz for a total fixed capacitance of 6 pF. The computed ideal case with no parasitics is also plotted for comparison. The fixed capacitance reduces the tuning range to less than $+2$ and -4 percent. The average VCO gain estimated from these values is $0.03\omega_o$ rad/(V·s). The maximum oscillation frequency was measured to be 350 MHz when a silver wire of 5 nH with a Q of 10 was employed as the external inductor. The oscillation frequency is essentially determined by the package inductance at these frequencies.

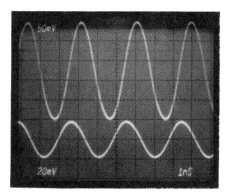

Fig. 13. PLL waveforms at 350 MHz (upper trace: VCO, lower trace: input).

TABLE I
MEASURED PLL CHARACTERISTICS

Supply Voltage	5 V
Power Consumption	270 mW
Maximum Frequency	350 MHz
Pull-in Range (at -23 dBm)	2 % at 350 MHz
DC Loop Gain	> $0.09\omega_o$ rad/sec
TC of Center Frequency	-100 ppm/°C
VC of Center Frequency	0.25 %/V
Total Offset	1.25 mV

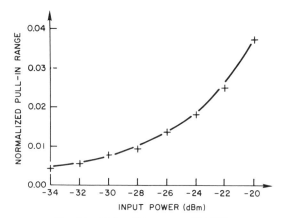

Fig. 14. PLL pull-in range at 350 MHz.

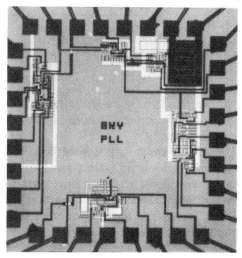

Fig. 15. Die photo of the test chip.

The complete closed-loop PLL was tested using a resistor ratio of 10 and a capacitor value of 10 μF in the loop filter to obtain a pole at 4.8 Hz and a zero at 53 Hz. A 10-nF chip capacitor was used for filtering the ripple which is at the VCO center frequency for a tristate phase detector. Fig. 13 shows the PLL waveforms at 350 MHz where the VCO output amplitude was 125 mV across 50 Ω. The clock amplifier provides more than ± 125 mV into the phase detector. The normalized pull-in range at a center frequency of 350 MHz is plotted as a function of the input signal level in Fig. 14. The pull-in range increases with the input power since the phase-detector gain is proportional to the input level. It is more than ± 1 percent for input levels greater than -23 dBm. In order to check the possibility of injection locking, the measurements were repeated by disabling the phase-detector current source. No injection locking was observed up to an input level of -10 dBm. The static phase error was measured as ± 0.1 nS at 350 MHz. The total pull-in range was 13 MHz at an input level of -20 dBm. This corresponds to a phase error of ± 0.22 rad or $\pm 12.6°$ for a pull-in range of ± 1.9 percent. Therefore, the dc-loop gain can be estimated as $0.09\omega_o$ rad/s under the closed-loop operating conditions. This is about half of what would be expected from a straightforward multiplication of the open-loop gains of each block. The difference is due partly to the 0.1-nS resolution limit of the oscilloscope used and partly to the small-signal assumptions made in defining the gain of each block.

Table I shows a summary of the performance characteristics of the monolithic PLL. The power consumption can be further reduced by lowering the bias current levels for the buffer stages. This will also improve the emitter-follower matching for differential outputs.

The die photo of the test chip is shown in Fig. 15. The active area is 0.5 mm². The minimum pad size and the number of pads determine the total chip area. The varactor diode is located on the upper right corner of the chip and occupies an area of 250×340 μm² with the metal lines. Substrate contacts are placed close to the critical devices in the high-frequency path such as the varactor diode.

V. CONCLUSIONS

Circuit design techniques for realizing high-frequency, low-power PLL's have been discussed. The frequency limitations of the main PLL building blocks have been investigated with a special interest in narrow-band filtering applications.

A varactor-tuned ECP-VCO and a single-balanced modulator have been designed as building blocks. The temperature and supply variations of the on-chip varactor have been compensated by a bandgap reference and a loop amplifier. A 2-μm oxide-isolated bipolar process has been used to fabricate the PLL test chip. The worst-case VCO center-frequency variation is on the order of ± 1 percent over a temperature range from 20°C to 100°C, for ± 1.25 mV total offset from the phase detector and loop amplifier, and for ± 5 percent supply voltage variation at 5 V. The measured pull-in range has been found to be adequate, allowing a small noise-bandwidth design for narrow-band applications ($Q_{PLL} > 100$) [7].

The maximum frequency of operation of the PLL is 350 MHz when housed in a 40-pin DIL package. Since the high-frequency performance is strongly affected by the package, the circuit is expected to achieve better performance with fewer pads and an optimum package. The test circuit dissipates relatively low power and occupies a small area and can therefore be integrated into a larger subsystem.

ACKNOWLEDGMENT

The authors acknowledge the assistance of W. Mack of Signetics Corporation for layout and characterization support and Signetics Corporation for die fabrication.

REFERENCES

[1] R. A. Blauschild and R. G. Meyer, "A low power, 5 V, 150 MHz PLL with improved linearity," in *IEEE ICCE Dig. Tech. Papers*, 1985, pp. 122–123.
[2] K. Kato *et al.*, "A low power dissipation PLL IC operating at 128 MHz clock," in *IEEE CICC Dig. Tech. Papers*, 1987, pp. 651–654.
[3] J. Tani *et al.*, "Parallel interface ICs for 120 Mb/s fiber optic links," in *IEEE ISSCC Dig. Tech. Papers*, 1987, 190–191.
[4] K. Matsumoto *et al.*, "A 700 MHz monolithic phase-locked demodulator," in *IEEE ISSCC Dig. Tech. Papers*, 1985, pp. 22–23.
[5] A. H. Neelen, "An integrated PLL FM demodulator for DSB satellite TV reception," in *IEEE ICCE Dig. Tech. Papers*, 1987, pp. 258–259.
[6] J. F. Ewen and D. L. Rogers, "Single chip, receiver/clock recovery circuit at 375 MHz," in *Optical Fiber Commun. Conf., Tech. Dig.*, 1986, pp. 116–117.
[7] E. Roza and P. W. Millenaar, "An experimental 560 Mbits/s repeater with integrated circuits," *IEEE Trans Commun.*, vol. COM-25, pp. 995–1004, Sept. 1977.
[8] E. Roza "Analysis of phase-locked timing extraction circuits for pulse code transmission," *IEEE Trans. Commun.*, vol. COM-22, pp. 1236–1249, Sept. 1974.
[9] R. R. Cordell *et al.*, "A 50 MHz phase- and frequency-locked loop," *IEEE J. Solid-State Circuits*, vol. SC-14, pp. 1003–1009, Dec. 1979.
[10] S. M. Sze, *Physics of Semiconductor Devices*, 2nd ed. New York: Wiley, 1981, ch. 2.
[11] M. Soyuer, "High-frequency monolithic phase-locked loops," University of California, Berkeley, Memo UCB/ERL M88/10, 1988.
[12] I. Getreu, *Modeling the Bipolar Transistor*, Tektronix, Inc., Beaverton, OR, 1976, ch. 2.
[13] J. F. Kukielka and R. G. Meyer, "A high-frequency temperature stable monolithic VCO," *IEEE J. Solid-State Circuits*, vol. SC-16, pp. 639–647, Dec. 1981.
[14] A. P. Brokaw, "A simple three-terminal bandgap reference," *IEEE J. Solid-State Circuits*, vol. SC-9, pp. 388–393, Dec. 1974.
[15] R. C. Halgren *et al.*, "Improved acquisition in PLLs with sawtooth phase detectors," *IEEE Trans. Commun.*, vol. COM-30, pp. 2364–2375, Oct. 1982.
[16] P. R. Gray and R. G. Meyer, *Analysis and Design of Analog Integrated Circuits*, 2nd ed. New York: Wiley, 1984, ch. 10.

A 6-GHz Integrated Phase-Locked Loop Using AlGaAs/GaAs Heterojunction Bipolar Transistors

Aaron W. Buchwald, *Member, IEEE*, Kenneth W. Martin, *Fellow, IEEE*,
Aaron K. Oki, *Member, IEEE*, and Kevin W. Kobayashi

Abstract—A fully integrated 6-GHz phase-locked-loop (PLL) fabricated using AlGaAs/GaAs heterojunction bipolar transistors (HBT's) is described. The PLL is intended for use in multigigabit-per-second clock recovery circuits for fiber-optic communication systems. The PLL circuit consists of a frequency quadrupling ring voltage-controlled oscillator (VCO), a balanced phase detector, and a lag-lead loop filter. The closed-loop bandwidth is approximately 150 MHz. The tracking range was measured to be greater than 750 MHz at zero steady-state phase error. The nonaided acquisition range is approximately 300 MHz. This circuit is the first monolithic HBT PLL and is the fastest yet reported using a digital output VCO. The minimum emitter area was 3 μm × 10 μm with f_t = 22 GHz and f_{max} = 30 GHz for a bias current of 2 mA. The speed of the PLL can be doubled by using 1-μm × 10-μm emitters in next-generation circuits. The chip occupies a die area of 2 mm × 3 mm and dissipates 800 mW with a supply voltage of −8 V.

I. INTRODUCTION

IN RECENT YEARS, there has been a significant research effort in the area of high-speed electronics for multigigabit-per-second communication. Higher speeds are required in order to take full advantage of the extremely broad-band capabilities of optical fibers. In particular, integrated solutions are sought for practical systems to reduce cost and improve reliability. One of the target bit rates for integrated fiber-optic receivers is 10 Gb/s, which is consistent with the SONET hierarchical specification [1]. Practical transmission systems at these data rates will open the way to unexplored territory in networking. Currently, the bandwidth (1400 GHz · km for 1.3-μm single-mode fibers) and low losses (0.15 dB/km) of optical fiber cannot be fully exploited. A bottleneck in system throughput exists due to speed limitations of the electronics in the receiver and transmitter. Several front-end circuits for receivers such as preamplifiers, postamplifiers, decision circuits, multiplexers, and

Manuscript received April 10, 1992; revised August 3, 1992. This work was supported by the University of California MICRO Program and TRW, Inc. under Grants 90-102 and 91-102.

A. W. Buchwald is with the Integrated Circuits and Systems Laboratory, University of California at Los Angeles, Los Angeles, CA 90024.

K. W. Martin is with the Department of Electrical Engineering, University of Toronto, Toronto, Canada M5S 1A4.

A. K. Oki and K. W. Kobayashi are with TRW Electronics Systems Group, Redondo Beach, CA 90278.

IEEE Log Number 9203913.

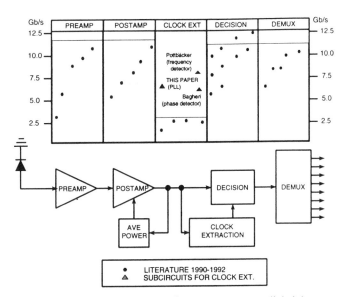

Fig. 1. Fiber-optic receiver status for nonreturn-to-zero digital data.

demultiplexers have been reported [2]–[11]. The maximum data rates of these circuits are illustrated graphically in Fig. 1. Although most of these circuits can process data at rates near 10 Gb/s, with some capable of handling rates greater than 20 Gb/s [12], [13], little has been reported on fully integrated clock extraction circuits above 2 or 3 Gb/s [14]. Clock recovery circuits presently limit the obtainable data rate of multigigabit-per-second integrated fiber-optic receivers. Current practical receivers that include methods for extracting the clock signal are limited to about 2.5 Gb/s, both for systems using a surface-acoustic-wave (SAW) filter for clock extraction [15], [16] and for systems using a phase-locked loop (PLL) [14]. However, various fundamental subcircuits for clock recovery at rates greater than 6 Gb/s have emerged recently [17]–[19]. This paper describes a PLL that may be used as a building block for clock recovery circuits in high-speed fiber-optic receivers to extend the maximum data rate of fully integrated receivers from 2.5 to 6 Gb/s.

II. BACKGROUND

Clock extraction circuits for nonreturn-to-zero (NRZ) data can be grouped into two main categories: open-loop

Reprinted from *IEEE Journal of Solid-State Circuits*, vol. SC-27, pp. 1752-1762, December 1992.

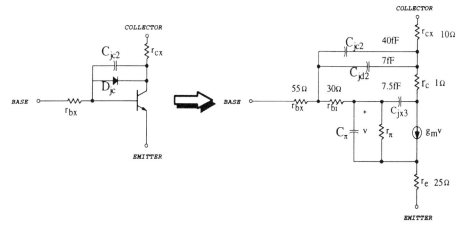

Fig. 2. HBT SPICE subcircuit and small-signal model parameter values.

filters, and closed-loop synchronizers. Previously, open-loop filters have been used almost exclusively in high bit-rate receivers. With this open-loop technique, the periodic timing information is extracted from the data by first using a nonlinear edge-enhancement circuit to generate a spectral line at the bit rate. The enhanced signal is then passed through a narrow-band filter centered at the bit-rate frequency. The filter must be highly selective (high Q) in order to minimize the phase jitter in the recovered clock signal. Typically, SAW filters have been used for this purpose; however, commercially available SAW filters are limited to frequencies less than 3 GHz [20].

The open-loop technique is attractive because it does not suffer from instabilities and nonlinear problems such as frequency acquisition and cycle slipping. However, open-loop systems usually need manual adjustment to center the clock edge in the bit interval. This one-time adjustment will not track phase offsets due to temperature variations and component aging. A SAW filter is also external to the receiver electronics and is bulky, leading to both packaging and interconnect problems.

In contrast with an open-loop filter, a closed-loop system can be integrated and can continually compensate for changes in the environment and the input bit rate. This phase-locking technique requires that a voltage-controlled oscillator (VCO) be tuned by a suitably filtered error signal to align its positive-going transitions to the center of the bit interval. Although the closed-loop feedback system has the desirable property of being self-adjusting, complications due to nonlinear frequency acquisition and tracking render the circuit difficult to design. The PLL described in this paper demonstrates functionality of key elements of a closed-loop clock-recovery circuit. A practical IC for extracting the clock from random NRZ data can be designed using the circuits described herein in conjunction with a nonlinear phase detector, such as those described by Enam and Abidi [21] and Bagheri et al. [18], and a frequency discriminator similar to circuits reported by Pottbäcker et al. [19] and Ransijn and O'Connor [14].

TABLE I
TYPICAL HBT DEVICE PARAMETERS

Device Size	$3 \mu m \times 10 \mu m$
dc Current	2 mA
Current Density	$67 \mu A/\mu m^2$
V_{be}	1.2–1.4 V
C_{je}	45 fF
Beta (β)	25–200
f_t	22 GHz
f_{max}	30 GHz

III. HETEROJUNCTION BIPOLAR TRANSISTORS

In order to achieve the high speed required in multi-gigabit fiber-optic data links, the PLL is fabricated in an AlGaAs/GaAs heterojunction bipolar transistor (HBT) process using 3-μm × 10-μm minimum-emitter-area devices [22]. This conservative minimum device dimension was chosen to maximize yield and ensure functionality in this first-generation PLL. An increase in speed by a factor of 2 can be obtained in second-generation circuits simply by substituting 1-μm × 10-μm devices for the existing transistors.

The key feature of an HBT, resulting in increased speed, is the formation of a heterojunction at the base–emitter interface such that the bandgap energy on the emitter side of the junction is larger than the energy gap on the base side. This energy difference blocks reverse charge-carrier injection from the base to emitter, resulting in near unity emitter injection efficiency independent of the doping levels. The freedom to optimize doping levels for wide-band performance, without suffering a degradation in current gain β, gives HBT's an approximate 2:1 speed advantage over comparable homojunction BJT's. The base doping of an HBT can be increased to lower base resistance and increase f_{max}; simultaneously, the emitter doping can be reduced, lowering the base-emitter junction capacitance C_{je}. Further improvements in speed result from using GaAs as the semiconductor material. High electron mobility reduces base-transit time τ_f, and the semi-insulating substrate reduces the collector–substrate capacitance C_{cs}. Typical device parameters for this process are given in Table I, and the small-signal SPICE model used in simulation is shown in Fig. 2. The process also has Schottky diodes available, as well as thin-film 100-Ω/sq NiCr resistors.

IV. CIRCUIT DESIGN

The PLL circuit is fully integrated, including a VCO, phase detector, and lag-lead loop filter that sets the noise bandwidth to approximately 200 MHz (or 1/30th of the VCO center frequency). An output buffer is also integrated as are three identical bias circuits. Each of the circuits composing the PLL will be described in the following sections.

A. Frequency Quadrupling Ring VCO

A frequency quadrupling ring VCO[1] was designed and fabricated separately from the PLL [23]. This VCO, which is illustrated in Fig. 3, has two quadrature outputs at twice the ring frequency and one output at four times the ring frequency. The core of this VCO is a four-stage ring oscillator. When an even number n of matched delay elements is used, then each pair of taps separated by $n/2$ stages will be 90° out of phase. For example, y_1 and y_3 are quadrature pairs as are y_2 and y_4. When each of these pairs are mixed, the resulting signals I and Q are at twice the ring frequency and are themselves in quadrature. Another level of frequency doubling can also be performed by mixing I and Q to obtain a signal X at four times the ring frequency.

The delay cell of the VCO core is shown in Fig. 4. This circuit uses a differential current steering input (*STEER*) for coarse adjustment of the VCO frequency and a control voltage (V_{CNTR}) of reverse-biased base–emitter junction capacitances for frequency fine tuning. Balanced differential design helps to minimize jitter due to common-mode noise and especially due to power supply coupling, which is a major source of jitter in high-frequency oscillators. Simulation results reveal that a single delay cell, which is terminated with a source resistor of value r_b and a load resistor of $C_\pi/C_\mu g_m$, achieves a delay time of approximately $1/f_{max}$, where f_{max} is the unity-power gain frequency of the transistor given approximately by

$$f_{max} \simeq \frac{1}{2}\sqrt{\frac{f_t}{2\pi C_\mu r_b}}. \tag{1}$$

To ensure oscillations, a gain greater than unity is required, and the load resistor must be increased accordingly. This increases the delay time of the ring oscillator cell, as do the emitter-follower buffer stages inserted before the frequency doubling mixers, resulting in an actual

[1] A patent application for this new VCO has been filed in cooperation with the UCLA Office of Intellectual Property Administration.

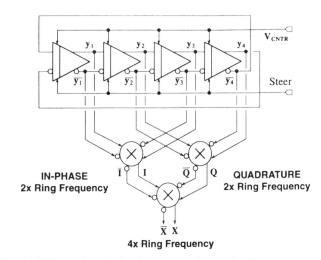

Fig. 3. VCO with I and Q, in-phase and quadrature, double frequency outputs, and a quadrupled frequency output X.

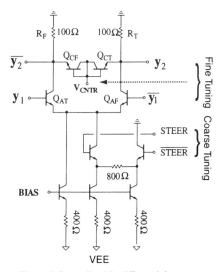

Fig. 4. Ring oscillator delay cell with differential current steering inputs for coarse tuning and a reverse-biased diode for fine tuning.

delay time of the loaded ring oscillator cell of between $1.5/f_{max}$ and $2/f_{max}$, depending on bias conditions. Therefore, the ring frequency f_1 is such that $f_{max}/16 < f_1 < f_{max}/12$, and the 4x signal achieves a maximum frequency in the range $f_{max}/4 < f_4 < f_{max}/3$.

Measured results of the VCO are summarized in Table II. The maximum obtainable frequency is 6.8 GHz. The tuning range is plotted in Fig. 5(a) as a function of the bias current per delay cell and in Fig. 5(b) as a function of the reverse-biased diode voltage. The VCO can be tuned by approximately 1 GHz by altering the bias current and by 500 MHz by modulating the load capacitance diode. A microphotograph of the VCO is shown in Fig. 6.

B. Fully Balanced Mixer

Frequency doubling and phase detection are performed by a fully symmetric circuit with the property of equal

TABLE II
MEASURED RESULTS OF VCO

Maximum Frequency	6.8 GHz
Power Dissipation	300–400 mW
Tuning Range (Steer)	6.25 GHz ± 400 MHz
Gain (Steer)	$2\pi(440\text{ MHz})/\text{mA}$
Tuning Range (V_{CNTR})	±200 MHz
Gain (V_{CNTR})	$2\pi(100\text{ MHz})/\text{V}$
Temperature Coefficient	1 MHz/°C (uncompensated)
Phase Jitter	<1° (rms)
Spectral Content	−100 dBc/Hz @ 100-kHz offset

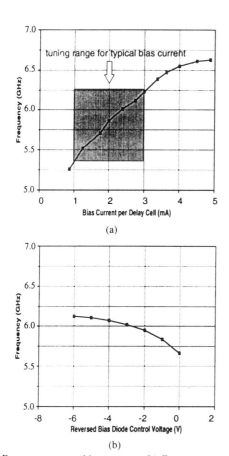

Fig. 5. (a) Frequency versus bias current. (b) Frequency versus control voltage.

delay paths for each input signal [24]. Half of this circuit is a Gilbert multiplier or, equivalently, a current-mode EXCLUSIVE-NOR gate, is shown in Fig. 7. When a single Gilbert multiplier is used as the complete mixer, differences in signal propagation delays between the top- and bottom-level input differential pairs result in an effective phase shift between the two signals being multiplied. This causes a steady-state phase error when the multiplier is used as a phase detector in a PLL, reducing both the tracking and acquisition ranges. This phase lag also gives rise to a dc offset voltage at the output of a frequency doubler when quadrature signals are multiplied. For this particular HBT process, the delay-time difference between a signal applied to the top differential pair and a signal applied to the bottom is on the order of 15 ps. This

corresponds to a phase-lag of 32° at 6 GHz, which is unacceptable.

By modeling the Gilbert multiplier of Fig. 7 as an ideal multiplier with an input phase difference, the circuit of Fig. 8 illustrates how two such mixers can be used in antiparallel to cancel the phase offset. Each mixer is identical, but their inputs are interchanged. Therefore, the resulting phase errors produced by the two mixers will be equal in magnitude but opposite in sign. Summing the result of each mixer, the phase error can be eliminated to the degree of matching accuracy of the two mixers. The fully symmetric circuit of Fig. 9 implements this phase-error compensation by summing the output current of the two Gilbert multipliers at the load resistor.

C. Loop Filter

The loop filter sets the PLL's closed-loop bandwidth as well as its dynamic response. Considerations in designing a loop filter are stability, frequency acquisition range, and phase-jitter suppression. The familiar linearized small-phase-error model of a PLL is shown in Fig. 10, where $F(s) = F_n(s)/F_D(s)$ is the transfer function of a loop filter with unity dc gain. K_d, K_f, and K_o are the gains of the phase detector, loop filter, and VCO, respectively. An FM input signal is also shown with a gain of K_{in}. The closed-loop transfer function of the PLL for a general loop filter is given by

$$H_\theta(s) = \frac{\theta_o(s)}{\theta_{\text{in}}(s)} = \frac{\dfrac{K_dK_fK_oF(s)}{s}}{1 + \dfrac{K_dK_oK_fF(s)}{s}}$$

$$= \frac{K_dK_fK_oF_N(s)}{sF_D(s) + K_dK_fK_oF_N(s)}. \quad (2)$$

Defining a gain $\Omega_k \triangleq K_dK_fK_o$ (rad/s), then for a lag-lead loop filter of the form

$$F(s) = \frac{1 + s\tau_z}{1 + s\tau_p} \quad (3)$$

the resulting closed-loop transfer function is second order and is given by

$$H_\theta(s) = \frac{\Omega_k(1 + s\tau_z)}{s^2\tau_p + s(1 + \Omega_k\tau_z) + \Omega_k}. \quad (4)$$

It is useful to express the loop parameters in terms of the undamped natural frequency $\omega_n = 2\pi f_n$ and the damping ratio ζ.

$$H_\theta(s) = \frac{1 + \dfrac{s}{\omega_n}\left(2\zeta - \dfrac{\omega_n}{\Omega_k}\right)}{1 + \dfrac{s}{\omega_n}2\zeta + \left(\dfrac{s}{\omega_n}\right)^2} \quad (5)$$

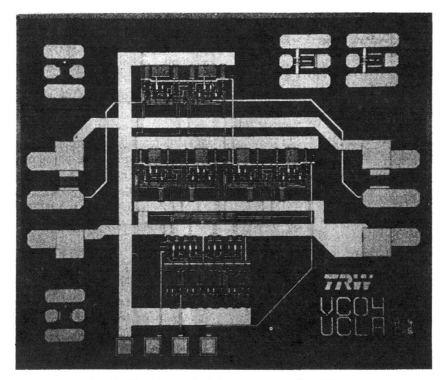

Fig. 6. Microphotograph of frequency quadrupling VCO.

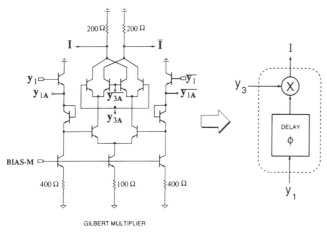

GILBERT MULTIPLIER

Fig. 7. Gilbert multiplier.

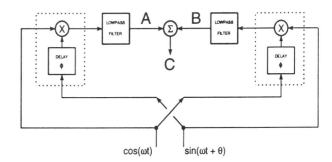

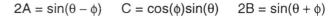

$2A = \sin(\theta - \phi)$ $C = \cos(\phi)\sin(\theta)$ $2B = \sin(\theta + \phi)$

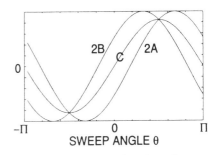

Fig. 8. Technique for compensating phase-lag using two matched Gilbert multipliers.

where

$$\omega_n^2 = \frac{\Omega_k}{\tau_p}$$

$$\zeta = \frac{1}{2}\left[\frac{\omega_n}{\Omega_k} + \omega_n \tau_z\right].$$

The transfer function for FM signals is identical to the phase-modulation transfer function except for a constant term:

$$H_{F_m}(s) = \frac{Fm_{\text{out}}(s)}{Fm_{\text{in}}(s)} = \frac{K_{\text{in}}}{K_o} H_\theta(s). \tag{6}$$

Another important transfer function relates the phase error $\phi(s)$ to the input phase:

$$H_\phi(s) = \frac{\phi(s)}{\theta_{\text{in}}(s)} = \frac{\dfrac{s}{\omega_n}\left(\dfrac{\omega_n}{\Omega_k}\right) + \left(\dfrac{s}{\omega_n}\right)^2}{1 + \dfrac{s}{\omega_n}(2\zeta) + \left(\dfrac{s}{\omega_n}\right)^2}. \tag{7}$$

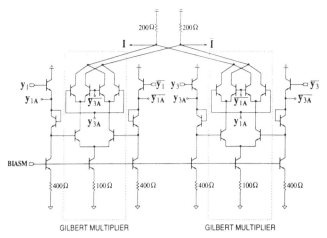

Fig. 9. Fully balanced mixer using two Gilbert multipliers in parallel.

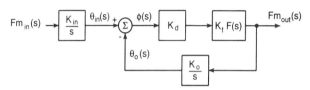

Fig. 10. Linearized small-phase-error model of PLL.

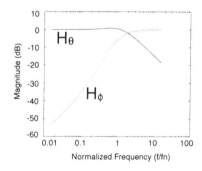

Fig. 11. Magnitude response of PLL closed-loop transfer functions H_θ and H_o.

The magnitudes of $H_\theta(s)$ and $H_\phi(s)$ are plotted in Fig. 11 as a function of the normalized frequency variable for the case of $\zeta = 1$ and $\Omega_k \gg \omega_n$. The loop filter has a limited bandwidth so that the PLL attenuates modulations of the carrier frequency above the undamped natural frequency of the loop f_n. The two transfer functions H_θ and H_ϕ have interesting interpretations regarding phase-jitter filtering. If we assume that the input to the PLL contains phase jitter but that the VCO of the PLL is jitter free, then the VCO output will be modulated by the input phase jitter. However, the original jitter will be filtered by the low-pass function H_θ. Therefore, to reduce the jitter of the PLL VCO, one should reduce the PLL's closed-loop bandwidth. Conversely, if the input signal is assumed to be jitter free and the PLL VCO has significant free-running phase jitter, then the negative feedback of the loop will act to modulate the VCO in such a way as to cancel its own phase jitter. The PLL will be able to track and suppress self-jitter within the loop bandwidth. The re-

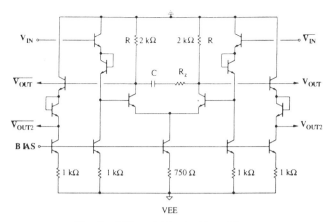

Fig. 12. Differential lag-lead loop filter.

sulting closed-loop VCO jitter will then be the original jitter filtered by the high-pass function of H_ϕ. In this case, jitter is reduced by increasing the loop bandwidth.

The circuit of Fig. 12 approximates a lag-lead characteristic. The small-signal transfer function for this filter, ignoring higher order poles due to parasitics, is given approximately by

$$\frac{\Delta V_{OUT}}{\Delta V_{IN}} = g_m R \left[\frac{1 + sCR_z}{1 + sC(R_z + 2R)} \right] = g_m R \left[\frac{1 + s\tau_z}{1 + s\tau_p} \right] \tag{8}$$

where

$$\tau_z = R_z C = \frac{2\zeta}{\omega_n} - \frac{1}{\Omega_k}$$

$$\tau_p = 2RC + \tau_z = \frac{\Omega_k}{\omega_n^2}.$$

Since the loop filter is integrated with the PLL, the maximum capacitor value is limited by area constraints to about 1 pF. The loop parameters and corresponding filter component values are given in Table III for the design goals of $f_n = 200$ MHz and $\zeta = 1$. The parasitic poles of the loop filter provide additional low-pass filtering of the 12-GHz double-frequency ripple from the output of the phase detector, reducing ripple-induced phase jitter. However, these higher order poles also add excess phase lag, which reduces the loop phase margin, and possibly causes ringing in the transient response. Simulations predict an overshoot in the step response of 5%, corresponding to an equivalent damping factor of $\zeta = 0.7$, which is approximately a two-pole Butterworth response.

D. Output Buffer

The output buffer is shown in Fig. 13. It consists of a pair of emitter-follower buffers, followed by a degenerated differential pair with 50-Ω on-chip load resistors. The nominal bias current is approximately 11 mA, which results in a maximum differential output voltage swing of 550 mV. Since the maximum anticipated differential input signal to the buffer is 2 V, a 300-Ω emitter degeneration

TABLE III
LOOP PARAMETERS AND COMPONENT VALUES

$K_d = 69$ (mV/rad)	$\tau_p = 5.5$ (ns)
$K_f = 25$	$\tau_z = 1.5$ (ns)
$K_o = 2\pi 800$ (Mrad/s/V)	$C = 1.0$ (pF)
$\Omega_k = 8685$ (Mrad/s)	$R = 2.0$ (kΩ)
	$R_z = 1.5$ (kΩ)

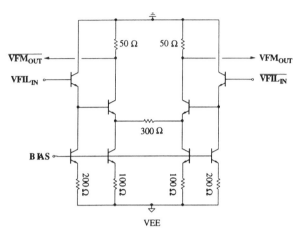

Fig. 13. Emitter degenerated output buffer with 50-Ω on-chip load resistors.

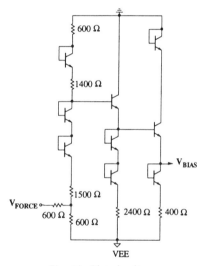

Fig. 14. Bias circuit.

resistor is used to accommodate a differential input signal of up to 3 V.

E. Bias Circuits

Three identical bias circuits are used, one of which is illustrated by Fig. 14. Separate circuits bias the mixers, the VCO-core delay cells, and the output stage. These bias circuits provide a nominal bias voltage of $V_{BE} + 550$ mV when V_{FORCE} is open circuited but can be altered from $V_{BE} + 400$ mV to $V_{BE} + 2.5$ V if V_{FORCE} varies from V_{EE} to GND.

V. RESULTS

A block diagram of the PLL circuit is shown in Fig. 15. A microphotograph of the complete PLL is shown in Fig. 16. To facilitate testing, an identical VCO was fabricated to provide an on-chip signal source. Testing of the chip was accomplished by frequency modulating the input VCO (*STEER*) signal and monitoring the buffered control voltage (FM_{OUT}) of the PLL VCO. These measurements were repeated for different values of V_{CNTR}, which adds stress to the loop by creating an initial frequency offset. The tracking range was measured by starting with the PLL in lock and slowly changing the FM input voltage until a loss of lock occurred. The acquisition range was measured by starting with the loop out of lock and varying the FM input until lock was established. The tracking and acquisition ranges are plotted in Fig. 17(a) for $V_{CNTR} = 0.0$ V. The tracking range for this condition is 750 MHz, and the acquisition range is approximately 300 MHz. Fig. 17(b) shows a plot of the same ranges for $V_{CNTR} = -0.3$ V, which adds a frequency offset and, therefore, a steady-state phase error to the loop. In this case, the tracking range is reduced to about 550 MHz, whereas the acquisition range is slightly less than 300 MHz. Fig. 18 shows a measured FM output waveform of the loop dynamically losing and regaining lock in response to modulation of V_{CNTR} by a 2.4-V peak-to-peak sine wave at 1 kHz. Gardner gives expressions for the maximum frequency deviation from the VCO center, Δf_p, that can be "pulled in" by the self-acquisition of the loop [25]. Expressed in terms of circuit parameters

$$|\Delta f_p| \simeq \frac{\Omega_k}{2\pi} \sqrt{2F(0)F(\infty)} = \frac{\Omega_k}{2\pi} \sqrt{2\tau_z/\tau_p} \qquad (9)$$

and in terms of loop parameters

$$|\Delta f_p| \simeq 2f_n \sqrt{\frac{\zeta \Omega_k}{\omega_n} - 1/2}. \qquad (10)$$

For the loop parameters given in Table III, $|\Delta f_p| \simeq 1.02$ GHz. Although (9) takes into account the sinusoidal phase-detector characteristic, it assumes that Ω_k is constant over the entire acquisition range. In this particular circuit, Ω_k results from a cascade of two differential pairs (the loop filter and the current steering VCO) and therefore has the functional form of a double-nested hyperbolic tangent, which reduces the gain substantially at the extremes of the tuning range. For an interval of 90% of the tuning range, the average gain $\overline{\Omega}_k$ is a factor of 4 less than Ω_k at the center frequency of the VCO. Replacing Ω_k in (9) with $\overline{\Omega}_k$ gives an acquisition range of ± 250 MHz, which is still significantly greater than the measured acquisition range ($|\Delta f_a| \simeq 150$ MHz). This discrepancy is due to offsets and noise in the actual circuit. In the presence of a large-frequency error, the dc value from the phase detector error signal is quite small and must be accumulated in the loop filter over several cycles, building up a voltage that tunes the VCO. Such a small error signal

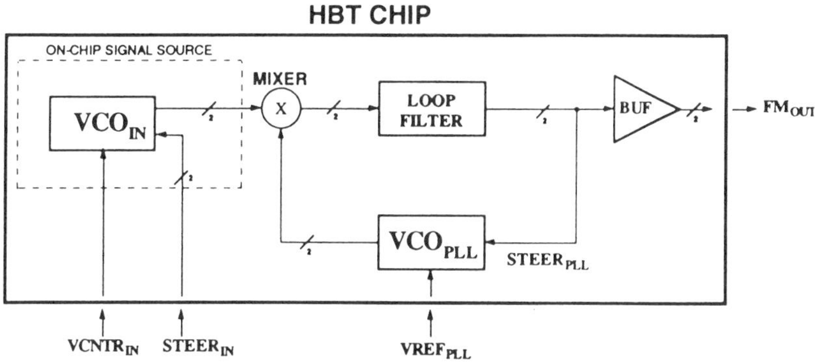

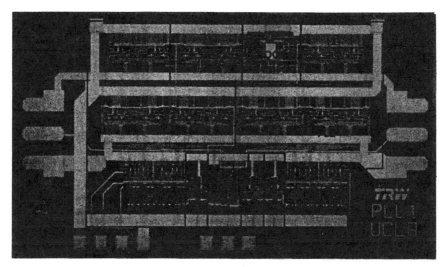

HBT CHIP

Fig. 15. Block diagram of 6-GHz HBT phase-locked loop.

Fig. 16. Microphotograph of 6-GHz HBT phase-locked loop.

is defeated by offsets and noise, and no tuning signal accumulates; as a result, the PLL cannot acquire.

The time required to "pull in" a frequency of Δf is given by

$$T_p(\Delta f) \simeq \frac{1}{2\pi f_n} \frac{1}{2\zeta} \left(\frac{\Delta f}{f_n}\right)^2 \qquad (11)$$

which shows that the acquisition time is proportional to the square of the initial frequency offset. For a frequency error equal to the theoretical limit of the acquisition range, $\Delta f = \Delta f_p$, and after substituting for f_n and ζ

$$T_p(\Delta f_p) \simeq 2\tau_p \left[\frac{\Omega_k \tau_z}{1 + \Omega_k \tau_z}\right]. \qquad (12)$$

For the usual case of $\Omega_k \tau_z \gg 1$

$$T_p(\Delta f) \simeq 2\tau_p \left(\frac{\Delta f}{\Delta f_p}\right)^2. \qquad (13)$$

This expression shows that the acquisition time depends only on the initial frequency error and the time constant of the dominant pole of the loop filter. For $\Delta f = 150$ MHz, $T_p = 0.25$ ns. However, (13) is not valid for small

frequency errors lying within the locking range of the PLL or for frequencies close to the edge of the acquisition range, as Fig. 19 illustrates. This plot shows the simulation results of frequency acquisition for Δf slightly less than Δf_p. For this case, the acquisition time is 60 ns, which is more than a factor of 10 greater than that predicted by (13). The phase-plane portrait for this simulation is shown in Fig. 20, where it can be seen that the loop settles to a steady-state phase offset of 32°, which is an artifact of the finite dc gain Ω_k.

$$\theta_{\text{steady-state}} = \frac{2\pi \Delta f}{\Omega_k}. \qquad (14)$$

The linear-tracking behavior and noise bandwidth can be determined by using small-signal modulations around the locking point. The measured closed-loop bandwidth varied from 100 to 200 MHz, depending on the steady-state phase error, with ζ ranging from 0.5 to 1.0. The change in closed-loop bandwidth is due to the compression nonlinearities mentioned previously. Loop gain is reduced in the presence of a steady-state phase error by the sinusoidal phase detector, the differential loop filter, and the current steering VCO control. In addition, there is

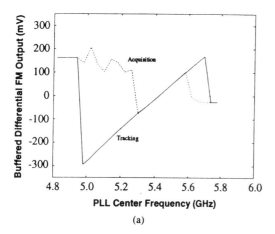

(a)

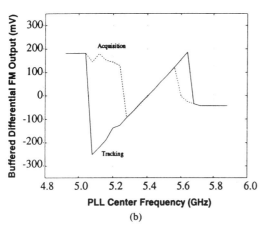

(b)

Fig. 17. DC tracking and acquisition ranges for V_{CNTR} = (a) 0.0 and (b) −0.3 V.

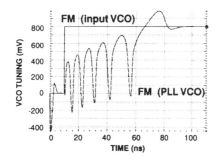

Fig. 18. Measured FM output showing PLL dynamically losing and reacquiring lock in response to a 2.4-V_{p-p} 1-kHz sine-wave modulation of V_{CNTR}.

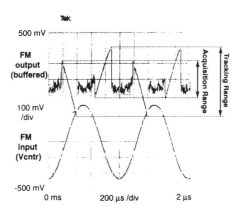

Fig. 19. Cycle-slipping behavior during frequency acquisition of the PLL simulated using SPICE.

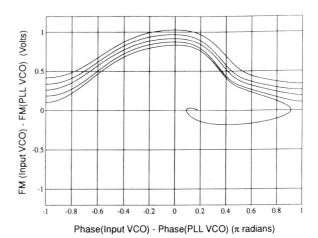

Fig. 20. Phase-plane portrait of the cycle-slipping behavior during frequency acquisition of the PLL simulated using SPICE.

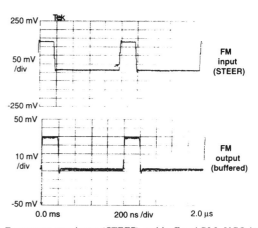

Fig. 21. Frequency step input (*STEER*) and buffered PLL VCO input signal (FM).

TABLE IV
SUMMARY OF MEASURED PLL RESULTS

Transistor Count	300
Die Area	2×3 mm
Supply Voltage	−8 V
Power Dissipation	800 mW
Maximum Center Frequency	6.8 GHz
Closed-Loop Bandwidth	100–200 MHz
Tracking Range	700 MHz
Acquisition Range	300 MHz
Acquisition Time Δf = 150 MHz	0.25 ns*

*Simulated

some amplitude modulation of the VCO with frequency, which also reduces the loop gain. Fig. 21 shows the PLL's pulse response for a 175-mV, 200-ns pulse to the positive current-steering FM input.

VI. SUMMARY

The measured results of the PLL are summarized in Table IV. A fully integrated PLL has been fabricated using AlGaAs/GaAs HBT's. The PLL is the fastest yet reported with a digital output VCO and is the first monolithic HBT PLL. The chip contains over 300 transistors.

A doubling of the speed of this PLL can be obtained in second-generation circuits by substituting 1-μm $\times$ 10-μm devices for the 3-μm $\times$ 10-μm minimum emitter-area transistors used. This PLL is a fundamental building block for multigigabit-per-second clock recovery circuits for use in fiber-optic communication systems.

References

[1] *Synchronous Optical Network (SONET) Transport Systems: Common Generic Criteria.* Morristown, NJ: Bellcore, Sept. 1990, 6th ed.

[2] M. Soda *et al.*, "A Si bipolar chip set for 10 Gb/s optical receiver," in *ISSCC Dig. Tech. Papers* (San Francisco, CA), Feb. 1992, pp. 100-101.

[3] H.-M. Rein, "Silicon bipolar integrated circuits for multigigabit-per-second lightwave communications," *J. Lightwave Technol.*, vol. 8, pp. 1371-1378, Sept. 1990.

[4] K. Runge *et al.*, "Silicon bipolar integrated circuits for multi-Gb/s optical communication systems," *IEEE J. Select. Areas Commun.*, vol. 9, pp. 636-644, June 1991.

[5] H. Hamano *et al.*, "High-speed Si-bipolar IC design for mulit-Gb/s optical receivers," *IEEE J. Select. Areas Commun.*, vol. 9, pp. 645-651, June 1991.

[6] J. N. Albers and H.-U. Schreiber, "A Si-bipolar technology for optical fiber transmission rates above 10 Gb/s," *IEEE J. Select. Areas Commun.*, vol. 9, pp. 652-655, June 1991.

[7] K. Hagimoto *et al.*, "Over 10 Gb/s regenerators using monolithic IC's for lightwave communication systems," *IEEE J. Select. Areas Commun.*, vol. 9, pp. 673-682, June 1991.

[8] K. Nakagawa and K. Iwashita, "High-speed optical transmission systems using advanced monolithic IC technologies," *IEEE J. Select. Areas Commun.*, vol. 9, pp. 683-688, June 1991.

[9] R. K. Montgomery *et al.*, "10 Gbit/s high sensitivity low error rate decision circuit implemented with C-doped AlGaAs/GaAs HBT's," *Electron. Lett.*, vol. 27, pp. 976-978, May 1991.

[10] J. Akagi *et al.*, "AlGaAs/GaAs HBT receiver ICs for a 10 Gbps optical communication system," in *Proc. IEEE GaAs IC Sym.* (New Orleans, LA), Oct. 1990, pp. 45-48.

[11] M. Tamamura *et al.*, "A 9.5 Gb/s Si-bipolar ECL array," in *ISSCC Dig. Tech. Papers* (San Francisco, CA), Feb. 1992, pp. 54-55.

[12] H.-M. Rein, J. Hauenschild, W. McFarland, and D. Pettengill, "23 Gbit/s Si bipolar decision circuit consisting of 24 Gbit/s MUX and DEMUX ICs," *Electron. Lett.*, vol. 27, pp. 974-976, May 1991.

[13] J. Hauenschild, H.-M. Rein, W. McFarland, J. Doernberg, and D. Pettengill, "Demonstration of retiming capability of silicon bipolar time-division multiplexor operating to 24 Gbit/s," *Electron. Lett.*, vol. 27, pp. 978-979, May 1991.

[14] H. Ransijn and P. O'Connor, "A 2.5 Gb/s GaAs clock and data regenerator IC," in *Proc. IEEE GaAs IC Symp.* (New Orleans, LA), Oct. 1990, pp. 57-60.

[15] B. Wedding, D. Schlump, E. Schlag, W. Pöhlmann, and B. Franz, "2.24-Gbit/s 151-km optical transmission system using high-speed integrated silicon circuits," *IEEE J. Select. Areas Commun.*, vol. 8, pp. 227-234, Feb. 1990.

[16] E. Schlag, B. Franz, and W. Pöhlmann, "Integrierte Si-bipolar schaltungen für ein optisches übertragungssystem von 2.4 Gbit/s," in *Proc. ITG Fachtagung Mikroelektronik für die Informationstechnik* (Stuttgart, Germany), Oct. 1989, pp. 221-226.

[17] A. W. Buchwald, K. W. Martin, A. K. Oki, and K. W. Kobayashi, "A 6GHz integrated phase-locked loop using AlGaAs/GaAs heterojunction bipolar transistors," in *ISSCC Dig. Tech. Papers* (San Francisco, CA), Feb. 1992, pp. 98-99.

[18] M. Bagheri *et al.*, "11.6 GHz 1:4 demultiplexer with bit-rotation control and 6.1 GHz auto-latching phase-aligner ICs," in *ISSCC Dig. Tech. Papers* (San Francisco, CA), Feb. 1992, pp. 94-95.

[19] A. Pottbäcker, U. Langmann, and H.-U. Schreiber, "A 8 Gb/s Si bipolar phase and frequency detector IC for clock extraction," in *ISSCC Dig. Tech. Papers* (San Francisco, CA), Feb. 1992, pp. 162-163.

[20] Z. Wang, U. Langmann, and B. Bosch, "Mulit-Gb/s silicon bipolar clock recovery IC optical receivers," *IEEE J. Select. Areas Commun.*, vol. 9, pp. 656-663, June 1991.

[21] S. K. Enam and A. A. Abidi, "Mos decision and clock-recovery circuits for Gb/s optical-fiber receivers," in *ISSCC Dig. Tech. Papers* (San Francisco, CA), Feb. 1992, pp. 96-97.

[22] M. E. Kim, A. K. Oki, G. M. Gorman, D. K. Umemoto, and J. B. Camou, "GaAs heterojunction bipolar transistor device and IC technology for high-performance analog and microwave applications," *IEEE Trans. Microwave Theory Tech.*, vol. 37, pp. 1286-1303, Sept. 1989.

[23] A. W. Buchwald and K. W. Martin, "A high-speed voltage-controlled oscillator with quadrature outputs," *Electron. Lett.*, vol. 27, pp. 309-310, Feb. 1991.

[24] L. Schmidt and H.-M. Rein, "New high-speed bipolar XOR gate with absolutely symmetrical circuit configuration," *Electron. Lett.*, vol. 26, pp. 430-431, 1990.

[25] F. Gardner, *Phaselock Techniques*, 2nd ed. New York: Wiley, 1979, ch. 5.

A 6 GHz 60 mW BiCMOS Phase-Locked Loop

Behzad Razavi, *Member, IEEE*, and JanMye James Sung

Abstract— The design of a 6 GHz fully monolithic phase-locked loop fabricated in a 1 μm, 20 GHz BiCMOS technology is described. The circuit incorporates a voltage-controlled oscillator that senses and combines the transitions in a ring oscillator to achieve a period equal to two ECL gate delays. A mixer topology is also used that exhibits full symmetry with respect to its inputs and operates with supply voltages as low as 1.5 V. Dissipating 60 mW from a 2 V supply, the circuit has a tracking range of 300 MHz, an rms jitter of 3.1 ps, and phase noise of -75 dBc/Hz at 1 kHz offset.

I. INTRODUCTION

HIGH-SPEED, low-power phase-locked loops (PLL's) find wide application in optical data links, ATM systems, and frequency synthesizers. In the multigigahertz range, most of PLL's and clock recovery circuits have been implemented in III-V technologies [1], [2], with silicon designs appearing only recently [3]. From system integration standpoint, it is desirable to design such circuits in mainstream VLSI technologies so that subsequent data processing can be performed on the same chip without incurring great power or yield penalty.

This paper describes the design of a 6 GHz BiCMOS phase-locked loop [4], the fastest reported in silicon technology. Fabricated in a 1 μm, 20 GHz process, the circuit requires no external components and dissipates 60 mW, a factor of 13 less than its counterpart in a AlGaAs/GaAs heterojunction bipolar technology [1]. Pushing the speed-power envelope of the process, the PLL employs a number of techniques to allow operation from supply voltages as low as 2 V.

The next section of the paper presents the PLL architecture and design issues. The building blocks are then described in sequence, starting with the voltage-controlled oscillator in Section III, the mixer in Section IV, and the pulse shaping circuit in Section V. Experimental results are summarized in Section VI.

II. PLL ARCHITECTURE

Fig. 1 shows the architecture of the phase-locked loop. It consists of a pulse-shaping circuit at the front end and a loop comprising a phase detector (PD), a low-pass filter (LPF), an error amplifier A_L, and a voltage-controlled oscillator (VCO). The VCO provides the main output in the form of current, which flows through external 50 Ω termination resistors. The control voltage of the VCO is also monitored and amplified by A_M so as to obtain a demodulated output when the input is frequency modulated.

Manuscript received July 5, 1994; revised July 30, 1994.
The authors are with AT&T Bell Laboratories, Holmdel, NJ 07733 USA.
IEEE Log Number 9406263.

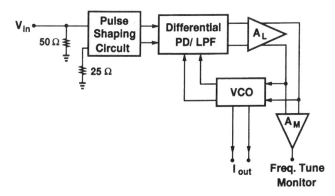

Fig. 1. PLL architecture.

The PLL utilizes differential signals in both the high-frequency path and the control of the VCO to suppress the effect of common-mode noise and also allow a robust design with low supply voltages.

The pulse shaping circuit serves two purposes. First, it converts the single-ended input to a differential signal having an amplitude equal to that of the VCO output. Second, it presents a driving impedance to the PD that is identical with the output impedance of the VCO. These precautions are necessary so as to lower the static phase error because, at 6 GHz, the mixer operates in small-signal regime and hence its phase error is sensitive to both the amplitude and the shape of its two inputs.

III. VOLTAGE-CONTROLLED OSCILLATOR

Several critical parameters of the PLL, such as speed, timing jitter, spectral purity, and power dissipation, strongly depend on the performance of the VCO.

Fig. 2 illustrates two conventional approaches to building monolithic oscillators. In the emitter-coupled oscillator of Fig. 2(a), the maximum frequency depends on the minimum value of capacitor C_1. As the value of this capacitor decreases and the oscillation frequency increases, the loop gain drops, the oscillation amplitude diminishes, and the circuit eventually fails to oscillate. Simulations indicate that this configuration or its variants do not attain a frequency of 6 GHz in our 20 GHz BiCMOS process.

In the ring oscillator of Fig. 2(b), the period of oscillation is given by twice the number of gate delays. However, it is difficult to ensure reliable oscillations if the number of stages is less than three because the total phase shift around the loop will not be sufficient to allow complete switching in each stage. Even regardless of these issues, a two-stage ring oscillator does not yield a frequency of 6 GHz in the process used here.

Reprinted from *IEEE Journal of Solid-State Circuits*, vol. SC-29, pp. 1560-1565, December 1994.

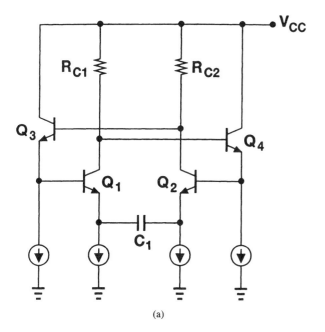

(a)

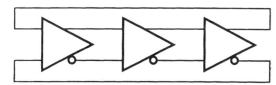

$$f = \frac{1}{6T_d}$$

(b)

Fig. 2. Conventional oscillator topologies. (a) Emitter-coupled oscillator; (b) ring oscillator.

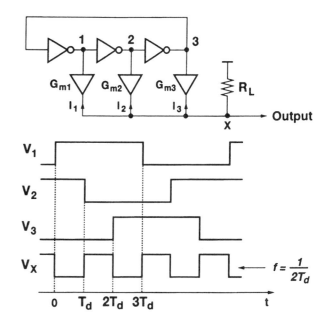

Fig. 3. Conceptual diagram of proposed VCO.

The VCO topology employed in this work senses and combines the transitions in consecutive stages of a ring oscillator so as to achieve a period equal to two gate delays. Fig. 3 shows a conceptual diagram of this technique. The circuit comprises a three-stage ring oscillator and transconductance amplifiers G_{m1}-G_{m3}, which sense the voltages at ports 1–3, respectively, and convert these voltages to current. The resulting currents are summed at node X and converted to voltage by R_L.

The circuit operates as follows: When the voltage at port 1 goes high (at $t = 0$), G_{m1} turns on and draws current from R_L, making V_X go low. After one gate delay, at $t = T_d$, the voltage at port 2 goes low and G_{m2} turns off, allowing V_X to go high. Similarly, after another gate delay, at $t = 2T_d$, the voltage at port 3 goes high and G_{m3} turns on, pulling V_X low again. Thus, for every transition at each port of the ring oscillator, one transition is generated at node X, thereby yielding an output period of $2T_d$ (as in a "one-stage ring oscillator"). Since the output frequency is independent of the number of inverters in the ring, three or more stages can be used to ensure sufficient phase shift around the loop.

To gain more insight into the circuit's operation, we note that if v_1, v_2, and v_3 were pure sinusoids and the G_m stages perfectly linear, the output frequency would *not* be three

times that of the ring, because frequency multiplication is not possible in a linear system. In fact, assuming

$$v_1 = A \sin \omega t \tag{1}$$
$$v_2 = A \sin(\omega t + 2\pi/3) \tag{2}$$
$$v_3 = A \sin(\omega t + 4\pi/3) \tag{3}$$

and $G_{m1} = G_{m2} = G_{m3} = G_m$, we have

$$v_X = $$
$$G_m R_L A[\sin \omega t + \sin(\omega t + 2\pi/3) + \sin(\omega t + 4\pi/3)] \tag{4}$$
$$= 0. \tag{5}$$

Thus, this technique requires nonlinear operation in the ring amplifiers or the G_m stages. More specifically, if the ring amplifiers are nonlinear differential circuits, their outputs contain odd-order harmonics, e.g.,

$$v_1 = A \sin \omega t + B \sin(3\omega t + \theta) \tag{6}$$
$$v_2 = A \sin(\omega t + 2\pi/3) + B \sin(3\omega t + \theta + 2\pi) \tag{7}$$
$$v_3 = A \sin(\omega t + 4\pi/3) + B \sin(3\omega t + \theta + 4\pi), \tag{8}$$

thereby giving

$$v_X = 3G_m R_L B \sin(3\omega t + \theta). \tag{9}$$

Similarly, if v_1, v_2, and v_3 are pure sinusoids but the G_m stages introduce nonlinearity, the fundamental is still suppressed while the third harmonic is enhanced. In essence, the G_m stages "sift" the third harmonic.

The VCO topology of Fig. 2 entails three design issues. First, the capacitive loading of the G_m stages tends to slow down the ring oscillator, making this technique more efficient in bipolar technology, where loading has less effect on the delay, than in CMOS. Second, the number of stages in the ring must be odd so that multiple levels do not occur in V_X. Third, any mismatch in the delay of these stages translates into

321

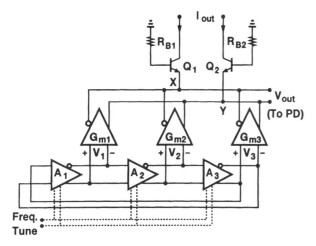

Fig. 4. VCO block diagram.

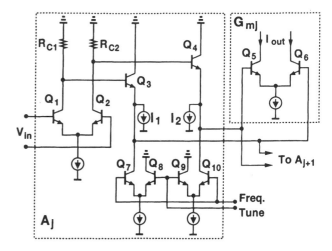

Fig. 5. Implementation of voltage and transconductance amplifiers.

jitter at the output. Thus, it is important to equalize the wiring capacitance seen at the output of each stage.

The actual implementation of the VCO is depicted in Fig. 4. Three differential voltage amplifiers A_1-A_3 constitute the main ring and their outputs are sensed by three differential transconductance amplifiers G_{m1}-G_{m3}. The frequency is tuned by varying the delay of each stage in the ring.

In combining the output currents of the G_m stages, the resistive load of Fig. 3 is replaced with common-base devices Q_1 and Q_2. This is because the capacitance at nodes X and Y is quite significant: it includes the capacitance seen at the output of each G_m stage and the input capacitance of the phase detector. With simple resistors connected to the summing nodes, the resulting time constant substantially attenuates the amplitude of the 6 GHz signal. The common-base devices, on the other hand, introduce an *inductive* component at nodes X and Y that is approximately equal to $(R_B + r_b)\tau_F$, where $R_B = R_{B1} = R_{B2}$, r_b is the base resistance, and τ_F is the base transit time. The resulting inductive peaking enhances the amplitude by approximately a factor of three. The value of R_{B1} and R_{B2} can vary by a factor of two with little effect on the output amplitude.

Fig. 5 depicts the circuit details of one stage of the VCO. Each of the amplifiers A_1-A_3 is implemented as a differential pair Q_1-Q_2 and two emitter followers Q_3-Q_4. Each G_m block simply consists of a current-steering pair Q_5-Q_6. Differential pairs Q_7-Q_8 and Q_9-Q_{10} adjust the bias current of the emitter followers to fine-tune the frequency of oscillation. Current sources I_1 and I_2 are used to avoid starving Q_3 and Q_4 during loop transients as well as provide a means for coarse frequency adjustment.

An important concern in the VCO design has been its low-voltage operation. Fig. 6 shows a section of the VCO along with voltage drops whose sum determines the minimum supply voltage. NMOS current sources prove useful here because, unlike their bipolar counterparts, they have negligible impact on speed even with voltage headrooms of a few hundred millivolts. Note that when the differential pair in Figure 6 switches, the common emitter node momentarily drops by about 150 mV, leaving only 250 mV across the current source and therefore precluding the use of bipolar devices here. In

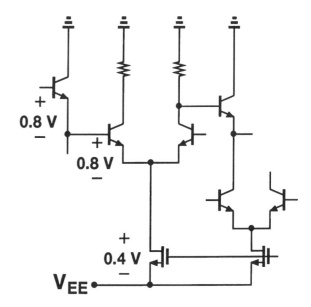

Fig. 6. Section of VCO circuit.

this design, the NMOS transistors are slightly in the triode region, but their current can still be controlled to compensate for temperature and supply variations.

IV. PHASE DETECTOR

Another critical building block of the PLL is the phase detector for it must mix two 6 GHz signals with reasonable gain and power dissipation. While the Gilbert cell is often utilized for mixing, it suffers from two drawbacks in this application. First, it employs stacked differential pairs, thus requiring a large voltage headroom. Second, it introduces substantial phase error at high frequencies because its input signals propagate through inherently different paths.

Fig. 7 shows a half-circuit equivalent of the phase detector/low-pass filter. The PD incorporates an exclusive OR gate comprising Q_1-Q_6 [5]. Since Q_3 is on only if both A and B are low, $I_{C3} \equiv \overline{A} \cdot \overline{B}$, where I_{C3} denotes the logical value of the Q_3 collector current. Similarly, $I_{C4} \equiv A \cdot B$. Therefore, the logical output is equal to $A \oplus B$.

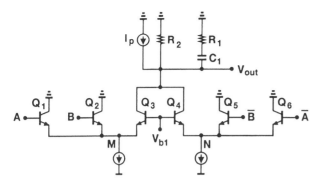

Fig. 7. Phase detector and low-pass filter.

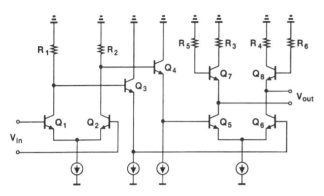

Fig. 8. Pulse shaping circuit.

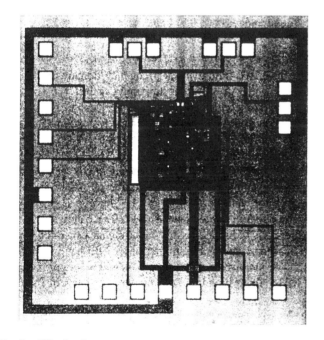

Fig. 9. PLL die photo.

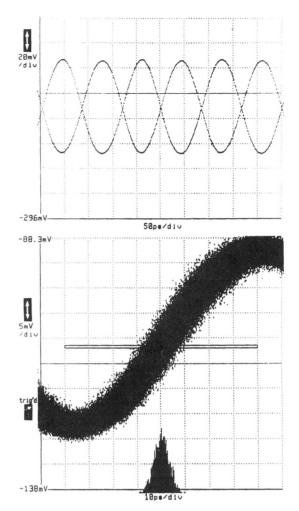

Fig. 10. Measured PLL output in time domain.

mode level of the inputs. Thus, V_{b1} is generated using a replica of the VCO output stage (i.e., common-base devices in Fig. 4).

The output current of the PD is directly low-pass filtered using the lead-network R_1, R_2, and C_1. Current source I_p is approximately equal to $0.75(I_{C3} + I_{C4})$, where I_{C3} and I_{C4} denote the collector bias currents of Q_3 and Q_4, respectively. This allows a larger R_2 and hence a higher gain for a given $I_{C3} + I_{C4}$. Note that M and N are the only high-speed nodes in this circuit.

The actual PD/LPF circuit utilizes two of the half circuits shown in Fig. 7, with A and $\overline{A}$ interchanged in one of the half circuits to generate fully differential outputs [5].

V. PULSE SHAPING CIRCUIT

The pulse-shaping circuit is shown in Fig. 8. This circuit provides a signal path identical to one stage of the VCO (namely, A_j and G_{mj} in Fig. 5) as well as a differential load identical to the impedances seen at nodes X and Y in Fig. 4. More specifically, in the pulse shaping circuit, Q_1-Q_6 replicate the role of Q_1-Q_6 in Fig. 5, whereas Q_7-Q_8 and R_3-R_6 emulate the load devices Q_1-Q_2 and R_{B1}-R_{B2} in Fig. 4. The devices and currents are scaled such that the output impedance is the same as that of the VCO.

In contrast with the conventional ECL XOR, the topology of Fig. 7 has two advantages: 1) it avoids stacked transistors and hence operates from a lower supply voltage, and 2) it is inherently symmetric with respect to inputs A and B, thereby providing equal phase shift for these signals and thus zero static phase error. Nevertheless, the value of V_{b1} should be accurately defined and controlled so that it tracks the common-

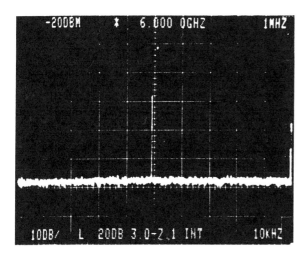

Horiz. 1 MHz/div
Vert. 10 dB/div

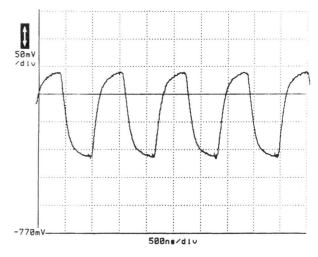

Fig. 12. Demodulated output for $\Delta f = \pm 10$ MHz at input.

TABLE I
PLL CHARACTERISTICS

Center Frequency	6 GHz
Tracking Range	300 MHz
Jitter	3.1 psec rms
Phase Noise	−75 dBc/Hz @ 1 kHz Offset
Power Dissipation	60 mW
Supply Voltage	2 V
Technology	20 GHz, 1-μm BiCMOS

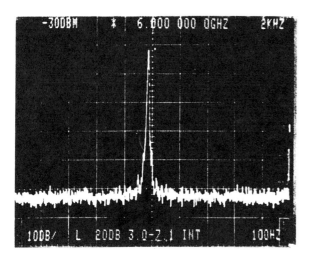

Horiz. 2 kHz/div
Vert. 10 dB/div

Fig. 11. PLL output spectrum.

VI. EXPERIMENTAL RESULTS

The phase-locked loop has been fabricated in a 1 μm, 20 GHz BiCMOS technology [6]. Shown in Fig. 9 is a photograph of the chip, whose active area measures approximately 500 μm × 500 μm. The circuit has been tested on wafer while running from a 2 V supply. High-speed Picoprobes from GGB Industries are used to apply the input and measure the output, while multicontact Cascade probes from Cascade Microtech provide power, bias, and ground connections. A ground ring on the chip establishes a low-inductance connection among the grounds of all the probes.

Fig. 10 shows the measured differential output and jitter histogram of the PLL. The circuit has a jitter of 3.1 ps rms and 30 ps peak-to-peak. The tracking range is 300 MHz and the center frequency can be varied by 700 MHz.

The measured output in the frequency domain is depicted in Fig. 11 with two different horizontal scales. The spectrum exhibits no coherent sidebands, and the center spectral line drops by 55 dB at 1 kHz offset when the resolution bandwidth is set to 100 Hz. This gives a phase noise of −75 dBc/Hz at 1 kHz offset.

The response of the PLL to a frequency-modulated input has also been examined. Fig. 12 depicts the measured frequency tune monitor voltage (output of A_M in Fig. 1) for a ±10 MHz modulation centered at 6 GHz. The bandwidth of this measurement is limited by the input signal generator, an HP 8341B, whose FM input amplifier has a 3 dB bandwidth of 10 MHz. The simulated closed-loop bandwidth is approximately 200 MHz. Table I summarizes the characteristics of the PLL

VII. CONCLUSION

High-speed, low-power circuit techniques make it possible to design high-performance phase-locked loops and clock recovery circuits in VLSI technologies. A PLL fabricated in a 1 μm, 20 GHz BiCMOS process has been presented that incorporates new VCO and mixer topologies. The circuit is suited to demodulation and frequency synthesis applications and can also be used in clock recovery with the addition of a frequency detector.

A VCO configuration has been introduced that achieves an oscillation period of two ECL gate delays by sensing and combining the transitions in a ring oscillator. Since the period is independent of the number of stages, the oscillator can be optimized for complete switching. A low-voltage exclusive OR gate has also been employed that, by virtue of its full symmetry, is free from systematic phase error. Using such techniques, the PLL operates at 6 GHz while dissipating 60 mW from a 2 V supply. It exhibits an rms jitter of 3.1 ps and phase noise of -75 dBc/Hz at 1 kHz offset.

ACKNOWLEDGMENT

The authors wish to thank R. G. Swartz for valuable comments and M. Tarsia for layout support.

REFERENCES

[1] A. Buchwald *et al.*, "A 6-GHz integrated phase-locked loop using AlGaAs/GaAs heterojunction bipolar transistors," *IEEE J. Solid-State Circ.*, vol. 27, pp. 1752–1762, Dec. 1992.

[2] H. Ransijn and P. O'Conner, "A PLL-based 2.5-Gb/s GaAs clock and data recovery IC," *IEEE J. Solid-State Circ.*, vol. 26, pp. 1345–1353, Oct. 1991.

[3] M. Soyuer, "A monolithic 2.3-Gb/s 100-mW clock and data recovery circuit in silicon bipolar technology," *IEEE J. Solid-State Circ.*, vol. 28, pp. 1310–1313, Dec. 1993.

[4] B. Razavi and J. Sung, "A 6-GHz 60-mW BiCMOS phase-locked loop with 2-V supply," *ISSCC Tech. Dig.*,, pp. 114–115, Feb. 1994.

[5] B. Razavi, Y. Ota, and R. G. Swartz, "Design techniques for low-voltage high-speed digital bipolar circuits," *IEEE J. Solid-State Circ.*, vol. 29, pp. 332–333, Mar. 1994.

[6] J. Sung *et al.*, "BEST2 - A high performance super self-aligned 3V/5V BiCMOS technology with extremely low parasitics for low-power mixed-signal applications," *Proc. IEEE CICC*, pp. 15–18, May 1994.

Design of PLL-Based Clock Generation Circuits

DEOG-KYOON JEONG, STUDENT MEMBER, IEEE, GAETANO BORRIELLO, STUDENT MEMBER, IEEE, DAVID A. HODGES, FELLOW, IEEE, AND RANDY H. KATZ, MEMBER, IEEE

Abstract —This paper describes the design of clock generation circuitry being used as a part of a high-performance microprocessor chip set. A self-calibrating tapped delay line is used to generate four nonoverlapping clock phases of a system clock. A charge-pump phase-locked loop (PLL) calibrates the delay per stage of the delay line. Using this technique, it is possible to obtain an accurate phase relationship between the off-chip reference clock and the internal clock signals. Experimental results show that required timing relations can be obtained with less than 2-ns clock skew for clock frequencies from 1 to 18 MHz.

I. INTRODUCTION

AS semiconductor processing technology advances, chip performance increases in two ways. One is the number of functional elements integrated onto a single die, and the other is the speed of operation of the circuits. Overall chip performance is jointly determined by these two quantities: the number of gates in one chip and the clock frequency. To fully exploit the increase in computational power of a chip, the interchip communication bandwidth must also be increased. However, with a synchronous communication protocol, it is impossible to increase the communication clock frequency without reducing clock skew between the chips. Clock skew results mostly from the different delays associated with the clock buffers integrated on chip. It varies with the loading on the clock buffer outputs, the process technology, process variations, and temperature variation from chip to chip. The derivation and distribution of clock signals is a critical problem to solve in order to improve overall system performance.

In the Symbolic Processing Using RISC's (SPUR) chip set [1], a four-phase nonoverlapping clock is used for internal communication, and stringent timing relations must be satisfied for interchip communication between the central processing unit (CPU) and the processor cache controller (PCC). There have been realizations [2]–[4] of accurate clock generators using phase-locked loops (PLL's). But in all of those cases, there is either no provision against interchip clock skew due to internal clock buffer delay or no control over nonoverlap time between clock

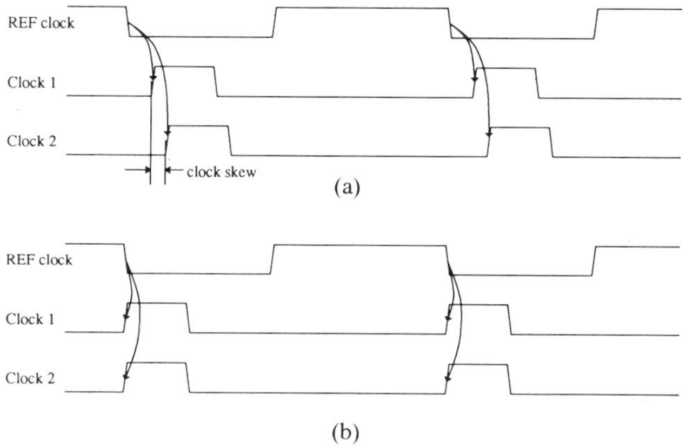

Manuscript received June 24, 1986; revised October 13, 1986. SPUR is supported by DARPA under Contract 482427-25840 by Navalex, California Micro, and Texas Instruments.

The authors are with the Department of Electrical Engineering and Computer Sciences, University of California, Berkeley, CA 94720.

IEEE Log Number 8613485.

Fig. 1. Clock generation from the external reference clock with (a) conventional scheme, and (b) new PLL-based scheme.

phases. In our design, by taking advantage of the extremely accurate phase tracking capability of charge-pump PLL's [5], [6], an edge of the internal clock is accurately aligned to an edge of the external clock (Fig. 1). This is accomplished by directly comparing the two phases through a sequential phase/frequency detector. Correct synchronization between chips is achieved regardless of the above-mentioned variations. All the sensitive circuit elements including clock buffers are within a negative feedback loop and the effect of the variations is tracked and removed by the PLL. The voltage-controlled oscillator (VCO) is composed of a multistage tapped delay line that is automatically calibrated to a precise delay per stage. The generation of arbitrary multiphase clocks is possible with proper decoding of the signals from the delay-line taps [7]–[10].

This paper describes the design and implementation of the SPUR clock generator in 2-μm n-well CMOS technology. Section II presents the design of each of the PLL circuit components. Section III considers the stability of the system. In Section IV, experimental results are provided to show the timing accuracy. Finally, Section V includes some closing remarks.

II. CLOCK GENERATION CIRCUITS

A charge-pump PLL is used to derive an on-chip four-phase nonoverlapping clock from an off-chip block. In addition to precisely determining timing relationships be-

Reprinted from *IEEE Journal of Solid-State Circuits*, vol. SC-22, pp. 255-261, April 1987.

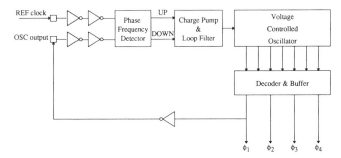

Fig. 2. Block diagram of the new clock generator based on the charge-pump PLL.

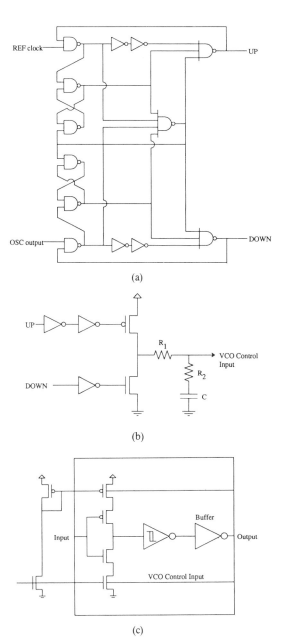

(a)

(b)

(c)

Fig. 3. Basic circuit elements. (a) Sequential logic PFD. (b) Charge pump and loop filter. (c) Delay cell.

tween internal clock phases, it is also used to eliminate clock skew between the reference (REF) clock and the internal clock. The rising edge of ϕ_1 is aligned to the falling edge of the REF clock to ensure correct synchronization throughout the chip set, i.e., all of the internal clocks in the different chips are in phase with respect to the falling edge of the REF clock. This requirement is hard to meet without using a charge-pump PLL due to the unavoidable process- and temperature-dependent delays of the clock buffers used to drive large on-chip capacitive loads. One important advantage of a charge-pump PLL is that it is capable of extremely accurate phase tracking with a passive RC loop filter. Nominal phase error can be practically zero regardless of its input frequency. On the other hand, conventional PLL's (for example, linear PLL's using an analog multiplier as a phase detector) have a finite phase error that is a function of input frequency [12]. The charge-pump PLL system shown in Fig. 2 eliminates clock skew due to the clock buffer simply by ensuring the same inverter delay between the ϕ_1-phase/frequency detector (PFD) path and the REF clock–PFD path. The remaining sources of clock skew are the mismatch of the inverter delay paths and the jitter of the PLL. Differences in the delay through the inverter paths can be minimized by using the same layout and orientation for the circuitry of both paths. The main causes of the jitter are finite input resistance, leakage current, and noise in the VCO. Since MOSFET devices are used to realize the VCO, the effect of the finite input resistance can be ignored. Junction leakage current in the picoampere range makes negligible phase jitter. Also, VCO noise due to the $1/f$ noise of the MOS transistors becomes negligible at the high clock frequencies typically used in high-performance digital circuits. Thus, by careful design of circuits and layout, clock skew between the REF clock and the internal clock can be made nominally zero.

A well-known sequential-logic PFD shown in Fig. 3(a) is used for phase/frequency detection [11], [12]. Since it has a memory to compare frequency as well as phase, the PFD is free from false locking to the second or third harmonics. Its outputs are UP and DOWN signals. When the falling edge of the REF clock leads the falling edge of the VCO output (OSC), UP is activated to low level until the falling edge of OSC arrives. Similarly, DOWN is activated when OSC leads REF. Both UP and DOWN are deactivated to high level when the loop is in a perfectly locked state. In no case are both of the signals activated. UP and DOWN are connected to the charge pump and loop filter of Fig. 3(b). These are comprised of three inverters, MOSFET switches, and a passive RC low-pass filter. The passive RC filter is provided off-chip so that a large capacitor can be used to guarantee stable operation over a wide frequency range and to provide flexibility in choosing the loop filter parameters.

The VCO is implemented as a simple ring oscillator. A series connection of delay cells forms a tapped delay line whose outputs are used to derive the four-phase nonoverlapping clock. Its oscillating frequency is determined by the delay time of the basic cell (Fig. 3(c)) and the number of stages. The delay time of each cell is determined by the amount of current supplied through the current source, the input capacitance, and the threshold of the Schmitt trigger. The two symmetric current sources are controlled by the

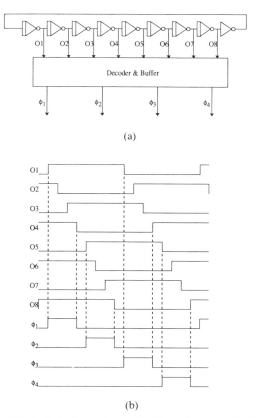

(a)

(b)

Fig. 4. Clock derivation scheme. (a) Block diagram. (b) Waveform generation.

VCO control voltage (i.e., the voltage across the loop filter capacitor in the steady state). To compensate for the asymmetry due to the difference in mobility of electrons and holes, a width ratio of 2:1 is maintained in all the clock generator circuitry. In applications where the timing accuracy is even more critical, a pair of delay cells can be used to obtain perfectly symmetric delays. The Schmitt trigger is included to achieve fast rising and falling outputs at low frequency. These signals become the inputs to the clock derivation circuitry.

The clock derivation scheme is depicted in Fig. 4. To get an odd number of inversions, one extra inverter is attached to the last stage. The amount of extra delay time due to the inverter is small compared to the total delay of the cells. Output waveforms of each of the delay cells can be assumed to be symmetric (50-percent duty cycle). ϕ_1 is derived by ANDing the outputs from the delay cell 1 and 4, ϕ_2 from delay cell 5 and 8, and so on. The ratio of clock high time to nonoverlapping time is 3:1 throughout the entire operating frequency regardless of the variations of process and temperature. Clock buffers consisting of cascaded inverters are used to drive on-chip capacitive loads of up to 3 pF with a rise/fall time of less than 2 ns. A separately buffered OSC output signal is derived to be fed back to the phase detector. By comparing the phases of the buffered clock with the REF clock rather than the original derived clock, internal clock edges are accurately aligned to the input REF clock edge regardless of the buffer delay time. This achieves correct synchronization across the chip set regardless of parameter variations as long as the time constant of the PLL is small enough to track the changes.

III. PLL STABILITY ANALYSIS

A complete analytical stability criterion for this particular type of PLL is difficult to derive because it has both linear and nonlinear elements and operates in the time-varying sampled-data domain. A simplified stability analysis for the second- and third-order PLL is presented in both the s and z domains in the literature [5]. The following analysis is, therefore, an extension of those in the literature. When we include the effect of logic delay in the simplified analysis of the second-order loop filter in the z domain, the stable operating condition becomes

$$K\tau_2 < \cfrac{1}{\cfrac{\pi}{\omega_i \tau_2}\left[\cfrac{\pi}{\omega_i \tau_2}+1-\cfrac{t_d}{\tau_2}\right]}$$

where $K = K_0 R_2 I_p$, K_0 = VCO gain in megahertz per volt,

$$I_p = \text{Max}\left[\frac{V_{cc}-V_x}{R_2}, \frac{V_x}{R_2}\right]$$

is maximum pumping current in amperes, V_x = average VCO input voltage, ω_i = input frequency in radians per second, $\tau_2 = R_2 C$, and t_d = logic delay time.

In the continuous-time domain, which assumes average time-continuous behavior, phase margin is calculated by

$$PM = \tan^{-1}\left[\frac{K\tau_2}{\sqrt{2}}\left\{1+\left[1+\left(\frac{2}{K\tau_2}\right)^2\right]^{0.5}\right\}^{0.5}\right]$$
$$-\frac{360°}{2\pi}\frac{K\tau_2}{\sqrt{2}}\left\{1+\left[1+\left(\frac{2}{K\tau_2}\right)^2\right]^{0.5}\right\}^{0.5}\frac{t_d}{\tau_2}.$$

Note that the last equation is not a function of input frequency. Since the first inequality is based on the z-domain analysis of transfer functions including the effect of a time-varying sampled-data characteristic, it is stricter than the s-domain criterion. Therefore the second criterion is valid only when the first condition is met. Derivation of the first relation is based on the z-plane pole-zero diagram shown in Fig. 5. When z_1 is less than 0, the loop becomes unconditionally *unstable*. Since z_1 is given by

$$\frac{\tau_2 - t_d}{T + \tau_2 + t_d}$$

where T = period of the input frequency, another necessary condition, $\tau_2 > t_d$, must be added for stability. The above criteria for stability are drawn in Fig. 6. Some important points are noticeable in the figure. The loop goes to a safer region as we increase the input frequency as long as the other parameters remain the same. Also, when we increase the phase margin by further increasing the loop gain K, the loop may go into an unstable region in

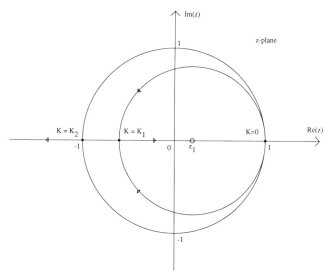

Fig. 5. Root locus of second-order PLL in z domain. Loop stability conditions are $z_1 > 0$ and $K < K_2$.

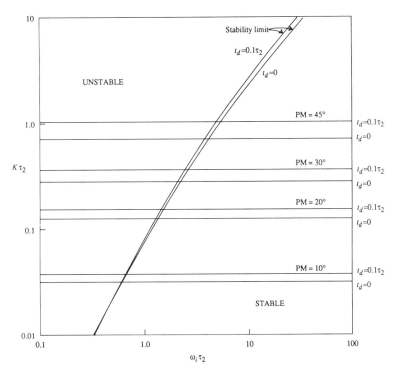

Fig. 6. Stability limit and phase margin.

the z domain. The most critical component affecting stability is R_2. As is shown in Fig. 7, when $R_2 = 0$ or ∞, the loop becomes unstable. There is a limited range of values for R_2 for stable operation. As we increase C, the loop goes into a more stable region as well as increasing the noise margin, at the expense of an increased start-up time. Logic delay time in the order of tens of nanoseconds does not degrade the overall stability much as long as $t_d \ll 0.1\tau_2$, which can be satisfied easily. In our application, efforts were made to get as wide an operating frequency range as possible so as to enable low-frequency testing of the chip. The capacitor was not integrated on chip so that a large value for C could be provided externally. Since the above

analysis ignores many parasitic effects such as the input capacitance of the VCO, VCO gain nonlinearity, and an asymmetric charge pump, we have to include a considerable margin for safe operation. The chosen loop filter parameters are $R_1 = 50$ kΩ, $R_2 = 100$ Ω, and $C = 0.1$ μF. With this large capacitance, we get only about 48° of phase margin, assuming $K_0 = 12$ MHz/V, $V_x = 1.5$ V, and $\omega_i = 2\pi 6.7$ MHz. A simulation program was written to plot the transient behavior and check the stability conditions. One sample output is shown in Fig. 8. With the above parameters, it takes about 16 000 cycles (2.4 ms) for the PLL to start up into the steady state of less than 1° of phase error at 6.7 MHz.

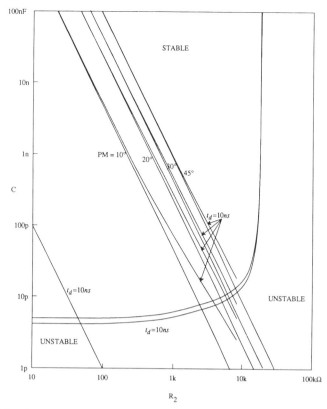

Fig. 7. Stability limit and phase margin as a function of R_2 and C ($\omega_i = 2\pi 6.7$ MHz, $I_p = 70$ μA, $K_0 = 12$ MHz).

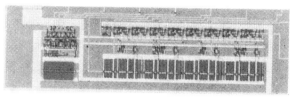

Fig. 9. Chip microphotograph.

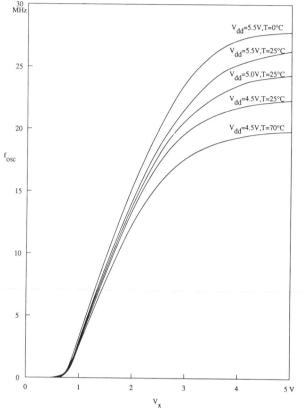

Fig. 10. Measured VCO characteristic.

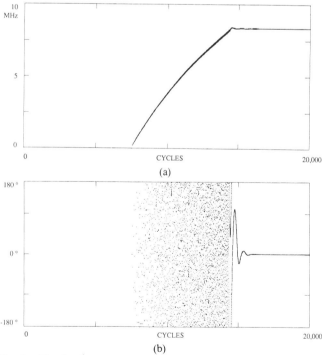

Fig. 8. Simulated start-up response ($\omega_i = 2\pi 6.7$ MHz, $R_1 = 50$ kΩ, $R_2 = 100$ Ω, $K_0 = 12$ MHz/V at $V_x = 1.5$ V). (a) Frequency response. (b) Phase response.

IV. EXPERIMENTAL RESULTS

The entire PLL system except for the loop filter was designed and fabricated in 2-μm n-well CMOS technology. Fig. 9 shows the chip microphotograph. The active area is about 0.4 mm^2 excluding pads. Fig. 10 shows the measured VCO characteristic. At room temperature, the maximum oscillating frequency is 24 MHz with $V_{dd} = 5.0$ V. The VCO control voltage V_x is 1.5 V at 6.7 MHz and its gain is about 12 MHz/V at this operating point. With the chosen loop filter parameters, it operates from 15 kHz up to 18 MHz. Fig. 11 shows the clock waveform operating at the maximum and minimum frequency. We may extend the maximum operating frequency by reducing the number of stages or eliminating the Schmitt trigger in the delay cell at the expense of degrading timing accuracy. At the minimum frequency and up to 100 kHz, nonnegligible jitter (less than 5°) was observed. Jitter is caused by VCO noise and loss of charge in the capacitor due to the leakage current of the junctions during the longer clock period. Also at low frequency it was found that ϕ_3 does not appear exactly in the middle of the clock period. This is believed to be due to inexact delay time of the delay cells at extremely small current levels. This effect can be reduced by including a capacitor at the input of the Schmitt trigger thereby increasing the overall current level at the expense of decreasing the maximum operating frequency. But for 1 MHz up to 18 MHz, no noticeable jitter was

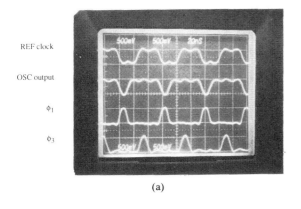

(a)

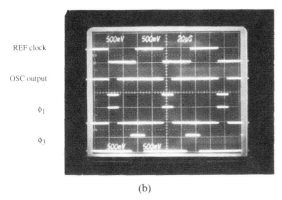

(b)

Fig. 11. Clock generator output (a) at maximum frequency (18 MHz), and (b) at minimum frequency (15 kHz). All the vertical scales must be multiplied by 10.

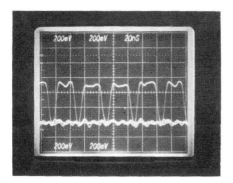

Fig. 12. Four-phase nonoverlapping clock operating at 6.7 MHz (vertical scales multiplied by 10).

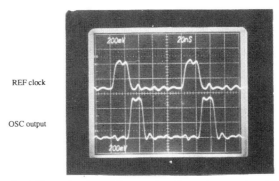

Fig. 13. Effect of a skewed input pulse. For an input pulse skewed by 20 ns, the resulting phase skew in the output is less than 1 ns (vertical scales multiplied by 10).

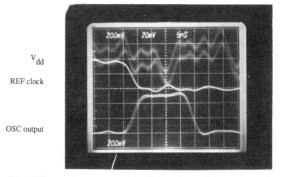

Fig. 14. Effect of power supply fluctuation. Upper trace shows the power supply fluctuation due to external and internal sources. Resulting jitter in the output is less than 2 ns (vertical scales multiplied by 10).

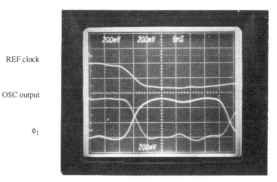

Fig. 15. Edge alignment between REF clock and ϕ_1 (vertical scales multiplied by 10).

found on the oscilloscope trace. Fig. 12 shows the generated four-phase nonoverlapping clock operating at 6.7 MHz, which is the nominal operating frequency of the SPUR chip set. Ratio of clock high time to nonoverlap time is maintained constant in all four phases. The effect of input noise can be seen in Fig. 13. One input pulse is skewed by 20 ns in the input pulse train. The resulting maximum skew of the output pulses appearing in the next cycle is less than 1 ns. This indirectly shows that the immunity to clock input noise is adequate for this application. Also, the effect of power supply fluctuation (for example, ripple from the switching power supply) can be seen in Fig. 14. The upper trace shows the V_{dd} fluctuation due to an externally applied 200-kHz 200-mV square-wave

signal and self-induced signal. The output pulse edge contains jitter of less than 2 ns. In case the power supply fluctuation is severe, a separate quiet supply line should be provided to prevent the interaction of the PLL with the power line signal. Fig. 15 shows the edge alignment, REF clock, and OSC output being the two inputs of the PFD. The rising edge of ϕ_1 is aligned to the falling edge of the REF clock by less than 2-ns skew, showing more than acceptable synchronization between the on-chip clock and the off-chip reference. Clock skew is less than 2 ns from 1 MHz up to the maximum operating frequency of 18 MHz. At high temperature (70°C), maximum operating frequency is reduced to about 14 MHz with $V_{dd} = 4.5$ V. No significant performance degradation was observed.

331

V. Conclusion

A clock generator using a charge-pump PLL was integrated in CMOS to generate a four-phase nonoverlapping clock. Experimental results show that precise timing relations with respect to an off-chip reference clock can be obtained through the use of a self-calibrating tapped delay line with less than 2-ns skew. In addition to generating arbitrary clock waveforms, the same design technique can be applied to the problem of communication between two systems with different clocks. By inserting two different frequency dividers at the input of the phase detector, it is possible to generate two clocks with accurate phase relationship, thereby avoiding the metastability that can cause synchronization failure in the system [13].

Acknowledgment

The authors would like to thank R. Allen at Xerox PARC for fabrication of the test chip, J. Gasbarro at Xerox PARC for valuable discussions and help, and K. Lutz for the setup of the test bench.

References

[1] M. D. Hill *et al.*, "SPUR: Multiprocessor workstation," *IEEE Computer*, vol. 19, no. 10, pp. 8–24, Nov. 1986.
[2] A. Iwata *et al.*, "Low power PCM CODEC and filter system," *IEEE J. Solid-State Circuits*, vol. SC-16, no. 2, pp. 73–79, Apr. 1981.
[3] R. Woudsma and J. M. Noteboom, "The modular design of clock-generator circuits in a CMOS building-block system," *IEEE J. Solid-State Circuits*, vol. SC-20, no. 3, pp. 770–774, June 1985.
[4] M. Bazes, "A novel precision MOS synchronous delay lines," *IEEE J. Solid-State Circuits*, vol. SC-20, no. 6, pp. 1265–1271, Dec. 1985.
[5] F. A. Gardner, "Charge-pump phase-locked loops," *IEEE Trans. Commun.*, vol. COM-28, pp. 1849–1858, Nov. 1980.
[6] F. A. Gardner, "Phase accuracy of charge pump PLL's," *IEEE Trans. Commun.*, vol. COM-30, pp. 2362–2363, Oct. 1982.
[7] A G. Bell and G. Borriello, "A single chip ethernet controller," in *ISSCC Dig. Tech. Papers*, Feb. 1983, pp. 70–72.
[8] T. D. Stetzler, "Clock circuit design considerations for high performance VLSI processors," M.S. thesis, Univ. of Calif., Berkeley, Sept. 1985.
[9] A. Bell, R. Lyon, and G. Borriello, "Self-calibrated clock and timing signal generator for MOS/VLSI circuitry," Patent 4 494 021, Jan. 15, 1985.
[10] G. Borriello, R. Lyon, and A. Bell, "Data and clock recovery system for data communication controller," Patent 4 513 427, May 1, 1985.
[11] H. Ebenhoech, "Make IC digital frequency comparators," *Electron. Des.*, vol. 15, no. 14, pp. 62–64, July 5, 1967.
[12] R. E. Best, *Phase-Locked Loops: Theory, Design, and Applications*. New York: McGraw-Hill, 1984.
[13] F. Anceau, "A synchronous approach for clocking VLSI systems," *IEEE J. Solid-State Circuits*, vol. SC-17, pp. 51–56, Feb. 1982.

A Variable Delay Line PLL for CPU–Coprocessor Synchronization

MARK G. JOHNSON, MEMBER, IEEE, AND EDWIN L. HUDSON, MEMBER, IEEE

Abstract —A fully integrated phase-locked loop (PLL) is used to time-align the hi-Z/low-Z transitions of a CMOS CPU and its floating-point coprocessor (FPC), resulting in minimum timing difference (skew) between the two devices at their shared data bus, and decreasing the bus cycle time. The PLL circuit abandons the traditional voltage-controlled oscillator (VCO) function, instead employing a CMOS voltage-controlled delay line (VCDL) to improve noise immunity, ease loop stabilization, and permit dynamically adjustable clock periods. With the PLL enabled, measured timing skew between the CPU and FPC is below 1 ns.

I. INTRODUCTION

MICROPROCESSORS are now operating at performance levels (10 MIPS or more) which demand high-bandwidth bus interfaces. External data-bus cycle times are often < 60 ns; otherwise the CPU will be forced to stall while waiting for data transfers to complete. The requirement for a small cycle time creates difficulties in the board-level design, as there is little available "dead time" to avoid contention on the data bus.

This paper describes an integrated phase-locked loop (PLL) for preventing bus contention between a CMOS CPU chip and its floating-point coprocessor (FPC). The PLL brings the hi-Z/low-Z transitions of the two devices into time alignment, ensuring that the CPU bus drivers are high impedance whenever the FP coprocessor's drivers are low impedance, and vice versa. By eliminating skew between the CPU and FPC, the PLL technique allows the cycle time to be shortened, yielding higher bandwidth data transfers and faster CPU performance. Time alignment of the CPU and FPC also ensures that the two devices clock their data-bus sampling latches coincidently, minimizing the bus setup and hold times.

II. THE PROBLEM: MANUFACTURING VARIANCES

Fig. 1 shows a simplified block diagram of the processor subsystem. An instruction cache and a data cache are connected to the CPU and the FPC by means of a shared

Manuscript received April 7, 1988; revised May 31, 1988.
The authors are with MIPS Computer Systems, Inc., Sunnyvale, CA 94086.
IEEE Log Number 8822547.

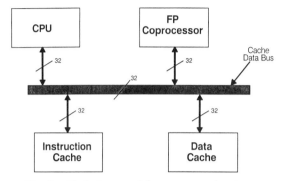

Fig. 1. Block diagram of the processor subsystem.

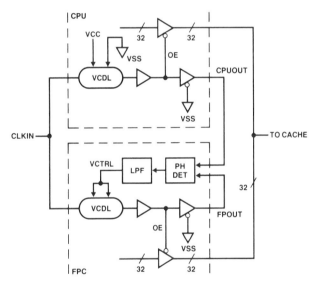

Fig. 2. Interconnection of the CPU and FPC.

three-state 32-bit data bus [1]. Any contention on this bus between the CPU and the FPC would result in large current peaks (injecting noise spikes onto the supplies and perhaps degrading long-term reliability), and would narrow the window of valid data (possibly violating hold time requirements).

The data-bus interconnections of the CPU and the FPC are shown in Fig. 2. The output-enable (OE) signals are controlled by an external reference clock input at the left of the figure. The delay from external clock in, to output

Reprinted from *IEEE Journal of Solid-State Circuits*, vol. SC-23, pp. 1218-1223, October 1988.

enable, and then to the pad-driver low-Z/hi-Z transition, is set by internal delays on each chip. CMOS fabrication parameters such as polysilicon line width, oxide thickness, and junction depth affect the internal delays, and since these process parameters are not perfectly constant in volume manufacturing, the low-Z/hi-Z transition times are (unfortunately) variable. It is not possible to guarantee that any given pair of CPU/FPC devices has received identical CMOS processing, since the two parts are built from different mask sets, in different manufacturing runs, and probably at different times. Therefore, due to the normal variability of CMOS volume fabrication, any given pair of CPU/FPC devices will not make simultaneous low-Z/hi-Z transitions, and there will be contention on the data bus.

III. ADJUSTABLE TIMING USING A PLL

To bring the CPU and FPC into time alignment, the output timing of the FPC is made adjustable, under control of a PLL. CPU timing is not adjustable, but is fixed by the particular CMOS processing parameters of each specific CPU chip. The PLL dynamically adjusts FPC timing until it matches the CPU.

Referring to Fig. 2, the CPU's output-enable signal controls 32 bus driver circuits, and also drives an off-chip clock called CPU_{OUT}. Identical circuitry on the FPC chip produces the FPC's own output enable and off-chip clock FP_{OUT}. The signals CPU_{OUT} and FP_{OUT} allow the PLL to *measure* the time that the two chips make hi-Z/low-Z transitions. The PLL then generates a control voltage *VCTRL* which varies the timing of the FPC's output enable such that FP_{OUT} occurs simultaneously with CPU_{OUT}. Ensuring that FP_{OUT} and CPU_{OUT} are coincident also ensures that the CPU and the FPC make simultaneous hi-Z/lo-Z transitions, thus CPU/FPC skew is eliminated. (The 33 pad drivers on each chip benefit from identical layout and monolithic fabrication, so that driver-to-driver variation on the same die is negligibly small.)

The PLL is integrated on the FP coprocessor chip, and consists of a variable timing element, a phase detector, and a loop filter. These circuit blocks are described in the following sections.

IV. ADJUSTABLE TIMING ELEMENT

Two major candidates for the adjustable timing element were identified: a voltage-controlled oscillator (VCO), and a voltage-controlled delay line (VCDL). In a VCO the output *frequency* is proportional to the input control voltage, and the transfer function contains a pole: $H(s) = K_1/s$. In a VCDL the output *phase* is proportional to the control voltage, and the transfer function is simply $H(s) = K_2$. Because its resultant loop contains fewer poles, the VCDL element was selected.

The VCDL loop's forward transfer function is zeroth order, $H(s) = K_2 K_3$ (where K_2 is the VCDL gain and

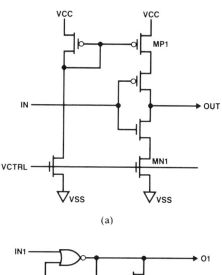

(a)

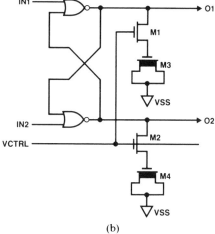

(b)

Fig. 3. (a) Current-starved inverter, after [2]. (b) Shunt capacitor delay stage, after [3].

K_3 is the phase detector gain), so the loop can be easily stabilized with a simple first-order filter. This permits the filter to be fabricated completely on-chip, which eliminates external noise sources such as pin-to-pin crosstalk and board/chip differential ground noise. Avoiding the VCO pole also enables the use of variable clock periods, as discussed in Section VII.

Variable delay elements in MOS technology can be built using two distinct approaches. In the first approach, the control voltage *VCTRL* is applied to a series-connected element which can "current starve" an inverter. A representative example of this approach appears in a CMOS clock generator [2] and is diagrammed in Fig. 3(a). *VCTRL* modulates the ON resistance of pull-down *MN*1, and, through a current mirror, pull-up *MP*1. These variable resistances control the current available to charge or discharge the load capacitance. Large values of *VCTRL* allow a large current to flow, producing a small resistance and a small delay.

In the second approach, the control element adjusts the resistance of a shunt transistor, which connects a large load capacitance to the output of a logic stage. Fig. 3(b) shows an example of this technique, from a DRAM controller chip built in NMOS [3]. (In the NMOS example cited, the data is sent differentially, but this is not an

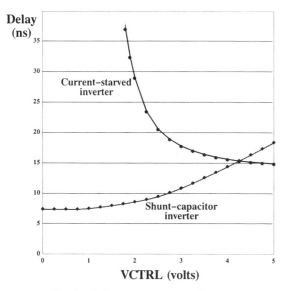

Fig. 4. Delay time versus control voltage.

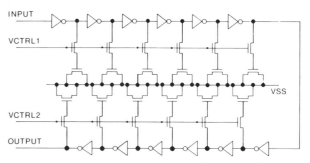

Fig. 5. CMOS VCDL used in the CPU and FPC.

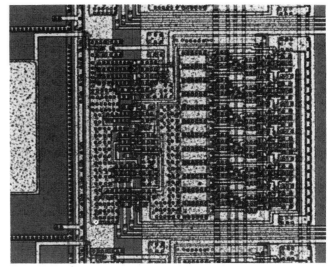

Fig. 6. Photograph of the VCDL.

essential feature of the circuit technique.) *VCTRL* modulates the resistance of shunt transistors $M1$ and $M2$, which connect the output signals $O1$ and $O2$ to MOS capacitors $M3$ and $M4$. The shunt transistors in essence control the amount of effective load capacitance "seen" by the driving gate. Large values of *VCTRL* decrease the resistance of the shunt transistor, so the effective capacitance at the logic gate output is large, producing a large delay.

Fig. 4 is a graph of delay time versus control voltage for the two circuit topologies of Fig. 3. These are plotted for 12 cascaded stages, operated at the worst-case conditions of low supply voltage and high junction temperature. Examining first the current-starved inverter circuit, Fig. 4 shows that it produces delays from 15 ns (at *VCTRL* = 5 V) on upward to near-infinite delays (where *VCTRL* $\leqslant$ V_{TN}). In the steep region of the curve, the starved transistors $MP1$ and $MN1$ operate slightly above threshold, where their impedance is quite high, making circuit nodes inside the delay line susceptible to crosstalk and noise injection. The steep slope also amplifies any noise present on the control signal *VCTRL* itself (see Section VIII). Finally, the input can be delayed by more than one full clock period, resulting in PLL operating in a very narrow band of *VCTRL* near V_{TN}. In practice it would be necessary to provide a mechanism that prevented *VCTRL* from falling below, say, 1.6 V_{TN}, or else force the PLL to always begin the phase capture process with *VCTRL* = V_{CC}, then release VCTRL to gradually decline until the (first) phase-lock capture point is encountered.

Delay of the shunt-capacitor circuit, also plotted in Fig. 4, has a range of about 12 ns and a slope of approximately 3 ns/V. Restricting the current-starved inverter circuit to a comparable 12-ns range (VCTRL between 2 and 5 V), the slope of its delay curve varies from approximately 1 ns/V to well over 20 ns/V. As discussed in Section VIII, larger slopes give rise to poor noise rejection, so the current-starved inverter topology was rejected and the shunt-

capacitor circuit was chosen. However, other workers have reported a VCDL implemented as a single-stage, current-starved inverter in NMOS technology [4].

Fig. 5 shows the single-ended CMOS implementation of the shunt capacitor inverter used to build the VCDL's in both the CPU and the FPC. Twelve identical stages are cascaded to give the desired delay adjustment range. The CPU's VCDL is set to the delay midpoint by connecting its first six stages to V_{CC} and its second six stages to ground, as indicated in Fig. 2. All 12 of the FPC's VCDL stages are connected to the PLL output voltage, *VCTRL*. An even number of stages cancels any difference between low-to-high and high-to-low propagation delays. Fig. 6 shows a photograph of a delay line as fabricated on the FPC chip.

V. PHASE DETECTOR

In a conventional VCO-based loop, the phase detector inputs differ in both frequency and phase. Commonly a "sequential-logic phase and frequency detector" is employed [2] to compare frequency as well as phase, permitting phase-lock acquisition over a range of frequencies. However, in this CPU/coprocessor design the synchronization signals CPU$_{OUT}$ and FP$_{OUT}$ are always at the same frequency (namely, that of the external reference clock), allowing the use of a simplified phase detection scheme.

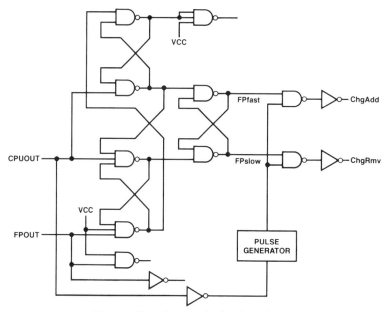

Fig. 7. Phase detector circuit schematic.

A straightforward edge-triggered *D* flip-flop is used for a phase detector, as shown in Fig. 7. The "clock" input of the flip-flop is connected to CPU_{OUT} and the "data" input is connected to FP_{OUT}. If CPU_{OUT} precedes FP_{OUT}, then the data input is logic ZERO when the clock rises, the flip-flop latches a ZERO, and the data-complement output *FPslow* is asserted. Similarly, if FP_{OUT} precedes CPU_{OUT}, the flip-flop latches a ONE and the data-true output *FPfast* is asserted.

The circuit of Fig. 7 is not the traditional implementation of an edge-triggered *D* flip-flop in CMOS technology. Usually a master–slave dynamic topology is invoked (having two transmission gates and two inverters), to lower the device count and reduce silicon area requirements. However, the larger circuit of Fig. 7 has several important advantages in this phase detector application. First, the NAND circuit has a symmetric sampling window: setup time is equal to hold time. The conventional CMOS circuit uses an input pass-transistor structure which gives a longer setup time than hold time, effectively introducing a time offset between CPU_{OUT} and FP_{OUT}. Second, the NAND circuit requires only single-polarity signals for clock and data, whereas the conventional CMOS circuit needs identically timed, complementary clock signals. Third, the NAND circuit has better resolution: its sampling window (= |setup time| + |hold time|) is below 50 ps, while the conventional CMOS topology requires about 400 ps for trouble-free operation (in the 2-μm microprocessor technology).

Capacitive loads are balanced on the input signals and on each of the cross-coupled pairs of gates in the flip-flop, to avoid systematic offsets. Actual NAND gates and inverters are used to match the gate oxide capacitances and their Miller-effect components. The flip-flop outputs are ANDed with a fixed-width pulse (about 10 ns wide), so the resulting signals *ChgAdd* and *ChgRmv* are active for a fixed time interval, independent of the clock frequency.

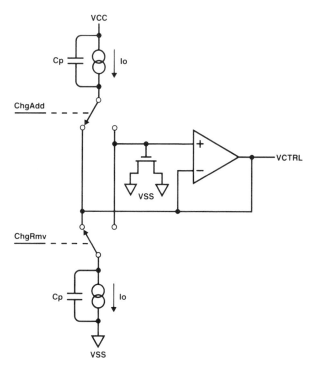

Fig. 8. Low-pass filter circuit schematic.

ChgAdd and *ChgRmv* are sent to the loop filter for time-averaging and smoothing.

VI. LOW-PASS FILTER

A charge-pump circuit [5] is used as the loop filter. The filter capacitor, a large n-channel transistor, is charged or discharged by a pair of matched switched current sources (Fig. 8). The switch control signals *ChgAdd* and *ChgRmv* are asserted for a fixed time, giving a fixed-size charge packet pumped into or out of the capacitor on each cycle.

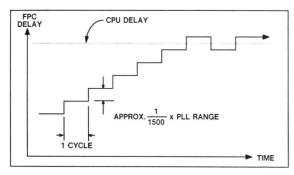

Fig. 9. Phase capture.

This ensures that the magnitude of each correction step is a fixed value, independent of the clock frequency. The pulse width, current-source sizes, and capacitor size are chosen so that approximately 1500 correction steps are required to slew the control voltage from one rail to the other. An operational amplifier connected as a unity-gain buffer provides a low-impedance version of the capacitor voltage, and its output *VCTRL* is sent to the VCDL's.

When a current source is connected to the filter capacitor, it sources or sinks a dc current I_O into or out of the capacitor. The unavoidable parasitic capacitance C_P at the current-source output also charge-shares with the filter capacitor, giving rise to a charge error term. To prevent charge-sharing errors, each parasitic capacitor C_P is clamped to *VCTRL* whenever its current source is not charging the filter capacitor. This guarantees that when C_P is later connected to the filter capacitor, very little ΔV is present (essentially, the offset voltage of the operational amplifier) and therefore very little charge-sharing can occur.

VII. PHASE CAPTURE

The range of delay through the VCDL sets the PLL's capture range. The design target was a range of approximately 12 ns at worst-case conditions (see Fig. 4) to enable FPC chips to phase lock onto CPU chips at opposite extremes of the process window. One possible phase-capture trajectory is depicted in Fig. 9. In this example, the FPC's native delay is less than the CPU's native delay, so the PLL should increase the delay through the FPC.

A fixed-size correction step is applied on each cycle, regardless of the magnitude of the phase error, so the PLL operates as a bang-bang control system. Eventually the phase error is reduced below the correction step size, and the loop applies alternating corrections ($\cdots, +Q, -Q, +Q, -Q, \cdots$) to maintain steady state. The correction step size is therefore designed to be quite small (1/1500 of the PLL range), to minimize phase jitter in steady state. Since 1500 corrections are needed to change the delay by 12 ns, each correction step is approximately (12 ns/1500) = 0.008 ns. The small step size forces a long capture time; however, in this application, phase capture need not occur quickly as it takes place only during the system reset interval. Once phase lock is acquired, the

CPU and FPC delays drift quite gradually (due to temperature and supply voltage drift), and are easily tracked by this fixed slope system.

Because the VCDL-based PLL controls time delay (phase) instead of frequency as in a VCO-based loop, it will operate correctly if the time-base reference clock has a time-varying period. This allows the external system to "pause" or stretch out the clock on a particular cycle, temporarily lengthening the clock period to permit some operation to complete. A VCO-based PLL such as [2] cannot track such a clock stretch operation, as it is required to apply a frequency step and maintain phase lock with zero settling time.

VIII. NOISE SENSITIVITY

Since the PLL is built on the same die as a 16.7-MHz CMOS coprocessor, high-frequency noise will be injected into the loop through the wells and through metallization capacitive coupling. The VCDL's small gain coefficient of 3 ns/V becomes advantageous, as it assists in noise rejection. For example, consider a hypothetical 100-mV noise step on the control voltage *VCTRL*, perhaps occurring when the first floating-point instruction is executed after a long quiescent interval of no FPC operations. The VCDL would produce only 0.1 V × 3 ns/V = 0.3 ns of jitter in the output clocks. In contrast, a VCO has a typical coefficient of 12 MHz/V [2], which would produce a much larger output error for the same hypothetical 100 mV of noise on the control voltage.

IX. RESULTS

Operation of the PLL is shown in Fig. 10. A slower-than-average CPU chip has been connected to a faster-than-average FPC chip. Fig. 10(a) shows the CPU$_{OUT}$ and FP$_{OUT}$ waveforms with the PLL disabled. When the PLL is enabled (Fig. 10(b)), the skew drops to about 0.4 ns as measured at the TTL trip point, 1.5 V. Nonzero residual phase error is attributed to phase detector input voltage offsets, supply-noise-induced VCDL variation, and imperfect board trace matching between CPU$_{OUT}$ and FP$_{OUT}$. Measurements on a range of fabricated devices indicate that a worst-case specification of 1-ns maximum skew has been achieved.

The PLL circuits have been implemented on a CMOS FPC using double-metal processing, 400-Å gate oxides, and 2.0-μm channel lengths. A photograph of the phase detector, filter capacitor, pulse generator, low-pass filter, and operational amplifier is shown in Fig. 11. These circuits are laid out in an area of 0.315 mm^2 (0.69 × 0.45 mm).

X. EXTENSION TO 25 MHz

Since these PLL circuits were first reported [6], the CPU and FPC chips have been fabricated in a more advanced CMOS technology, with 0.8-μm channel lengths and 225-Å

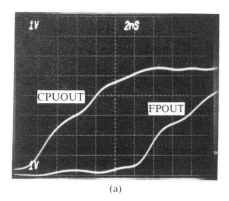

(a)

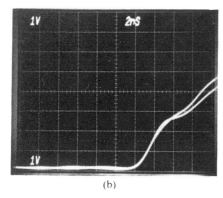

(b)

Fig. 10. (a) Output timing with PLL disabled. (b) Output timing with PLL enabled.

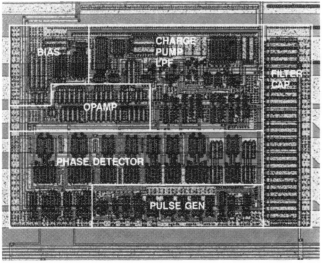

Fig. 11. Photograph of phase detector and low-pass filter.

up from the 12 stages used for 16.7-MHz operation. This adds more tuning range, in gate delays, but the total range in nanoseconds has actually decreased due to the faster gate speeds obtained with 0.8-μm channel lengths. Second, the bias current was increased in the operational amplifier, yielding a lower output impedance for driving *VCTRL*. This is intended to ameliorate the expected increase in injected noise at higher clock rates.

ACKNOWLEDGMENT

The authors thank J. Kinsel, T. Vo, D. Freitas, R. Kunita, and T. Riordan for their technical discussions and suggestions, and C. Kyle for mask layout.

REFERENCES

[1] J. Moussouris *et al.*, "A CMOS RISC processor with integrated system functions," in *Proc. 1986 COMPCON, IEEE*, pp. 126–131.
[2] D-K. Jeong, G. Borriello, D. Hodges, and R. Katz, "Design of PLL-based clock generation circuits," *IEEE J. Solid-State Circuits*, vol. SC-22, no. 2, pp. 255–261, Apr. 1987.
[3] M. Bazes, "A novel precision MOS synchronous delay line," *IEEE J. Solid-State Circuits*, vol. SC-20, no. 6, pp. 1265–1271, Dec. 1985.
[4] M. Forsyth, W. Jaffe, D. Tanksalvala, J. Wheeler, and J. Yetter, "A 32-bit VLSI CPU with 15-MIPS peak performance," *IEEE J. Solid-State Circuits*, vol. SC-22, no. 5, pp. 768–775, Oct. 1987.
[5] F. Gardner, "Charge-pump phase-lock loops," *IEEE Trans. Commun.*, vol. COM-28, pp. 1849–1858, Nov. 1980.
[6] M. Johnson and E. Hudson, "A variable delay line phase locked loop for CPU-coprocessor synchronization," in *ISSCC Dig. Tech. Papers*, Feb. 1988, pp. 142–143.

gate oxides. These newer chips, which represent a linear scaling of the 2-μm layout databases, operate at 25 MHz and above. The PLL circuits were modified in two locations to accommodate the new, faster process. First, the CPU and the FPC VCDL's were lengthened to 16 stages,

A PLL Clock Generator with 5 to 110 MHz of Lock Range for Microprocessors

Ian A. Young, *Member, IEEE*, Jeffrey K. Greason, and Keng L. Wong, *Member, IEEE*

Abstract—A microprocessor clock generator based upon an analog phase-locked loop (PLL) is described for deskewing the internal logic control clock to an external system clock. This PLL is fully integrated onto a 1.2-million-transistor microprocessor in 0.8-μm CMOS technology without the need for external components. It operates with a lock range from 5 up to 110 MHz. The clock skew is less than 0.1 ns, with a peak-to-peak jitter of less than 0.3 ns for a 50-MHz system clock frequency.

I. Introduction

A FULLY integrated phase-locked loop (PLL) has been designed for application and integration on high-clock-frequency microprocessors. The motivation for doing this comes from three areas.

First, as microprocessors increase clock frequency to 50 MHz and higher, it is necessary to eliminate the delay between external and internal clock (clock skew) caused by the on-chip clock driver delay (Fig. 1(a)). As the microprocessor increases in size to 1 million transistors and beyond, the capacitive load from the logic circuits on the clock driver (Fig. 1(b)) has grown to values of several nanofarads. Therefore, the delay through the clock driver can be 2 ns or more. This delay causes large setup and hold times for the input/output signals and is a limitation on the design of systems at high clock frequencies. An on-chip PLL can eliminate this clock skew.

Also, many microprocessors use logic that requires a precise 50% duty cycle clock. To generate this clock, a clock source at twice the logic operating frequency is needed. Instead of generating this high-frequency signal on the system board and bringing it into the chip, an on-chip PLL can synthesize this multiple of the external clock.

Finally, clocking a microprocessor internally faster than the external bus is an option that is attractive for system integration. A clock generator based upon a PLL can synthesize the internal clocking frequency to be greater than the external frequency, for example, twice the external frequency.

However, integrating a PLL on a microprocessor chip is difficult because of noise. This involves integrating an analog circuit (that is required to generate precision timing) on a die that has a large amount of generated digital noise.

This paper describes a PLL-based deskewed clock generator that has been fully integrated on a microprocessor [1], i.e., no external components are used. This PLL circuit has a wide range of frequency of lock, from 5 to 110 MHz, and minimum sensitivity to process, supply, and temperature changes. This paper also describes what was done to reject the digital noise to achieve a skew of less than 0.1 ns with peak-to-peak jitter of 0.3 ns.

II. The Phase-Locked-Loop Clock Generator

A block diagram of the deskewed clock generator is shown in Fig. 2. The external clock goes through an input buffer to the phase–frequency detector (PFD). The feedback of internal clock is compared to the external clock for phase and frequency error. The input offset phase error of the PFD determines the skew between internal and external clock.

The PFD generates UP/DOWN pulsed signals to the charge pump which is followed by a loop filter. The loop filter stabilizes the PLL even with component variation due to the manufacturing process. The output of the loop filter is a control voltage for the voltage-controlled oscillator (VCO). The VCO is followed by a divide-by-2 circuit that generates an accurate 50% duty cycle clock waveform for the microprocessor. The VCO output frequency is, therefore, operating at twice the microprocessor clock

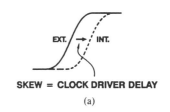

SKEW = CLOCK DRIVER DELAY

(a)

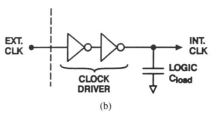

(b)

Fig. 1. Clock skew definition.

Manuscript received April 23, 1992; revised July 15, 1992.

The authors are with Portland Technology Development, Intel Corporation, Hillsboro, OR 97124.

IEEE Log Number 9203397.

Reprinted from *IEEE Journal of Solid-State Circuits*, vol. SC-27, pp. 1599-1607, November 1992.

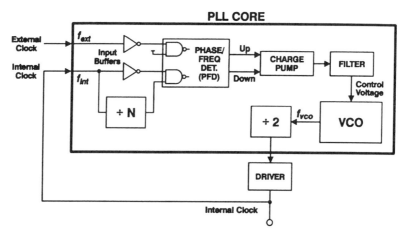

Fig. 2. Functional block diagram.

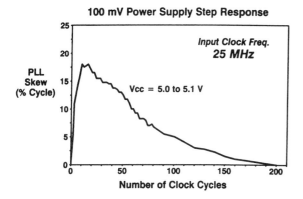

Fig. 3. Conventional PLL clock generator.

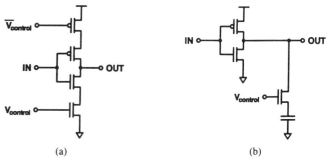

Fig. 4. Conventional delay elements. (a) Current-starved inverter. (b) Variable C_{load}.

frequency. The driver circuit buffers the clock signal to drive the logic of the microprocessor.

The divide-by-N circuit allows the feedback internal clock to be N times the external clock. For example, $N = 2$ makes the internal clock twice the external clock frequency. The divide-by-N circuit creates a signal that enables only one out of N feedback internal clock pulses to propagate into the PFD. This circuit, since it simply masks out clock cycles, does not introduce any additional delay in the feedback clock. The largest value for N is a function of the minimum stable operating input frequency for the PLL, and the maximum VCO output frequency.

Since the PLL is on the same chip as the microprocessor, it is difficult to isolate the PLL from the digital noise generated by the core logic and output buffers. The performance of a conventional PLL clock generator when it experiences a 100-mV step voltage on the power supply is shown in Fig. 3. This figure shows the skew between internal and external clock over time measured in clock cycles after the step voltage. The measurement shows that the skew or phase error accumulates to as much as 18% of the clock cycle (65°) before the PLL has time to correct this error through the feedback. Clearly this is unacceptable since at least 100 mV of power supply noise is very likely to occur in a microprocessor. Over an order of magnitude improvement in the power supply noise rejection of this conventional PLL clock generator is needed to achieve the acceptable level of less than 2% of cycle

(0.72°). Fig. 4 shows some commonly found delay elements used in conventional ring-oscillator-based VCO designs used for PLL's. These show the high sensitivity to power supply noise of Fig. 3.

III. PLL COMPONENTS

A. Phase–Frequency Detector

The conventional sequential frequency and phase detecting logic circuit is used as shown in Fig. 5. This circuit has a monotonic phase error transfer characteristic over the range of input phase error up to ±1 cycle (360°). It also is insensitive to duty cycle. This circuit can operate up to very high frequency (approximately 400 MHz in this process), since the critical path is limited by just three gate delays: two from the cross-coupled, two-input NAND's, and one from the four-input RESET NAND.

B. The Charge Pump

The charge pump which is based on one described in [3] is shown in Fig. 6. The UP and DOWN signals switch current sources I_{up} and I_{dn} onto node $V_{control}$, thus delivering a charge to move $V_{control}$ UP or DOWN. I_{up} and I_{down} need to be equal. When nodes $N1$ and $N2$ are not switched to $V_{control}$ they are biased by the unity-gain amplifier. This suppresses any charge sharing from the parasitic capacitance on $N1$ or $N2$ that can cause mismatch between the UP and DOWN current sources.

The PFD and charge-pump circuits provide a transfer function of input phase error to output charge per clock

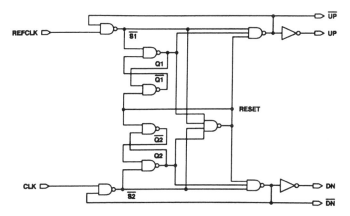

Fig. 5. Phase–frequency detector.

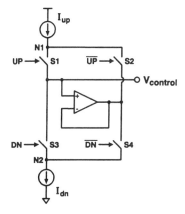

Fig. 6. Charge pump.

cycle. This transfer characteristic must be controlled over supply, temperature, and process variations since it determines the clock skew.

C. The Loop Filter

The loop filter shown in Fig. 7 consists of two capacitors made with MOSFET gate oxide and one resistor made with the n-well implant. For low-jitter PLL design, the frequency jumps inherent to simple RC low-pass filter compensation are a concern [5]. By adding capacitor C_3 in parallel to the series RC impedance, we form a second-order filter with impedance of

$$Z(s) = \left(\frac{b-1}{b}\right) \frac{(s\tau_2 + 1)}{sC_1 \left(\frac{s\tau_2}{b} - + 1\right)} \quad (1)$$

where $b = 1 + C_1/C_3$ and $\tau_2 = R_2 C_1$.

This filter realizes two poles and one zero, which makes the PLL a third-order system with a continuous-time transfer function of

$$H(s) = \frac{K\left(\frac{b-1}{b}\right)\left(s + \frac{1}{\tau_2}\right)}{\frac{s^3 \tau_2}{b_2} + s^2 + K\left(\frac{b-1}{b}\right)s + \frac{K(b-1)}{b\tau_2}} \quad (2)$$

where K is the open-loop gain of the PLL.

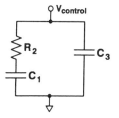

Fig. 7. Loop filter.

Stability of the system is maintained even with the process variation of these on-chip components, which is approximately $\pm 10\%$ for both the capacitance and resistance. While the large-size capacitor C_1 was around 200 pF, this was not a limitation on die size since the total area occupied by this PLL was 0.4% of the total microprocessor die area.

IV. Voltage-Controlled Oscillator (VCO) Issues

A. Power Supply Coupling

Fig. 8 shows what the power supply noise looks like if displayed on a time scale of many clock cycles. The noise consists of high-frequency noise at the clock frequency or higher. This noise comes from cycle-to-cycle switching of large-capacitance nodes within the chip, i.e,. nodes switching every cycle. There is a second component of noise, this one with a lower repetition rate that occurs at over an order of magnitude lower frequency. However, this noise changes the power supply voltage from one cycle to the next in a step-like manner. The cause of this noise is the variation of the circuit activity within the microprocessor. This is a function of the software program that is running. The processor can have intervals when there is heavy circuit activity in switching large amounts of capacitance within the chip, and intervals when there is very little circuit activity. This noise appears as steps or impulses on the power supply of the PLL. If the rejection of the noise coupling into the PLL is not high, then the PLL clock skew (phase error) will display a transient impulse response in Fig. 8(b). The time constant of the settling response is determined by the loop gain of the PLL and can be on the order of microseconds. Note that the actual peak-to-peak jitter in this case becomes dominated by the peaks in the impulse transient noise response, instead of the high-frequency cycle-to-cycle dithering jitter. The jitter would be determined by the latter if there was no supply variation due to the processor activity.

Power supply noise can directly change the VCO output frequency and phase. This is a function of the VCO's stability with changes in the supply (the power supply rejection). Fig. 9 shows how power supply sensitivity of the VCO couples noise into the PLL closed-loop system. Sensitivity of the VCO to power supply changes will cause a direct change in the VCO output frequency, represented as an error frequency f_n away from the VCO operating frequency f_{vco}. Note that the VCO has the transfer function $1/s$, hence the frequency noise source has to be added

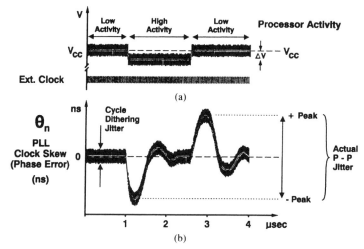

(a)

(b)

Fig. 8. Power supply coupling.

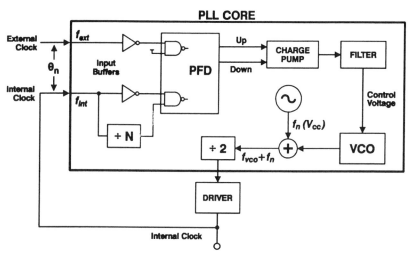

Fig. 9. Power supply noise coupling.

at the output of the VCO and not its input. Since the VCO output frequency divided by 2 is the processor core operating frequency, the processor has to operate at the frequency $[f_{vco} + f_n]/2$.

At the input to the PFD circuit, this internal clock frequency signal is compared to the external clock frequency for phase error. Initially, on the first cycle after the noise step in the power supply voltage, this frequency error causes a phase error (delta) = $[f_n/f_{vco}] \cdot 360°$ of cycle. Since the PLL system has a long time constant compared to the clock period, it takes a number of clock cycles before the feedback can adjust the control voltage to correct the VCO output frequency and phase error. This accumulation of phase error by the PLL over a number of cycles before the maximum phase error occurs makes it a requirement that the initial frequency error f_n, due to the VCO sensitivity to power supply voltage, must be minimized. Therefore, if the power supply rejection of the VCO circuit is not high enough, it can be the dominant source of PLL jitter.

Note also that without any external filtering of the VCC onto the chip, noise on the VCC coming from the motherboard also influences the PLL jitter.

B. VCO Frequency versus Control Voltage Characteristic

When choosing a VCO circuit approach it is important to consider the control voltage versus frequency (V–f) characteristic. A linear characteristic will minimize the VCO sensitivity (i.e., characteristic slope) variation as a function of control voltage or operating frequency, providing PLL stability over the widest possible frequency range.

Also, it is desirable for the V–f characteristic to have a positive slope as shown in Fig. 10. This means that the maximum operating frequency is achieved at maximum control voltage, which is only limited by the maximum VCC applied. If the slope is negative then the VCO minimum frequency is limited by the maximum VCC, i.e., the minimum frequency actually is increased as the applied VCC is reduced. A VCO implemented with the delay element in Fig. 4(b) has this undesirable negative slope V–f characteristic.

The VCO circuit used in this PLL has a linear, positive slope V–f characteristic. This VCO uses a current-controlled differential delay element and is described in the next section.

342

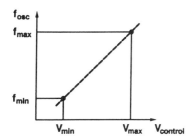

Fig. 10. Oscillator frequency versus control voltage.

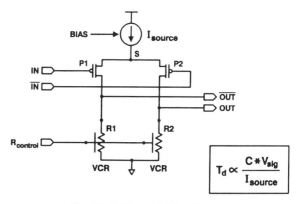

$$T_d \propto \frac{C * V_{sig}}{I_{source}}$$

Fig. 11. Differential delay element.

V. VCO CIRCUIT DESIGN

A. Current-Controlled Differential Delay Element

The VCO is based on a five-stage ring oscillator where each stage is a current-controlled differential delay cell as shown in Fig. 11.

The delay cell is based on a P-MOSFET source-coupled pair with voltage-controlled resistor (VCR) load elements. The n-well containing the P-MOSFET devices is biased at a voltage referenced to V_{SS}. This eliminates noise coupling from the power supply to the signal nodes. The use of P-MOSFET devices makes the signals referenced to V_{SS}, eliminating frequency shifts due to nonlinear capacitances. Tail current sources are cascoded for high impedance to V_{CC}.

Delay through this cell is a function of the tail current I_{source}, the differential voltage swing, and the capacitance at OUT and OUT#. By controlling signal $R_{control}$ to the VCR load, the voltage swing is held constant and independent of supply voltage. If the capacitance is constant, then the delay is inversely proportional to the variable I_{source}. The frequency of a ring-oscillator VCO is, therefore, directly proportional to I_{source}, with the desirable positive linear slope.

Note that for high rejection of power supply noise the differential signal is referenced to V_{SS} and the variable current source is cascoded for high impedance to V_{CC} supply.

B. Voltage-Controlled Resistor (VCR)

Fig. 12 describes the circuit used as a VCR. The voltage applied to $R_{control}$ controls the impedance of MOSFET devices $M1$ and $M3$ and thus the current into node "IN" for a given voltage at "IN." For a V_{IN} range from 0 to 1.0 V, varying $R_{control}$ voltage produces a family of I_{IN} versus V_{IN} curves shown in Fig. 12(b). These I–V characteristics provide the load for the differential delay element.

The VCR is a key component in the delay cell. What is required is an element that is suitable as a load resistor for source/drain voltages between 0 and 0.8 V (V_{ref}) and having the widest possible range of resistance to maximize the VCO frequency range of operation. As shown in Fig. 12, the VCR uses a combination of three N-MOSFET's to provide a more linear I_d versus V_d characteristic than a single MOSFET. This VCR circuit configuration has been able to achieve higher dynamic range

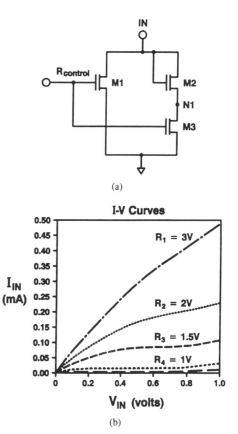

(a)

I-V Curves

(b)

Fig. 12. Voltage-controlled resistor.

than another VCR reported recently [4], due to the series $M3$ MOSFET. With the techniques described, the delay cell improved the power supply noise rejection of the VCO by six times to 0.7%/V compared to 4.5%/V previously reported [2].

C. V–I Converter

Fig. 13 shows the additional circuits for the VCO using the differential delay element. To control the frequency with a voltage input signal, a voltage-to-current converter is realized using an NMOS source follower, applying the voltage across an n-well resistor. This transconductance is very linear and has a positive slope. The current sources are cascoded.

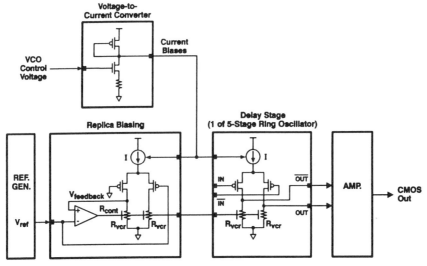

Fig. 13. Differential delay element VCO.

D. Replica Biasing

The signal amplitude is held constant by means of replica biasing and a power supply independent reference generator (V_{ref}). The replica biasing circuit uses an op amp and a copy of the delay cell to generate the appropriate bias to the VCR [2]. The replica biasing uses the same delay cell as the VCO, together with a differential amplifier. One input to the source-coupled p-channel pair is grounded (left input), while the other input is connected to V_{ref}. All the current I flows on the left side of the differential pair. This is the maximum signal condition. The amplifier gain is high enough to cause the feedback (drain of the VCR) to be equal to V_{ref}. Since $R_{control}$ is also used to bias the resistors in the VCO delay stages, the signal amplitude in the VCO is equal to V_{ref}. The V_{ref} voltage is about 0.8 V. The amplifier at the output of the ring oscillator converts the small-signal swing to CMOS levels.

The voltage swing reference generator (V_{ref}) in Fig. 13 is a self-biasing supply-independent MOSFET current source, based on the mobility difference between two N-MOSFET's with different current density. This current source is mirrored into a diode-connected N-MOSFET to generate a voltage bias that tracks with the N-MOSFET based VCR's process variation.

VI. CMOS Process Technology

This PLL clock generator was realized using a 0.8-μm n-well CMOS technology. The features of this process are shown in Table I. The PLL circuit has been realized in both two- and three-layer metal versions of this process.

The die photograph of the PLL located on a 1.2-million-transistor microprocessor die, using the 0.8-μm CMOS three-layer metal process, is shown in Fig. 14(a). The area occupied by the PLL circuit is 0.31 mm^2, less than 0.4% of this 82-mm^2 processor die [1]. Fig. 14(b) shows the enlarged micrograph of the PLL circuit. It can be seen

TABLE I
0.8-μm CMOS Parameter Summary

MOSFET Parameters		
	NMOS	PMOS
T_{ox}	150 Å	150 Å
S/D	LDD	LDD
L_{eff}	0.6 μm	0.6 μm
V_t	0.6 V	0.6 V
I_{dsat}	0.4 mA/μm	0.2 mA/μm

Process Parameter Silicided Source/Drain & Poly	
Metal 1 Pitch	2.0 μm
Metal 2 Pitch	2.4 μm
Metal 3 Pitch	2.4 μm—*optional

that digital signals are routed over the circuit in the third layer of metal.

VII. Measured Results

Table II shows the measured performance of the clock generator when functioning with the microprocessor running at 50 MHz. Table III summarizes the measured characteristics of the VCO circuit used in this PLL.

To characterize the PLL rejection of power supply noise, Fig. 15 shows the transient phase error response of the PLL to a positive or negative 500-mV step on the V_{CC} level (5.75 to 6.25 V for worst case). The peak-to-peak jitter was found to scale with the size of the step. The same step response over the range of input clock periods is given in Fig. 16.

Fig. 17 shows that the clock skew (of internal clock relative to the external clock) is 88 ps at a clock frequency of 66 MHz. Fig. 18 is a digital oscilloscope display of the rising edge of the internal clock superimposed with a histogram of its jitter. It compares the jitter when the microprocessor is running a quiet pattern to the jitter when it is

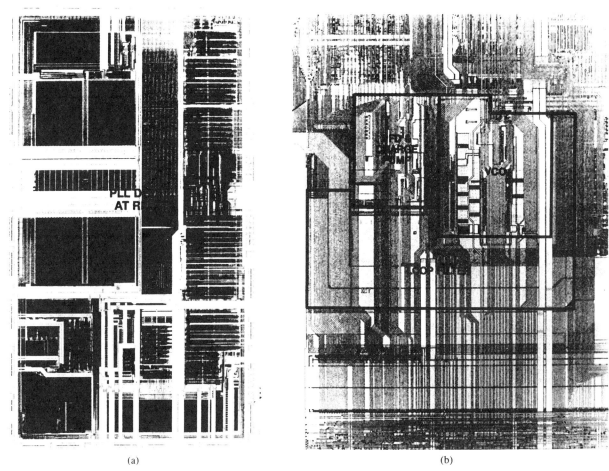

(a)   (b)

Fig. 14. (a) Microprocessor micrograph. (b) PLL detail photograph.

TABLE II
MICROPROCESSOR RUNNING TEST PROGRAM AT 50 MHz

Peak-to-Peak Jitter	<0.3 ns
Phase Error (Skew)	<0.1 ns @ 50 MHz
PLL Lock-in Time	75 μs
PLL Power Dissipation	16 mW
Silicon Area of PLL (w/o clock driver)	0.31 mm^2/500 mil^2

TABLE III
VCO MEASURED CHARACTERISTICS (5 V, 25°C)

Frequency Range	10 to 220 MHz
Sensitivity Range	43 MHz/V @ 120°C
	77 MHz/V @ 0°C
Supply Sensitivity	0.7%/V
Temperature Sensitivity	4700 ppm/°C

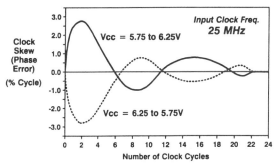

Fig. 15. 500-mV power supply step response.

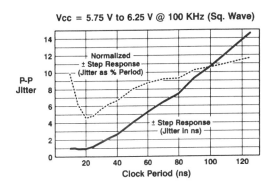

Fig. 16. Power supply rejection test.

running the noisy test pattern. It can be seen that there is not a significant increase in the level of jitter between these two test cases; the peak-to-peak jitter increases from 244 to 280 ps and rms jitter (σ) increases from 36 to 39.3 ps.

The measured schmoo diagram of the PLL clock generator's functional operating frequency versus supply voltage is shown in Fig. 19 (circuit operating with N =

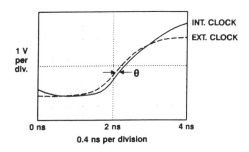

Reset (quiet) @ 66 MHz
θ_{skew} = 88 ps

Fig. 17. Clock skew measured results.

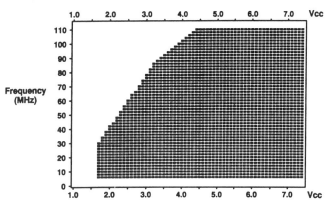

Reset (quiet) pattern
running @ 50 MHz
P-P jitter = 244 ps
Sigma (jitter) = 36 ps

Noisy test pattern
running @ 50 MHz
P-P jitter = 280 ps
Sigma (jitter) = 39.3 ps

Fig. 18. Jitter measured results.

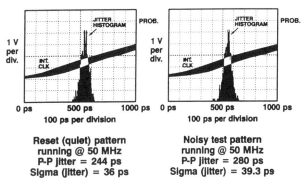

Fig. 19. PLL schmoo plot.

1). The operation of the PLL with external input frequency as low as 5 MHz and as high as 100 MHz has been demonstrated. Since this PLL implementation provides a 50% duty cycle clock regardless of input duty cycle, the PLL VCO is operating as high as 220 MHz.

VIII. CONCLUSION

This circuit has demonstrated that a PLL analog circuit can be integrated into a large microprocessor chip and provide the deskewing accuracy required for clock frequency synthesis up to more than 100 MHz, with clock skew of less than 0.1 ns and peak-to-peak jitter of less than 0.3 ns (with an rms jitter of less than 40 ps). To achieve this, a PLL with a significantly improved power supply rejection was realized, realizing a 30 times improvement factor compared to conventional PLL implementations.

Low generated internal clock skew and jitter was achieved without the need for any external components. This PLL operates in lock over a wide operating frequency range from 5 to 110 MHz, more than a factor of 20. Internal to the PLL, the VCO output frequency has achieved high-frequency operation, operating up to 220 MHz.

ACKNOWLEDGMENT

The authors wish to acknowledge the excellent work done by J. Smith and M. Wegman in making the electrical measurements on the PLL circuit.

REFERENCES

[1] J. Schutz, "A 100 MHz microprocessor," in *ISSCC 1991 Dig. Tech. Papers*, pp. 90–91.
[2] B. Kim and P. Gray, "A 30 MHz hybrid analog/digital clock recovery circuit in 2-μm CMOS," *IEEE J. Solid-State Circuits*, vol. 25, pp. 1385–1394, Dec. 1990.
[3] M. Johnson and E. Hudson, "A variable delay line PLL for CPU-coprocessor synchronization," *IEEE J. Solid-State Circuits*, vol. 23, pp. 1218–1223, Oct. 1988.
[4] G. Moon, M. Zaghloul, and R. Newcomb, "An enhancement-mode voltage-controlled linear resistor with large dynamic range," *IEEE Trans. Circuit Syst.*, vol. 37, p. 1284, Oct. 1990.
[5] F. A. Gardner, "Charge pump phase locked loops," *IEEE Trans. Commun.*, vol. COM-28, pp. 1849–1858, Nov. 1980.

A Wide-Bandwidth Low-Voltage PLL for PowerPC™ Microprocessors

Jose Alvarez, *Member, IEEE,* Hector Sanchez, Gianfranco Gerosa, *Member, IEEE,* and Roger Countryman

Abstract—A 3.3 V Phase-Locked-Loop (PLL) clock synthesizer implemented in 0.5 μm CMOS technology is described. The PLL supports internal to external clock frequency ratios of 1, 1.5, 2, 3, and 4 as well as numerous static power down modes for PowerPC™ microprocessors.[1] The CPU clock lock range spans from 6 to 175 MHz. Lock times below 15 μs, PLL power dissipation below 10mW as well as phase error and jitter below $\pm$100 ps have been measured. The total area of the PLL is 0.52 mm^2.

I. INTRODUCTION

THIS paper describes a fully integrated clock synthesizer analog PLL for PowerPC™ RISC microprocessors [1]. The major functional blocks for the PLL are shown in Fig. 1. The PLL generates an internal CPU clock with 50% duty cycle and low jitter from an external system clock (BUSCLK) [2], [3]. The PLL supports CPU to BUSCLK ratios of 1, 1.5, 2, 3, and 4 while maintaining proper edge alignment. The noninteger 1.5× mode allows the BUS to run at its maximum frequency of 66 MHz while the CPU clock runs at 100 MHz.

This design supports BUSCLK frequencies between 16.6–100 MHz and does not require an accurately controlled duty cycle. The required bandwidth of the Voltage-Controlled-Oscillator (VCO) is reduced from 167 MHz to 100 MHz by using a novel dual-divider circuit in the feedback [4]. The VCO gain is lowered as a direct result of the reduced bandwidth, which improves its noise rejection characteristics. Good noise rejection results in low cycle to cycle jitter, which maximizes the usable cycle time of the CPU clock.

A programmable charge-pump is used in conjunction with two dividers in the feedback path to reduce the variation of damping coefficient and natural frequency by 58%. Good control of both stability parameters is critical for systems which use a board-level PLL to generate BUSCLK. Systems with cascaded PLLs must have well controlled stability parameters to minimize the cumulative phase error from BUSCLK to CPU clock due to input phase modulation. The stability parameter adjustment is implemented in the charge pump by decoding the divider settings and adjusting the reference current to either 50 μA for low divider settings or 100 μA for higher ones.

The PLL lock time is also affected by the stability parameters and is important for power management because the input clock could be stopped and restarted frequently in order to minimize power dissipation. The PLL is shut down when the microprocessor is in one of the static power saving modes and can be restarted in less than 15 μs. The on-chip *RC* filter allows for a fast lock-time as well as for improved reliability, lower system cost, and clock jitter.

This design is implemented in a 4 layer metal CMOS process with 0.5 μm feature sizes; total area is 0.52 mm^2 including the on-chip *RC* filter.

The process technology will be described in Section II followed by more detailed discussions of the phase detector (Section III), charge pump (Section IV), VCO circuits (Section V), and dividers (Section VI). Finally, the clock distribution (Section VII), jitter (Section VIII) and simulation strategy (Section IX) will be described.

II. CMOS PROCESS TECHNOLOGY

A 0.5-μm 3.3-V Nwell CMOS technology with four layers of metal is used in this PLL design. Both NFET and PFET transistors have silicided gates and salicided source/drain junctions. Effective channel lengths of 0.45 and 0.5 μm produce saturated device currents of 300 and 130 μA/μm for NFET and PFET transistors respectively. Sheet resistances for both polysilicon and diffusions are kept below 5 ohms/square. A p– epi layer over p+ substrate is used for CMOS latchup robustness.

Four layers of metal are available to the design in which the last metal is very thick and is exclusively used for power, clock, and critical signal routing. 0.8 μm wide metal1 and 0.9 μm wide metal2 and metal3 are used for local as well as global routing. 0.5 μm contacts and 0.9 μm via1 and via2 are filled with tungsten plugs; the last via between metal3 and last metal is a 1.0 μm wide standard tapered contact. As a result, very aggressive contacted pitches of 1.8 μm are obtained for metal1 through metal3, with a conservative 4.8 μm pitch for last metal. No metals are allowed to be routed over sensitive areas of the PLL such as the on-chip filter, charge pump and VCO. A separate clean VDD is supplied for these circuits. Table I lists the key process parameters and design rules employed in this design. A die photo of the PLL is shown in Fig. 2.

III. PHASE/FREQUENCY DETECTOR CIRCUIT

A. General Operation of Phase Detector

A digital phase and frequency detector (PFD) monitors phase as well as frequency differences between BUSCLK and

[1] PowerPC, PowerPC 603, and PowerPC 604 are trademarks of International Business Machines Corporation.

Manuscript received September 26, 1994; revised January 23, 1995.

The authors are with Motorola, Somerset Design Center, Austin, TX 78758 USA.

IEEE Log Number 9409887.

Reprinted from *IEEE Journal of Solid-State Circuits*, vol. SC-30, pp. 383-391, April 1995.

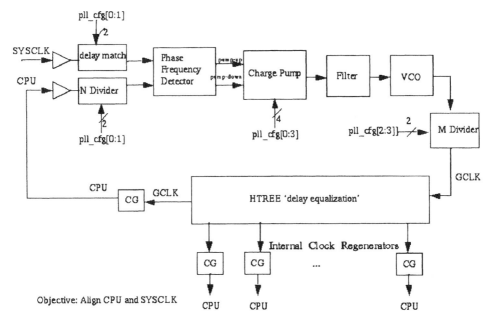

Fig. 1.   Block diagram for PowerPC PLL which shows all the cells, control signals, and input and output clocks.

TABLE I
PROCESS FEATURES

TECHNOLOGY	0.5 μm N-well CMOS with p- epi on p+ substrate
gate oxide:	13.5 nm
Leffn / Leffp:	0.45 / 0.50 μm
Vtn / Vtp:	0.60 / -0.80 volts
Idsatn / Idsatp:	300 / 130 μA/μm
METALS (AL)	width / contacted pitch / resistance (mohms/square) @ T=25C
metal1	0.8 / 1.4 / 80.0
metal2, metal3	0.9 / 1.8 / 55.0
last metal	2.4 / 4.8 / 18.0

the feedback clock. The PFD generates pump-down or pump-up signals when the feedback clock leads or lags BUSCLK, respectively. These pulse-width modulated signals control the switched current sources in the charge pump.

B. Circuit Implementation

A pair of fast latches with reset detect the falling edges of the input clocks. Since the D-inputs are connected to VDD, a logic '1' is latched after the falling edge of each clock. The latches are reset after both inputs arrive, which completes a pump-up or pump-down cycle. The delay through the reset path, which was set to approximately 1 ns, produces a narrow pulse at the outputs of the PFD. Narrow pulses are sent to the charge pump even after the input signals are phase locked. This technique minimizes the PFD dead-band, a low gain region, because an additional input phase error just widens one of the already-existing phase detector output pulses. A Set/Reset latch in the reset path guarantees that both latches have been reset properly before the reset signal is removed.

IV. CHARGE PUMP CIRCUIT

A. Functional Description

The charge pump circuit consists of a pair of switched current sources which either sink or source current pulses to an

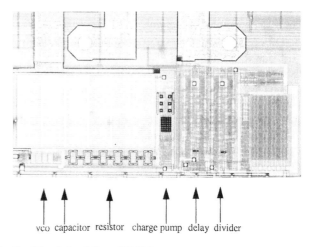

Fig. 2.   Die photo of PowerPC PLL.

on-chip *RC* filter. The control signals for the switches, UPB and DOWN are generated by the PFD circuit. Their pulse-width, along with the selected current level, determines the amount of charge that the charge pump will supply to the filter capacitor. The dc voltage across the filter capacitor controls the frequency, while the transient voltage across the filter resistor controls the phase of the VCO.

The primary function of the charge pump (Fig. 3) is to inject a predetermined amount of charge on the filter capacitor if the feedback clock has a lower frequency than the reference clock or if it phase-lags the reference clock. Similarly, the circuit discharges the capacitor when the frequency of the feedback clock is too fast or if it phase-leads the reference. The output current is automatically selected based on the frequency divider selection, which allows better control of the damping coefficient and natural frequency of the PLL system for different feedback divider selections. Both the damping coefficient and the natural frequency of the PLL are a function of the ratio $I_p/(N * M)$, where N is the phase detector

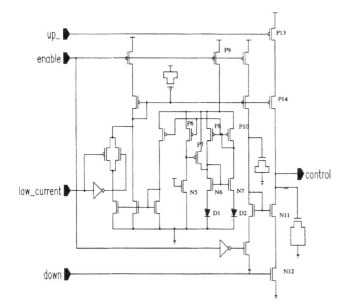

Fig. 3. Schematic diagram for programmable charge pump circuit showing the control signals, current reference circuit, output switches, and passive on-chip filter.

divider while M is the VCO divider. Therefore, to maintain the same system stability parameters for a given temperature and supply voltage, I_p is adjusted so that $I_p/(N*M)$ is constant for varying N and M. The stability parameter adjustment is implemented in the charge pump by decoding the divider settings and adjusting the reference current to 50 μA for division rates of 2, 3, and 4 and to 100 μA for division rates of 6 and 8. A slightly underdamped system, with a damping coefficient less than 1, was targeted in order to obtain fast lock times but still be able to respond properly to input phase modulation.

B. Current Reference

The current reference circuit consists of a pair of ratioed p+ to nwell diodes (D1 and D2 in Fig. 3), a pair of ratioed NMOS transistors (N6 and N7), a PMOS load (P8 and P10) and a start-up circuit (P6, P7 and N5). This circuit has the following advantages: first, it does not require resistors. The polysilicon and source/drain junctions are silicided, which makes large resistors impractical. Second, the current reference has a positive temperature coefficient which helps compensate for the reduced VCO gain at high temperature. This partial temperature compensation results in a consistent amount of phase correction for a given phase error at the input. Third, it can be easily turned off with PMOS transistor P9 to minimize power dissipation. And finally, static power dissipation is reduced by keeping the reference current level to only 25 μA.

The following simplified design equations describe the operation of the current reference circuit.

For Diodes:

$$I_d = A \times I_s \times e\left(\frac{V_d}{V_{th}}\right), \tag{1}$$

for NMOS transistor in saturation:

$$I_{ds} = \frac{W}{L} \times K \times (V_{gs} - V_t)^2 \tag{2}$$

where
  W transistor width
  L transistor length
  K mobility $\times C_{ox}/2$
  V_d diode voltage
  V_{th} thermal voltage, KT/q
  I_s saturation current
  A diode area
  D1, D2 ratioed diodes
  N6, N7 NMOS transistors connected to the anode terminal
of the forward biased diodes.

$$V_{gs}(\text{N6}) + V_d(\text{D1}) = V_{gs}(\text{N7}) + V_d(\text{D2}) \tag{3}$$

$$V_{gs}(\text{N6}) - V_{gs}(\text{N7}) = V_d(\text{D2}) - V_d(\text{D1}) \tag{4}$$

$$V_{gs}(\text{N6}) - V_{gs}(\text{N7}) = V_{th}\left[\ln\left\{\frac{I}{A(\text{D2}) \times I_s}\right\} - \ln\left\{\frac{I}{A(\text{D1}) \times I_s}\right\}\right] \tag{5}$$

$$\left(\sqrt{\frac{I}{k}}\right) \times \left\{\left(\sqrt{\frac{L}{W(\text{N6})}}\right) - \left(\sqrt{\frac{L}{W(\text{N7})}}\right)\right\}$$
$$= V_{th} \times \ln\left\{\frac{A(\text{D1})}{A(\text{D2})}\right\} \tag{6}$$

$$I = \left(\frac{(\sqrt{K}) \times V_{th} \times \ln\left\{\frac{A(\text{D1})}{A(\text{D2})}\right\}}{\left(\sqrt{\frac{L}{W(\text{N6})}}\right) - \left(\sqrt{\frac{L}{W(\text{N7})}}\right)}\right)^2. \tag{7}$$

Because this circuit has two stable operating points, a start-up circuit is required to guarantee that it always settles at $I_{ref} = 25\,\mu$A. A long channel PMOS (P7 in Fig. 3) injects 2 μA into the current reference. Once current starts flowing in the reference circuit, the voltage at the gate of the long channel PMOS device is raised to VDD, which removes the start-up current. This simple analysis neglects the body effect of N6 and N7 which will slightly increase the reference current.

All the transistors are relatively wide for better matching and have long channels to minimize short channel effects and maximize output conductance. High output conductance allows the transistors in the current mirrors to operate as ideal current sources when biased to the saturation region.

By taking the partial derivative of (7) with respect to temperature, one can verify that it has a positive temperature coefficient (TC). Also, it can be easily seen that I_{ref} varies by KV_{th}^2, where V_{th}^2 increases at a faster rate than the corresponding decrease of K with temperature. This results in a net increase in current with temperature and therefore a positive TC.

The properly scaled reference current is mirrored to the output stage, which sources or sinks current into the filter. Annular gates are used for the output current mirrors (N11

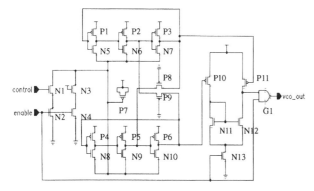

Fig. 4. Schematic diagram for Voltage Controlled Oscillator circuit showing the synchronized ring oscillators, control device, and differential stage.

and P14) to minimize their source capacitance and improve the transition times of the output switches (N12 and P13). The current reference circuit can also be shut off during the power saving modes.

C. On-Chip RC Filter

The on-chip *RC* filter consists of a single NMOS transistor where the gate forms the top plate while the grounded source and drain form the lower plate. The resistor is made of silicided poly. The resistance and capacitance values, 1.5 Kohm and 300 pF, respectively, are chosen to obtain optimum damping coefficient and natural frequency.

V. VOLTAGE CONTROLLED OSCILLATOR CIRCUIT

A. Functional Description

The VCO (Fig. 4) consists of two synchronized, current-controlled, ring oscillators with an NMOS current source. A differential stage (transistors P10-P11, N11-N13, G1) converts the output of the cross coupled rings to single ended and restores the VCO voltage swing to full rail. This circuit topology has excellent noise rejection, in the order of 1 ps/mV of power supply noise, and wide bandwidth, over 100 MHz, even for supply voltages less than 3 V. A dedicated analog supply pin is provided for the VCO and charge pump to minimize power-supply-induced clock jitter. The VCO is designed so that its gain curve falls within the range of interest for all process corners and conditions resulting in a gain of 130 MHz/V (Fig. 5).

B. Ring Oscillator Circuits

The oscillators were designed to cover the full frequency range over process, temperature (0–105°C), and supply voltage (3–3.6 V). The channel length, which determines the maximum frequency of operation, is as long as possible in order to minimize sensitivity to process variations and yet meet the maximum frequency specification for worse case process corners and conditions. The cross coupled transistors (transistors P8, P9) synchronize the two oscillators (transistors P1–P6, N5–N10) so that they run at 180° out of phase. The devices conduct enough current to synchronize the oscillators immediately after power up as well as to keep them 180° out of

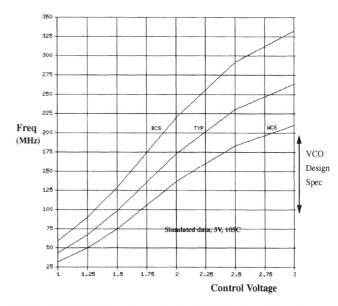

Fig. 5. VCO Gain Curve from simulated data for three different process corners. It shows the design meets the frequency specification over full process corners.

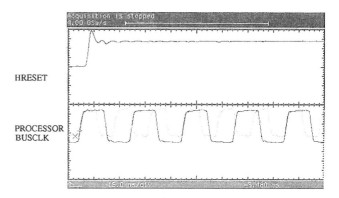

Fig. 6. Oscilloscope picture from CPU and BUS clocks in 1.5× mode after HRESET is deasserted.

phase afterwards. The outputs of the oscillators are connected to a differential stage which restores full rail-to-rail swing.

An NMOS transistor is used to convert the reference voltage from the *RC* filter into pull-down current for the VCO (transistor N1). The W/L is selected to maximize the control voltage range while the length was made long to minimize short channel effects. The ground reference of N1 is the same as the one of the filter capacitor, which helps to ensure that the V_{gs} and I_d of N1 are unaffected by ground bounce. A parallel current path is provided to ensure oscillation for low control voltages. A long channel transistor (transistor N3), connected in parallel with the control NMOS (transistor N1), conducts a small amount of dc current in order to guarantee that the VCO oscillates at a minimum frequency even when the control device operates in the subthreshold region.

An additional capacitor (transistor P7), connected from the virtual ground node to analog VDD, limits the voltage swing on the virtual ground node and therefore enhances the noise rejection of the VCO. The oscillators can also be stopped during power management modes by cutting off the DC current paths to ground (transistors N2, N4, and N13).

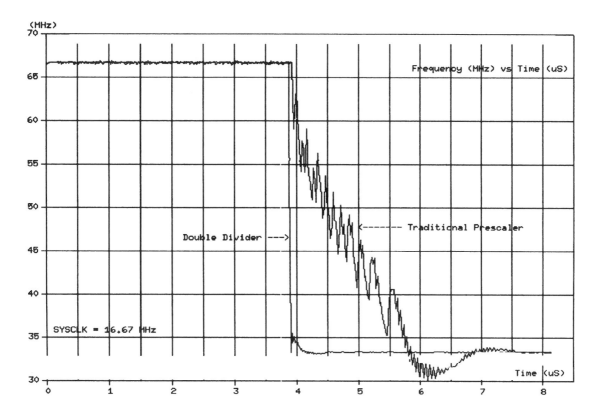

Fig. 7. Frequency versus time measurements from silicon which show the response of the dual divider and prescaler PLL's to a change in multiplier factor. The chart illustrates the faster relock time of the dual divider implementation.

VI. DIVIDER CIRCUITS

A. VCO Divider

The output of the VCO is divided down by a high speed dynamic divider (M divider) which, in conjunction with a second divider (N divider) in the feedback loop, sets the frequency of the processor clock. The VCO divider restores the duty cycle of the global clock to 50% and can be programmed to divide by 2, 4, or 8.

B. Phase Detector Divider

The programmable phase detector divider (N divider) can be configured to divide by 1, 2, 3, 4, and 1.5. The division rate of this circuit determines the final processor to BUS clock frequency ratio. The output of the divider circuit is connected to the feedback input of the phase detector. Once the PLL is locked, the output of the N divider oscillates at the BUS frequency for any of the selected processor to BUS ratios. However, the input of the divider, which is the processor clock, oscillates at a frequency of BUS*N divider. Therefore, the different processor to BUS ratios are obtained by setting the division rate N to the desired rate. Division factors of 1, 2, 3, and 4 were implemented using traditional counter circuits. However, the noninteger divide-by-1.5 required a different circuit. This divider consists of a clock chopper circuit, which generates a constant-width pulse for every rising or falling edge that is detected and results in a multiplication factor of 2×. A divide-by-3 circuit completes the signal conditioning for this mode. In the 1.5× mode the unqualified clock doubling

makes it possible to obtain either one of the two CPU-to-BUS clock phase relationships. For multiprocessor systems to operate properly, all PLL's should have the CPU and BUS clocks phase locked identically at all times. A circuit is included to ensure the CPU clocks generated from all the different PLL's have identical phases when RESET is deasserted. Fig. 6 shows the microprocessor and BUS clocks after RESET with the PLL in 1.5× mode.

C. Delay Matching Circuit

The programmable delay element is almost identical to the divider circuit, differing only in selected metal connections. It is included in the reference clock path in order to guarantee proper phase alignment of the CPU and BUS clocks. The delay element consists of the slave portion of the master-slave (M/S) flip-flop used in the divider. It was designed to propagate BUSCLK through the same devices that determine the CLK to Q delay of the dividers. To achieve similar delay through the divider and delay element over process corners, all the diffusions and polysilicon patterns were drawn and oriented identically. To further improve matching, the layouts for the delay matching and divider circuits were placed as close as possible to each other.

D. Advantages of Dual Divider Implementation

The dual-divider topology minimizes the required VCO bandwidth, which results in lower gain and improved noise rejection. Furthermore, a narrow bandwidth requirement results in higher manufacturing yields since a larger amount of process

351

TABLE II
COMPARISON OF VCO BANDWIDTH FOR PLL WITH PRESCALER VERSUS ONE WITH DUAL DIVIDER CIRCUITS. THE DUAL DIVIDER IMPLEMENTATION REQUIRES 99 MHz LESS VCO BANDWIDTH THAN THE PRESCALER, WHILE ALLOWING CHANGES TO THE CPU/BUSCLK RATIO WITHOUT LOSING FREQUENCY LOCK

| | | | Prescaler | | | Double Divider | | |
mode	BUSCLK	f_{cpu}	N	M	f_{vco}	N	M	f_{vco}
1X	16.5	16.5	1	2	33	1	8	132
2X	16.5	33	2	2	66	2	4	132
4X	16.5	66	4	2	132	4	2	132
1X	33	33	1	2	66	1	4	132
1X	66	66	1	2	132	1	2	132
1X	100	100	1	2	200	1	2	200

TABLE III
VCO BANDWIDTH AND OPERATING POINT COMPARISON FOR PRESCALER VS. DUAL DIVIDER PLL. IF THE PRODUCT OF THE $N * M$ DIVIDERS IS KEPT CONSTANT, THE FREQUENCY OF THE CPU CLOCK CAN BE DOUBLED WITHOUT CHANGING THE VCO FREQUENCY, WHICH RESULTS IN MUCH FASTER RELOCK TIME

| | | | Prescaler | | | Dual Divider | | |
Mode	f_{bus}	f_{cpu}	N	M	f_{vco}	N	M	f_{vco}
2X	16.5	33	2	2	66	2	4	132
4X	16.5	66	4	2	132	4	2	132

variation can be tolerated. The required VCO bandwidth for the dual-divider PLL and for the traditional N divider and fixed prescaler PLL is presented in Table II. For the same input frequency range, the dual-divider topology requires a VCO with 99 MHz less bandwidth. Table III shows that the PLL can switch between 1, 2, and 4× modes without losing frequency lock since the VCO frequency is kept constant.

Fig. 7 shows that the dual-divider topology requires only 480 ns to phase-relock after PLL is reconfigured from 4× to 2× and the CPU frequency is changed from 66 to 33 MHz accordingly. Since the product of the N and M dividers was unchanged, the VCO remained at the same frequency and the PLL did not need to regain frequency lock. Only a small amount of time was required to realign the phase of the feedback clock. A similar mode change to the traditional synthesizer resulted in a change of VCO frequency and therefore frequency relock. Measurements showed a relock time of 3.7 μs.

VII. CLOCK DISTRIBUTION MODELING AND DESIGN

A. Skew Control

A high performance microprocessor with an on-chip PLL strives to minimize the skew between the BUS clock and the internal clocks that drive the latches. It is therefore very important to design a low-skew clock distribution network to take full advantage of the PLL's natural phase alignment operation. Furthermore, since the delay through the clock distribution is unimportant for proper phase alignment, there is no need for high power-consuming clock distribution schemes such as clock-wire grids or extremely large drivers with very wide clock distribution lines. Therefore, an H-Tree style balance clock distribution was chosen because it minimizes the clock distribution power consumption and lessens the effect of RC's.

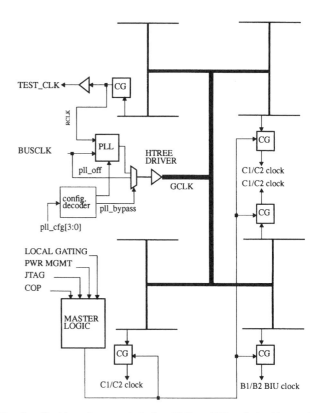

Fig. 8. Clocking elements including H-Tree, PLL, clock driver, clock regenerators, test, and power management logic.

B. H-Tree Design

Given a set of metal lines of width W and length L, the delay of a signal starting at the middle of a clock distribution wire in the shape of an H to the top and bottom endpoints of the H is electrically the same. That is, the RC and the C seen from the middle to every endpoint is equivalent. A clock distribution based on an H-Tree replicates as many H's as necessary to cover the entire clock distribution area (Fig. 8). This is done in a uniform fashion extending from the endpoints of each preceding H. The design priorities for the clock distribution are minimum skew, sharp rise and fall times, 50% duty-cycle, and low power consumption. Rise and fall times below 1 ns are typically the design target for a 10 ns CPU clock period. This is important to minimize the sensitivity of the global clock to power supply noise, which could change the trip point of the clock regenerator circuits and produce clock skew.

The initial stage of designing the clock distribution and the clock driver requires an approximation of the total number of loads on the H-Tree, their input capacitance, and an approximate total area of coverage. This allows the designer to perform SPICE simulations with RC network representations for each branch of each H in an H-Tree. These RC networks, or pi-networks, have process-dependent R's and C's with embedded variations for width, thickness, and metal to metal spacing.

At this stage of the design, the clock driver size is chosen based on the criteria previously outlined. It is important to properly characterize the driver delay and rise/fall times with proper parasitics extracted from layout. Furthermore, it is necessary to ensure proper power supply routing to the driver, since it conducts a significant amount of instantaneous current which can cause noticeable *IR* drops. Fully contacted diffusions are usually necessary in order to meet electromigration design rules. Liberal use of bar vias is very much recommended to minimize localized *RC* drops inside the driver. Guard rings are desirable around the driver to ensure proper protection against CMOS latch-up. The clock distribution is designed for 1 ns rise/fall times and 250 ps typical delay. An automated routing tool connects the last branches of the H-Tree to the clock inputs of the clock regenerators. A layout-based timing analysis is performed to verify the skew control targets. This tool predicts 290 ps delay to the clock regenerator feeding the PLL. The measured delay on the PowerPC 603™ microprocessor is 230 ps. A total of 135 ps of skew at the inputs of the H-Tree targets is also attained. The clock distribution for 266 clock regenerators with 80 fF of input capacitance per regenerator dissipates 120 mW at 80 MHz including the PLL and clock driver circuits [5].

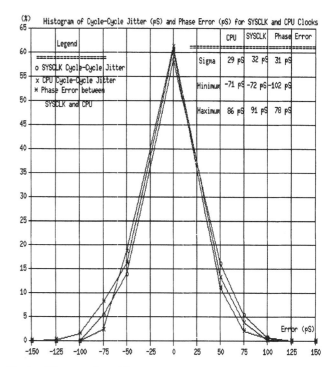

Fig. 9. Histogram from silicon measurements which shows CPU and BUS-CLK jitter as well as clock skew.

VIII. JITTER AND LOCK RANGE MEASUREMENTS

A. Jitter in Quiet Environment

Thorough characterization of the PLL is performed in a test card environment that permits full control of the level of activity in the PowerPC 604™ and PowerPC 603™ microprocessors [6], [7]. With the microprocessor in an inactive state, running only the PLL and H-Tree, at 3.3V, room temperature, 1× mode, and 50 MHz BUSCLK, the CPU clock shows a $-71/+86$ ps peak deviation from the mean measured period with a standard deviation (sd) of 29 ps, and a peak phase error of -102 ps/$+78$ ps with a standard deviation of 31 ps (Fig. 9). These measurements include 163 ps of peak-to-peak pulse generator jitter.

B. Jitter with I/O Activity

It is important to characterize the PLL clock stability under extreme conditions of I/O activity since large instantaneous voltage drops can occur which will cause jitter. A worst-case I/O activity test in which the cache is disabled and the microprocessor is forced to write to memory continuously while switching as much as 64 Data I/O's from high to low shows a ±180 ps peak deviation from the mean period with a standard deviation of 78 ps and a peak phase error between $+143$ps and -230 ps with a standard deviation of 75 ps. It should be noted that about half the total measured jitter is seen at the output of the VCO, indicating that the rest of the jitter is primarily due to the digital circuits in the feedback path to the phase detector. Furthermore, if the test is repeated with the PLL disabled while bypassing BUSCLK to the H-Tree, the measured peak jitter in the CPU clock is only 50 ps less, which indicates most of the jitter is due to variation in the propagation delay of the feedback clock and not from the VCO. Similar jitter and skew data was obtained for the 604 microprocessor.

C. Frequency Lock Range

The PLL lock range was measured for each of the four CPU-to-BUS ratios. To test this, the frequency of the reference clock to the PLL was swept from 16.6–100 MHz while monitoring the CPU clock. Fig. 10 shows the valid frequency range of the feedback clock for each mode. The chart also shows that the measured lock range for all modes exceeded the design specifications. Voltage (Fig. 11) and temperature schmoo plots recently obtained from silicon show good functionality at voltages significantly lower than 3 V and temperatures higher than 105°C.

IX. SIMULATION STRATEGY

The basic building blocks of the PLL are designed and simulated individually, with PLL system simulations providing necessary feedback to adjust the design. Each building block is characterized in SPICE for the 0.5 μm, 3.3 V, CMOS processes. PLL simulations were performed initially with SPICE macro models, followed by schematics, and ultimately layout extracted parasitics. The design is robust over a wide range of supply voltage (3–3.6 V), temperature (0–105°C), and process variations (worse-case to best-case process parameters).

X. CONCLUSION

The fully integrated analog Phase-Locked-Loop described in this paper was successfully implemented in Motorola and IBM's 0.5 μm, 3.3 V, four layer metal, CMOS processes. The design consists of a 1) digital phase/frequency detector with low deadband; 2) charge pump with programmable current level for improved control of stability parameters;

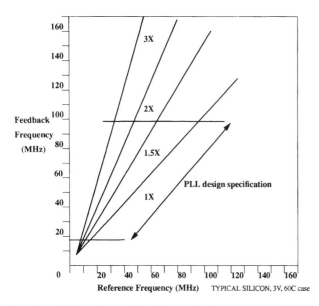

Fig. 10. Measured Lock Range from Silicon for 604 PLL. The measured input frequency range for each clock multiplier setting is shown as well as the design specification.

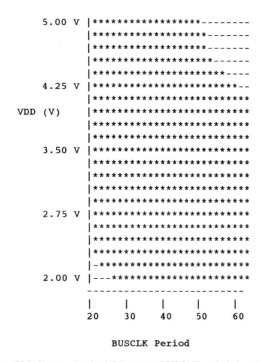

BUSCLK Period

Fig. 11. PLL Shmoo plot for VDD versus BUSCLK period (in ns) at 25°C with PLL in 2× mode.

3) fully integrated *RC* filter; 4) current-controlled VCO with enhanced noise rejection; 5) programmable VCO divider for wide bandwidth and duty cycle control; and 6) programmable phase detector divider to support five different processor to bus frequency ratios. Novel circuit topologies were developed to increase the lock range, reduce sensitivity to input phase modulation, improve manufacturing yields, and reduce power dissipation.

Silicon measurements show full functionality for all the supported CPU to BUS clock ratios, including 1×, 1.5×,

2×, 3×, and 4×. The measurements also showed excellent performance; 3 sigma skew of 87 ps, 3 sigma jitter of 93 ps, and power supply rejection of approximately 1 ps of jitter per millivolt of power supply noise.

ACKNOWLEDGMENT

The authors would like to thank D. Bearden, K. Burch, M. Conner, B. Cornelius, V. Deems, B. Gulliver, C. Hanke, G. Menoskey, J. Montanaro, P. Parkinson, J. Prado, J. Reyes, and S. Thadasina.

REFERENCES

[1] J. Alvarez *et al.*, "A wide-bandwidth, low voltage PLL for PowerPC microprocessors," in *Symp. VLSI Circuits*, June 1994, pp. 37–38.
[2] F. M. Gardner, "Charge-pump phase-lock loops," *IEEE Trans. Commun.*, vol. COM-28, no. 11, pp. 1849–1858, Nov. 1980.
[3] K. M. Ware, H. S. Lee, and C. G. Sodini, "A 200-MHz CMOS phase-locked loop with dual phase detectors," *IEEE J. Solid-State Circuits*, vol. 24, no. 6, pp. 1560–1568, Dec. 1989.
[4] I. A. Young, J. K.Greason, and K. L. Wong, "A PLL clock generator with 5–110 MHz lock range for microprocessors," *IEEE J. Solid-State Circuits*, vol. 27, no. 11, pp. 1599–1607, Nov. 1992
[5] D. Pham *et al.*, "A 3.0W, 75SPEC92int, 85SPEC92fp superscalar RISC microprocessor," in *ISSCC*, Feb. 1994, pp. 112–113.
[6] S. P. Song *et al.*, "The PowerPC 604 RISC microprocessor," *IEEE Micro*, pp. 8–17, Oct. 1994.
[7] G. Gerosa *et al.*, "A 2.2 watt, 80 MHz superscalar RISC microprocessor," *IEEE J. Solid-State Circuits*, vol. 29, no. 12, pp. 1440–1454, Dec. 1994.

A 30 – 128 MHz Frequency Synthesizer Standard Cell

Ricky F. Bitting, William P. Repasky

NCR Microelectronics Products Division, 2001 Danfield Court
Fort Collins, Colorado, 80525

ABSTRACT

A frequency synthesizer standard cell, with an output frequency range of 30 – 128 MHz, has been implemented using a 1.0 µm (drawn), double level metal, double level poly, n–well CMOS process. The cell area is about 2100 mil². The synthesizer is implemented using a programmable, third order, charge–pump phase–locked loop with a sequential logic phase/frequency detector. An on–chip loop filter is used to eliminate any required off–chip components. A jitter standard deviation of 50 pS was measured at an output frequency of 100 MHz, using an external 10 MHz reference. The supply current required by the synthesizer cell alone was 9.0 mA, with a total current of 15.0 mA when the feedback, reference, and output counters are included.

INTRODUCTION

Frequency synthesizer standard parts have recently started to displace multiple crystal oscillators in applications which require multiple clock frequencies, such as computers and computer peripherals[1]. Using these parts permits equipment manufacturers to reduce board space, save power, and easily reprogram required clock frequencies without changing components. As levels of integration and clock frequencies increase, it will become advantageous to use synthesizers within larger digital ASICs as well, for many of the same reasons.

Current generation digital CMOS ASICs generally operate at clock frequencies in the 25 to 50 MHz range. The required clocks are usually generated using crystal oscillator standard cells with external crystals and passive components. Above 30 MHz, 3rd or 5th overtone oscillators are necessary. These oscillators are harder to design than fundamental mode oscillators, and require an external inductor, 3 capacitors, and sometimes a resistor. The supply current required by a typical standard cell crystal oscillator increases roughly as frequency squared. At 10 to 15 MHz (which is typical of synthesizer reference frequencies), a supply current of only about 2 mA is required. At 100 MHz, a crystal oscillator requires 50 to 75 mA of supply current. In contrast, a synthesizer running at 100 MHz from a 10 MHz crystal oscillator reference would require only 17 mA. Finally, if multiple, independent output frequencies or programmability are required, the synthesizer is clearly the best choice.

CELL ARCHITECTURE

BLOCK DIAGRAM

A block diagram of the frequency synthesizer cell, and associated digital counters, is shown in Figure 1. The major functional blocks consist of a current-controlled oscillator, voltage-to-current converter, an on-chip loop filter, loop control logic, charge-pump, phase/frequency detector, a duty cycle buffer circuit (DUTY BUF) for generating an output waveform with improved duty cycle (FOUTD) from the 7-phase ICO outputs (FOUT(7:1)), and a current reference (not shown) for supplying required bias currents to all of the synthesizer subcells.

The digital counters (shown as ÷M, ÷N, ÷P) required to realize a frequency synthesizer with a non-unity frequency transfer function, and also programmability (if desired), are added by the user through use of a digital standard cell library. This approach permits the user to choose the divider configurations required (length, fixed or programmable, binary or divide-by-n). Typically 2 counters are used – the feedback counter (÷M), and the reference counter (÷N). An additional counter (÷P) can also be added to one of the cell outputs (such as FOUTD) if clock frequencies of less than 30 MHz, or frequency steps of less than 1 MHz, are needed.

The charge-pump current, and loop filter damping resistor value, are programmable to maintain loop stability and fast aquisition time over a wide range of ICO and reference frequencies. The optimum settings of the charge-pump current, and loop filter damping resistor, are determined by the loop control logic as a function of the feedback counter divide ratio (M). The loop filter and charge-pump can also be externally controlled, if desired, by the FILCON and M(8:2) pins. The use of an external loop filter is also supported, via the CPSEL, EXTFILT, and VCOIN pins. Output ISOUT provides a current which mirrors the frequency controlling current supplied to the ICO. It can be used to drive a secondary slaved ICO, or a precision delay line. The cell also has a power-down mode.

Reprinted from *Proc. CICC*, pp. 24.1.1 - 24.1.6, May 1992.

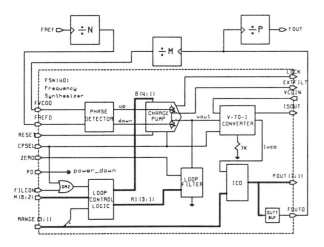

Figure 1 Frequency Synthesizer Cell Block Diagram

THEORY OF OPERATION

The frequency synthesizer cell is implemented using a programmable, third order, charge-pump phase-locked loop[2-7]. A type IV sequential logic phase/frequency detector is used to eliminate the possiblity of harmonic locking. This loop configuration has a theoretically infinite capture range, which in practice is limited by the charge-pump voltage swing and the ICO frequency range. If the ICO frequency is less than the reference frequency, the phase detector will generate finite width pulses on the **up** output while the **down** output will always be low. These pulses will be converted into equal width positive current pulses by the charge pump. The current pulses are smoothed by the loop filter smoothing capacitor (C_2), to reduce the ICO cycle-to-cycle jitter, and further integrated by loop filter components R_1 and C_1 (see figure 5 for a schematic diagram of the loop filter). The time average output voltage, **vout**, will slowly increase, increasing the frequency of the ICO. This process continues until the FVCOD and FREFD inputs of the phase detector are in frequency and phase lock. Once the loop is locked, the charge-pump will only need to deliver extremely narrow output current pulses to correct for voltage droop in the loop filter due to leakage[5]. If the ICO frequency is greater than the reference, a similar process occurs but with pulses on the phase detector **down** output. The synthesizer output frequency is given by

$$F_{out} = F_{ref} \frac{M}{NP} \qquad (1)$$

where F_{ref} is the reference frequency, and M, N, and P are the divisors of the feedback, reference, and output counters, respectively.

LOOP DYNAMICS

The dynamic performance of the frequency synthesizer phase-locked loop is primarily that of a second order system, since the pole due to the smoothing capacitor is usually placed at least an order of magnitude above the loop crossover frequency.

Given this assumption, the loop natural frequency (F_n), the damping factor (ς), and the loop crossover frequency (F_c), are give by

$$F_n = \frac{(I_P K_0 / M C_1)^{1/2}}{2\pi} \qquad (2)$$

$$\varsigma = \pi F_n R_1 C_1 \qquad (3)$$

$$F_c = 2\varsigma F_n \qquad (4)$$

where I_P is the charge-pump output current, K_0 is the VCO gain in Hz/V (ICO gain × V-I converter gain), and R_1 and C_1 are the values of the loop filter components. The value of the loop filter capacitor C_1 has been fixed at 100 pf. R_1 is programmable from 5 KΩ to 40 KΩ in 5 KΩ steps. I_P is programmable from 25 µA to 375 µA in 25 µA steps. With the value for C_1 substituted, equations 2 and 3 become

$$F_n = 1.59 \times 10^4 (I_P K_0 / M)^{1/2} \qquad (5)$$

$$\varsigma = 3.14 \times 10^{-10} F_n R_1 \qquad (6)$$

The minimum loop bandwidth is limited by capacitor values that can be reasonably integrated on-chip, the minimum charge-pump current that can be used, and the maximum allowable loop aquisition time. The maximum loop bandwidth is limited by the amount of jitter that can be tolerated. The presence of higher frequency parasitic poles in the loop response also limit the maximum loop bandwidth for which stable operation is possible. F_n is typically in the range of 50 to 200 KHz, and can be as high as 500 KHz for feedback counter divisors of 4 or less. ς should always be in the range of 0.5 to 1.5 to assure optimum loop transient response and guarantee stability. Higher frequency parasitic poles, including the one introduced by the smoothing capacitor, will tend to reduce the phase margin and increase ringing above that predicted by the above second order equations[3].

Due to the discrete-time (or sampled-data) nature of the charge-pump PLL, one additional criterion needs to be satisfied to assure stability

$$F_0 / F_c > 0.637 \varsigma^2 [(1 + 1/\varsigma^2)^{1/2} - 1] \qquad (7)$$

where F_0 is the phase detector input frequency (at FREFD)[4]. This equation defines the minimum phase detector input frequency that can be used for a given loop crossover frequency (F_c). The ratio F_0 / F_c is typically 3 to 5. Loop parameters were chosen to assure that F_0 is always greater than 1 MHz for M greater than 16.

DETAILED CIRCUIT DESCRIPTION

CURRENT-CONTROLLED OSCILLATOR

The ICO, which is implemented using a 7-stage current-controlled ring oscillator, is shown in Figure 2[8]. Transistors MN1 and MP1 form the basic current-controlled delay stage. The delay stage has been designed to operate with a bias current of 100 to 500 μA. The voltage on each stage of the ring is sensed by the source follower MN2–MN3. The following two inverter stages of the buffer level shift and square up the output waveform. This configuration was used to minimize capacitive loading and thus maximize the operating frequency of the ring oscillator.

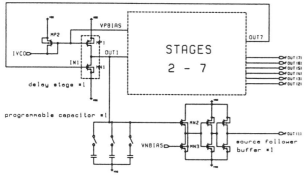

Figure 2 Simplified Schematic Diagram of ICO

Five overlapping frequency ranges are used to accommodate expected process variations without requiring the use of trimming. The frequency ranges are defined as follows:

RANGE(3:1)	Min. Frequency	Max. Frequency
0	30 MHz	40 MHz
1	40 MHz	55 MHz
2	55 MHz	75 MHz
3	75 MHz	100 MHz
4	100 MHz	128 MHz

Ranges 0 to 3 are realized by adding 3, 2, 1, or 0 capacitors, respectively, to each of the 7 stages of the ring. It was necessary to use a separate ring oscillator for the highest frequency range to eliminate the parasitic capacitive loading of the 3 transistor switches.

The 7 ICO outputs, FOUT(7:1), are phase shifted with respect to each other by 1/7 of the ICO period (for clock edges of like polarity). A plot of these output waveforms, as well as FOUTD, is shown in Figure 3. These outputs can be used for generating multi-phase clocks or for controlling the   relative phase shift of 2 clocks by multiplexing the output tap used for one   of the clock signals.

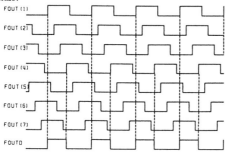

Figure 3 Typical FOUT(7:1) and FOUTD waveforms

The FOUTD output is generated in a way to guarantee that its duty cycle will always be between 3/7 and 4/7 (or about 43% to 57%) independent of the duty cycle of the FOUT(7:1) waveforms. Up to 7 different FOUTD waveforms could be generated in like manor, with well controlled relative phase shifts of 1/7 of the ICO period.

CHARGE–PUMP

The proper design of the charge-pump is key to obtaining low jitter with an on-chip loop filter, since realizable capacitor values are obviously limited. Output glitch energy needs to be minimized. Because of this requirement, design techniques developed for high speed video DACs were employed[9,10]. Mirror image p-channel and n-channel 4-bit current steering DACs were used to realize the current sourcing and sinking capability required of a charge pump, as well as a programmable output current.

A schematic diagram of the current switch architecture used in the DACs in shown in Figure 4. A triple ratioed cascode biasing scheme is used to develop the bias voltages **vbp**, **vpcas**, and **vbp2**[14]. This assures that the output voltage compliance of the current switch will be maximized, as well as minimizing the voltage swing on the gates of MP4 and MP5 to minimize clock feedthrough to the output. The charge-pump output voltage swing is specified at 0.7 to V_{DD}–1 volts. This results in a minimum 5 to 1 voltage swing ratio (0.7 to 3.5 V) at V_{DD} = 4.5V, which is consistent with the ICO current range of 100 to 500 μA.

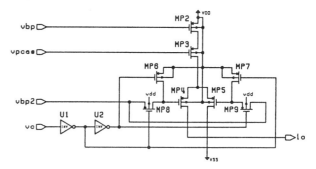

Figure 4 Charge-pump P-channel Current Switch

LOOP FILTER

A schematic diagram of the loop filter is shown in Figure 5. Relatively narrow p+ diffused resistors were used in order to minimize distributed capacitance. The distributed capacitance of the resistors also limited the maximum resistor values that could be used without deviating significantly from the ideal filter impedance verses frequency characteristics. N-wells connected to analog V_{DD} were placed under the double-poly capacitors to minimize noise coupling from the substrate. The p+ resistors were also fabricated in N-wells connected to analog V_{DD}. Transistor MN1 is used to zero the loop filter voltage on power-up to insure that the ICO does not initialize at a frequency that is too high for the feedback counter to operate properly.

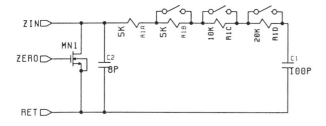

Figure 5 Loop Filter

PHASE DETECTOR

The D flip-flop and RS latch implementations of the Type IV phase detector were simulated and considered for use in the design[11]. The desired maximum operating frequency for the phase detector was 64 MHz. Any "dead-zone" in the transfer characteristic was considered unacceptable.

Both phase detector implementations are susceptible to cross-over distortion in the region of zero phase difference between the REFIN and VCOIN input ports. A "dead-zone" or region where the small signal gain is zero, can lead to low frequency jitter[12]. The feedback loop is essentially broken when the small signal gain is zero. The high impedance node of the charge-pump/loop filter will leak off charge until the phase difference of the inputs becomes large enough for the phase detector to exit the "dead zone" and turn on the charge pump to correct the phase error. In some applications the "dead zone" can actually be an asset since the charge pump currents do not have to be exactly matched to provide low steady-state phase error. However, this is generally not true for a frequency synthesizer PLL.

To realistically evaluate the performance of a phase detector in an actual PLL circuit, it is best to simulate the composite response of the phase detector driving the charge-pump. This is necessary since parameters such as switching threshold and input capacitance of the charge-pump UP and DOWN inputs, as well as charge-pump dynamic response, are significant in determining the composite phase detector/charge-pump transfer characteristics. Thus the composite transistor level circuits were simulated using SABER[TM](Analogy®, Inc.) for both phase detector implementations. A 10 pF capacitive load was connected to the charge-pump output. The transfer curves shown in Figures 7 and 9 are obtained by varying the delay between the REFIN and VCOIN positive going edges, in multiple transient simulations, while monitoring the final charge pump output voltage.

The RS latch implementation of the phase detector is shown in Figure 6 and the resulting transfer characteristics in Figure 7. The results are shown for additional load capacitances of 0.25 pF (rslat1), 1.0 pF (rslat2) and 2.5 pF (rslat3) on the UP and DOWN lines. Any added load capacitance tends to slow down the output rise and fall times. The results show that the UP and DOWN outputs do not get above the threshold of the charge pump for small input phase differences. This results in a "dead zone" in the transfer characteristics. It can be shown that the susceptibility to cross-over distortion of the RS latch implementation is primarilily due to the relationship between the

propagation delay of the internal gates versus the rise and fall time of the UP and DOWN outputs[13].

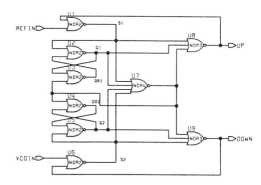

Figure 6 RS Latch Implementation of the Type IV Phase Detector

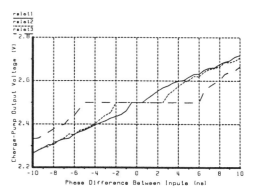

Figure 7 RS Latch Phase Detector Transfer Function

The DFF implementation of the phase detector is shown in Figure 8 and the resulting transfer characteristics in Figure 9. Simulations clearly show that this phase detector has much more margin for extra load capacitance, using our CMOS process and digital standard cells, than the RS latch implementation of the phase detector. The results were obtained using the same test circuit as before but for load capacitances of 0 (dpfd1), 2.5 pF (dpfd2), and 5.0pF (dpfd3). There is negligble crossover distortion for small capacitance values. The output of the charge pump will remain non-zero given that the capacitance on the UP and DOWN lines is low enough to permit these signals to transition through the switching thresholds of the charge-pump inputs. The non-linearity in the curve results from the UP and DOWN outputs not reaching the rails as discussed in reference 13.

Care was taken in the design of the charge-pump to match the source and sink currents to assure a low steady-state phase error. Simulation show a typical steady-state phase error of 25-50 pS. Measurement results indicate a phase error of typically less than 200 pS.

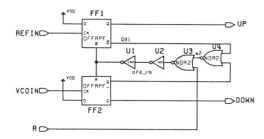

Figure 8 D Flip-flop Implementation of the Type IV
Phase Detector

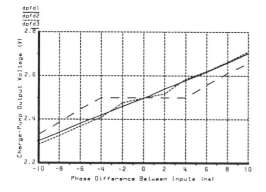

Figure 9 D Flip-Flop Phase Detector Transfer
Function

DESIGN METHODOLOGY

A top-down design methodology was used for developing the frequency synthesizer cell. Ideal behavioral level models were written for all of the basic PLL functional blocks using SABER^{TM} and their MAST^{TM} modeling language. Behavioral simulations were then run to help determine required subcell specifications. Following the design of a transistor level block, the transistor level description was substituted for its equivalent behavioral function and re-simulated. Any performance differences could then be attributed to non-ideal behavior in the transistor level block.

To verify the top level synthesizer cell following the completion of the subcell design, more detailed pin-for-pin compatible SABER^{TM} subcell models were developed. SABER^{TM} digital models were used for all of the logic circuits for improved efficiency. All of the programmable features of the loop were modeled, including the charge-pump/loop filter programming verses M (feedback counter divisor) value. Simulations were then run to analyze loop frequency acquisition transient response for the full range of M values, using a behavioral divide-by-n counter. The loop was found to be unstable for certain M values at phase detector input frequencies greater than 1 MHz as predicted by equation 7. These results prompted us to re-design the loop control logic to meet the minimum $F_0 \leq 1$ MHz criterion. A plot of the simulated frequency acquisition transient response, for an M value of 10 and a reference frequency step of 4 to 10 MHz, is shown in Figure 10. This simulation took 38 CPU minutes to run on an Apollo DN5500.

A digital (QUICKSIM^{TM}) behavioral level model of the cell was also developed. This model was used for final test chip simulations and is also intended for customer use to simulate ASIC designs which utilize the synthesizer cell. This function is a unique case of being able to develop a reasonably accurate (in terms of functionality) digital BLM for a mixed-signal circuit

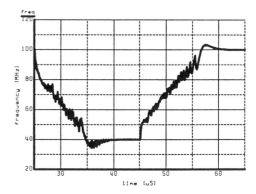

Figure 10 Simulated Synthesizer Frequency
Acquisition Response

TEST CHIPS

Two different synthesizer test chips were constructed. The first one includes all of the synthesizer subcells. The second chip consists of 2 identical synthesizers complete with 8-bit divide-by-n feedback and reference counters and divide by 1, 2, 4, or 8 output counters. Isolated power buses were used for each synthesizer for the analog and digital sections, and the output pad cells (total of 12 power pins) to reduce jitter as well as cross-talk between the 2 synthesizers. The synthesizers are programmed via a microprocessor compatible serial interface. The area of this chip is 19,800 mil^2. Both test chips were completely functional and met target specifications on the first pass.

MEASUREMENT RESULTS

Characterization of the frequency synthesizer cell is currently in progress. At the time this paper was written, (January 1992) only typical measurement results on a small sample of parts were available.

An HP5372A Time Interval Analyzer was used to measure duty cycle, and acquisition time. Output jitter was measured with both the HP5372A and an HP54120 oscilloscope. This was necessary since the noise floor of the HP5372A is about 70 pS (1σ), where as the oscilloscope has a jitter resolution of about 5 pS and can also generate histograms. A typical jitter histogram (measured with the HP54120) is shown in Figure 11. Output jitter does not vary significantly over operating conditions and typically has a standard deviation of 50 to 100 pS, with a maximum peak-to-peak amplitude of 250 to 500 pS.

A measured frequency verses time plot is shown in Figure 12. The reference is being switched between 20 and 50 MHz with a reference counter divisor of N=5. All of the conditions for this measurement are identical to those used for the simulation shown in Figure 10. Note the similarity of the simulated and measured responses.

Measurement of the synthesizer output duty cycle yielded values ranging from 45 to 50% over the full frequency range. The duty cycle actually improves at the upper end of the frequency range due to the design of the DUTY BUF circuit.

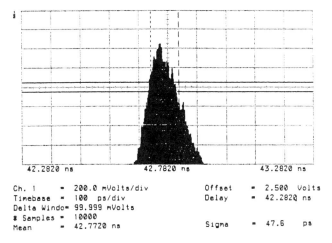

Ch. 1   = 200.0 mVolts/div   Offset   = 2.500 Volts
Timebase = 100 ps/div   Delay   = 42.2820 ns
Delta Windo= 99.999 mVolts
Samples = 10000
Mean   = 42.7720 ns   Sigma   = 47.6   ps

Figure 11 Typical Measured Output Jitter Histogram

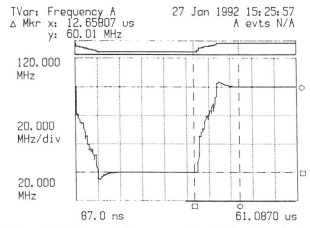

TVar: Frequency A   27 Jan 1992 15:25:57
Δ Mkr x: 12.65807 us   A evts N/A
       y: 60.01 MHz

120.000
MHz

20.000
MHz/div

20.000
MHz
       87.0 ns   61.0870 us

Figure 12 Measured Frequency Acquisition Transient Response

CONCLUSION

A frequency synthesizer standard cell with a 30 to 128 MHz output frequency range was described. It has been shown that acceptable performance in terms of output jitter and duty cycle can be obtained.

REFERENCES

1. Richard S. Miller and Paul F. Beard, "Improve Clock Synthesis in Laptops with a Frequency Generator," *Electronic Design*, pp. 111-120, September 12, 1991.

2. Roland E. Best, *Phase-Locked Loops*, McGraw-Hill, New York, 1984.

3. Gerd Ascheid and Heinrich Meyr, *Synchronization in Digital Communications*, John Wiley, New York, pp. 79-95, 1990.

4. Floyd M. Gardner, "Charge-Pump Phase-Lock Loops," *IEEE Trans. Commun.*, vol. COM-28, pp. 1849-1858, Nov. 1980.

5. ____ , "Phase Accuracy of Charge-Pump PLLs," *IEEE Trans. Commun.*, vol. COM-30, pp. 2362-2363, Oct. 1980.

6. Jerrell P. Hein and Jeffrey W. Scott, "z-Domain Model for Discrete-Time PLL's," *IEEE Transactions on Circuits and Systems*, vol. 35, pp. 1393-1400, Nov. 1988.

7. Deog-Kyoon Jeong *et al.*, "Design of PLL-Based Clock Generation Circuits," *IEEE J. of Solid State Circuits*, vol. SC-22, no. 2, pp. 255-261, April 1987.

8. R.J. Baumert *et al.*, "A Monolithic 50 - 200 MHz CMOS Clock Recovery and Retiming Circuit," *Proceedings of the 1989 Custom Integrated Circuits Conference*, pp. 14.5.1-14.5.4.

9. Takahiro Miki *et al.*, "An 80-MHz 8-bit CMOS D/A Converter," *IEEE J. of Solid State Circuits*, vol. SC-21, no. 6, pp. 983-988, Dec. 1986.

10. Kang K. Chi *et al.*, "A CMOS Triple 80-Mbit/s Video D/A Converter with Shift Register and Color Map," *IEEE J. of Solid State Circuits*, vol. SC-21, no. 6, pp. 989-996, Dec. 1986.

11. Mehmet Soyuer and Robert G. Meyer, "Frequency Limitations of a Conventional Phase-Frequency Detector," *IEEE J. of Solid State Circuits*, vol. 25, no. 4, pp. 1019-1022, Aug. 1990.

12. Eric Breeze, "A New Design Technique for Digital PLL Synthesizers," *IEEE Transactions on Consumer Electronics*, vol. CE-24, no. 1, pp. 24-33, Feb. 1978.

13. Dave Gavin and Ronald Hickling, "A PLL Synthesizer Utilizing a New GaAs Phase Frequency Comparator," *GigaBit Logic 1991 GaAs IC Data Book & Designer's Guide*, pp. 8-95 to 8-110, August 1990, Reprinted from RF EXPO '88 Proceedings.

14. Phillip Allen and Douglas Holberg, *CMOS Analog Circuit Design*, Holt, Rinehart and Winston, New York, pp. 225-226, 1987.

Cell-Based Fully Integrated CMOS Frequency Synthesizers

Dejan Mijuskovic, *Member, IEEE,* Martin Bayer, *Member, IEEE,* Thecla Chomicz, Nitin Garg, *Member, IEEE,* Frederick James, Philip McEntarfer, and Jeff Porter, *Member, IEEE*

Abstract— A family of standard cells for phase-locked loop (PLL) applications is presented. The applications are processed using a 1.5 μm, n-well, double-polysilicon, double-layer metal CMOS process. Applications include frequency synthesis for computer clock generation, disk drives, and pixel clock generators for computer monitors, with maximum frequencies up to 80 MHz. The synthesizers require no external components since the loop filter and oscillator are on chip with the phase ferquency detector and the charge pump. Special voltage and current reference cells are discussed. Analysis of noise sources in the PLL demonstrates the need for reducing the phase noise of the system. A low phase noise is achieved through supply rejection techniques and by placing the oscillator in a high-gain feedback loop to minimize its noise contributions. Laboratory measurements of completed silicon show synthesizers with exceptionally linear gain, as well as transient responses and phase noise similar to predicted results.

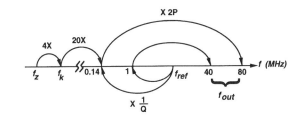

Fig. 1.   Relationship of critical frequencies.

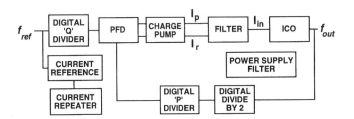

Fig. 2.   PLL block diagram.

I. INTRODUCTION

TO meet the requirements of computer and pixel clock frequency synthesizers, five key design goals are outlined. First, the PLL architecture must be fully integrated and include the oscillator and loop filter. Second, the maximum output frequency must be at least 80 MHz. Third, frequency synthesis with 50 kHz resolution is needed. This results in a phase detector input frequency on the order 140 kHz and forces the PLL bandwidth to be as low as 7 kHz (Fig. 1). Fourth, phase noise must be low, as dictated by the pixel clock applications for high-resolution computer monitors. This requires distinguishing between period jitter and low-frequency phase noise. Recently reported work [1], [2] in this area concentrates solely on reducing period jitter. Lastly, the circuit must be capable of performing in the noisy on-chip environment of mixed signal integrated circuits. Since these design goals produce a conflicting set of requirements, special design solutions are used. For example, the low PLL bandwidth necessary for high-resolution synthesis conflicts with the low phase noise specification. The low bandwidth also requires large loop filter time constants, which are difficult to realize on chip.

The basic PLL architecture is represented in Fig. 2. The output frequency is given by

$$f_{\text{out}} = f_{\text{ref}} \frac{2P}{Q} \qquad (1)$$

where P is the modulus of the feedback counter, Q is the modulus of the input divider, and all frequencies (f_x) are in hertz (Hz). These digital dividers are constructed for each application using a digital cell library. The phase frequency detector (PFD), charge pump, loop filter, and current-controlled oscillator (ICO) can be modified depending on the desired output frequency range, frequency synthesis resolution, and loop bandwidth. Since the loop is actually a sampled time loop, and hand calculations are performed assuming a continuous time loop, the calculated design parameters are verified using a mixed mode behavioral simulator. Time domain simulation of the PLL is impractical using a purely analog circuit simulator. Behavioral simulations are performed using models written at the cell level, not the transistor level, and the high-frequency portions of the PLL are modeled digitally. This mixed-mode modeling greatly reduces the number of time steps that the analog portion of the simulation requires, and therefore offers run time savings of two to three orders of magnitude over transistor level simulations.

Manuscript received July 21, 1993; revised October 19, 1993.

D. Mijuskovic, M. Bayer, T. Chomicz, F. James, P. McEntarfer, and J. Porter are with the Semicustom Operation, High Performance Microprocessor Division, Motorola, Inc., Chandler, AZ 85224.

N. Garg was with the Semicustom Operation, High Performance Microprocessor Division, Motorola, Inc., Chandler, AZ 85224. He is now with the Integrated Circuit Laboratory, David Sarnoff Research Center, Princeton, NJ 08540.

IEEE Log Number 9214794.

Reprinted from *IEEE Journal of Solid-State Circuits,* vol. SC-29, pp. 271-279, March 1994.

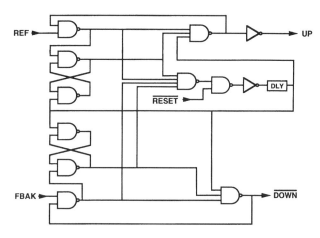

Fig. 3.   Phase frequency detector block diagram.

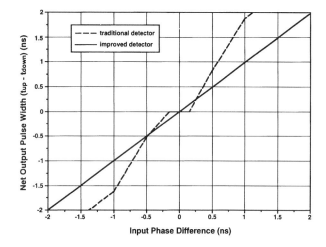

Fig. 4.   Phase frequency detector transfer characteristic.

II. PLL CELLS

The individual analog cells are designed and independently verified using a SPICE-like analog circuit simulator. Cell performance is simulated over process, voltage, and temperature variations (PVT). Process variations are simulated by using extracted and skewed models of the process. Supply voltage and temperature variations are typically specified to be between 4.5 and 5.5 V and from 0 to 70°C (T_A).

A. Phase Frequency Detector

The digital PFD is a type-four detector [3], [4], and is designed to eliminate the region of low gain near phase lock (Fig. 3). The elimination of this "dead zone" is accomplished by producing an "up" current pulse and a "down" current pulse during each cycle. In lock, both pulses are coincident and have a designed minimum width. The transfer characteristic of the PFD is shown in Fig. 4. One version of the detector can be held reset to allow switching between reference frequency sources without disturbing the loop.

B. The Charge Pump and Loop Filter

The charge pump cell [5] contains two charge pump circuits which are driven by the same digital signals, UP, DWN, and

their complements (Fig. 5). These pumps produce two similar output currents, I_p and I_r. The current I_p drives the integrator portion of the loop filter, while I_r bypasses the integrator and provides a stabilizing zero in the transfer function [6]. These currents are on the order 1–10 μA, where the minimum charge pump current is limited by the switching speed requirement. In order to avoid the dead zone, the final current values must be reached during the UP/DWN pulse. By using these small current values, the loop filter can be placed on chip. However, careful design and layout techniques must be used to minimize the capacitive coupling, thereby reducing charge injection at the output of the charge pump. The voltage swing at the charge pump control inputs is the minimum swing required to completely turn off the switching devices. The output of the I_p charge pump is held at virtual V_{ref} by the amplifier in the loop filter. Since the output of the filter is held at V_{ref} by the operational transconductance amplifier (OTA) in the ICO and no dc current flows through $R3$ while the loop is in lock, the output of the I_r charge pump is also held at V_{ref}. In order to minimize unwanted charge transfer, nodes $DP1$ and $DR1$ are also held at V_{ref}. The third pole in the open-loop transfer function, which reduces ripple at the ICO input, is realized in the I_r path by $R3$ and $C3$. The resistor R_T accomplishes the voltage-to-current conversion required to drive the ICO.

The PLL open-loop gain is

$$A_{ol}(s) = \frac{K_i I_p}{2\pi 2P} \cdot \frac{1 + \left(R_3 C_3 + \frac{I_r}{I_p}R_T C\right)s}{R_T C s^2 (1 + R_3 C_3 s)} \tag{2}$$

where K_i, in rad/(A ·s), is the gain of the ICO. In addition to two poles at the origin, this function exhibits a stabilizing zero at

$$f_z = \frac{1}{2\pi\left(R_3 C_3 + \frac{I_r}{I_p}R_T C\right)} \approx \frac{1}{2\pi \frac{I_r}{I_p}R_T C} \tag{3}$$

Making $I_r/I_p > 1$ helps reduce the on-chip size of components R_T and C. The third pole is located at

$$f_{p3} = \frac{1}{2\pi R_3 C_3}. \tag{4}$$

The PLL bandwidth is found from $|A_{ol}(j2\pi f_k)| = 1$ as

$$f_k = \frac{K_i I_r}{4\pi^2 2P}. \tag{5}$$

The Bode plot of $|A_{ol}|$ is shown in Fig. 6.

C. The Current-Controlled Oscillator

The ICO consists of an OTA, a ring oscillator, a frequency-to-current converter, and a clean supply generator (Fig. 5). The ICO frequency range of one octave was chosen to be 40–80 MHz since all lower octaves can be obtained by means of digital division. A ring oscillator structure is used because of its high-frequency capability. However, from previous experience and first-order analysis, the noise in the ring oscillator is too high for these applications. Attempts to reduce the noise within the ring lead to unacceptably high penalties in power and size. The large transistor gate areas required

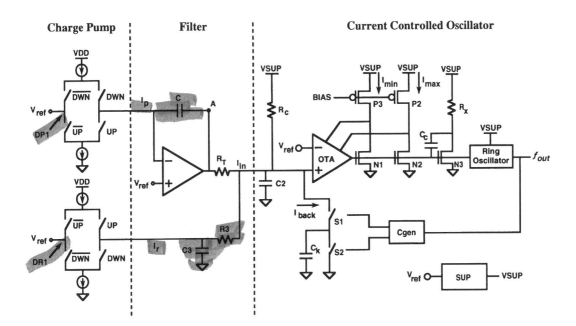

Fig. 5. PLL charge pump, loop filter, and current-controlled oscillator.

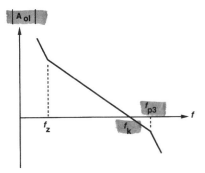

Fig. 6. Bode plot of PLL open-loop gain.

for low noise conflict with the high-frequency requirements. Therefore, the inherently high phase noise of the ring oscillator is suppressed by a high gain feedback loop in which the output frequency f_{out} is converted to current I_{back} and compared with the ICO input current I_{in}. The resistor R_c provides the center current to the ICO. The ring oscillator control current is limited to the range defined by I_{min} and I_{max}. When the control current mirrored by $N2$ exceeds I_{max}, the drain of $P2$ is pulled toward ground, turning on a device in the OTA which limits further increase in control current. A similar mechanism using $N1$ and $P3$ controls excessive decreases in control current. The components $N3, R_x$, and C_c compensate the ICO loop. The block SUP uses V_{ref} to generate a clean supply, $VSUP$, for the ICO.

The ring oscillator is a three-stage differential circuit [Fig. 7(a)]. The output of the third stage is converted into a single-ended signal by the differential to single converter, $D2S$, and subsequently ac-coupled to level shift and preserve a 50% duty cycle. Biasing circuitry for the ring oscillator and one stage are shown in Fig. 7(b). Transistors $N1, N2, N3, P1$, and $P2$ form the differential stage. The maximum voltage of the

oscillator waveforms at OP and OM is limited by $P4$ and $P3$, respectively. Similarly, $N4$ and $N5$ limit the minimum voltage at these nodes. During their limiting action, these four devices conduct the variable control current. The gate voltages of these devices must vary with the control current to maintain the desired clamping levels. Transistor $P5$ conducts current proportional to the control current, and provides it to $P6$ which is matched to $P3$ and $P4$. The operational transconductance amplifier (OTAP) develops V_{tp1} needed to keep the source of $P6$ at V_{tp}. This value for V_{tp1} ensures that the maximum voltage at OP and OM is clamped at V_{tp} and is independent of the control current. Precision clamping is possible because $\partial V/\partial t = 0$ at peak values of the ring oscillator waveform. Since no current is delivered to capacitive loads at these instants, the peak voltage values are determined by transistor dc characteristics and bias voltages.

Clamping the waveform to constant amplitude does not eliminate the nonlinearity in the ring oscillator. As the control current increases, the voltage overdrive needed at IP and IM to switch the current increases. This results in an increased switching time at higher currents (frequencies) and would cause severe nonlinearity of the ring oscillator gain. Waveforms at IP and IM are linear during the transition through the switching region. Therefore, the switching time is proportional to the excess voltage V_{sat} of the switching transistor, $N2$ or $N3$. This excess voltage is a result of current $2I_{\text{control}}$ flowing through the device. It is assumed that the time lost in the switching process also tracks V_{sat}. Reducing the signal swing by an amount proportional to V_{sat} of the switching device compensates for this lost time. While $P6$ provides a constant clamping level, $N7$ is weaker than $N4$ and $N5$ such that the minimum voltage at OP and OM slightly increases as the control current increases. The result is a smaller waveform amplitude at higher control currents which compensates for a large portion of ring oscillator nonlinearity.

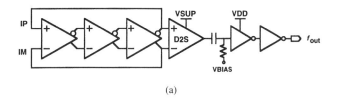

(a)

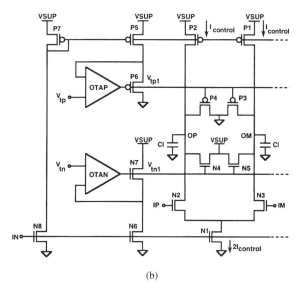

(b)

Fig. 7. (a) Ring oscillator. (b) Ring oscillator detail.

By using this variable clamping technique, the frequency and the gain of the ring oscillator from Fig. 7(a) and (b) are given by

$$f_{rng} = S \frac{I_{\text{control}}}{C_l(V_{tp} - V_{tn})} \qquad (6)$$

and

$$G_{rng} = \frac{\partial f_{rng}}{\partial I_{\text{control}}} = \frac{S}{C_l(V_{tp} - V_{tn})} \qquad (7)$$

where V_{tp} and V_{tn} track V_{ref} and S is a proportionality constant. The load C_l consists of 70% double poly capacitance and 30% parasitics. Simulations using I_{control} proportional to the product of V_{ref} and double polysilicon capacitance C_{dp} yield frequency variations of $\pm 3\%$ over PVT. The gain is verified by observing the product $G_{rng}V_{\text{ref}}C_{dp}$, which varies $\pm 11\%$ over the frequency range and PVT. While differential delay stages based on triode-biased loads [7], [8] are simpler and potentially faster, the resulting oscillator gain varies more over PVT and control current. At a given load current and voltage, the resistive component of a triode-biased load varies over PVT, changing the output waveform and stage delay.

Frequency-to-current conversion is accomplished by means of the switched capacitor resistor formed by $C_k, S1, S2$, and the nonoverlapping clock generator C_{gen} (Fig. 5). Two such converters, driven by opposite polarity clocks, are used in the actual circuit and their outputs are connected in parallel. Each converter contains switched capacitance $C_k/2$. This technique improves the stability of the ICO loop in two ways. First, the value of $C2$ needed for filtering of the switched current I_{back} is reduced by two. Since $C2$ is responsible for the lowest

nondominant pole in the ICO loop, the frequency of that pole is increased by two. Second, the phase shift due to the digital nature of this loop is halved. The gain K_i of the ICO is a function of the feedback circuit, and is

$$K_i \equiv 2\pi \frac{\partial f_{\text{out}}}{\partial I_{\text{in}}} = \frac{2\pi}{C_k V_{\text{ref}}} \qquad (8)$$

since $I_{\text{in}} + I_{RC} = f_{\text{out}} C_k V_{\text{ref}}$.

Evaluating the transfer function of the converter yields

$$K_f \equiv \frac{\partial V_{\text{out}}}{\partial f_{\text{in}}} = \frac{-C_k V_{\text{ref}}}{C_k f_{\text{out}} + \frac{R_c + R_T}{R_c R_T}} \approx -C_k V_{\text{ref}} \frac{R_c R_T}{R_c + R_T}. \qquad (9)$$

Using (7) and (9) and selecting the size of $N3$ to conduct I_{control}, the unity-gain bandwidth of the ICO loop is

$$f_u = \frac{S}{2\pi} \cdot \frac{g_m}{C_c} \cdot \frac{R_c R_T}{R_x(R_c + R_T)} \cdot \frac{V_{\text{ref}}}{V_{tp} - V_{tn}} \cdot \frac{C_k}{C_l}. \qquad (10)$$

Here, g_m is the transconductance of the OTA in the ICO. Resistors R_c, R_T, and R_x track each other, as do voltages V_{ref}, V_{tp} and V_{tn}. As previously mentioned, two thirds of C_l tracks C_k, leaving most of the bandwidth variations in the g_m/C_c term. The bandwidth of the ICO loop is therefore proportional to the bandwidth of the op amp, and is consequently well defined and nominally 1 MHz. The ICO loop introduces a pole into the main loop at approximately 1 MHz. Since this pole is well above the unity-gain bandwidth of the main loop, its impact on the stability of the latter is negligible. Noise generated within the ICO loop is suppressed by the open-loop gain. The OTA has a folded cascode configuration in order to provide high loop gain at low frequencies, which is important for the suppression of flicker noise.

Since each application may require different output frequency ranges, frequency synthesis resolution, and loop bandwidth, the loop parameters can be optimized by recalculating the I_r/I_p ratio, R_T, R_c, R_x, R_3, C_3 and C. In some cases, it may also be necessary to adjust the limiting currents I_{min} and I_{max}. The result of these recalculations is slight modifications of the charge pump, the loop filter, and the ICO.

III. BIASING CELLS

In order to meet the low noise requirements, three cells are used to bias the PLL. Since frequency components in the power supply can modulate the ICO frequency, good supply rejection is crucial in achieving low phase noise. A two-pole low-pass filter cell with a cutoff frequency of 30 Hz is utilized (Fig. 8). The diodes represent appropriately connected MOS transistors. $D1 - D4$ are matched. $D5$ and $D6$ are matched, and their W/L is about 1000 times larger than the W/L of $D1 - D4$. The small current I_0 keeps all diodes in the subthreshold region. The high impedance of $D1$ and $D4$, along with C, determine the pole location. An OTA forces V_{back} to be equal to $VDD/2$. By virtue of diode matching, the output reference voltage is also $VDD/2$. Source and drain diffusions connected to V_{ref} and V_{back} have minimum areas, resulting in a worst case leakage current on the order 1 pA. This leakage current develops a voltage drop of only 0.4 mV

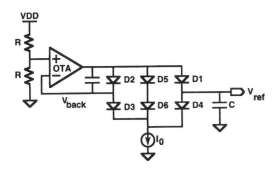

Fig. 8. Low-pass filter.

on the 400 MΩ impedance of these nodes. In addition, leakage phenomena at V_{ref} and V_{back} tend to cancel. Due to device characteristics in the subthreshold region, the variation of the current through $D1$ and $D4$ is larger than that of saturated transistors, resulting in a large but tolerable spread of the pole frequency. Two filter sections in series form the desired second-order filter. The low-pass filter solution is preferred to an untrimmed bandgap reference since the output voltage can be defined more accurately, resulting in a higher value of the clean supply voltage, $VSUP = 4.4$ V. The power supply filter does not provide protection below 30 Hz, but the open-loop gain of the PLL does. The loss of the PLL gain with increasing frequency is compensated by the increasing power supply rejection of the filter.

A current reference cell generates a bias current I_b proportional to V_{ref}, f_{ref}, and on-chip capacitance by means of a switched capacitor circuit similar to the one used in the frequency-to-current converter from Fig. 5. The bias current is then filtered and replicated in a separate cell and used in other cells as needed. The bias current is given by

$$I_b = C_b V_{\text{ref}} f_{\text{ref}} \quad (11)$$

where C_b is the value of the switched capacitor. Charge pump currents, I_p and I_r and frequency-limiting currents $I_{\min}$ and $I_{\max}$ are proportional to I_b. I_r is then

$$I_r = DI_b = DC_b V_{\text{ref}} f_{\text{ref}}. \quad (12)$$

D is a constant of proportionality.

Substituting (8) and (12) into (5) results in

$$f_k = \frac{D}{2\pi 2P} \cdot \frac{C_b}{C_k} \cdot f_{\text{ref}} \quad (13)$$

showing that f_k varies only with f_{ref} and the feedback counter modulus P. Limiting currents $I_{\min}$ and $I_{\max}$ result in ring oscillator frequencies $f_{\min}$ and $f_{\max}$ which are a function only of f_{ref} since the voltage and capacitance variations again cancel. This can be verified by inspecting (6) and (11). The worst case stability of the PLL occurs at the lowest f_k, i.e., when f_k is closest to f_z. At these very low frequencies, the impact of the third and other nondominant poles is negligible. Consequently, at low frequencies, the loop can be treated as a second-order system [5], [6] with a damping factor ξ, which can be represented as

$$\xi = \frac{1}{2}\sqrt{\frac{f_k}{f_z}}. \quad (14)$$

For $f_z = 3$ kHz, $f_k = 7$ kHz, and the worst case spread of the ratio f_k/f_z, the damping factor is 0.623.

IV. NOISE ANALYSIS

A noise source with frequency f_m causes frequency deviation Δf_{out} and phase deviation $\Delta\theta_{\text{out}}$ at the output of the PLL. They are related [4] by

$$\Delta\theta_{\text{out}} = \frac{\Delta f_{\text{out}}}{f_m}. \quad (15)$$

For example, if a noise source with an f_m of 10 kHz changes the frequency of a 50 MHz ICO by 0.02%, the resulting phase deviation is 1 rad, an unacceptably high value. Low-frequency and high-frequency noise are different in terms of their origins, methods of suppression, and importance in various applications. Low-frequency phase noise is caused by power supply noise, random resistor and transistor thermal noise, and random transistor flicker noise. High-frequency phase noise is mainly due to digital switching, and can be considered as predominantly deterministic. High-frequency noise yields low values for $\Delta\theta_{\text{out}}$ due to large f_m. For instance, at frequencies above the PLL bandwidth, the power supply rejection is mainly provided by the op amps in the circuit. The PSRR of an op amp decreases at frequencies above its dominant pole, typically with the slope of -20 dB/dec. This results in Δf_{out} increasing in proportion to f_m. Therefore, $\Delta\theta_{\text{out}}$ does not increase due to the loss of PSRR in the op amps.

Period jitter, or edge-to-edge jitter, can be used to characterize the high-frequency noise. Period jitter is, in fact, a measure of the frequency deviation since

$$\frac{\Delta T_{\text{out}}}{T_{\text{out}}} = -\frac{\Delta f_{\text{out}}}{f_{\text{out}}} \quad (16)$$

holds for $\Delta f_{\text{out}} \ll f_{\text{out}}$. In combination with (15), this yields the relationship between the phase deviation and the period jitter as

$$\Delta\theta_{\text{out}} = -\frac{\Delta T_{\text{out}} \cdot f_{\text{out}}^2}{f_m}. \quad (17)$$

This suggests that it is meaningful to observe the phase deviation for low values of f_m and use the period jitter when f_m is large.

The fundamental method for the treatment of noise in PLL's and oscillators can be found in [4]. Accurate formulas for some cases of noise exist in [4] and [9]. However, these expressions cannot be applied to all cases required by this design. Therefore, accuracy is traded for insight by developing a set of simplified approximate expressions for the phase noise. The phase noise power spectrum density (PSD) at the output of the PLL in the open-loop configuration is given by

$$\phi_{ol} = G_{ol}^2(f_m) \cdot \phi_{\text{in}}(f_m) \quad (18)$$

where $\phi_{\text{in}}(f_m)$ is the voltage or current PSD of some noise source, $G_{ol}(f_m)$ is the modulus of the corresponding transfer function, and $f_m = \omega/2\pi$ is used as the variable, as is often done in noise calculations. The output phase PSD with the loop closed is then

$$\phi_{\text{out}}(f_m) = \frac{\phi_{ol}(f_m)}{|1 + A_{ol}(j2\pi f_m)|^2} = \frac{G_{ol}^2(f_m)\phi_{\text{in}}(f_m)}{|1 + A_{ol}(j2\pi f_m)|^2} \quad (19)$$

where $A_{ol}(j2\pi f_m)$ is the PLL open-loop gain. The mean-square value of the phase noise is obtained by integrating

$$\overline{\theta_{\text{out}}^2} = \int_0^\infty \phi_{\text{out}}(f_m) df_m. \quad (20)$$

In order to simplify this integration, an approximation based on the Bode diagram from Fig. 6 is used:

$$\frac{1}{|1 + A_{ol}(j2\pi f_m)|^2} \approx \begin{cases} \dfrac{f_m^4}{f_z^2 \cdot f_k^2} & \text{for } f_m \leq f_z \\ \dfrac{f_m^2}{f_k^2} & \text{for } f_z < f_m \leq f_k \\ 1 & \text{for } f_m > f_k. \end{cases} \quad (21)$$

Besides the obvious inaccuracies at the segment boundaries, the approximation neglects the nondominant effects on the loop. This is justified by the very small device noise contribution at frequencies above 1 MHz. The infinite integration limit for $\overline{\theta^2}$ is kept for simplicity.

$G_{ol}^2(f_m)$ always contributes the term f_m^{-2} due to the frequency-to-phase transition at the output of the PLL. An additional f_m^{-2} terms exists if the noise is low-pass filtered with one pole close to the origin. Finally, in the case of flicker noise, $\phi_{\text{in}}(f_m)$ has a term f_m^{-1}. Therefore, most noise sources in this design result in an open-loop phase noise PSD at the PLL output of the form

$$\phi_{ol}(f_m) = C \cdot f_m^{-n}. \quad (22)$$

This result yields (23) at the bottom of this page. The integral I_n for $n = 2, 3, 4$ is then evaluated as

$$I_2 = \frac{2}{f_k}\left(1 - \frac{f_z}{3f_k}\right) \quad (24)$$

$$I_3 = \frac{1}{f_k^2}\left(1 + \ln\frac{f_k}{f_z}\right) \quad (25)$$

and

$$I_4 = \frac{2}{f_k^3}\left(\frac{f_k}{f_z} - \frac{1}{3}\right). \quad (26)$$

For values of $n > 4$, the integral I_n does not converge.

The most pronounced noise source is the flicker noise generated by the OTA in the ICO, and is mainly due to the noise in the differential pair and load devices. This noise source can be represented as an equivalent voltage source in series with the negative input terminal of the OTA, and has the following transfer function to the PLL output phase:

$$G_{ol}^2(f_m) = \frac{\left[f_{\text{out}} + \dfrac{(R_c + R_T)}{C_k R_c R_T}\right]^2}{V_{\text{ref}}^2} \cdot \frac{1}{f_m^2} = \frac{B}{f_m^2}. \quad (27)$$

The input noise PSD takes the form

$$\phi_{\text{in}}(f_m) = \frac{A}{f_m} \quad (28)$$

and the open-loop output phase PSD becomes

$$\phi_{ol}(f_m) = \frac{AB}{f_m^3}. \quad (29)$$

The expression for $\overline{\theta^2}$ can be written directly as

$$\overline{\theta^2} = \theta_{\text{rms}}^2 = ABI_3 = AB\frac{1}{f_k^2}\left(1 + \ln\frac{f_k}{f_z}\right). \quad (30)$$

Substitution of numerical values with $f_k = 15.57$ kHz (midrange) results in $\theta_{\text{rms}} = 3.3°$ versus the simulated result of $3.1°$. To aid the suppression of noise, the OTA is designed with large area transistors which, as previously stated, are impossible to use inside the high-frequency ring oscillator.

The PSD of the noise current injected into the ICO input by the centering resistor R_c is $\phi_{\text{in}} = 4kT/R_c$, and the corresponding output phase PSD becomes

$$\phi_{ol}(f_m) = \frac{4kT}{R_c} \cdot \frac{K_i^2}{4\pi^2} \cdot \frac{1}{f_m^2}. \quad (31)$$

Since (31) is of the form f_m^{-2}, I_2 is used:

$$\overline{\theta^2} = \frac{4kT}{R_c} \cdot \frac{K_i^2}{4\pi^2} \cdot \frac{2}{f_k}\left(1 - \frac{f_z}{3f_k}\right). \quad (32)$$

θ_{rms} is calculated to be $0.62°$, and simulation of θ_{rms} yields $0.59°$.

In some PLL designs [1], [2], the gain of the oscillator is purposely made low in order to reduce the noise due to other PLL components. The resulting reduction of the frequency control range usually requires a coarse control of the center frequency. This approach is justified if the oscillator is not the dominant noise source, and if the mechanism defining the center frequency has better noise performance than the rest of the PLL. Neither is the case in this design; therefore, this approach is not used.

It should be noted that defining the center current of the ICO with a transistor current source increases θ_{rms} of the PLL three to five times. Flicker noise in that and other devices in the bias chain is eliminated when a resistor is used. Additionally, the thermal noise current PSD of the transistor is about five times larger than that of the resistor for device parameters within the design limits.

The thermal noise generated by the $R_F = 400$ MΩ output impedance of the VDD filter also deserves attention. The second filter section generates a white PSD which is subjected to a single pole roll-off at $f_p = 30$ Hz:

$$\phi_{\text{in}}(f_m) \approx 4kTR_F\frac{f_p^2}{f_m^2}. \quad (33)$$

$$\overline{\theta_{\text{out}}^2} = \int_0^\infty \phi_{\text{out}}(f_m) df_m = C\left(\int_0^{f_z} f_m^{-n} \cdot \frac{f_m^4}{f_z^2 \cdot f_k^2} df_m + \int_{f_z}^{f_k} f_m^{-n} \cdot \frac{f_m^2}{f_k^2} df_m + \int_{f_k}^\infty f_m^{-n} df_m\right) = CI_n \quad (23)$$

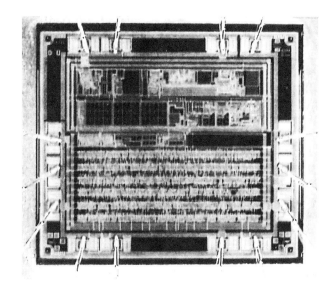

Fig. 9.   Photomicrograph of a cell-based PLL application.

By taking into account the impact of V_{ref} on V_{sup}, the ICO, and the loop filter, $G_{ol}^2(f_m)$ is found to be

$$G_{ol}^2(f_m) = \frac{(f_c - f_{\mathrm{out}})^2}{V_{\mathrm{ref}}^2} \cdot \frac{1}{f_m^2} \qquad (34)$$

where f_c is the center frequency of the ICO and f_{out} is the ICO output frequency. Since ϕ_{ol} contains an f_m^{-4} term, this case calls for I_4:

$$\overline{\theta^2} = 4kTR_F f_p^2 \frac{(f_c - f_{\mathrm{out}})^2}{V_{\mathrm{ref}}^2} \cdot \frac{2}{f_k^3}\left(\frac{f_k}{f_z} - \frac{1}{3}\right). \qquad (35)$$

This expression yields $\theta_{\mathrm{rms}} = 0.063°$ versus a simulated $0.057°$.

The remaining noise sources are analyzed using the same approach. Transistor switches $S1$ and $S2$ in the ICO only contribute sampled thermal noise since their flicker noise is heavily attenuated in the linear region of operation. The contribution of the sampled thermal noise can be modeled with a resistor

$$R \propto \frac{1}{C_k f_{\mathrm{out}}} \qquad (36)$$

connected between the ICO input and ground. The resulting phase noise is lower than that due to R_c or R_T. Sampled noise calculation principles can be found in [10]. The op amp and voltage-to-current conversion resistor are the main noise contributors in the loop filter (Fig. 5). Therefore, the op amp is designed for low noise. Charge pump noise is negligible due to the very low output duty cycle when the loop is in lock.

Contributions of all noise sources are calculated, and then simulated using a special transistor level noise model of the PLL developed in support of this design. Individual noise sources are modeled at the transistor level and linked to the ideal, noise-free model of the PLL by means of dependent

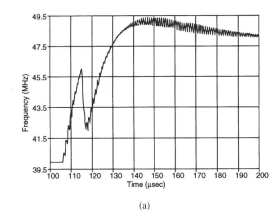

(a)

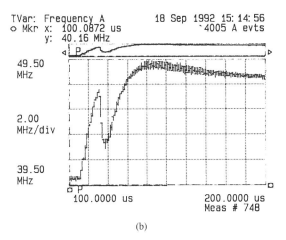

(b)

Fig. 10.   Simulated versus measured transient step response of PLL.

voltage and current sources. Discrepancies between analytic expressions and simulation for this design are within 28% for $\overline{\theta^2}$ and 14% for θ_{rms}. While the simulation results are accurate for a limited number of points within the multidimensional parameter space, analytic expressions provide orientation and enable decision making at early stages of the design.

V. RESULTS

Five applications are complete. A photomicrograph of one of the applications containing a frequency synthesizer is shown in Fig. 9. The analog area of the chip is approximately 2.36 mm^2. Characterization is performed in both the time and frequency domains. Measured PLL behavior is close to the predicted values in all cases. The measured frequency step response (Fig. 10) is very similar to the response predicted by simulation. Samples from a production lot of silicon demonstrate that the gain of the ICO is exceptionally linear and slightly higher than predicted, but well within the specification. The measured transfer characteristic of Fig. 11 depicts the lot mean, with the error bars representing one standard deviation of the data. These measurements are accomplished by observing the voltage at node A (Fig. 5) for a number of programmed output frequencies. Seen from node A, the oscillator behaves like a voltage-controlled oscillator (VCO).

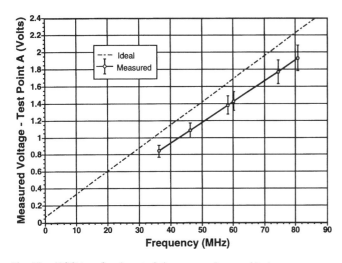

Fig. 11. VCO transfer characteristic: measured versus ideal.

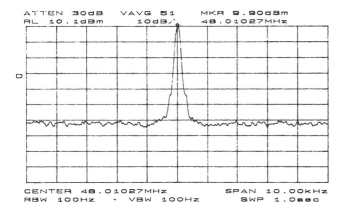

CENTER 48.01027MHz   SPAN 10.00kHz
RBW 100Hz · VBW 100Hz   SWP 1.0sec

Fig. 12. Spectral analysis of synthesized frequency.

TABLE I
SUMMARY OF PLL FUNCTIONALITY AND PERFORMANCE

Maximum output frequency	80 MHz
Reference frequency	14MHz–18MHz
Frequency programming resolution	∼0.1%
PLL bandwidth	7kHz–26kHz
Loop filter	Internal
DC power dissipation	30 mW
Total power dissipation	125 mW
(f_{out} =80 MHz, C_{load} = 25 pF)	
Period jitter	100 ps
Phase noise	4°–7.5° rms
(Bandwidth dependent)	

The phase noise and period jitter are measured using a time interval analyzer. The low-frequency phase noise is a combination of flicker and thermal noise. Measurements yield rms values between 4° and 7.5°, depending upon the actual PLL bandwidth. The period jitter is measured at 100 ps. The results of the spectral analysis are shown in Fig. 12. For a summary of results, see Table I.

VI. CONCLUSIONS

The development of a family of standard cells for PLL applications has been presented. The loop cells, which include the PFD, the charge pump, the filter, and the ICO, are all on chip and are designed to reduce phase noise and process parameter dependence. If necessary, the loop characteristics can be adapted to each application with a minimum of modifications to these cells. Three biasing cells: a power supply filter, a capacitive bias generator, and a current repeater, supplement the loop cells. Because the applications using these PLL's require low phase noise, special emphasis is placed on the analysis of the noise sources and their contributions to the overall noise performance of the PLL. Results from laboratory measurement of completed silicon are consistent with hand calculations and both behavioral and transistor level simulations.

ACKNOWLEDGMENT

The authors would like to thank the following people for their contributions in the development of these synthesizers: G. Goodman for digital circuit design, C. Bellman for C program development, A. Harris for cell model development, D. Davis, L. Fischer, S. Lewis, C. Threatt, C. Yu, and D. Zyriek for place and route and test development, and C. Clevenger, B. Harris, D. Holland, and R. Palmer for cell layout and mask preparation.

REFERENCES

[1] R. F. Bitting *et al.,* "A 30–128MHz frequency synthesizer standard cell," in *Proc. IEEE 1992 Custom Integrated Circuits Conf.,* pp. 24.1.1–24.1.6.
[2] R. Shariatdoust *et al.,* "A low jitter 5 MHz to 180 MHz clock synthesizer for video graphics," in *Proc. IEEE 1992 Custom Integrated Circuits Conf.,* pp. 24.2.1–24.1.5.
[3] R. E. Best, *Phase-Locked Loops Theory, Design, & Applications.* New York: McGraw-Hill, 1984, p. 8.
[4] D. H. Wolaver, *Phase-Locked Loop Circuit Design.* Englewood Cliffs, NJ: Prentice-Hall, 1991.
[5] F. M. Gardner, "Charge-pump phase-lock loops," *IEEE Trans. Commun.,* vol. COM-28, pp. 1849–1858, Nov. 1980.
[6] F. M. Gardner,, *Phaselock Techniques,* 2nd ed. New York: Wiley, 1979.
[7] B. Kim, D. N. Helman, and P. R. Gray, "A 30-MHz hybrid analog/digital clock recovery circuit in 2-μm CMOS," *IEEE J. Solid-State Circuits,* vol. 25, pp. 1385–1394, Dec. 1990.
[8] I. A. Young, J. K. Greason, and K. L. Wong, "A PLL clock generator with 5 to 110 MHz of lock range for microprocessors," *IEEE J. Solid-State Circuits,* vol. 27, pp. 1599–1607, Nov. 1992.
[9] A. Blanchard, *Phase-Locked Loops: Application to Coherent Receiver Design,* reprint ed. FL: Krieger, 1992.
[10] R. Gregorian and G. Temes, *Analog MOS Integrated Circuits for Signal Processing.* New York: Wiley, 1986.

Fully Integrated CMOS Phase-Locked Loop with 15 to 240 MHz Locking Range and ±50 ps Jitter

Ilya Novof, *Member, IEEE*, John Austin, Ram Kelkar, *Member, IEEE*, Don Strayer, and Steve Wyatt

Abstract— A fully integrated phase-locked loop (PLL) in a digital 0.5 um CMOS technology is described. The PLL has a locking range of 15 to 240 MHz. The static phase error is less than ±100 ps with a peak-to-peak jitter of ±50 ps at a 100 MHz output frequency. The PLL has a resistorless architecture achieved by the implementation of feedforward current injection into the current controlled oscillator.

I. INTRODUCTION

PHASE-locked loops (PLL) are widely used for clock-phase synchronization, frequency synthesis, and clock distribution in ASIC and microprocessor applications [1]–[5]. A typical PLL application in ASIC is shown in Fig. 1; the internal clock needs to be distributed on a large chip through the clock distribution tree. The PLL can perform either phase synchronization or frequency synthesis or both simultaneously. The PLL synchronizes the clock phase at the output of the clock distribution tree (feedback clock) with the external clock (deskew the clock) and synthesizes the higher internal chip frequency from the external clock, which simplifies the system board-level design.

The timing diagram in Fig. 1 shows that both the feedback and external clock rising edges are synchronized. The feedback clock frequency is divided internally to match that of the external clock.

It is highly desirable that the standard digital CMOS process be used in the PLL design because process modifications to accommodate the PLL increase product cost. Other desirable features include minimal sensitivity to noise and a fully integrated design. A standard digital CMOS process is used to produce differential circuits that are designed to reduce their sensitivity to substrate and supply noise. To further reduce PLL jitter due to power supply noise, the analog and digital PLL portions are wired to two separate die power-supply pads. The PLL function includes multiplication of frequency and synchronization of input and output clock phases. The architecture is unique because resistors are not needed for PLL loop stabilization. The PLL was designed with a locking range from 15–240 MHz (worst case) and a static phase error of less than ±100 ps with a peak-to-peak cycle-to-cycle jitter of ±50 ps in typical applications.

Manuscript received May 2, 1995; revised September 14, 1995.
I. Novof, J. Austin, R. Kelkar, and S. Wyatt are with IBM Microelectronics Division, Essex Junction, VT 05452 USA.
D. Strayer was with IBM Microelectronics Division, Essex Junction, VT 05452 USA. He is now with the Systran Corporation, Dayton, OH.
IEEE Log Number 9415858.

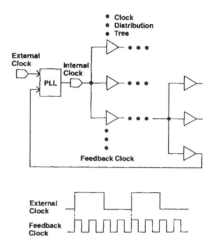

Fig. 1. Typical PLL applications in ASIC.

II. PLL ARCHITECTURE

Fig. 2 shows a block diagram of the differential resistorless PLL. The PLL phase-frequency detector is tri-state sequential and without a "dead zone," or a region of low gain near phase-lock. Its inputs are the reference clock, PLL output clock, and the output of the feedback divider/ pulse generator. A pulse-masking technique is used to nullify the effect of the feedback divider delay on the PLL static phase error. A forward divider extends the operational PLL frequency range from 15–240 MHz. The programmable feedback divider/pulse generator divides the feedback clock, establishing the PLL frequency multiplication factor (from 1–10). The initialization circuit precharges the loop filter during power-up or system-reset to guarantee the proper PLL startup sequence. The decoder programs the auxiliary charge pump output or forward current as well as setting the forward and feedback divider ratios. The PLL includes a test controller that provides full PLL test capability and reduces PLL test time and PLL pin count.

III. PLL LOOP DESIGN

In conventional loop designs, a second-order filter (R_{ser} and C_{ser} in series and a second smaller C_{pr} in parallel across the series RC) is fed by current pulses from the charge pump. The action of the charge pump can be best explained by examining the voltage across the filter $v(s)$ where

$$v(s) = i(s) \times R_{ser} + \frac{i(s)}{sC_{ser}} \qquad (1)$$

Reprinted from *ISSCC Dig. Tech. Papers*, pp. 112-113, February 1995.

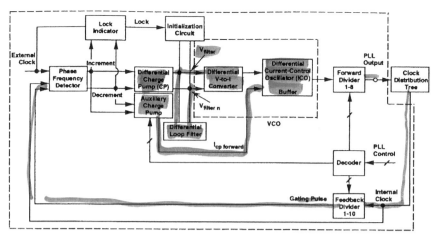

Fig. 2. PLL block diagram.

and $i(s)$ is that portion of the charge pump output that flows through the series RC structure. The filter output $v(s)$ is fed into a voltage-to-current converter whose output i_1 is fed into a current-controlled oscillator. Denoting the gain of the voltage-to-current converter as g, we have

$$i_1 = g \times v(s)$$
$$= g \times R_{ser} \times i(s) + g \times \frac{i(s)}{sC_{ser}}. \quad (2)$$

It is easy to see that the first term in (2), $g \times R_{ser} \times i(s) = i_d$ is really the current multiplied by a gain factor $g \times R_{ser}$, while the second term represents the integral of the injected charge. The current fed into the oscillator is thus the sum of two components.

The open loop gain GH in conventional designs is of the form

$$GH = \left[\frac{KI(s + w_z)}{s^2(s + w_p)} \right] \quad (3)$$

where

$$w_z = \left(\frac{1}{R_{ser} \times C_{ser}} \right)$$

and

$$w_p = \left(\frac{1}{R_{ser} \times C_{pr}} \right).$$

The added zero (w_z) improves the phase margin and ensures that, at unity-gain frequency, the slope of the transfer function corresponds to a single pole. The zero is thus positioned at a frequency lower than unity-gain frequency, while the pole (w_p) is positioned above the unity-gain frequency. The zero in the numerator is controlled by the series components, while the parallel capacitor controls the pole; the two track each other through the series resistor, which affects both.

In this PLL loop architecture, only a single capacitor C_{ser} is needed rather than the three components (two capacitors and one resistor) in conventional designs. The auxiliary charge pump in Fig. 2 creates the zero, while the pole in the ICO (current-controlled oscillator) circuit substitutes for the filter pole. In our implementation, the total injected current is created by summing the two terms in (2). The auxiliary

charge pump creates the first term i_d by applying the correct gain to the current, while the main charge pump implements the capacitive integration through the filter and the voltage-to-current converter. Therefore, changing the applied gain effectively changes the value of the equivalent filter resistance. The two components of the total current are then summed at the current-controlled oscillator input. A key benefit of this resistorless implementation is the space and cost saving achieved because a resistor is not needed; also the gain $g \times R_{ser}$ can be easily changed to accommodate a wide range of input- and output-clock operating frequencies (changing the gain changes the equivalent resistor value which moves the location of the zero; this can be used to compensate for a wide range of input and output frequencies).

In [4], the authors mention the use of an auxiliary charge pump to realize the zero; the approach detailed above differs from the one in the reference in implementation and in the use of variable gain to easily change the value of the resistor and thus the position of the zero. Also, our approach results in fewer filter components, whereas the one referenced uses an extra R and C to create the pole.

The loop transfer function is derived using the following gain definitions:

1) Charge Pump gain: $I_{cp}/(2\pi)$ mA, where I_{cp} is the dc level of the charge pump output.
2) Filter gain: $1/sC_1$ (C_1 is filter capacitor)
3) V–I gain: $g_1 e^{-sT_v}$ mA/V
4) V–I dc gain: g mA/V
5) ICO gain: $2\pi K_1/s(1 + sT_1)$, where K_1 is the dc gain in GHz/mA, and T_1 is the ICO internal pole.
6) Forward path gain: $A_{g1} e^{-sT_g}$
7) Forward path dc gain: A_g
8) Clock Tree delay: T_d ns
9) Forward divider: D_f
10) Feedback divider: D_b

where the forward path is defined as the path through the auxiliary charge pump and the V–I converter.

The V–I and forward path gains shown above include time constants T_v and T_g to model intrinsic poles resulting from the circuit implementation. Adding the delay terms e^{-sT_g} and

e^{-sT_v} and modifying the dc gains as shown in (4) provides a way to model the poles (at T_g and T_v) in the forward path as well as in the V–I converter.

$$g_1 = \frac{g}{\sqrt{[1 + (2\pi f T_v)^2]}}$$

$$A_{g1} = \frac{A_g}{\sqrt{[1 + (2\pi f T_g)^2]}}. \quad (4)$$

The action of the pole, when modifying the amplitude, is captured by (4), while the phase delay provided by the pole is approximated by the delay term. Modeling the poles as shown results in approximate (even pessimistic) phase delays, but the technique is easy to use, and in cases where the pole locations are above the unity-gain frequency, it leads to accurate Bode plots.

Using the gain expressions as defined above, it is easy to write the loop transfer function GH as

$$GH = \left\{ \frac{I_{cp} K_1 g_1 \left[1 + s C_1 \left(\dfrac{A_{g1}}{g_1}\right) e^{-s(T_g - T_v)}\right]}{D_f D_b C_1 s^2 (1 + s T_1)} \right\}$$
$$\cdot\, e^{-s(T_d + T_v)}. \quad (5)$$

The loop transfer function here takes the same form as that provided in conventional designs (3), with one zero and three poles (two of which are at the origin). The zero is created by the filter capacitor and the equivalent resistor A_{g1}/g_1 and the pole is the internal ICO pole. The internal ICO pole is formed by adding a capacitor at the input to the oscillator. Loop design proceeds by first measuring the intrinsic pole location and then adding capacitance to position the pole in the desired location. The value of the equivalent resistance can be easily changed by varying the value of the gain A_{g1}.

In the conventional second-order filter, the pole and zero locations are both controlled by the value of the resistance, which provides tracking between the pole and the zero. In our implementation, the pole is formed by the internal ICO pole, which is independent of the resistance value; thus, the pole and the zero do not track as the resistor value is changed. This is not a limitation because nearby devices track each other well. On the other hand, this scheme saves both a resistor and capacitor because the ICO pole is an internal pole resulting in a smaller filter.

Another key advantage of the present implementation is the ease with which the equivalent resistor value can be changed via the gain A_g to accommodate a wide range of forward and feedback divider ratios. Stability analysis using (4) revealed that only three values of gain A_g would be sufficient to realize a stable design for the following forward and feedback divider ratios:

$$\text{For} \quad PM = 2 \cdots 6, \qquad A_g = 0.7$$
$$PM = 7 \cdots 16, \qquad A_g = 1.0 \quad \text{and}$$
$$PM = 18 \cdots 40, \qquad A_g = 1.7$$

where $PM = (D_f \times D_b)$ is the product of the forward and feedback dividers.

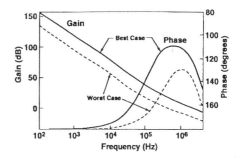

Fig. 3. PLL bode plot.

The gain of the forward path A_g is set in the auxiliary charge pump through the decoder circuit (Fig. 2).

Fig. 3 shows the minimum and maximum gain and phase for A_g set to 1.0; the phase plot begins at $-180°$, corresponding to the double pole at the origin in (4), and the zero pulls the phase up. The high frequency pole then pulls the phase down; thus, the action of the zero in providing phase margin can be clearly seen. The predicted minimum phase margin is 45°, and the loop bandwidth is 491 KHz from Fig. 3 (the phase margin is calculated by adding 180 to the phase at any particular frequency). Similar values of phase margin were obtained for $A_g = 1.7$ and $A_g = 0.7$; the loop bandwidth varied from 195–537 KHz, respectively.

Monte-Carlo analysis was conducted on the loop design to check phase and gain margins for each of the three gain values. Tolerance values used in the analysis included the effects of process variation, circuit tolerance, as well as temperature and power supply. A time-domain simulation model composed of behavioral macro-models of the PLL key elements was built and used to confirm the loop design. Monte-Carlo analysis of the time-domain model was conducted for each of the three gain values using the above tolerance values. The analysis showed that the loop was stable and the output clock locked in to the input clock for initial oscillator frequencies both above and below the output frequency. The settling time predicted was 4 μs minimum to 9 μs maximum.

IV. PLL Components

A. Phase-Frequency Detector

The circuit diagram of the gated phase-frequency detector is shown in Fig. 4(a). The PLL is designed to synchronize the rising edges of the external and feedback clocks. The detector is designed as a falling-edge detector because there is an inverting receiver between the die pad, to which the external clock is applied, and the detector input.

A matching inverting receiver was introduced in the PLL feedback path to compensate for the delay through the input receiver. The sequential phase and frequency detector has a monotonic phase error transfer characteristic over the full clock cycle independent of the duty cycle. A conventional detector [4] has a region of low gain near phase lock or the "dead zone." To minimize the PLL static phase error, it is desirable to use a detector without this dead zone. To eliminate the detector dead zone, a delay element was added, which consists of inverters $G1$ and $G2$ in Fig. 4(a). This creates

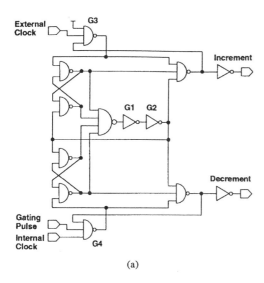

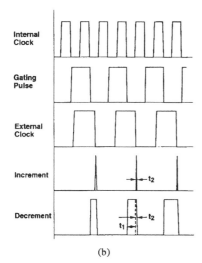

Fig. 4. PLL phase-frequency gated detector. (a) circuit diagram. (b) timing diagram.

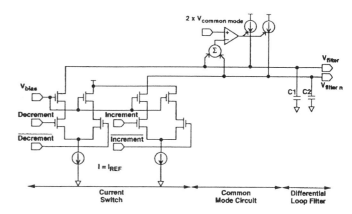

Fig. 5. Differential charge pump and loop filter.

minimum-width increment and decrement pulses when the input signals are in phase. Because the widths of the increment and decrement minimum pulses are the same, the net charge added to the loop filter by the charge pump is zero. Gates $G3$ and $G4$ in Fig. 4(a) are used to gate the feedback clock edges by the feedback divider/pulse generator narrow pulses nullifying the effect of the feedback divider delay on the PLL static phase error. Phase-frequency gated detector timing diagrams, when the internal clock phase leads the external one, are shown in Fig. 4(b). The external clock and gating pulse are of the same frequency, however, the internal clock is of higher frequency. The detector output decrement pulse consists of $t1$ and $t2$ portions, however, the increment portion has only $t2$. Time interval $t1$ corresponds to a phase difference between external and internal clocks; $t2$ is defined by the delay through $G1$ and $G2$.

B. Differential Charge Pump and Loop Filter

The primary differential charge pump and loop filter are shown in Fig. 5. This charge pump has common-mode feed-

back that increases the output dynamic range by maintaining constant common-mode voltage at the loop filter capacitor. The phase-frequency detector was simulated together with the charge pump to compensate for the slight delay between the inverted and noninverted outputs of the phase-frequency detector.

The loop filter is fully differential and consists of two 250-pF capacitors. The capacitor structures are created as segmented NFET's in an N-well to improve capacitor linearity and time constant. A feature of the PLL design is illustrated in Fig. 6, which shows a loop filter capacitor and current reference resistor. This structure consists of a MOS capacitor and first-level-metal resistor used for the bandgap reference described below. The resistor is stacked on top of the capacitor to minimize the chip area used. The charge pump current switch consists of two differential stages that are controlled by the decrement and increment outputs of the phase-frequency detector. The decrement controls the current injection in the $C1$ capacitor of the differential loop filter. The increment controls the current injection in the $C2$ capacitor of the differential loop filter. The charge pump current is set at $I = I_{ref} = 35\,\mu\mathrm{A}$.

C. Bandgap Reference

The current reference circuit is based on a standard bandgap voltage reference circuit [5]. A 1.2 KΩ metal resistor is added between one of the p$^+$ to n-well diodes and NFET current mirror to tighten the output current statistical distribution.

D. Auxiliary Charge Pump

An auxiliary charge pump is used to create a zero in the PLL transfer characteristic to ensure PLL loop stability. It injects current directly into the current-controlled oscillator (ICO) input, bypassing the differential loop filter and voltage-to-current ($V-I$) converter. The auxiliary charge pump circuit diagram is shown in Fig. 7; the charge pump current value changes in conjunction with the programmable value of the feedback and forward dividers ($I1 = 0.7I_{ref}$, $I2 = I_{ref}$, $I3 = 1.7I_{ref}$). Differential switches $T1$ and $T2$ are controlled by the increment and $T5$ and $T6$ by decrement outputs of the phase-frequency detector. The differential switches control the net

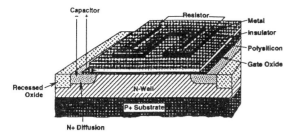

Fig. 6. Loop filter capacitor and current reference resistor structure.

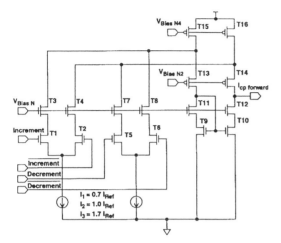

Fig. 7. Auxiliary charge pump circuit diagram.

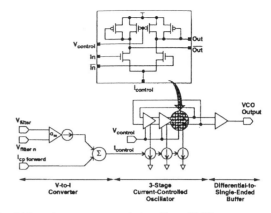

Fig. 8. Differential voltage-controlled oscillator (VCO).

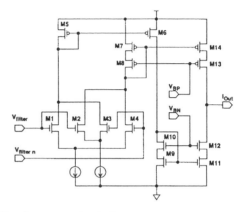

Fig. 9. V-to-I converter circuit diagram.

current flow through the current mirrors $T9, T10, T15, T16$. Cascaded FET's $T3, T4, T7, T8$, and $T11–T14$ are added to increase the output impedances.

E. Differential Voltage-Controlled Oscillator (VCO)

Fig. 8 shows the voltage-controlled oscillator (VCO), which consists of a V–I converter, current-controlled oscillator (ICO), and differential-to-single-ended buffer. The ICO is controlled by the sum of the V–I converter and auxiliary charge-pump currents. The current summation is achieved by driving both V–I converter and auxiliary charge-pump currents into the same current mirror and mirroring the resulting current into ICO.

A V–I converter circuit diagram is shown in Fig. 9. It is a transconductance stage with a differential input and converts the differential loop filter voltage into current. It consists of cross-coupled differential pairs $M1$–$M4$ and cascaded current mirrors $M5, M6, M7, M9, M11, M14$. Cascaded FET's $M8, M10, M12, M13$ increase output impedances. Differential pair cross-coupling extends the transconductance circuit linear operational region. A basic CMOS differential stage has an S-shaped transfer function. In our circuit, a second differential pair is connected in such a way that its output current adds to the current from the main pair but in opposite phase. The tail current and device sizes of this pair are chosen so that this pair swings through its active region over a smaller input voltage range than the main pair does. The second pair, therefore, works against the main pair in the region where

the main pair's gain is greatest and has little effect at more extreme input voltages where the main pair has less gain. The current ratio of the main and supplemental differential pairs is $I1/I2 = 2.5$.

The ICO circuit was designed to have a minimum power supply and substrate noise sensitivity and, at the same time, provide a wide and linear frequency operational range; VCO power-supply sensitivity is 0.01% per 1% power supply-change. Fig. 10 illustrates the VCO transfer function; it has a linear range of at least 45–240 MHz. VCO nominal gain is 178 MHz/V. The ICO is a three-stage ring oscillator; each stage is fully differential and consists of a source-coupled NFET pair and symmetrical load elements comprising a diode-connected PFET and a variable PFET current source. The oscillator control current is mirrored into the current sources in such a way that the output-voltage swing across the load remains constant over a wide range of control currents and output frequencies. This increases the ICO linear range of operation.

Fig. 11 shows the differential-to-single-ended buffer circuit diagram. It consists of two source followers, a differential stage, and two inverters. The source followers are used to increase the differential-to-single-ended buffer input impedance and minimize its loading effect on the ICO output. The source followers consist of NFET's $K3$, $K6$, and current sources. The differential stage converts the differential signal into a

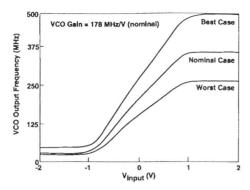

Fig. 10. VCO transfer function.

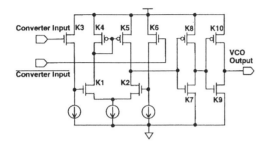

Fig. 11. Differential-to-single-ended buffer circuit diagram.

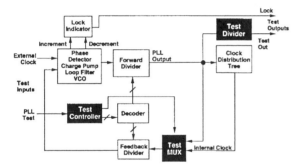

Fig. 12. PLL functional-test block diagram.

single-ended one. It consists of NFET differential pair $K1$, $K2$, and PFET load $K4$, $K5$. Two CMOS inverter stages consisting of FET's $K7–K10$ are added to provide sufficient drive capability and improve the duty cycle. The differential-to-single-ended converter converts the differential ICO output into single-ended CMOS levels with a 50% $\pm3\%$ duty cycle.

V. PLL DESIGN METHODOLOGY

An in-house program was developed to plot PLL open-loop Bode plots, optimize loop parameters, and to assure loop stability for all PLL operation modes. Statistical analysis was performed for various parameter changes such as process, temperature, power-supply voltage deviations. A closed-loop behavioral time-domain PLL model was developed and requires just a fraction of CPU simulation time compared to the device-level simulation. It made the PLL detailed statistical simulations of the PLL closed loop behavior practical, which

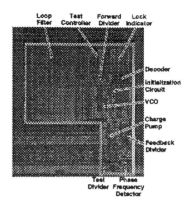

Fig. 13. PLL micrograph.

permitted PLL optimization for settlement time, jitter, and static phase error.

VI. PLL FUNCTIONAL TEST

Fig. 12 is a PLL functional-test block diagram; the PLL test support circuits are all integrated on-chip. The phase-detector outputs are examined by a lock indicator that indicates the PLL locked condition when the output clock is in phase with the input. Pulses are applied to the test-controller input to force the forward and feedback dividers through all the programmable states. The test divider reduces the PLL output frequency by a factor of 8, a value that can be measured by the production tester. PLL lock conditions and output frequencies are tested to ensure coverage of all combinations of the forward and feedback dividers. This approach reduces test costs because it uses a low-frequency production tester to ac test high-frequency PLL's. The test controller circuit minimizes test time since it does not require time-consuming scan operations to load the values of the PLL dividers.

VII. MEASURED RESULTS

Fig. 13 shows a micrograph of the PLL that occupies 0.82 mm^2 area and uses only two layers of metal for inclusion in an ASIC family. Module-level tests were performed on a typical ASIC chip with internal and I/O buffer switching activity. Fig. 14 shows the output clock at 100 MHz with an input clock of 25 MHz; the forward and feedback divider ratios were set to 2 and 4, respectively. The jitter histogram of the output is also shown. Table II shows the measured PLL performance at 3.3 V power supply and ambient temperature. The PLL power-supply noise sensitivity was measured by applying a square wave of various frequencies and amplitudes to the analog power-supply pin. As expected, the PLL jitter sensitivity was higher at a lower power-supply noise frequency. At a 100 kHz power-supply noise frequency, the PLL jitter sensitivity was 82 ps per 1% power-supply change versus only 9 ps at 100 MHz. Fig. 15 shows PLL output clock jitter dependence on the external clock jitter. Measurements were performed at two different PLL output clock frequencies: 90 MHz and 120 MHz. The external clock was modulated with a jitter frequency of 100 kHz; this low-frequency jitter is not well attenuated and propagates from the PLL input to output. The typical power

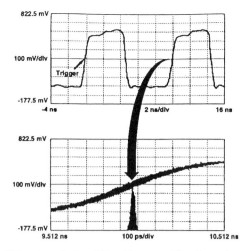

Fig. 14. PLL output clock and jitter histogram. Experimental; peak-to-peak jitter = 90 ps.

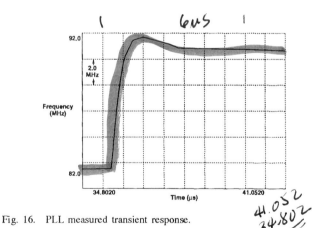

Fig. 15. PLL jitter measurements. PLL external clock jitter frequency = 100 kHz.

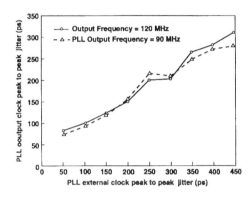

Fig. 16. PLL measured transient response.

TABLE I
3.6-V CMOS 5L PARAMETER SUMMARY

MOSFET Parameters		
	NFET	PFET
V_t	0.57 V	-0.71 V
L_{eff}	0.46 µm	0.51 µm
I_{dsat}	0.346 mA/µm	0.188 mA/µm
T_{ox}	13.5 nm	13.5 nm
Junction Depth	0.25 µm	0.30 µm
S/D	LDD	LDD
Process Parameters		
Minimal Lithographic Image	500 µm	
Metal 1 Pitch	1.8 µm	
Metal 2 Pitch	1.8 µm	
Metal 3 Pitch	1.8 µm	
Metal 4 Pitch	1.8 µm	
Metal 5 Pitch	1.8 µm	

TABLE II
PLL PERFORMANCE SUMMARY

PLL Area (including all support circuits, 2-level metal)	0.82 mm^2
Output Clock Cycle-to-Cycle Jitter (peak-to-peak)	Less than 100-ps Jitter
Static Phase Error	±100 ps
Lock-In Time	15 µs
Power Dissipation	33 mW
Bandwidth	537 kHz
Phase Margin	50°
VCO Frequency DC Power-Supply Sensitivity	0.01% per 1% Power-Supply Change
PLL Jitter AC Power-Supply Sensitivity	9 ps per 1% Power-Supply Change at 100-MHz Square Wave
	82 ps per 1% Power-Supply Change at 100-kHz Square Wave

measured PLL output cycle-to-cycle jitter is less than ±100 ps (peak-to-peak).

VIII. CONCLUSION

A manufacturable 15 to 240 MHz PLL has been developed in a digital CMOS technology. It can operate as a part of a noisy logic CMOS chip and is designed to address a wide range of applications. The PLL is fully integrated with built-in test support circuits and has demonstrated clock static phase error of less than ±100 ps and peak-to-peak jitter of ±50 ps.

ACKNOWLEDGMENT

The authors thank K. Short, T. Frank, M. Styduhar, E. Schneider, R. Chmela, A. Debrita, and R. Smith for their support in the design, layout, and characterization of this PLL design.

consumption was measured at 33 mW (85 mW worst-case) at a power supply of 3.0–3.8 V (3.3-V nominal). The operational temperature range is 0–100°C, and the loop is yielding an over-damped system with a phase margin greater than 45° under all conditions. The PLL measured transient response is shown on Fig. 16. It is within 10% of the simulated response.

The most demanding PLL application is on a large microprocessor with a maximum amount of internal and I/O buffer circuit switching activity. Under these conditions, the

REFERENCES

[1] Young *et al.*, "PLL clock generator with 5–110 MHz of lock range," *IEEE J. Solid-State Circuits*, vol. 27, pp. 1599–1607, Nov. 1992.

[2] M. Franz *et al.*, "A 240 MHz phase-locked loop circuit implemented as a standard macro on CMOS SOG gate arrays," in *Proc. IEEE 1992 Custom Integrated Circuits Conf.*, May 1992, pp. 25.1.1–25.1.4.

[3] I. Novof *et al.*, "Fully-integrated CMOS phase-locked loop with 15–240 MHz locking range and ± 50 ps jitter," in *IEEE 1995 Int. Solid-State Circuit Conf.*, pp. 112–113.

[4] D. Mijuskovic *et al.*, "Cell-based fully integrated CMOS frequency synthesizers," *IEEE J. Solid-State Circuits*, vol. 29, pp. 271–280, Mar. 1994.

[5] J. Alvarez *et al.*, "A wide-bandwidth low-voltage PLL for power PC microprocessors," *IEEE J. Solid-State Circuits*, vol. 30, pp. 383–392, Apr. 1995.

PLL Design for a 500 MB/s Interface

Mark Horowitz, Andy Chan, Joe Cobrunson, Jim Gasbarro, Thomas Lee, Wing Leung, Wayne Richardson, Tim Thrush, Yasuhiro Fujii[1]

Rambus Incorporated, Mountain View, CA/[1]Fujitsu Limited, Kawasaki, Japan

When operating pins at high data rates the key problem is to control timing skews (both on- and off-chip) so data on the pins can be read in a short time. The problem of external skews is solved by a clocking scheme where clock and data signals travel the same distance between the sender and receiver so there is little skew on the $600mV_{pp}$ external signals [1]. There are two clocks: one for incoming data (RxClk) and one for outgoing data (TxClk). In this bus protocol, data is sampled on each clock edge. For maximum setup and hold margins, each transmitted bit is centered between clock edges (Figure 1).

A PLL generates the properly-skewed internal clocks to operate the bus. Figure 2 is a block diagram of the PLL, consisting of one main loop and two fine loops (one each for the receive and transmit clocks). The main loop is a VCO-based second-order loop using a 6-stage, small-swing, differential ring oscillator VCO, and is locked to the incoming RxClk after it is amplified to full CMOS levels. VCO and input clock frequency are halved to allow the phase/frequency detector more time to settle.

The main loop provides twelve clock outputs dividing the period into as many signals spaced at 30° intervals. While the spacing is uniform, the absolute phase shift between these internal clocks and the external clock is not well determined, since the delay of the clock amplifier depends on the clock input level. All twelve outputs of the VCO feed two digital 'fine' loops (only one shown in detail in the Figure) that align precisely the phases of transmit and receive internal clocks.

To correct for skew as much as possible, the type of circuit used as input sampler functions as phase detector for the fine loop. The fine loop delays the internal clock (relative to the input RxClk) by an amount that causes the sampler to produce high and low outputs with precisely equal frequency, thereby compensating for sampler setup time.

The fine loop is a first-order digital loop and uses the output of the input sampler to control an up/down counter. The counter output feeds a phase selector that chooses an adjacent pair of clock signals from the output of the VCO. Since the 30° VCO output resolution is not fine enough, the low-order counter bits control a phase interpolator giving the digital loop 15 steps between adjacent phases. Resolution of the digital loop is 2°, or slightly over 20ps with a 4ns cycle.

Figure 3 shows phase selection/interpolation in more detail. The first stage (the phase selector) generates a pair of phases (odd and even) to interpolate between, while the phase interpolator generates output phase between these two input signals. The odd phase selector changes its output (the input to the phase interpolator) only when the phase interpolator is set all the way to the even phase, and vice versa. Since the output clock does not depend on the even phase input when it is changed, the change cannot cause a glitch. A small digital FSM keeps track of which phase input is earlier, and ensures that a high output of the clock sampler (signal Late) always causes the phase of the output to decrease. Two shift registers control odd and even phase selectors, and an up/down counter and current-mode DAC control the phase interpolator.

Drive clock generation uses two sets of phase selectors/interpolators. The output of the first interpolator feeds a dummy output driver (to compensate for its delay), whose output in turn is sampled to control the loop. This creates a clock that would generate an output aligned to the external clock. To generate the real transmit clock, the second phase interpolator uses the same shift registers and counter output, but renumbers the output phases from the VCO by three. This shift produces a clock that is 90° out of phase with the clock driving the dummy driver. Thus when this clock is used in an output driver, the data appears centered on the external clock.

This two-level loop can be partially powered down. As long as the main loop is left on continuously, the digital loops and the internal clocks need to be turned on only occasionally to track drifts, a power saving of approximately 70%. Interpolating between ring oscillator outputs means that the digital loops have no edge conditions, in contrast with delay-locked loops that employ finite delay lines. There are no start-up issues, deadbands, or large jitter regions in this design.

Like previous designs to reduce jitter in the clock system, the VCO, phase selectors, and phase interpolators use fully differential circuits with high power-supply rejection [2]. A schematic of the VCO buffer cell is shown in Figure 4. The pMOS current sources with the cross-coupled diode loads provide a differential load impedance independent of common-mode voltage, making cell delay insensitive to common-mode noise. The outputs of the VCO are buffered by source followers before they leave the buffer cell to isolate the VCO from the fine loops.

The phase interpolator uses the same structure as the VCO buffer cell to maintain noise properties (Figure 5). The nMOS differential pair is split into two, one pair driven by the output of the odd phase selector and the other, by the output of the even. Low-order counter bits control the relative current to the differential pairs. When the output should not depend on the even phase, current in that differential pair is set to be zero, while current in the odd differential pair is i. Since total gate current is constant, the output transition is roughly a fixed delay from the weighted average of input arrival times.

This PLL occupies 500x2500mm² on a 4.5Mb DRAM using a 0.8µm technology [3]. Figure 6 is a PLL micrograph. The composite loop draws 160mA from a single 5V supply at 250MHz when fully active, dominated by the dynamic power dissipated driving the capacitive clock loads. This decreases to approximately 50mA during standby.

The loop and interface operate from 50MHz to over 330MHz (660Mb/s), limited by test equipment. Figure 7 is a histogram showing regenerated sampling-clock jitter less than 36ps, and peak-to-peak jitter of 184ps. Figure 8 shows generation of data centered on the bus clock within 100ps of ideal position.

References

[1] Kushiyama et al., "500 MByte/sec Data Rate 512 Kbits x 9 DRAM Using a Novel I/O Interface", Sym. on VLSI Circuits, Digest of Technical Papers, pp. 66-67, June 1992.

[2] Young, I. et al., "A PLL Clock Generator with 5 to 110MHz Lock Range for Microprocessors", ISSCC DIGEST OF TECHNICAL PAPERS, p. 50, Feb. 1992.

[3] The Fujitsu MB814953.

Reprinted from *ISSCC Dig. Tech. Papers*, pp. 160-161, February 1993.

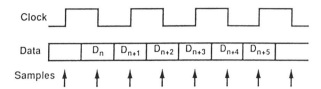

Figure 1: Timing of output signals, clocks and internal sample points.

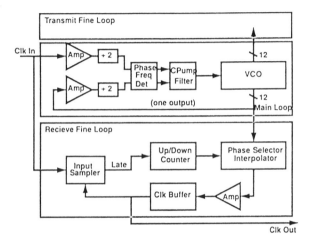

Figure 2: Block diagram of receive clock PLL.

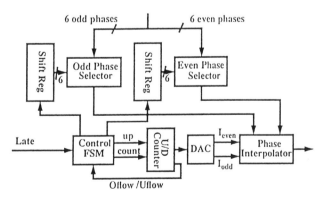

Figure 3: Receive fine loop.

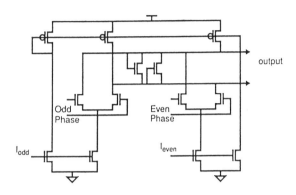

Figure 4: Schematic of the VCO buffer design.

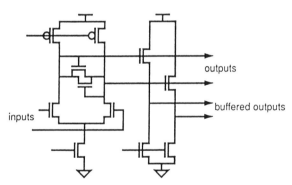

Figure 5: Schematic of the phase interpolator.

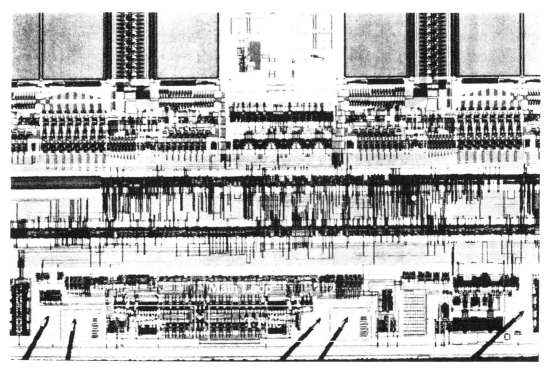

Figure 6: Micrograph showing PLL components.

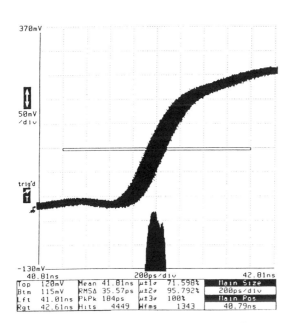

Top	120mV	Mean	41.81ns	μ±1σ	71.598%	Main Size	
Btm	115mV	RMSΔ	35.57ps	μ±2σ	95.792%	200ps/div	
Lft	41.81ns	PkPk	184ps	μ±3σ	100%	Main Pos	
Rgt	42.61ns	Hits	4449	Wfms	1343	40.79ns	

Figure 7: PLL jitter on the internal clock.

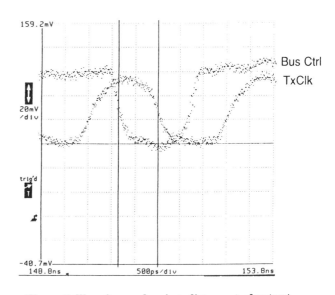

**Figure 8: Waveforms showing alignment of output
BusCtrl, and external TxClk at 660Mb/s.**

379

Part 5
Clock and Data Recovery Circuits

IN this part, a number of papers on the subject of clock and data recovery have been collected. The first three papers introduce implementations in CMOS and BiCMOS technologies for Integrated Services Digital Network (ISDN) and disk-drive applications, while the next three papers target the SONET OC-3 standard (156 Mb/s).

The next four papers demonstrate rates from 480 Mb/s to 2.3 Gb/s, employing vastly different architectures and capture characteristics. The last five papers are based on the "quadricorrelator" architecture, achieving speeds as high as 8 Gb/s.

Additional References

J. Sonntag and R. Leonowich, "A monolithic CMOS 10 MHz DPLL for burst-mode data retiming," *ISSCC Dig. Tech. Papers,* San Francisco, CA, pp. 194–195, February 1990.

R. J. Baumert et al., "A monolithic 50–200 MHz CMOS clock recovery and retiming circuit," *Proc. CICC,* Santa Clara, CA, pp. 14.5.1–14.5.4, 1989.

Z.-G. Wang et al., "19-GHz monolithic integrated clock recovery using PLL and 0.3-μm FTE-length quantum-well HEMTs," *ISSCC Dig. Tech. Papers,* San Francisco, CA, pp. 118–119, February 1994.

An Analog PLL-Based Clock and Data Recovery Circuit with High Input Jitter Tolerance

SAM YINSHANG SUN, MEMBER, IEEE

Abstract — A clock and data recovery circuit for a T1 network is described. A fully integrated PLL extracts the carrier signal embedded in the data. Two trimming DAC's simultaneously bring the VCO center frequency and the PLL closed-loop bandwidth to their specified values. A triple sampler captures the jittering data and aligns them with the recovered clock. The input jitter tolerance of this circuit is three times more than previously reported PLL-based circuits.

I. Introduction

RECENT advances in the Integrated Service Digital Network (ISDN) have made its realization for the general public a likely prospect in the near future. For its primary rate interface, 1.544 Mbit/s in the U.S. and 2.048 Mbit/s in Europe, a higher input jitter tolerance requirement has been newly imposed demanding better performance of its line interface unit (LIU). This requirement challenges the performance of the key circuit block in the LIU, namely the clock and data recovery circuit, to a greater level.

Clock and data recovery circuits can be categorized into three types: *LC* tuned tank, digital PLL-based, and analog PLL-based circuits. The *LC* tank requires tuning of the external components and is sensitive to the data pattern. The digital PLL suffers high-speed phase "hit" on its output which is inherent to the scheme. The analog PLL can be fully integrated and provides a smooth clock output. Nevertheless, the performance of the analog PLL-based circuit needs to be optimized. Designed to be highly tolerant of input jitter, this paper describes a fully integrated analog PLL-based clock and data recovery circuit in a monolithic LIU using novel techniques to meet the performance challenge.

II. Circuit Description

The circuit shown in Fig. 1 is constructed with a signal slicer to convert the alternate mark inversion (AMI) signal into a return-to-zero (RZ) signal, a charge-pump PLL to extract the carrier clock embedded in the input data stream, and a triple sampler to capture the jittering data and align them with the recovered clock (RCLK). A frequency comparator is also included to monitor the PLL lock-in status.

A. Charge-Pump Phase-Locked Loop

As illustrated in Fig. 2 the PLL consists of a data-activating frequency/phase detector (FPD), a charge-pump loop filter (LF), and a current-controlled relaxation oscillator (ICO) [1]. Two current DAC's are used to trim the VCO center frequency and the PLL closed-loop bandwidth respectively. The FPD compares the phase difference between the RZ data and VCO output RCLK and steers the LF current sources to charge or discharge the large capacitor C. The voltage change on C will then drive the VCO output frequency to a direction to reduce the phase difference. The small capacitor C_2 is used to suppress the ripple in the charge-pump LF.

The capture range of this PLL is approximately the same as the VCO frequency range because the frequency/phase detector shown in Fig. 3 compares not only the phase difference, but also the frequency difference. When a pulse is present on RZ data this FPD will be activated as follows: if the phase of RZ data leads the RCLK phase, the UP output will be ON for a duration proportional to the phase difference; if RZ data lags RCLK, the DOWN output will be ON in the same fashion. The output logic prevents the UP/DOWN outputs from being ON simultaneously. This FPD has 1 bit of memory to keep the last comparison result that will eliminate an output phase jump at the beat frequency when the two input signals have different frequencies. When no pulse appears on RZ data, these two outputs will be deactivated and the VCO is free running until the next pulse comes.

The VCO consists of two parts. One is a relaxation ICO; its output frequency is controlled by its input current I_p as shown in Fig. 4. Two identical holding capacitors and auto-reset comparators provide near 50-percent duty cycle on clock output without a dividing counter. The VCO center frequency is switchable from 1.544 to 2.048 MHz by changing the trigger level of the comparators. The other part of the VCO is a 4-bit current DAC with VCO input

Manuscript received September 2, 1988; revised December 13, 1988.

The author is with the Semiconductor Product Division, Rockwell International, Newport Beach, CA 92660.

IEEE Log Number 8826245.

Reprinted from *IEEE Journal of Solid-State Circuits*, vol. SC-24, pp. 325-330, April 1989.

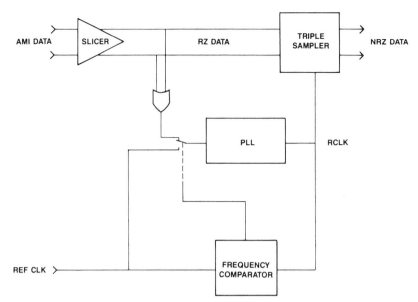

Fig. 1. Functional block diagram.

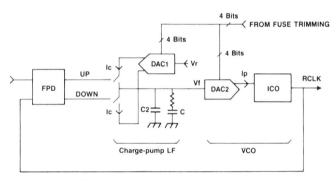

Fig. 2. Charge-pump PLL block diagram.

voltage V_f as its reference voltage. This DAC functions as a programmable voltage-to-current converter to adjust the transfer curve between the input voltage V_f and output current I_p to the ICO under different digital input codes. The VCO center frequency f_o and VCO gain K_o are delineated in the expressions below:

$$f_o = \frac{1}{2V_T} \cdot \frac{I_p}{C_h} \qquad (1)$$

$$K_o = 0.8\pi \cdot f_o. \qquad (2)$$

The other DAC is used in the charge-pump loop filter with a preset reference voltage V_r to provide a suitable injecting current I_c to determine the PLL closed-loop bandwidth which can be expressed by (3) [1]:

$$f_n^2 = \frac{K_o}{8\pi^3} \cdot \frac{I_c}{C} \qquad (3)$$

$$\xi = \pi R C f_n. \qquad (4)$$

The PLL is a second-order loop of which the damping factor ξ is determined in (3) by the resistor R, capacitor C, and PLL closed-loop bandwidth f_n. The two DAC's are made identically in the circuit and layout with the intention to adjust these key circuit parameters.

Two inherently conflicting aspects were considered when deciding the PLL closed-loop bandwidth. Stable locking to a random data stream needs a narrow bandwidth while a reliable data recovery demands a wide bandwidth to enable RCLK to quickly track the jittering data phase instance. The key performance characteristic of a clock/data recovery circuit is its input jitter tolerance. The input jitter tolerance is the maximum amount of phase jittering on an input data stream tolerable by the circuit that still recovers the clock and data from input. Fig. 5 shows the input jitter tolerance of a conventional PLL-based circuit with different closed-loop bandwidths. It is apparent that the wide loop bandwidth yields better performance in low jitter frequencies below 10 kHz while the narrow loop bandwidth produces a better result in higher frequencies.

B. Triple Sampling Data Recovery

To alleviate the dilemma of choosing the PLL closed-loop bandwidth, a triple sampler for data recovery was developed in an attempt to relax the phase tracking requirement. Instead of sampling the RZ signal only at 90° as the conventional approach, the triple sampler samples at 0°, 90°, and 180° as shown in Fig. 6. Whichever sampler detects a ONE, the output will be ONE due the characteristic of the RZ signal. This scheme makes the acceptable instantaneous phase error between the RZ signal and RCLK as high as 180° in contrast to the 90° from the conventional approach. The closed-loop bandwidth was then chosen to be 8 kHz for optimal jitter performance. Fig. 7 shows the measured PLL phase jitter frequency response (i.e., PLL phase-transfer function).

III. CIRCUIT ADJUSTMENT

The circuit integration on silicon demands adjustments for the VCO center frequency and the PLL closed-loop bandwidth to compensate for the large component varia-

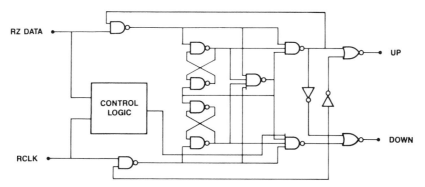

Fig. 3. Schematic of data activating frequency/phase detector.

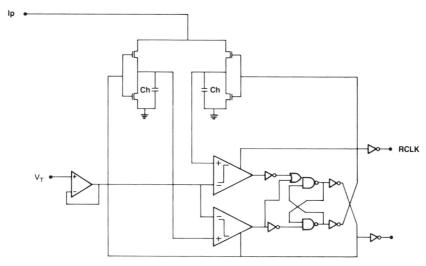

Fig. 4. Schematic of current-controlled oscillator (ICO).

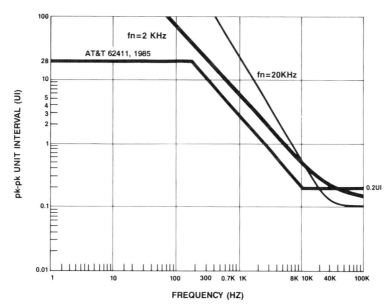

Fig. 5. Input jitter tolerance of the prototype with different PLL closed-loop bandwidth.

tions in MOS fabrication process. In mass production measuring the VCO center frequency is fast but measuring the PLL closed-loop bandwidth takes a relatively long time. Instead of calibrating them separately, a new technique was developed to complete the two tasks with one

trimming. Two identical current DAC's were used to trim these two key parameters. At the wafer test stage, the on-chip polysilicon fuses were blown according to the VCO frequency measurement. The resulting digital code controlled the VCO current DAC to adjust the transfer

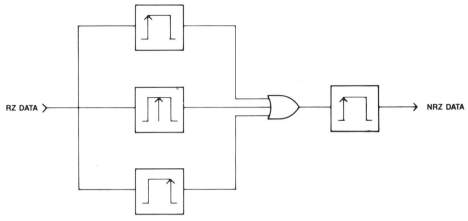

Fig. 6. Triple sampler data recovery.

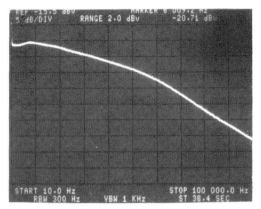

Fig. 7. Measured PLL phase jitter frequency response to a random data input.

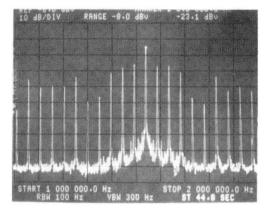

Fig. 8. Frequency spectrum of a repetitive data pattern.

curve between the VCO input voltage and the charging current I_p which allows the VCO to oscillate at the desired frequency. The same digital code was also used to change the current DAC of the charge-pump LF to adjust the charging current I_c which, in turn, centered the PLL closed-loop bandwidth to a predetermined frequency.

The correlation between these two adjustments is apparent when examining the mechanisms of the VCO frequency and the formation of the PLL closed-loop bandwidth. As expressed in (1) the VCO center frequency f_o is proportional to its charging current I_p divided by capacitance C_h. The PLL closed-loop bandwidth f_n^2 (see (2)) is also proportional to its charging current I_c divided by holding capacitance C. The input voltages of the two identical DAC's in the VCO and LF are equal by design and by definition. With the same digital input code the ratio of the DAC's output current I_p to I_c is approximately constant within the precision of component matching in the process. That is,

$$I_p/I_c = A \qquad (5)$$

where A is a constant. Therefore, the PLL closed-loop bandwidth is derived as

$$f_n^2 = \frac{0.2V_T}{A\pi^2} \cdot \frac{C_h}{C} \cdot f_o^2. \qquad (6)$$

Thus, trimming I_p by a certain percentage to let f_o reach a desired value will also change I_c by the same percentage, consequently making f_n come near the predetermined bandwidth. This is accomplished by the simultaneous control of the two identical DAC's used in the VCO and the charge-pump LF circuits.

IV. ANTI-HARMONIC LOCK TECHNIQUE

One inherent danger in the analog PLL-based clock and data recovery circuit is the PLL being locked onto the subharmonic of the data carrier frequency, especially when the data pattern is repetitive. The spectrum of a repetitive data pattern shown in Fig. 8 has local peaks at a subharmonic frequency near the peak of the carrier frequency. A commonly used scheme is to limit the VCO frequency range to within 6 or 7 percent of the center frequency to avoid locking onto the frequency outside the VCO range [2]. This scheme provides only limited protection to subharmonic locking and is useless to subharmonic locking within the VCO range. The nearest local peak on Fig. 8 is about 3 percent away from the carrier frequency, where the PLL was observed to be locked onto it in the prototype experiment. Further reduction of the VCO range creates problems under different operating conditions and temperatures.

386

Fig. 9.   Die photo of the clock and data recovery circuit.

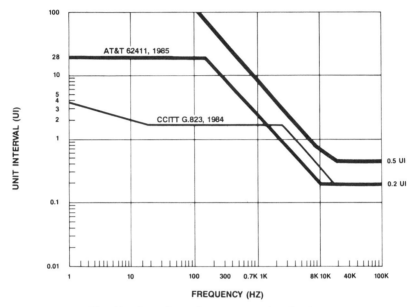

Fig. 10.   Input jitter tolerance to a random data input.

A unique technique was developed to detect and correct potential subharmonic locking. As shown in Fig. 1 a frequency comparator was added to monitor the recovered clock RCLK against a precision reference clock also derived from the data stream. If the RCLK frequency deviates from the reference frequency, it will be declared abnormal and the PLL input will be switched to the reference clock for training and then switched back to the RZ data stream. Since the trained VCO clock is very close to the carrier frequency of the data it will easily lock onto it. Thus, instead of resisting the subharmonic locking, this circuit detects the subharmonic locking and corrects it.

V.   EXPERIMENTAL RESULT

This circuit was realized with a 3-μm bulk CMOS process and occupied 5500 mil^2. A die photo of this circuit is shown in Fig. 9. The input jitter tolerance given in Fig. 10 exceeds the requirement of the newly revised AT&T pub. 62411 "High Capacity Digital Service Interface Specification" [3]. The high jitter frequency tolerance is at least

three times better than previously reported PLL clock and data recovery circuits [2]. The performance is summarized in Table I.

VI.   CONCLUSION

A high input jitter tolerance was achieved in this analog PLL-based clock and data recovery circuit. The task of making the circuit system stable and fast data phase cap-

TABLE I
DEVICE CHARACTERISTICS

Technology	CMOS 3 micro double poly
Power Supply	5 V ± 10%
Power Consumption	25 mW (typical)
Silicon Area	5500 mil^2
Allowable Instant Phase Error	± 0.47 UI
PLL loop Bandwidth	8 KHz ± 1 KHz
Pull-in Time	< 200 us
Tolerance to "0" stream	300 "0"s (typical)
Intrinsic Jitter Output	0.01 UI (All "1" input) 0.015 UI (Random data)
Max. Static Phase Error	15º

turing was accomplished by adjusting the key PLL parameters and using a novel data recovery circuit. Full integration in CMOS technology makes this circuit a useful functional block which can be incorporated easily into VLSI systems.

ACKNOWLEDGMENT

The author would like to express his gratitude to S. Hao and Dr. K.-L. Lee for their suggestions and encouragement. The author would also like to thank R. R. Co for the initial work on the PLL architecture.

REFERENCES

[1] F. M. Gardner, "Charge-pump phase-lock loops," *IEEE Trans. Commun.*, vol. COM-28, pp. 1849–1858, Nov. 1980.
[2] K. J. Stern, N. S. Sooch, D. J. Knapp, and M. A. Nix, "A monolithic line interface circuit for T1 terminals," in *ISSCC Dig. Tech. Papers*, 1987, pp. 292–293.
[3] AT&T PUB 62411, "High capacity digital service interface specification," 1985.
[4] S. Y. Sun, "A PLL based clock and data recovery circuit with high input jitter tolerance," in *Proc. IEEE CICC*, 1988, pp. 9.7.1–9.7.3.

A 30-MHz Hybrid Analog/Digital Clock Recovery Circuit in 2-μm CMOS

BEOMSUP KIM, STUDENT MEMBER, IEEE, DAVID N. HELMAN, STUDENT MEMBER, IEEE,
AND PAUL R. GRAY, FELLOW, IEEE

Abstract —A high-speed hybrid clock recovery circuit composed of an analog phase-locked loop (PLL) and a digital PLL (DPLL) for disk drive applications is described. The chip operates at a maximum data rate of 33 MHz from a single 5-V power supply and achieves fast acquisition, a decode window of 94% of full window width, effective sampling jitter of 100-ps rms, and an effective input sampling rate of 1 GHz. The ring oscillator in the analog PLL shows a 62 ppm/°C temperature coefficient (TC) and 4.5%/V supply sensitivity of free-running frequency. The total power dissipation is about 600 mW and the active area is 30 000 mil^2 in a 2-μm single-poly double-metal n-well CMOS process.

I. INTRODUCTION

IN HIGH-SPEED magnetic recording data separators and in certain data communication applications [1]–[4], clock recovery from received data must be performed with stringent requirements on rapid phase acquisition, static phase offset, sensitivity of decode error to phase jitter, and programming capability. Currently, most high-speed clock recovery circuits make use of an analog phase-locked loop (PLL) implemented mostly in bipolar technology [1], [2] using either an emitter-coupled multivibrator or a starved ring-oscillator VCO, and utilize techniques such as zero phase start and PLL time constant gear shifting to achieve fast acquisition at the start of the data record. These techniques are limited in their ability to implement more sophisticated fast-acquisition algorithms at high speed, and also are not well-suited to the CMOS implementation needed to achieve higher levels of integration and lower power dissipation.

One of the most challenging areas in the design of clock recovery circuits is the disk-drive read channel. Fig. 1 shows a typical disk-drive read channel using a pulse detection scheme. Here, the disk head reads the data stored in the magnetic medium through magnetic flux

changes and sends small electric signals to the preamplifier. The output of the preamplifier is amplified through an AGC to a proper voltage level and filtered by a pulse slimming filter. The peak detector detects the peaks of the signal and generates pulses corresponding to these peaks. Then, the pulses coming from the peak detector are used to derive the clock information and to detect data through a clock recovery circuit called a data separator.

There are several requirements for this data separator. When the disk drive begins reading a new data record, the data clock recovery circuit must align its local clock with the incoming data bit stream as rapidly as possible. Techniques of zero phase start and variable loop time constant are used to speed up the acquisition time, but an acquisition time of 20–40-b cycles is still required for the current implementations. Since the acquisition time is wasted time, it is desirable to reduce it. In the steady state, the clock recovery circuit must decode data correctly in the presence of a large amount of jitter and other circuit impairments. Therefore, the static phase offset should be reduced and the decoding window width should be increased. For high-performance disk drives using varying bit rates for constant areal density, the data separator is required to track multiple-frequency data bit streams. Also, for system-level tests and diagnostics, the data separator should be able to provide variable window width and offset under software control.

One of the key performance measures of the data separator is the maximum effective decoding window width, as shown in Fig. 2(a). Here the data separator generates a sequence of time windows within which a data transition can occur. The solid lines in the top diagram correspond to the window boundaries when the local clock does not have jitter, and the solid lines in the second diagram are data transition edges when the input data contains no jitter. Improper decoding can result from a combination of jitter on the data transition edge, jitter on the local clock giving window boundary uncertainty, and static phase offset of the data in a decoding window. This aspect of performance can be characterized by the bit error rate as a function of pulse location in the decoding window as shown in Fig. 2(b). In typical disk

Manuscript received May 10, 1990; revised August 8, 1990. This work was supported by the National Science Foundation under Grant MIP-8801013, the State of California MICRO program, National Semiconductor, Level One Communications, Texas Instruments Incorporated, and Xerox Corporation. Chip fabrication was by MOSIS.

B. Kim and D. N. Helman were with the Department of Electrical Engineering and Computer Sciences, University of California, Berkeley, CA. They are now with Chips and Technologies, Inc., San Jose, CA 95134.

P. R. Gray is with the Department of Electrical Engineering and Computer Sciences, University of California, Berkeley, CA 94720.

IEEE Log Number 9039478.

Reprinted from *IEEE Journal of Solid-State Circuits*, vol. SC-25, pp. 1385-1394, December 1990.

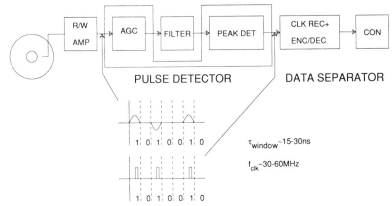

Fig. 1.   Typical disk-drive read channel.

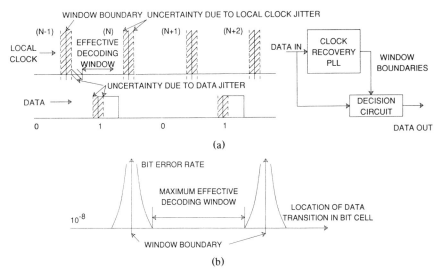

(a)

(b)

Fig. 2.   (a) Maximum effective decoding window in the presence of jitter. (b) Bit error rate as a function of pulse location in a decoding window.

drives, a maximum effective decoding window width of 90% or more is required at the 10^{-8} bit error rate.

Virtually all existing clock recovery circuits for disk drives make use of an analog PLL implemented in bipolar technology with zero phase start and gear shifting techniques. There are several disadvantages to these approaches. The static and dynamic phase error depends on the precision matching of analog devices and path delays, and fast phase acquisition requires complex analog design and often requires many external components. Phase acquisition selection algorithms are limited to those that can be practically implemented in analog PLL's. Finally, it is difficult to implement sophisticated techniques such as window shift and shrinks needed for high-level system tests and diagnostics within the disk-drive system. Traditional digital PLL circuits can alleviate these problems, but they are not practically usable because of large phase jumps normally associated with them, i.e., granularity in the time domain is limited to a minimum of one clock interval by the DPLL architecture. Finally, since the speed of CMOS circuits is currently slower than that of bipolar circuits, most data separators are implemented in bipolar technology and are not well-suited to the CMOS technol-

ogy needed to achieve a higher level of integration and lower power dissipation in disk-drive read channel electronics.

The hybrid analog/digital clock recovery circuit presented in this paper can overcome many of these limitations. The hybrid clock recovery circuit uses an analog PLL to generate multiple clocks for a 1-GHz effective sampling rate in a 2-μm technology, and a digital PLL (actually a digital signal processor) to process edge detection, acquisition, tracking, and programming.

Section II describes the hybrid clock recovery circuit concept, which embodies several desirable attributes of analog and digital PLL's. In Section III, several circuit issues are addressed to realize high-performance hybrid clock recovery circuits. Section IV discusses the experimental results from a prototype chip. Finally, conclusions are given in Section V.

II.   SYSTEM DESCRIPTION

The basic concept of the clock recovery system is shown in Fig. 3. A ring oscillator composed of a long chain of inverters is permanently locked onto a reference clock at

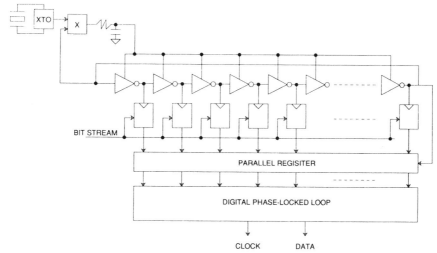

Fig. 3. Hybrid analog/digital clock recovery architecture.

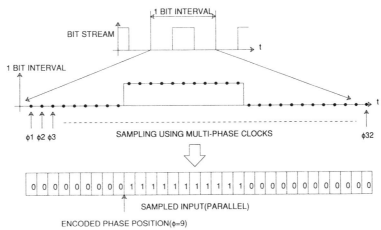

Fig. 4. Multiphase sampling.

the data window rate (the read reference clock in the case of a disk-drive data separator). In the prototype described later, the ring oscillator composed of 16 differential stages having 32 taps on the ring is locked onto the 30-MHz reference clock with a conventional phase/frequency detector, an on-chip charge pump, and a loop filter. Since the dynamics of this loop do not affect phase acquisition or tracking in the digital part of the system, the analog PLL design is not critical. Each tap spaced one gate delay (about 1 ns) apart is used to latch the data samples into one of the 32 latches, so that at the end of one round-trip (one bit interval), 32 samples spaced 1-ns delay apart are stored in the 32 data latches (Fig. 4). In a 2-μm technology, this gives an effective sampling rate of 1 GHz; in 1-μm technology it is about 4 GHz. However, since these samples are sequential, the following logic circuits would need to process the 1-GHz sampled data serially. This is very difficult using CMOS logic circuits. Therefore, a serial-to-parallel translator is necessary to reduce the data processing rate.

The parallel phase sampler accomplishes the data processing rate translation. First, the incoming data are sampled by 32 data latches clocked using 32 taps from the

ring oscillator. Since the latching occurs sequentially through the 32 latches, the outputs of the latches are also available in a sequential manner. The next registers, as shown in Fig. 5, consist of two sets of latches. Each set contains 32 latches: the first 16 latches in the left set are clocked by ϕ_a, and the second 16 latches in the same set are clocked by ϕ_b, a delayed version (16 taps later) of ϕ_a. The first 16 latches in the right set are clocked by ϕ_c, and the second 16 latches in the same set are clocked by ϕ_d, a delayed version (16 taps later) of ϕ_c. Here, ϕ_c is the inverted version of ϕ_a. A four-phase clock generator produces ϕ_a, ϕ_b, ϕ_c, and ϕ_d. Fig. 6 shows the clock diagram.

Next a multiplexer takes the 32-b data from the two registers alternatively. For the odd cycle, the multiplexer takes the left register outputs (32 b), and for the even cycle, the multiplexer takes the right register outputs (32 b). Then the outputs of the multiplexer are latched into another register. Now the data rate is 1/32 (32 b at 30 MHz) of the sampling rate (1 b at 1 GHz).

After subsequently being moved to the holding register, the bit pattern is evaluated to determine the location of valid data transitions in the decode window using digital

Fig. 5. Parallel phase sampler architecture.

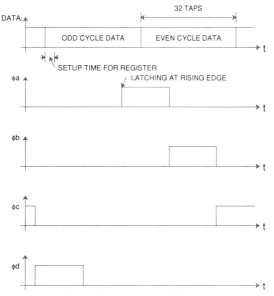

Fig. 6. Clock diagram for parallel phase sampler.

transition detectors of complexity appropriate to the application. Noise filtering is easily applied to eliminate the effects of isolated noise pulses and transition sampling noise. The location of the data transition is then encoded into a binary number through a binary encoder. If the data sequence is such that two transitions can occur in a decode window, encoder complexity must be increased to handle that case.

The following PLL operation is purely digital and is based on the location of detected transitions. The center of the current decode window is held in a current phase tap register, and this register is updated by the digitally low-pass filtered phase error signal. The bandwidth of the low-pass filter is easily controlled by a multiplying constant K as shown in Fig. 7. The phase error signal is simply the difference in tap location between the current window center and the occurrence of a valid data transition. When the window center is not at the center of the 32-b data latches, the window over which valid transitions

are accepted as valid within that particular time slot can extend into the previously stored 32 samples or the next, necessitating a simultaneous pipeline storage of transition locations from three sets of samples in order to evaluate symbol values. Zero phase start, variable loop time constant, and other modes are implemented digitally. Other specialized disk-drive functions, such as variable window width and offset, pulse pair compensation, and so forth can also be implemented directly in the digital domain. Also the loop dynamics are determined digitally avoiding critical design of the analog PLL normally used to lock and track the incoming data in clock recovery circuits. This approach can achieve virtually full window-width decoding since the window width depends only on phase resolution. Here the phase resolution is about 1 ns in a 2-μm process. In data communication applications such as Ethernet, the more sophisticated symbol decoding allowed by this approach allows the jitter margin to be extended from the ± 18 ns, typical of current Ethernet serial interface chips sampling at the $1/4$ point, to a value that approaches the window half-width (50 ns).

The loop as described here is first order, but in the event that frequency differences can exist between the reference frequency and the data frequency, it is straightforward to add a frequency offset register to form a second-order loop with zero static phase error in the presence of frequency offsets. In summary, Table I explains the algorithm of the digital PLL for the first-order case.

III. Circuit Description

Realization of a low-jitter, high-performance clock recovery circuit using hybrid analog/digital architecture requires careful attention to certain aspects of the circuit design. In this section, several circuit design issues are addressed. The first issue is the phase noise in the data sampling process, which in turn arises from a number of sources including ring-oscillator phase noise due to the

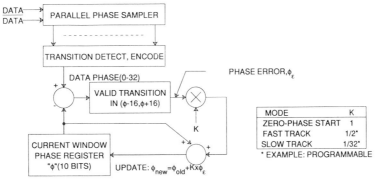

Fig. 7. Digital PLL implementation for the first-order loop.

TABLE I
DPLL ALGORITHM FOR THE FIRST-ORDER CASE

Step I - Compute Current Window Phase Error ϕ_e

- Measure Current Window Phase Error $\phi_i(nT)$

    if Transition Exists

- Load Feedback Clock Phase $\phi_o(nT)$

- Compute Phase Difference $\phi_\varepsilon(nT) = \phi_i(nT) - \phi_o(nT)$

    if Transition Exists, otherwise Put $\phi_\varepsilon(nT) = 0$.

Step II - Update Current Feedback Clock Phase ϕ_o

- Take Current Feedback Clock Phase $\phi_o(nT)$

- Calculate $\phi_o((n+1)T) = \phi_o(nT) + K\phi_\varepsilon(nT)$

    where in Zero Phase Start Case, $K = 1$

    in Fast Mode, $K = 1/4$

    in Slow Mode, $K = 1/32$

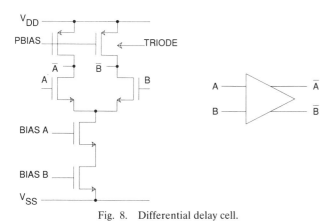

Fig. 8. Differential delay cell.

inherent thermal noise in the inverters themselves, noise due to the power supply noise injection, and sampling noise occurring in the data latches used to sample the data.

A first-order analysis of thermal noise accumulation indicates that it is related to the thermal noise at the input of each inverter stage, inverter rise time, signal swing, and loop bandwidth. For a given technology and set of design parameters, the total effective mean-square thermal noise at the input of each inverter when the

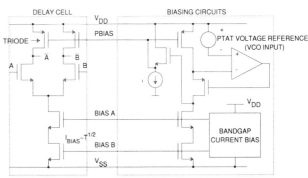

Fig. 9. Biasing circuit for delay cell.

inverter is in its active region is related to $\sqrt{kT/C_{gs}}$ where C_{gs} is the gate capacitance of the inverter stage. A sampling jitter of 0.05-ns rms was chosen as a target, which dictated the use of relatively large devices (160/2) in the inverter chain as a result of thermal noise consideration.

A second major issue is coupling of supply noise. Traditionally, CMOS ring-oscillator based PLL's have used starved inverters together with, for example, an on-chip regulator to reduce supply sensitivity and noise injection [5]. In this instance, however, rejection of high-frequency supply noise was a very critical requirement in order to avoid degradation of sampling jitter. Furthermore, in order to latch data at every inverter delay time, the ring-oscillator stage must be able to produce a positive-going edge for each unit delay time. In order to meet these objectives, an ECL-like logic implementation shown in Fig. 8 was used in the inverter chain. The load elements are triode-region PMOS devices. This logic form was carried through the data latches giving completely balanced signals throughout. (A schematic diagram of the dual-latch data path from ring-oscillator tap to latched CMOS output is shown in Fig. 11.)

The primary advantage of the differential delay cell is lower susceptibility to power supply noise because the inherent differential structure rejects the power supply noise. Another merit is the capability of the oscillator to utilize an even number of stages. In general, the single-ended ring oscillators require an odd number of stages to oscillate. However, an even number of stages may be used in a differential ring oscillator if the last-stage outputs are

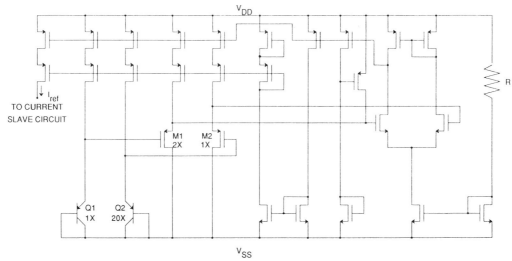

Fig. 10. Bandgap current bias circuit.

crossed and connected to the first-stage inputs. Finally, the differential delay cell has two outputs, therefore each cell generates two taps of the ring oscillator. If a 16-stage differential ring oscillator is used, a total of 32 taps are available, i.e., each delay cell gives an output and its complement. In the starved ring-oscillator case, if 32 taps are required, a 32-stage ring oscillator is necessary. (Actually 33 stages would be required to make the ring oscillator oscillate.)

There are two problems associated with the differential delay cell. The first problem is longer gate delay. In CMOS technology, the square-law MOSFET drain current characteristic makes the delay inversely proportional to the voltage swing. Since the voltage swing is limited by biasing considerations to about 1 V in the differential delay cell, the delay is longer than that of the inverter delay cell. However, since the differential delay cell has two outputs, tap-to-tap spacing (for example, the time delay between one rising edge and the next rising edge) is only one gate delay. In the case of the starved ring oscillator, two stages are required to generate two successive rising edges. Therefore, the time resolution of the inverter delay cell is two gate delays while the time resolution of the differential delay cell is one gate delay. Since the differential delay cell is about 30–50% slower than the inverter delay cell, this delay penalty is compensated.

The second problem is the control of the delay cell bias point. Since the PMOS loads should be in the triode region, a method is required to control both the cell bias current and the PMOS drain-to-source voltage in order to prevent the PMOS loads from leaving the triode region. A replica biasing scheme has been developed which maintains the PMOS devices in the triode region and which also compensates the free-running frequency variation due to temperature drift. Fig. 9 shows the biasing scheme.

Temperature- and supply-independent biasing of these circuits is important in order to achieve the goal of supply noise rejection and good control of the free-running fre-quency. This is accomplished by controlling the cell bias current and voltage swing of the delay cell using a replica biasing scheme. The bandgap current biasing circuit generates the cell bias current which is proportional to $T^{1/2}$. Also, a replica of the delay cell is used to force the PBIAS voltage line to a value that gives a voltage swing of the delay cell equal to that of the on-chip PTAT voltage reference. Nominal voltage swing is set to 1 V. The combination of voltage and current biasing gives a propagation delay which is first-order independent of temperature and power supply and makes the process dependence of free-running frequency depend only on mobility and on L^2. An average free-running frequency temperature coefficient (TC) of 62 ppm/°C and supply sensitivity of 4.5%/V were measured over a sample of devices taken from two runs. The analog PLL is controlled by a phase-frequency detector and a charge-pump circuit. Fig. 10 shows the bandgap current bias circuit.

Since the hybrid analog/digital clock recovery circuit relies heavily on the sampling of the incoming data, the problem of metastability in the data latches must be considered. A dual-latch data path is used to reduce the probability of metastability-induced data errors. The second latch is placed in the regenerative mode by a delayed version of the clock taken from an inverter several stages down the ring, allowing it enough time to make its decision. Fig. 11 shows the dual-latch data path.

The digital PLL (DPLL) consists of several arithmetic units. Here, a subtractor, a multiplier, and an accumulator with an adder take the roles of a phase detector, a loop filter, and a VCO, respectively, in an analog PLL (refer to Table I). Since the DPLL updates the phase in one window time and has a feedback path, the total time taken to process the data through the feedback path should be less than one window cycle. Since the window time is small, the total data-path delay can be too long to allow complete processing within a single window cycle, requiring pipelining of the computation in the feedback path. However, two important factors should be consid-

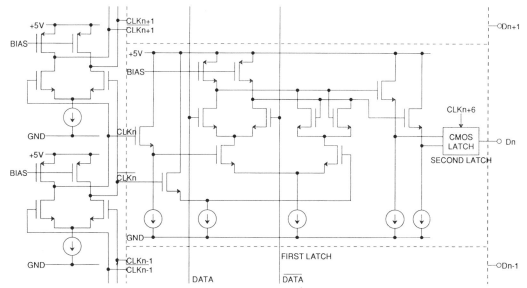

Fig. 11. Dual-latch parallel data path.

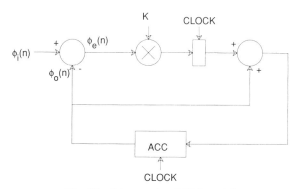

Fig. 12. Discrete-time DPLL model.

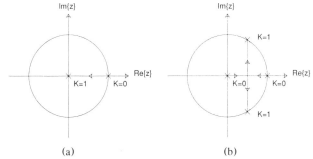

(a)   (b)

Fig. 13. Pole–zero diagram for DPLL's: (a) nonpipelined, and (b) pipelined.

ered when pipelining is used. First, the stability of the pipelined DPLL is degraded since the pipelining adds an additional pole to the loop transfer function. Second, static phase offset is introduced in the DPLL since the pipelining also adds one more zero to the loop transfer function. Fig. 12 shows the simplified discrete-time model of the DPLL. Here, a pipeline buffer is used between the multiplier and adder. When there is no pipelining, the buffer is replaced by a direct connection. The loop transfer function in the z domain for the nonpipelined DPLL is then represented by

$$\frac{\Phi_o(z)}{\Phi_i(z)} = \frac{K}{z - (1 - K)}. \tag{1}$$

Also the error transfer function is represented by

$$\frac{\Phi_e(z)}{\Phi_i(z)} = \frac{(z - 1)}{z - (1 - K)}. \tag{2}$$

From (2), it is guaranteed that the loop has zero static phase offset. The location of the pole in both equations determines the stability of the loop. Fig. 13(a) shows the pole–zero diagram of the transfer function. From this figure, it can be seen that the location of the pole always

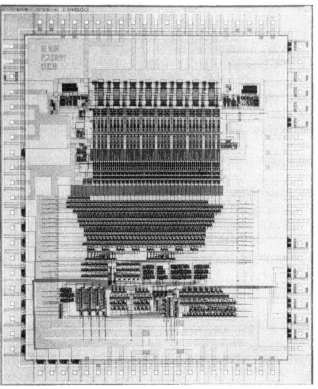

Fig. 14. Hybrid clock recovery chip photograph.

395

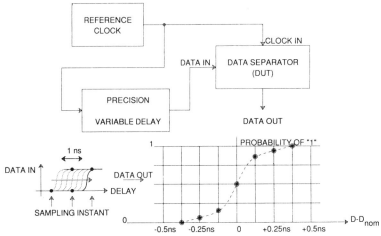

Fig. 15. Sampling jitter measurement.

remains inside of the unit circle as long as $0 < K < 1$. Therefore, the DPLL is always stable.

In the pipelined DPLL case, the pipeline buffer is active, and it delays the output of the multiplier by one window cycle. The transfer function in this case is represented by

$$\frac{\Phi_o(z)}{\Phi_i(z)} = \frac{K}{z^2 - z + K} \tag{3}$$

and the error transfer function is represented by

$$\frac{\Phi_e(z)}{\Phi_i(z)} = \frac{z(z-1)}{z^2 - z + K}. \tag{4}$$

In (4), the $(z-1)$ term appears again in the numerator. Therefore, the loop has zero static phase error again. To determine the stability of the loop, a pole–zero diagram for the transfer function is shown in Fig. 13(b). Again, the two poles remain inside of the unit circle as long as $0 < K < 1$, and the DPLL with pipelining is stable.

The above calculation for the zero static phase error is based on the linear analysis without the input phase quantization effect. Since the hybrid clock recovery circuit samples the input at discrete time intervals spaced one gate delay apart, the input phase (the location of the transition) is obtained in the form of a binary number with a finite length, which results in quantization errors in the phase domain. This is analogous to quantization errors introduced in the voltage domain by the A/D converter as it samples the input at discrete voltage intervals. This phase quantization noise can be modeled as an additive noise in the input and represented by

$$\sigma_N^2 = \frac{\Delta^2}{12} = \frac{2^{-2b}}{12} \tag{5}$$

where Δ is the time interval and b is the number of digits representing the input phase. Since the quantization phase noise disturbs the loop settling, zero static phase error cannot be achieved. Instead, a small phase fluctuation centered at zero exists in $\phi_e(n)$. For small values of K,

the phase error variation is almost the same as the input phase quantization noise power (σ_N^2).

IV. Experimental Results

A prototype clock recovery circuit was fabricated and tested in a 2-μm single-poly, double-metal, n-well CMOS process. The test circuit consisted of the 32-stage analog PLL and parallel phase sampler, and a complete programmable digital PLL of 4000-gate complexity. Fig. 14 shows a chip photograph of the prototype hybrid analog/ digital clock recovery circuit.

The ring-oscillator sampling jitter measurement is shown in Fig. 15. Here, an analog PLL in the data separator is permanently locked into the reference clock, which is running at 30 MHz. A precision variable delay was used to move a ZERO-TO-ONE data transition in time through the sampling interval of one of the 32 latches. Here, in each sampling instance, the probability of a ONE for each delayed version of data transitions was measured and a distribution function has been plotted (Fig. 15). The width of the transition region from 100% probability of a latched ZERO to 100% probability of a latched ONE gives an indication of the effective jitter in the data sampling process. From this measurement, about 100-ps rms jitter was obtained from the chip. Here, the rms jitter includes the input jitter, which is controlled to be negligible.

Bit error rate in the complete data detector was measured as a function of data transition location in the window. Here an isolated transition was moved within a long chain of evenly spaced RZ pulses. The effective measured decode window was 31 ns, or 94% of the full window width of 33 ns. This is limited by a combination of the jitter mentioned above and the sampling resolution (phase quantization noise). Fig. 16 shows the measured bit error rate versus pulse location in a decode window.

Because of the zero phase start algorithm, phase acquisition is instantaneous, limited only by the jitter on the first detected transition. Initial acquisition is followed by a period of PLL operation with a short time constant to

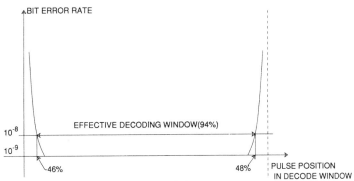

Fig. 16. Decode window measurement.

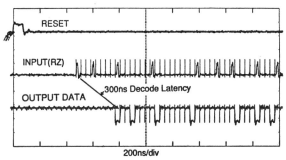

Fig. 17. Typical phase acquisition sequence.

TABLE II
MEASURED DATA, 5 V, 25°C UNLESS INDICATED

Ring Oscillator and Phase Sampler	
Free Running Freq	30MHz
Std Deviation of Free-Run Freq, (2 runs, 20 devices)	0.84MHz (2.9%)
TC of Free-Running Freq	62ppm/deg C
Supply Sens of Free-Running Freq	4.5%/volt
Power Dissipation	390mW*
Silicon Area, 2μ	10K square mils
Silicon Area, 1μ**	4K square mils
Complete Analog/Digital PLL	
Silicon Area, 2μ	30K square mils
Power Dissipation	600mW*

*Idiosyncratic of particular level-shift implementation; easily reduced to 300 mW by straightforward modification of one level shifter in the data path.
**Taken from 1-μm prototype layout, now in fabrication.

allow decay of phase error due to jitter on the first transition. Then the loop time constant is shifted to a large value for normal operation. Fig. 17 displays the typical acquisition sequence showing error-free data recovery with no preamble. Here the incoming data and the local clock are at the same frequency but out of phase. The vertical lines are added to indicate the window boundaries. Each data transition contains all information of the data and clock. From this picture, for each input pulse, the corresponding correctly positioned synchronized output pulse is generated from the data separator chip. In this example, the first pulse was intentionally given 25% jitter to inject error in the acquisition process.

The temperature coefficient of the free-running frequency of the ring oscillator was measured from 0 to 70°C using single 5-V power supply. The measured TC is 62 ppm/°C. The VCO gain of the ring oscillator is 6 MHz/V. On the other hand, the power supply sensitivity of the free-running frequency is 4.5%/V. The active area for the whole circuitry is 30 000 mil^2 in 2 μm. The power dissipation of the chip is 600 mW. Here, some of this power consumption is due to the particular design used in one of the level shifters. A straightforward improvement would give a value of about 500 mW. Table II summarizes the experimental results.

V. CONCLUSION

A clock recovery architecture using a hybrid analog/digital PLL in CMOS technology for disk-drive read applications has been explained. Currently most high-speed disk drive applications make use of an analog PLL implemented in bipolar technology. These techniques are limited in their ability to implement more sophisticated algorithms at high speed and are not well-suited to the implementation of CMOS technology needed to achieve higher levels of integration and lower power dissipation in disk-drive read channel electronics. Using the hybrid approach, the above limitations are overcome. Also, fast acquisition, large jitter rejection, high-speed sampling, a large effective decode window, and flexible programming capabilities are achieved.

When future scaled technologies are considered, one important issue, phase jitter, should be addressed again. As device sizes are scaled down, the absolute value of the phase jitter decreases. However, the tap-to-tap delay decreases more rapidly. Hence, the ratio of the phase jitter to tap delay increases, resulting in degraded phase resolution. One possible method of reducing the phase jitter is to use a delay-locked loop instead of a phase-locked loop to generate the multiphase clocks [8]. Since the delay-locked loop does not have the problem of phase accumulation, the phase jitter is much smaller than that of the phase-locked loop.

REFERENCES

[1] J. Kellis and S. Mehrotra, "A 15 Mb/s data separator and write compensation circuit for Winchester disk drives," in *ISSCC Dig. Tech. Papers*, vol. 29, Feb. 1984, pp. 232–233.

[2] W. Llewellyn, M. Wong, G. Tietz, and P. Tucci, "A 33 Mb/s data synchronizing phase-locked loop circuit," in *ISSCC Dig. Tech. Papers*, vol. 31, Feb. 1988, pp. 12–13.

[3] A. Bell and G. Boriello, "A single chip NMOS Ethernet controller," in *ISSCC Dig. Tech. Papers*, vol. 26, Feb. 1983, pp. 70–71.

[4] H. Haung, D. Banatao, G. Perlegos, T. Wu, and T. Chiu, "A CMOS Ethernet serial interface chip," in *ISSCC Dig. Tech. Papers*, vol. 27, Feb. 1984, pp. 184–185.

[5] K. Ware, H. Lee, and C. Sodini, "A 200 MHz CMOS PLL with dual phase detectors," in *ISSCC Dig. Tech. Papers*, vol. 32, Feb. 1989, pp. 192–193.

[6] "Digital PLL decoder," U.S. Patent 4 584 695, National Semiconductor, 1980.

[7] "Data communication system and bit timing circuit," U.S. Patent 4 189 622, NCR Corp., 1977.

[8] J. Sonntag and R. Leonowich, "A monolithic CMOS 10 MHz DPLL for burst-mode data retiming," in *ISSCC Dig. Tech. Papers*, vol. 33, Feb. 1990, pp. 194–195.

A BiCMOS PLL-Based Data Separator Circuit with High Stability and Accuracy

Shyoichi Miyazawa, *Member, IEEE*, Ryutaro Horita, *Member, IEEE*, Kenichi Hase, *Member, IEEE*,
Kazuo Kato, *Associate Member, IEEE*, and Shinichi Kojima

Abstract —This article describes a data separator which can work in Winchester disk drives at a read/write speed of up to 30 Mb/s. To realize high stability and accuracy in reproducing data in high-speed transfers, not only a digital synchronization field detector is used, but also an analog dual-mode phase-locked loop (PLL) which has a new phase detector. The new phase detector has constant gain in the data field, independent of pattern. In addition to the above circuit, the IC incorporates a run-length-limited (RLL) 2–7 code encoder/decoder and a write compensator. By using the 2-μm BiCMOS process, the total power consumption is kept as low as 400 mW even at the high transfer rate of 30 Mb/s.

I. Introduction

A DATA separator has been developed to meet the requirements of the Winchester disk drive industry which has increased capacity and decreased size.

As capacities increase, read/write circuit performance becomes critical. To meet this requirement, a monolithic read/write IC has been developed [1].

The write circuit consists of an encoder using the run-length-limited 2–7 (RLL 2–7) code, which is a form of data compression code, and a write compensation circuit.

The read circuit consists of a data separator incorporating an analog phase-locked loop (PLL), a sync field detector, and a decoder of the RLL 2–7 code. The output signals from the magnetic disk contain large jitter because of the S/N characteristics of the disk, head, and amplifier, and bit-to-bit interference due to peak shift. The data separator therefore is required to separate data and clock timing from such signals.

In a conventional PLL circuit, we could not design the filter of the PLL optimally to cancel the effects of jitter. The reason is that, because data are random in the data field and the output signal includes frequency deviations, the gain of the PLL other than the filter varies due to frequency deviation. Furthermore, in the sync field, the PLL must lock quickly without erroneous locking to the output signal of the fixed pattern.

Thus the demand is for the development of a data separator that locks quickly in the sync field and makes the gain of the circuit, except the filter, fixed in the data field where frequency deviation occurs.

Additionally, as data records are written over old ones on

Manuscript received March 8, 1990; revised October 10, 1990.
S. Miyazawa, R. Horita, and K. Hase are with the Microelectronics Products Development Laboratory, Hitachi, Ltd., 292 Yoshida-cho, Totsuka-ku, Yokohama 244, Japan.
K. Kato is with the Hitachi Research Laboratory, Hitachi, Ltd., Ibaraki, Japan.
S. Kojima is with Takasaki Works, Hitachi, Ltd., Gunma, Japan.
IEEE Log Number 9041213.

the magnetic disk, write splicing occurs as the speed of rotation constantly fluctuates. If the data separator starts to lock in the write splice area, the wrong signal coming from the write splice may cause the PLL to run out of control.

Therefore, the data separator must lock in the sync field and, for this reason, a sync field detector is required. The conventional sync field detector was an analog circuit comprising passive devices like resistors and capacitors [4]. In such systems, individual tolerances and the temperature coefficients of components affect the accuracy of detection. So, demand arose for a digital sync field detector which would be temperature stable and accurate.

This article describes a data separator which will lock quickly, precisely, and reliably, combined with a digital sync field detector with high accuracy.

II. Study and Design of the PLL

A. Conventional PLL Used in Magnetic Disk Units and Its Problems

The conventional PLL used to read data from magnetic disks consists of a phase detector (PD), loop filter (LF), and VCO. The input and output are digital signals. In order to realize high-speed locking in the sync field and stability against jitter in the data field, natural angular frequency ω_n and damping factor ζ are optimized by switching the output current of the phase detector and the constant of the loop filter. By making use of the fact that signals with a fixed frequency are recorded in the sync field, the phase detector compares the frequency and phase in the sync field to attain a wide capture range and prevent erroneous locking. In the data field, it detects the phase difference of the VCO clock and the input data whose frequency ranges from 1/3 to 1/8 of the VCO clock.

However, the phase detector detects a phase difference only when there is input data, and produces a pulse current of a fixed intensity I_p for a duration that is proportional to the phase difference. So, the open-loop transfer function $G(s)$, natural angular frequency ω_n, and damping factor depend on the ratio N of frequency of the VCO clock to that of the input data as follows:

$$G(s) = \frac{I_p \cdot K_o\{1 + S(C1 + C2)R\}}{2\pi N \cdot S^2 C1(1 + SC2R)} \tag{1}$$

$$\omega_n = \sqrt{\frac{I_p}{2\pi} \cdot \frac{K_o}{C1 \cdot N}} \tag{2}$$

$$\zeta = \frac{(C1 + C2)R}{2}\sqrt{\frac{I_p \cdot K_o}{2\pi C1 \cdot N}} \qquad (3 \leqslant N \leqslant 8). \tag{3}$$

Reprinted from *IEEE Journal of Solid-State Circuits*, vol. SC-26, pp. 116-121, February 1991.

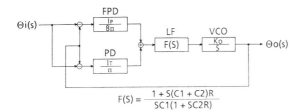

$$F(S) = \frac{1 + S(C1 + C2)R}{SC1(1 + SC2R)}$$

Fig. 1.   PLL block diagram.

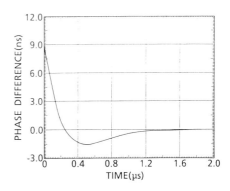

Fig. 2.   Measured phase step response using frequency phase detector.

For this reason, the open-loop frequency response varies in the data field where the frequency ranges from $1/3$ to $1/8$ of the VCO clock frequency, and optimization is difficult. The system producing pulse current with a fixed intensity I_p will produce a current pulse every time it compares the phase, even after the completion of phase synchronization. So, the output voltage waveform of the loop filter that controls the VCO varies due to the current pulse, and the VCO clock includes jitter.

The VCO clock is used to produce the detection window the decoder uses to read data, and jitter reduces the decoding margin and degrades the data read margin of the device.

B. Solution to Problems and Circuit Design

1) The PLL shown in Fig. 1 is proposed to solve the above problems. A frequency phase detector (FPD) is used in the sync field where data are written with the same frequency, and the output is a conventional pulse current with a fixed intensity I_p. This realizes a wider capture range, preventing erroneous locking and faster phase locking. In the data field where the frequency ranges from $1/3$ to $1/8$ of the VCO clock frequency, a phase detector is used. The output is a dc current which is proportional to the phase difference of the input data and VCO clock and lasts until the next time–phase difference is detected. It realizes a frequency response that does not depend on the frequency ratio N of the VCO clock to the input data, and reduces the degradation of the data read margin caused by VCO jitter during synchronization.

2) A frequency phase detector compares the VCO clock divided by four with the rising edge of the read data (RAWRD) and, if the VCO clock is earlier, allows current with a fixed intensity I_p to flow from the loop filter during the time interval just as long as the phase difference.

If the phase difference is zero, no current flows to or from the loop filter. If the VCO clock is later, it allows current with a fixed intensity I_p to flow to the loop filter during the time interval just as long as the phase difference. This circuitry presents a wide range of linear comparison and a wide capture range (actually limited by the dynamic range of VCO output and parameters of the circuit), and is suitable for the synchronization when phase synchronization is required to be complete in a few microseconds. Fig. 2 shows the resulting measurement of phase step response at 30 Mb/s (acquisition time: less than 2 μs).

3) Fig. 3 shows the schematic of a PLL using a phase detector and Fig. 4 shows its timing diagram. The signal T_c is generated between the rising edge of the read data and the falling edge of the next VCO clock. The signal T_s is generated between the falling edge of the VCO clock and the rising edge of the next VCO clock. The signal T_D is generated between the rising edge of the VCO clock and the

falling edge of the next VCO clock. While the signal T_c is coming out, SW1 is closed and current I_c flows from the capacitor C_c. Then the signal T_s causes SW2 to close and the output voltage V_{op} of the operational amplifier is sampled and held by the capacitor C_s.

A differential amplifier converts the voltage V_{op} to current I_D. While the signal T_D is coming out, SW3 is closed and current I_D flows to the capacitor C_c. We analyze this process quantitatively as follows. Suppose we sample the phase difference n times. Then $I_{D(n)}$ and $V_{op(n)}$ at the nth sampling are

$$I_{D(n)} = g_m \cdot V_{op(n)} \tag{4}$$

$$V_{op(n)} = V_{op(n-1)} + \frac{T_c \cdot I_c - T_D \cdot I_{d(n-1)}}{C_c} \tag{5}$$

where g_m is the gain of the differential amplifier. These equations may be rewritten as follows:

$$I_{D(n)} = \frac{T_c \cdot I_c}{C_c} g_m + \left(1 - \frac{T_d \cdot g_m}{C_c}\right) \cdot I_{D(n-1)}. \tag{6}$$

This recursive equation can be solved as follows:

$$I_{D(n)} = \frac{T_c}{T_D} I_c \cdot \left\{ 1 - \left(1 - \frac{T_D \cdot g_m}{C_c}\right)^{n-1} \right\}$$

$$+ \left(1 - \frac{T_D \cdot g_m}{C_c}\right)^{n-1} \cdot I_{D(1)}. \tag{7}$$

This indicates that if the condition

$$\frac{T_D \cdot g_m}{C_c} \geqslant 2 \tag{8}$$

holds, the expression of current I_D will diverge; otherwise it will converge. If the condition

$$\frac{T_D \cdot g_m}{C_c} = 1 \tag{9}$$

holds, $I_{D(n)}$ becomes as follows:

$$I_D = I_{D(n)} = \frac{T_c}{T_D} I_c. \tag{10}$$

So, if the circuit is designed so that the output current I_T meets the following:

$$I_T = I_D - I_c \tag{11}$$

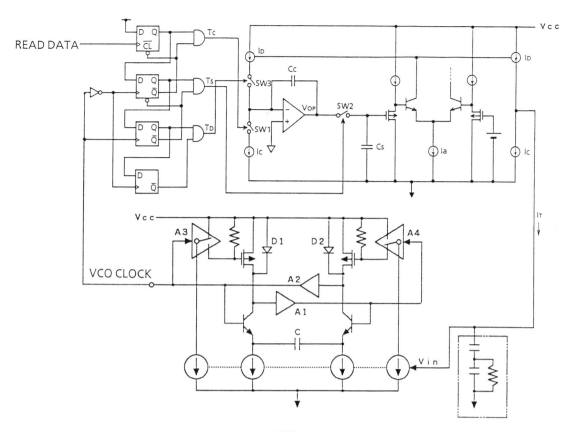

Fig. 3.   Schematic of PLL using phase detector.

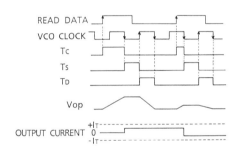

Fig. 4.   Timing diagram for phase detector.

the output current I_T is

$$I_T = \frac{T_c - T_D}{T_D} I_c. \qquad (12)$$

As the phase difference $\Delta\phi$ is given as

$$\Delta\phi = T_c - T_D \qquad (13)$$

the output current I_T is dc current which is proportional to the phase difference $\Delta\phi$.

Next we examine (9) used as a condition for the quantitative analysis of the phase detector. T_D corresponds to the half period of the VCO clock and remains fixed if the VCO clock frequency is fixed and its duty is controlled. Deviation of the capacitance C_c can be reduced, even considering the mask precision, the thickness of spin-on glass (SOG) + phosphosilicate glass (PSG), and its dielectric constant. The gain g_m of the differential amplifier is given as follows:

$$g_m = \frac{I_a}{V_T} = \frac{qI_a}{kT} \qquad (14)$$

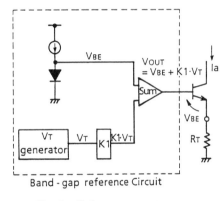

Fig. 5.   Reference current source.

where

q electric charge of an electron,
k Boltzman constant,
T absolute temperature (Kelvin).

This has a temperature dependence of about -3300 ppm/°C if I_a is fixed. To cancel this temperature dependence, a bandgap reference circuit as shown in Fig. 5 is used. Its output voltage is applied to resistor R_T through an emitter follower. Then the current I_a of the differential amplifier is

$$I_a = \frac{V_{\text{out}} - V_{BE}}{R_T} = \frac{K_1 \cdot V_T}{R_T} \qquad (15)$$

where K_1 is a constant. So, the gain g_m of the differential

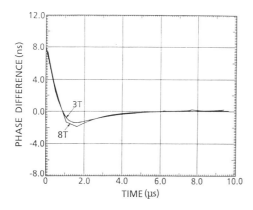

Fig. 6. Measured phase step response using phase detector.

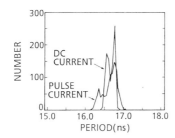

Fig. 7. Measured VCO period.

amplifier is

$$g_m = \frac{1}{V_T} \frac{K_1 \cdot V_T}{R_T} = \frac{K_1}{R_T}. \qquad (16)$$

K_1, which is given as a ratio of two resistors, does not vary with temperature. Therefore it is possible to make the gain g_m independent of temperature by connecting the resistor R_T externally. It is also possible to make (9) independent of temperature and to cause the phase detector to produce a dc current that is proportional to the phase difference $\Delta\phi$.

When we use the circuitry described above, the open-loop transfer function $G(s)$, characteristic frequency ω_n, and damping factor ζ of the system are as follows in the data field:

$$G(s) = \frac{S(C1+C2)R \cdot I_T \cdot K_o + I_T \cdot K_o}{S^2 C1 \cdot \pi (1 + SC2R)} \qquad (17)$$

$$\omega_n = \sqrt{\frac{I_T \cdot K_o}{\pi \cdot C1}} \qquad (18)$$

$$\zeta = \frac{(C1+C2)R}{2} \cdot \sqrt{\frac{I_T \cdot K_o}{\pi \cdot C1}}. \qquad (19)$$

They do not depend on the ratio N of frequency of the VCO clock to that of the read data. So, you need not consider the effect of N and can set the system in an optimal condition. Fig. 6 shows two curves of phase step response measured with an experimental circuit when $N = 3$ and 8. These curves are almost equal, though differences of the sampling intervals exist. Fig. 7 compares VCO jitter at the time of phase synchronization with the conventional current pulse system and the dc current system. VCO jitter is lower by 500 ps with the dc current system, and the decode margin can be increased by the same amount.

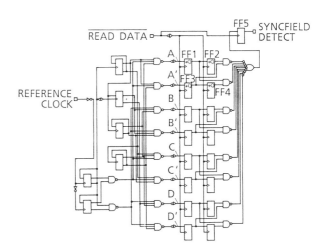

Fig. 8. Circuit diagram of sync field detector.

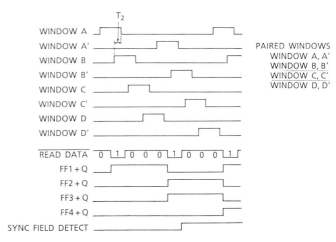

Fig. 9. Timing diagram of sync field detector.

III. DIGITAL SYNC FIELD DETECTOR

The sync field detector was conventionally an analog circuit composed of a monostable multivibrator, and it has the following problems:

1) in the analog circuit, individual differences of components are great (IC's, C, R), so adjustment is needed for accurate operation;
2) error due to temperature drift cannot be ignored (measured as $600 \sim 1000$ ppm) and the design must allow a sufficient margin;
3) if the pattern written in the sync field is not the highest frequency of the data code, the conventional system will fail since it will detect all frequency patterns that are higher than that frequency as sync fields.

The digital sync field detector has been developed to solve these problems.

Fig. 8 is the circuit diagram of the sync field detector and Fig. 9 is a timing diagram. Eight-phase window pulses generated by the reference clock are classified into four pairs so that the difference in phase of each pair is $4T_0$ (T_0: the period of reference clock). To detect the pattern of sync fields, the interval of 1's is detected when two neighboring 1's in the input data exist in a paired window. Fig. 9 shows

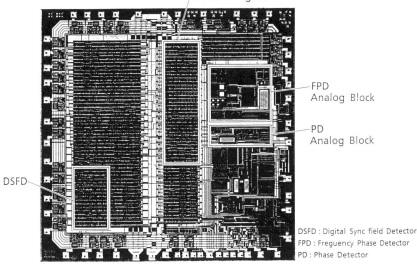

FPD&PD Digital Block

FPD
Analog Block

PD
Analog Block

DSFD

DSFD : Digital Sync field Detector
FPD : Frequency Phase Detector
PD : Phase Detector

Fig. 10.   Photograph of the data recovery circuit.

the eight-phase window pulses paired into four, and the timing of the case where the read data is a $4T$ pattern (repeat of "1000"). The figure shows that the sync field detect signal works when a $4T$ pattern is detected.

The window pulses generated from the reference clock are asynchronous with the read data and they have different frequencies and phases. The data read from a disk includes much jitter. To secure operation whatever the phase relation, arrangement is made so that windows are continuous and neighboring windows overlap. Fig. 9 shows windows A and B overlap. Permissible jitter T_1 is within $\pm 50\%$ of the data bit length T as it is determined from the capability of the following data separator. Overlap T_2 of windows and permissible jitter T_1 are as follows:

$$T_2 = |T_1|. \tag{20}$$

On the other hand,

$$T_1 = \pm 0.5T. \tag{21}$$

So, overlap $T2$ is as follows:

$$T_2 = 0.5T. \tag{22}$$

At this time, permissible frequency deviation $\Delta f\theta$ is found to be 12.5% from

$$\Delta f\theta = \frac{T_2}{4T} \cdot \frac{1}{T} \tag{23}$$

Though this circuit is designed to detect the $4T$ pattern, it may detect a $3T$ or $5T$ pattern incorrectly due to the overlap T_2. But this happens only when the phase relation is such that data enter in the specific 50% area of windows for the time of 6 or 8 bytes, and this does not actually happen because of the frequency deviation and jitter actually involved in read data.

Since this is a digital circuit, adjustment is not necessary and a very accurate sync field detector can be realized with the effects of temperature drift eliminated.

IV. Operation of PLL

Operation of the PLL is controlled through the READ GATE (RG) and PHASE SYNC (PS) terminals. The PLL follows the reference clock (REFCLK) when RG is negated.

TABLE I
Device Characteristics

General	Technology	2-μm BiCMOS
	Supply Voltage	5 V, $\pm 5\%$
	Power Consumption	400 mW (typical)
	Chip Size	4.7×4.9 mm^2
	Data Transfer Rate	30 Mbs
PLL	Acquisition Time	2 μs
	Jitter Output	500 ps
	Capture range	$\pm 10\%$
	Lock Range	$\pm 10\%$
Sync Field Detect	Tolerance to Input Jitter	$\pm 0.5T$

When PS is asserted according to the sync field detect (SYNCDETECT) output of the sync field detector or an external signal while RG is asserted, the PLL starts comparing the frequency and phase of the read data (RAWRD) and locking with the charge pump in a high-gain state. After a fixed time of 6 or 8 bytes when PS is disasserted, the PLL switches from the frequency phase comparison mode to the phase comparison mode and operates for T/I conversion at normal gain. When RG is disasserted again, the PLL returns to follow REFCLK. The NRZRD terminal remains at Hi-Z when RG is disasserted, produces L after RG is asserted as long as the high-gain state continues and, in the normal-gain state, produces read data decoded by the conversion rules of the RLL 2–7 code.

V. Results

This IC using the 2-μm BiCMOS process incorporates the functions of a PLL, sync field detector, RLL 2–7 encoder/decoder, and is realized as a chip of 4.9×4.7 mm^2. Fig. 10 shows a photo of the IC.

The PLL in this IC takes 2 μs until it locks by using a frequency phase detector. This is twice as fast as conventional data synchronizer/2-7RLL ENDEC; jitter is also halved in the stable state. The sync field detector is adjustment-free because it is a digital circuit. The power consumption of the IC is reduced to 400 mW thanks to the $2\text{-}\mu\text{m}$ BiCMOS process.

Table I summarizes the performance of the IC.

VI. Conclusion

A high-speed read/write IC suitable for magnetic disk drives has been developed using a dual-mode analog PLL and digital sync field detector.

The dual-mode analog PLL has a wide decode margin, locks up quickly, and operates stably without being affected by the frequency deviation of data. The digital sync field detector is adjustment-free and detects sync fields very accurately.

The results of this research ensure the feasibility of the development of very large mixed analog–digital LSI incorporating the hard disk control function, analog signal amplifiers, and a read/write circuit using the BiCMOS process.

Acknowledgment

The authors wish to thank Dr. S. Kawamura and N. Tanaka for their encouragement, and H. Sato and T. Sase for their technical suggestions.

References

[1] J. T. Kellis and S. Mehrozra, "A 15-Mb/s data separator and write compensation circuit for Winchester disk drives," in *ISSCC Dig. Tech. Papers*, Feb. 1986, pp. 232–233.
[2] K. Kato *et al.*, "A low power dissipation PLL IC operating at 128 MHz clock," in *IEEE 1987 CICC Conf. Rec.*, May 1987, pp. 651–654.
[3] R. W. Wood, "Magnetic recording systems," *Proc. IEEE*, vol. 74, no. 11, pp. 1557–1569, Nov. 1986.
[4] "Data synchronizer/2, 7 RLL ENDEC with write precompensation SSI 32D535," product manual, Silicon Systems.

A Versatile Clock Recovery Architecture and Monolithic Implementation

Invited Paper

Lawrence M. DeVito

Abstract—A family of monolithic phase-locked loops recover clock and retime NRZ data. At 155 MHz, random plus pattern jitter with a 2^7 code is 1.8 degrees rms, and static phase error is 4 degrees. Devices fabricated on both junction-isolated and dielectric-isolated bipolar processes are described. Measurement techniques to verify compliance with international telecommunication standards are also described.

1. Introduction

Fiber-optic serial digital data communication networks are finding increased application in both traditional telecommunications with the synchronous optical network (SONET) and now in datacom local area networks such as asynchronous transfer mode (ATM). This increased demand creates a need for small and easy-to-use fiber-optic receivers, key elements of which are the recovery of the implicit clock signal embedded in the non-return-to-zero (NRZ) serial data stream, and re-establishing the synchronous timing of the data using the recovered clock as the reference. This paper describes several integrated-circuit devices for clock recovery and data retiming. One circuit for a data rate of 52 MHz is described, and two different devices at a data rate of 155 MHz are also described. All three devices are fully compliant with the SONET/SDH (Synchronous Digital Hierarchy) specifications and CCITT G.958.

Motivation for this work is to produce a generic clock recovery core function that is easy to use and provides very high performance in a wide variety of real systems. Other systems where the device may be used are in electrical physical media such as recently emerged ATM on category 5 unshielded twisted pair (UTP) transmission lines. These systems benefit technologically from sharing specifications with SONET, and can also benefit economically with the availability of cost-effective clock recovery devices. For a single device (or, more likely, a reusable architecture) to glide easily across different systems and even different physical media, minimal constraints may be placed on its application. For example, use of a reference frequency as an acquisition aid may be acceptable in a switch application, but places an unacceptable cost and size burden on an optical receiver module. Thus, the trick is to find cost-effective techniques to allow a stand-alone clock recovery function.

The devices described in this paper [1,2] are fabricated on two different processes. Devices at 52 MHz and 155 MHz are fabricated on a 3.5-GHz junction-isolated bipolar process including laser-trimmed silicon-chromium thin-film resistors. Power consumption is 140 mA from a single 5-V supply. A newer device is fabricated on a dielectric-isolated bonded wafer complementary bipolar process [3,4] with both npn and pnp f_T of 4 GHz. This process also includes laser-trimmed silicon-chromium thin-film resistors. Power consumption of the newer device is 30 mA from a single 5-V supply. The new device also has a ring oscillator VCO [5,6] specifically developed for lower jitter generation as compared to the multivibrator in the original devices.

A complete schematic of the clock recovery application is shown in Figure 1. The device uses standard ECL levels at input and output and the data has an NRZ format. Only one external component is required: a capacitor to set the phase-locked loop damping factor. It is important to note the simplicity of this application. Unlike prior art, the device does not require a crystal or a ringing tank [7–10]; no external reference frequency is required [11–14]. There are no control functions to manipulate between acquisition and tracking modes, and no data preamble for acquisition aid is required. The device simply locks to random data, and a lock-detect indication is given when acquisition is complete. Further, unlike prior art, this device is monolithic [8], and the jitter is specified [15]. Separate analog and digital power and substrate pins allow for ease of use in a realistic situation with inevitable noise on the power supply.

Figure 2 is a summary of the device performance. The most important performance specifications to note are the jitter and static phase offset. With a 2^7 pseudorandom code, the jitter at 52 MHz is only 2 degrees rms. At 155 MHz, the jitter of the junction-isolated device is 4 degrees rms, and that of the dielectric-isolated device is 2 degrees rms. The static phase offset, or window centering, is 3 degrees at 52 MHz and either 18 degrees or 4 degrees for the two devices at 155 MHz.

The loop bandwidth with random data is approximately 0.1% of the clock rate, consistent with applications such as SONET and fiber distributed data interface (FDDI). For other applications, different loop bandwidths are needed. The bandwidth is mask programmable over a range of 0.01% to 1%. Clearly, the center frequency is also mask programmable. The damping factor is user programmable with an external capacitor over a range of 1 to 10. The acquisition time is a function of damping factor.

Edge jitter on retimed data output is shown in Figure 3 for a 155-MHz clock rate with both maximum data density and 2^7 pseudorandom code. Note the well-formed Gaussian distribution even with the pseudorandom code. Note also that the increase in jitter when going from maximum data density (2.3 degrees rms) to pseudorandom code (3.6 degrees rms) is explained entirely by the doubling of the jitter bandwidth with doubling data density. The self-noise of the voltage-controlled oscillator (VCO) in a phase-locked loop is reduced in proportion

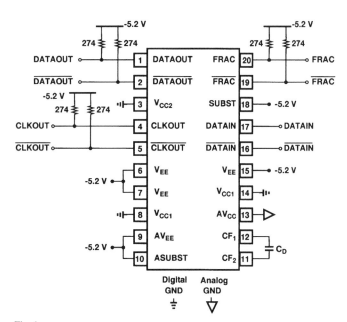

Fig. 1

Clock Rate	52 MHz	155 MHz	155 MHz
Process	Junction-Isol.	Junction-Isol.	Dielectric-Isol.
Jitter (deg. rms)			
Max Data	1.4	2.3	
2^7 PRN	1.9	3.6	1.8
Static Phase (deg.)	3	18	4
Loop BW (kHz)	52	100	92
Damping Factor	1-10	1-10	1-10
Acquistion Time			
Damping = 1 (us)	160	50	50
Damping = 10 (ms)	16	5	5
Power (from 5 V) (mW)	625	700	150
Die Size (kmil^2)	20	20	10

Fig. 2

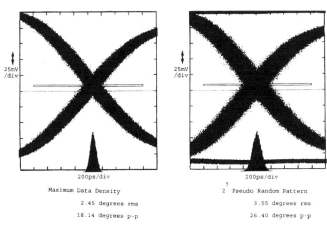

Maximum Data Density
2.45 degrees rms
18.14 degrees p-p

2^7 Pseudo Random Pattern
3.55 degrees rms
26.40 degrees p-p

Fig. 3

to the square root of the jitter bandwidth. This means there is essentially no pattern jitter in this clock recovery device. Elimination of pattern jitter is attributed to a combination of a newly developed phase detector and judicious placement of higher order poles in the loop.

In an application as a regenerator, the clock recovery device is called upon to meet both receive-jitter tolerance and transmit-jitter generation requirements. In addition, bridging these two sets are the SONET OC3 and CCITT G.958 STM-1 Type A requirements for jitter transfer: 0.1-dB jitter peaking, and the 130-kHz low-pass jitter bandwidth. One of the devices described in this paper has internal oscillator self-jitter of 2 degrees rms, and thus easily meets the transmit-jitter generation requirement of rms unit interval (UI) = 0.01, or 3.6 degrees rms. It is interesting to note that jitter bandwidth and jitter generation must be traded off against each other. Some clock recovery devices [16,17] raise the jitter bandwidth of the phase-locked loop so the noise of the internal VCO can be made to look smaller due to the increased loop gain over a wider range of frequencies. However, the penalty is failure to meet the jitter transfer specifications of 130-kHz low-pass required of a regenerator. The self-noise of the VCO in a phase-locked loop is reduced in proportion to the square root of the jitter bandwidth. Thus, if one erroneously chooses a high-jitter bandwidth, then loop jitter can be lowered. For example, the jitter bandwidth of the devices described in this paper is 110 kHz, easily meeting the regenerator requirement. However, if the jitter bandwidth were mistakenly increased to 1 MHz, the loop jitter would be decreased from 2 degrees rms to 0.6 degrees rms, or 11 ps. But that would be wrong.

2. ARCHITECTURE

2.1 Two-Loop Architecture

A two-loop architecture is employed in these clock recovery devices, as seen in Figure 4. A frequency detector guarantees that the device locks to the proper data frequency, and a phase

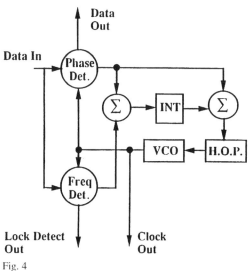

Fig. 4

detector then aligns the clock edges with the input data edges. Two loops are necessary because a phase-locked loop alone has a slow and unreliable frequency acquisition process. First, the rotational frequency detector [15,18] reduces the error in frequency between the input data and the VCO. Once the frequency error is sufficiently small, the phase detector takes over and aligns the phase of the clock to match the input data. During this final stage of operation, the output of the frequency detector is identically zero, and no longer affects the operation of the circuit. No control functions are needed to initiate acquisition or change mode of operation after acquisition.

False lock to data sidebands is completely eliminated by this two-loop architecture. Such false lock is a problem found in more ordinary single-loop PLL clock recovery devices. In single-loop architectures, an initial frequency acquisition aid is needed because a phase detector alone cannot be relied on to achieve frequency acquisition. Sometimes a "pilot tone" is used to first lock the clock recovery to a center frequency reference, and then control is switched to the input data. Besides the extra burden and expense of the crystal oscillator needed for the pilot tone and the control circuitry, this method still suffers from the possibility of false lock to data sidebands because the data is ultimately tracked by a phase-sensitive-only detector. Another method of using the pilot tone is to have two complete phase-locked loops: for example, one may be used as a transmit synthesizer and the other is the receive clock recovery. The synthesizer is then locked to the pilot tone and the clock recovery loop uses an identical VCO with a center tuning voltage derived from the synthesizer. Even in this case, the data is tracked only by a phase-sensitive detector and false lock to data sidebands is possible. Also, in many applications for fiber-optic receivers, this transmit synthesizer is not needed, and this represents an extra expense. For example, in a regenerator, the receive signal is the only frequency of interest and no independent transmit frequency reference is needed. In fact, it is undesirable to have any extraneous frequencies in the module because there is always the possibility of unintentional locking caused by stray coupling.

Detail of the phase and frequency detectors is shown in Figure 5. No control functions are used: both the phase detector and the frequency detector are always active. During frequency acquisition, higher gain allows the frequency loop to dominate. The frequency detector senses cycle slips across the *BC* quadrant boundary. During the frequency acquisition process, cycle slips between the VCO and the data are indicated at the lock detect (called FRAC here) output. The phase loop locks data transitions to the *AD* quadrant boundary, effectively disabling the frequency loop.

Note that the frequency loop is first order and requires no transmission zero for stability as does the second-order PLL. The zero is implemented in the phase loop only by shunting the phase detector path around the shared integrator. Higher order poles (HOPs) in the loop transmission to filter the VCO control signal are needed for two reasons: first, to increase tolerance to high-frequency jitter on the data input; and second, to reduce the residual internal pattern jitter caused by the phase detector. Two real poles at seven times the loop bandwidth reduce the worst-case pattern jitter by a factor of 50. Without these higher order poles, the fast shunt path from the phase detector into the VCO would produce 5 degrees of peak-to-peak jitter.

2.2 Higher Order Poles

Placement of the higher order poles is explained with Figure 6. With a constraint that the crossover frequency remains constant, we want to move the two HOPs to as low a frequency as possible without compromising stability. The least stable case is with a damping factor of unity and with maximum data density as shown in the figure. On the root locus, we start with the two usual closed loop poles. Then the HOPs are added to the loop transmission as two real poles at some very high frequency, so that they have no effect on the location of the lower frequency closed loop singularities. The HOPs contribute two new closed loop poles in the root locus diagram. Then the HOPs are moved to a lower frequency, which causes one of the usual low-frequency closed loop poles to move to a higher frequency, and one of the newly added HOPs to move to a lower frequency. The HOPs are moved in to the point where the lower frequency HOPs and the

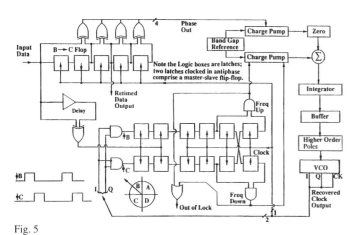

Fig. 5

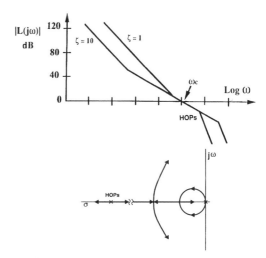

Fig. 6

higher frequency usual pole just meet. In this extreme situation, the closed loop bandwidth is increased dramatically, but in the more realistic case of random data, the closed loop bandwidth is not changed much by the presence of the HOPs. And in the much more representative situation of damping factor of 5 to 10, the HOPs have negligible effect on the details of the low-frequency jitter transfer function.

Measured jitter transfer functions of two different devices, one with and one without HOPs, are given in Figure 7. The entire passband of the loop is contained in the first two data points, below 100 kHz. Beyond the tracking bandwidth of the loop, the response falls off. The two devices are similar up to 600 kHz. Beyond this point, the HOPs suppress jitter in one device, but in the other, the direct path from the phase detector to the VCO allows unwanted frequency modulation of the VCO by the input jitter.

JITTER TRANSFER FUNCTION

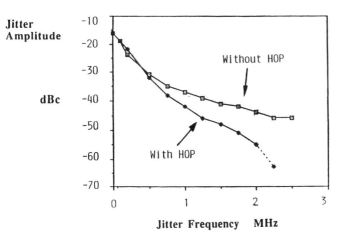

Fig. 7

2.3 Jitter Tolerance Theory

Jitter tolerance is a measure of the ability of the clock recovery device to track a jittered input data signal. Jitter on the input data, X in Figure 8, is best thought of as phase modulation, and is usually specified in unit intervals. The PLL must provide a clock signal, Y in Figure 8, that tracks this phase modulation in order to accurately retime jittered data. For the VCO output to have a phase modulation that tracks the input jitter, some modulation signal must be generated at the output of the phase de-

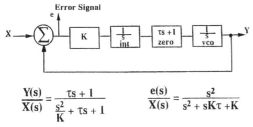

$$\frac{Y(s)}{X(s)} = \frac{\tau s + 1}{\frac{s^2}{K} + \tau s + 1} \qquad \frac{e(s)}{X(s)} = \frac{s^2}{s^2 + sK\tau + K}$$

Fig. 8

tector; this is the error signal, *e* in Figure 8. The modulation output from the phase detector can only be produced by a phase error between the data input and the clock input. Hence, the PLL can never perfectly track jittered data. However, the magnitude of the phase error depends on the gain around the loop. At low frequencies, the integrator provides very high gain, and thus very large jitter can be tracked with small phase errors between input data and recovered clock. At frequencies closer to the loop bandwidth, the gain of the integrator is much smaller, and thus less input jitter can be tolerated. These features are seen in Figure 9, the theoretical jitter tolerance for the clock recovery phase-locked loop. Note that there are two curves, one each for loop damping of 1 and 10. Also indicated are slew limits set by finite VCO tuning range of 1 and 10%.

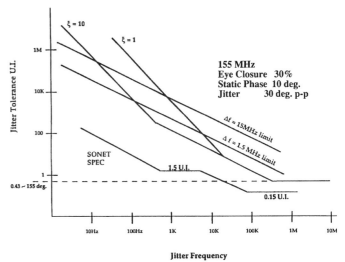

Fig. 9

If the magnitude of the error signal, *e* in Figure 8, is greater than the eye opening of the input data signal, then data retiming errors are made. This is the limit of "jitter tolerance." The error transfer function (*e/X* in Figure 8) is a high-pass filter. Thus, we expect to tolerate large amounts of jitter at low frequencies, because the high-pass filter attenuates. At higher frequency, the high-pass transmits all the input jitter and only small amounts of jitter may be tolerated. The corner frequency of this high-pass is the same as the -3-dB frequency of the low-pass jitter transfer function, *Y/X*, shown in Figure 8. Thus, jitter tolerance can be determined in two different ways. An input signal with increasing amounts of jitter can be tracked until errors are detected, or the bandwidth of the jitter transfer function can be measured. Finally, now, the jitter transfer function of the phase-locked loop is shown in Figure 10. The -3-dB bandwidth of the PLL is about s_{high}.

Knowing the bandwidth of the jitter transfer function is important for another reason. There is a specification in SONET that the bandwidth of the jitter transfer function be less than 130 kHz for the OC3 155-MHz repeater. If this bandwidth is too large, jitter in a string of repeaters will accumulate quickly and limit the number of repeaters. Thus, we see conflicting requirements on the jitter transfer function: first, it should be wideband

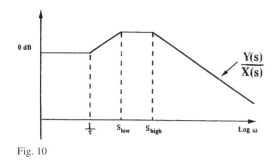

Fig. 10

to accommodate lots of jitter; second, it should be narrowband to filter jitter and prevent accumulation in a string of repeaters.

2.4 Jitter Peaking Theory

The SONET specifications for an OC-3 regenerator require a low-pass jitter transfer function with a bandwidth less than 130 kHz and a maximum gain in the passband of less than 0.1 dB. This gain in the jitter transfer function is commonly called *jitter peaking,* and contributes to the accumulation of jitter in a string of regenerators [19]. From Figure 10 it is seen that the zero in the closed loop transfer function occurs at a lower frequency than the first closed loop pole. This results in jitter peaking that can never be eliminated; but the peaking can be reduced to negligible levels by overdamping the loop. However, overdamping has some deleterious effects, such as long acquisition time and awkwardly large capacitor values in the loop filter. Thus, it is necessary to construct an analytical model for jitter peaking in order to do engineering trade-offs in the system design of the PLL. (It is interesting here to note that jitter peaking can be fundamentally eliminated by an architectural change in the PLL [20].)

Referring to Figure 8, the open loop transmission is given by Eq. (1).

$$L(s) = \frac{OD(RCs + 1)}{s^2 C} \tag{1}$$

where O is the gain of the oscillator in units of radians per volt-second; D is the gain of the phase detector in units of amperes per radian; R and C are series elements in the loop filter comprising an integrator with a zero. The closed loop transfer function Y/X is thus as shown in Eq. (2).

$$\frac{Y(s)}{X(s)} = \frac{OD(RCs + 1)}{s^2 C + ODRCs + OD} \tag{2}$$

When factored, the denominator of Eq. (2) gives two roots as closed loop poles, given in Eqs. (3) and (4).

$$s_{low} = \frac{ODR}{2}\left(-1+\sqrt{1-\frac{4}{CODR^2}}\right) \tag{3}$$

$$s_{high} = \frac{ODR}{2}\left(-1-\sqrt{1-\frac{4}{CODR^2}}\right) \tag{4}$$

The jitter peaking can be approximated as the ratio of frequencies of the pole, s_{low}, in Eq. (3), and the closed loop zero from Eq. (2). This is given in Eq. (5).

$$\frac{s_{low}}{s_z} = \frac{CODR^2}{2}\left(1-\sqrt{1-\frac{4}{CODR^2}}\right) \tag{5}$$

Simplifying this expression requires approximation of the square root shown in Eq. (6), giving better than 0.1% accuracy for values of x less than 0.2.

$$\sqrt{1-x} \approx 1 - \frac{x}{2} - \frac{x^2}{8} \tag{6}$$

Finally, now, jitter peaking is given in Eq. (7).

$$JP \approx \frac{s_{low}}{s_z} \approx 1 + \frac{1}{CODR^2} \tag{7}$$

It is convenient to express this jitter peaking in dB, as shown in Eq. (8).

$$JP_{dB} = \frac{8.686}{CODR^2} \ dB \tag{8}$$

The jitter bandwidth is easily determined by finding the purely imaginary frequency at which the value of the squared modulus of the transfer function of Eq. (2) is equal to one-half. The unilluminating result is given in Eq. (9).

$$w_{-3dB} = \sqrt{\frac{OD}{C}} \sqrt{1 + \frac{CODR^2}{2} + \sqrt{\left(1+\frac{CODR^2}{2}\right)^2 + 1}} \tag{9}$$

For practical values of the variables, all the square roots may be approximated by neglecting the "one plus" terms, giving the useful result of Eq. (10):

$$w_{-3dB} = ROD \tag{10}$$

For representative values: $O = 446 \times 10^6$ rad/(V-sec), $D = 796 \times 10^{-9}$ A/rad, $R^2 = 1.588 \times 10^6 \ \Omega^2$, Figure 11 shows the analytical model approximations for jitter peaking as compared to the values obtained by exact evaluation. The approximations are close enough to allow useful engineering trade-off between circuit elements. For example, if in fabrication of a monolithic PLL, the VCO gain, O, is higher than expected, the jitter bandwidth can most easily be restored to design value by proportionately lowering the series resistor, R, in the loop filter. However, it is clear from Eq. (8) that the loop filter integration capacitor, C, must scale as the square of the resistor to preserve jitter peaking.

C	CODR^2	Eq. 8	Eq. 2	Eq. 10	Eq. 9
47 nF	26.5	0.33	0.26	71.2 kHz	74 kHz
100 nF	56.4	0.15	0.13	71.2 kHz	73 kHz
220 nF	124	0.07	0.062	71.2 kHz	72 kHz
330 nF	186	0.047	0.042	71.2 kHz	72 kHz

Fig. 11

3. PHASE DETECTOR

3.1 Operation of Phase Detector

A simplified schematic of the phase detector [21] is shown in Figure 12. The logic elements are latches: two latches in cascade comprise a master-slave flipflop. The XOR outputs control the charge pump. X1 controls the first up-ramp section; X2 controls the first down-ramp; X3 controls the second down-ramp; and finally, X4 controls the second up-ramp. The first pump signal is initiated by a data transition and terminated by a clock edge. The remaining pump signals are both initiated and terminated by clock edges. The arrival time of a data transition controls the width of the X1 pulse, giving the phase measurement. As the data bit travels down the shift register, X2,3,4 give equal-width pulses. A single phase measurement persists for two clock cycles. Data transitions on subsequent clock periods simply produce overlapping charge pumps with the first half of the second pump sequence canceling the second half of the first pump sequence. The phase measurement process is still active during the overlap time. The four phases of charge pump provide two benefits: first, static phase error caused by duty cycle distortion in the clock signal is reduced; and second, the triwave transient created in the charge pump has a net area of zero. The zero area, in combination with the higher order poles, has a benefit of reducing pattern jitter.

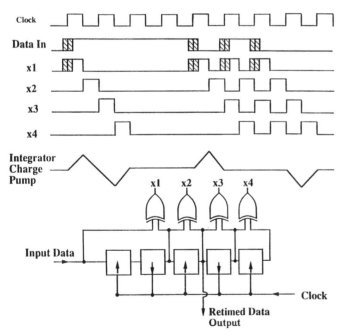

Fig. 12

3.2 Pattern Jitter

The phase detector produces transients on the charge-pump output, as shown in Figure 13, in response to a section of dense data and a section of sparse data. Also, the transient produced by

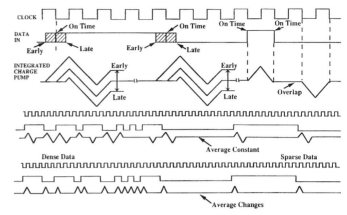

Fig. 13

a prior art phase detector [22] is shown in the figure as well. A data transition arriving on time produces a transient with zero net area: thus, the VCO phase (responsive to volt-seconds on its control line) is unchanged. Pattern jitter is eliminated by this property because the accumulation of phase by the VCO is rendered independent of data density. For random data with a density of 0.5, the net area of the charge-pump transient is still zero. Although the balance is delayed half a clock cycle on average, pattern jitter suffers little. Figure 13 shows that the average level of the charge-pump output is independent of the data density in contrast to the prior art phase detector. Note that, because the area of the transient is zero, the average value of this waveform does not change with data density. But because the transient created by the prior art phase detector has net area, the average value of the charge-pump waveform does change with data density. Since the VCO frequency responds to the average value of the charge-pump voltage, the change in data density incorrectly produces a change in VCO frequency. Since the frequency of the data is constant, this type of phase detector will create phase errors to equilibrate the average level for dense and sparse data. It is these phase errors that are pattern jitter. This source of pattern jitter is completely eliminated by the phase detector used in this circuit, as shown in Figure 12.

When the PLL is locked, the phase detector controls the loop by comparing the input data signal to the clock signal produced by the VCO. Clearly, characteristics of the phase detector are crucial to meeting the SONET OC$_3$ and CCITT G.958 STM-1 requirements. While these specifications are quite explicit on items such as jitter peaking and jitter transfer bandwidth, there are other more subtle issues that must also be correctly analyzed to produce a device that is truly fit for use in the intended application. These more subtle issues include pattern jitter and consecutive identical digits (CIDs). Pattern jitter is especially important in a string of regenerators because each regenerator sees an identical pattern, and contributes its own correlated jitter that each subsequent stage will then need to accommodate. In the limiting case of very small jitter peaking and up to 50 regenerators in cascade, correlated pattern jitter contributes twice as much to the accumulated jitter as does random jitter [19,

p. 62]. Problematic as this pattern jitter seems, random jitter must also be tightly controlled. Consider a worst-case system with the 0.1-dB allowable spec limit of jitter peaking in each of 50 regenerators. Here, alignment jitter due to random noise actually accumulates three times faster than does the correlated pattern jitter [19, p. 99]. A real system, of course, will be somewhere between these two extremes of performance, necessitating exemplary performance in both pattern jitter and random jitter.

3.3 Consecutive Identical Digits

Another requirement that is placed on the clock recovery is to continue to produce a valid clock output signal even if there is a long string of transitionless data or consecutive identical digits. This is described in Appendix I to CCITT G.958. This is the fatal performance flaw of clock recovery techniques based on surface acoustic wave (SAW) filters on high-Q tanks [8]. Because a SAW is a narrow-bandpass filter, it cannot produce an output with no input. When input transitions cease, the filter will continue to produce a clock signal as it rings down; but eventually the clock signal will cease. A high Q for the bandpass will allow the clock to not fade away, but such a high Q places unreasonable demands on center frequency accuracy. In a SAW filter, if the center frequency is not correct, then static phase error results. While static phase must always be compensated in the retiming path of the SAW, the temperature stability of static phase due to center frequency drift of the bandpass is more difficult to accommodate. So the SAW bandpass Q must be limited to a value that allows premature fading of the clock with a long string of transitionless data. A phase-locked loop, on the other hand, has no fundamental limitation on the length of transitionless data strings that can be tolerated. This is not to say that a phase-locked loop will guarantee adequate performance: careful attention to detail is necessary to achieve good performance. One such detail is in the phase detector.

The phase-locked loop clock recovery device described in this work will coast over more than 1000 CIDs. (It is interesting that we do not know exactly how many CIDs can be tolerated, because the test equipment that inserts blocks of zeros into data pattern is limited to 1000 zeros maximum.) In addition to this extremely good CID performance, the device has virtually no pattern jitter. Both of these features are attributable in part to the newly developed phase detector. When data density changes, as with data patterns, or when the data ceases, as with CID, the average value of the output of the phase detector does not change. Because the VCO responds to this average value of the phase detector output, the VCO continues to oscillate at the same frequency when data density changes or during a CID. This unchanged frequency of oscillation means that accurate timing information is provided by the clock. By contrast, if the VCO frequency was changed ever so slightly by the transition into a CID, then the incorrect frequency would cause the phase of the recovered clock to drift away from the ideal phase.

4. FREQUENCY DETECTOR

4.1 Operation of Frequency Detector

The frequency detector is shown in Figure 14; here, the logic elements are D-type flipflops. This is a rotational frequency detector [5,18], essentially a digital implementation of a quadricorrelator [10, 23]. In general, if there is a frequency difference between two signals, then their phase relationship will change with time at a rate proportional to the frequency difference. The circular diagram in Figure 14 is a phasor representation of the clock signal created by the VCO: four phases of the clock signal are represented by A,B,C,D: a ray from the origin rotates counterclockwise at an angular rate of ω_c. The phasor ω_d represents the data frequency: this ray also rotates counterclockwise. If the clock and the data frequencies are not equal, then the two phasors rotate at different speeds. Typically, however, the clock is used as a phase reference and only the data phasor is thought of as moving around the circle. But now, the data phasor can rotate in either direction, depending on the relative frequencies. When the data frequency is lower than the clock frequency, the data phasor rotates counterclockwise. The circuit detects cycle slips when the data phasor moves from B to C, producing frequency down pulses, or from C to B, producing frequency up pulses. When the PLL has acquired phase lock, the data phasor is locked to the AD quadrant boundary, and the frequency detector produces no further outputs. The In-Phase and Quadrature signals are easily obtained from the VCO.

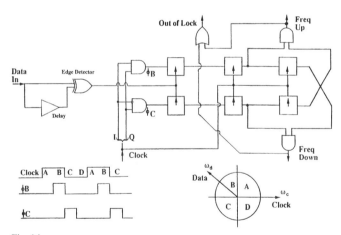

Fig. 14

The frequency detector takes samples, or "snapshots," of this clock phasor at the instants of occurrence of input data transitions. The frequency difference information is contained in the rate and direction of rotation of these phasor samples over time. Imagine an analogy to a bicycle wheel with a red ribbon tied to one of the spokes. If the wheel is spinning and a strobe light flashes at the same frequency as the rotation rate of the wheel, the ribbon will appear to be fixed in angular position. If the strobe flashes at a slightly different frequency, then the ribbon will rotate in a direction determined by the polarity of the frequency difference, and at a rate determined by the magnitude of the frequency difference. The apparent rotation of the ribbon is the frequency error phasor, ω_e. The frequency of revolution of the ribbon is exactly equal to the frequency difference between the bicycle wheel (clock) and the strobe (data). Every complete revolution of the ribbon is called a *cycle slip.* In this manner, the edges of the data input sample or "strobe" the clock phasor.

Note the clock period is divided into four quadrants labeled *A,B,C,D* in the figure. The frequency detector observes the error phasor ω_e as it rotates past the boundary between the *B* and *C* quadrants in the phasor diagram. For example, if a data transition strobes the clock while it is in the *B* quadrant and then later strobes the clock while it is in the *C* quadrant, then we say a *B* to *C* transition has occurred and the clock frequency is faster than the data frequency. Conversely, if a data transition is first detected in *C* and then *B,* the clock is then known to be faster than the data. This method is feasible if the frequency error is small enough so the change in angular position of the ribbon is less than half a revolution on consecutive strobe flashes: this means the frequency error must be less than half the data rate. This restriction is easily accommodated in the device by using laser-wafer trimming of the center frequency of the VCO. The other requirement is that the center frequency drift be small enough to keep the clock within an octave of the data frequency. This is most easily achieved, as the center frequency drift of both the multivibrator and the ring oscillator VCO in the two different devices is less than 1000 ppm/C. This drift results in less than 10% shift in frequency over temperature.

The transitions of the error phasor between the *B* and *C* quadrants occur once per revolution of the ribbon. Thus, for larger frequency error between the clock and data there are more frequent boundary crossings, also called "cycle slips." The frequency detector delivers pulses of current to the charge pump to either raise (*C* to *B* transition) or lower (*B* to *C* transition) the frequency of the clock. The higher frequency of cycle slips gives more pulses of current and thus a larger magnitude of the output signal from the frequency detector.

The process of frequency detection is complicated by the fact that the input data are random; a transition may not occur in any given data period. Returning to the bicycle wheel analogy, the strobe is intermittent and flashes, on average, half of the time. One can easily imagine that for small enough frequency differences it is still possible to determine the apparent angular velocity (magnitude and direction) ω_e of the strobed ribbon. A careful analysis shows that, with random data, the range of the frequency detector is in excess of $\pm 25\%$ of the data rate. With a maximum density data pattern (1010 . . .) every cycle slip produces a pulse at the frequency detector output. However, with random data not every cycle slip produces a pulse. The density of pulses at the frequency detector output increases with the density of data transitions. The probability that a cycle slip will produce an output pulse increases as the frequency error approaches zero. After the frequency error has been reduced to a small value, the phased-locked loop takes control of the error phasor and maintains coincidence between the data transitions and the *A-D* boundary of the clock. Because the frequency detector is sensitive only to crossings of the *B-C* boundary, no further output pulses are produced. It is in this manner that the frequency detector output becomes identically zero, and no longer affects the operation of the circuit. No control functions are needed to initiate acquisition or change mode of operation after acquisition.

4.2 False Lock

There exists a theoretical possibility that the frequency detector described here can fail to achieve lock through a pathological interaction with the phase detector. This possibility exists only with a short repetitive data pattern, as shown in Figure 15. Here, data transitions are marked as dots on the phase error curve. The essential feature of this data pattern is the long gap of no data transitions that can straddle the point of cycle slip, rendering the frequency detector mute. Further, the phase relation between the remaining data transitions and the VCO are all accurately measured by the phase detector. These phase measurements can all sum to zero, giving a stable operating point for the loop. If the VCO frequency changes, then the phase of the beat note will change in such a way as to produce a nonzero sum of phase of the data transitions. This nonzero sum provides a restoring force to maintain control of the loops in the false locked state.

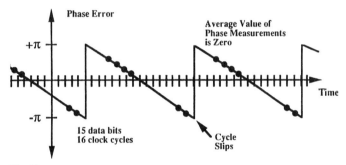

Fig. 15

This phenomenon has not been observed experimentally. Further, in practical systems, scrambling is used to prevent short patterns. Also, with long patterns, this phenomenon cannot happen because the phase loop will force acquisition during the span of zero sum phase. Further, this effect cannot happen with random data even if there are long gaps with no data transitions. With random data, the gaps with no transitions will not always straddle the cycle slip, and thus the frequency detector will provide the frequency feedback signal as intended.

5. CHARGE PUMP

5.1 Charge-Pump Circuit

Figure 16 shows the combined phase and frequency charge pump and loop filter. The phase detector controls the switches labeled "exor", and the frequency detector controls the switches labeled "freq up" and "freq down." The phase charge pump is a simple differential current. The currents flow through both the resistor and the capacitor, forming the transmission zero to stabilize the phase loop. The frequency charge pump consists of two

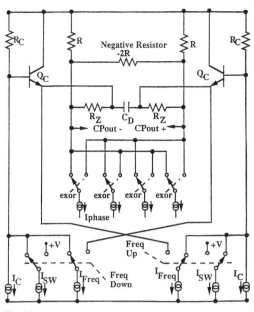

Fig. 16

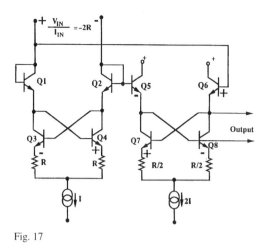

Fig. 17

single-ended currents with voltage clamps. The frequency pump current flows only through the capacitor, and not through the resistors, thus there is no transmission zero in the frequency loop. When the frequency pump is not active, these clamp devices are reverse biased. When the frequency pump is active, one side of the capacitor is clamped to the common-mode voltage of the phase pump and current is drawn from the other side of the capacitor. The common-mode voltage is set by I_{phase} and R. The clamp voltage of the frequency pump is set by R_c, I_c, and the V_{be} of Q_c. The voltage output is taken differentially from CP_{out}. The zero to compensate the loop is formed by resistors R_z in series with capacitor C_D. The resistors R_z (internal mask programmable SiCr thin film) set the loop bandwidth. Capacitor C_D is external; user selection of the capacitor controls the damping ratio.

5.2 Negative Resistor

The negative resistor in Figure 16 is needed to achieve sufficient dc gain in the charge-pump integrator. The implementation shown in Figure 17 is a circuit with a negative real driving point impedance. The parallel combination of equal magnitude positive and negative resistors is ideally infinite, yielding a perfect, lossless integrator as the loop filter. The parallel combination used in this charge pump affords a minimum multiplication factor of 500 of the magnitude of R. This minimum is given by component matching of: 1 mV of base-emitter voltage mismatch, 10% of β mismatch, and 0.1% resistor mismatch. The input voltage is impressed across the resistors, R, with polarity shown for emphasis. But the differential current is coupled back to the input terminals with opposite polarity. The devices Q5-Q8 compensate for errors caused by finite β in the primary cell, Q1-Q4. The output from the emitters Q5-Q6 may drive a load that is large compared to the product of β and R without degrading performance. This negative resistor scheme allowed the charge pump to fit easily into the constraints of a 5-V power supply. The

usual approach to get high impedance charge pump in bipolar technology is to use degenerated pnp current sources. But that would require about 1 V more headroom than this circuit.

5.3 Balanced Differential Circuits

Several design features combine to allow for easily meeting the jitter transfer specifications. The clock recovery PLL, and especially the charge pump, uses balanced differential circuit structures exclusively. The benefit of the balanced differential circuits is elimination of unwanted feedthrough and cross talk on the chip. Unwanted feedthrough and cross talk can provide alternate signal paths within the PLL that can alter the phase response characteristics. For example, with a damping factor of 5, the jitter peaking is 0.08 dB, and the reactance of the 0.047-μF damping capacitor is 30 Ω at the 100-kHz bandwidth of the loop. Because the phase detector full-scale output current is 5 μA for 180°, we see that a 150-μV signal would be generated by a full-scale output current from the phase detector flowing through the damping capacitor. Thus, in order for the phase-locked loop to behave as expected, it is necessary for the stray coupling around the damping capacitor to be much less than 150 μV. This very low stray coupling is achieved with the balanced differential circuits.

6. VOLTAGE-CONTROLLED OSCILLATOR

6.1 Choice of VCO

As mentioned in Section 1, clock recovery devices were built using two different types of VCO: multivibrator and ring oscillator. The original circuit [1] used a multivibrator that was discovered to have too much jitter generation. As seen from Figures 2 and 3, the jitter on the recovered clock is about 4 degrees. Because the SONET requirement for transmit jitter is 0.01 UI, or 3.6 degrees, the device is disallowed from use as a regenerator. The theory of jitter in the multivibrator [24] could not accurately predict performance. Subsequent theoretical developments [25] indicated fundamental flaws in the multivibrator related to the infinite gain during the positive feedback

413

switching transient. For this reason, the multivibrator was abandoned. Theoretical development of jitter in ring oscillators [5,6,26] allowed realization of a new clock recovery device [2,27] with 1.9 degrees of jitter.

6.2 Frequency Stability and Tuning

Guaranteed acquisition of the loop requires a compatibility between the VCO and the frequency detector. The frequency of the VCO must never be displaced from the nominal center frequency by an amount larger than the detection range of the frequency detector. In addition, the tuning range of the VCO must be large enough to allow locking even when temperature variations have caused the center frequency to drift away from nominal. However, the tuning range cannot be made arbitrarily large: start-up transients could cause the VCO initial condition to be at an extreme of its tuning range. Because the range of the frequency detector has been shown in Section 4 to be ±25%, acceptable performance of the VCO is to have a center frequency stability better than 10% over the temperature range and a tuning range larger than ±10%. Note that the center frequency stability encompasses both initial frequency accuracy and temperature drift. The solution chosen here is laser-wafer trimming of silicon-chromium thin-film resistors to set the initial frequency of the VCO to within 1% of nominal.

Often the interplay among center frequency accuracy, temperature drift, and loop acquisition leads to consideration of a crystal-controlled oscillator (VCXO). If a VCXO is chosen, a major performance liability to address is having enough tuning range to meet jitter tolerance requirements. As seen in Figure 9, the VCO must have a 0.02% minimum tuning range to satisfy these requirements. Another serious detraction of a crystal-controlled oscillator is the large size and high cost: thus, one is ineluctably drawn to a monolithic VCO.

6.3 Multivibrator

One VCO is an emitter-coupled multivibrator with collector clamps [28]. A translinear voltage-to-current converter (Figure 18) takes the input voltage from the charge pump and causes a

differential current to flow through diodes D3-D4. The tuning currents in the three differential pairs, Q9-Q14, are a linear representation of the input voltage, giving a ±12% tuning range. The differential signal current mimicked by Q5-Q6 is coupled back to the input devices Q1,2. Polarity of this feedback is chosen opposite the signal flow through the resistors and the diodes, giving an infinite differential input impedance. Devices Q5-Q6 also act as buffers to drive Q7-Q8.

The amplitude of the multivibrator triwave is set by the voltage across the resistor R_a, and the temperature drift of the VCO center frequency is compensated with a temperature-dependent current. The power supply rejection of the VCO is 1% per volt. This high level of performance ensures the device will operate properly in a real system with noise on the power supplies. A wide range of process variation can be tolerated because the VCO center frequency is laser-wafer trimmed to within 0.5% of nominal using SiCr thin-film resistors. The in-phase and quadrature signals required by the frequency detector are easily obtained from the multivibrator. The in-phase output is taken from collectors of Q9-Q12 and the quadrature output is taken from collectors of Q10-Q11.

6.4 Substrate Displacement

The multivibrator has a common-mode step on the timing capacitor once per half cycle. Displacement current into the substrate through the pocket capacitance would flow into the positive supply and out the substrate connection, causing voltage transients on bond-wire and leadframe inductances. These transients can cause jitter by modulating the VCO frequency and also by coupling noise to other parts of the circuit. To prevent this, internal bypass is used to terminate the substrate displacement. Also, ac power supply rejection of the VCO is improved through bypassing of a few critical nodes.

Figure 19 is a very simplified schematic of the multivibrator; the inductors model the leadframe and bond wires. The timing capacitor consists of two symmetrical halves, and there is a total of 10 pF parasitic capacitance from the emitter nodes V1-V2 to the substrate. The waveform at the emitters has 400 mV common-mode steps with 1-ns edges; the displacement current into

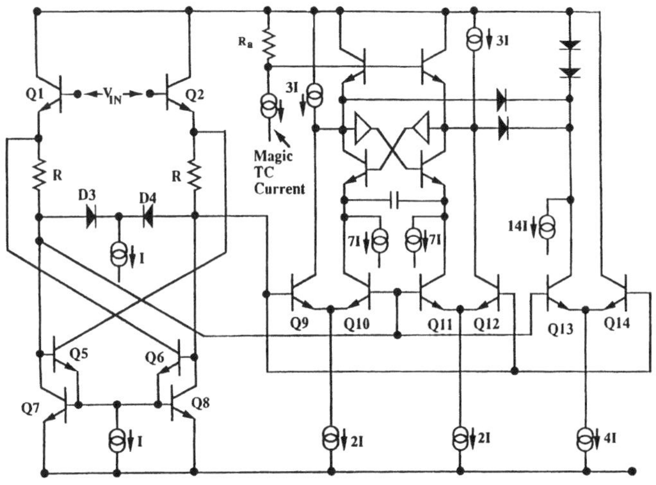

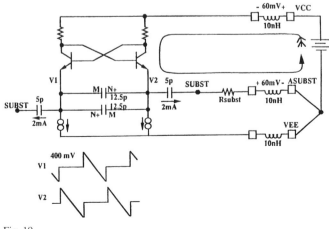

Fig. 18

Fig. 19

the substrate is 4 mA. If this current is allowed to flow in the bond wires, voltage transients will develop.

The 100 pF bypass capacitor in Figure 20 forces the displacement current to flow in a tight loop through the bypass and through the transistors. The current is thus prevented from flowing in the bond wires. For this bypass structure to be effective, the series resistance must be small. This is accomplished in the physical layout by surrounding the timing capacitor with the bypass structure. Another requirement of the physical layout is that enough substrate resistance be included to prevent the bypass capacitor and lead inductance from ringing.

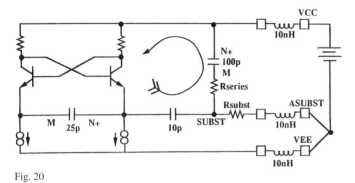

Fig. 20

A cross section of the bypass capacitor, the timing capacitor, and 10-pF parasitic-to-substrate is shown in Figure 21. Note that the bypass path is through the N+ , metal, and ISO diffusion, so that this series resistor is a small value. This substrate resistor must be large enough to prevent ringing. This is easily arranged by contacting the substrate some distance away from this structure.

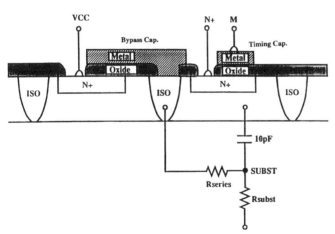

Fig. 21

6.5 Ring Oscillator

The ring oscillator and detailed calculation of jitter are adequately discussed in the literature [2,5,6,26] and will not be duplicated here. It could be mentioned that the ring oscillator has improved immunity to noise on the power supply due to exclusive use of balanced differential circuit structures. And, in addi-

tion, noise immunity is further improved by special bypass structures [29]. These bypass structures couple samples of the power supply noise to internal bias circuits and thus supply noiseless currents to the delay elements of the ring oscillator. In this manner, several hundred millivolts of power supply noise can be totally rejected and the clock recovery device will continue to produce a low-jitter clock signal. Figure 22 shows this compensation for power supply noise. As expected, the impedance of the current source is very high at low frequency. However, as the emitters of Q4 track power supply variations, the displacement current through the collector-base capacitance of the current source would lower the impedance, and the displacement current would flow into emitters of Q4 and modify the delay element. However, in Figure 22, the bypass capacitor supplies the displacement current demanded by the collector-base capacitance, and the series resistor in the emitters of Q4 maintains a high impedance of the current source to high frequencies. The emitters of Q4 track power supply variations at the top of the resistor. The bypass capacitor forms a voltage divider with the collector-base capacitance, forcing about 90% of the power supply noise to appear at the bottom of the resistor. Thus, the impedance of the current source at high frequency is about ten times the value of the

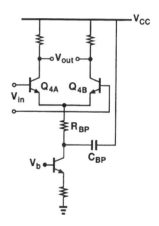

Supply Rejection

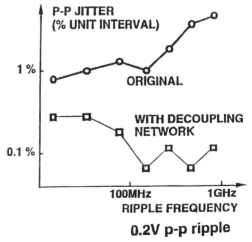

Fig. 22

415

resistor. Very little dynamic current related to power supply noise flows through the delay element.

7. MEASUREMENTS

7.1 Jitter Tolerance Measurement

The test setup used to measure jitter tolerance of the clock recovery devices is shown in Figure 23. This is a fairly straightforward application of the bit-error-rate (BER) test sets except the transmitter is used with an external clock. This external clock has jitter on it in the form of frequency modulation. The measurement is made by increasing the amplitude of jitter at a fixed frequency and monitoring the bit error rate. When the BER is 10^{-10}, this is called the *limit* of the jitter tolerance. The measured jitter tolerance is shown in Figure 24 and compared to the SONET specification.

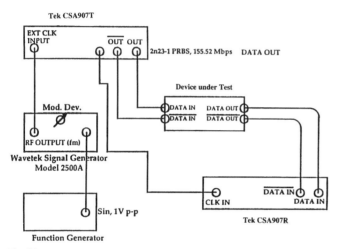

Fig. 23

In this measurement setup, it is necessary to use a frequency-modulated carrier to generate enough jitter at low frequencies to fully test the device. Most jitter generators will not create a test

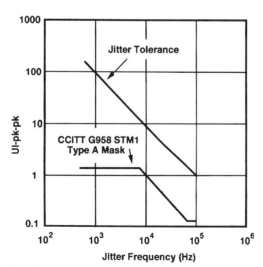

Fig. 24

signal with a large number of unit intervals of jitter. Signal generators capable of external frequency modulation are commonly calibrated in zero-to-peak frequency deviation for a 1-V zero-to-peak modulation input. The ratio of the peak frequency deviation to the modulation frequency is known in radio argot as deviation ratio (DevRat). The peak-to-peak UI of jitter is then given by Eq. (11).

$$DevRat = \pi UI_{pp} \tag{11}$$

This is most easily seen from the extended concept of frequency as the instantaneous derivative of phase [30]. A frequency-modulated clock is represented as

$$c(t) = \cos\left(pt + \frac{\Delta\omega}{m}\sin mt\right) \tag{12}$$

where p is the clock rate, m is the jitter modulation frequency, and $\Delta\omega$ is the zero-to-peak frequency deviation. The instantaneous frequency of this modulated clock is simply the derivative of the argument of the cosine:

$$f(t) = p + \Delta\omega \cos mt \tag{13}$$

Thus, Eq. (13) verifies the peak frequency deviation is $\Delta\omega$; and $\Delta\omega/m$ is recognized as the deviation ratio. Returning to Eq. (12), the peak-to-peak phase deviation in radians is

$$\Delta Rad_{pp} = \frac{2\Delta\omega}{m} = 2DevRat \tag{14}$$

Finally, now, this phase deviation must be divided by 2π to yield jitter in UI_{pp}, resulting in the expression of Eq. (11).

7.2 Jitter Transfer Function Measurement

Maximum bandwidth requirements for a regenerator necessitate measurement of the jitter transfer function of a clock recovery device. Also, characterization of jitter tolerance is sometimes aided by jitter transfer measurement. Figure 25 shows a test fixture to phase modulate a reference carrier with the excitation from a network analyzer, and then the modulated carrier clocks a data generator driving the device under test (DUT). The phase of the recovered clock is then compared to the original unmodulated reference carrier, and the result is detected and displayed by the network analyzer as a transfer function.

The jitter transfer function of a clock recovery device [1] with various loop-damping ratios has been measured. For damping ratios of 1, 2, 5, and 10, the circuit exhibits jitter peaking of 2,

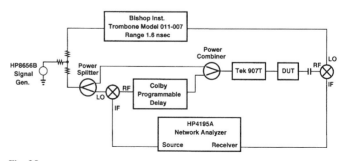

Fig. 25

416

0.5, 0.08, and 0.02 dB, respectively, and − 3-dB bandwidths of 130, 100, 90, and 90 kHz, respectively. Because the SONET specification for OC3 is maximum bandwidth of 130 kHz and maximum jitter peaking of 0.1 dB, we are required to use a damping of 5 or greater.

Figure 26 shows measured jitter transfer function of another clock recovery device [2] with different damping capacitors. The family of curves here agrees remarkably well with the theoretical result given in Figure 11.

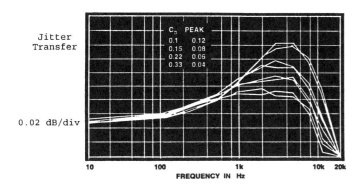

Fig. 26

Figure 27 shows the general operating principle of the phase modulator used in the test setup of Figure 25. The circuit is an Armstrong phase modulator [31]: a single carrier is split into two paths and then recombined. One of the two paths contains a mixer, so the output of that path can be amplitude modulated. The two paths also differ in length by a quarter of a wavelength of the carrier. The phase modulator can be tuned to different carrier frequencies by adjusting the path difference with the programmable delay. This modulator is capable of very high modulation rates, but is limited to producing about 0.5 UI maximum phase shifts.

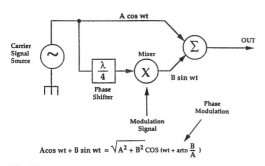

$$Acos \, wt + B \sin wt = \sqrt{A^2 + B^2} \, COS \, (wt + artn \frac{B}{A} \,)$$

Fig. 27

7.3 Bit Error Rate versus Signal-to-Noise Ratio

The clock recovery device is intended to operate with standard ECL signal levels at the data input. Although not recommended as an application of the device, smaller input signals are useful for exploring the robustness of the clock recovery architecture. Figure 28 shows the bit error rate versus input signal-to-noise ratio (SNR) for input signal amplitudes of full 900-mV

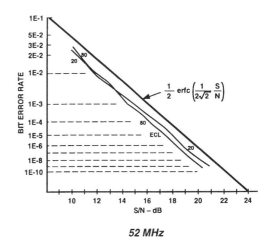

52 MHz

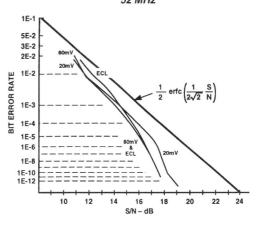

155 MHz

Fig. 28

ECL, and decreased amplitudes of 80 mV and 20 mV. Wideband amplitude noise is summed with the data signals, as shown in Figure 29. The full ECL and 80-mV signals give virtually indistinguishable results. The 20-mV signals also provide adequate performance when in lock, but signal acquisition may be impaired.

Note that the data of Figure 28 are plotted on special graph paper that renders the theoretical result as a straight line, which

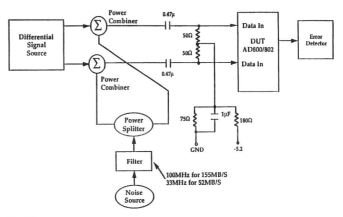

Fig. 29

417

is labeled as the complementary error function. The data all fall close to the line, but more importantly, the slope of the data line matches the theoretical result. The fact that the data indicate bit error rates better than theoretical is simply a case of an over-simplified (i.e., wrong) theory. The data at 155 MHz do not match the ideal slope as well as the data at 52 MHz. This is because the noise generator did not supply truly Gaussian noise. In fact, the generator is known to clip extreme peaks of the noise. Consider the situation of high signal-to-noise ratio and low bit error rates. In this case, it is the extreme peaks of the noise that cause errors to be made, so it is reasonable to expect the measurement to be sensitive to the exact characteristics of the noise peaks supplied by the generator.

For the calculation of BER versus SNR for a binary signal with additive white Gaussian noise, we note that if the amplitude of the binary signal is *A* and decisions between 0's and 1's are made by a comparator threshold at half the amplitude, *A/2*, then errors can be made if a noise peak is greater than *A/2*. Two types of errors can occur: a 0 can be mistaken for a 1, and a 1 for a 0. The total error rate is then found by summing the areas in tails of the two Gaussian distribution curves, giving the result:

$$\text{BER} = \frac{1}{2}\,\text{erfc}\left(\frac{1}{2\sqrt{2}}\frac{S}{N}\right) \qquad (15)$$

where erfc(.) denotes the complementary error function and *S/N* is the raw signal-to-noise ratio.

Figure 28 shows the logarithm of the BER plotted as a function of the SNR in dB. Note that the complementary error function is a very steeply descending curve. Any data plotted on these axes would need to be compared to this curve. It is difficult to fit such a curve to a few data points. Also, it is often desirable to make BER measurements at high error rates and extrapolate performance to lower error rates. This extrapolation is especially difficult to do if the ideal result is a steeply descending curve such as erfc. So, what is needed is a special graph paper that has the BER axis scrunched up at the bottom to make the erfc a straight line. Then, comparing data to theory and extrapolating from a few points is easy. Figure 30 shows a graphical technique for constructing such a graph paper. First, plot the erfc as shown and then run horizontal lines on the BER decades to the curve. Next, run perpendicular vertical lines: these are the new BER axis scale. Now, turn the paper sideways and arbitrarily pick a point on the 1E-12 BER line. Call this

point 23 dB. From this point, draw a straight line of any convenient slope. Wherever this line crosses the 1E-1 BER line is the 8.2-dB point on the SNR axis.

The test setup for measuring BER versus SNR is shown in Figure 31. The Tektronix CSA 907T and 907R are used to generate data and measure BER in an ordinary way. The data output from the generator is attenuated and mixed with noise to create any desired SNR at the input to the DUT. The hybrid combiners are needed for two reasons. First, they provide isolation between the *A* and *B* input ports. Second, one of the combiners inverts the phase of the noise signal to be added to the already differential data from the generator. Each of the combiners is separately terminated in 50 Ω at the input of the DUT.

Detailed schematics of the hybrid combiners are shown in Figures 32 and 33. On both the inverting and non-inverting cir-

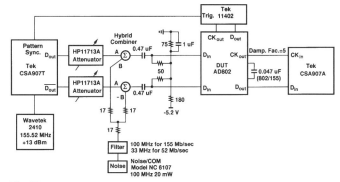

Fig. 31

cuit, the *B* output has about 12 ns more delay than the *A* output. This is not a problem as long as the proper *A* and *B* connections are observed as in Figure 31. The transformers in these circuits are wound on iron powder toroids [32].

Figure 34 is the filter used in Figure 31. The noise must be low-pass filtered with a corner frequency at 0.7 times the bit rate to simulate the dynamics of a fiber-optic receiver. These fiber-optic receivers generally have an analog bandwidth of 0.7 of the bit rate as a compromise between intersymbol interference and noise. The filter is not used for measurements at 155 MHz because the noise source itself is bandlimited to 100 MHz. It is interesting to note that the averaging of the noise signal by the

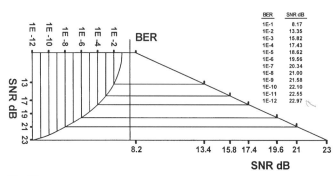

BER	SNR dB
1E-1	8.17
1E-2	13.35
1E-3	15.82
1E-4	17.43
1E-5	18.62
1E-6	19.56
1E-7	20.34
1E-8	21.00
1E-9	21.58
1E-10	22.10
1E-11	22.55
1E-12	22.97

Fig. 30

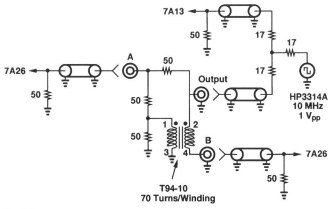

Fig. 32

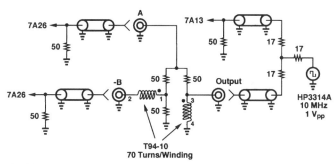

Fig. 33

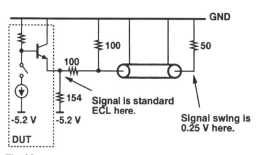

C = 87.5 pF

L = 437 nH (T80-12, 14 turns
            bus wire w/ teflon spaghetti)

R = 50 Ω

Fig. 34

33-MHz filter makes the noise more Gaussian. The clipping of extreme peaks causes the unexpected slope of the 155-MHz data in Figure 28. But the 52-MHz data lie more closely to theoretical expectations.

7.4 Back-Terminated ECL

Figure 35 shows the back-terminated ECL circuit used on all clock recovery evaluation boards. The whole point is to provide

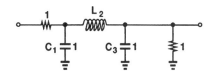

Fig. 35

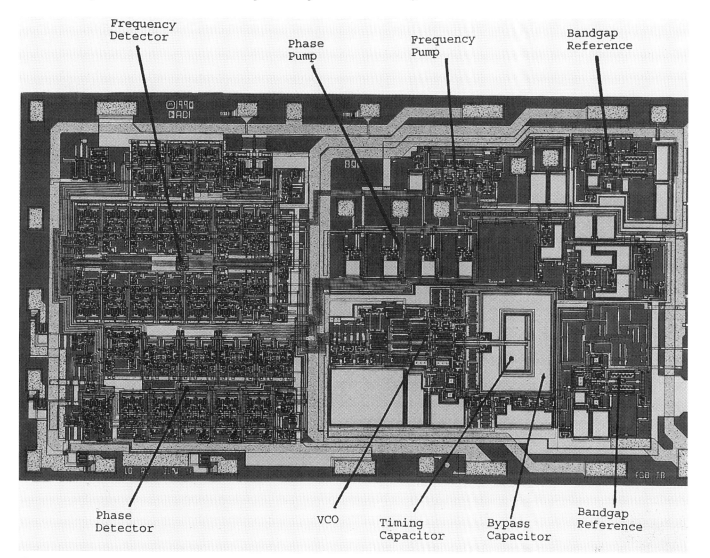

Fig. 36 (*See discussion on p. 420.*)

the output signal from a 50-Ω source so that a 50-Ω load can be driven. These demo boards can be connected directly to the 50-Ω-to-ground input on a scope. Because in any real bench setup there are discontinuities in the transmission lines due to connectors and other imperfections, it is very helpful to have the signal come from a terminated source to eliminate reflections. These reflections, if not terminated, may lead to confusing results, usually in the form of excess jitter.

8. CONCLUSION

8.1 Die Photograph

A die photo is shown in Figure 36 with the various sections of the device denoted. Note the bypass structure that wraps around the timing capacitor in the multivibrator VCO. Note also the enlarged bond pads to accept multiple bond wires on power, ground, and substrate connections to reduce lead inductance.

8.2 Acknowledgments

The author would like to acknowledge the invaluable contributions of his colleagues: R. Croughwell, A. Gusinov, J. Newton, J. Bulzacchelli, and J. McNeill for detailed engineering; B. Surette, F. Benkley, E. Ferrari, D. Williams, and M. Fazio for test and characterization measurements; and T. Freitas for layout.

References

[1] L. DeVito et al., "A 52MHz and 155MHz clock recovery PLL," *1991 ISSCC Dig. Tech. Papers,* p. 142.

[2] J. A. McNeill, R. Croughwell, L. DeVito, and A. Gusinov, "A 150 mW, 155 MHz phase locked loop with low jitter VCO," *Proc. IEEE Int. Symp. Circuits and Syst.* (ISCAS), vol. 3, pp. 49–52, May, 1994, London, U.K.

[3] S. Feindt et al., "XFCB: A high speed complementary bipolar process on bonded SOI," *Proc. IEEE Bipolar Circuit and Techn. Meet.,* pp. 264–267, 1992, Minneapolis, MN.

[4] S. Feindt et al., "A complementary bipolar process on bonded wafers," *Proc. Spring Electro-Chemical Soc. Meet.,* pp. 189–196, 1993, Santa Clara, CA.

[5] J. A. McNeill, "Jitter in ring oscillators," *Proc. IEEE Int. Symp. Circuits and Syst.* (ISCAS), vol. 6, pp. 201–204, May 1994, London, U.K.

[6] J. A. McNeill, "Jitter in ring oscillators," Ph.D. thesis, Boston University, Boston MA, 1994.

[7] B. Kim et al., "A 30MHz high-speed analog/digital PLL in 2μm CMOS," *1990 ISSCC Dig. Tech. Papers,* p. 104, San Francisco, CA.

[8] G. Andrews et al., "A 300Mb/s clock recovery and data retiming system," *1987 ISSCC Dig. Tech. Papers,* San Fransisco, CA, p. 188.

[9] J. Tani et al., "Parallel interface ICs for 120Mb/s fiber optic links," *1987 ISSCC Dig. Tech. Papers,* San Fransisco, CA, p. 190.

[10] R. Cordell et al., "A 50MHz phase- and frequency-locked loop" *IEEE J. Solid-State Circuits,* vol. SC-14, no. 6, pp. 1003–1009, December 1979.

[11] J. Sonntag and R. Leonowich, "A monolithic CMOS 10MHZ DPLL for burst-mode data retiming," *1990 ISSCC Dig. Tech. Papers,* p. 194, San Francisco, CA.

[12] J. Scott et al., "A 16Mb/s data detector and timing recovery circuit for token ring LAN," *1989 ISSCC Dig. Tech. Papers,* p. 150, San Fransisco, CA.

[13] W. Llewellyn et al., "A 33Mb/s data synchronizing phase-locked loop circuit," *1988 ISSCC Dig. Tech. Papers,* p. 12, San Fransisco, CA.

[14] B. Thompson, H. S. Lee, and L. DeVito, "A 300MHz BiCMOS serial data transceiver," *IEEE J. Solid-State Circuits,* vol. SC-29 no.3, pp. 185–192, March 1994.

[15] R. Leonowich and J. Steininger, "A 45MHz CMOS phase/frequency-locked loop timing recovery circuit," *1988 ISSCC Dig. Tech. Papers,* San Fransisco, CA, p. 14.

[16] N. Ishihara and Y. Akazawa, "A monolithic 156 Mb/s clock and data recovery PLL circuit using the sample-and-hold technique," *ISSCC Dig. Tech. Papers,* pp. 110–111, February 1994, San Fransisco, CA.

[17] Cypress Semiconductor Corp., "Data sheet CY7B951," San Jose CA, 1994.

[18] M. Soyuer, "High-frequency monolithic phase-locked loops," Ph.D. thesis, University of California, Berkeley, UCB/ERL M88/10, 9 February 1988.

[19] P. Trischitta and E. Varma, *Jitter in Digital Transmission Systems.* Norwood MA: Artech House, 1989.

[20] T. Lee and J. Bulzacchelli, "A 155MHz clock recovery delay and phase-locked loop," *IEEE J. Solid-State Circuits,* vol. 27, no. 12, pp. 1736–1745, December 1992.

[21] L. DeVito, U.S. Patent 5,027,085, "Phase detector for phase-locked loop clock recovery system," 7 May 1990.

[22] C. Hogge, "A Self-correcting Clock Recovery Circuit," *IEEE J. Lightwave Technology,* vol. LT-3, no. 6, pp. 1312–1314, December 1985.

[23] D. Wolaver. *Phase-Locked Loop Circuit Design.* Englewood Cliffs, NJ: Prentice Hall, 1991.

[24] A. A. Abidi and R. G. Meyer, "Noise in relaxation oscillators," *IEEE J. Solid-State Circuits,* vol. SC-18, December 1983.

[25] C. Verhoeven, "First order oscillators," Ph.D. thesis, Delft University, 1990.

[26] T. C. Weigandt, B. Kim, and P. R. Gray, "Analysis of timing jitter in CMOS ring oscillators," *Proc. IEEE Int. Symp. Circuits and Syst.* vol. 4, pp. 27–30, June 1994, London, UK.

[27] Analog Devices, Inc., "Data Sheet AD807," Norwood MA, 1994.

[28] B. Gilbert, "A versatile monolithic voltage-to-frequency converter," *IEEE J. Solid State Circuits,* vol. SC-11, no. 6, pp. 852–864, December, 1976.

[29] L. DeVito and J. McNeill, U.S. Patent 5,418,498, "Low jitter ring oscillators," 23 May 1995.

[30] L. Arguimbau. *Vacuum-Tube Circuits and Transistors,* New York: John Wiley, p. 473, 1956.

[31] E. H. Armstrong, "A method of reducing disturbances in radio signaling by a system of frequency modulation," *Proc. IRE,* vol. 24, no. 5, pp. 689–740, May 1936.

[32] Micrometals, Inc., "Iron powder toroidal cores," Catalog 3, Anaheim, CA.

A 155-MHz Clock Recovery Delay- and Phase-Locked Loop

Thomas H. Lee, *Member, IEEE*, and John F. Bulzacchelli, *Student Member, IEEE*

Abstract—This paper describes a completely monolithic delay-locked loop (DLL) that may be used either by itself as a deskewing element, or in conjunction with an external voltage-controlled crystal oscillator (VCXO) to form a delay- and phase-locked loop (D/PLL). By phase shifting the input data rather than the clock, the DLL and D/PLL provide jitter-peaking-free clock recovery. Additionally, the jitter transfer function of the D/PLL has a low bandwidth for good jitter filtering without compromising acquisition speed. The D/PLL described here exhibits less than 1° rms jitter on the recovered clock, independent of the input data density. No jitter peaking is observed over the 40-kHz jitter bandwidth.

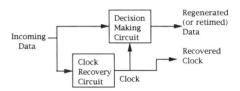

Fig. 1. Simplified block diagram of a digital receiver.

I. INTRODUCTION

THIS paper describes the performance of a completely monolithic delay-locked loop (DLL) that may be used either by itself as a deskewing element, or in conjunction with an external voltage-controlled crystal oscillator (VCXO) to form a delay- and phase-locked loop (D/PLL) [1]) that enables jitter-peaking-free clock recovery at the SONET OC-3 frequency of 155.52 MHz. In addition, it will be shown that the D/PLL also may possess a low jitter bandwidth to provide jitter filtering without compromising acquisition performance.

Section II provides a general background on the problem of clock recovery, while Section III examines the properties of DLL's and D/PLL's. Section IV presents specific implementation details for the DLL, Section V presents experimental results, and Section VI concludes with a summary.

II. BACKGROUND

The ability to regenerate binary data is an inherent advantage of digital transmission of information. To perform this regeneration with the fewest bit errors, the received data must be sampled at the optimum instants of time. Since it is generally impractical to transmit the requisite sampling clock signal separately from the data, the timing information is usually derived from the incoming data itself. The extraction of this implicit signal is called clock recovery, and its general role in digital receivers is illustrated in Fig. 1.

When the incoming data signal has spectral energy at the clock frequency, a synchronous clock can be obtained simply by passing the incoming data through a bandpass filter, often realized either as an *LC* tank or surface-acoustic wave (SAW) device, tuned to the nominal clock frequency. Because of bandwidth restrictions, however, in most signaling formats the incoming data signal has no spectral energy at the clock frequency, complicating clock recovery. Such signals must first undergo appropriate nonlinear preprocessing ahead of the resonator.

While these clock recovery circuits generally offer rapid acquisition (typically in the hundreds of clock cycles), they operate essentially at a fixed clock frequency, and provide no output in the absence of an input. Furthermore, the phase of the resonator's output may not necessarily be correct to provide optimally timed sampling for the decision circuit and thus generally requires compensation. Proper phase alignment of data and clock must be provided on an open-loop basis and is therefore difficult to maintain over temperature, supply, and process variations. Finally, neither *LC*- nor SAW-based clock recovery circuits are practically realizable in monolithic form.

Another classical approach employs phase-locked loops (PLL's), as illustrated in Fig. 2. Because of its amenability to monolithic construction, a PLL is an attractive alternative to tuned circuit clock recovery. Furthermore, conventional PLL's offer a comparatively wide tuning range, and provide an output of some kind at all times, in contrast with resonant approaches. However, because the desired loop bandwidths are often much smaller than the tuning range, acquisition is generally comparatively slow, even with frequency-acquisition aids.

To appreciate a more subtle difficulty with conventional PLL realizations, consider the linearized block diagram of a second-order PLL shown in Fig. 3, and the associated root locus diagram of Fig. 4. The input–output phase

Manuscript received April 27, 1992; revised July 15, 1992.

T. H. Lee was with Analog Devices, Wilmington, MA 01887. He is now with Rambus Inc., Mountain View, CA 94040.

J. F. Bulzacchelli is with the Department of Electrical Engineering, Massachusetts Institute of Technology, Cambridge, MA 02139.

IEEE Log Number 9203623.

Reprinted from *IEEE Journal of Solid-State Circuits*, vol. SC-27, pp. 1736-1746, December 1992.

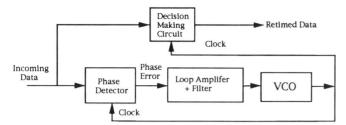

Fig. 2. PLL-based clock recovery system.

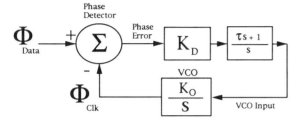

Fig. 3. Linearized block diagram of second-order PLL.

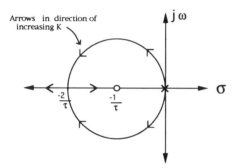

Fig. 4. Root locus for second-order PLL.

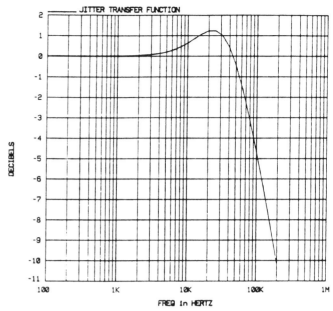

Fig. 5. Jitter transfer function of critically damped second-order PLL.

transfer function (also known as the jitter transfer function) for this system is as follows:

$$H(s) = \frac{\phi_{\text{clk}}(s)}{\phi_{\text{data}}(s)} = \frac{K(1 + \tau s)}{s^2 + K\tau s + K}$$

where $K = K_D K_O$.

As seen in the Bode diagram of Fig. 5, the magnitude of the jitter transfer function of such a system exceeds unity over some range of frequencies because of the presence of the closed-loop zero at a frequency lower than that of the poles. This gain in excess of unity is known as jitter peaking and is particularly objectionable in systems that employ many clock recovery units in cascade (such as in repeaters) because of the resulting exponential growth of jitter [3].

The traditional solution is to increase the damping ratio to reduce the spacing between the zero and the lowest frequency pole. As can be inferred from the root locus, however, the amount of jitter peaking can be reduced but never completely eliminated since the zero always occurs at a lower frequency than that of the poles. Additionally, frequency acquisition speed often suffers as damping increases in PLL's that employ acquisition aids, such as that described in [2]. Hence, simply increasing the damping ratio is frequently unsatisfactory.

Note that the problem of jitter peaking disappears if a way can be found to move the necessary loop-stabilizing zero to a frequency higher than that of the lowest frequency pole. One possibility is to employ a loop of different order. As shown in the root locus of Fig. 6, a third-order loop, for example, permits placement of the zero as desired. Unfortunately, the conditional stability of such loops makes them difficult to use.

Another option is to retain a second-order loop, but to move the necessary loop-stabilizing zero out of the forward path so that there is then no closed-loop zero. The next section explores methods for accomplishing this goal.

III. THE DELAY- AND PHASE-LOCKED LOOPS

Traditional PLL's adjust the phase of the clock to obtain phase alignment with the incoming data. However, significant advantages accrue if one instead shifts the input data to align with the clock, as in the two clock recovery schemes shown in Fig. 7.

A. The Delay-Locked Loop (DLL)

If a clock signal of precisely the correct frequency is available, the DLL connection shown in solid lines may be used. As seen, the DLL employs a voltage-controlled phase shifter (VCPS) driven under loop control to align data with the incoming clock. The loop is first-order, as shown in the linearized block diagram of Fig. 8, with the VCPS characterized by a gain constant K_ϕ, with units of radians/volt. A single loop integrator suffices to drive the steady-state phase error to zero.

Note that the jitter transfer function of the DLL simply equals zero; the DLL cannot transfer jitter to the clock since the same (external) clock both generates and retimes the data. Although this fact does not imply that the DLL's decision circuit then functions with improved bit-error rate

422

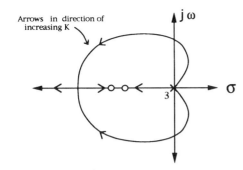

Fig. 6. One possible root locus for a third-order PLL.

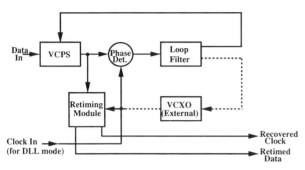

Fig. 7. Block diagram of DLL and D/PLL.

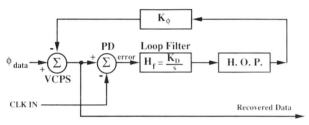

Fig. 8. Linearized block diagram of DLL. (The block labeled H.O.P. represents higher order poles and is discussed in Section IV-A.)

(BER) as a consequence, the BER of *subsequent* clock recovery stages would improve.

Phase errors are nulled out with a speed commensurate with the loop bandwidth $K_D K_\phi$ so that increasing the loop bandwidth makes the DLL acquire more quickly. Furthermore, because only phase acquisition needs to take place, acquisition can be faster than that of PLL's or even of resonant filter approaches. And, unlike the case of a PLL, jitter filtering is independent of the loop bandwidth, so that only acquisition speed and loop stability considerations bound the loop bandwidth.

It should be noted that, in general, any real VCPS will have a finite range. As a consequence, it is possible for the VCPS to be driven to the end of its control range under certain circumstances. Once this occurs, acquisition is prematurely halted, and data can be lost.

This predicament can be avoided by resetting the DLL. Before acquisition begins, the integrator can be initialized to the midpoint of the phase shifter's range. Since the initial phase error must lie between $-\pi$ and π radians, the phase shifter will not be driven more than $\pm\pi$ radians ($= \pm 0.5$ *unit intervals*, or clock periods) from the middle of its range. Hence, in principle, acquisition will always be

successful as long as the total range of the VCPS exceeds 2π radians.

In practice, a range considerably larger than 2π radians is desirable, for jitter on the incoming data can temporarily drive the phase shifter more than $\pm\pi$ radians from the middle of its range. Additional range is also needed if the integrator is not initialized to the exact midpoint of the control range. For most applications, a total range of approximately 3π radians is sufficient.

While this type of DLL exhibits no jitter peaking (because the jitter transfer function is zero), it suffers from the need for an exact external frequency reference. In most cases of practical interest, no such reference is available local to the clock recovery module.

B. The Delay- and Phase-Locked Loop (D/PLL)

A loop that does not require an external frequency reference is the combined D/PLL comprising all of the elements shown in Fig. 7. As shown, the D/PLL contains two parallel control loops. The phase detector, loop amplifier, and VCXO form the core of a PLL, while the phase detector, loop amplifier, and VCPS form the core of a DLL, and is effectively summed with the first loop.

The two loops in the D/PLL act in concert to null out phase errors as follows: if the clock is behind the data, the phase detector drives the VCXO to a higher frequency and simultaneously increases the delay through the VCPS. Both of these actions serve to reduce the initial phase error since the faster clock picks up phase, while the delayed data loses phase. Eventually, the initial phase error is reduced to zero.

To investigate the loop dynamics, consider the linearized block diagram of Fig. 9 and the associated loop transmission magnitude plot of Fig. 10. As can be seen, the low-frequency loop transmission is controlled by the second-order PLL path, while the high-frequency loop behavior is controlled by the first-order DLL path.

The advantages of this connection derive from the manner in which loop stabilization is accomplished. Rather than realizing the zero in the loop filter as is customary (e.g., with a resistor in series with an integrating capacitor), the *phase shifter* provides the necessary zero, as is readily verified by considering the loop transmission behavior. The first-order DLL path sums with the second-order PLL path to yield a loop transmission that inflects at the intersection (at $\omega = K_O/K_\phi$) of the two components. Hence the zero location (the inflection point) is set by a ratio of gain constants, in contrast with prior art.

To establish that the arrangement of Fig. 9 also evades jitter peaking, let us derive explicitly the jitter transfer function for the D/PLL.

From the block diagram of Fig. 9, the control voltage v_1 may be expressed as follows:

$$v_1 = \frac{K_D}{s}(\phi_{\text{data}} - K_\phi v_1 - \phi_{\text{clk}}).$$

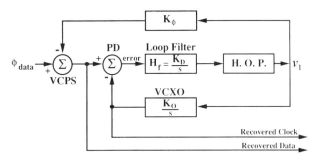

Fig. 9. D/PLL linearized block diagram.

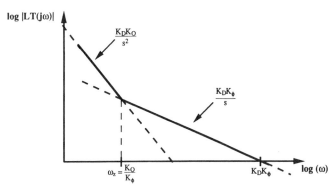

Fig. 10. D/PLL loop transmission.

Now, we may also write:

$$\phi_{\text{clk}} = v_1 \frac{K_O}{s}.$$

Eliminating v_1 from these two expressions leads to the desired jitter transfer function:

$$H(s) = \frac{\phi_{\text{clk}}(s)}{\phi_{\text{data}}(s)} = \frac{K}{s^2 + K\tau s + K}$$

where $K = K_D K_O$ and $\tau = K_\phi / K_O$. Note that this jitter transfer function is all-pole, with the same closed-loop poles as a standard second-order PLL (and hence the same stability), but there is no closed-loop zero. Thus, as long as one picks K and τ so that the loop possesses a damping ratio ζ greater than or equal to 0.707, this D/PLL exhibits no jitter peaking. This condition is readily satisfied by locating the zero well below loop crossover.

Elimination of the closed-loop zero also lowers the bandwidth of the jitter transfer function, perhaps dramatically. In a conventional overdamped PLL, the jitter transfer function has a bandwidth nearly equal to the location of the higher frequency pole because the closed-loop zero nearly cancels the low-frequency pole. In this D/PLL, however, it is the lower frequency pole that sets the closed-loop bandwidth because the loop transmission zero does not appear in the closed-loop transfer function. If the D/PLL is overdamped, the low-frequency pole resides near the loop transmission zero, as easily seen from the root locus. Thus, if a PLL and D/PLL possess identical pole locations, the D/PLL can have a much smaller jitter bandwidth. This difference becomes most prominent at high damping ratios, where the two poles are widely separated in frequency. The implication is that for the same loop crossover frequencies (and hence the same acquisition speed), the D/PLL can provide much more jitter filtering.

C. Acquisition Behavior of the D/PLL

The acquisition behavior of the D/PLL can be studied most conveniently by considering separately the response of the loop to an initial phase error and the response to an initial frequency error. Assuming that the system is linear, the total response can be obtained by superposition. We will later articulate the conditions under which the assumption of linearity holds.

Let us first assume that the frequency of the D/PLL's VCXO equals the bit rate of the incoming data, but that an initial phase error ϕ_O exists. The D/PLL's response to this initial phase error has two distinct stages. During the first stage, the control voltage is adjusted to a new value with a time constant ($= 1/K_D K_\phi$) of the DLL portion of the loop (recall that the DLL controls the loop behavior at high frequencies). Simultaneously, the phase-shifter delay and the VCXO frequency are adjusted. Since the frequency of the VCXO cannot be pulled very far from the bit rate of the incoming data, the phase of the VCXO does not change very rapidly. Consequently, the VCXO's phase may be considered stationary during the first stage of acquisition. As a result, the phase error is reduced toward zero as fast as the DLL bandwidth $K_D K_\phi$ during this first stage.

During the second stage of acquisition, the control voltage decays slowly (with the time constant of the slow pole) from its new value back to its original value, for at the start of the second stage, the VCXO's frequency does not match the bit rate of the incoming data. (The VCXO's frequency only matches the bit rate of the incoming data when the control voltage is at its original value.) Because of this frequency error, the phase of the VCXO with respect to the incoming data changes (albeit slowly). As the phase of the VCXO changes, the delay of the phase shifter is adjusted so that the phase of the delayed data tracks that of the VCXO. This process continues until the control voltage reaches its original value, at which point the VCXO's frequency equals the bit rate of the data, and acquisition is complete.

It must be stressed that the phase error between the delayed data and the clock may be insignificant during the second stage of acquisition. Examination of the block diagram of Fig. 9 readily shows that the phase error of the D/PLL is proportional to the time derivative of the control voltage. Since the control voltage is changing during the second stage of acquisition, a nonzero phase error can persist for a long time. However, if the control voltage changes slowly, the magnitude of this slow tail can be negligible. Overdamping the loop is advantageous in this context since the time constant of this slow mode is that of the low-frequency pole, which in turn is nearly the location of the loop transmission zero. Overdamping moves the zero, and hence the frequency of the slow mode, to lower frequencies and thereby reduces the magnitude of the error tail. Under such conditions, the second stage of acquisition can be ignored, for the data can be successfully regenerated as soon as the first stage of acquisition is completed.

Let us now consider the response of the loop to an initial frequency error of $\Delta \omega$. We may simplify the situation considerably by recalling that the second stage of phase acquisition just studied is essentially one of frequency acquisition. The frequency error at the beginning of the second stage is equal to ϕ_O / τ, while the phase error is negligibly small (if the loop is sufficiently well damped). Hence, by analogy, we conclude that the phase error due to an initial frequency error decays away with the time constant of the low-frequency pole, which is nearly that of the zero. A VCXO with its excellent center frequency accuracy and narrow tuning range is generally required to ensure that this phase error tail is negligible.

Unlike the second-order PLL, the D/PLL realizes rapid acquisition without compromising jitter filtering. While phase errors are nulled out as quickly as the DLL bandwidth $K_D K_\phi$, the jitter transfer function's bandwidth is predominantly controlled by the low-frequency pole at (very nearly) K_O / K_ϕ. Increasing K_D makes the D/PLL acquire more rapidly, but does not diminish the D/PLL's ability to filter jitter.

In practice, the D/PLL's ability to filter jitter is limited by the finite range of the phase shifter, the variability of the incoming data's bit rate, and the instability or uncertainty of the VCXO's center frequency. Together, the variability of the bit rate and the instability of the VCXO's center frequency determine the VCXO's required tuning range, as the tuning range must be large enough to ensure that the VCXO's frequency can always be pulled to the bit rate of the incoming data.

An important consideration is that the entire tuning range of the VCXO is not necessarily usable, for one must guarantee that the phase shifter is never driven out of range. (Recall that the VCXO and the phase shifter share the same control voltage.) As explained earlier, the phase shifter stabilizes the D/PLL by implementing the necessary loop transmission zero. If the phase shifter is ever driven to the end of its range, the loop transmission zero disappears, and the D/PLL becomes unstable. Hence, signal swings must be arranged to guarantee that the VCXO's control range is a *subset* of the phase shifter's control range.

If the VCXO's control range is a subset of that of the phase shifter, it follows that

$$\frac{\Delta \omega}{K_O} \leq \frac{\Delta \phi}{K_\phi}$$

where $\Delta \omega$ equals the VCXO's required tuning range (in radians per second), and $\Delta \phi$ equals the phase shifter's to-

tal range (in radians). Rearranging terms, one obtains

$$\frac{K_O}{K_\phi} \geq \frac{\Delta \omega}{\Delta \phi}.$$

This last equation expresses a lower limit on the bandwidth of the jitter transfer function, as the bandwidth of the jitter transfer function is approximately K_O / K_ϕ.

As a numerical example, consider a VCXO with a tuning range of 200 ppm of the bit rate, and a phase shift range of 3π radians. The lower limit on the bandwidth is then only 21 ppm of the bit rate, which accommodates exceptionally good jitter filtering.

On the other hand, use of a relaxation-type oscillator with its poorer center frequency stability greatly diminishes the D/PLL's ability to filter jitter. Suppose $\Delta \omega$ were 50% of the bit rate, and that the phase shifter range were still 3π radians. In this instance, the lower limit on the jitter bandwidth would be 2500 times larger than in the VCXO example. Because its narrow tuning range improves jitter filtering, a VCXO is highly desirable in a D/PLL.

A final consideration is that the D/PLL may require resetting if rapid acquisition is required because of the finite range of the voltage-controlled phase shifter. Normally, initial phase errors are rapidly nulled out during the first stage of phase acquisition. However, the first stage of acquisition, which is basically the acquisition process of the DLL, can halt prematurely if the VCPS is driven to the end of its range.

As in the DLL, initializing the D/PLL's integrator to the midpoint of the phase shifter's range eliminates this problem, provided that the total range of the phase shifter is sufficient. As before, a total range of approximately 3π radians is sufficient.

If rapid acquisition is unimportant, then resetting is unnecessary. As long as the bit rate of the incoming data lies within the tuning range of the VCXO, the D/PLL eventually acquires. If the phase shifter is driven to the end of its range, the first stage of acquisition terminates prematurely and a significant phase error may remain. Now, if the phase shifter is at the end of its range, then the VCXO must also be as well, since the VCXO's control range is a subset of that of the phase shifter. The resulting frequency error slews the phase error toward zero in an open-loop manner. Once the phase error is zero, closed-loop behavior resumes and the loop ultimately settles to a condition of zero phase and frequency error.

IV. Implementation of the DLL

This section describes the realization of a monolithic DLL in a 3-GHz silicon bipolar process, beginning with some background material on phase detectors.

A. Phase Detectors

Before describing the phase detector actually used in this particular DLL, it is useful first to examine briefly the properties of two related phase detectors. The first of

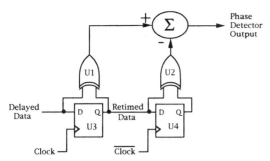

Fig. 11. Hogge phase detector.

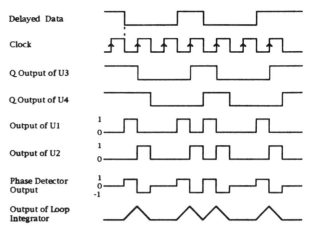

Fig. 12. Waveforms of Hogge's detector with clock and data aligned.

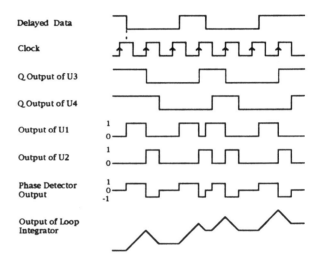

Fig. 13. Waveforms of Hogge's detector with data ahead of clock.

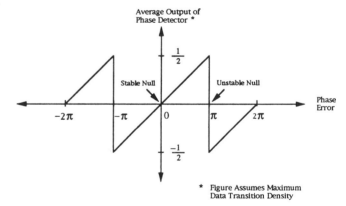

Fig. 14. Transfer characteristic of Hogge's detector.

these, due to Hogge [4], is shown in Fig. 11. This circuit directly compares the phases of the delayed data and the clock in the following manner. After a change in the state of the delayed data, the D input and Q output of D-type flip-flop $U3$ are no longer equal, causing the output of XOR gate $U1$ to go high. The output of $U1$ remains high until the next rising edge of the clock, at which time the delayed data's new state is clocked through $U3$, eliminating the inequality between the D and Q lines of $U3$. At the same time, XOR gate $U2$ raises its output high because the D and Q lines of $U4$ are now unequal. The output of $U2$ remains high until the next falling edge of the clock, at which time the delayed data's new state is clocked through $U4$.

If we assume that the clock has a 50% duty cycle, $U2$'s output is a positive pulse with a width equal to half the clock period for each data transition. $U1$'s output is also a positive pulse for each data transition, but its width depends on the phase error between the delayed data and the clock; its width equals half a clock period when the delayed data and the clock are optimally aligned. Hence, the phase error can be obtained by comparing the widths of the pulses out of $U1$ and $U2$.

Fig. 12 is a timing diagram for this detector with the delayed data and clock optimally aligned (in this case, with the falling edge of the clock). In this case, the output of the phase detector has zero average value, and there is no net change in the loop integrator's output.

If the delayed data are ahead of the clock, the output of the phase detector has a positive average value, as shown in Fig. 13. As a result, the loop integrator's output exhibits a net increase. Conversely, if the delayed data were behind the clock, the phase detector's output would have a negative average value, and the loop integrator's output would exhibit a net decrease.

Plotting the phase detector's average output (assuming maximum data transition density) as a function of phase error yields the sawtooth characteristic shown in Fig. 14. Consistent with Fig. 12, the phase detector's average output equals zero when the phase error between the delayed data and the clock is zero.

While one noteworthy feature of this phase detector is that the decision-making circuit is an integral component of the phase detector (for the output of flip-flop $U3$ is the retimed data), this detector does suffer from a sensitivity to the data transition density. Since each triangular pulse on the output of the loop integrator has positive net area (see Fig. 12), the presence or absence of such a pulse affects the average output of the loop integrator. The data-dependent jitter thus introduced is often large enough to be objectionable.

The phase detector shown in Fig. 15 greatly reduces this problem by replacing the triangular correction pulses (which have net area, even when the delayed data and clock are properly aligned) with "triwaves" whose net area is zero when clock and data are aligned.

426

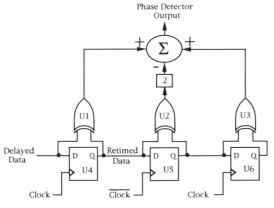

Fig. 15. Triwave phase detector.

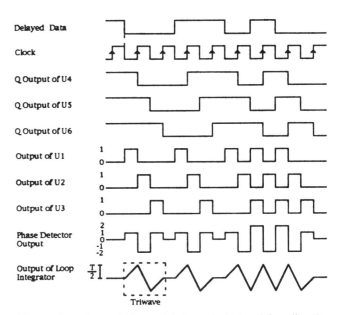

Fig. 16. Waveforms of triwave detector with clock and data aligned.

As in Hogge's detector, the width of $U1$'s output is dependent on the phase error between the delayed data and the clock, while the outputs of $U2$ and $U3$ are always a half clock cycle wide (assuming that the clock possesses a 50% duty cycle). The phase error can thus be obtained by comparing the variable width pulse from $U1$ with the fixed width pulses from $U2$ and $U3$. Note that the pulses out of $U1$ and $U3$ are weighted by 1, while the pulse out of $U2$ is weighted by -2.

Fig. 16 is the timing diagram for the triwave detector with the delayed data and the clock optimally aligned. Note that each data transition initiates a three-sectioned transient (the triwave) on the output of the loop integrator, and that this triwave has zero area. Therefore, its presence or absence does not change the average output of the loop integrator. Hence, the triwave detector exhibits a much reduced sensitivity to data transition density.

The triwave detector is somewhat more sensitive to duty cycle distortion in the clock signal, however, than Hogge's implementation, because of the unequal weightings used [5]. This sensitivity to duty cycle can be restored to that of Hogge's implementation with the simple modification shown in Fig. 17. The modified triwave detector uses two distinct down-integration intervals clocked on opposite edges of the clock, rather than a single down-integration of twice the strength clocked on a single clock edge. As a consequence, duty cycle effects are attenuated.

Filtering of the high-frequency ripple present in the output of any of these phase detectors can be provided by the addition of poles placed beyond loop crossover. These "higher order poles" (HOP's), shown in Figs. 8 and 9, can reduce the jitter induced by this ripple to insignificant levels.

The low clock duty cycle sensitivity, small data-dependent jitter, and the integral decision/retiming property of the modified triwave detector are the reasons the present DLL design uses this particular form of phase detector.

In the actual implementation of the phase detector, the necessary summations are performed by steering currents into and out of a capacitor with a differential charge pump,

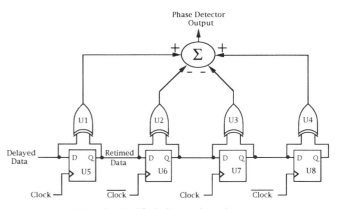

Fig. 17. Modified triwave phase detector.

as shown in Fig. 18. The current switches are connected to the outputs of the XOR gates shown in Fig. 17.

No reset function is provided in this particular implementation. Hence, although acquisition is typically fast (40 clock cycles by design), it is possible for acquisition to terminate prematurely, as described in Section III-C. Acquisition takes considerably longer in such cases.

Because of headroom limitations, resistive loads are used in the charge pump, as seen in Fig. 18. To achieve a sufficiently high dc gain (for low static phase error), a negative resistance is placed in parallel with the resistive loads, as shown in simplified form in Fig. 19. To understand how the negative resistor functions, assume that $Q1$ and $Q2$ act as ideal voltage followers so that $v_e = v_{in}$. Because of the cross-connections of the bases, the current i_{in} is equal to $-v_{in}/2R$, so that $R_{in} = -2R$.

Fig. 20 shows an improved negative resistor that compensates for the nonzero output impedance of the emitter followers by adding diode-connected transistors $Q1$ and $Q2$. If we assume that the transistors possess infinite β and infinite Early voltage, the collector currents of $Q1$

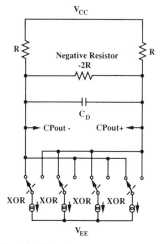

Fig. 18. Phase-detector charge pump.

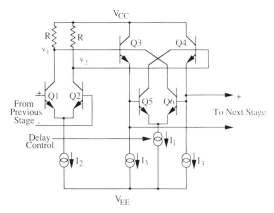

Fig. 21. Delay cell.

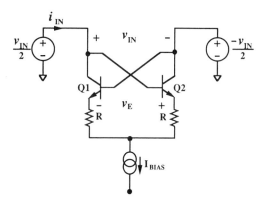

Fig. 19. Simple negative resistance circuit.

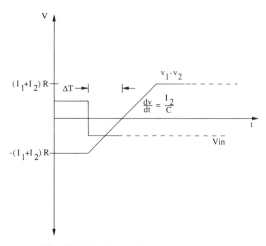

Fig. 22. Idealized delay cell waveforms.

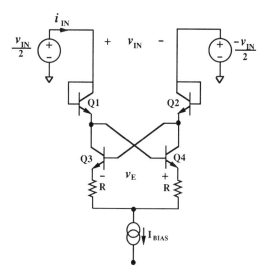

Fig. 20. Improved negative resistance circuit.

and $Q3$ will be equal, as will those of $Q2$ and $Q4$. Hence, $v_{BE1} = v_{BE3}$ and $v_{BE4} = v_{BE2}$. As a result

$$v_{BE1} + v_{BE4} = v_{BE2} + v_{BE3}.$$

That is, the voltage drop between the base of $Q1$ and the emitter of $Q4$ equals the voltage drop between the base of $Q2$ and the emitter of $Q3$. Therefore v_e equals v_{in} exactly.

In practice, finite β and Early voltage degrade the performance of the negative resistor somewhat; these effects may also be cancelled if desired [3]. Even without such additional compensation, however, increases of a factor of 50 in the effective load resistance are routinely achievable, assuming a resistor match of 0.1% and a β of 100.

B. Phase Shifter

The phase shifter is based on the cascadable current-starved differential-amplifier cell shown in Fig. 21. The slew-limited outputs of $Q1$ and $Q2$ drive a differential comparator formed by $Q3$–$Q6$. Control of the delay is by variation of the total differential voltage (by varying I_1) that the input pair must slew before the comparator changes state, leading to an approximately linear delay-versus-control characteristic. Additionally, the positive

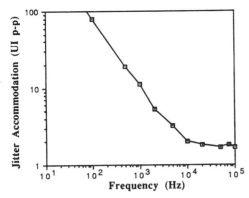

Fig. 23. Measured D/PLL jitter accommodation.

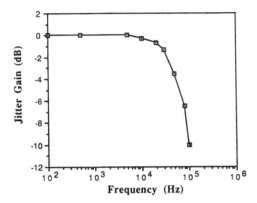

Fig. 24. Measured D/PLL jitter transfer function.

feedback action of the comparator ensures rapid switching once the thresholds are reached, reducing both the window of time over which noise could induce jitter and any pattern sensitivity that could arise if node capacitances were not restored to consistent initial conditions.

Fig. 22 shows idealized waveforms for the delay cell, where exponential transients have been coarsely approximated as straight-line segments. If we further assume that the comparator switches as soon as the differential voltage applied to it becomes zero, the delay can be expressed as follows:

$$\Delta T \approx \frac{RC}{2} \left(1 + \frac{I_1}{I_2} \right)$$

where C is the effective capacitance seen at each collector of the input pair.

As stated in Section III-A, the phase shifter must be able to provide at least ± 0.5 unit intervals (UI's) of shift range, and a somewhat greater value is desirable to improve acquisition in the presence of jitter as well as to absorb inevitable system offsets. This DLL employs a delay chain that provides a typical measured total phase-shift range of approximately 2.2 UI. This value is also consistent with the requirements of telecommunication standards such as SONET OC-3 specifications.

V. Experimental Results

In addition to avoiding jitter peaking, a practical clock recovery circuit must be able to accommodate the jitter that is always present on real input signals. Jitter accommodation in the D/PLL is governed by different factors in three distinct frequency regimes [5].

Fig. 23 shows the measured jitter accommodation of the D/PLL formed by combining the DLL with an external VCXO that possesses a useable tuning range of approximately 335 ppm. As can be seen, the low-frequency characteristic has a $1/f$ shape. In this region of operation, the VCXO's tuning range controls the jitter accommodation. The $1/f$ shape is due to the integral relationship between phase and frequency, which may be expressed as

$$\#UI_{pp} = \frac{\Delta f_{pk}}{\pi f_{\text{jitter}}}$$

where Δf is the peak tuning deviation from center of the VCXO in hertz and f_{jitter} is the frequency of the sinusoidal input phase modulation.

At midrange frequencies (the DLL regime of operation), bounded roughly at the lower end by where the accommodation of the VCXO equals the range of the phase shifter and at the higher end by the loop crossover frequency, the phase shifter controls jitter accommodation. Hence, the midrange accommodation is roughly constant with frequency and equal to the phase shift range of 2.2 UI. In this design, the midrange extends to approximately 3 MHz to accommodate nulling of phase errors in approximately 40 clock periods.

At frequencies above the loop crossover frequency, jitter accommodation necessarily drops below unity, to a value determined by the width of the eye opening and the static phase error. Because of test limitations, measurements in this regime are absent from Fig. 23.

Fig. 24 shows the measured jitter transfer gain of the D/PLL. There is no evidence of jitter peaking anywhere within the 40-kHz bandwidth. Note that this closed-loop bandwidth is only approximately 250 ppm of the bit rate, while the loop crosses over at approximately 3 MHz, in contrast with conventional loops where both bandwidths must be the same.

Fig. 25 is a histogram of jitter on the recovered clock for the D/PLL. With an input that possesses the maximum data transition density, the rms jitter is less than 1°. The rms jitter changes insignificantly even when the D/PLL is driven with a $2^{23}-1$ pseudorandom bit sequence (PRBS), indicating that there is negligible data-dependent jitter.

The 2.5×4.3-mm^2 die shown in Fig. 26 draws 70 mA from a standard ECL -5.2-V power supply.

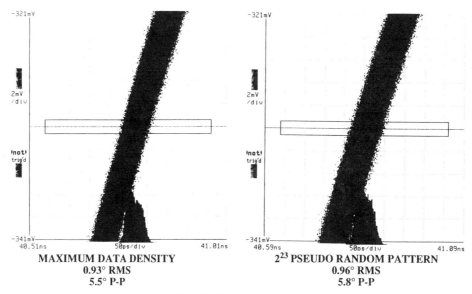

MAXIMUM DATA DENSITY
0.93° RMS
5.5° P-P

2²³ PSEUDO RANDOM PATTERN
0.96° RMS
5.8° P-P

Fig. 25. Jitter histograms.

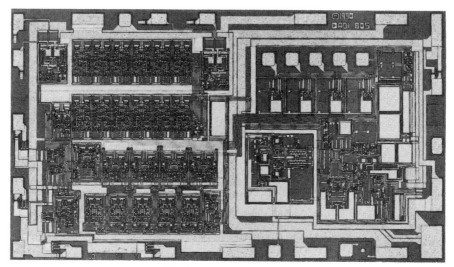

Fig. 26. DLL die photo.

VI. SUMMARY

By shifting the phase of the input data relative to the clock with a voltage-controlled phase shifter, the DLL and D/PLL avoid many of the trade-offs that limit the performance of conventional clock recovery circuits. Rapid acquisition can be achieved without compromising jitter filtering, and neither the DLL nor the D/PLL exhibits jitter peaking. These architectures thus offer a combination of jitter filtering and rapid acquisition not achievable with conventional clock recovery circuits.

ACKNOWLEDGMENT

The authors are grateful to B. Surette for assistance in testing; to T. Freitas for expertise in layout; to R. Croughwell, E. Ferrari, and L. DeVito for invaluable help and enlightening discussions; and to S. Hubbard for aid in manuscript preparation.

REFERENCES

[1] J. Bulzacchelli, U.S. Patent 5 036 298, July 1991.
[2] P. Trischitta and E. Varma, *Jitter in Digital Transmission Systems.* Dedham, MA: Artech House, 1989.
[3] L. DeVito *et al.,* "A 52MHz and 155MHz clock-recovery PLL," in *ISSCC Dig. Tech. Papers,* Feb. 1991, pp. 142–143.
[4] C. R. Hogge, "A self-correcting clock recovery circuit," *J. Lightwave Technol.,* vol. LT-3, no. 6, pp. 1312–1314, 1985.
[5] J. Bulzacchelli, "A delay-locked loop for clock recovery and data synchronization," Master's thesis, Massachusetts Inst. of Technology, Cambridge, June 1990.
[6] T. H. Lee and J. F. Bulzacchelli, "A 155MHz clock recovery delay-and phase-locked loop," in *ISSCC Dig. Tech. Papers,* Feb. 1992, pp. 160–161.

A Monolithic 156 Mb/s Clock and Data Recovery PLL Circuit Using the Sample-and-Hold Technique

Noboru Ishihara, *Member, IEEE* and Yukio Akazawa, *Member, IEEE*

Abstract—The PLL circuit described here performs the function of data and clock recovery for random data patterns by using a sample-and-hold technique, and four component circuits (a phase comparator, a delay circuit, a voltage-controlled oscillator, and a S/H switch with a low-pass-filter) were specially designed to further stabilize the PLL operation. A test chip fabricated using Si bipolar process technology demonstrated error-free operation with an input of $2^{23}-1$ PRBS data at 156 Mb/s. The rms data pattern jitter was reduced to only 1.2 degrees with only an external power supply bypass capacitor.

I. INTRODUCTION

ALTHOUGH optical-fiber communication systems have been used mainly for high-speed, high-density, long-distance communications, the advantages of optical-fiber transmission in local systems are now being explored in applications, such as Local Area and Wide Area Network (LAN, WAN) systems for multimedia, Fiber To The Home (FTTH) [1], and the board-to-board interconnections in exchange systems and computers [2]. Emphasis is shifting from increasing the operating speed of components to reducing their size, power consumption, and cost and to eliminating the need for adjustment and trimming. These goals can be achieved by using fully monolithic circuits.

A clock and data recovery circuit is one of the key components of optical transmission receivers. Besides the appropriate timing operation between data and the recovered clock, it must be stable for random input patterns which include consecutive data. To maintain the clock signal during the input of consecutive data bits, a conventional nonlinear clock recovery circuit needs a high Q-filter (such as a SAW or tank filter). Such a filter is difficult to integrate in monolithic form with the other electrical circuits. A circuits using a Phase Lock Loop (PLL) are easily implemented in a fully monolithic form. But data and clock outputs have large jitter, and the loop easily loses lock for random input data patterns. Efforts to create new PLL structures suitable for data and clock recovery are therefore ongoing at laboratories around the globe [3]–[7]. A 156-Mb/s monolithic circuit with an 3.5 degree rms pattern jitter for 2^7-1 data has already been reported [3].

The present paper describes a PLL circuit which determines appropriate timing between data and the recovered clock automatically by using a new phase-comparing technique and which maintains lock for random pattern data by using the

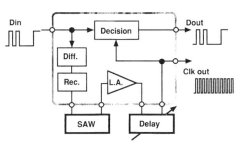

Fig. 1. Conventional clock and data recovery structure using a nonlinear extraction technique.

sample-and-hold (S/H) technique. To further stabilize PLL operation, four key component circuits were investigated. For $2^{23}-1$ pseudo random bit sequence (PRBS) data input at 156 Mb/s, a chip fabricated using 16-GHz Si bipolar process technology (SST-1A [8]) demonstrated error free operation and a root-mean-square data pattern jitter of 1.2 degrees. It's only external component was a power supply bypass capacitor.

II. PROBLEMS IN CONVENTIONAL CLOCK AND DATA RECOVERY

A. Nonlinear Extraction Technique

In a conventional clock and data recovery circuit using a nonlinear clock extraction technique (Fig. 1), a clock signal is extracted from NRZ input data by differentiating and rectifying circuits whose combined function is similar to that of a frequency doubler. This technique is used because NRZ data does not include a portion of its clock frequency. To maintain the clock signal during consecutive input ONES or ZEROS, a high-Q filter and a high-gain limiting amplifier are used.

The input data is retimed in the decision circuit, where a tunable delay generator is used to precisely adjust the clock timing manually. The tunable delay generator may be integrated monolithically, but it is difficult to estimate the range to be adjustment because this way depends on the characteristics of the parasitic impedance effects of the external hybrid circuit.

A circuit that is stable when the data includes more than 100 consecutive ONES or ZEROS has been built, but because it needs special devices, such as a SAW filter or a tunable delay generator, its components cannot be easily integrated on a single chip.

B. PLL Circuit Technique

A PLL circuit is shown in Fig. 2, and a timing chart of the PLL operation is illustrated in Fig. 3. A phase comparator compares the input data phase (a) with the voltage-controlled

Manuscript received May 23, 1994; revised, August 5, 1994.

The authors are with NTT LSI Laboratories, 3-1 Morinosato Wakamiya, Atsugi-shi, Kanagawa Pref., 243–01 Japan

IEEE Log Number 9406260.

Reprinted from *IEEE Journal of Solid-State Circuits*, vol. SC-29, pp. 1566-1571, December 1994.

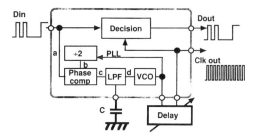

Fig. 2. Conventional clock and data recovery structure using the PLL technique.

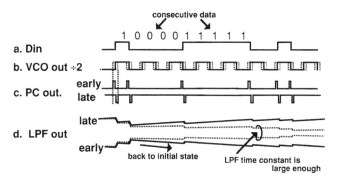

Fig. 3. Timing chart of a conventional PLL.

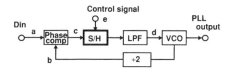

Fig. 4. PLL structure using a sample-and-hold switch.

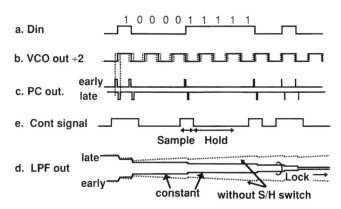

Fig. 5. Timing chart of the PLL using a sample-and-hold switch.

oscillator output divided by 2 (b), and it outputs a signal pulse (c) including phase error. In this comparator, when the input data phase is earlier than the VCO output signal divided by 2, an upper pulse is obtained. When the phase is later, an lower pulse is obtained. The phase error signal (d) is extracted by the LPF and controls the VCO to the appropriate frequency, stabilizing the PLL. But with random pattern data input, the phase comparator output pulses cannot be obtained during bit periods of no transitions. During these periods the VCO control voltage (LPF output) goes back to its initial state, and the VCO oscillation is independent of the input data. Thus, for random input data, there are two stabilizing forces in the PLL. One is synchronizes the phase of the input data with the phase of the VCO output, and the other is pulls the PLL back to its initial state during the input of consecutive data bits. The latter force becomes stronger as the number of consecutive data bits increases, making the PLL more unstable and resulting in larger output jitter. To prevent this, either the time constant of the LPF should be large enough as to not to go back to the initial state, or the VCO oscillation frequency should be initially set to synchronize the input data as close as possible. This means that an external component such as a high-Q filter or a large capacitor—either of which is difficult to integrate with the other electrical circuit—become necessary in addition to a power supply bypass capacitor. A clock recovery PLL circuit also needs to have high-Q characteristics equivalent to those of the nonlinear extraction circuit. And because it is necessary to adjust the clock timing, there is still the problem of a tunable delay generator.

III. ENHANCING CLOCK AND DATA RECOVERY PLL FUNCTION

A. Equivalent High-Q Method

To stabilize PLL operation with small output jitter, it is important to get enough margin for the tolerance to consecutive input data bits, and we studied the possibility of doing this by using an active circuit structure rather than high-Q device or an LPF with a large time constant. As shown in Fig. 4, for equivalent high-Q operation, a S/H switch is inserted so that the PLL does not to go back to the initial state during the input of consecutive data bits. The timing chart is illustrated in Fig. 5. In this circuit, the phase output signal can be transferred to the LPF only when the S/H switch is in the sample mode. By setting the S/H switch in the hold mode during the consecutive data, the control voltage for the VCO (LPF output) can be kept constant and the VCO control voltage can be prevented from going back to its initial state. Thus even for the random input data, the PLL can be operated with only one force synchronizing the input data phase with the VCO output phase. Furthermore, because the large-time-constant LPF or high-Q devices become unnecessary, it also becomes easier to implement the PLL in a fully monolithic form. This means that the equivalent Q of the PLL can be high.

Tolerance for the number of consecutive data bits depends on the S/H switch performance. The longer the S/H switch maintains the output hold level, the more consecutive data bits can be tolerated. To use this structure without manual adjustment, a circuit that detects consecutive data is necessary in order to control the S/H switch automatically. But there are no problems in producing an appropriate control signal. It can be obtained automatically by using the input data. Details of the circuit for this are given in following subsection C.

B. Automatic Clock Timing Adjustment

Automatic clock timing adjustment is basically done by using a flip-flop (F/F) function. A block diagram of the circuit is shown in Fig. 6, and the timing chart is shown in Fig. 7. The F/F function is ordinarily used as the decision circuit, and ideally its output is delayed 90 degrees from the input data when the clock timing is appropriate. The phase difference (e)

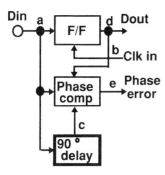

Fig. 6. Phase comparison for automatic clock timing adjustment.

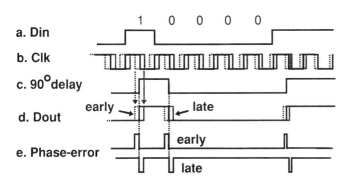

Fig. 7. Timing chart of phase comparison for automatic clock timing adjustment.

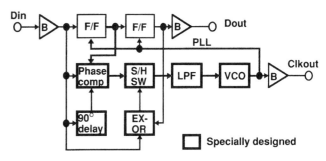

Fig. 8. Structure of the whole clock and data recovery PLL.

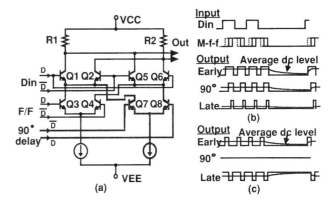

Fig. 9. Phase comparator. (a) Circuit. (b) Timing chart of one-multiplier circuit. (c) Timing chart of two-multiplier circuit.

between the real clock and appropriate clock for the input data can therefore be obtained by comparing the output phase of the F/F output (d) with the ideal 90-degree-delayed phase for the input data (c). Both the recovered clock frequency and the phase timing for the input data can be adjusted automatically by using this phase difference signal in the PLL.

Another feature of this structure is that the PLL is resistant to duty cycle variation or wave form distortion in the input data, since the clock timing is always adjusted by a feedback effect at the average time point of rise and fall input data edges. Still another advantage is that high accuracy for the 90-degree delay is not necessary because the F/F function generally has a phase margin of more than 200 degrees. The circuit can be operated even if the delay differs from 90 degrees. Tolerance for the delay range is determined by the F/F phase margin and a capture range of the phase comparator. This is an advantage for integrating the circuit with high yield. Furthermore, when this structure is used in the PLL, the function of division by 2 (shown in Fig. 2), can be replaced by the F/F function. This is because the frequency portion of the F/F output is half the frequency of the clock signal.

C. PLL Block Diagram

A complete block diagram of the PLL is shown in Fig. 8. The first F/F circuit performs a function equivalent to the division by 2 shown in Fig. 2. The control signal for the S/H switch is automatically generated by an Exclusive OR of the input data and the output of a second F/F circuit. The output of the phase comparator is sent to the LPF only when the EX-OR output signal (S/H switch control signal) is high. For

consecutive data the control signal is low and, as illustrated in Fig. 5, this keeps the S/H switch in the hold mode.

IV. CIRCUIT DESIGN

To verify the effectiveness of this concept, we designed a circuit to operate at a data rate of 156 Mb/s. And to further improve the PLL stability, we specially designed several key component circuits.

A. Phase Comparator

In a conventional phase comparator indicated with solid lines in Fig. 9(a), Two inputs are compared by using a multiplication operation. A problem, however, is that, as illustrated in Fig. 9(b), the average dc level of the output signal changes depending on the data mark ratio even when the phase difference between the two inputs is ideal (90 degrees).

To solve this problem, two multiplier circuits are combined as shown in Fig. 9(a). The first multiplier circuit consisting of $Q1$ to $Q6$ and $R1$ and $R2$, compares the phase of the input data with that of the F/F output. The second multiplier circuit, consisting of $Q1$ and $Q2$, $Q5$ to $Q8$, and $R1$ and $R2$, compares the phase of the input data and the 90-degree delayed input data. In this circuit, the change in the average output dc level is due only to the data mark ratio. Therefore, by subtracting the output of the second circuit from that of the first, the effect of the mark ratio can be reduced and the VCO frequency can be kept stable. As shown in Fig. 9(c), the output is independent of the mark ratio when the phase difference is the ideal 90 degrees.

433

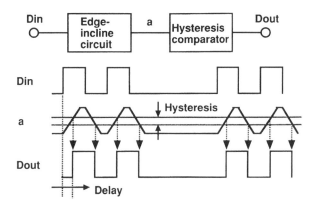

Fig. 10. Principle of delay circuit.

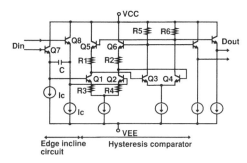

Fig. 11. Delay circuit.

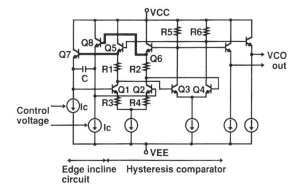

Fig. 12. Voltage-controlled oscillator.

B. Delay Circuit

A delay circuit that includes an edge-incline circuit and a hysteresis comparator are used for the 90-degree delay and the voltage-controlled oscillator. Fig. 10 shows the basic principle of the delay circuit for the pulse data. The rise and fall edges of the input data are inclined by the incline circuit and the hysteresis comparator reproduces the data edges when the input signal crosses the threshold voltage. Since there is hysteresis in the comparator, a delay also occurs at the comparator. The total delay is the sum of all the individual delays, and the maximum is about 1 bit period. This is twice that of a conventional RC delay circuit and easily covers a 90-degree delay.

Fig. 11 shows the detail of the delay circuit. The data edge is inclined by the capacitor C and the charging current I_c. Positive feedback is used to produce hysteresis and renew the waveform in the comparator. The amount of hysteresis is

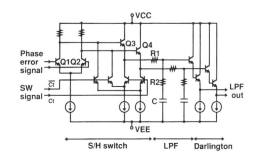

Fig. 13. Sample-and-hold switch with a low-pass filter.

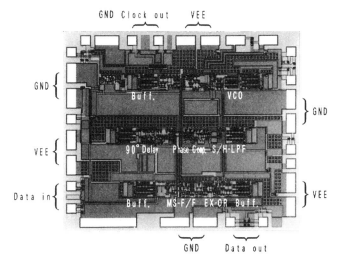

Fig. 14. Microphotograph of PLL IC.

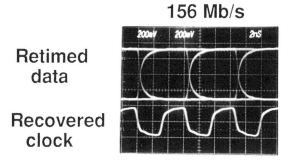

Fig. 15. Output waveform.

determined by the gain of the differential amplifier consisting of $Q1$, $Q2$, and $R1$ to $R4$. The comparator operates as a Schmitt trigger circuit. The value of C, I_c, and the hysteresis voltage are chosen so that data is delayed by 90 degrees: C is 4 pF, and I_c is 1 mA.

C. Voltage-Controlled Oscillator

A similar delay circuit technique is used in the VCO. An oscillation can be obtained by channeling positive feedback from the output of the comparator back to its inputs through

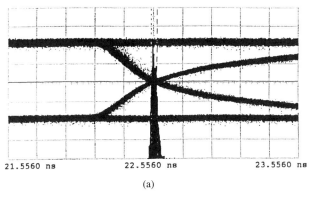 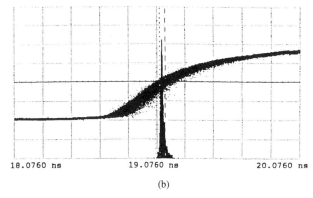

| (a) | (b) |

Fig. 16. Measured output jitter of data and clock output. (a) Retimed data. Rms jitter: 1.2 degrees (20 ps). H: 200 ps/div. V: 150 mV/div. (b) Recovered clock. Rms jitter: 1.2 degrees (20 ps). H: 200 ps/div. V: 60 mV/div.

the delay circuit as shown in Fig. 12. The frequency can be varied by changing the charging current I_c. In a conventional multivibrator-type VCO that uses diode loads, the limiting caused by the diode loads is thought to produce a large jitter and to limit the frequency range. This circuit, in contrast, has a small output jitter and a wide oscillation range. Based on the worst case analysis considering fabrication tolerance, the free-running frequency of the VCO is designed to be around 156 MHz. The controllable frequency range is designed to be more than 50 MHz. The value of C is 16 pF and a typical I_c value is 1 mA.

D. S/H Switch with Low-Pass Filter

The S/H switch with a LPF is shown in Fig. 13. A differential-type circuit is used to prevent common-mode noise from affecting the VCO. When the control signal C_t is high, the input signal, which includes phase error, is sent to the LPF. When it is low, $Q3$ and $Q4$ turn off and the output of the LPF is constant. The switch gain is designed to be high (14 dB) in order to change the mode immediately. The LPF is a lag-lead type. To allow operation with more than 24 consecutive data bits, the low-pass resistance $R1$, capacitance C and damping resistance $R2$, were optimized by a circuit simulator to be 2000 Ω, 30 pF, and 1000 Ω, respectively. And to retain the hold state accurately, a Darlington circuit was inserted between the filter and the VCO.

E. Whole PLL Circuit

The other circuits shown in Fig. 8 are conventional differential-type circuits. An input buffer was added to increase the input amplitude sensitivity. The gain and the sensitivity are designed to be 14 dB and less than 20 mV. The output buffer consisting of the differential amplifier and emitter followers, are capable of driving 50 Ω loads. The F/F and EX-OR functions are provided by conventional logic circuits, and the logic swing is designed to be 800 mV. A band-gap voltage reference circuit is also included to guard against variations in the supply voltage or temperature.

To stabilize the operation during random input data, the PLL parameters are optimized as follows. The transfer gain of the phase comparator including the S/H switch is 0.5 V/rd, and the transfer gain of the VCO is 0.1 MHz/mV. The open loop gain is therefore about 3×10^8. From these parameters and

TABLE I
IC CHARACTERISTICS: MEASURED VALUES FOR PRBS 2^{23}–1 DATA

Parameter	Value	
Bit rate	155.52 Mb/s	
External LPF capacitor	nonconnect	2200 pF
Natural angle freq.	7.5×10^7 rad/s	8.7×10^6 rad/s
Dumping factor	1.2	9.6
Capture range	> ± 10 MHz	> ± 10 MHz
Lock range	> ± 15 MHz	> ± 15 MHz
Rms data jitter	1.2° (20 ps)	0.95° (17 ps)
Rms clock jitter		
Consecutive data	23 bits	> 320 bits
Supply voltage	-5.2 V and -2 V	
Power consumption	320 mW	

the LPF constants described in the previous section, we can calculate that the natural angle frequency and damping factor are 7.5×10^7 rad/s and 1.2.

V. FABRICATION RESULTS

A PLL circuit was fabricated using Si-bipolar super-self-aligned process technology (SST-1A[8]). The cut-off frequency of the transistor at the typical bias current is 16 GHz at $V_{ce} = 2$ V. Transistors used in the circuit are 0.8×8 and 0.5×16 μm^2 in emitter size. A relatively small bias current (less than half the typical value) was supplied for those transistors in order to reduce the power consumption and the chip layout was designed with special consideration of the line capacitance for the delay and the VCO part. A microphotograph of the IC is shown in Fig. 14. It measures 3×2.5 mm^2 and contains two 30-pF capacitors for the low-pass filters. To prevent coupling between the circuit blocks due to the effect of the parasitic impedance of the packaging, separate power supply lines are used for each block.

A. Measurement System

The PLL IC was mounted in a 7-mm-square ceramic package and evaluated. A random pattern was input by the pattern generator, and the error rate was measured with an error detector using the PLL IC's clock output signal. Waveforms and jitter were observed with an oscilloscope.

B. Measured Characteristics

The output waveforms obtained when a 2^{23}–1 PRBS data are shown in Fig. 15. Error-free operation (less than 10^{-11}) was

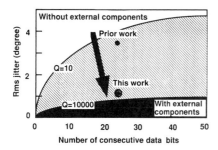

Fig. 17. Relation between jitter and number of consecutive data bits.

obtained at a data rate of 156 Mb/s. The eye opening of the output data was sufficiently wide, and the clock extraction was very precise. The minimum sensitivity for the input amplitude was 20 mV, keeping error-free operation with an rms output jitter of 1.2 degrees.

Jitter histograms measured with a sampling oscilloscope are shown in Fig. 16. The edges of the output signal are expanded. Both the retimed data and recovered clock have an rms jitter of 1.2 degrees.

The IC characteristics are summarized in Table I. With only an external power supply bypass capacitor, the IC can be operated with $2^{23}-1$ PRBS input data that includes up to 23 consecutive bits. The capture-and-lock range is more than ± 10 MHz. This wide range is due to the equivalent high-Q operation. The chip consumes 320 mW at supply voltages of -5.2 V and -2 V. After the PLL is locked, it can keep the lock status even if the power supply voltage changes from -4 to -6 V. The free-running frequency is 158 MHz. For the temperature variation from 0°C to 80°C, the free running frequency varies from 156 to 163 MHz and the capture-and-lock range is more than ± 10 MHz. Jitter decreases as the temperature increase, but from 0°C to 80°C, the difference is less than 5%. By adding an external LPF capacitor, the rms jitter can be reduced to less than 1 degree and the tolerance for the number of consecutive data bits can of course be increased.

The relationship between jitter and the number of consecutive data bits for the conventional nonlinear circuit with a Q device is plotted in Fig. 17, and the PLL-IC data is also shown on this plot. The shaded area is for circuits needing external components such as a SAW or a large LPF capacitor. To keep the output jitter small when the data includes a large number of consecutive bits, external components are necessary. The limit above which external components are necessary had generally been thought to be around an equivalent Q of 10, but our work has pushed the limit to an equivalent Q of more than 10,000, which means that the performance has been improved 1000-fold.

VI. CONCLUSION

To create a fully monolithic, adjustment-free clock and data recovery PLL circuit without sacrificing performance, we investigated the following new circuit techniques:

1) An equivalent high-Q operation using a S/H switch.
2) A timing clock adjusted automatically by using a reference delay.

3) Four kinds of component circuits (a phase comparator, a delay circuit, a VCO, and a S/H switch with an LPF) that stabilize the operation.

The effectiveness of these techniques was demonstrated by fabricating an IC implementing a 156-Mb/s clock and data recovery PLL-IC with only an external bypass capacitor.

ACKNOWLEDGMENT

We are grateful to S. Horiguchi for discussions, and to Y. Inabe, S. Fujita, J. Sekine, and K. Takiguchi of NTT Electronics Technology Corp. for help in fabricating and evaluating the IC.

REFERENCES

[1] N. Miki and K. Okada, "Access flexibility with passive double star systems," *IEEE 5th Conf. Opt./Hybrid Access Networks Proc.*, 1993.
[2] K. Yukimatsu and Y. Shimazu, "Optical interconnections in switching systems," *IEICE Trans. Electron.*, vol. E77-C, no. 1, pp. 2–8, Jan. 1994.
[3] L. DeVito, J. Newton, R. Croughwell, J. Bulzacchelli, and F. Benkley, "A 52-MHz and 155-MHz Clock-Recovery PLL," in *ISSCC Dig. Tech. Papers*, Feb., 1991, pp. 142–143.
[4] Analog Devices Data Sheet, AD-802.
[5] J. T. Wu and R. C. Walker, "A bipolar 1.5Gb/s monolithic phase-locked loop for clock and data extraction," in *1992 Symp. VLSI Circuits Dig. Tech. Papers*, 1992, pp. 70–71.
[6] M. Soyer, "A monolithic 2.3Gb/s 100mW clock and data recovery circuit in silicon bipolar technology," *IEEE J. Solid-State Circuits*, vol. 28, pp. 1310–1313, Dec. 1993.
[7] H. Ransijin and P. O'Connor, "A PLL-based 2.5Gb/s GaAs clock and data regenerator IC," *IEEE J. Solid State Circuits*, vol. 26, pp. 1345–1353, Oct. 1991.
[8] S. Konaka, Y. Yamamoto, and T. Sakai, "A 30ps Si bipolar IC using self-aligned process technology," in *Conf. Solid State Devices and Mat.*, Extended Abstract, 1984, pp. 209–212.

A Monolithic 480 Mb/s Parallel AGC/Decision/Clock-Recovery Circuit in 1.2-μm CMOS

Timothy H. Hu, *Student Member, IEEE,* and Paul R. Gray, *Fellow, IEEE*

Abstract—A parallel architecture for high-data-rate AGC/ decision/clock-recovery circuit, recovering digital NRZ data in optical-fiber receivers, is described. Improvement over traditional architecture in throughput is achieved through the use of parallel signal paths. An experimental prototype, fabricated in a 1.2-μm double-poly double-metal n-well CMOS process, achieves a maximum bit rate of 480 Mb/s. The chip contains variable gain amplifiers, clock recovery, and demultiplexing circuits. It yields a BER of 10^{-11} with an 18 mV $_{\mathrm{p-p}}$ differential input signal. The power consumption is 900 mW from a single 5 V supply.

I. INTRODUCTION

LIGHTWAVE communications receivers with data rates in the 500 Mb/s range and above will be very important for SONET-based telecommunications, ATM-based LAN's, and other high speed data communication applications. A major problem limiting the deployment of optical-fiber communication systems in this range is the cost of the optical components and the terminal electronics attributed to the high cost technology, package, and assembly required.

The traditional architecture of a digital optical-fiber receiver is shown in Fig. 1. It consists of a photodetector and a low-noise preamplifier that generate a low-level voltage proportional to photon flux, a main amplifier, decision circuit, and a clock recovery block. Following the decision function, the digital data bit stream is usually deserialized back to its parallel form by a demultiplexer, using clocking provided by the frequency divider, and passed to a digital data link control for high level manipulation of data received.

The analog components of the receiver, including the main amplifier, decision circuit, and clock recovery circuit, require analog bandwidths on the order of the data rate. As a result, high speed technologies such as GaAs and high speed bipolar have typically been used to implement receivers in the 500 Mb/s range and above. These technologies are characterized by higher cost per function and lower integration levels than, for example, digital CMOS technology [3]–[6]. This paper describes a modified architecture for the amplifier, decision, and clock recovery sections of a data communications receiver

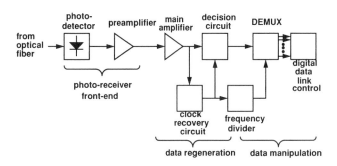

Fig. 1. Traditional digital optical-fiber receiver architecture.

which utilizes parallel analog data paths in order to reduce the required bandwidth of operation. This in turn potentially allows the use of digital CMOS in the implementation, resulting in the possibility of higher levels of integration and lower cost.

II. PARALLEL RECEIVER ARCHITECTURE

The basic approach taken in this work is to demultiplex the high-data-rate bit stream of analog symbols as early as possible in the data path, immediately following the preamplifier in this case. The remaining analog processing operations of gain, decision, and timing recovery are done in lower speed, parallel paths, with each parallel channel having its own amplifier and decision circuit. The output of each parallel channel is then passed into the data link control in parallel. A multiphase clock is used to strobe the input demultiplexer and the decision circuits in each parallel channel, and the phase of this sampling is controlled by a decision-directed clock recovery system to be described later. A block diagram of the parallel receiver is shown in Fig. 2.

Operation of the gain and decision functions are similar to a conventional implementation, except that in this case, these functions are operating on a sampled-data basis where settling time is more important than small-signal bandwidth. While the speed requirements on these functions are reduced by approximately the demultiplexing ratio, additional time must be allowed for recovery and reacquisition of the next data value, resulting in a gain that in fact is more like half of that ratio. Because of the low signal level of the input, input offset mismatches can result in a fixed pattern noise that detracts directly from the dynamic range of the signal. As a result, AC coupling is required on a per-channel basis if low-level signals are to be recovered at low bit error rates.

Manuscript received May 24, 1993; revised September 3, 1993. This work was supported by the National Science Foundation under Grant MIP-8801013, the California Micro Program, Texas Instruments Corporation, National Semiconductor, and Level One Communications.

The authors are with the Department of Electrical Engineering and Computer Sciences, University of California, Berkeley, Berkeley, CA 94720.

IEEE Log Number 9213191.

Reprinted from *IEEE Journal of Solid-State Circuits,* vol. SC-28, pp. 1314-1320, December 1993.

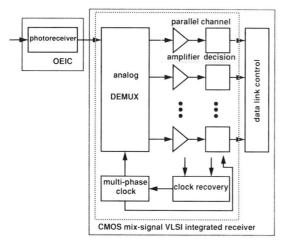

Fig. 2. Block diagram of a parallel receiver.

In the prototype described in this paper, this DC decoupling is accomplished using a DC feedback path with on-chip low-pass filter in each of the channels. The use of AC coupling introduces the requirement that the line code used be inherently DC balanced. Considerations of offset cancellation and clock recovery, which make use of the analog information in the waveform, led to the use of nonlimiting AGC amplifiers in the prototype described here. AGC feedback information is provided to the channel amplifiers by comparing received data amplitude with a fixed reference.

Since the signal is inherently sampled at the input, clock recovery is more complex in a parallel receiver implementation than in conventional receivers. One approach is to place additional sampling channels in the parallel structure which sample the data waveform at time points halfway between the two adjacent data channel sampling points; a point at which the input waveform should be passing through zero if the data alignment with clock is correct. Using this information together with the recovered data value, a timing function can be generated. The number of timing channels required depends on the line code used and the speed of timing acquisition that is required. One timing channel per data channel would result in very rapid acquisition and compatibility with any line code at the cost of hardware. Fewer numbers of timing channels are likely to be required in most systems. In the prototype described here, only one timing channel was used for simplicity.

The maximum bit rate that can be achieved with the parallel architecture is ultimately limited by errors in the input analog sampling and demultiplexing circuit. The analog input demultiplexing is most easily implemented by an array of differential MOS sample-and-hold (S/H) circuits, controlled by multiphase clock edges separated by exactly one bit period of the input data. The sampling process in these circuits introduces sampling jitter, DC offsets that can be different in different channels, intersymbol interference (ISI) and pulse distortion due to limited sample-mode bandwidth, and non-linearity due to signal-dependent charge injection. Since DC offsets are removed in later circuitry, sample-mode bandwidth can be made very large by using large MOS switch device sizes relative to sampling capacitor size, which results in sample

mode bandwidths significantly larger than could be achieved using the same MOS device in a linear amplifier. The potential advantage of the parallel approach rests on the wide bandwidth achievable in this way.

Since the sampling jitter of the demultiplexer depends primarily on the clock source, a low-noise, multiphase clock generator is required. A low-noise differential ring oscillator was used for this purpose in the prototype described here, with multiphase clock edges taken from taps of the ring oscillator.

The disadvantage of this architecture is that more hardware is needed, resulting in larger chip size and higher power dissipation. Nevertheless, because of the use of highly integrated CMOS technology with most of the circuits operating at a much lower rate, the chip area and power consumption are comparable with implementations in other high speed technologies.

III. PARALLEL RECEIVER IMPLEMENTATION

A prototype receiver with 8 parallel data channels and one timing channel, implementing the functions in the dotted box in Fig. 2, was designed and fabricated in a 1.2-μm double-poly, double-metal, n-well CMOS process. The prototype with typical output waveform from the photoreceiver is shown in Fig. 3. In the following sections, the major design aspects of the prototype are described.

A. Analog Demultiplexing

The analog demultiplexer consists of an array of differential NMOS switches and sampling capacitors, shown as simplified single-ended form in Fig. 3. As mentioned earlier, the principle factor limiting the throughput of the parallel receiver is errors in these input sample-and-hold circuits. The key factor enabling parallel receiver implementation to improve throughput is the very high sample mode bandwidth achievable in MOS sample-and-hold functions, if DC errors due to charge injection can be controlled. In this particular instance, for example, the sampling capacitor values were 1pF and the NMOS switch sizes were 100/1.2 with 1.5 V gate drive over threshold in the on state. The result is a sampling bandwidth of 1 GHz, and a total charge injection offset at zero signal, including both overlap and channel charge, assuming a 10% mismatch between the differential inputs, is approximately 27 mV. This input sample-mode bandwidth can support an input data rate up to 500 Mb/s, assuming that intersymbol interference must be kept less than 5%. As mentioned earlier, the DC term of the offset is taken out later, and the signal dependent term can be viewed as a gain error term and corrected by the gain-control circuitry. Both the sample mode bandwidth and the offset error improved with scaled technologies.

Typical output levels for optical fiber preamplifiers are in the range of 5 to 15 mV minimum, with maximum values at high signal levels in the hundreds of millivolts. To achieve the bit error rates normally required, a SNR of 25–30 dB is needed. As a result, kT/C noise considerations dictates a minimum sampling capacitor size in the range of a few tenths of a picofarad. In this case, a convervative value of 1 pF was used. A more important problem is digital noise coupling from

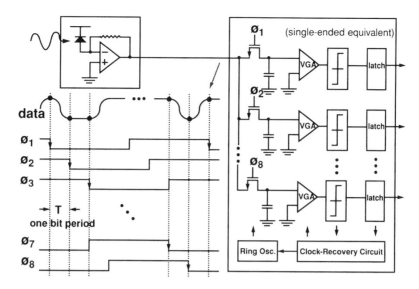

Fig. 3. Parallel receiver implemented with 8 parallel channels.

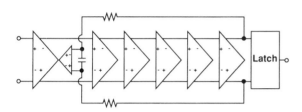

Fig. 4. Variable gain channel amplifier with latch.

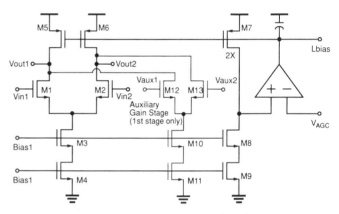

Fig. 5. MOS basic variable gain unit.

adjacent channels and from other digital circuitry on the same chip. This was minimized in the prototype by a differential signal path through the sample and hold (and in fact through the entire receiver), and careful attention to layout. A fully differential data path with simple sample-and-hold stages as shown before, is used to minimize the DC offset error. On-chip input signal line termination is also used to reduce the effect of extra phase-shift due to the bonding pads and bonding wires. Once the signal is on chip, individual source follower buffers are used to drive the samplers in each channel, so that loading on the inputs is minimized, and channel-to-channel coupling is also minimized.

B. Channel Amplifier

The range of signal amplitudes applied to the receiver dictated a channel amplifier dynamic range of 40 dB with a maximum gain of 100. An offset cancellation scheme must be used to lower the total input error voltage due to input offset voltage of the channel amplifier and the mismatch in the sampling switches of the analog demultiplexer.

The variable gain amplifier consists of a cascade of variable gain units. Using earlier results on optimum configurations in multistage wide band amplifiers [1], [2], an amplifier with 9 stages and a gain per stage of 1.8 is optimum for a gain of 100 in this technology. Because of the broad minimum in the optimum gain per stage, 6 stages are used in this work with a maximum gain per stage of 2.15 (6.7 dB) to minimize power and area.

A critical problem in the design of the channel amplifier was achieving the required offset cancellation without excessive degradation of bandwidth, since most input offset cancellation schemes add parasitics to the signal path. For this prototype, input offset cancellation is done by a continuous feedback path from the outputs back to the inputs through an auxiliary stage as shown in Fig. 4. The auxiliary stage is outside the signal path and adds very little parasitic to the first input stage so as to maintain the maximum bandwidth possible for the technology. Each channel has its own feedback network, so the capacitor cannot be too large because of area occupied. Since the absolute accuracy of the RC product is not important, the large-value resistors used in the feedback loop were implemented by long channel triode-mode MOSFET's to get the high resistance required. The corner frequency of the RC network in this case was 10 KHz.

As shown in Fig. 5, the variable gain load resistors are implemented by two PMOS devices (M5-M6) biased in the triode region. The resistance is controlled by a replica biasing circuit (M7-M9) that creates the correct gate voltage to bias the output common mode voltage of the differential pair. In this circuit, the biasing current in the differential pair is kept constant and the gain value is varied by changing

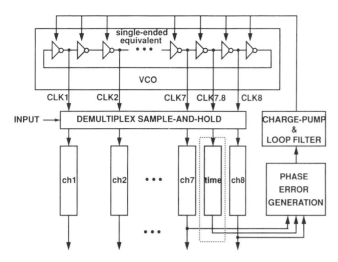

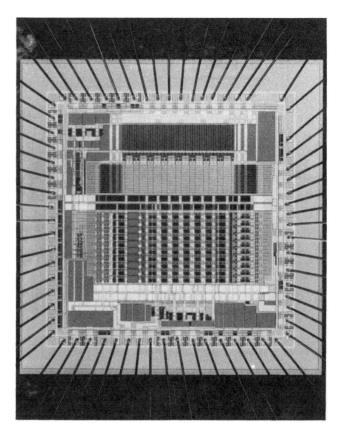

Fig. 6. Block diagram for clock recovery.

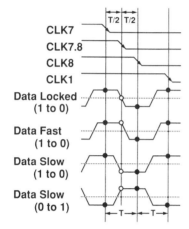

Fig. 7. Examples of decision directly phase error detection.

Fig. 8. Chip photograph of experimental prototype.

the load resistance value, which in turn is controlled by the feedback circuit that controls the DC voltage across the load resistors with the replica biasing circuit. The range of gain values achievable with this simple circuit is adequate for this application, but is limited by the fact that at the low gain extreme, the gate voltage applied to the PMOS loads reaches the negative supply and at the high-gain extreme the PMOS loads approach the saturation region on large signal swings. As a result, the gain range of each stage is only about 1 : 3 with a single 5 V supply. From simulation, the settling time to 95% point for a 5 mV input with a gain value of 100 is about 8 ns. The power dissipated in each channel amplifier is about 25 mW.

C. Clock Recovery

As mentioned earlier, the data channels sample the bit stream, resulting in a baud-rate sampled system. While it is possible to generate a timing function and perform clock recovery in a baud-rate sampled system [7], for reasons of system robustness, an additional timing sampling channel was inserted in this prototype. Clock recovery was implemented using a decision directed minimum-likelihood algorithm [8] and a charge-pump PLL. The clock recovery loop is shown

in block diagram form in Fig. 6. In this prototype a single extra timing channel was inserted between two adjacent data channels, although more timing channels may be required depending on the line code used and the speed of timing acquisition required. For the timing channel, a sampling clock edge exactly in between the sampling clocks for the two adjacent data channels is needed. This was obtained by tapping off an intermediate point in the ring oscillator. The phase error information is then passed to a charge-pump and then a loop filter to control the ring oscillator and complete the PLL. The sampling instants of the multiphase clock edges from the ring oscillator is adjusted to force the output of the timing channel to be zero. The differential loop filter (2R's and 2C's) was implemented externally in this prototype.

As shown in Fig. 7, the phase error is generated by defining the timing error voltage as the output voltage of the timing channel. From the first waveform shown, the clock is locked onto the input data waveform. Channel 7 and channel 8 sampling is in the center of the bit cell and the error voltage is zero. The following two waveforms correspond to positive and negative phase error, and the error voltage is positive and negative, respectively. However, in the last waveform, the phase error is negative but the error voltage is positive because of the fact that the polarity of the transition itself is negative instead of positive. The sense of the phase error signal must be adjusted in polarity accordingly, requiring that the phase detection be decision directed.

Initial acquisition of phase upon power up or arrival of data must be provided for in such a receiver. Requirements vary

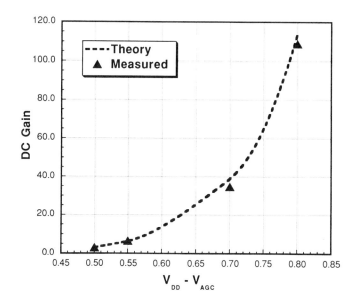

Fig. 9. DC gain versus ($V_{\mathrm{DD}} - V_{\mathrm{AGC}}$) of channel amplifier.

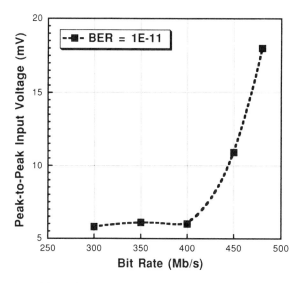

Fig. 10. Minimum input voltage versus input bit rate.

widely on the speed with which this must be accomplished. The choice of just one timing channel for phase detection requires a narrow bandwidth for the PLL results in a narrow capture range for the charge-pump PLL. In this case, initial acquisition would be achieved by alternate means, such as locking the PLL to a reference crystal and transferring control to the data detector upon the arrival of data. Alternatively, use of a large number of timing channels, such as one per data channel, would allow direct rapid acquisition using timing channel information.

Another consideration in the choice of number of timing channels is implication for requirements on receiver channel offset and on pulse distortion in the analog circuitry preceding sampling. Input offset and pulse distortion effectively shift the zero crossing point on the leading or trailing pulse edge. As a result, when one polarity of transition appears consecutively many times at the input of the timing channel, the clock edge is shifted as a transient phase error because the PLL tries to move the edges of the clock to force a zero output from the timing channel. When the opposite kind of transition occurs, the phase error generator detects an error voltage in the opposite direction. This results in random wandering of the sampling edge for the timing channel results in a pattern dependent jitter. The combination of density of timing channels and choice of line code must be such that the probability of long time periods of detection of only one polarity of transition for timing recovery is low if the receiver is to be insensitive to offset and pulse distortion of the incoming data.

IV. EXPERIMENTAL RESULTS

The experimental prototype shown in Fig. 8 is implemented in a 1.2-μm double-poly, double-metal n-well CMOS technology. The chip area is 4 mm $\times$ 4 mm with an active area of 3 mm $\times$ 3 mm. It consists of 10 parallel channels—8 data channels, one timing channel, and one extra dummy channel for testing. It is packaged in a 68-pin LCC package. The

prototype operates with a single 5 V supply. All experiments were done at room temperature.

The DC gain of the channel amplifier was first measured. It agrees well with the calculated values and is shown in Fig. 9.

The bit error rate (BER) performance of the receiver was measured using an HP71600 series tester. The input used for testing was a $2^{15}-1$ pseudorandom bit sequence (PRBS) code. An important aspect of a terminal receiver is the minimum peak-to-peak input voltage required to achieve a fixed error rate at different bit rates. A plot of this is shown in Fig. 10. In order to achieve a BER of 10^{-11}, a minimum signal of about 6 mV$_{\mathrm{p-p}}$ is required for input bit rate of 400 Mb/s and below. This minimum input signal level appears to be limited by the overall noise floor of the receiver. Operation above 400 Mb/s at the same error rates requires progressively larger signal input amplitude because the sample-mode bandwidth and amplifier speed limitations. The maximum bit rate that can be achieved is at 480 Mb/s with an input voltage of 18 mV$_{\mathrm{p-p}}$.

A jitter tolerance measurement was made to evaluate the performance of the clock-recovery circuit. A reference input voltage of 18mV was set to give a BER of 10^{-9} with no external jitter. The input voltage was then increased by 0.5 dB, and external jitter with different jitter frequencies was inserted on the input waveform. The peak-to-peak magnitude of the jitter, at each jitter frequency, was adjusted until the BER was 10^{-9} again. The results for different jitter frequencies are shown in Fig. 11. It is believed that the jitter performance was degraded significantly by a process-related large offset voltage in the channel amplifiers, the result of which is that the DC offset correction circuit could not correct all of the offset voltage. Two taps of the oscillator are available off-chip through a digital output pad driver. The measured rms jitter of the recovered clock at 480 Mb/s with an 18 mV$_{\mathrm{p-p}}$ input was 78 ps. With a 100 mV$_{\mathrm{p-p}}$ input to reduce the effect of the large DC offset, the measured rms jitter was 24 ps.

The capture range of the parallel receiver was about ±500 kHz while the lock range was about ±1 MHz for an input bit rate of 480 Mb/s using random data. When using a training sequence of 1111111100000000, giving alternating 0-to-1 and

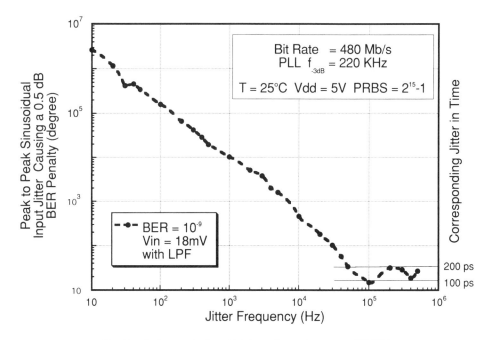

Fig. 11.   Peak-to-peak sinusoidal input jitter tolerance at 480 Mb/s.

T = 25 °C, V$_{DD}$ = 5V, PRBS = 2^{15}-1

Max. Input Bit Rate	480 Mb/s
Min. Input Amplitude for BER = 10^{-11} @ 480 Mb/s @ 400 Mb/s	18 mV$_{p-p}$ 6 mV$_{p-p}$
Output Data Amplitude	5 V$_{p-p}$ CMOS level
Power Dissipation	900 mW
AGC Gain Range	0dB to 40 dB
Chip Size	4 mm X 4 mm
Active Area	3 mm X 3 mm
Technology	1.2 μm CMOS
Package	68-pin LCC

Fig. 12.   Performance summary.

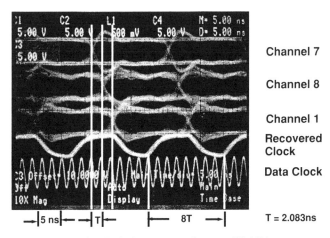

Fig. 13.   Typical output waveforms at 480 Mb/s.

1-to-0 transitions each time for clock recovery, the capture range increases to ±5 MHz because of the increase in effective "gain" of the phase detector due to the higher density of timing transitions.

The overall performance of the experimental prototype is listed in Fig. 12. An oscilloscope picture of some typical output waveforms are shown in Fig. 13. The bottom waveform is the reference clock at 480 MHz used by the pattern generator to transmit the PRBS data. On top of it is the recovered clock from one tap of the ring oscillator, used to synchronize the oscilloscope and also used as the input clock for the bit error rate tester. Note that the oscillation frequency is exactly 1/8 of that of the input data rate. The top three waveforms are typical eye diagrams from channel 7, 8 and 1. The output amplitude is a 0 to 5 volt full CMOS level. The eye opening lasted for

a full 8 input bit period and the eye crossings are staggered by exactly one input bit period.

V.   SUMMARY AND CONCLUSIONS

It has been demonstrated that the parallel architecture can be used for the implementation of high-data-rate optical-fiber receivers with relatively low speed technology. Operation at 480 Mb/s is demonstrated with a 1.2-μm commercially available CMOS technology [9]. Comparison of this result with results of more conventional circuit approaches is difficult because of the different speed capabilities of the wide spectrum of technologies used to implement receivers reported to date.

One way to make relative comparison of different circuit implementations is to normalize the data rate achieved with the device f_T of the technology. This is inherently an imprecise comparison in the case of the MOS technologies because the inherent device speed, as measured by device f_T, is itself a function of biasing point and the way the device is used. However, because most high speed circuits use the NMOS devices in a near velocity-saturated mode where f_T is not a very strong function of bias, and because practical considerations usually dictate the use of device $V_{gs} - V_t$ on the order of 0.5 V, it is possible to assign an approximate f_T value to different linewidth technologies. Using this approach, it is possible to approximately compare the normalized speed performance of published clock recovery circuits. Only clock recovery circuits with on-chip VCO, using phase-locked loops (PLL's) were included in this comparison and they may still have different system requirements.

For virutally all implementations described to date [10]–[16], a device f_T to data rate ratio of at least $12 : 1$ has been used. In one instance, the figure is $8 : 1$, for a highly optimized high-speed NMOS implementation [10]. By comparison, the work reported here achieves a $5 : 1$ ratio of device f_T to data throughput rate. This level of performance demonstrates the potential of parallel approaches for achieving high throughput rates in amplifier/clock recovery functions using parallel data channel architectures.

REFERENCES

[1] R. P. Jindal, "Gigahertz-band high-gain low-noise AGC amplifier in fine-line NMOS," *IEEE J. Solid-State Circuit*, vol. SC-22, no. 4, pp. 512–521, Aug. 1987.
[2] D. Soo, "High-frequency voltage amplification and comparison in a one-micron NMOS technology," PhD dissertation, University of California, Berkeley, 1985.
[3] D. M. Pietruszynski, J. M. Steininger, and E. J. Swanson, "A 50-Mb/s CMOS monolithic optical receiver," *IEEE J. Solid-State Circuits*, vol. 23, no. 6, pp. 1426–1433, Dec. 1988.
[4] Y. Akazawa *et al.*, "A design and packaging techniques for a high-gain gigahertz-band single-chip amplifier," *IEEE J. Solid-State Circuits*, vol. SC-21, no. 3, pp. 417–423, June 1986.
[5] R. Reimann and H. M. Rein, "A single-chip bipolar AGC amplifier with large dynamic range for optical-fiber receivers operating up to 3 Gb/s," *IEEE J. Solid-State Circuits*, vol. 24, no. 6, pp. 1744–1748, Dec. 1989.
[6] Y. Imai *et al.*, "A high-gain GaAs amplifier with AGC function," *IEEE Electron Device Lett.*, vol. EDL-5, no. 10, pp. 415–417, 1984.
[7] K. H. Mueller and M. Muller, "Timing recovery in digital synchronous data receivers," *IEEE Trans. Commun.*, vol. COM-24, pp. 516–531, May 1976.
[8] J. H. Chiu and L. S. Lee, "The minimum likelihood— a new concept for bit synchronization," *IEEE Trans. Commun.*, vol. COM-35, no. 5, May 1987.
[9] T. H. Hu and P. R. Gray, "A monolithic 480 Mb/s parallel AGC/decision/clock-recovery circuit in 1.2-μm CMOS technology," in *ISSCC Dig. Techn. Pap.*, 1993, pp. 98–99.
[10] S. K. Enam and A. A. Abidi, "NMOS IC's for clock and data regeneration in gigabit-per-second optical-fiber receivers," *IEEE J. Solid-State Circuits*, vol. 27, no. 12, pp. 1763–1774, Dec. 1991.
[11] B. Lai and R. C. Walker, "A monolithic extraction data retiming circuit," in *ISSCC Dig. Tech. Pap.*, Feb. 1991, pp. 144–145.
[12] L. DeVito *et al.*, "A 52 MHz and 155 MHz clock-recovery PLL," in *ISSCC Dig. Techn. Pap.*, Feb. 1991, pp. 142–143.
[13] H. Ransijn and P. O'Connor, "A PLL-biased 2.5-Gb/s clock and data regenerator IC," *IEEE J. Solid-State Circuits*, vol. 26, no. 10, pp. 1345–1353, Oct. 1991.
[14] J. Tani, D. Crandall, J. Corcoran, and T. Hornak, "Parallel interface IC's for 120 Mb/s fiber optic links," in *ISSCC Dig. Tech. Pap.*, Feb. 1987, pp. 190–191.
[15] M. Soyuer and H. A. Ainspan, "A monolithic 2.3 Gb/s 100 mW clock and data recovery circuit," in *ISSCC Dig. Techn. Pap.*, Feb. 1993, pp. 158–159.
[16] D-L. Chen and R. Waldron, "A single-chip 266 Mb/s CMOS transmitter/receiver for serial data communications," in *ISSCC Dig. Techn. Pap.*, Feb. 1993, pp. 100–101.

A Monolithic 622Mb/s Clock Extraction Data Retiming Circuit

Benny Lai, Richard C. Walker

In high-speed digital transmission systems, a digital waveform must be regenerated to insure data integrity. This is usually done by extracting a clock directly from the input data stream using a non-linear function with a band-pass filter, and then using this clock to trigger a decision circuit which reshapes the data stream. The band-pass filter is typically a SAW filter or PLL. This method is accepted as the norm for systems above 200Mb/s. However, the design of such a system requires careful alignment of the data relative to the clock over temperature, supply variation, process variations, and age. Also, this approach is relatively expensive since the components are difficult to integrate.

In this report, an architecture based on a different technique is described. This retiming circuit incorporates all the functions of clock recovery, data regeneration, and clock to data alignment. Except for two external capacitors, the circuit is fully integrated. The targeted bit rate is centered at 622.08Mb/s, the standard rate for SONET OC-12.

The overall circuit is comprised of three main blocks, as shown in Figure 1. The first block is a phase/frequency detector, which compares NRZ data to the clock, and produces a binary output. In addition, a built-in decision circuit retimes the data in the center of the data eye. The second block is a charge pump which integrates the detector output. The clock is generated by the third block, a special VCO with one input which sets the center frequency, and another which toggles the VCO between two small but discrete frequency offsets. The output of the detector directly controls the toggling. The integrator output sets the center frequency of the VCO.

The block diagram of the detector is shown in Figure 2. This phase detector is similar to one proposed by Alexander.[1] The input is first split by a low-noise amplifier into flip-flops **B** & **T**. The output of **B** feeds flip-flop **A**. Both **A** & **B** are toggled by CLK, while **T** is toggled by $\overline{\text{CLK}}$. This results in the sampling of the input data at three distinct points; prior to, in the vicinity of, and following each potential transition. If a transition is present, the phase relationship of the data and the clock can be deduced to be fast or slow. The phase detector output is combined with that of a frequency detector, which yields the final detector output. Note that under locked conditions, **T** is always at the edge of the data eye, and that flip-flops **A** & **B** are always sampling the center of the data eye. Therefore, the output of flip flop **A** could be used as the retimed data output, emulating the function of a decision circuit.

The VCO is designed with a ring oscillator, as shown in Figure 3.[2] This ring consists of three variable-delay cells plus a bang-bang delay cell which has an additional small delay inserted on command. The variable delay cell is a continuous interpolation between two cells of different delays. The analog inputs to these cells form the center frequency control for the VCO. The bang-bang delay cell is designed with three sections. The digital input is sliced down to a lower level. This voltage is then converted into current with a transconductance cell, which modulates the bias current of buffer biased below the peak f_T current.

The overall ring is designed to have a nominal delay of about 800ps, corresponding to 622MHz, with an extended range of ±20%. The bang-bang time is about 3ps. The response time of the bang-bang modulation is within 700ps, much less than one clock cycle at 622 Mb/s.

The circuit was laid out using a quick-turnaround prototyping and production tool for gigahertz ICs and was fabricated using the a silicon bipolar process with a 10GHz peak f_T.[3] The chip measures 2.8x2.8mm, and is mounted in a microwave surface-mount package.[4] External 39nF monoblock capacitors are used for the integrator and supply bypassing. Die and package photos are shown in Figure 4.

With V_{EE} = -5.2V and V_{TT} = -2.0V, the total power dissipation is 1.25W. With a PRBS=(2^{23}-1) input at 622Mb/s, the acquisition of the retiming circuit is measured with a time frequency analyzer as shown in Figure 5. The VCO starts at its upper frequency and acquires lock in less than 1ms. The eye diagrams of the input and regenerated output, as well as the recovered clock, are shown in Figure 6. The jitter of the recovered data measured at the eye crossing is shown in Figure 7. The measured rms jitter generated by the bang-bang is 7.4ps, which is less than 0.5% of the eye at 622Mb/s. An integrated clock extraction data retiming circuit operating at 622Mb/s has advantages over traditional designs: tolerance to environmental variations, autocentering of the clock to the input eye, and the lack of adjustments in production. The level of integration also yields the benefits of reduction in size, improved reliability, and the potential for major cost reduction. This architecture can be extended to other bit rates.

Acknowledgments:

The authors thank D. Lee, G. Flower, R. Dugan, J. Wu, J. Chang, C. Zee and V. Ho for valuable help and discussions, and P. Petruno, J. Aukland, C. Yen and T. Hornak for guidance and support. Appreciation is extended to the staff at HP-BDC, especially C. Bacon and G. Smith for processing and CAD support.

References

[1]Alexander, J. D. H., "Clock Recovery from Random Binary Signals", IEEE Electrons Letters, vol 11, pp. 541-542, October 1975

[2]Walker, R. C., "A Fully Integrated High-Speed Voltage Controlled Ring Oscillator", November 1989. U.S. Patent Number 4884041.

[3]Flower, G. et al., "MASTERSLICE II: A Quick Turnaround Prototyping and Production Tool for Gigahertz ICs", in Proc. 1989 Bipolar Circuits and Technology Meeting, pp. 23-26, September 1989

[4]Ellenberger, J., "Packaging Faster Silicon Circuits", Microwave and RF, vol. 27, pp. 121-124, August 1988

Reprinted from *ISSCC Dig. Tech. Papers*, pp. 144-145, February 1991.

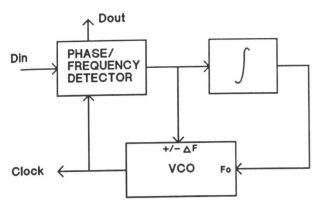

Figure 1: Block diagram of retiming circuit

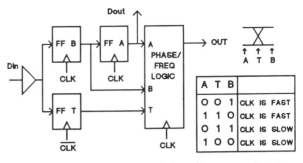

Figure 2: Block diagram of phase/frequency detector

A	T	B	
0	0	1	CLK IS FAST
1	1	0	CLK IS FAST
0	1	1	CLK IS SLOW
1	0	0	CLK IS SLOW

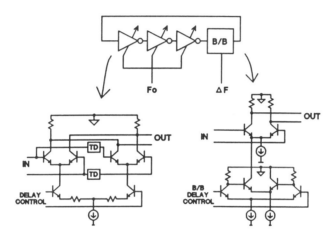

Figure 3: Voltage-controlled ring oscillator

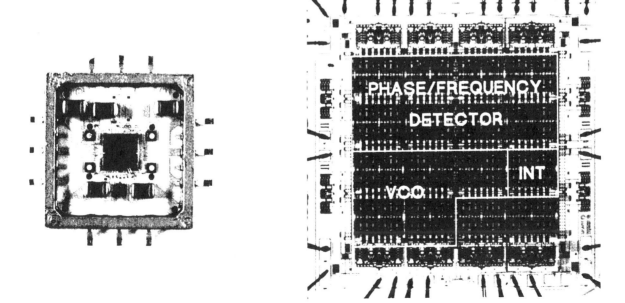

Figure 4: Package and die photos of retiming circuit

445

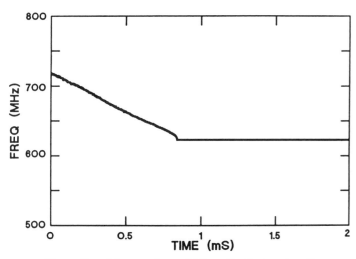

Figure 5:   Measured acquisition of retiming circuit

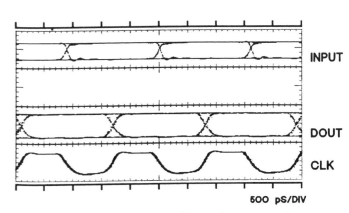

Figure 6:   Eye diagrams of input & output, and
             extracted clock

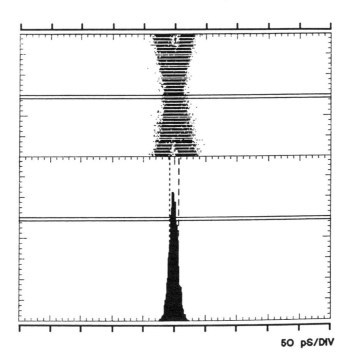

Figure 7:   Jitter histogram of regenerated data

A 660Mb/s CMOS Clock Recovery Circuit with Instantaneous Locking for NRZ Data and Burst-Mode Transmission

Mihai Banu, Alfred Dunlop

Recently, there has been a growing need for high-speed digital communication channels that can handle asynchronous data packets. These burst-mode channels are critical components for an emerging new generation of high-performance systems such as asynchronous transfer mode (ATM) networks, or superfast combinatorial crossbar switches. This high-speed, low-power, fully-integrated clock recovery circuit for these systems locks instantaneously to the first arriving NRZ data transition and at the same time handles data with low-transition densities. This performance is obtained by a broad-band open-loop approach based on matched gated oscillators [1]. The relative standing of this technique compared to conventional nonoversampling methods is summarized in Table 1.

The circuit block diagram is shown in Figure 1. Two matched gated square-wave oscillators (A and B) are started and stopped successively by the data signal. The recovered clock signal is obtained by combining their outputs. This memory-less system has the important property that any accumulating phase errors due to oscillator-frequency/transmission-rate mismatch are discarded every time the oscillators are stopped. The recovered clock signal does inherit all data transition jitter, but it provides the timing information necessary to pass the data to an elastic store or other memory-based interface. Automatic tuning of the two oscillators is accomplished indirectly by allowing a third matched oscillator (C) to run continuously and phase-locking it to an external reference signal.

The scheme is implemented in standard 0.9μm CMOS digital technology (no capacitors available) and is designed for operation at 622Mb/s. The high-speed gated oscillator, based on a five-stage inverter ring, is shown in Figure 2. This circuit has a 450MHz-690MHz output frequency range over all process and commercial temperature variations. The output duty cycle is optimized for increased immunity to input signal jitter. A conventional charge-pump PLL with sequential phase/frequency detector (P/FD) and passive loop filter is sufficient to provide the tuning function. The loop filter is implemented only with transistors used either as resistors (triode region) or as capacitors. For high operating speed, the P/FD digital circuit and the latches are built with pseudo-nMOS gates and dynamic shift registers. Except for the conceptual aspects, the chip design was done automatically on a workstation with almost no manual intervention [2]. A typical design iteration, consisting of layout generation, circuit extraction including parasitics, and circuit simulation, takes a few minutes.

A photograph of the monolithic clock recovery circuit whose area is only 0.3mm² is shown in Figure 3. The chip contains additional 50Ω ECL/CMOS and CMOS/ECL (100K) level translators and drivers (not shown) described elsewhere [3]. At room temperature, the circuit operates properly in the 450-690Mb/s range with a power dissipation of 600mW (at -5V supply, in/out buffers not included). The minimum operating power supply voltage at 666Mb/s is 3.25V. Typical waveforms and eye diagrams at 660Mb/s are shown in Figure 4 and Figure 5. The instantaneous locking to an arbitrary data transition is demonstrated in Figure 6. Here, a sudden input phase shift of

216° was intentionally applied. Oscillator matching to about 99.8% is accomplished as illustrated by Figure 7 where 64 "ones" followed by 64 "zeros" are properly handled. The transition density of this data pattern is only 1.56%. Proper operation is maintained with ±0.35% reference clock frequency error.

References

[1] Banu, M., A. E. Dunlop, "Clock Recovery Circuits with Instantaneous Locking", Electronics Letters, vol. 28, no. 23, pp. 2127-2130, Nov. 1992.

[2] Dunlop, A. E., J. P. Fishburn, D. D. Hill, D. D. Shugard, "Experiments Using Automatic Physical Design Techniques for Optimizing Circuit Performance", Proceedings, IEEE Int. Symp. Circuits Syst., pp. 847-851, May 1990.

[3] Gabara, T. J., "600 MHz 100K ECL Output Buffer Fabricated in 0.9 μm CMOS", Proceedings, IEEE TENCON'92, Nov. 1992.

Narrow-band Systems	Broad-band Systems
Open-loop (e.g. SAW filters)	Open-loop (This work)
-simple, not fully integrated	-simple, fully-integrated, small
-many transitions to lock	-locks on first data transition
-handles only occasional low transition density	-handles low transition density
-high jitter rejection	-no jitter rejection
Closed-loop (e.g. narrow-band PLL)	Closed-loop (e.g. broad-band PLL)
-complex design, circuits difficult full integration	-moderate complexity, can be fully integrated
-many transitions to lock	->5 to 10 transitions to lock
-handles low transition density	-difficulty with low transition density
-high jitter rejection	-low jitter rejection

Table 1: Non-oversampling clock-recovery methods

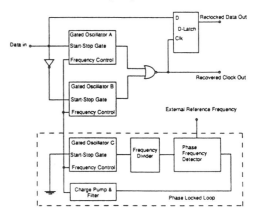

Figure 1: Circuit block diagram.

Reprinted from *ISSCC Dig. Tech. Papers*, pp. 102-103, February 1993.

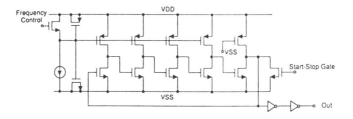

Figure 2: Gated oscillator with frequency control.

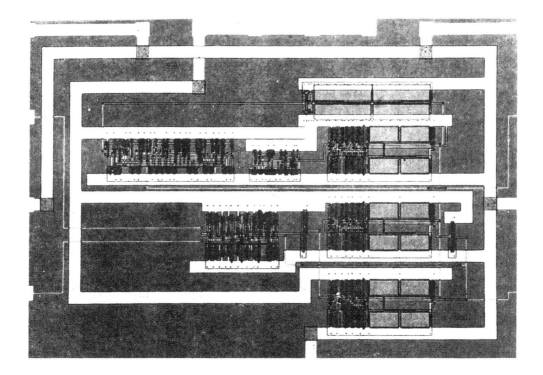

Figure 3: Chip micrograph.

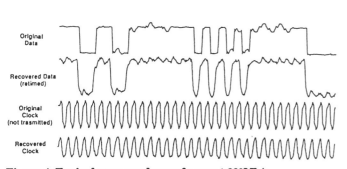

Figure 4: Typical measured waveforms at 660Mb/s.

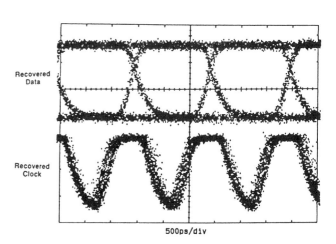

500ps/div

Figure 5: Typical eye diagrams at 660Mb/s
for a 2^{31} -1 pseudo-random input sequence.

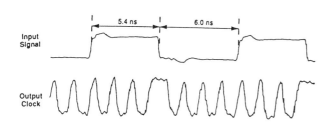

Figure 6: System response for sudden
0.9ns input edge jump; correct period is 1.5ns.

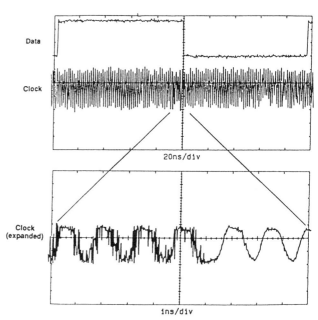

Figure 7: Output signals at 660Mb/s with input 64
consecutive "1s" and 64 consecutive "0s".

A Monolithic 2.3-Gb/s 100-mW Clock and Data Recovery Circuit in Silicon Bipolar Technology

Mehmet Soyuer, *Member, IEEE*

Abstract—A monolithic clock and data recovery PLL circuit is implemented in a digital silicon bipolar technology without modification. The only external component used is the loop filter capacitor. A self-aligned data recovery architecture combined with a novel phase-detector design eliminates the need for nonlinear processing and phase shifter stages. This enables a simpler design with low power and reduced dependence on the bit rate. At 2.3 Gb/s, the test chip consumes 100 mW from a -3.6-V supply, excluding the input and output buffers. The worst-case rms jitter of the recovered clock is less than 14 ps with $2^{23} - 1$ pseudorandom bit sequence.

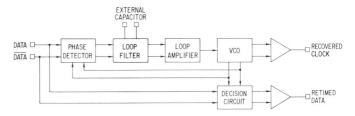

Fig. 1.   Clock and data recovery block diagram.

I. INTRODUCTION

THERE is an increasing need for low-power monolithic circuits that can perform the functions required by multi-gigabit front-end subsystems for fiber and wireless communications. Monolithic implementation of clock and data recovery functions for high-speed fiber-optic data links has been difficult above 1 Gb/s. A recent silicon bipolar receiver chip with integrated PLL reached a 1.5-Gb/s data rate [1], whereas a GaAs IC combined with a silicon bipolar chip performed the clock and data recovery functions at bit rates exceeding 2.5 Gb/s [2]. The power dissipation of these chips is greater than 1 W. A mature silicon technology, such as an existing high-performance digital bipolar VLSI technology, is an excellent candidate for the integration of such mixed-signal functions in low-power, multigigabit receivers and transmitters.

This paper presents a single-chip clock and data recovery PLL implemented in a standard digital silicon bipolar technology without modification. The only external component is the loop filter capacitor. At 2.3 Gb/s, the chip consumes 100 mW from a -3.6-V supply, excluding the input and output buffers. This enables integration of these functions into a larger receiver subsystem while keeping the power dissipation low. To the best of our knowledge, this is the fastest single-chip implementation of clock and data recovery reported to date.

II. CIRCUIT DESIGN

The PLL-based clock and data recovery chip is intended for serial fiber data-link applications above 2 Gb/s. The block diagram of the overall circuit is shown in Fig. 1. It is important to note that the nonlinear processing stage (e.g., a transition

Manuscript received May 24, 1993; revised July 28, 1993.
The author is with the IBM Research Division, T. J. Watson Research Center, Yorktown Heights, NY 10598.
IEEE Log Number 9212566.

detector device) and the phase shifter of conventional schemes (see Fig. 1 in [2]) are absent in this block diagram. This is crucial to designing a simple architecture with reduced dependency on the bit rate and achieving low power at multigigabit/s data rates. The chip is designed to receive NRZ data from a postamplifier, which generates logic level signals from the low-level input. The NRZ input signal enters the chip differentially and is level shifted by a pair of emitter followers. The inputs to these followers are terminated by on-chip 50-Ω resistors. The chip outputs are the retimed data and the recovered clock synchronized to the NRZ input. The PLL portion consists of a relaxation VCO with on-chip timing capacitors, a dual master/slave flip-flop phase detector, on-chip loop filter resistors and a loop amplifier. Another master/slave flip-flop is used for the decision circuit. As already mentioned, this architecture does not require a nonlinear stage preceding the PLL or a phase shifter following the PLL. The nonlinear processing function, which is needed to create a discrete clock frequency from the input spectrum, is inherent to the phase detector design. This enables a simpler design with low power and reduced dependency on the bit rate.

The phase detector and decision circuit designs are based on the same master/slave flip-flop macro circuit. The phase detector uses two of these macros, as shown in Fig. 2. Because both rising and falling data transitions are used, the VCO output jitter is less than that of a design with a single master/slave flip-flop. The static phase error is also smaller when a dual flip-flop is used as the phase detector. This is verified by computer simulations with behavioral models substituted for the PLL blocks. The details of this approach are further discussed in the following section. The phase detector design is actually part of a dual-edge sensitive phase-frequency detector circuit proposed previously [3]. Therefore, with the addition of more logic circuits, it can be converted from a phase and transition detector device into a phase-frequency and transition detector device that still works with random data. The frequency detectors that can work with random data are discussed in [4].

Reprinted from *IEEE Journal of Solid-State Circuits*, vol. SC-28, pp. 1310-1313, December 1993.

450

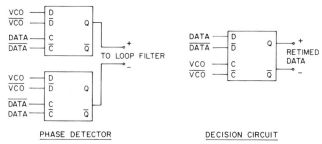

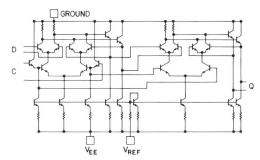

PHASE DETECTOR   DECISION CIRCUIT

Fig. 2.   Phase detector and decision circuit block diagram.

Fig. 3.   Edge-triggered D-type flip-flop.

In this implementation, the falling transitions of the VCO waveform are phase locked to the transitions of the DATA input. Thus, rising VCO transitions can be directly used to sample the DATA waveform in the decision circuit without any phase adjustments. This phase detector and decision circuit combination provides optimum sampling of DATA by centering one VCO edge between DATA transitions.

Differential logic with two-level current steering is used throughout the design of the edge-triggered D-type flip-flops, as shown in Fig. 3. Each master/slave flip-flop macro comprises nine current sources; two for tail currents, six for emitter followers and one for the reference branch. For a typical bias current of 0.8 mA, the power dissipation is 25 mW per D-type flip-flop. The nominal logic swing is 300 mV. The decision circuit outputs feed an open-collector output buffer stage that employs a pair of large devices and can deliver up to 16 mA to a 50-Ω load. A similar buffer stage is also used to measure one of the VCO outputs across a 50-Ω oscilloscope input impedance. Hence, the total power dissipation of the test chip, including all the buffers, can be as high as 250 mW, depending on the test conditions. However, since our intention is to use this architecture as part of a larger receiver chip rather than a stand-alone clock-recovery module, this additional power dissipation can be reduced in a highly integrated receiver design. Therefore, one of our design goals in this work is to minimize the contribution of the clock and data recovery core portion to the total receiver power dissipation so that total power dissipation levels of less than 1 W are feasible.

Fig. 4 shows the loop filter and amplifier. A resistive divider in the passive loop filter reduces the phase detector output by a factor of three. This decreases VCO jitter due to the ripple waveform at the filter output. The dc-to-ac attenuation ratio provided by the filter is 36. Thus, the phase detector output at high frequencies is attenuated by a factor of 108. For an external capacitor of 100 pF, the filter pole and zero are at 0.448 and 16 MHz, respectively.

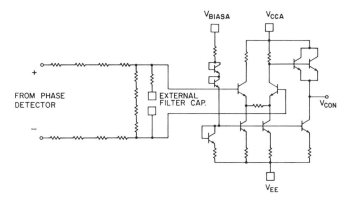

Fig. 4.   Loop filter and amplifier.

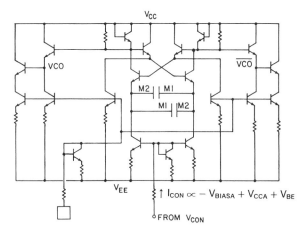

Fig. 5.   Voltage-controlled oscillator circuit.

The loop amplifier is a single-stage differential-to-single ended converter with emitter degeneration. Its output controls the VCO frequency and provides partial compensation against frequency variations caused by supply and temperature fluctuations [5]. The VCO circuit is a relaxation-type multivibrator as shown in Fig. 5. The timing capacitors are implemented with first- and second-level metal as electrodes. The VCO circuit has a maximum tuning range of $\pm 20\%$ when used with the loop amplifier. Its center frequency is adjusted externally by varying the bias current of the loop amplifier. The center-frequency drift with supply and temperature variations is estimated to be less than $\pm 7\%$ over $\pm 5\%$ in V_{EE} and 30–100°C. A frequency-acquisition aid circuit is necessary in the final design to cover the VCO center-frequency variations without frequency trimming [3].

III. CLOSED-LOOP PERFORMANCE AND SIMULATIONS

Phase-detector analysis and design is crucial to the understanding of the closed-loop operation and determining its performance limits. Contrary to the operation of most analog and digital phase detectors with a well-defined linear range around their stable operating point, the digital phase detector used in this design functions as a coarse quantizer. Therefore, additional care must be taken to control the jitter created by the phase-detector output waveform. Some properties of phase detectors with binary quantization have been discussed in the literature [6]–[8]. A simple binary phase detector can easily be implemented using a single

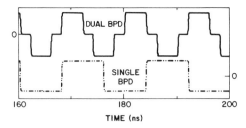

Fig. 6. Phase-detector outputs with three-level (top) and binary (bottom) quantization.

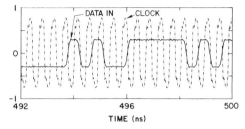

Fig. 7. Simulated PLL waveforms at 2.3 Gb/s.

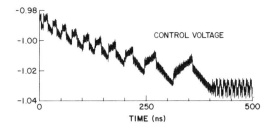

Fig. 8. Simulated capture transient at 2.3 Gb/s.

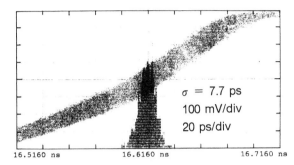

Fig. 9. Jitter histogram of recovered clock at 2.3 Gb/s (8B/10B idle sequence).

D-type flip-flop, as discussed in [6]. Ideally, a phase detector of this type results in square-wave phase characteristics with discontinuities at multiples of π radians. PLL's that use such phase detectors are sometimes referred to as bang-bang loops since they actually do not have any linear region of operation. The loop is highly nonlinear and time varying; hence, time-domain analysis is usually preferred to gain a better insight into its operation. Our experience suggests that an approximate linear analysis (e.g. assuming the phase detector gain is finite but very large) will be of limited value in evaluating such bang-bang loops for most applications.

In general, a single D-flip-flop implementation can be made sensitive to either the rising or the falling edge of the DATA waveform by sampling the VCO with DATA. Since the phase detector output in our design is taken differentially from the two D-type flip-flops sampled by opposite edges of the input data, the phase error signal can have one of three discrete levels, as opposed to two levels used in the single flip-flop implementation. However, the phase detector operation is still highly nonlinear, and a linearized approach is not very useful for closed-loop analysis, similar to the binary phase detector case [6]. This leads to extensive use of computer simulation programs like ASTAP, in which FORTRAN functions can be used for behavioral modeling for the PLL components as mentioned previously. Fig. 6 illustrates some of the points made so far using such models. In this example, a periodic data input of 1111100000 is used at 2.5 Gb/s. The solid and dashed lines show the dual and single D-type phase detector outputs, respectively, when the loop is locked. The VCO and the filter/amplifier characteristics are kept the same in both simulations. It is clear from these outputs that the three-level quantization provided by the dual D-type design will create less pattern-dependent jitter. Actually, in this example, the VCO output jitter is reduced by a factor of two by using the dual D-type design.

Behavioral-model analysis of the closed loop substantially reduces the computer simulation time during the initial phases of the design. However, in order to obtain more precise results,

several device-level simulations were also run on a mainframe computer. Two outputs from a device-level simulation are shown in Figs. 7 and 8. Fig. 7 shows the data and clock waveforms at phase-lock for an 8B/10B code idle sequence at 2.5 Gb/s [9]. The simulated static phase error is 18 ps. This is the average phase offset between the falling VCO edge and the DATA transitions in Fig. 7. The capture transient on the control voltage is shown in Fig. 8. The frequency lock is achieved around 400 ns for an initial offset of about 1%. The simulated capture range is less than $\pm 2\%$. Finally, it is worth noting that the proper operation of this loop does not require the existence of a master or frame transition in DATA, and therefore it is not restricted to a special coding scheme, as opposed to another class of bang-bang loops discussed in the recent literature [1].

IV. EXPERIMENTAL RESULTS

Measurements are made at wafer level. The loop filter capacitor and the bypass capacitors are mounted on a high-frequency probe card. The maximum VCO center frequency is 2.3 GHz when a single -3.6-V supply is used. Measured VCO cycle-to-cycle jitter is 1650 ppm when free running at 23 GHz.

Both pseudorandom data inputs and 8B/10B code patterns are used in the measurements. The frequency acquisition range is found to be close to the estimated range of $\pm 2\%$ with 8B/10B idle sequence. Measured rms jitter of the recovered clock is less than 14 ps with $2^{23} - 1$ pseudorandom data and less than 8 ps with an idle sequence of the 8B/10B code up to 2.3 Gb/s (Fig. 9). The jitter goes down to 2 ps with a 1010 periodic input. Worst-case PLL differential input sensitivity is 85 mV$_{pp}$ for a BER less than 10^{-12} with 8B/10B code patterns. The VCO center-frequency range is extendable to 2.6 GHz using $+0.3$- and -3.6-V supplies across the loop amplifier with negligible increase in the power dissipation. This enables the clock and data recovery circuit to function

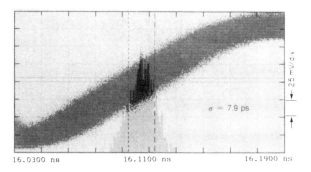

Fig. 10. Jitter histogram of recovered clock at 2.6 Gb/s ($2^7 - 1$ PRBS).

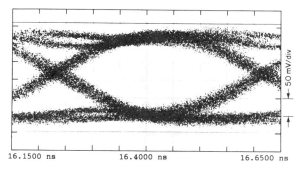

Fig. 11. Eye diagram of retimed data at 2.6 Gb/s ($2^{23} - 1$ PRBS).

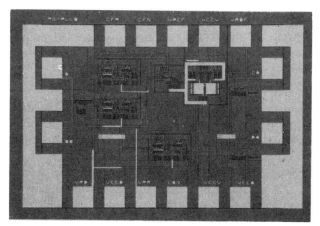

Fig. 12. Test chip micrograph.

up to 2.6 Gb/s. Recovered clock jitter at 2.6 Gb/s is shown in Fig. 10. The rms clock jitter is 8 ps with $2^7 - 1$ pseudorandom data. The data eye diagram when the PLL is locked to $2^{23} - 1$ pseudorandom data input at 2.6 Gb/s is shown in Fig. 11. The peak-to-peak jitter is about 60 ps, with a corresponding recovered clock jitter of less than 14 ps.

The circuit is fabricated using a digital bipolar VLSI process similar to that described in [9]. The ground rules used in this work have resulted in an NPN device with 0.7-μm effective emitter width and peak f_T of about 30 GHz. Table I summarizes the chip characteristics. The chip micrograph is shown in Fig. 12. The test circuit occupies 0.85×1.25 mm^2 and is pad-limited for flexibility in testing.

V. CONCLUSIONS

A high-speed low-power clock and data recovery PLL in a digital bipolar VLSI technology has been presented. The

TABLE I
CHIP CHARACTERISTICS

Data Rate	2.3 Gb/s @ 3.6 V
	2.6 Gb/s @ 3.6/3.9 V
Core Power	100 mW
Recovered Clock Jitter	14 ps rms ($2^{23} - 1$ PRBS)
Retimed Data Jitter	60 ps p–p ($2^{23} - 1$ PRBS)
Diff. Input Sensitivity	85 mV p–p (10^{-12} BER)
Technology	30 GHz Silicon Bipolar

circuit does not need any external high-frequency components and can be combined with more functions in a single chip. It has good jitter performance both with coded and uncoded data inputs. This design approach can be applied to higher bit rates with the existing high-performance digital bipolar VLSI technology. It has been demonstrated that high-speed mixed-signal functions can be realized with a digital bipolar VLSI technology, thereby allowing the integration of high-speed, low-power front-end data communication circuits.

ACKNOWLEDGMENT

The author would like to thank H. A. Ainspan, B. M. Fleischer, and E. Panagiotopoulos for their help, J. F. Ewen, K. Y. Toh, M. Arienzo, and J. E. Kelly III for encouragement and support, and the IBM East Fishkill ASTC for chip fabrication.

REFERENCES

[1] R. Walker *et al.*, "A 2-chip 1.5 Gb/s bus-oriented serial link interface," *ISSCC Dig. Tech. Papers*, pp. 226–227, Feb. 1992.
[2] H. Ransijn and P. O'Connor, "A PLL-based 2.5-Gb/s GaAs clock and data regenerator IC," *IEEE J. Solid-State Circuits*, vol. 26, pp. 1345–1353, Oct. 1991.
[3] M. Soyuer, "Phase-frequency and transition detector device," *IBM Tech. Discl. Bull.*, vol. 33, no. 12, pp. 346–348, May 1991.
[4] D. G. Messerschmitt, "Frequency detectors for PLL acquisition in timing and carrier recovery," *IEEE Trans. Commun.*, vol. 27, pp. 1288–1295, Sept. 1979.
[5] M. Soyuer and J. D. Warnock, "Multigigahertz voltage-controlled oscillators in advanced silicon bipolar technology," *IEEE J. Solid-State Circuits*, vol. 27, pp. 668–670, Apr. 1992.
[6] N. R. Aulet *et al.*, "IBM enterprise systems multimode fiber optic technology," *IBM J. Res. Develop.*, pp. 553–576, July 1992.
[7] J. F. Oberst, "Pull-in range of a phase-locked loop with a binary phase comparator," *Bell Syst. Tech. J.*, pp. 2289–2302, Nov. 1970.
[8] J. R. Cessna and D. M. Levy, "Phase noise and transient times for a binary quantized digital phase-locked loop in white Gaussian noise," *IEEE Trans. Commun.*, vol. 20, no. 2, pp. 94–104, Apr. 1972.
[9] A. X. Widmer and P. A. Franaszek, "A dc-balanced, partioned-block, 8B/10B transmission code," *IBM J. Res. Develop.*, pp. 440–451, Sept. 1983.
[10] T. C. Chen *et al.*, "A submicron high performance bipolar technology," *Dig. Tech. Papers, Symp. VLSI Technology*, pp. 87–88, 1989.

A 50 MHz Phase- and Frequency-Locked Loop

ROBERT R. CORDELL, MEMBER, IEEE, J. B. FORNEY, CHARLES N. DUNN, MEMBER, IEEE, AND
WILLIAM G. GARRETT, MEMBER, IEEE

Abstract—A monolithic phase/frequency-locked loop has been developed for operation at up to 50 MHz. The loop combines wide capture range and narrow bandwidth, making it ideal for timing recovery in digital transmission systems. The 24-pin device features an electronically-tuned voltage-controlled *LC* oscillator and includes the input differentiation and full-wave rectification circuitry required for clock recovery from unipolar nonreturn-to-zero (NRZ) data.

I. INTRODUCTION

A well-known limitation of conventional phase-locked loops (PLL's) is the fact that the capture range is not much larger than the noise bandwidth [1]–[3]. In applications requiring very narrow-band filtering, such as timing recovery in synchronous digital communication systems, this often necessitates the use of an expensive crystal-controlled voltage-controlled oscillator (VCO). The phase- and frequency-locked loop (P/FLL) to be described here provides an economical solution to this lingering problem. Another noteworthy feature of the device is that it is capable of operating at speeds in excess of 50 MHz. It was initially designed for timing recovery in a digital regenerator, so a brief look at this function will put the task in perspective.

II. TIMING RECOVERY IN DIGITAL REGENERATORS

Digital signals are widely used in telecommunications systems for transmission of 24 to over 4000 multiplexed PCM voice channels over twisted pair, coaxial cable, or optical fiber media. These digital transmission systems rely on regularly spaced regenerators to reconstruct the bit stream, making up for loss, dispersion, and noise in the transmission medium. For the purposes of this discussion, we will consider regenerators designed for unipolar nonreturn-to-zero (NRZ) data (i.e., simple binary). The incoming signal is first amplified and equalized, and is then applied to a decision circuit which generally consists of a comparator and a *D*-type flip-flop. A timing recovery circuit extracts timing information from the equalized data signal and supplies a stable high-level clock signal to the decision circuit so that decisions are made at the center of each time slot. Since the retiming information is extracted from the data, such regenerators are called self-timed regenerators. Our discussion here will focus on the timing recovery circuit.

The timing recovery process for unipolar NRZ signaling is illustrated by the sequence of waveforms shown in Fig. 1 [4]. The spectrum of such a signal has a null at the baud frequency. However, reliable timing information exists in the incoming digital signal as the locations in time of the data transitions. If the incoming data is differentiated and full-wave rectified,

then so called "transition pulses" occur whenever a transition in the data occurs. Random data with a mean transition density of 50 percent will produce a transition pulse waveform that looks like a sine wave at the baud frequency with missing portions, and such a pulse stream will have a very strong spectral component at the baud frequency. With the addition of proper bandpass filtering, amplification, and phase shift, the baud frequency component in the transition pulse spectrum can be processed into the required clock signal.

The required filtering operation is often performed with a high-*Q* tank or crystal filter. The pulses excite the filter, and it continues to ring and produce clock even when many pulses are absent.

However, much improved timing recovery performance can be achieved if a PLL is used for the filtering operation. The PLL is a simple feedback system in which phase is controlled. Choice of the loop low-pass filter (LPF) sets the overall bandwidth (i.e., the loop's gain crossover frequency), and very narrow bandwidths are easily achieved by choosing an LPF with a low cutoff frequency.

Since a PLL can lock onto a very small spectral component at the baud frequency, the data coding employed can be essentially unrestricted. Long strings of zeros or ones can be tolerated with almost no effect on timing phase or amplitude. The PLL can easily be designed for effective *Q* factors of greater than 1000, while at the same time tolerating large frequency offsets without producing static phase offset.

While providing superior performance once phase-lock is

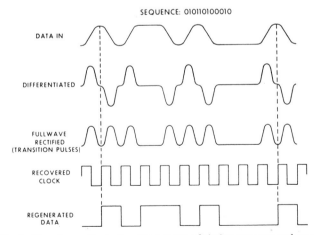

SEQUENCE: 010110100010

DATA IN

DIFFERENTIATED

FULLWAVE RECTIFIED (TRANSITION PULSES)

RECOVERED CLOCK

REGENERATED DATA

Fig. 1. Waveforms illustrating the process of timing recovery and regeneration for unipolar NRZ data. A narrow-band PLL is used to derive the recovered clock from the transition pulses.

Reprinted from *IEEE Journal of Solid-State Circuits*, vol. SC-14, pp. 1003-1009, December 1979.

achieved, the conventional PLL is seriously limited by its small capture range, which is generally less than ten times its closed loop bandwidth. For a bandwidth of 0.1 percent (Q of 1000), this means a capture range of less than ±1 percent. As a result, a VCO with a very precise center frequency is required, such as a crystal-controlled VCO, and these oscillators are expensive.

III. PHASE- AND FREQUENCY-LOCKED LOOPS

If an ideal frequency difference detector (FDD) could be added to a conventional PLL, it could be used to generate an error signal that would tune the VCO toward the incoming frequency and thus enable acquisition to occur over a fairly wide range. The PLL dynamics could then be designed without regard to capture range considerations. Such an arrangement is shown in Fig. 2. Off-the-shelf phase/frequency detectors do exist, but they utilize sequential logic which operates on transitions of the two input signals, and thus assumes sine wave or square wave inputs [5]. Data, by their very nature, have missing transitions and thus will confuse these detectors.

A frequency difference detector which does not utilize sequential logic and is not confused by data has previously been developed at Bell Labs [6]. A second phase detector, called a quadrature phase detector (QPD), is added to the PLL and is driven with the input signal and a quadrature VCO signal, as shown in Fig. 3. When the loop is out of lock, the beat signals generated by the phase detectors will be 90° out of phase, one proportional to the cosine of the frequency difference, the other to the sine. One will lead the other by 90° if the VCO frequency is too high. The reverse will be true if the VCO frequency is too low. This reversal in the phase relationship can be detected by a third phase detector (multiplier) and used to produce a positive or negative signal of the correct polarity to decrease the VCO frequency error.

In practice, both phase detector outputs are quantized, and the one originating from the QPD is also differentiated. The differentiator produces a positive and a negative pulse each

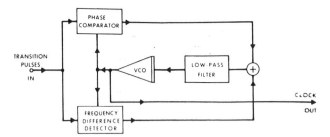
Fig. 2. Block diagram of a P/FLL.

time a cycle is "slipped" (i.e., one cycle of the beat signal). These pulses are then multiplied by the quantized output of the conventional phase detector to produce a unipolar pulse train whose polarity is indicative of the sign of the frequency difference. Two pulses are produced for each cycle slip, so the integrated output is proportional to frequency difference. After lock is achieved, the beat signals disappear and the output goes identically to zero because one input to the multiplier (from the differentiator) is zero. Thus, the FDD is completely out of the circuit after lock is achieved and cannot detract from PLL performance.

Discrete P/FLL's using this circuit have been built and are in use, but they are fairly large and expensive, requiring eight DIP's and over 100 passive components. Integration, however, makes the circuit very attractive.

IV. AN INTEGRATED P/FLL

Our objective was to integrate all of the active circuitry required for timing recovery, including the differentiator and full-wave rectifier (DIFF/FWR) on a single chip. All important operating parameters are set by external passive components, so that a variety of applications between low frequencies and 50 MHz can be accommodated.

Fig. 4 is a schematic block diagram showing the device connected for timing recovery. Operating at a center frequency

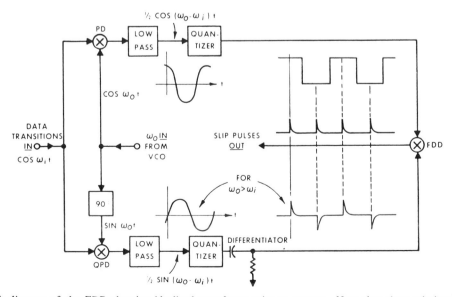
Fig. 3. Block diagram of the FDD showing idealized waveforms prior to capture. Note that phase relationship of beat frequency signals depends on sense of frequency difference.

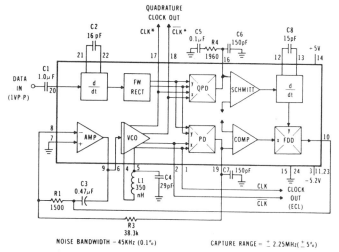

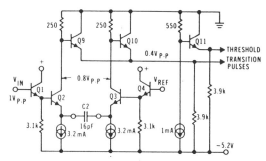

Fig. 4. A 44.7 MHz timing recovery circuit showing the 24-pin DIP P/FLL and its associated passive components.

Fig. 5. The DIFF/FWR circuit, including external differentiating capacitor C_2.

of 44.7 MHz, the loop as shown provides a noise bandwidth of 0.1 percent (44.7 kHz) and jitter peaking of less than 0.04 dB. The capture range is a full ±5 percent (±2.25 MHz). The 24-pin device operates from the standard logic power supplies of +5 and −5.2 V. A 1 V pp data signal is ac-coupled to the self-biased input of the DIFF/FWR, while the output is a balanced emitter-coupled logic (ECL) signal. The VCO center frequency is set by an external LC tank.

The two phase comparators each provide an output at a 2 kΩ impedance level, allowing capacitors C_6 and C_7 to provide the initial low-pass filtering. The lag–lead combination of C_5 and R_4 is optional; the nominal charge on C_5 provides increased immunity to unwanted FDD activity during data intervals containing very few transitions.

The FDD multiplier is arranged to produce current pulses at its output. These pulses are fed directly to the integrating capacitor C_3. The output of the conventional PLL phase detector is fed through R_3 to the inverting input of the op amp. Notice that the op amp is connected as an integrator with a zero added to its transfer function by R_1. The large dc gain of the integrator forces the static phase offset to be essentially that of the phase detector, independent of VCO mistuning. The zero in the transfer function allows the loop to act essentially as a first-order loop in the vicinity of the gain crossover frequency, so as to provide very small jitter peaking.

The Differentiator/Full-Wave Rectifier

Since differentiation and full-wave rectification of a 44.7 Mbit/s digital signal implies the need for very good high frequency response, care was taken to employ a circuit which would provide the function with few stages and without the need for significant amounts of amplification.

The circuit chosen is shown in Fig. 5. Transistors Q_2 and Q_3 form a differential pair whose emitters are connected by the external differentiating capacitor C_2. The input data signal appears at the emitter of Q_2, and a reference voltage appears at the emitter of Q_3. Since the data signal appears across C_2, the current through C_2 is a differentiated version of the data signal. Q_2 and Q_3 thus produce a balanced differentiated version of the data signal at their collectors.

This signal is directly applied to emitter followers Q_9 and Q_{10}, whose connected emitters provide the full-wave rectification. Thus, a positive-going transition pulse is generated at every transition of the input data. A slicing reference in the middle of the transition pulse waveform is provided by Q_{11}. An additional pair of emitter followers and level shift diodes which provide proper input levels for the phase detectors are not shown.

The Phase Detectors

The operation of the two identical phase detectors is similar in principle to that of the popular double-balanced phase detector found in most monolithic PLL's [3]. However, in order to minimize clock feedthrough during periods when there are no data transition pulses, a modification has been made which makes the phase detector go out of the circuit during these periods and results in what is called a tristate phase detector.

The outputs of the conventional phase detector and the tristate phase detector are illustrated in Fig. 6 for idealized data transition and clock inputs. The condition representing zero static phase offset, resulting in zero average dc output from the phase detector, is shown. The double-balanced modulator performs essentially a logical EXCLUSIVE NOR function. This results in direct clock feedthrough when there are no data transition pulses. One undesirable result of this is that any asymmetry in the clock waveform can result in some dc offset at the output of the phase detector, even when no data transitions are occurring.

The output of the tristate phase detector also has positive and negative states, but in addition it has a zero output condition, which occurs when there are no transition pulses. The greatly reduced activity at the output during periods without data transitions is clearly evident.

The schematic of the phase detector is shown in Fig. 7. The actual phase detector consists of Q_{20-23}, which implements a double balanced modulator with the exception that one of the upper differential pairs is replaced by a pair of resistors ($R_{28,29}$). When the transition pulse input is low, this resistor pair routes the current from Q_{23} equally to both output collector circuits, so that zero output results.

The remainder of the phase detector consists of an arrangement of two p-n-p current mirrors and one n-p-n current mirror which implements a balanced to single-ended current conversion. The resulting single-ended current output is then applied to R_{34} to form a voltage.

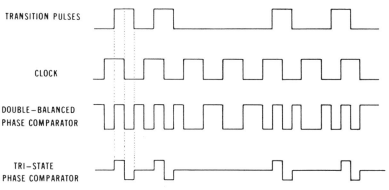

TRANSITION PULSES

CLOCK

DOUBLE−BALANCED
PHASE COMPARATOR

TRI−STATE
PHASE COMPARATOR

Fig. 6. Idealized waveforms illustrating operation of the double-balanced and tristate phase detectors under conditions of zero static phase error (i.e., inputs are in perfect quadrature).

Quantizers and Frequency Difference Detector

For proper FDD operation during acquisition, the beat signal outputs from the phase detectors must each be converted into a rectangular waveform by a quantizer.

As shown in Fig. 8, the output from the phase detector is quantized by a simple comparator realized from a differential pair.

The quantized output of the QPD must feed a differentiator, and this makes fairly sharp transitions on the rectangular waveform desirable so as to achieve good differentiator efficiency. The desired fast transitions are achieved by passing the QPD output through a Schmitt trigger instead of an ordinary comparator. The Schmitt trigger function is achieved by utilizing a differential pair as a comparator and providing a small amount of positive feedback to its "reference" input at the base of Q_{44}. The positive feedback provides fairly fast edges with virtually no hysteresis. Too much hysteresis would make the circuit unable to respond to small beat notes from the QPD. The single-ended output of the circuit is buffered by an emitter-follower (Q_{46}) before being applied to the external differentiator capacitor.

The basic function required of the frequency difference detector is to multiply the differentiated signal from the QPD by the rectangular balanced signal from the phase detector quantizer.

In theory, the FDD output goes identically to zero after acquisition, because there are no slip pulses being differentiated and thus one input to the "multiplier" is zero. However, inevitable dc offsets would ordinarily cause some output under these conditions. In order to avoid the degrading effects of such offsets, the input of the multiplier receiving the differentiated pulses is provided with a "dead zone" centered about zero. In essence, the multiplier is completely shut down unless pulses are present which exceed the boundary of the dead zone. The dc offsets in the multiplier thus cannot introduce any error into the loop.

The circuit has much in common with a traditional double-balanced modulator, however the bottom differential pair found in the conventional design has been replaced by a pair of differential stages acting as comparators, one for positive input excursions and one for negative input excursions. The "reference" input of each of the comparators is set at ± 0.15 V, respectively. The peak value of the differentiated pulses is 1 V. If the voltage at the bases of Q_{57} and Q_{60} is substantially less than 0.15 V in either direction, Q_{58} and Q_{60} will stay "on" and little or no current will be supplied to the upper portion of the circuit (Q_{53-56}). The FDD will thus be essentially shut down.

The balanced output of the multiplier is converted to a single-ended current by three current mirrors as in the phase detectors. A 5K resistor provides a load for this current.

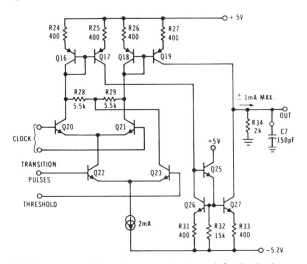

Fig. 7. The tristate phase detector circuit used for both the normal (PD) and quadrature (QPD) phase detectors in the P/FLL.

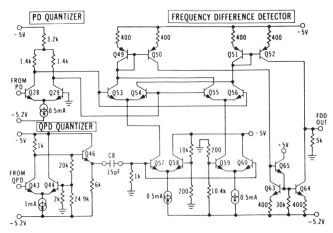

Fig. 8. Circuit diagram of the quantizers and frequency difference detector.

457

Efficient operation at 50 MHz suggests the use of an *LC* oscillator, as opposed to the more conventional emitter-coupled multivibrator [7], [8]. However, voltage control of these *LC* oscillators usually requires the use of a discrete varactor diode. A fully integrated approach to voltage control of an *LC* oscillator has been chosen for this device.

The concept of the voltage-controlled *LC* oscillator is really quite simple. Rather than modifying the center frequency of the tank with a varactor, frequency changes are induced by forcing the oscillator to operate at frequencies away from the center frequency of the tank. This is done by varying the phase shift in the active part of the oscillator, and depends on the fact that an oscillator operates at the frequency where the phase around the positive feedback loop is exactly 0° [9]. If, for example, lagging phase shift is added to the active half of the oscillator, the system will oscillate at a frequency slightly below the center frequency of the tank, causing the parallel-resonant tank to look inductive and, thus, generate an off-setting leading phase shift in the passive half of the system.

A conceptual diagram showing the operation of the oscillator is shown in Fig. 9. A conventional oscillator loop is formed by the path from V_{out}, through amplifier A, and through the parallel-resonant tank consisting of R, L, and C. The active portion of the oscillator is from V_{out} to V_s.

Leading or lagging phase shift is introduced into the active half of the loop by inserting a quadrature signal, controlled in magnitude and polarity by the four quadrant multiplier, back into the loop at the summer. For the moment, the 90° phase shift element can be thought of as a separate network. The frequency control signal, which can be thought of as dc, is applied to the "Y" input of the high-speed four quadrant multiplier. We thus have

$$V_s = A \cdot V_{out} + V_c \cdot jV_{out} = (A + jV_c) \cdot V_{out}$$

$$\theta = \arctan(V_c/A)$$

where θ is the phase shift through the active half of the oscillating loop. If the maximum magnitude of $V_c \cdot jV_{out}$ equals $A \cdot V_{out}$, then a ±45° phase shift can be achieved. Since it is well known that a tank circuit generates ±45° of phase shift at its 3 dB frequencies, it follows that, for this case, the oscillator frequency can be shifted over the range of frequencies within the 3 dB bandwidth of the tank. This corresponds to about ±5 percent in this design, where the tank Q is 10.

Notice that the 90° phase-shifted version of the oscillator output serves the dual purpose of driving the QPD as well as the frequency-control multiplier.

The circuit of the oscillator is shown in Fig. 10. The portion labeled "frequency control" is really just a fairly conventional four quadrant transconductance multiplier. The single-ended control input is applied to the base of Q_{87}, while the single-ended quadrature signal and a dc reference are applied to the two cross-coupled differential pairs. The output of the multiplier, proportional to the product of the quadrature signal and the control voltage, appears at the collectors of Q_{95} and Q_{99}, and is added to the in-phase signal at the collector of Q_{96}.

The portion of the figure labeled "oscillator" consists of a straightforward oscillator combined with circuitry which gen-

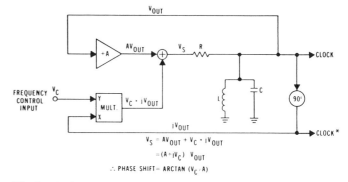

Fig. 9. Block diagram illustrating electronic voltage control of the *LC* oscillator by means of phase shift in the active portion of the oscillator.

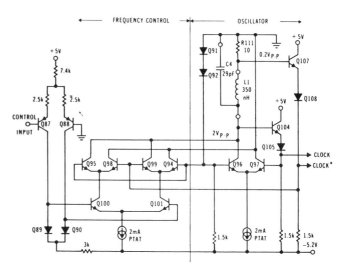

Fig. 10. Simplified circuit of the voltage-controlled *LC* oscillator. ECL output circuitry (clock drivers) is not shown.

erates the quadrature signal for the frequency control circuit. The oscillator consists of a differential pair, Q_{96} and Q_{97}, driving a parallel resonant *LC* tank circuit in the collector circuit of Q_{96}. The oscillating loop is completed by an emitter follower (Q_{104}) and a level-shift diode (Q_{105}), which buffer and route the tank voltage back to the input of the differential pair.

A signal in quadrature with the above-mentioned main oscillator signal appears across R_{111}. This phase shift is achieved because the circulating current in a tank at resonance is 90° out of phase with the voltage across the tank. Because resistor R_{111} sets the tank Q at about 10, the circulating current is approximately ten times greater than the current supplied by the active devices. The voltage produced across the tank is about 2 V pp, while the quadrature signal developed across R_{111} is on the order of 200 mV pp. The quadrature signal is buffered and level-shifted by Q_{107} and Q_{108}, and is then applied to the frequency control multiplier.

In order to provide uniform oscillator performance over temperature, it is desirable that the transconductance of the amplifying devices in the main and quadrature signal paths not be a function of temperature. To achieve a constant transconductance, the current sources have been arranged to provide a bias current which is proportional to absolute temperature (PTAT). Not shown are the clock drivers, which

TABLE I
PERFORMANCE CHARACTERISTICS OF THE P/FLL IC.
NOISE BANDWIDTH IS FOR THE CONNECTION SHOWN IN FIG. 4.

f_{MAX}	> 50 MHz
f_0 TOLERANCE	$< \pm 1\%$
T.C. (f_0)	$< \pm 100$ PPM/°C
CAPTURE RANGE	$\pm 5\%$
NOISE BANDWIDTH	0.1%
SUPPLY VOLTAGE	+5, −5.2 V
POWER DISSIPATION	< 600 mW

merely consist of differential pairs driving emitter followers so as to produce the ECL clock signals.

V. PERFORMANCE

Table I shows some key performance characteristics of the P/FLL. In particular, notice the good oscillator tolerance and temperature coefficient. These are directly attributable to the use of an *LC* oscillator design; a multivibrator simply cannot perform this well as these frequencies [8]. The capture range is limited by the ±5 percent sweep range of the VCO; however, the FDD will produce a reliable error output for frequency differences of greater than ±15 percent. Although the device is rated for a maximum operating frequency of 50 MHz, reliable operation at 70 MHz is typical. The DIFF/FWR appears to be the present limitation on top frequency. There is no fundamental low-frequency limitation, but the oscillator tank inductor becomes a significant expense at frequencies below about 1 MHz. The bandwidth shown is for the connections illustrated in the timing recovery application (Fig. 4), and is set by external components.

Fig. 11 shows the complete process of timing recovery and regeneration using this chip, a comparator, and a *D*-type flip-flop. The circuit is being fed a fixed word pattern at 44.7 Mbits/s. From top to bottom we see the input data, which has been passed through a simulated equalized line network, the differentiated and full-wave rectified data, the recovered clock, and the regenerated data.

VI. TECHNOLOGY

The device is made with a 300 MHz complementary bipolar process. The vertical p-n-p's available in this process provided an advantage in the phase detector current mirrors, where beat frequency signals as high as 5 MHz must be passed, and in the op amp where a small unity-gain compensating capacitor could be used.

Fig. 12 is a photomicrograph of the P/FLL integrated circuit. The chip is beam-leaded, with a total of 40 beams, and is approximately 101 mils on a side. Great care was taken in the layout to optimize high-frequency performance. In particular, extra isolations and wide power buses are used to minimize crosstalk. Also, most transistors in critical high-frequency paths use a double base contact to reduce base resistance and assure good high-frequency performance. Both ion-implanted 200 Ω/square base and 2 kΩ/square resistors are used.

The ceramic interconnect pattern to which the beam-leaded chip is bonded provides an interesting advantage of the beam-leaded technology: we were able to provide many of the clock,

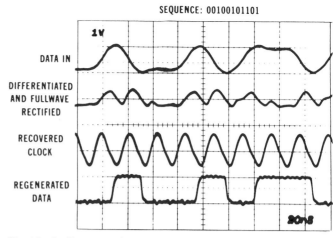

Fig. 11. Oscilloscope photograph illustrating timing recovery and regeneration from a fixed-word pattern at 44.7 Mbits/s using the P/FLL IC in combination with a comparator and a *D*-type flip-flop.

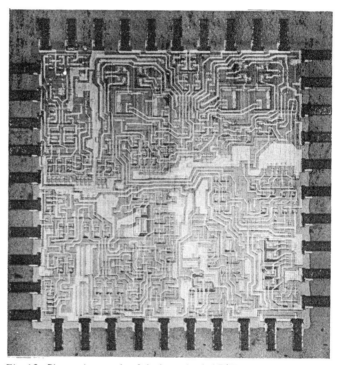

Fig. 12. Photomicrograph of the beam-leaded P/FLL integrated circuit.

data, and power lead interconnections on the ceramic instead of on the chip. This greatly eased the layout.

VIII. CONCLUSION

The design of a monolithic P/FLL capable of operating at frequencies in excess of 50 MHz has been presented. The use of a unique frequency difference detector, made economically attractive by integration, allows the device to provide an inexpensive solution to the lingering problem of limited capture range in narrow-band PLL applications. The P/FLL will find initial application as a timing recovery circuit in an optical fiber digital transmission system operating at 44.736 Mbits/s [4].

References

[1] F. M. Gardner, *Phaselock Techniques*. New York: Wiley, 1966.

[2] A. J. Viterbi, *Principles of Coherent Communication*. New York: McGraw-Hill, 1966.

[3] Signetics, *Phase-Locked Loop Applications Book*. Signetics, 1972.

[4] T. L. Maione, D. D. Sell, and D. H. Wolaver, "Practical 45 Mb/s regenerator for lightwave transmission," *Bell Syst. Tech. J.*, vol. 57, pp. 1837–1855, July–Aug. 1978.

[5] MC404 Phase-Frequency Detector Data Sheet, Motorola Inc., 1972.

[6] J. A. Bellisio, "A new phase-locked loop timing recovery method for digital regenerators," in *IEEE Int. Conf. Communications Rec.*, vol. 1, June 1976, pp. 10–17, patent 4 015 083.

[7] R. R. Cordell and W. G. Garrett, "A highly stable VCO for application in monolithic phase-locked loops," *IEEE J. Solid-State Circuits*, vol. SC-10, pp. 480–485, Dec. 1975.

[8] E. N. Murthi, "A monolithic phase-locked loop with post detection processor," *IEEE J. Solid-State Circuits*, vol. SC-14, pp. 155–161, Feb. 1979.

[9] L. A. Harwood, "An integrated one-chip processor for color TV receivers," *IEEE Trans. Consumer Electron.*, vol. CE-23, pp. 300–310, Aug. 1977.

NMOS IC's for Clock and Data Regeneration in Gigabit-per-Second Optical-Fiber Receivers

S. Khursheed Enam and Asad A. Abidi, *Member, IEEE*

Abstract—The design and performance of two essential analog circuits in optical-fiber receivers is described. A time-interleaved decision circuit is capable of regenerating 35-mV nonreturn-to-zero (NRZ) data inputs to full logic levels at 1.1 Gb/s with 10^{-11} bit error rate (BER), and a phase-locked loop (PLL) extracts the clock from a 2^{23} long pseudorandom sequence at 1.5 Gb/s with 13-ps rms jitter. The two circuits have been implemented as 1-μm NMOS IC's, and in their core area dissipate 200 and 350 mW, respectively.

I. INTRODUCTION

HIGH cost has held back the widespread use of optical-fiber links for the serial transmission of data at gigabit-per-second rates. Although fiber cable itself is relatively cheap, connectors, couplers, and the interface electronics remain expensive. The high cost of electronics may partly be attributed to expensive IC fabrication processes, and the remainder to the high-frequency packaging and assembly of the many parts required in an interface chip set. These costs are expected to come down when similar advances to those realized in telecommunications IC's are obtained, namely high levels of integration that reduce the number of chips and eliminate the required discrete components, and the use of widespread, low-cost IC processes that can easily attain economies of scale.

Much of the work in recent years on optical-fiber receivers and transmitter circuits has used high-speed compound semiconductor or silicon bipolar IC technologies to demonstrate operation at gigabit-per-second data rates. In most of this work, performance at the ultimate speed took precedence over cost. Although exotic technologies will always be able to attain higher frequencies of operation than more mature production-oriented processes such as silicon MOS, for operation at around 1-Gb/s data rates where all these technologies may compete, the latter may ultimately have advantages in density of integration and simplicity of use. Motivated by this, and building on a steady accumulation of demonstrations (mostly by AT&T Bell Laboratories) of the use of scaled NMOS for optical-fiber interfaces, we demonstrate in this work new

Manuscript received May 21, 1992; revised August 27, 1992. This work was supported by AT&T Bell Laboratories and the State of California MICRO Program.

The authors are with the Integrated Circuits and Systems Laboratory, Electrical Engineering Department, University of California, Los Angeles, CA 90024-1594.

IEEE Log Number 9204141.

architectures and NMOS realizations of circuits for regeneration of clock and data signals in gigabit-per-second optical-fiber receiver electronics [1]. These have so far remained the most challenging analog circuits in this application, and have only been demonstrated as integrated solutions in GaAs [2] and silicon bipolar [3] technologies.

II. GIGABIT-PER-SECOND CIRCUITS IN SCALED MOS

It continues to surprise many even today that silicon MOS circuits are capable of operation at gigahertz bandwidths. This section gives a perspective on the development of scaled MOS technology for high-speed applications, and summarizes some key circuit results pertaining to the optical-fiber interface.

A. Technology

Efforts to scale down the dimensions of MOSFET's in the 1970's to attain higher densities of integration also led to the realization that the frequency of operation could be scaled up beyond gigahertz at channel lengths approaching 1 μm. The transit time of a carrier from source to drain in a MOSFET determines the maximum frequency at which voltage gain may be attained, based on transconductance charging intrinsic FET capacitance. Electrons move at drift saturation velocity ($v_{dsat} \cong 10^7$ cm/s) across most of the inversion layer when FET's with a channel length of about 1 μm are subject to $V_{DS} > 1$ V. The transit time is then 10 ps, implying an f_T of 16 GHz.

A comprehensive technology development program was undertaken to realize ''fine-line'' MOS devices and circuits [4]. This included advances in lithography, etching, device design, and so on. The major parasitics in fine-line MOSFET's remain the capacitances of the source and drain junctions to substrate. The measured f_T of fabricated 1-μm NMOSFET's was found to be 8 GHz, within a factor of 2 of the prediction from a simple model assuming an entirely velocity saturated channel. This gives rise to the prospect of silicon MOS analog and digital circuits operating with bandwidths and clock rates exceeding 1 GHz.

B. Previous Work

Aside from the customary ring oscillators used to measure the ultimate switching speed of a logic inverter, several high-speed NMOS circuits have been built for the

Reprinted from *IEEE Journal of Solid-State Circuits*, vol. SC-27, pp. 1763-1774, December 1992.

optical-fiber application. A low-noise transresistance amplifier with a bandwidth of almost 1 GHz was developed for photocurrent amplification, accompanied by a variable gain amplifier with a 30-dB range at the same bandwidth [5]. A data regeneration circuit capable of operation up to 750 MS/s was described [6]. A multiplexer/demultiplexer set operating at 3 Gb/s was implemented [7]. Laser driver IC's operating at rates exceeding 1 Gb/s have also been demonstrated [8]. These circuits together constitute most of the high-speed analog portions of optical-fiber interfaces.

The maximum speed of operation depends on whether the circuit is driven by large or small signals. In a laser driver or data multiplexer, for example, higher toggle rates are possible than the −3-dB bandwidth of a small-signal amplifier because these circuits are driven by large signals, often with voltage swings approaching the power supply. Overdriven circuits such as these may often be required to operate at a near-unity voltage gain. A data communications receiver consists of a mixture of small- and large-signal circuits. Although built in the same technology, the highest frequency of operation of the large-signal portions, such as the multiplexer or laser driver, may exceed the highest possible frequency of the small-signal circuits with high gain, such as the wide-band photocurrent amplifier.

III. CLOCK RECOVERY

A. The Problem

A digital communications receiver must retime the recovered data in the received waveform. The recovered clock is then used to synchronize the data regenerator. In a serial data link, a periodic signal must be recovered from a succession of arriving pulses spaced apart by random multiples of a unit bit period. Passive resonators tuned to the known data rates of transmission may be used to recover the clock. At data rates of 1 Gb/s, SAW resonators are most often used to extract the underlying clock from the data stream. The data pulses stimulate resonant oscillations at the tuned frequency.

A clock recovery scheme relying on a high-Q passive resonator results in a low jitter in the recovered clock, but it also suffers from some disadvantages. A large power is dissipated in driving signals off-chip and into the low resonator impedance. Waveforms at the full data rate can cause stray coupling into nearby signal and power supply lines. The resonator typically has a large unknown group delay, which must be compensated with a manually adjusted off-chip delay line to obtain the optimum clock edge to strobe the data regenerator. This scheme cannot accommodate drifts in the group delay with temperature and aging, or clock recovery over a range of data rates.

A frequency- and phase-locked loop (FPLL), on the other hand, can circumvent all these problems [9]. The signal can be kept entirely on chip, and the local voltage-controlled oscillator (VCO) does not require an external

tuning element, such as a resonator, inductor, or varactor. The clock may be recovered from data rates spanning the entire frequency range of the VCO, which can also absorb drifts in the clock recovery circuits with temperature and power supply. A phase-locking action may be used to deliver the optimal strobe to a data regenerator on chip, because propagation delays over interconnects on chip are usually a negligible portion of a clock cycle.

There are several design challenges in implementing an on-chip FPLL. A high-frequency VCO is required. A VCO without external tuning elements will inevitably have a large phase noise, particularly due to flicker noise in FET's, so means must be developed to prevent this noise from producing a large jitter in the recovered clock. A phase detector to function with random data must be developed that suppresses pattern-induced jitter in the recovered clock. Finally, a phase alignment scheme must be included to provide the optimum strobe to the data regenerator.

B. Architecture

Once lock has been acquired to the data stream, an FPLL must minimize the jitter induced in the clock by the random data stream. There are two aspects to this problem.

1) Stable Lock to NRZ Data: Data in an optical fiber channel are most often encoded in nonreturn to zero (NRZ) format to obtain the highest throughput within a given channel bandwidth. A PLL using a conventional phase detector, or for that matter a passive resonator, cannot stably recover the clock from this waveform because NRZ data contains no spectral energy at the baud rate. The waveform must be rectified, or somehow subject to nonlinear processing, to induce a spectral component at the baud. The resonator or VCO will lock to this frequency. In the context of a PLL with a conventional phase detector, transitions in the unrectified waveform may induce jumps from a stable to unstable equilibrium and *vice versa* (Fig. 1). A jump to an unstable equilibrium will cause the phase to slip by a cycle to return to the stable equilibrium. The data pattern thus induces jitter, or perhaps even loss of lock. The same data presented to the loop after rectification, or in RZ format, will always result in lock at a stable equilibrium (Fig 2).

Rectification was obtained in this implementation by an analog multiplication of the data waveform by itself, which has the effect of creating unipolar pulses at every transition of the NRZ data. The rationale for this choice will be described in Subsection C.

2) Maintaining Stable Lock in Absence of Transitions: In response to a string of data 0's that produces no transitions at the PLL input, the phase detector will also produce *zero* output. If now in the presence of data transitions the phase detector must produce a *nonzero* output to lock the VCO, a pattern-induced jitter will appear in the recovered clock. To suppress this jitter, a loop filter requiring a zero phase-detector output to lock the

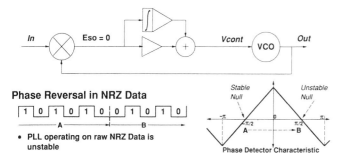

Phase Reversal in NRZ Data

| 1 | 0 | 1 | 0 | 1 | 0 | 0 | 1 | 0 | 1 | 0 |

- PLL operating on raw NRZ Data is unstable

Fig. 1. NRZ data applied to PLL input causes pattern-dependent jumps from stable to unstable operating points of feedback loop.

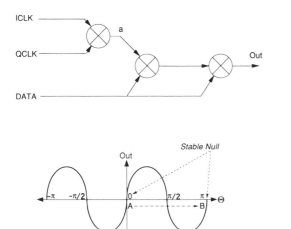

Fig. 2. Data rectified after multiplication with itself, when compared with a rectified clock, maintains loop operation at a stable operating point.

VCO frequency to the data clock must be used to maintain the same steady-state conditions in the presence or absence of data transmissions. If the low-pass loop filter maintains a constant input to the VCO, the loop will produce the same output in both the absence or presence of data transitions, and a steady clock will be recovered.

Loop lock with zero phase-detector output over the entire VCO range may only be guaranteed with the use of an integrator in the loop filter. An integrator can hold a constant output with zero input. A direct path will also be required in parallel with the integrator to let feedback action in the loop correct for noise or short-term drift in the VCO. Therefore, a proportional plus integral (PPI) loop filter is best suited to this application.

A possible source of offset, and thus jitter, arises if the duty cycle of the VCO oscillation is not equal to 50%. In the absence of data transitions, this may result in a nonzero phase-detector output. It is desirable, therefore, to shape the VCO waveform applied to the phase detector such that the output of that detector is guaranteed zero in the absence of data transitions. Since the data are multiplied by themselves at one input of the phase detector, the VCO output must be similarly multiplied by itself to produce a double frequency. If this multiplication is carried out on two VCO waveforms in *exact quadrature*, the output will be zero in the absence of data. A voltage-con-

trolled phase shifter enclosed in a master loop containing a replica of the main phase detector and an integrator may be used to drive the two VCO outputs into precise quadrature (Fig. 3). The other inputs to the master phase detector are connected to zero signal voltage. The VCO clock phases in the actual PLL are then slaved to this master loop.

C. Circuit Design

1) Voltage-Controlled Oscillator: A gigabit-per-second clock regeneration PLL requires a VCO capable of operation to frequencies near 1 GHz. The VCO need not have a large range of oscillation frequency, but its f–V transfer function should be fairly linear to maintain a constant loop gain in the PLL over its lock range. MOS VCO's capable of operation at gigahertz frequency have been described in the past. Notable is a tuned VCO employing an off-chip inductor [10] with varactor tuning, and a relaxation-type VCO employing parasitic capacitance as the timing element [11]. The LC tuned oscillator has the potential of very high frequencies of oscillation, but suffers from the disadvantages that the high-frequency signal must be taken off chip, and that the sweep range is limited by the characteristics of the varactor. The relaxation oscillator affords a wide sweep range, but its f–V characteristics may be unacceptably nonlinear, and it may not offer the signal swings desired for the subsequent circuits.

We have developed a new class of voltage-controlled ring oscillator that exploits some features unique to FET's for frequency control, although a similar technique in the current mode has been described recently for tuning an LC bipolar transistor oscillator [12]. Fixed frequency ring oscillators consisting of a delay line of inverters in negative feedback are of course well known in the circuit art; they have been converted to VCO's by using RC delays [13] or pull-up and pull-down current sources [14] controlled by a voltage. This technique, however, tends to compromise the maximum attainable oscillation frequency, because the voltage-controlled elements introduce a capacitive load per stage in the delay line. The circuit presented here may be called a *delay interpolating VCO* [15], [16]. The outputs at two taps from an otherwise unloaded inverter delay line are interpolated with MOSFET's used as voltage-controlled resistors to obtain the feedback signal (Fig. 4). On the one extreme, with $M1$ fully ON and $M2$ OFF, the shortest delay is obtained in the ring oscillator, which defines the maximum oscillation frequency; on the other, with $M1$ OFF and $M2$ ON, the largest delay results in the minimum frequency. Proper operation requires that the first tap be taken after an odd number of inverters in the delay line, followed by a second tap after an even number, and that the delay between the first and second tap $(T_2 - T_1)$ should not exceed the delay from the input to the first tap (T_1). The first condition, by forcing an odd number of inverter delays in the feedback loop at the two extremes of interpolation, ensures that the circuit always oscillates, while the second

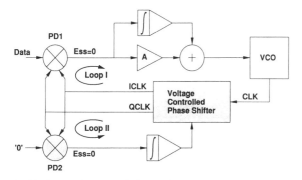

Fig. 3. Clock waveforms applied to main PLL (Loop I) input are maintained in precise quadrature through phase control exerted by master loop (Loop II).

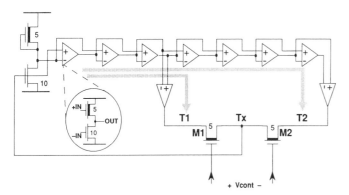

Fig. 4. Delay interpolating voltage-controlled ring oscillator. Feedback signal (T_x) is derived by interpolation of signals tapped from inner loop (T_1) and outer loop (T_2).

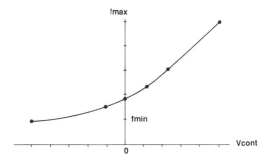

Fig. 5. Typical frequency–voltage characteristic of delay interpolating VCO.

condition suppresses multiple modes of oscillation coexisting in the loop. A delay variable by somewhat less than 2 : 1 is obtained, implying the same range of variable frequency. Push–pull NMOS inverters are used throughout in the inverter delay line, and depletion MOSFET's which do not turn off at large voltage swings as the voltage-controlled interpolation resistors. A differential voltage controls the interpolation resistors. The maximum frequency of oscillation is set by five inverter delays, the minimum by nine delays (Fig. 5). A higher range of frequencies may be obtained by using a shorter delay line, and the requirement for an odd number of inverter delays may be circumvented by tapping antiphase signals from even numbers of differential delay elements.

A delay-line oscillator also affords the possibility of tapping off multiple phases of the oscillation. This is best illustrated by the following circuit.

2) Voltage-Controlled Phase Shifter: The voltage-controlled phase shifter resembles the VCO in that it contains a delay line of NMOS inverters, here implemented as quasi-differential circuits, with voltage-controlled interpolation between three taps on the delay line (Fig. 6). The VCO output is injected into one end of the delay line. The phase difference of the two interpolated outputs from the taps may be continuously swept with the differential control voltage until exact quadrature is obtained, as sensed by the phase detector in the master control loop.

These are referred to as the in-phase (I) and quadrature (Q) clocks.

3) Phase Detector: The phase detector plays a critical role in determining the purity of the clock recovered from the received data. It must foremost be capable of operation at twice the data bandwidth, as is required to recover the clock from an NRZ data stream. At the intended data rates, this rules out the use of a digital or regenerative phase detector. An analog phase detector in a wide-band configuration must be used.

The NRZ data are doubled in frequency, or equivalently, rectified, by multiplication of the analog waveform by itself. Then, this is compared in phase with the frequency doubled clock obtained by multiplying the I and Q clock waveforms emerging from the phase shifter. The phase comparison is itself an analog multiplication. Thus, a quadruple frequency component emerges at the output of the phase detector, only whose average value is of interest. This repeated doubling of frequencies may be sustained in a current-mode circuit operating at a low-impedance level, where node capacitances do not substantially attenuate the high-frequency signal. An NMOS version of a four input analog multiplier has been used here (Fig. 7) to implement this previously described signal flow (Fig. 2).

The fully balanced multiplier consists of a stack of six FET's, all of which operate in saturation in a 5-V power supply owing to short-channel effects. The lower two levels are driven by the two large-signal VCO outputs in quadrature, and the upper two by level shifted copies of the small data signal. The differential current waveform at every level is the product of the signals applied to that level and all the ones below it. The differential current at the fourth level, I_{out}, is converted to a voltage at the high-impedance-load FET's, and as the average value is of interest, the low-pass filtering by node capacitance at this point is only desirable. The two clock phases applied to the lower levels of the phase detector are in exact quadrature when the average value of the differential output voltage is zero in the absence of data transitions. Thus, a replica phase detector consisting of only these two levels driven by the I and Q clock phases is used in the master control loop (Fig. 8). Its output after integration servos the phase shifter to produce a steady-state average value of zero. The FET's in the replica detector are biased at

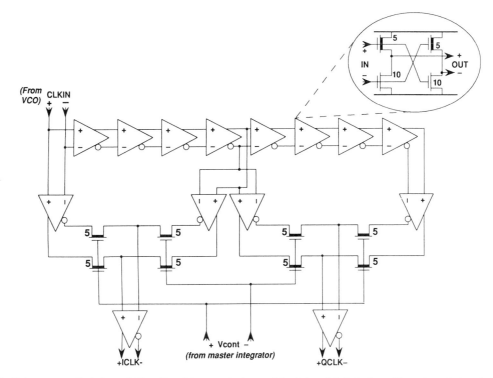

Fig. 6. Voltage-controlled phase shifter. V_{cont} drives apart phases of the interpolated clocks until exact quadrature is obtained.

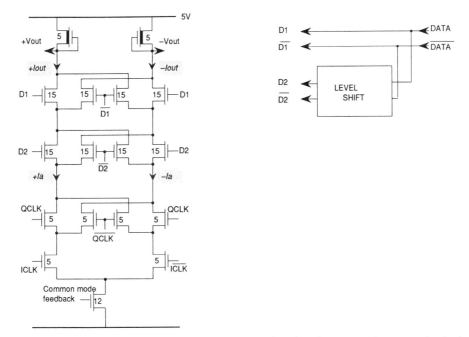

Fig. 7. Analog phase detector. Inputs D_1 and D_2 are level-shifted copies of the input data. Phase correction is obtained from the average value of V_{out}.

the same common-mode voltage as the FET's in the main detector to match the gain and signal delays in both circuits.

The exact gain of the phase detector depends on the transition wave shapes and amplitudes of the input signals, and cannot be expressed by a simple analytic expression. Roughly speaking, if all input signals to the detector are large enough to switch the current fully from one leg of a differential pair to the other, the output signal will

consist of the tail bias current switched with a duty cycle determined by the relative phases of the input and clock waveforms, and by their transition times. Loop gain was predicted during design using approximations and simulations.

A separate frequency detector was implemented on-chip with I and Q channels following a well-known configuration [17] to aid in the acquisition of an input frequency away from the free-running VCO frequency. The detector

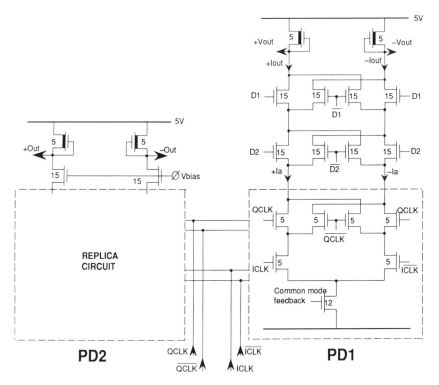

Fig. 8. Replica phase detector (PD2) used in master loop consists of the lower stages of the main phase detector (PD1), and is similarly biased.

D. Loop Dynamics

produces pulses with a polarity which drives the VCO towards the input frequency, and once lock is acquired the detector produces zero output. The outputs from the phase and frequency detector are then summed to drive the VCO.

D. Loop Dynamics

The dynamics of the loop determine its tracking characteristics in the presence of noise, and are mainly determined by the loop filter.

The PPI filter was implemented by two parallel paths in the circuit. The phase-detector output voltage was applied to an off-chip op-amp-based integrator, and also fedforward on-chip for the wide-band proportional path to detect rapid changes. The two differential voltages were summed together on-chip in a differentially driven NMOS unity-gain push–pull stage. The integrator in the master loop was an identical op-amp-based circuit. Thus, the master and slave loops will differ only to the extent of op-amp offsets, and mismatch between the main and replica phase detectors.

The closed-loop bandwidth of the PLL was set by a compromise between conflicting requirements for low-noise clock recovery. On the one hand, the loop must be narrowband when tracking to reject pattern-dependent jitter produced by intersymbol interference in the data. On the other hand, as the PLL uses a MOS local oscillator, a wide closed-loop bandwidth will assist in suppressing broad-band phase noise in the oscillator. A compromise was struck by setting the -3-dB closed-loop bandwidth

at about 10 MHz. Finally, peaking in the closed-loop frequency response at the band edge is undesirable because it will tend to enhance pattern-induced jitter at those frequencies, which can accumulate in a chain of transceivers. In this second-order loop containing two integrators, one in the filter and one due to the integrating action of the VCO in converting frequency to phase, peaking may be produced by parasitic poles near the band edge.

The voltage-controlled phase shifter has some useful consequences in the small-signal loop dynamics. The integrator in the master servo circuit will hold the delay in the phase shifter constant over small-signal variations of the input. Thus, fluctuations in the VCO frequency will produce a departure from quadrature in the phase of the I and Q clock signals, making the phase detector into a frequency-to-phase converter. This action may be modeled by a feedforward branch with gain K_p across the VCO integration block, which after a straightforward linear feedback analysis of the loop results in a left-half s-plane zero at $-1/K_p$ in the loop gain (Fig. 9(a)). The feedforward signal sums into the return path of the feedback loop, and therefore does *not* introduce a zero in the closed loop transfer function. Owing to the large dc loop gain introduced by the integrator, the poles of the closed loop system will come to rest on the real axis of the s-plane (Fig. 9(b)), and the feedforward zero will tend to suppress high-frequency peaking due to parasitic poles. The loop bandwidth will be set by the second, higher frequency closed-loop pole, and any frequency difference between the first pole and the low-frequency zero (z) will produce a slight peaking in the loop passband. Thus, a wide-band loop is

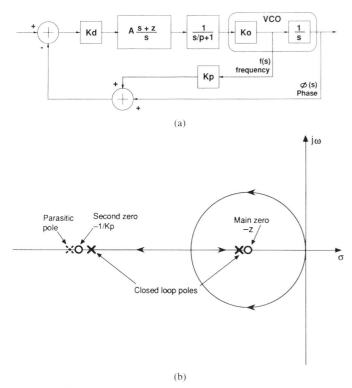

(a)

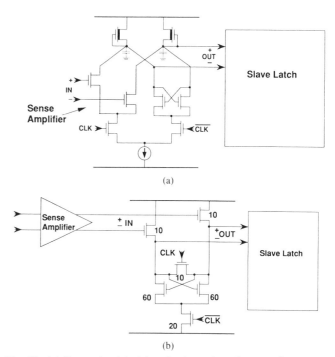

(a)

(b)

Fig. 9. (a) Small-signal model of PLL, including phase shifter. (b) Root locus of this feedback loop, showing final positions of closed-loop poles and zeros.

Fig. 10. (a) Conventional decision circuit consists of a merged sense amplifier and latch. (b) Improved decision circuit continuously couples amplified signal, and improves resetting with shorting switch.

obtained with negligibly small jitter peaking. A similar idea implemented slightly differently has been described to suppress jitter peaking in a narrow-band clock recovery loop [17].

IV. DATA REGENERATION

A. The Problem

The analog waveform after preamplification and AGC must be converted into data bits. The timing of these bits occupy is determined by the recovered clock. The analog waveform may be converted to logic levels either by a regenerative latch strobed by the recovered clock, or in a limiting amplifier whose output is sampled with the appropriately delayed recovered clock. The former method is most common, and has also been employed in this work.

B. Architecture

The combination of a sense amplifier and regenerative latch is most often used as a data regenerator, or *decision circuit* (Fig. 10(a)). The shared nodes are a convenient means to couple the analog signal into the latch. The sense amplifier and latch are enabled on alternate phases of the clock cycle. While the sense amplifier samples a new bit, the latch is reset to erase memory of the previous bit. It will then regenerate the amplified sample to a logic level.

This work has dealt with improvements in the topology of the combination of sense amplifier and latch, and with a parallel architecture to speed up the throughput rate of a MOS data regenerator to attain gigabit-per-second per-

formance. The sense amplifier and latch are decoupled in a way that neither is capacitively loaded by the other, and that their individual speeds may be optimized independently. A further doubling of throughput rate is then obtained by time interleaving the operation of two data regeneration channels [19]. A single wide-band sense amplifier distributes data to the two channels, then a current-mode multiplexer combines the data into a single output stream at the full rate.

C. Circuit Design

1) Improved Regenerative Latch: The conventional merged sense amplifier and latch has been implemented in 1-μm NMOS and shown to operate at data rates of up to 750 MS/s [6]. To improve its sensitivity and extend the maximum data rate of operation, this circuit must be modified to overcome some fundamental limitations.

The merging of the conventional sense amplifier and latch compromises the maximum clock rate of operation in two ways. First, when the amplifier is activated, the latch, although not biased, still imposes a capacitive loading at the output nodes; *vice versa*, the amplifier imposes a capacitive load when the latch is activated. Second, the input signal is isolated from the latch while it is regenerating. An input data bit can actually aid regeneration of the latch because its polarity does not change until the next bit period. Thus, in the improved design reported here, the amplifier continues to drive the latch through a buffer during the regeneration period [20].

The NMOS latch uses ratioed source-follower FET's as the load devices (Fig. 10(b)). A separate amplifier designed specifically for wide-band operation drives the

467

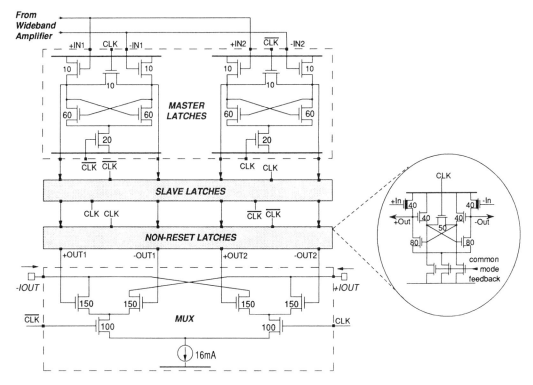

Fig. 11. Two-channel time-interleaved decision circuit. Data signal is distributed to the two channels by one wide-band sense amplifier, and the channel outputs are multiplexed to the full data rate.

source followers. The amplifier is not clocked, but operates continuously. While the latch is being reset, the load FET's act as source followers to imbalance the latch nodes with the amplified data signal. On the next half clock cycle, this imbalance is regenerated to full logic levels. The amplified signal coupled into the latch during regeneration improves the sensitivity, and increases the tolerance to deviations in the latch strobe clock edge from its ideal position in time.

2) Time-Interleaved Architecture: The maximum throughput rate of the latch was doubled by employing two channels in parallel operating time interleaved (Fig. 11). Data were distributed simultaneously to the source-follower inputs of the two banks of latches, and sampled on alternate clock cycles. Two latches were cascaded in each bank in a master–slave arrangement to reduce the input window of metastability. These latches produce a return to zero output due to the reset cycle. A third latch in each bank produced a nonreset output by acting as a source follower while the previous stage regenerated, and as a latch when the previous stage was reset. Thus, each bank consisted of three latches in cascade.

Two sets of differentially driven current switches sampled the outputs of the two banks at the full data rate, and steered a current source into a differential output driving an off-chip 50-Ω load. This accomplished a multiplexing in current mode of the data emerging from the two channels of data regenerators.

Deviations from a 50% duty cycle in the recovered clock will cause the two banks of latches to operate unequally spaced apart in time, possibly leading to a deg-

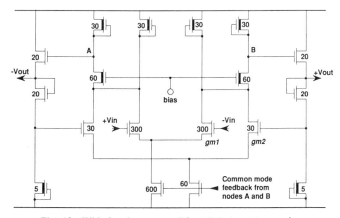

Fig. 12. Wide-band sense amplifier. Gain is set by g_{m1}/g_{m2}.

radation in the bit error rate of the recovered data. Push–pull NMOS inverters were used in the VCO and as the buffers to force the duty cycle of the oscillation close to the desired 50%. Any remaining errors in the duty cycle had little effect on the error rate, owing to the large clock phase margin (CPM) of this latch, as described in Section V. Once correctly regenerated, the multiplexed data stream may, if so desired, be retimed at the full data rate by an external clock.

3) Wide-Band Amplifier: An all NMOS balanced differential amplifier consisted of an input and a feedback differential pair connected in a shunt feedback topology (Fig. 12). Cascode FET's enhanced the open-loop gain. The overall gain was set to about 10 by the ratio of the g_m of the input stage to that of the feedback stage. The -3-dB bandwidth was about 1 GHz.

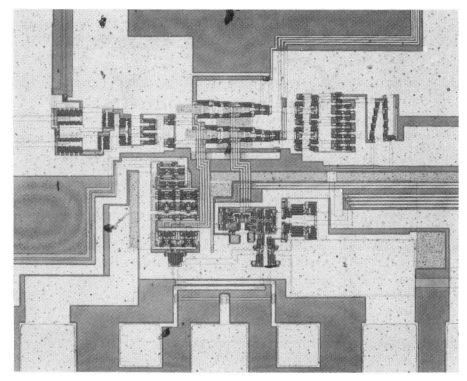

Fig. 13. Chip microphotograph of clock regeneration IC.

V. EXPERIMENTAL RESULTS

The clock and data regeneration circuits were evaluated separately. The IC's were mounted in high-frequency packages with stripline connections leading up to the wirebonded chips, with nearby decoupling capacitors to a ground plane on the supply and bias lines. Pseudorandom data patterns up to 1.8 Gb/s were supplied by a HP 71604A pattern generator, and error rate measurements were made on a HP 71603A error performance analyzer. Time-domain evaluations of the waveforms, including eye patterns, were obtained from a Tektronix CSA803 communications signal analyzer. Dispersion was deliberately introduced in the data waveforms by transmitting them through several meters of coaxial cable.

A. Clock Regeneration

The clock regeneration IC was fabricated in a 1-μm enhancement–depletion NMOS technology with an active area of 625×400 μm, and its core dissipated 350 mW from a 5-V power supply (Fig. 13). The clock was successfully recovered from pseudorandom data received at rates from 1.4 to 1.7 Gb/s. The VCO range was somewhat lower than predicted, probably because of incomplete turn-off of the depletion-mode FET's used in the voltage-controlled interpolator at the extremes of the range. An output swing of almost 4-V peak to peak was obtained into 50 Ω, with rise times somewhat greater than 200 ps. The jitter in the clock waveform due to noise in the VCO was first calibrated by applying an undispersed, jitter-free periodic 0011 $\cdots$ pattern at the input. The

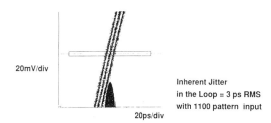

20mV/div

20ps/div

Inherent Jitter
in the Loop = 3 ps RMS
with 1100 pattern input

Fig. 14. Measured overlay of zero crossings of clock recovered from jitter-free periodic input. Histogram below.

measured jitter was 3-ps rms (Fig. 14). This remarkably low value is due to the suppression of all flicker noise and most thermal noise in the NMOS VCO by the wide-band negative feedback in the PLL.

The capability of the clock regeneration circuit to suppress pattern dependent jitter was then measured by applying a 1.8-Gb/s data waveform to the input, 100-mV peak to peak, consisting of a $2^{23} - 1$ long pseudorandom sequence. Dispersion through the cable introduced a 77-ps rms jitter into the zero crossings of the data pattern. A stable clock with a jitter of only 13-ps rms was successfully recovered from the data (Fig. 15). As the jitter in the input pattern decreases with the pseudorandom sequence length, so does the jitter in the recovered clock. The high quality of recovered clock may be attributed to the phase detector design and master–slave arrangement for VCO phase alignment.

The effectiveness of the proportional plus integral loop filter was separately verified. During normal operation with the integrator in place, the rms jitter in the recovered clock remained constant at 13-ps rms for $2^{23} - 1$ long

469

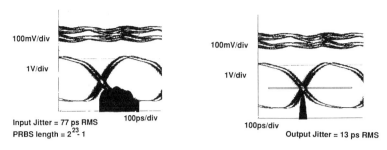

Fig. 15. Measured jitter in data and recovered clock. Histogram at left for data, at right for clock.

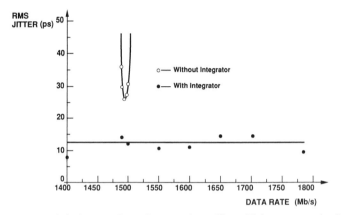

Fig. 16. Measured jitter in recovered clock versus input data rate. Loop filter with integrator maintains low jitter across entire range.

pseudorandom sequences across the *entire* lock range of the PLL. With the integrator removed, the jitter deteriorated rapidly as the input data rate moved away from the center frequency of the VCO (Fig. 16). These data experimentally verify that low-jitter clock recovery over a wide range of clock rates absolutely requires a second-order loop with large dc loop gain. More interestingly, the jitter in the clock recovered with the integrator present is almost half the lowest attainable jitter without it, which implies that the higher loop gain at low frequencies usefully suppresses undesirable fluctuations in the loop, such as caused by power supply hum leaking in through the VCO and by MOSFET flicker noise. These sources of noise appeared in the measured frequency spectra of the recovered clock without the integrator.

B. Data Regeneration

The data regeneration IC occupied an active area of 560 × 300 μm on a 1-μm NMOS IC, and its core dissipated 200 mW from a 5-V power supply (Fig. 17). Pseudorandom input data waveforms of 35 mV amplitude could be resolved at 10^{-11} bit error rate (BER) up to a data rate of 1.1 Gb/s (Fig. 17). The sensitivity rapidly decreased at higher data rates, owing we believe to clock kickback from the strobe pulses which inhibits the latches from resetting completely. This effect, too, was responsible for

the constant sensitivity as the data rate is lowered, contrary to the expected increase in sensitivity of a flip-flop at lower toggle rates.

A smaller maximum data rate of operation and a worsened sensitivity in the data regenerator constitute two of the few discrepancies between the simulated performance of the circuits and experimental observation. Simulations were carrried out using an early release of the BSIM model in SPICE, capable of modeling short-channel effects. For lack of adequately measured BSIM parameters for the 1-μm NMOS process at the time of design, the FET capacitance C_{gs} was underestimated. This resulted in the larger than predicted clock kickback through the source followers.

The clock phase margin (CPM) is another measure of the quality of the data regenerator. This specifies the maximum deviation of a clock strobe edge from the center of the input data eye which will not degrade the BER of the output data. A large CPM means that the data regenerator may tolerate phase misalignments in the clock recovery, and also some routing delay in the recovered clock used as its strobe. CPM is often specified as the degrees of phase misalignment from the center of the data eye, with one data bit being 360°. The CPM of our data regenerator was measured as 650 ps at 1.1 Gb/s, corresponding to 234° (Fig. 19). This large CPM may principally be attributed to the continuous amplifier aid presented to the regenerative latch.

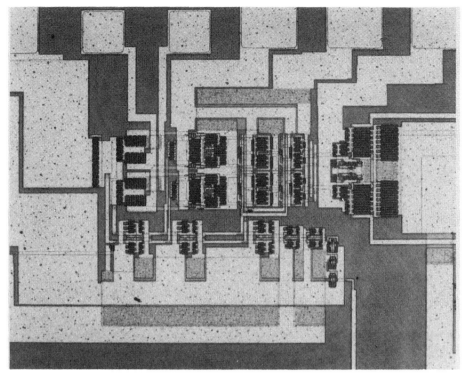

Fig. 17. Chip microphotograph of data regeneration IC.

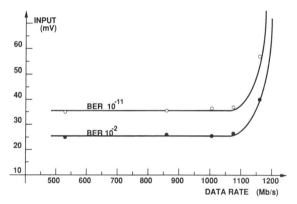

Fig. 18. Measured input amplitude at two BER levels versus data rate. Internal clock feedthrough is thought to limit high-frequency operation and sensitivity at low frequencies.

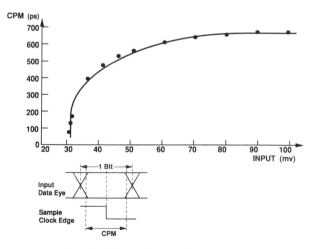

Fig. 19. Measured clock phase margin.

VI. CONCLUSIONS

New architectures and circuit techniques have been described for the design of circuits to regenerate data and clock waveforms from the amplified optical signals in optical-fiber links. These circuits have demonstrated a performance and speed heretofore not thought possible in a conventional silicon MOS realization. Aside from some low-performance off-chip components which in principle could also be integrated with the high-speed portions, the circuits represent the final building blocks for the realization of an all-MOS optical-fiber receiver/transmitter chip set capable of operation beyond 1 Gb/s. With its high levels of integration and low cost, a silicon MOS implementation of the interface electronics should contribute to a more widespread use of gigabit-per-second optical-fiber links.

ACKNOWLEDGMENT

The authors are grateful to many present and past members of AT&T Bell Laboratories, Murray Hill, NJ, particularly G. E. Smith, L. W. Nagel, J. A. Michejda, D. C. Dennis, and S-C. Fang, who were responsible for the successful fabrication of the IC's in the very last run of an experimental process. Testing and evaluation were carried out at the facilities of Gigabit Logic, Inc., Newbury Park, CA.

REFERENCES

[1] S. K. Enam and A. A. Abidi, "MOS decision and clock recovery circuits for Gb/s optical fiber receivers," in *ISSCC Dig. Tech. Papers* (San Francisco), Feb. 1992, pp. 96–97.

[2] J. F. Ewen, D. L. Rogers, A. X. Widmer, F. Gfeller, and C. J. Anderson, "Gb/s fiber optic link adapter chip set," in *GaAs IC Symp. Tech. Dig.*, Nov. 1988, pp. 11–14.

[3] R. C. Walker, T. Hornak, C-S. Yen, J. Doernberg, and K. H. Springer, "A 1.5-Gb/s link interface chipset for computer data transmission," *IEEE J. Selected Areas Commun.*, vol. 9, pp. 698–703, June 1991.

[4] M. P. Lepselter *et al.*, "A systems approach to 1 micron NMOS," *Proc. IEEE*, vol. 71, p. 640, May 1983.

[5] A. A. Abidi, "Gigahertz transresistance amplifiers in fine line NMOS," *IEEE J. Solid-State Circuits*, vol. SC-19, pp. 986–944, Dec. 1984.

[6] D. Soo *et al.*, "A 750 MS/s latched comparator," in *ISSCC Dig. Tech. Papers* (New York), 1985, pp. 146–147.

[7] R. J. Bayruns *et al.* "A 3 GHz 12-channel time-division multiplexer demultiplexer chip set," in *ISSCC Dig. Tech. Papers*, 1986, pp. 192–193.

[8] K. R. Shastri, K. N. Wong, and K. A. Yanushefski, "1.7 Gb/s NMOS laser driver," in *Proc. IEEE Custom IC Conf.* (Rochester, NY), May 1988, pp. 5.1.1–5.1.4.

[9] D. H. Wolaver, *Phase-Locked Loop Circuit Design.* Englewood Cliffs, NJ: Prentice-Hall, 1991.

[10] T. Yamada *et al.*, "A 1.2 GHz single chip NMOS PLL," in *ISSCC Dig. Tech. Papers*, vol. 32, 1985, pp. 24–25.

[11] M. Banu, "100 kHz–1 GHz NMOS variable frequency oscillator with analog and digital control," in *ISSCC Dig. Tech. Papers* (San Francisco), vol. 31, Feb. 1988, pp. 20–21.

[12] N. M. Nguyen and R. G. Meyer, "A 1.8 GHz monolithic LC voltage controlled oscillator," in *ISSCC Dig. of Tech. Papers*, pp. 158–159 (San Francisco), Feb. 1992,

[13] A. G. Bell and G. Borriello, "A single chip NMOS ethernet controller," in *ISSCC Dig. Tech. Papers*, vol. 30, Feb. 1983, pp. 70–71.

[14] K. M. Ware, H.-S. Lee, and C. G. Sodini, "A 200-MHz CMOS phase-locked loop with dual phase detectors," *IEEE J. Solid-State Circuits*, vol. 24, pp. 1560–1568, Dec. 1989.

[15] S. K. Enam and A. A. Abidi, "A gigahertz voltage controlled ring oscillator," *Electron Lett.*, vol. 22, pp. 677–679, June 5, 1986.

[16] S. K. Enam and A. A. Abidi, "A 300-MHz CMOS voltage-controlled ring oscillator," *IEEE J. Solid-State Circuits*, vol. 25, pp. 312–315, Feb. 1990.

[17] J. A. Bellisio, "New phase locked loop timing recovery method for digital regenerators," in *Int. Commun. Conf. Rec.*, (Philadelphia), June 1976, pp. 10–17.

[18] T. H. Lee and J. F. Bulzachelli, "A 155 MHz clock recovery delay and phase locked loop," in *ISSC Dig. Tech. Papers* (San Francisco), Feb. 1992, pp. 160–161.

[19] D. Clawin and U. Langmann, "Multigigabit/second decision circuit," in *ISSCC Dig. Tech. Papers*, 1985, pp. 222–223.

[20] S. K. Enam and A. A. Abidi, "Decision and clock recovery circuits for gigahertz optical fiber receivers in silicon NMOS," *J. Lightwave Technol.*, vol. LT-5, pp. 367–372, Mar. 1987.

A PLL-Based 2.5-Gb/s GaAs Clock and Data Regenerator IC

Hans Ransijn and Paul O'Connor, *Member, IEEE*

Abstract —A GaAs IC that performs clock recovery and data retiming functions in 2.5-Gb/s fiber-optic communication systems is presented. Rather than using surface acoustic wave (SAW) filter technology, the IC employs a frequency- and phase-lock loop (FPLL) to recover a stable clock from pseudo-random NRZ data. The IC is mounted on a 1-in × 1-in ceramic substrate along with a companion Si bipolar chip that contains a loop filter and acquisition circuitry. At the SONET OC-48 rate of 2.488 Gb/s, the circuit meets requirements for jitter tolerance, jitter transfer, and jitter generation. The data input ambiguity is 25 mV while the recovered clock has less than 2° rms edge jitter. The circuit functions up to 4 Gb/s with a 40-mV input ambiguity and 2° rms clock jitter. Total current consumption from a single 5.2-V supply is 250 mA.

I. INTRODUCTION

MANY circuit functions required in high-speed digital fiber-optic communications systems have been realized in monolithic form. Single-chip implementations of transimpedance amplifiers, main amplifiers, laser drivers, decision circuits, and multiplexers–demultiplexers have been demonstrated at speeds up to at least 2.5 Gb/s [1]–[5]. As in any electronic system, obvious advantages of integration are reduction of cost, physical size, and power dissipation.

The data regeneration function has so far mainly been implemented with discrete components, primarily because of the difficulty of integrating stable high-Q filters used for the clock extraction function at these high frequencies. As Fig. 1 illustrates, the data coming from the main amplifier are typically split into two paths. One leads to a monolithic decision circuit, which uses a comparator to discriminate between logic levels and a flip-flop to sample and retime the data in the center of the bits. The other path uses a nonlinear element to extract a spectral component at the baud (the nonreturn-to-zero (NRZ) data spectrum itself has a null at the bit frequency), and a narrow-band filter such as a surface acoustic wave (SAW) filter [6] or a dielectric resonator filter [7] to extract a stable clock signal. Because of the insertion loss of the filters (20 dB typically for SAW's) and the strong modulation of the clock's amplitude by the data pattern, a

Manuscript received April 16, 1991; revised May 23, 1991.
H. Ransijn is with AT&T Bell Laboratories, Reading, PA 19609.
P. O'Connor is with the Instrumentation Division, Brookhaven National Laboratory, Upton, NY 11973.
IEEE Log Number 9102414.

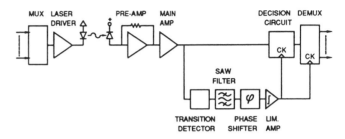

Fig. 1. High-bit-rate fiber-optic communication system with conventional clock and data regeneration.

high-gain limiting amplifier is required to make the clock signal usable for the decision circuit. Still another component, i.e., a phase shifter, is needed to align the clock edge in the center of the data bit. This alignment must be accurately maintained over temperature and aging, a task which is complicated by the large differences in power levels along the clock path, requiring components to be separated to prevent unwanted coupling.

It is clear that a dramatic reduction of regenerator cost and complexity can be obtained through monolithic implementation of the clock recovery function and its integration with a decision circuit. Thus far, however, reporting of such circuits has been limited mainly to applications in short-span optical data links at bit rates not exceeding 1.25 Gb/s [8].

This paper reports on a GaAs E/D HFET IC, based on a frequency- and phase-lock loop, that can regenerate data and clock at bit rates exceeding 2.5 Gb/s with a performance that matches that of SAW-filter-based regenerators, and that can be applied in optical data links as well as in repeatered optical transmission systems [9]. Section II discusses the circuit's architecture. Design details are given in Section III, while the process technology and packaging scheme are dealt with in Section IV. In Section V we will show the results of extensive characterization of the device and their relevance to SONET standards for the OC-48 bit rate of 2.48832 Gb/s [10]. Conclusions are presented in Section VI.

II. ARCHITECTURE

There are several considerations determining the preferred architecture for the clock and data regenerator (CDR). The first one deals with the need for a frequency detector as part of the phase-locked loop. An on-chip

Reprinted from *IEEE Journal of Solid-State Circuits*, vol. SC-26, pp. 1337-1344, October 1991.

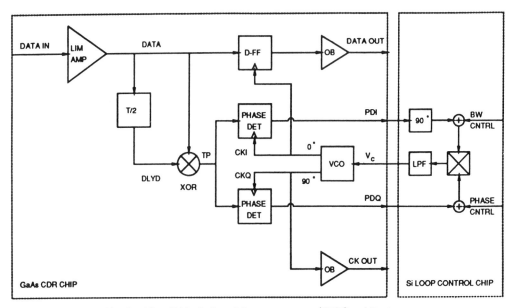

Fig. 2. Monolithic clock and data regenerator (CDR) architecture.

VCO is likely to show some sensitivity to temperature variations and should therefore have a sufficiently large frequency range to accommodate for temperature drift. Since a practical phase-lock loop has a limited capture range, a frequency detector or some other kind of acquisition aid is indispensable.

Sequential and rotational phase–frequency detectors [11], [12] are attractive because they can be implemented by means of digital logic, but they are limited in speed. The type of phase–frequency detector used in this design, Richman's "quadricorrelator," dating back to the early days of color television [13], was first proposed by Bellisio [14] to be used in a clock recovery circuit and was successfully implemented by Cordell *et al.* in a 50-MHz circuit [15]. Its only high-speed components are two simple analog phase detectors that produce low-frequency outputs. The speed of the entire circuit is therefore limited only by the decision circuit.

The alignment of clock and data at the input of the decision circuit is critical. The fact that the incoming data require preprocessing tends to complicate this alignment. An elegant solution to this problem is offered by Hogge [16], by *post*-processing the inputs and outputs of cascaded flip-flops, but it is hard to fit a frequency detector into this scheme. The quadricorrelator, on the other hand, fits in seamlessly with the conventional approach taken here, which simply relies on the minuteness of delays experienced on-chip as compared to a board-level circuit.

Fig. 2 shows the resulting architecture of the CDR. The decision circuit function is performed by a D-type flip-flop, which is preceded by a high-gain limiting amplifier and followed by an output buffer, providing 50-Ω complementary outputs. A half-bit delay generator and EXCLUSIVE-OR circuit form transition pulses [17] that provide both the required spectral content at the bit frequency and an accurate timing signal at the reference inputs of the two quadricorrelator phase detectors.

Since the phase detectors are driven by quadrature clock phases CKI and CKQ, their outputs PDI and PDQ are in quadrature as well. Whether the clock is faster or slower than the bit frequency determines which phase detector output will lead or lag. By shifting the two output signals another 90° with respect to one another, and feeding them to a four-quadrant multiplier, a signal results with a dc component, whose sign depends on that of the frequency error. Once frequency lock has been established, and the multiplier output becomes zero, an offset at the PDI input of the multiplier enables the PDQ phase detector to drive the loop filter, which closes the frequency- and phase-locked loop (FPLL). As the loop forces PDQ to zero, thus centering CKQ with respect to the half-bit transition pulses, the 0° clock phase CKI lines up with the center of the bit period and ensures that the flip-flop can sample the data at the optimum decision time. A timing diagram of the phase-locked situation is shown in Fig. 3. Since the phase detectors and the flip-flop are both implemented with SCFL-type circuits, with the lowest logic level driven by the clock, good tracking of remaining delay differences is ensured.

The low-frequency functions of the FPLL are taken care of by a semicustom Si bipolar IC. Although the speed of this part of the circuit is noncritical, matching of devices determines accuracy of clock and data alignment. A Si bipolar process was therefore chosen as the most appropriate technology to implement this function.

III. DESIGN

Beside the performance aspects of the design, most of our attention has been focused on suppression of unwanted behavior of the circuit. Although parameters such as input ambiguity, clock, and attainable bit rate are prime objectives, the real challenges in a circuit such as this, with its various types of signals, are in finding ways to

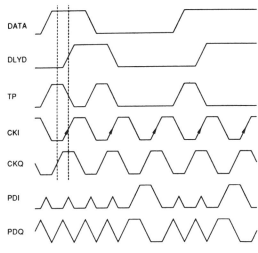

Fig. 3. CDR timing diagram.

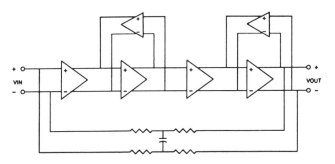

Fig. 4. Input limiting amplifier block diagram.

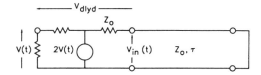

Fig. 5. Half-bit delay generator principle.

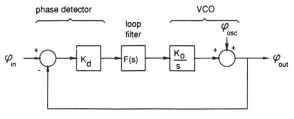

Fig. 6. Linear PLL model, accounting for VCO phase noise.

route the high-speed signals and bypass the bias signals without introducing crosstalk and interference that could easily result in reduced sensitivity, or worse, injection locking of the PLL. The physical layout of the chip as well as its environment are as important as the electrical design.

Differential circuitry is used throughout the entire chip, to prevent any type of coupling through common-mode signals and to simplify power-supply bypassing. Careful design and accurate device models are applied to ensure stable operating points.

A. Input Limiting Amplifier

As is true for a simple decision circuit, the input limiting amplifier of the CDR largely determines the circuit's ability to resolve the incoming data eye vertically and discriminate between logic levels, corrupted by noise, and intersymbol interference. Important performance aspects include gain and bandwidth, but at least as important is maintaining pulse integrity under large input signal conditions.

The amplifier, whose block diagram is shown in Fig. 4, is made up of two sections, each of which employs a limiting transconductance stage followed by an active feedback transimpedance stage to extend the bandwidth. In view of the inherent high offsets in FET input stages, overall dc feedback is applied around the amplifier to

ensure that the slicing level stays in the vertical center of the data eye.

B. Half-Bit Delay Generator

The accuracy and stability of the half-bit delay generator is critical to the alignment of clock and data at the flip-flop inputs. Although a fixed off-chip delay line would provide the desired stability, a separate board or package design for every bit rate would be required, as well as, most likely, several iterations to fine-tune the electrical length of the line. To circumvent these problems, we have used the delay properties of a shorted line stub (see Fig. 5).

A pulse $V(t)$ is launched onto this line stub by a signal source whose internal impedance equals the line's characteristic impedance. At the shorted end of the line the pulse is reflected back to the source and returns at the source's terminals undistorted, but with a negative sign and delayed by twice the electrical length of the line. The total voltage across the terminals is now the superposition of the original pulse and the (negative and delayed) reflected pulse:

$$V_{in}(t) = V(t) - V(t - 2\tau).$$

By subtracting the terminal voltage from the original pulse, a delayed version V_{dlyd} of the latter is obtained. The delay can be adjusted for any given data rate by placing the short on the line stub at the appropriate distance from the chip. A more detailed device-level description of the delay generator can be found in [18].

C. Voltage-Controlled Oscillator (VCO)

Special demands are placed on the spectral purity of the local oscillator in a PLL, used for timing recovery, as can be illustrated using the linear PLL model of Fig. 6. By examining the loop equation below, it is easily shown that within the loop bandwidth, where the loop transfer function $H(s)$ has a magnitude equal to one, the phase fluctuations from the transition detector, $\varphi_{in}(s)$, are passed and those generated by the VCO, $\varphi_{osc}(s)$, are suppressed.

Outside the loop bandwidth the pattern jitter is rejected, but the VCO phase noise goes unchecked:

$$\varphi_{\text{out}}(s) = H(s) \cdot \varphi_{\text{in}}(s) + \{1 - H(s)\} \cdot \varphi_{\text{osc}}(s).$$

Although long-term stability of VCO phase is not essential, at offset frequencies where the loop transfer function starts rolling off, the VCO phase sidebands should be down far enough so as not to contribute to the overall clock jitter.

The VCO uses an external shorted stub line tank to ensure short-term stability. The tank is tuned by on-chip diode varactors and, like the delay generator, it can be adjusted for the desired bit rate by shorting the stub at the right distance from the chip. The active part of the VCO is based on a simple differential pair that, if configured in a positive feedback arrangement, presents a non-linear negative resistance to the tank resulting in a self-limiting oscillator [19].

The oscillator, including the tank circuit, is completely balanced, which not only prevents injection locking, but also helps reduce upconversion of $1/f$ noise [20], [21].

D. Si Bipolar Loop Control Chip

We chose an in-house complementary-bipolar linear array [22] to implement the loop control functions of the CDR. The vertical p-n-p transistors available in this process allow for low-power operation and simplify loop filter design.

The 90° phase shift of the PDI phase detector output (Fig. 1) is in reality implemented as a relative 90° shift between the PDI and PDQ branches. A simple RC low-pass filter in the PDQ branch and an CR high-pass filter in the PDI branch ensure a quadrature phase relationship over the required range of beat frequencies (approximately dc to 40 MHz).

The four-quadrant multiplier uses a Gilbert architecture [23]. One of the properties of this type of multiplier is that it preserves the modulation index of the input signal, in this case the phase-detector gain, which means a constant PLL bandwidth, independent of bias current.

Rather than using an op-amp/integrator structure for the loop filter, we chose to use high-output-impedance complementary current mirrors, driving a passive grounded loop filter. Latch-up is eliminated this way, and power-supply noise is reduced to a minimum.

E. Phase-Locked-Loop Dimensioning

As shown in Section III-C, there is a lower limit for the PLL bandwidth with respect to the PLL's ability to suppress VCO phase noise. The upper limit is dictated by the total allowed amount of pattern jitter (0.01 UI or 3.6° rms for SONET OC-48). Even if a perfectly clean oscillator signal were available, the bandwidth could not be chosen arbitrarily small. An important performance measure for a regenerator in a transmission system is its jitter tolerance [24], defined as the ability of a regenerator to tolerate incoming jitter, or timing fluctuations of the

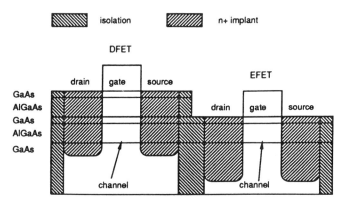

Fig. 7. Enhancement/depletion heterojunction FET cross section.

incoming data. Any relative phase shift between data and recovered clock exceeding the bit period would result in errors. At low jitter frequencies, where these fluctuations can span several bit periods, the regenerator must therefore follow the data jitter. At higher jitter frequencies the PLL's low-pass jitter transfer function will reject most of the incoming jitter, and the decision-circuit phase margin combined with the peak-to-peak jitter of the recovered clock determines the jitter tolerance. As a result the PLL bandwidth is a compromise between the desired jitter tolerance template and the allowed pattern jitter.

IV. IMPLEMENTATION

The CDR IC was implemented in AT&T's 0.9-μm self-aligned refractory gate IC (SARGIC) E/D heterojunction FET (HFET) process. A cross section of the two device types, MBE-grown enhancement and depletion-mode HFET's, is shown in Fig. 7. The active area of the devices is formed by the interface of a doped AlGaAs layer grown on top of an undoped GaAs buffer. A "two-dimensional gas" of high mobility electrons forms the channel. The W/WSi$_x$ refractory gate is used as a mask for the n$^+$ Si implants that form the source and drain of the HFET. A layer of undoped GaAs on top of the AlGaAs layer places the gate at the appropriate distance from the channel to define the E-FET. The complete structure includes two more layers, one AlGaAs and one GaAs layer, that space the D-FET gate further away from the channel and that are selectively removed to make the EFET. The devices show a g_m of 170 and 200 mS/mm and an f_T of 24 and 28 GHz for the DFET and the EFET, respectively. The DFET I_{DSS} is 80 mA at $V_{GS} = 0$ V, while the EFET I_{DSS} amounts to 60 mA at $V_{GS} = 0.5$ V. The devices are isolated by oxygen implantation and interconnected with two levels of Ti/Pt/Au metal, using 2-μm lines and spaces. TaN resistors and MIM capacitors are available for termination and bypassing, respectively.

Fig. 8 shows a photomicrograph of the chip, which measures 1.5 × 2.34 mm. A 1-in × 1-in alumina substrate was designed that accepts the GaAs CDR chip as well as the Si loop control chip (Fig. 9). It contains the two coplanar strip stub lines for the VCO and the delay generator, respectively. The high-speed I/O lines begin

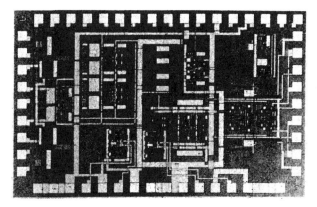

Fig. 8. CDR chip photomicrograph.

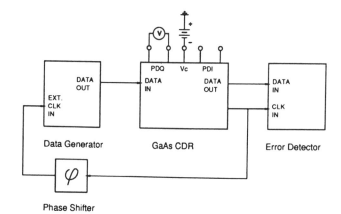

Fig. 10. Open-loop wafer and device test.

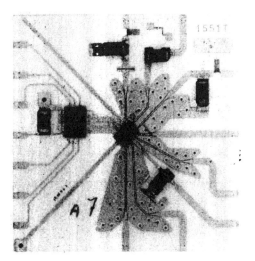

Fig. 9. GaAs CDR chip (center) and Si control chip (left) on 1-in × 1-in alumina substrate.

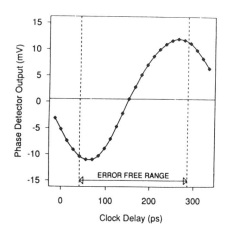

Fig. 11. Phase-detector output alignment.

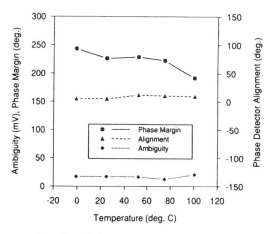

Fig. 12. Performance versus temperature.

as microstrip lines at the edge of the substrate, but change into coplanar lines to allow close proximity to the chip edge for bonding. Excellent high-speed performance has been observed, obtaining a return loss for a 50-Ω terminated line of 25 dB up to 4 GHz and an insertion loss of two cascaded lines of less than 1.5 dB at 4 GHz.

V. RESULTS

The CDR is tested at the wafer level for full speed functionality using a $2^{23} - 1$ pseudorandom data pattern. The high-speed probe card employs two shorted stubs for the delay generator and VCO ports, which are optimized for 2.5-Gb/s operation. During wafer test, the VCO is allowed to free run, driving the external clock input of a data pattern generator through an external delay line that in turn drives the data input of the CDR (see Fig. 10). Beside VCO functionality, the decision-circuit phase margin and the phase-detector alignment can be tested in this manner. This test is repeated after mounting the GaAs die on the alumina substrate. At this time the circuit is characterized extensively over temperature, supply voltage, data rate, and input voltage.

Fig. 11 shows the alignment of the phase-detector output with respect to the clock phase margin. Phase margin, phase-detector alignment, and ambiguity level are plotted in Fig. 12 against a fixture temperature ranging from 0 to 100°C.

Another aspect of the open-loop behavior of the CDR is VCO performance. In Fig. 13 the VCO frequency is plotted versus the control voltage for temperatures between 0 and 100°C. Since the PLL bandwidth is propor-

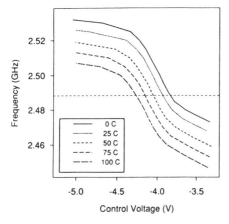

Fig. 13.   VCO tuning curve versus temperature.

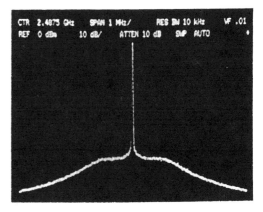

Fig. 16.   Phase-locked VCO spectrum.

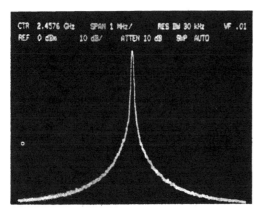

Fig. 14.   Free-running VCO spectrum.

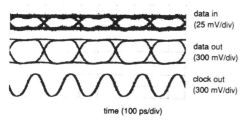

data in
(25 mV/div)

data out
(300 mV/div)

clock out
(300 mV/div)

time (100 ps/div)

Fig. 15.   CDR eye diagram at 2.5 Gb/s with 25-mV$_{p-p}$ data input.

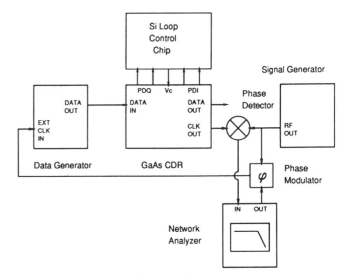

Fig. 17.   Jitter transfer measurement.

tional to VCO gain, the target bit rate (dashed line) should remain within the high-gain region. The plot illustrates that a minimum VCO tuning range is required which exceeds the temperature drift of the tuning curve, hence the need for a frequency detector as stated in Section II. A photograph of the free-running VCO spectrum is shown in Fig. 14.

Next, the Si loop control chip is mounted and full closed-loop performance of the CDR can be verified. The total supply current averages 250 mA, of which 10 mA is consumed by the Si chip. An eye diagram of the CDR, locked to a 2.488-Gb/s pseudorandom pattern of 25 mV$_{p-p}$ and performing error free (BER $< 10^{-12}$), is shown in Fig. 15. Output rise and fall times are under 100 ps. The locked VCO spectrum under nominal input voltage

(600 mV$_{p-p}$) is shown in Fig. 16. The data sidebands can be seen to follow the PLL bandpass response.

A more accurate characterization of this filter response involves phase modulation of the data pattern and subsequent demodulation of the recovered clock as depicted in Fig. 17. A low-frequency network analyzer can display the equivalent low-pass response, generally referred to as jitter transfer function. As Fig. 18 shows, a flat response, within 0.1 dB, is obtained, with a -3-dB bandwidth of 1.2 MHz, giving an equivalent Q of the PLL filter of 1000. The SONET OC-48 interface standard specifies a jitter transfer function with peaking less than 0.1 dB within a 2-MHz bandwidth.

The clock jitter of the CDR is measured in a similar fashion. By demodulating the recovered clock, using the clock of the pattern generator as a reference, and normalizing the measured rms mixer output with respect to the mixer conversion gain, a measure for the rms jitter in radians is obtained. Conversion to degrees gives a typical jitter number of 2° rms, with best results as low as 1.6° rms, again using a $2^{23} - 1$ pseudorandom pattern.

By offsetting the PDQ phase detector input to the Si chip, we can shift the retiming phase of the clock some-

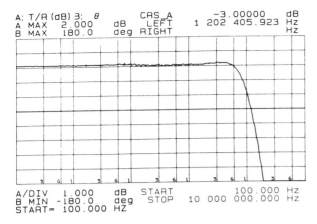

Fig. 18. Jitter transfer function at 2.5 Gb/s.

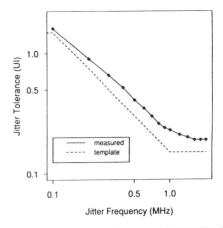

Fig. 20. Jitter tolerance. Measured curve and SONET OC-48 template.

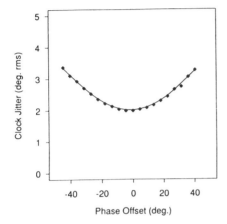

Fig. 19. Jitter versus phase offset.

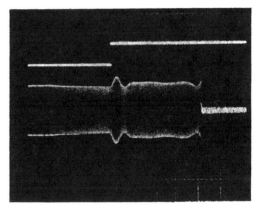

Fig. 21. Acquisition behavior. Top trace: blanking signal (low-data off, high-data on); bottom trace: demodulated CDR clock. Time scale: 1 ms/div.

what, with the purpose of verifying the correct decision point under stressed input conditions. Although a slight penalty is paid in terms of pattern jitter, as shown in Fig. 19, this method can provide a useful diagnostic tool in regenerator troubleshooting.

We measured the CDR's jitter tolerance by frequency modulating an external clock source, driving the pattern generator. In lieu of an optical test, we took the peak-to-peak phase modulation, which caused errors, as the measured value. SONET OC-48 calls for a jitter tolerance template of 1.5 UI up to 100 kHz, a 20-dB/decade rolloff to 1 MHz, from where the template remains constant. This template is shown in Fig. 20, along with measured data on a typical device, showing performance better than required.

The FPLL's lockup time is determined in principle by the output of the frequency detector, but is in practice limited by the time constant of the ac-coupled input. We measured the acquisition time by periodically interrupting the input data to the device and monitoring the output of the mixer, demodulating the recovered clock. In Fig. 21 the top trace shows the data blanking signal (30-Hz square wave), while the bottom trace shows the mixer output, going from a high-frequency beat signal to a dc level, indicating phase lock, in approximately 4 ms.

Although the CDR is intended to function at the SONET OC-48 rate, the circuit is useful at much higher bit rates. The parts that limit the speed of the device were designed to be the limiting amplifier and D-flip-flop, rather than the delay generator, VCO, or phase detectors. The architecture will therefore be useful at any speed at which a stand-alone decision circuit can be made to work. Fig. 22 shows eye diagrams of the CDR running error free at 4 Gb/s, both at the ambiguity input level of 40 mV and at the nominal 600-mV input. Phase margin at this data rate is somewhat reduced at 150°. The clock spectrum of Fig. 23 is similar to the 2.5-Gb/s case. The clock jitter is 2° rms.

VI. Conclusions

We have developed a circuit that integrates clock and data regeneration functions for 2.5-Gb/s fiber-optic communication systems and that is useful up to 4 Gb/s. Based on analog frequency- and phase-locked loop techniques, this circuit is limited in performance only by the speed of its core decision circuit function. Its performance easily matches that of SAW-filter or dielectric resonator-based circuits, but requires none of the critical

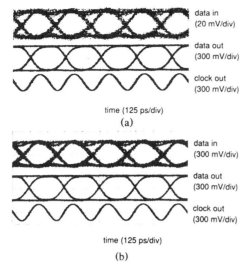

data in (20 mV/div)

data out (300 mV/div)

clock out (300 mV/div)

time (125 ps/div)

(a)

data in (300 mV/div)

data out (300 mV/div)

clock out (300 mV/div)

time (125 ps/div)

(b)

Fig. 22.   CDR eye diagram at 4 Gb/s: (a) 40-mV$_{p-p}$ input, and (b) 600-mV$_{p-p}$ input.

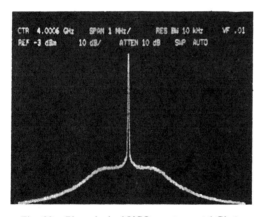

Fig. 23.   Phase-locked VCO spectrum at 4 Gb/s.

temperature compensations and avoids costly board-level assembly. While the 0.9-μm GaAs HFET technology assures margin at high speed, an accurate filter response is obtained by applying a low-speed Si bipolar process for the loop filter function.

ACKNOWLEDGMENT

The authors wish to thank all who contributed to this project, in particular S. Allen and J. Brown for IC layout, D. Johnson for wafer process overview, D. Daugherty and L. Snowden for wafer test, E. Tobias for assembly, J. Beccone, B. Crispell, and G. Salvador for test fixture design, B. Fulmer and C. Miller for automating device tests, and D. Butherus and P. Dorman for their support.

REFERENCES

[1] B. L. Kasper, J. C. Campbell, J. R. Talman, and A. H. Gnauck, "An 8 Gbit/s optical receiver using an InGaAs avalanche photodiode and a GaAs preamplifier," in *CLEO'87 Proc.*, 1987, p. 302.

[2] R. Reimann and H. M. Rein, "A 4 Gb/s limiting amplifier for optical-fiber receivers," in *ISSCC Dig. Tech. Papers*, Feb. 1987, pp. 172–173.

[3] F. S. Chen and F. Bosch, "A 4 Gbit/s GaAs MESFET laser-driver IC," *Electron. Lett.*, vol. 22, no. 18, pp. 932–933, Aug. 28, 1986.

[4] P. O'Connor *et al.*, "Monolithic multigigabit/s GaAs decision circuit for lightwave system applications," *IEEE Electron Device Lett.*, vol. EDL-5, pp. 226–227, July 1984.

[5] M. Ida, N. Kato, and T. Takada, "A 4-Gb/s GaAs 16:1 multiplexer/1:16 demultiplexer LSI chip," *IEEE J. Solid-State Circuits*, vol. 24, pp. 928–932, no. 4, Aug. 1989.

[6] B. Fleischmann, W. Ruile, and G. Riha, "Rayleigh-mode SAW filters on quartz for timing recovery at frequencies above 1 GHz," in *Ultrasonics Symp. Conf. Dig.* (Williamsburg, VA), July 1984, pp. 163–167.

[7] W. I. Way *et al.*, "High speed circuit technology for multigigabit/sec optimal communications systems," in *IEEE Int. Conf. Commun. Conf. Rec.* (Philadelphia, PA), June 1988, pp. 313–318.

[8] J. F. Ewen, "Fully integrated Gbit/s clock recovery circuit in GaAs," in *Optical Fiber Commun. Conf. Tech. Dig. Series*, vol. 1 (New Orleans, LA), Feb. 1988, p. 116.

[9] H. Ransijn and P. O'Connor, "A 2.5 Gb/s GaAs clock and data regenerator IC," in *IEEE GaAs IC Symp. Dig.*, 1985, pp. 57–60.

[10] "Synchronous Optical Network (SONET) transport systems: Common generic criteria," Bellcore, Tech. Advisory TA-NWT-000253, Issue 6, Sept. 1990.

[11] I. Shahriary *et al.*, "GaAs monolithic digital phase/frequency discriminator," in *IEEE GaAs IC Symp. Dig.*, 1985, pp. 183–186.

[12] D. G. Messerschmitt, "Frequency detectors for PLL acquisition in timing and carrier recovery," *IEEE Trans. Commun.*, vol. COM-27, pp. 1288–1295, Sept. 1979.

[13] D. Richman, "Color-carrier reference phase synchronization accuracy in NTSC color television," *Proc. IRE*, vol. 42, pp. 106–133, Jan. 1954.

[14] J. A. Bellisio, "A new phase-locked loop timing recovery method for digital regenerators," in *IEEE Int. Conf. Commun. Rec.*, June 1976, pp. 10–17.

[15] R. R. Cordell *et al.*, "A 50 MHz phase- and frequency-locked loop," *IEEE J. Solid-State Circuits*, vol. SC-14, no. 6, pp. 1003–1010, Dec. 1979.

[16] C. R. Hogge, Jr., "A self correcting clock recovery circuit," *J. Lightwave Technol.*, vol. LT-3, pp. 1312–1314, Dec. 1985.

[17] D. J. Millicker, R. D. Standley, and K. Runge, "Delay-and-multiply timing recovery circuit for lightwave transmission systems using NRZ format," in *OFC Conf. Dig.* (San Diego, CA), 1985, pp. 38–40.

[18] J. G. Ransijn, "Delay generator," U.S. Patent 5 014 286, May 7, 1991.

[19] M. G. Crosby, "Two-terminal oscillator," *Electronics*, pp. 136–137, May 1946.

[20] H. B. Chen, A. van der Ziel, and K. Amberiadis, "Oscillator with odd-symmetrical characteristics eliminates low-frequency noise sidebands," *IEEE Trans. Circuits Syst.*, vol. CAS-31, pp. 807–809, Sept. 1984.

[21] A. N. Riddle and R. J. Trew, "A novel GaAs FET oscillator with low phase noise," in *IEEE MTT-S Dig.*, 1985, pp. 257–260.

[22] "ALA201/202 UHF semicustom linear arrays," AT&T, data sheet, June 1989.

[23] B. Gilbert, "A dc-500 MHz amplifier/multiplier principle," in *ISSCC Dig. Tech. Papers*, 1968, pp. 114–115.

[24] P. R. Trischitta and P. Sannuti, "The jitter tolerance of fiber optic regenerators," *IEEE Trans. Commun.*, vol. COM-35, pp. 1303–1308, Dec. 1987.

A 2.5-Gb/sec 15-mW BiCMOS Clock Recovery Circuit

Behzad Razavi and James Sung

AT&T Bell Laboratories, Holmdel, NJ 07733, USA

High-speed low-power clock recovery circuits find wide application in high-performance communication systems [1]. This paper describes the design of a 2.5-Gb/sec 15-mW clock recovery circuit (CRC) fabricated in a 20-GHz 1-μm BiCMOS technology [2]. Employing a modified version of the "quadricorrelator" architecture [3, 4], the circuit extracts the clock from a non-return-to-zero (NRZ) data sequence using both phase and frequency detection.

Shown in Fig. 1, the original quadricorrelator architecture consists of a front-end operator (differentiator and full-wave rectifier) followed by a phase/frequency-locked loop including in-phase and quadrature mixers and a voltage-controlled oscillator (VCO) with quadrature outputs. The combination of Loop 1 and Loop 2 serves as a frequency-locked loop while Loop 3 functions as a phase-locked loop. The circuit operates as follows. The front-end operator creates a frequency component at $\omega_1 = 2.5$ GHz from an NRZ sequence at 2.5 Gb/sec. This component is mixed with the VCO outputs and low-pass filtered, yielding quadrature signals $\sin(\omega_1 - \omega_2)t$ and $\cos(\omega_1 - \omega_2)t$. One of these signals is differentiated and mixed with the other and the result is low-pass filtered, thereby producing a dc signal whose magnitude and polarity correspond to those of $\omega_1 - \omega_2$. This signal drives ω_2 closer to ω_1, and when $\omega_1 - \omega_2$ is sufficiently small, Loop 3 begins to dominate, thus acquiring the phase lock.

To obtain a compact, low-power implementation, the quadricorrelator architecture can be modified as shown in Fig. 2. Here, the front-end operator is moved to each arm such that it can be merged with the subsequent mixer and low-pass filter. The resulting circuit performs differentiation, rectification, mixing, and low-pass filtering, and is denoted as DRML. Similarly, the second differentiator, mixer, and low-pass filter can be combined (DML in Fig. 2). Note that the high-speed path now includes only three blocks, allowing significant savings in power dissipation. The architecture is fully differential to suppress common-mode noise and provide robust operation from a 3-V supply.

The DRML implementation is depicted in Fig. 3. A capacitively-degenerated differential pair, Q_3-Q_4, approximates the differentiator, and two Gilbert multipliers, Q_5-Q_8 and Q_9-Q_{12}, perform full-wave rectification and mixing, respectively. The circuit operates as follows. When D_{in} goes high, Q_4 turns off momentarily, disabling Q_7 and Q_8. At the same time, Q_6 and Q_{11}-Q_{12} are also off. Thus, the entire tail current $I_{EE1} + I_{EE2}$ flows from $Q_3, Q_5,$ and Q_9-Q_{10}, and is therefore multiplied by the VCO output. Due to symmetry, the same result is obtained when D_{in} goes low. Consequently, on every data transition the instantaneous value of the VCO output is sampled and deposited as charge on C_1.

The interface between the VCO and the DRML circuit entails an important issue: the feedthrough of data transitions from the input of DRML to the output of the VCO. Illustrated in Fig. 4, this effect originates from the capacitive path between these two ports and disturbs the VCO whenever D_{in} changes. While differential operation reduces the feedthrough, capacitance nonlinearities still require that the VCO internal circuit be isolated from this disturbance.

The VCO is realized as a two-stage ring oscillator so as to provide quadrature outputs (Fig. 5). We note that each stage in the ring must drive three circuits: the other stage, a mixer, and an output buffer. (Both stages are loaded with buffers to maintain symmetry.) In the VCO of Fig. 5, these outputs are taken from different ports to distribute the loading as well as ensure isolation of the VCO from the feedthrough noise.

Since a simple two-stage ECL ring oscillator suffers from both small-amplitude oscillations due to insufficient phase shift and narrow tuning range due to limited gain, the VCO of Fig. 5 incorporates additional phase/gain elements. Shown in Fig. 6 is the implementation of each stage, consisting of emitter followers Q_1-Q_2, a cross-coupled pair Q_3-Q_4, and two differential pairs Q_5-Q_6 and Q_7-Q_8. At high frequencies, the total capacitance seen at nodes X and Y allows the cross-coupled pair to contribute substantial gain (even at low bias currents) and phase shift. As a result, the VCO center frequency can be varied by approximately 1 GHz with little degradation in the voltage swings.

The signal driving each mixer is taken from the *collectors* of Q_1 and Q_2 and buffered by the pair Q_7-Q_8. At 2.5 GHz, the voltage gain from V_{in} to nodes M and N is approximately equal to the total capacitance seen at X and Y divided by that at M and N - roughly unity. This technique effectively isolates the internal VCO signal path from the DRML feedthrough noise.

All the current sources in the CRC are implemented with NMOS devices. For the current levels used here, NMOSFETs introduce less capacitance and less thermal noise and consume less voltage headroom than their bipolar counterparts.

The fabricated prototype has been tested on wafer with a 3-V supply. The circuit (excluding the I/O buffers) dissipates 15 mW. Fig. 7 shows the measured input and output waveforms of the CRC for a 2.5-Gb/sec pseudorandom NRZ sequence of length $2^{20} - 1$. The rms jitter is 9.5 psec and the capture range is 300 MHz. Fig. 8 depicts the output spectrum. The circuit exhibits a phase noise of -80 dBc/Hz at 50 kHz offset.

References

[1] M. Soyuer, "A Monolithic 2.3-Gb/s 100-mW Clock and Data Recovery Circuit in Silicon Bipolar Technology," *IEEE Journal of Solid-State Circuits*, vol. SC-28, pp. 1310-1313, Dec. 1993.

[2] J. Sung *et al*, "BEST2 - A High Performance Super Self-Aligned 3V/5V BiCMOS Technology with Extremely Low Parasitics for Low-Power Mixed-Signal Applications," *Proc. IEEE CICC*, pp. 15-18, May 1994.

[3] D. Richman, "Color-Carrier Reference Phase Synchronization Accuracy in NTSC Color Television," *Proc. of IRE*, vol. 42, pp. 106-133, Jan. 1954.

[4] J. A. Bellisio, "A New Phase-Locked Loop Timing Recovery Method for Digital Regenerators," *IEEE Int. Conference Rec.*, vol.1, June 1976, pp. 10-17.

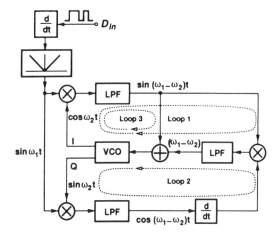

Fig. 1. Operation of quadricorrelator.

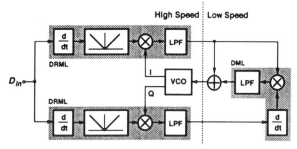

Fig. 2. Clock recovery architecture.

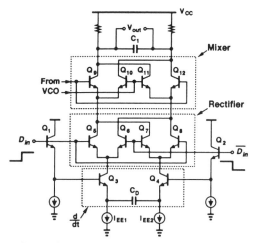

Fig. 3. Differentiator, rectifier, mixer, and LPF.

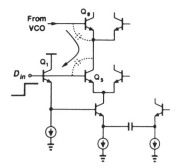

Fig. 4. Feedthrough from D_{in} to VCO.

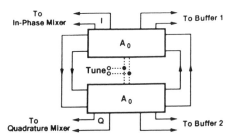

Fig. 5. VCO block diagram.

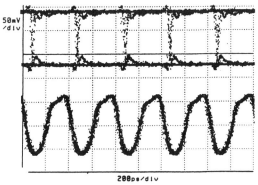

Fig. 6. Implementation of one stage of VCO.

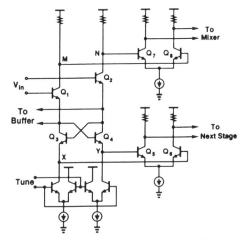

Fig. 7. Measured input/output waveforms.

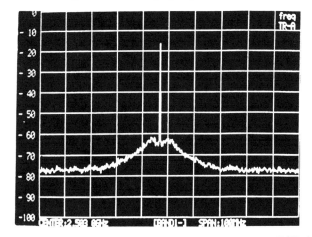

Fig. 8. Output spectrum (Horiz. 10 MHz/div., Vert. 10 dB/div.)

482

An 8 GHz Silicon Bipolar Clock-Recovery and Data-Regenerator IC

Ansgar Pottbäcker and Ulrich Langmann, *Senior Member, IEEE*

Abstract—A silicon bipolar IC for data regeneration and clock recovery which includes a phase/frequency detector (PFD), a quadrature voltage controlled oscillator (VCO), and an MS *D*-flipflop (DFF) is presented. The VCO is based on a modified two-stage ring oscillator approach and presents a wide tuning range of 2-to-9 GHz. Data regeneration at 8 Gb/s (with the on-chip VCO) and PFD operation up to 15 Gb/s (with an external VCO) are demonstrated. The IC for clock and data regeneration was fabricated with a 25 GHz f_T 0.4 μm emitter width bipolar process. The power dissipation is 2.25 W.

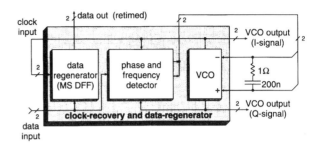

Fig. 1.   Block diagram of clock-recovery and data-regenerator.

I. INTRODUCTION

AT THE receiving end of an optical fiber transmission link, the clock signal is needed for data regeneration and/or demultiplexing. Therefore, the clock has to be extracted from the received binary data themselves, for which, normally, NRZ format is chosen. Regarding the factors of cost and reliability, an approach employing a single monolithic IC for clock-recovery and data-regeneration is superior to multichip solution.

In contrast to the data-regenerator which can be simply realized by an edge-triggered *D*-FF, the clock-recovery circuit needs much more attention. Currently there are two main concepts for clock-recovery circuits. The first concept is based on a passive high-Q resonator type filter [1],[2], whereas the second clock-recovery approach uses a phase locked loop (PLL) [3]. The major advantages of using a PLL instead of a passive resonator type filter are:

- A PLL can be implemented as a monolithic integrated circuit and
- the phase between the extracted clock and the received data is locked by a PLL.

In clock-recovery designs based on passive filters, often an additional phase-alignment circuitry is needed. However, pull-in can be a painfully slow process in a conventional PLL, and the pull-in range is not larger than the noise bandwidth. Therefore, an acquisition aid is indispensable for a PLL since a very narrow-band filtering is required. This leads to a clock-recovery approach based on a phase and frequency locked loop (PFLL).

Manuscript received May 23, 1994; revised, August 15, 1994. This work was supported in part by the Research Center Darmstadt of the Deutsche Bundespost Telekom.

A. Pottbäcker was with Ruhr University Bochum, Germany. He is now with SICAN GmbH, Hanover, Germany.

U. Langmann is with the Ruhr University Bochum, Department of Electrical Engineering, D–44780 Bochum, F. R. Germany.

IEEE Log Number 9406262.

II. ARCHITECTURE OF THE CLOCK AND DATA REGENERATOR IC

The block diagram of the clock and data regenerator (CDR) IC shown in Fig. 1 comprises three main processing blocks: a master-slave-D-flipflop which acts as a data regenerator, a phase and frequency detector (PFD) and VCO. The circuit is designed fully differential in order to optimize the operating speed and to minimize the problem of cross talk. The VCO provides both an in-phase and a quadrature output over the full tuning range. All inputs are terminated on-chip with 50 Ω resistors and the outputs are driven by 8 mA open collector output buffers. This simplifies the design of the loop filter and therefore only two external components—a resistor and a capacitor—are required. These form, together with the internal 50 Ω resistors of the tuning input, a passive lead-lag loop filter. Due to the phase and frequency locked operation of the regenerator the pull-in range is independent of the loop bandwidth and the low pass filter time constant can be made very large. In this way the loop remains locked, even if a longer bit sequence with a constant logic level occurs in the input data stream. Since the maximum operating speed of the regenerator is limited by the VCO, the option of connecting an external VCO is provided.

III. THE PHASE AND FREQUENCY DETECTOR

The block diagram of the PFD and the corresponding output signals are shown in Fig. 2. The detector is composed of two identical sample and hold cells which serve as phase detector, denoted as PD, and quadrature phase detector, denoted as QPD [4]. At every transition of the input data stream the in-phase signal and the quadrature signal of the VCO are sampled. This operation provides beat notes at the outputs Q_{PD} and Q_{QPD} if the frequency of the VCO and the bit-rate frequency are unequal. Depending on the sign of the frequency difference the signal of the quadrature phase detector leads or lags relative to

Reprinted from *IEEE Journal of Solid-State Circuits*, vol. SC-29, pp. 1572-1576, December 1994.

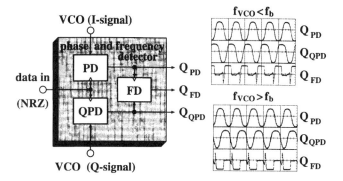

Fig. 2. Block diagram of phase and frequency detector.

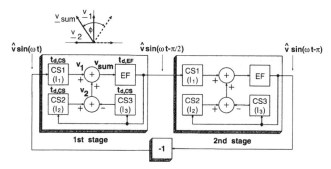

Fig. 3. Block diagram of VCO (modified 2-stage ring oscillator).

the signal of the phase detector. This phase shift is analyzed by the frequency detector, denoted as FD which delivers a signal with a strong dc component at the output. This signal is superimposed to the phase detector output and ensures that the closed loop is driven towards lock. Once lock is achieved, the beat notes vanish and the frequency detector is switched off. Therefore no additional phase noise is added by the frequency detector in the locked state. A more detailed description of the operation principle and a simplified circuit diagram of the PFD can be found in [4].

If the phase shift between the in-phase and quadrature signal of the VCO doesn't exactly equal $90°$, the phase and frequency detector still works; the pull-in range, however, becomes asymmetrical in this case, referring to the free-running frequency of the VCO. Therefore, a VCO which provides inphase and quadrature outputs over the full tuning range is desired.

IV. THE VOLTAGE-CONTROLLED OSCILLATOR

There are two basic concepts for VCO's that generate wideband in-phase and quadrature signals. The first concept is based on an oscillator running at double the frequency. The in-phase and quadrature signals are subsequently generated by a master-slave D-FF which is wired as a static frequency divider. The second approach takes advantage of the signal phases in a 2-stage ring oscillator, if the output of the second stage is connected to the inverting input of the first stage. However, both concepts show an inherent speed penalty of about a factor of 2 in comparison with VCO's without quadrature outputs. In order to reduce this performance gap we used a *modified* 2-stage ring oscillator [5] in our regenerator design (cf., Fig. 3). The oscillator is designed fully differential and the required phase shift of $180°$—indicated here by the minus one—is realized by connecting the output of the second stage to the inverting input of the first stage. Only two simple circuit elements are used in the oscillator: a current switch, in Fig. 3 denoted as CS, and an emitter follower stage, denoted as EF.

Frequency tuning is achieved by controlling the signal phase in the ring oscillator. This is done by two additional current switches CS2 and CS3, which provide a positive and a negative feedback within each stage. The resulting sign and value of the inner-stage feedback is controlled by the gain of the current switches, CS2 and CS3. If we assure that the delay times $t_{d,CS}$ of the three current switches are nearly identical, we get the

vector diagram shown above the first stage in Fig. 3. The voltage $v_1(t)$ and $v_2(t)$ may be expressed as

$$v_1(t) = R_c I_1 \sin\left(\omega(t - t_{d,CS})\right) \quad (1)$$

$$v_2(t) = R_c(I_2 - I_3)\sin\left(\omega(t - t_{d,CS}) - \frac{\pi}{2}\right) \quad (2)$$

where R_c denotes the common collector resistance and I_1 to I_3 the current through the current switches (CS); $t_{d,CS}$ is the delay caused by a current switch. Based on the vector representation of v_1 and v_2 the resulting voltage sum v_{sum} is

$$v_{sum}(t) = R_c\sqrt{I_1^2 + (I_2 - I_3)^2}\sin[\omega(t - t_{d,CS}) - \Phi];$$
$$\Phi = \arctan\left(\frac{I_2 - I_3}{I_1}\right). \quad (3)$$

For a steady-state oscillation the phase at the output of the first stage must have a phase shift of $90°$ (c.f., Fig. 3):

$$v_{sum}(t - t_{d,EF}) = \hat{v}\sin\left(\omega t - \frac{\pi}{2}\right).$$

This leads to the following expression for the oscillation frequency of the ring oscillator:

$$f_{VCO} = \frac{\frac{\pi}{2} - \arctan\left(\frac{I_2 - I_3}{I_1}\right)}{2\pi t_d}; \quad t_d = t_{d,CS} + t_{d,EF}. \quad (4)$$

This equation describes how the oscillation frequency f_{VCO} can be adjusted by controlling the currents I_1 to I_3 of the current switches.

In case of a positive feedback within the stages (i.e., $I_2 > I_3$) the overall negative feedback of the connected two stages must prevent a latching of the dc coupled ring oscillator. Therefore, the lower limit of the oscillation frequency is given by $I_1 > I_2 - I_3$. In case of a strong negative feedback within the stages (i.e., $I_2 \ll I_3$) each stage forms a conventional 1-stage oscillator, phase-locked to the other by the current switch CS1. Therefore, the upper limit of the oscillation frequency is the frequency of a 1-stage oscillator.

To summarize, the theoretical tuning range of the presented dc coupled ring oscillator structure is:

$$\frac{1}{8t_d} < f_{VCO} < \frac{1}{2t_d}. \quad (5)$$

However, in a real circuit the factor between the upper and lower limit is closer to 2 than to 4, but it is possible to increase the tuning range by additionally controlling the delay time t_d.

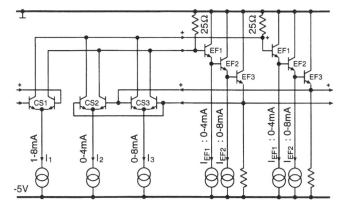

Fig. 4.   Circuit diagram of one ring oscillator stage.

The simplified circuit diagram of one stage is shown in Fig. 4. Besides a reduction of cross talk the differential design permits the use of a smaller internal voltage swing resulting in a higher oscillation frequency. The sum of currents I_1 to I_3 of the current switches is kept constant, which provides a nearly constant voltage swing of about 230 mV across the load resistance of 25 Ω. The tuning range of the ring oscillator is further increased by additionally controlling the current of the first two emitter follower stages. For controlling these currents an extra tuning input is provided. The VCO needs only a single -5 V power supply, and the input and output are compatible to standard current mode logic.

Note that there is a trade-off between a large tuning range and a low phase noise of the oscillator. In applications where a low phase noise and a stable center frequency are more important than a wide tuning range, the load resistance in the RO may be replaced by short stub transmission lines or on chip spiral inductors. As shown by simulations the tuning range remains about 10% of the center frequency in such cases.

V. FABRICATION AND MEASURED PERFORMANCE OF THE CLOCK- AND DATA-REGENERATOR IC

Fig. 5 is the chip photomicrograph of the generator IC. This chip was fabricated with a self-aligned bipolar technology which features a transit frequency of about 25 GHz and an emitter stripe width of 0.4 μm [6]. The three main processing blocks are the data regenerator, the phase and frequency detector and the two stage ring oscillator. Special attention was paid to reduce the interconnection delay in the ring oscillator. The signal path of one stage is about 200 μm.

In Fig. 6 the measured tuning characteristics of the VCO are shown, if the same tuning voltage is applied to both tuning inputs—one for controlling the currents in the current switches and the other for controlling the emitter follower currents. This diagram shows that the VCO can be tuned from 2 to 9 GHz by a differential 600 mv voltage swing. Both the in-phase and the quadrature signal are provided over the full tuning range. Fig. 7 displays the temperature dependence of the frequency of the free-running VCO. The frequency shift due to temperature change is much less than the VCO tuning range and remains within the pull-in range of the PFLL.

In order to demonstrate that the maximum operating frequency of the regenerator is only limited by the on-chip

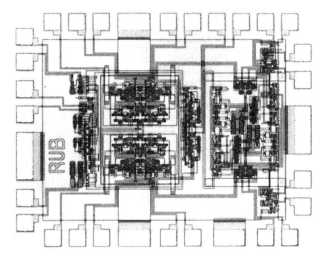

Fig. 5.   Chip photomicrograph.

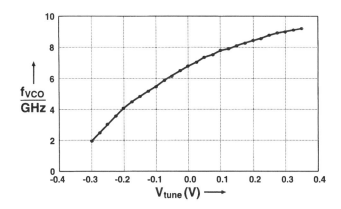

Fig. 6.   Measured tuning characteristics of the VCO.

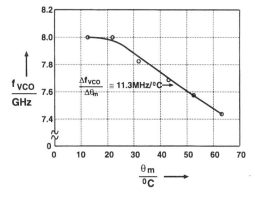

Fig. 7.   Measured temperature dependence of the free-running VCO frequency.

VCO, open-loop measurement of the regenerator with an external 15 GHz sine-wave generator was performed. The generator was biased for a fixed frequency offset of ± 200 MHz. The output signals of the detector are shown in the upper diagram of Fig. 8, with a 7.5 GHz sine wave signal, representing a periodic 15 Gb/s zero/one bit-sequence, applied to the data input. The corresponding output signals for a PRBS input signal can be seen in the lower diagrams. The superimposed noise is suppressed by the loop filter and does not deteriorate the operation of the detector. The dc component

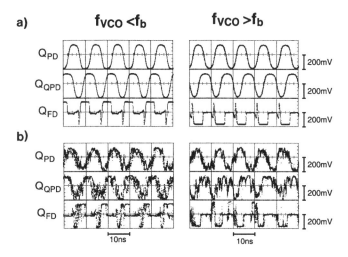

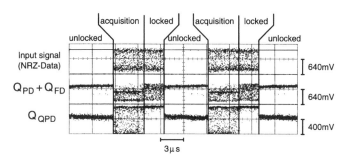

Fig. 8. Measured PFD output signals at 15 Gb/s: PFLL Open. Input signal: a) 7.5 GHz sine wave representing a periodic 15 Gb/s zero/one sequence; b) PRBS signal of 15 Gb/s.

Fig. 9. Measured acquisition of the PFLL at 8 Gb/s.

of the frequency detector output changes with the sign of the frequency difference and ensures that the closed loop could be driven towards lock.

In Fig. 9 the acquisition process is demonstrated at a bit rate of 8 Gb/s. The oscillator is biased for a frequency offset of about 200 MHz. In order to get a periodic representation of the acquisition process, the data signal, which is shown in the upper signal trace, is switched from a constant logic 1 to a PRBS and vice versa. Since a constant logic level in the NRZ format carries no clock information the loop starts in an unlocked state. After switching to a PRBS signal, the acquisition process starts. The signal of the quadrature output toggles with the beat frequency between the low and high level having no clear dc component, in contrast to the summed output of the phase detector and the frequency detector. This signal is now responsible for driving the loop towards lock. After an acquisition time of about 4 μs, this is roughly two times the low pass filter time constant of the loop filter, lock is achieved. Now the output of the quadrature detector remains high and the frequency detector is switched off. Finally, the data input is switched back to a constant logic 1 and the loop falls back to the unlocked state, since no further clock information is supplied.

The data regeneration at 8 Gb/s is demonstrated in Fig. 10. The regenerator is locked to a PRBS of length of $2^{11}-1$. The upper diagram presents the signal traces and the lower diagram the corresponding eye diagram representations. The incoming

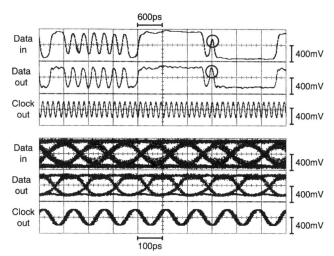

Fig. 10. Measured regenerator operation at 8 Gb/s: loop locked.

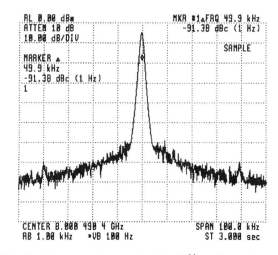

Fig. 11. Signal spectrum of extracted clock: $(2^{11}-1)b$ PRBS.

data signal, the regenerated data signal and the extracted clock are displayed here. The single bit circled in this figure and the eye diagrams demonstrate the regenerating effect of the data regenerator.

The measured signal spectrum of the extracted clock, when locked to an 8 Gb/s PRBS data signal, is presented as Fig. 11. The measured single side phase noise of the extracted clock is about -90 dBc/Hz at 50 kHz offset from carrier.

The jitter histogram of the extracted clock in the time domain is displayed in Fig. 12. The measured rms time jitter, as observed on the sampling scope, is about 1.6 ps when a PRBS signal of length of $2^{11}-1$ is applied. In the free-running case, we measured an rms phase jitter of about 1.9 ps. Note that the systematic rms time jitter of the measuring setup is about 1 ps.

Table I summarizes the most important chip data. The maximum experimental bit rate of the regenerator chip is 8 Gb/s with the on-chip VCO, and 15 Gb/s with an external VCO. The tuning range of the on-chip VCO is 2-to-9 GHz. The measured rms phase jitter of the extracted clock is 1.6 ps, when locked to a PRBS of length of $2^{11}-1$. The single side phase noise in the frequency domain is -91 dBc/Hz at 50 kHz offset from carrier. The total power dissipation of the

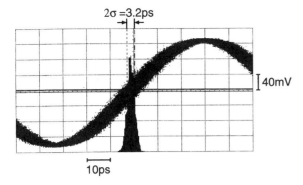

2σ =3.2ps

40mV

10ps

Fig. 12.   Jitter-histogram of extracted clock: $(2^{11} - 1)b$ PRBS.

TABLE I
CHARACTERISTICS OF THE CLOCK-RECOVERY AND DATA-REGENERATOR

Max. bit rate with	on-chip VCO	8 Gb/s
	external VCO	15 Gb/s
Tuning range of on-chip VCO		**2-9 GHz**
RMS phase jitter at 8 Gb/s		**1.6ps**
Phase noise at 50 kHz offset		**-91 dBc/Hz**
Total power dissipation		**2.25 W**
Transistor count		**297**
Chip size		$1.53 \times 2.0\text{mm}^2$

chip is about 2.3 W, when connected to a single -5 V power supply. The transistor count is about 300 and the chip size is about 1.5×2.0 mm^2.

VI. CONCLUSION

In conclusion, we have demonstrated the design and the performance of a clock and data regenerator IC, which possess the following main features:

The chip directly processes the incoming NRZ data signal and needs no signal preprocessing. Only two external components—a capacitor and a resistor—are required. Due to a phase and frequency locked-loop, a fast acquisition is achieved and the pull-in range is independent of the loop-bandwidth. The VCO features a wide tuning range and provides in-phase and quadrature outputs over the full tuning range.

ACKNOWLEDGMENT

We gratefully acknowledge the fabrication of the experimental CDR IC by Hewlett Packard Co., and assistance by B. Wüppermann and W. J. Hillery of HP ICBD Bipolar Business Unit R&D, Palo Alto, CA. We also thank Z. Lao of the Ruhr-Universität Bochum for the temperature measurements of the VCO.

REFERENCES

[1] Z. Wang, U. Langmann, and B. G. Bosch, "Multi-Gb/s silicon bipolar clock recovery IC," *IEEE J. Selected Areas Commun.*, vol. 9, no. 5, pp. 656–663, 1991.
[2] D. Briggmann, G. Hanke, U. Langmann, and A. Pottbäcker, "Clock recovery circuits up to 20 Gbit/s for optical transmission systems," *1994 IEEE MTT-S Int. Microwave Symp., Dig. Tech. Papers*, pp. 1093–1096.
[3] Z. Wang *et al.*, "19 GHz monolithic integrated clock recovery using PLL and 0.3 μ gatelength quantum-well HEMTs," *1994 IEEE Int. Solid-State Circuit Conf.(ISSCC)*, Feb. 1994, *Dig. Tech. Papers.*, pp. 118–119.
[4] A. Pottbäcker, U. Langmann, and H.-U. Schreiber, "A Si bipolar phase and frequency detector IC for clock extraction up to 8 Gb/s," *IEEE J. Solid-State Circuits*, vol. 27, pp. 1747–1751, Dec. 1992.
[5] A. Pottbäcker and U. Langmann, "An 8 GHz Silicon bipolar clock-recovery and data-regenerator IC," *1994 IEEE Int. Solid-State Circuits Conf. (ISSCC)*, Feb. 1994, *Dig. Tech. Pap.*, pp. 116–117.
[6] W. M. Huang *et al.*, "A high-speed bipolar technology featuring self-aligned single-poly base and submicrometer emitter contacts," *IEEE Electron. Device Lett.*, vol. 11, no. 9, pp. 412–414, 1990.

Author Index

Subject Index

Editor's Biography

BEHZAD RAZAVI received the B.Sc. degree in electrical engineering from Tehran (Sharif) University of Technology, Tehran, Iran, in 1985, and the M.Sc. and Ph.D. degrees in electrical engineering from Stanford University, Stanford, CA, in 1988 and 1991, respectively. Since December 1991 he has been a Member of the Technical Staff at AT&T Bell Laboratories, Holmdel, NJ, where his research involves integrated circuit design for communication systems. His current interests include wireless transceivers, data conversion, clock recovery, frequency synthesis, and low-voltage low-power circuits.

Dr. Razavi has been a Visiting Lecturer at Princeton University, Princeton, NJ, and Stanford University. He is also a member of the Technical Program Committee of the International Solid-State Circuits Conference (ISSCC). He has served as Guest Editor to the IEEE Journal of Solid-State Circuits (JSSC) and International Journal of High Speed Electronics and is currently an Associate Editor of JSSC.

Dr. Razavi received the Beatrice Winner Award for Editorial Excellence at the 1994 ISSCC, the best paper award at the 1994 European Solid-State Circuits Conference, and the best panel award at the 1995 ISSCC. He is the author of the book "Principles of Data Conversion System Design," also published by IEEE Press.